ENVIRONMENTAL SCIENCE

FIFTEENTH EDITION

ABOUT THE COVER PHOTO

In 2005, nature journalist Richard Louv hypothesized that many people, especially children, have experienced *nature-deficit disorder,* a series of problems resulting from their spending increasingly less time in the natural world. Many children and young adults spend most of their free time indoors watching TV and using smart phones, computers, and other electronic devices. Evidence indicates that such isolation from nature could be contributing to stress, anxiety, depression, irritability, difficulty in dealing with change, and excessive body weight. In the United States, according to the Centers for Disease Control and Prevention, about 33% of all children and 69% of all adults over age 20 are overweight or obese. Also, the indoor air in U.S. homes and buildings is typically 2 to 5 times more polluted than outdoor air, according to the U.S. Environmental Protection Agency, which could be contributing to the increasing incidence of certain lung ailments.

Nature-deficit disorder is partly an effect of urbanization. More than half of the world's people now live in urban areas, many of which do not have enough parks and recreational areas to make it easy for people to get out. Cities also have higher crime rates than do rural areas, and the continuous news cycle along with social networking keep people hyper-informed about crime and other threats. Thus, many people are afraid to venture out.

Research indicates that children and adults can gain many benefits by playing and exploring outdoors, hiking, jogging, snorkeling (see cover photo), fishing, gardening, and bird-watching. Such activities can foster better health, reduce stress, improve mental abilities, and stimulate imagination and creativity. Experiencing nature can also provide a sense of wonder and connection to life on Earth, which keeps us alive and supports our economies.

Environmental scientists have identified this increasing isolation from nature as one of the five major causes of the environmental problems we face. Without an understanding of our utter dependence on nature for food, shelter, clean air, clean water, and many other natural resources and services, we become more likely to degrade our environment. With such an understanding, we will be more likely to reverse such degradation and to contribute positively to the environment and thus to our own well-being.

Jason Edwards/National Geographic Creative

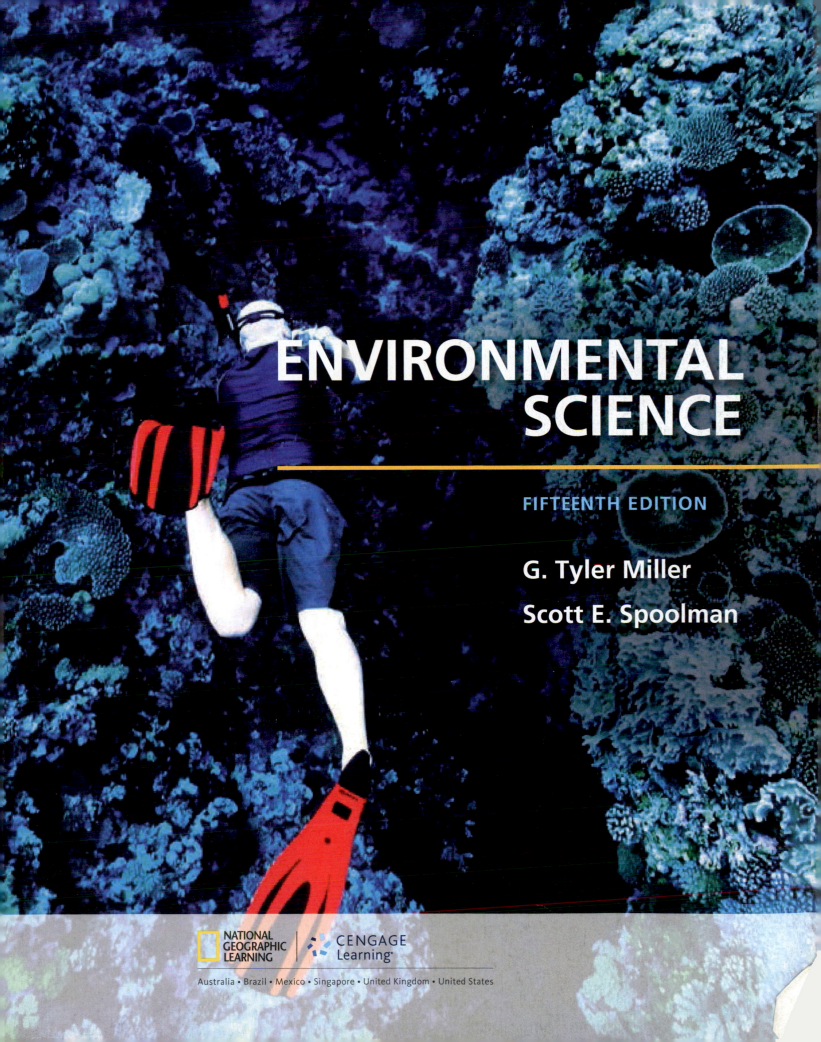

ENVIRONMENTAL SCIENCE

FIFTEENTH EDITION

G. Tyler Miller

Scott E. Spoolman

NATIONAL GEOGRAPHIC LEARNING | CENGAGE Learning

Australia • Brazil • Mexico • Singapore • United Kingdom • United States

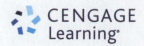

Environmental Science, **Fifteenth Edition**
G. Tyler Miller, Scott E. Spoolman

Product Director: Mary Finch

Senior Product Team Manager: Yolanda Cossio

Senior Content Developer, Media Developer:
Jake Warde

Associate Content Developers: Kellie
Petruzzelli; Casey Lozier

Product Assistant: Chelsea Joy

Market Development Manager: Julie Schuster

Content Project Manager: Hal Humphrey

Senior Art Director: Pamela Galbreath

Manufacturing Planner: Karen Hunt

Production Service: Dan Fitzgerald, Graphic
World Inc.

Photo Researcher: Carly Bergey, PreMedia
Global

Text Researcher: Kavitha Balasundaram,
PreMedia Global

Copy Editor: Graphic World Inc.

Illustrator: ScEYEnce Studios

Text and Cover Designer: Hespenheide Design

Cover Image: © Jason Edwards/National
Geographic Creative

Compositor: Graphic World Inc.

For product information and technology assistance, contact us at
Cengage Learning Customer & Sales Support, 1-800-354-9706.

For permission to use material from this text or product,
submit all requests online at **www.cengage.com/permissions.**
Further permissions questions can be e-mailed to
permissionrequest@cengage.com.

Library of Congress Control Number: 2014946046

Student Edition:

ISBN-13: 978-1-305-09044-6

ISBN-10: 1-305-09044-6

Loose-leaf Edition:

ISBN-13: 978-1-305-25750-4

ISBN-10: 1-305-25750-2

Cengage Learning
20 Channel Center Street
Boston, MA 02210
USA

Cengage Learning is a leading provider of customized learning solutions with
office locations around the globe, including Singapore, the United Kingdom,
Australia, Mexico, Brazil, and Japan. Locate your local office at
www.cengage.com/global.

Cengage Learning products are represented in Canada by Nelson Education, Ltd.

To learn more about Cengage Learning Solutions, visit **www.cengage.com.**

Purchase any of our products at your local college store or at our preferred
online store **www.cengagebrain.com.**

Printed in Canada
Print Number: 01 Print Year: 2014

FSC
www.fsc.org
MIX
Paper from
responsible sources
FSC® C011825

BRIEF CONTENTS

Michael Nichols/National Geographic Creative

CONTENTS

Dudarev Mikhail/Shutterstock.com

JENS SCHLUETER/AFP/Getty Images

Darlyne A. Murawski/National Geographic Creative

Robert King/Shutterstock.com

Kirsten Wahlquist | Dreamstime.com

Ken Hawkins/Alamy

James Forte/National Geographic Stock

Seatraveler/Dreamstime.com

Dick Ercken/Shutterstock.com

Joel Sartore/National Geographic Creative

Jim West/Age Fotostock

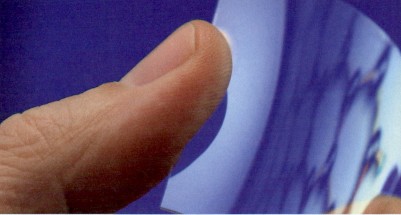
Vincenzo Lombardo/Getty Images

ssuaphotos/Shutterstock.com

Mark Thiessen/National Geographic Creative

DIANE COOK/LEN JENSHEL/National Geography Creative

David Grossman/Alamy

For Instructors

We wrote this book to help instructors achieve three important goals: *first,* to explain to their students the basics of environmental science; *second,* to help their students in using this scientific foundation to understand the environmental problems that we face and to evaluate possible solutions to them; and *third,* to inspire their students to make a difference in how we treat the earth on which our lives and economies depend, and thus in how we treat ourselves and our descendants.

We view environmental problems and possible solutions to them through the lens of *sustainability*—the integrating theme of this book. We believe that most people will still be able to live comfortable and fulfilling lives, and that societies will be more prosperous and peaceful, when sustainability becomes one of the chief measures by which personal choices and public policies are made. We consistently challenge students to work toward attaining such a future.

For this reason, we are happy to be working with the National Geographic Society in the production of this book. This partnership has allowed us to include many stunning and informative photographs, numerous maps, and many new stories of National Geographic Explorers and other researchers who have received funding from National Geographic—people who are making a positive difference in the world. With these new tools, we continue to tell of the good news from various fields of environmental science, hoping to inspire young people to commit themselves to making our world a more sustainable place to live for their own and future generations.

What's New in This Edition?

- *Our partnership with National Geographic* has given us access to hundreds of amazing photographs, numerous maps, and inspiring stories of *National Geographic Explorers and grantees*—people who are leading the way in environmental science, education, or entrepreneurial enterprises.

- A *stunning new design* with a National Geographic look that enhances visual learning.

- *New Core Case Studies* for 10 of the book's 17 chapters bring important real-world stories to the forefront for use in applying those chapters' concepts and principles.

- New Supplement 6, Geologic and Biological Time Scale, that locates major developments related to life on Earth, including the mass extinctions, within the earth's geologic time scale.

Sustainability Is the Integrating Theme of This Book

Sustainability is the overarching theme of this textbook. You can see the sustainability emphasis by looking at the Brief Contents (p. v).

Six principles of sustainability play a major role in carrying out this book's sustainability theme. These principles are introduced in Chapter 1. They are depicted in Figure 1.2 (p. 6) and Figure 1.5 (p. 9) and summarized in Supplement 7 (pp. S50–S51), and we apply them throughout the book, with each reference marked in the margin as shown here (see pp. 50 and 219).

We use the following five major subthemes to integrate material throughout this book:

- ***Natural capital.*** Sustainability depends on the natural resources and ecosystem services that support all life and economies. See Figures 1.3, p. 7, and 9.4, p. 189.

- ***Natural capital degradation.*** We describe how human activities can degrade natural capital. See Figures 1.7, p. 10, and 7.17, p. 146.

- ***Solutions.*** We present existing and proposed solutions to environmental problems in a balanced manner and challenge students to use critical thinking to evaluate them. See Figures 9.14, p. 195, and 11.11, p. 257.

- ***Trade-offs.*** The search for solutions involves trade-offs, because any solution requires weighing advantages against disadvantages. Our Trade-offs diagrams located in several chapters present the benefits and drawbacks of various environmental technologies and solutions to environmental problems. See Figures 13.9, p. 322, and 16.11, p. 439.

- ***Individuals Matter.*** Throughout the book, Individuals Matter boxes and some of the Case Studies describe what various scientists and concerned citizens (including several National Geographic Explorers) have done to help us work toward sustainability (see pp. 17, 209, and 237). Also, a number of What Can You Do? diagrams describe how readers can deal with the problems we face (see Figures 11.21, p. 267, and 13.44, p. 353). Eight especially important steps that people can take are summarized in Figure 17.21 (p. 478).

Other Successful Features of This Textbook

- ***Up-to-Date Coverage.*** Our textbooks have been widely praised for keeping users up to date in the rapidly changing field of environmental science. We have used thousands of articles and reports published in 2011–2014 to update the information in this book. Major new or updated topics include planetary boundaries and ecological tipping points (Science Focus 3.3, p. 58), hydraulic fracturing, or fracking (Science Focus 13.1, p. 318), and the rising threat of ocean acidification (Science Focus 9.3, p. 206), along with dozens of other important topics.

- **Concept-Centered Approach.** To help students focus on the main ideas, we built each major chapter section around a key question and one to three key concepts, which state the section's most important take-away lessons. In each chapter, all key questions are listed at the front of the chapter, and each chapter section begins with its key question and concepts (see pp. 187 and 262). Also, the concept applications are highlighted and referenced throughout each chapter.

- **Science-Based Approach.** Chapters 2–7 cover scientific principles important to the course and discuss how scientists work (see Brief Contents, p. v). Important environmental science topics are explored in depth in Science Focus boxes distributed among the chapters throughout the book (see pp. 206 and 408). We also integrate science coverage throughout the book in various Case Studies (see pp. 175 and 203) and in numerous figures.

- **Global Coverage.** This book provides a global perspective, first on the ecological level, revealing how all the world's life is connected and sustained within the biosphere, and second, through the use of information and images from around the world. This includes more than 40 maps in the basic text and in Supplement 4. At the end of each chapter is a Global Environment Watch Exercise that applies this global perspective.

- **Core Case Studies.** Each chapter opens with a Core Case Study (see pp. 162 and 216), which is applied throughout the chapter. These applications are indicated by the notation **Core Case Study** wherever they occur (see pp. 171 and 240). Each chapter ends with a *Tying It All Together* box (see pp. 181 and 244) that connects the Core Case Study and other material in the chapter to some or all of the principles of sustainability.

- **Case Studies.** In addition to the 17 Core Case Studies, some 42 additional Case Studies (see pp. 220, 259, and 322) appear throughout the book (and are listed in the Detailed Contents, pp. vi–xiii). Each of these provides an in-depth look at specific environmental problems and their possible solutions.

- **Critical Thinking.** The Note to Students (p. xxiii) describes critical thinking skills, and specific critical

thinking exercises are used throughout the book in several ways:

- In dozens of *Thinking About* exercises that ask students to analyze material immediately after it is presented (see pp. 117 and 263).
- In all *Science Focus* boxes.
- In dozens of *Connections* boxes that stimulate critical thinking by exploring often surprising connections related to environmental problems (see pp. 166 and 396).
- In the captions of many of the book's figures (see Figures 8.4, p. 166, and 11.13, p. 260).
- In end-of-chapter *Critical Thinking* questions (see pp. 126 and 356).

- **Visual Learning.** With a new design heavily influenced by material from National Geographic and more than 440 photographs, many of them from the archives of National Geographic, this is the most visually arresting environmental science textbook available (see Figures 5.9, p. 92, 7.16, p. 145, and 9.13, p. 195). Add in the more than 130 diagrams, each designed to present complex ideas in understandable ways relating to the real world (see Figures 3.3, p. 44, and 4.2, p. 65), and you also have one of the most visually informative textbooks available.

- **Flexibility.** To meet the diverse needs of hundreds of widely varying environmental science courses, we have designed a highly flexible book that allows instructors to vary the order of chapters without exposing students to terms and concepts that could confuse them. We recommend that instructors start with Chapter 1, which defines basic terms and gives an overview of sustainability, population, pollution, resources, and economic development issues that are discussed throughout the book. This provides a springboard for instructors to use the other chapters in almost any order. One often-used strategy is to follow Chapter 1 with Chapters 2–7, which introduce basic science and ecological concepts. Instructors can then use the remaining chapters in any order desired. Some instructors follow Chapter 1 with Chapter 17 on environmental economics, politics, and worldviews, respectively, before proceeding to the chapters on basic science and ecological concepts. We provide a second level of flexibility in seven Supplements (see p. xiii in the Detailed Contents), which instructors can assign as desired to meet their needs. Examples include Some Basic Chemistry (Supplement 3), Maps and Map Analysis (Supplement 4), Environmental Data and Data

Analysis (Supplement 5), and a new Supplement 6 showing a Geologic and Biological Time Scale.

- **In-Text Study Aids.** Each chapter begins with a list of *Key Questions* showing how the chapter is organized (see pp. 312–313). When a new *key term* is introduced and defined, it is printed in boldface type and all such terms are summarized in the glossary at the end of the book. In most chapters, *Thinking About* exercises reinforce learning by asking students to think critically about the implications of various environmental issues and solutions immediately after they are discussed in the text (see p. 320). The captions of many figures contain similar questions that get students to think about the figure content (see Figures 13.5, p. 320, and 13.34, p. 345). In their reading, students also encounter *Connections* boxes, which briefly describe connections between human activities and environmental consequences, environmental and social issues, and environmental issues and solutions (see pp. 347 and 349). Finally, the text of each chapter wraps up with three *Big Ideas* (see p. 353), which summarize and reinforce three of the major take-away lessons from each chapter, and a *Tying It All Together* section that relates the Core Case Study and other chapter content to the principles of sustainability (see p. 354).

Each chapter ends with a *Chapter Review* section containing a detailed set of review questions that include all of the chapter's key terms in bold type; *Critical Thinking* questions that encourage students to think about and apply what they have learned to their lives; *Doing Environmental Science*—an exercise that will help students to experience the work of various environmental scientists; a *Global Environment Watch* exercise taking students to Cengage's GREENR site where they can use this tool for interesting research related to chapter content; and a *Data Analysis* or *Ecological Footprint Analysis* problem built around ecological footprint data or some other environmental data set. (See pp. 127 and 357.)

Online Solutions and Resources

- **MindTap.** MindTap is a new approach to highly personalized online learning. Beyond an e-Book, homework solution, digital supplement, or premium website, MindTap is a digital learning platform that works alongside your campus Learning Management System (LMS) to deliver course curriculum across the range of electronic devices in your life. MindTap is built on an "app" model allowing enhanced digital collaboration and delivery of engaging content across a spectrum of Cengage and non-Cengage resources. Visit the Instructor's Companion Site for tips on maximizing your MindTap course.

- **Aplia.** Aplia™ for Environmental Science is an online interactive learning solution that improves comprehension and outcomes by increasing student effort and engagement. Aplia provides automatically graded assignments that were written to make the most of the web medium and contain detailed, immediate explanations on every question. Students come to class prepared and ready to participate. Diverse types of questions aim to reinforce, extend, and apply key concepts by focusing on case studies, data analysis, real-world applications, global perspectives, and more. Aplia homework is also available via MindTap.

- **Instructor's Companion Site.** Everything you need for your course in one place! This collection of book-specific lecture and class tools is available online via www.cengage.com/login. Access and download PowerPoint presentations, images, instructor's manual, videos, and more.

- **Cognero Test Bank.** Available to adopters. Cengage Learning Testing Powered by Cognero is a flexible, online system that allows you to author, edit, and manage test bank content from multiple Cengage Learning solutions; create multiple test versions in an instant; and deliver tests from your LMS, your classroom, or wherever you want.

- **BBC Videos for Environmental Science.** These short, informative video clips cover current news stories on environmental issues from around the world. These clips are a great way to start a lecture or spark a discussion. Available on the Instructor's Companion Site and within MindTap.

Help Us Improve This Book or Its Supplements

Let us know how you think this book can be improved. If you find any errors, bias, or confusing explanations, please e-mail us about them at:

- **mtg89@hotmail.com**
- **spoolman@tds.net**

Most errors can be corrected in subsequent printings of this edition, as well as in future editions.

Acknowledgments

We wish to thank the many students and teachers who have responded so favorably to the 14 previous editions of *Environmental Science*, the 18 editions of *Living in the Envi-*

ronment, the 11 editions of *Sustaining the Earth,* and the 7 editions of *Essentials of Ecology,* and who have corrected errors and offered many helpful suggestions for improvement. We are also deeply indebted to the more than 400 reviewers, who pointed out errors and suggested many important improvements in the various editions of these three books.

It takes a village to produce a textbook, and the members of the talented production team, listed on the copyright page, have made vital contributions. Our special thanks go to Senior Content Developer Jake Warde, Senior Content Project Manager Hal Humphrey, production editor Dan Fitzgerald, copy editor Chris DeVito, chapter layout specialist Cheryl Whitley, Senior Art Director Pam Galbreath, photo researcher Carly Bergey, artist Patrick Lane, Product Development Manager Alexandria Brady, assistant editor Chelsea Joy, Assistant Content Developers Kellie Petruzzelli and Casey Lozier, and Cengage Learning's hard-working sales staff. Finally, we are very fortunate to have the guidance, inspiration, and unfailing support of life sciences Senior Product Team Manager Yolanda Cossio and her dedicated team of highly talented people who have made this and our other book projects such a pleasure to work on.

G. Tyler Miller

Scott E. Spoolman

Guest Essayists

Guest essays by the following authors are available as assignable activities via MindTap: **M. Kat Anderson**, ethnoecologist with the National Plant Center of the USDA's Natural Resource Conservation Center; **Lester R. Brown**, president, Earth Policy Institute; **Alberto Ruz Buenfil**, environmental activist, writer, and performer; **Robert D. Bullard**, professor of sociology and director of the Environmental Justice Resource Center at Clark Atlanta University; **Michael Cain**, ecologist and adjunct professor at Bowdoin College; **Herman E. Daly**, senior research scholar at the School of Public Affairs, University of Maryland; **Lois Marie Gibbs**, director, Center for Health, Environment, and Justice; **Garrett Hardin**, professor emeritus (now deceased) of human ecology, University of California–Santa Barbara; **John Harte**, professor of energy and resources, University of California–Berkeley; **Paul G. Hawken**, environmental author and business leader; **Jane Heinze-Fry**, environmental educator; **Paul F. Kamitsuja**, infectious disease expert and physician; **Amory B. Lovins**, energy policy consultant and director of research, Rocky Mountain Institute; **Bobbi S. Low**,

professor of resource ecology, University of Michigan; **John J. Magnuson**, Director Emeritus of the Center for Limnology, University of Wisconsin–Madison; **Lester W. Milbrath**, director of the research program in environment and society, State University of New York–Buffalo; **Peter Montague**, director, Environmental Research Foundation; **Norman Myers**, tropical ecologist and consultant in environment and development; **David W. Orr**, professor of environmental studies, Oberlin College; **Noel Perrin**, adjunct professor of environmental studies, Dartmouth College; **John Pichtel**, Ball State University; **David Pimentel**, professor of insect ecology and agricultural sciences, Cornell University; **Andrew C. Revkin**, environmental author and environmental reporter for the *New York Times;* **Vandana Shiva**, physicist, educator, environmental consultant; **Nancy Wicks**, ecopioneer and director of Round Mountain Organics; and **Donald Worster**, environmental historian and professor of American history, University of Kansas.

Pedagogy Contributors

Dr. Dean Goodwin and his colleagues, Berry Cobb, Deborah Stevens, Jeannette Adkins, Jim Lehner, Judy Treharne, Lonnie Miller, and Tom Mowbray, provided excellent contributions to the Data Analysis and Ecological Footprint Analysis exercises. Mary Jo Burchart of Oakland Community College wrote the in-text Global Environment Watch Exercises.

Cumulative List of Reviewers

Barbara J. Abraham, Hampton College; Donald D. Adams, State University of New York at Plattsburgh; Larry G. Allen, California State University–Northridge; Susan Allen-Gil, Ithaca College; James R. Anderson, U.S. Geological Survey; Mark W. Anderson, University of Maine; Kenneth B. Armitage, University of Kansas; Samuel Arthur, Bowling Green State University; Gary J. Atchison, Iowa State University; Thomas W. H. Backman, Lewis-Clark State College; Marvin W. Baker, Jr., University of Oklahoma; Virgil R. Baker, Arizona State University; Stephen W. Banks, Louisiana State University in Shreveport; Ian G. Barbour, Carleton College; Albert J. Beck, California State University–Chico; Eugene C. Beckham, Northwood University; Diane B. Beechinor, Northeast Lakeview College; W. Behan, Northern Arizona University; David Belt, Johnson County Community College; Keith L. Bildstein, Winthrop College; Andrea Bixler, Clarke College; Jeff Bland, University of Puget Sound; Roger G. Bland, Central Michigan University; Grady

Blount II, Texas A&M University–Corpus Christi; Lisa K. Bonneau, University of Missouri–Kansas City; Georg Borgstrom, Michigan State University; Arthur C. Borror, University of New Hampshire; John H. Bounds, Sam Houston State University; Leon F. Bouvier, Population Reference Bureau; Daniel J. Bovin, Université Laval; Jan Boyle, University of Great Falls; James A. Brenneman, University of Evansville; Michael F. Brewer, Resources for the Future, Inc.; Mark M. Brinson, East Carolina University; Dale Brown, University of Hartford; Patrick E. Brunelle, Contra Costa College; Terrence J. Burgess, Saddleback College North; David Byman, Pennsylvania State University, Worthington–Scranton; Michael L. Cain, Bowdoin College; Lynton K. Caldwell, Indiana University; Faith Thompson Campbell, Natural Resources Defense Council, Inc.; John S. Campbell, Northwest College; Ray Canterbery, Florida State University; Deborah L. Carr, Texas Tech University; Ted J. Case, University of San Diego; Ann Causey, Auburn University; Richard A. Cellarius, Evergreen State University; William U. Chandler, Worldwatch Institute; F. Christman, University of North Carolina–Chapel Hill; Lu Anne Clark, Lansing Community College; Preston Cloud, University of California–Santa Barbara; Bernard C. Cohen, University of Pittsburgh; Richard A. Cooley, University of California–Santa Cruz; Dennis J. Corrigan; George Cox, San Diego–State University; John D. Cunningham, Keene State College; Herman E. Daly, University of Maryland; Raymond F. Dasmann, University of California–Santa Cruz; Kingsley Davis, Hoover Institution; Edward E. DeMartini, University of California–Santa Barbara; James Demastes, University of Northern Iowa; Robert L. Dennison, Heartland Community College; Charles E. DePoe, Northeast Louisiana University; Thomas R. Detwyler, University of Wisconsin; Bruce DeVantier, Southern Illinois University at Carbondale; Peter H. Diage, University of California, Riverside; Stephanie Dockstader, Monroe Community College; Lon D. Drake, University of Iowa; Michael Draney, University of Wisconsin–Green Bay; David DuBose, Shasta College; Dietrich Earnhart, University of Kansas; Robert East, Washington & Jefferson College; T. Edmonson, University of Washington; Thomas Eisner, Cornell University; Michael Esler, Southern Illinois University; David E. Fairbrothers, Rutgers University; Paul P. Feeny, Cornell University; Richard S. Feldman, Marist College; Vicki Fella-Pleier, La Salle University; Nancy Field, Bellevue Community College; Allan Fitzsimmons, University of Kentucky; Andrew J. Friedland, Dartmouth College; Kenneth O. Fulgham, Humboldt State University; Lowell L. Getz, University of Illinois at Urbana–Champaign; Frederick F. Gilbert, Washington State University; Jay Glassman, Los Angeles Valley College; Harold Goetz, North Dakota State University; Srikanth Gogineni, Axia College of University of Phoenix; Jeffery J. Gordon, Bowling Green State University; Eville Gorham, University of Minnesota; Michael Gough, Resources for the Future; Ernest M. Gould, Jr., Harvard University; Peter Green, Golden West College; Katharine B. Gregg, West Virginia Wesleyan College; Stelian Grigoras, Northwood University; Paul K. Grogger, University of Colorado at Colorado Springs; L. Guernsey, Indiana State University; Ralph Guzman, University of California–Santa Cruz; Raymond Hames, University of Nebraska–Lincoln; Robert Hamilton IV, Kent State University, Stark Campus; Raymond E. Hampton, Central Michigan University; Ted L. Hanes, California State University–Fullerton; William S. Hardenbergh, Southern Illinois University at Carbondale; John P. Harley, Eastern Kentucky University; Neil A. Harriman, University of Wisconsin–Oshkosh; Grant A. Harris, Washington State University; Harry S. Hass, San Jose City College; Arthur N. Haupt, Population Reference Bureau; Denis A. Hayes, environmental consultant; Stephen Heard, University of Iowa; Gene Heinze-Fry, Department of Utilities, Commonwealth of Massachusetts; Jane Heinze-Fry, environmental educator; Keith R. Hench, Kirkwood Community College; John G. Hewston, Humboldt State University; David L. Hicks, Whitworth College; Kenneth M. Hinkel, University of Cincinnati; Eric Hirst, Oak Ridge National Laboratory; Doug Hix, University of Hartford; Kelley Hodges, Gulf Coast State College; S. Holling, University of British Columbia; Sue Holt, Cabrillo College; Donald Holtgrieve, California State University–Hayward; Michelle Homan, Gannon University; Michael H. Horn, California State University–Fullerton; Mark A. Hornberger, Bloomsburg University; Marilyn Houck, Pennsylvania State University; Richard D. Houk, Winthrop College; Robert J. Huggett, College of William and Mary; Donald Huisingh, North Carolina State University; Catherine Hurlbut, Florida Community College at Jacksonville; Marlene K. Hutt, IBM; David R. Inglis, University of Massachusetts; Robert Janiskee, University of South Carolina; Hugo H. John, University of Connecticut; Brian A. Johnson, University of Pennsylvania–Bloomsburg; David I. Johnson, Michigan State University; Mark Jonasson, Crafton Hills College; Zoghlul Kabir, Rutgers, New Brunswick; Agnes Kadar, Nassau Community College; Thomas L. Keefe, Eastern Kentucky University; David Kelley, University of St. Thomas; William E. Kelso, Louisiana State University; Nathan Keyfitz, Harvard University; David Kidd, University of New Mexico; Pamela S. Kimbrough; Jesse Klingebiel, Kent School; Edward J. Kormondy, University of Hawaii–Hilo/West Oahu College;

John V. Krutilla, Resources for the Future, Inc.; Judith Kunofsky, Sierra Club; E. Kurtz; Theodore Kury, State University of New York at Buffalo; Troy A. Ladine, East Texas Baptist University; Steve Ladochy, University of Winnipeg; Anna J. Lang, Weber State University; Mark B. Lapping, Kansas State University; Michael L. Larsen, Campbell University; Linda Lee, University of Connecticut; Tom Leege, Idaho Department of Fish and Game; Maureen Leupold, Genesee Community College; William S. Lindsay, Monterey Peninsula College; E. S. Lindstrom, Pennsylvania State University; M. Lippiman, New York University Medical Center; Valerie A. Liston, University of Minnesota; Dennis Livingston, Rensselaer Polytechnic Institute; James P. Lodge, air pollution consultant; Raymond C. Loehr, University of Texas at Austin; Ruth Logan, Santa Monica City College; Robert D. Loring, DePauw University; Paul F. Love, Angelo State University; Thomas Lovering, University of California–Santa Barbara; Amory B. Lovins, Rocky Mountain Institute; Hunter Lovins, Rocky Mountain Institute; Gene A. Lucas, Drake University; Claudia Luke, University of California–Berkeley; David Lynn; Timothy F. Lyon, Ball State University; Stephen Malcolm, Western Michigan University; Melvin G. Marcus, Arizona State University; Gordon E. Matzke, Oregon State University; Parker Mauldin, Rockefeller Foundation; Marie McClune, The Agnes Irwin School (Rosemont, Pennsylvania); Theodore R. McDowell, California State University; Vincent E. McKelvey, U.S. Geological Survey; Robert T. McMaster, Smith College; John G. Merriam, Bowling Green State University; A. Steven Messenger, Northern Illinois University; John Meyers, Middlesex Community College; Raymond W. Miller, Utah State University; Arthur B. Millman, University of Massachusetts, Boston; Sheila Miracle, Southeast Kentucky Community & Technical College; Fred Montague, University of Utah; Rolf Monteen, California Polytechnic State University; Debbie Moore, Troy University Dothan Campus; Michael K. Moore, Mercer University; Ralph Morris, Brock University, St. Catherine's, Ontario, Canada; Angela Morrow, Auburn University; William W. Murdoch, University of California, Santa Barbara; Norman Myers, environmental consultant; Brian C. Myres, Cypress College; A. Neale, Illinois State University; Duane Nellis, Kansas State University; Jan Newhouse, University of Hawaii–Manoa; Jim Norwine, Texas A&M University–Kingsville; John E. Oliver, Indiana State University; Mark Olsen, University of Notre Dame; Carol Page, copy editor; Bill Paletski, Penn State University; Eric Pallant, Allegheny College; Charles F. Park, Stanford University; Richard J. Pedersen, U.S. Department of Agriculture, Forest Service; David Pelliam, Bureau of Land Management, U.S. Department of the Interior; Murray Paton Pendarvis, Southeastern Louisiana University; Dave Perault, Lynchburg College; Barry Perlmutter, College of Southern Nevada; Carolyn J. Peters, Spoon River College; Rodney Peterson, Colorado State University; Julie Phillips, De Anza College; John Pichtel, Ball State University; William S. Pierce, Case Western Reserve University; David Pimentel, Cornell University; Peter Pizor, Northwest Community College; Mark D. Plunkett, Bellevue Community College; Grace L. Powell, University of Akron; James H. Price, Oklahoma College; Marian E. Reeve, Merritt College; Carl H. Reidel, University of Vermont; Charles C. Reith, Tulane University; Erin C. Rempala, San Diego City College; Roger Revelle, California State University–San Diego; L. Reynolds, University of Central Arkansas; Ronald R. Rhein, Kutztown University of Pennsylvania; Charles Rhyne, Jackson State University; Robert A. Richardson, University of Wisconsin; Benjamin F. Richason III, St. Cloud State University; Jennifer Rivers, Northeastern University; Ronald Robberecht, University of Idaho; William Van B. Robertson, School of Medicine, Stanford University; C. Lee Rockett, Bowling Green State University; Terry D. Roelofs, Humboldt State University; Daniel Ropek, Columbia George Community College; Christopher Rose, California Polytechnic State University; Richard G. Rose, West Valley College; Stephen T. Ross, University of Southern Mississippi; Robert E. Roth, Ohio State University; Dorna Sakurai, Santa Monica College; Arthur N. Samel, Bowling Green State University; Shamili Sandiford, College of DuPage; Floyd Sanford, Coe College; David Satterthwaite, I.E.E.D., London; Stephen W. Sawyer, University of Maryland; Arnold Schecter, State University of New York; Frank Schiavo, San Jose State University; William H. Schlesinger, Ecological Society of America; Stephen H. Schneider, National Center for Atmospheric Research; Clarence A. Schoenfeld, University of Wisconsin–Madison; Madeline Schreiber, Virginia Polytechnic Institute; Henry A. Schroeder, Dartmouth Medical School; Lauren A. Schroeder, Youngstown State University; Norman B. Schwartz, University of Delaware; George Sessions, Sierra College; David J. Severn, Clement Associates; Don Sheets, Gardner-Webb University; Paul Shepard, Pitzer College and Claremont Graduate School; Michael P. Shields, Southern Illinois University at Carbondale; Kenneth Shiovitz; F. Siewert, Ball State University; E. K. Silbergold, Environmental Defense Fund; Joseph L. Simon, University of South Florida; William E. Sloey, University of Wisconsin–Oshkosh; Robert L. Smith, West Virginia University; Val Smith, University of Kansas; Howard M. Smolkin, U.S. Environmental Protection Agency; Patricia M. Sparks, Glassboro State College; John E. Stanley, University of Virginia; Mel Stanley, California State Polytechnic University, Pomona; Richard Stevens, Monroe Community College; Norman R. Stewart, University of Wisconsin–Milwaukee; Frank E. Studnicka, University of Wisconsin–Platteville; Chris Tarp, Contra Costa College; Roger E. Thibault, Bowling Green State University; Nathan E. Thomas, University of South Dakota; William L. Thomas, California State University–Hayward; Shari Turney, copy editor; John D. Usis, Youngstown State University; Tinco E. A. van Hylckama, Texas Tech University; Robert R. Van Kirk, Humboldt State University; Donald E. Van Meter, Ball State University; Rick Van Schoik, San Diego State University; Gary Varner, Texas A&M University; John D. Vitek, Oklahoma State University; Harry A. Wagner, Victoria College; Lee B. Waian, Saddleback College; Warren C. Walker, Stephen F. Austin State University; Thomas D. Warner, South Dakota State University; Kenneth E. F. Watt, University of California, Davis; Alvin M. Weinberg, Institute of Energy Analysis, Oak Ridge Associated Universities; Brian Weiss; Margery Weitkamp, James Monroe High School (Granada Hills, California); Anthony Weston, State University of New York at Stony Brook; Raymond White, San Francisco City College; Douglas Wickum, University of Wisconsin–Stout; Charles G. Wilber, Colorado State University; Nancy Lee Wilkinson, San Francisco State University; John C. Williams, College of San Mateo; Ray Williams, Rio Hondo College; Roberta Williams, University of Nevada–Las Vegas; Samuel J. Williamson, New York University; Dwina Willis, Freed-Hardeman University; Ted L. Willrich, Oregon State University; James Winsor, Pennsylvania State University; Fred Witzig, University of Minnesota at Duluth; Martha Wolfe, Elizabethtown Community and Technical College; George M. Woodwell, Woods Hole Research Center; Peggy J. Wright, Columbia College; Todd Yetter, University of the Cumberlands; Robert Yoerg, Belmont Hills Hospital; Hideo Yonenaka, San Francisco State University; Brenda Young, Daemen College; Anita Závodská, Barry University; Malcolm J. Zwolinski, University of Arizona.

G. TYLER MILLER

G. Tyler Miller has written 62 textbooks for introductory courses in environmental science, basic ecology, energy, and environmental chemistry. Since 1975, Miller's books have been the most widely used textbooks for environmental science in the United States and throughout the world. They have been used by almost 3 million students and have been translated into eight languages.

Miller has a professional background in chemistry, physics, and ecology. He has a PhD from the University of Virginia and has received two honorary doctoral degrees for his contributions to environmental education. He taught college for 20 years, developed one of the nation's first environmental studies programs, and developed an innovative interdisciplinary undergraduate science program before deciding to write environmental science textbooks full time in 1975. Currently, he is the president of Earth Education and Research, devoted to improving environmental education.

He describes his hopes for the future as follows:

If I had to pick a time to be alive, it would be the next 75 years. Why? First, there is overwhelming scientific evidence that we are in the process of seriously degrading our own life-support system. In other words, we are living unsustainably. Second, within your lifetime we have the opportunity to learn how to live more sustainably by working with the rest of nature, as described in this book.

I am fortunate to have three smart, talented, and wonderful sons—Greg, David, and Bill. I am especially privileged to have Kathleen as my wife, best friend, and research associate. It is inspiring to have a brilliant, beautiful (inside and out), and strong woman who cares deeply about nature as a lifemate. She is my hero. I dedicate this book to her and to the earth.

SCOTT E. SPOOLMAN

Scott Spoolman is a writer and textbook editor with more than 30 years of experience in educational publishing. He has worked with Tyler Miller since 2003 as a contributing editor and lately as coauthor of *Living in the Environment, Environmental Science,* and *Sustaining the Earth.* With Norman Myers, he coauthored *Environmental Issues and Solutions: A Modular Approach.*

Spoolman holds a master's degree in science journalism from the University of Minnesota. He has authored numerous articles in the fields of science, environmental engineering, politics, and business. He worked as an acquisitions editor on a series of college forestry textbooks. He has also worked as a consulting editor in the development of over 70 college and high school textbooks in fields of the natural and social sciences.

In his free time, he enjoys exploring the forests and waters of his native Wisconsin along with his family—his wife, environmental educator Gail Martinelli, and his children, Will and Katie.

Spoolman has the following to say about his collaboration with Tyler Miller:

I am honored to be working with Tyler Miller as a coauthor to continue the Miller tradition of thorough, clear, and engaging writing about the vast and complex field of environmental science. I share Tyler Miller's passion for ensuring that these textbooks and their multimedia supplements will be valuable tools for students and instructors. To that end, we strive to introduce this interdisciplinary field in ways that will be informative and sobering, but also tantalizing and motivational.

If the flip side of any problem is indeed an opportunity, then this truly is one of the most exciting times in history for students to start an environmental career. Environmental problems are numerous, serious, and daunting, but their possible solutions generate exciting new career opportunities. We place high priorities on inspiring students with these possibilities, challenging them to maintain a scientific focus, pointing them toward rewarding and fulfilling careers, and in doing so, working to help sustain life on the earth.

My Environmental Journey — G. Tyler Miller

My environmental journey began in 1966 when I heard a lecture on population and pollution problems by Dean Cowie, a biophysicist with the U.S. Geological Survey. It changed my life. I told him that if even half of what he said was valid, I would feel ethically obligated to spend the rest of my career teaching and writing to help students learn about the basics of environmental science. After spending six months studying the environmental literature, I concluded that he had greatly underestimated the seriousness of these problems.

I developed an undergraduate environmental studies program and in 1971 published my first introductory environmental science book, an interdisciplinary study of the connections between energy laws (thermodynamics), chemistry, and ecology. In 1975, I published the first edition of *Living in the Environment*. Since then, I have completed multiple editions of this textbook, and of three others derived from it, along with other books.

Beginning in 1985, I spent ten years in the deep woods living in an adapted school bus that I used as an environmental science laboratory and writing environmental science textbooks. I evaluated the use of passive solar energy design to heat the structure; buried earth tubes to bring in air cooled by the earth (geothermal cooling) at a cost of about $1 per summer; set up active and passive systems to provide hot water; installed an energy-efficient instant hot water heater powered by LPG; installed energy-efficient windows and appliances and a composting (waterless) toilet; employed biological pest control; composted food wastes; used natural planting (no grass or lawnmowers); gardened organically; and experimented with a host of other potential solutions to major environmental problems that we face.

I also used this time to learn and think about how nature works by studying the plants and animals around me. My experience from living in nature is reflected in much of the material in this book. It also helped me to develop the six simple principles of sustainability that serve as the integrating theme for this textbook and to apply these principles to living my life more sustainably.

I came out of the woods in 1995 to learn about how to live more sustainably in an urban setting where most people live. Since then, I have lived in two urban villages, one in a small town and one within a large metropolitan area.

Since 1970, my goal has been to use a car as little as possible. Since I work at home, I have a "low-pollute commute" from my bedroom to a chair and a laptop computer. I usually take one airplane trip a year to visit my sister and my publisher.

As you will learn in this book, life involves a series of environmental trade-offs. Like most people, I still have a large environmental impact, but I continue to struggle to reduce it. I hope you will join me in striving to live more sustainably and sharing what you learn with others. It is not always easy, but it sure is fun.

Cengage Learning's Commitment to Sustainable Practices

We the authors of this textbook and Cengage Learning, the publisher, are committed to making the publishing process as sustainable as possible. This involves four basic strategies:

Using sustainably produced paper. The book publishing industry is committed to increasing the use of recycled fibers, and Cengage Learning is always looking for ways to increase this content. Cengage Learning works with paper suppliers to maximize the use of paper that contains only wood fibers that are certified as sustainably produced, from the growing and cutting of trees all the way through paper production.

Reducing resources used per book. The publisher has an ongoing program to reduce the amount of wood pulp, virgin fibers, and other materials that go into each sheet of paper used. New, specially designed printing presses also reduce the amount of scrap paper produced per book.

Recycling. Printers recycle the scrap paper that is produced as part of the printing process. Cengage Learning also recycles waste cardboard from shipping cartons, along with other materials used in the publishing process.

Process improvements. In years past, publishing has involved using a great deal of paper and ink for the writing and editing of manuscripts, copyediting, reviewing page proofs, and creating illustrations. Almost all of these materials are now saved through use of electronic files. Very little paper and ink were used in the preparation of this textbook.

Students who can begin early in their lives to think of things as connected, even if they revise their views every year, have begun the life of learning.

Mark Van Doren

Why Is It Important to Study Environmental Science?

Welcome to **environmental science**—an *interdisciplinary* study of how the earth works, how we interact with the earth, and how we can deal with the environmental problems we face. Because environmental issues affect every part of your life, the concepts, information, and issues discussed in this book and the course you are taking will be useful to you now and throughout your life.

Understandably, we are biased, but *we strongly believe that environmental science is the single most important course that you could take.* What could be more important than learning about the earth's life-support system, how our choices and activities affect it, and how we can reduce our growing environmental impact? Evidence indicates strongly that we will have to learn to live more sustainably by reducing our degradation of the planet's life-support system. We hope this book will inspire you to become involved in this change in the way we view and treat the earth, which sustains us, our economies, and all other living things.

You Can Improve Your Study and Learning Skills

Maximizing your ability to learn involves trying to *improve your study and learning skills.* Here are some suggestions for doing so:

Develop a passion for learning. A passion for learning will serve you well while studying environmental science and in whatever career you choose.

Get organized. Planning is a key life skill.

Make daily to-do lists. Put items in order of importance, focus on the most important tasks, and assign a time to work on these items. Shift your schedule as needed to accomplish the most important items.

Set up a study routine in a distraction-free environment. Develop a daily study schedule and stick to it. Study in a quiet, well-lit space. Take breaks every hour or so. During each break, take several deep breaths and move around; this will help you to stay more alert and focused.

Avoid procrastination. Do not fall behind on your reading and other assignments. Set aside a particular time for studying each day and make it a part of your daily routine.

Make molehills out of mountains. It can be difficult to read an entire chapter or book, write a paper, or cram for a test within a short period of time. Instead, break these large tasks (mountains) down into a series of small tasks (molehills). Each day, read a few pages of the assigned book or chapter, write a few paragraphs of the paper, and review what you have studied and learned.

Ask and answer questions as you read. For example, "What is the main point of a particular subsection or paragraph?" Relate your own questions to the key questions and key concepts addressed in each major chapter section.

Focus on key terms. Use the glossary in your textbook to look up the meaning of terms or words you do not understand. This book shows all key terms in **bold** type and lesser, but still important, terms in *italicized* type. The *Chapter Review* questions at the end of each chapter also include the chapter's key terms in bold. Flash cards for testing your mastery of key terms for each chapter are available within MindTap, or you can make your own.

Interact with what you read. You could mark key sentences and paragraphs with a highlighter or pen or with asterisks and notes in the margin or electronically if you are using MindTap (which may be synced with an Evernote account). You might also mark important pages that you want to return to by using Post-it notes or by folding down page corners.

Review to reinforce learning. Before each class session, review the material you learned in the previous session and read the assigned material.

Become a good note taker. Learn to write down the main points and key information from any lecture. Review, fill in, and organize your notes as soon as possible after each class.

Check what you have learned. At the end of each chapter, you will find review questions that cover all of the key material in each chapter section. We suggest that you try to answer each of these questions after studying each chapter section. Waiting to do this for the entire chapter after you complete it can be overwhelming.

Write out answers to questions to focus and reinforce learning. Write down your answers to the critical thinking questions found in the *Thinking About* boxes throughout the chapters, in many figure captions, and at the end of each chapter. These questions are designed to inspire you to think critically about key ideas and connect them to other ideas and to your own life. Also, write down your answers to all chapter-ending review questions. Mind-Tap has additional questions for each chapter. Save your answers for review and test preparation.

Use the buddy system. Study with a friend or become a member of a study group to compare notes, review material, and prepare for tests. Explaining something to someone else is a great way to focus your thoughts and reinforce your learning. Attend any review sessions offered by instructors or teaching assistants.

Learn your instructor's test style. Does your instructor emphasize multiple-choice, fill-in-the-blank, true-or-false, factual, or essay questions? How much of the test will come from the textbook and how much from lecture material? Adapt your learning and studying methods to this style.

Become a good test taker. Avoid cramming. Eat well and get plenty of sleep before a test. Arrive on time or early. Calm yourself and increase your oxygen intake by taking several deep breaths. (Do this also about every 10–15 minutes while taking the test.) Look over the test and answer the questions you know well first. Then work on the harder ones. Use the process of elimination to narrow down the choices for multiple-choice questions. For essay questions, organize your thoughts before you start writing. If you have no idea what a question means, make an educated guess. You might earn some partial credit and avoid getting a zero. Another strategy for getting some credit is to show your knowledge and reasoning by writing something like this: "If this question means so and so, then my answer is _____."

Develop an optimistic but realistic outlook. Try to be a "glass is half-full" rather than a "glass is half-empty" person. Pessimism, fear, anxiety, and excessive worrying (especially over things you cannot control) are destructive and lead to inaction.

Take time to enjoy life. Every day, take time to laugh and enjoy nature, beauty, and friendship.

You Can Improve Your Critical Thinking Skills

Critical thinking involves developing skills to analyze information and ideas, judge their validity, and make decisions. Critical thinking helps you to distinguish between facts and opinions, evaluate evidence and arguments, and take and defend informed positions on issues. It also helps you to integrate information, to see relationships, and to apply your knowledge to dealing with various problems and decisions. Here are some basic skills for learning how to think more critically.

Question everything and everybody. Be skeptical, as any good scientist is. Do not believe everything you hear and read, including the content of this textbook, without evaluating the information you receive. Seek other sources and opinions.

Identify and evaluate your personal biases and beliefs. Each of us has biases and beliefs taught to us by our parents, teachers, friends, role models, and our own experience. What are your basic beliefs, values, and biases? Where did they come from? What assumptions are they based on? How sure are you that your beliefs, values, and assumptions are right and why? According to the American psychologist and philosopher William James, "A great many people think they are thinking when they are merely rearranging their prejudices."

Be open-minded and flexible. Be open to considering different points of view. Suspend judgment until you gather more evidence, and be willing to change your mind. Recognize that there may be a number of useful and acceptable solutions to a problem, and that very few issues are either black or white. Try to take the viewpoints of those you disagree with in order to better understand their thinking. There are trade-offs involved in dealing with any environmental issue, as you will learn in this book.

Be humble about what you know. Some people are so confident in what they know that they stop thinking and questioning. To paraphrase American writer Mark Twain, "It's what we know is true, but just ain't so, that hurts us."

Find out how the information related to an issue was obtained. Are the statements you heard or read based on firsthand knowledge and research or on hearsay? Are unnamed sources used? Is the information based on reproducible and widely accepted scientific studies or on preliminary scientific results that may be valid but need further testing? Is the information based on a few isolated stories or experiences or on carefully controlled studies that have been reviewed by experts in the field involved? Is it based on unsubstantiated and dubious scientific information or beliefs?

Question the evidence and conclusions presented. What are the conclusions or claims based on the information you're considering? What evidence is presented to support them? Does the evidence support them? Is there a need to gather more evidence to test the conclusions? Are there other, more reasonable conclusions?

Try to uncover differences in basic beliefs and assumptions. On the surface, most arguments or disagreements involve differences of opinion about the validity or meaning of certain facts or conclusions. Scratch a little deeper and you will find that many disagreements are based on different (and often hidden) basic assumptions concerning how we look at and interpret the world around us. Uncovering these basic differences can allow the parties involved to understand one another's viewpoints and to agree to disagree about their basic assumptions, beliefs, or principles.

Try to identify and assess any motives on the part of those presenting evidence and drawing conclusions. What is their expertise in this area? Do they have any unstated assumptions, beliefs, biases, or values? Do they have a personal agenda? Can they benefit financially or politically from acceptance of their evidence and conclusions? Would investigators with different basic assumptions or beliefs take the same data and come to different conclusions?

Expect and tolerate uncertainty. Recognize that scientists cannot establish absolute proof or certainty about anything. However, the results of science have varying degrees of certainty.

Check the arguments you hear and read for logical fallacies and debating tricks. Here are six of many examples of such debating tricks. *First,* attack the presenter of an argument rather than the argument itself. *Second,* appeal to emotion rather than facts and logic. *Third,* claim that if one piece of evidence or one conclusion is false, then all other related pieces of evidence and conclusions are false. *Fourth,* say that a conclusion is false because it has not been scientifically proven. (Scientists never prove anything absolutely, but they can often establish high degrees of certainty.) *Fifth,* inject irrelevant or misleading information to divert attention from important points. *Sixth,* present only either/or alternatives when there may be a number of options.

Do not believe everything you read on the Internet. The Internet is a wonderful and easily accessible source of information that includes alternative explanations and opinions on almost any subject or issue—much of it not available in the mainstream media and scholarly articles. Blogs of all sorts have become a major source of information, more important than standard news media for some people. However, because the Internet is so open, anyone can post anything they want to some blogs and other websites with no editorial control or review by experts. As a result, evaluating information on the Internet is one of the best ways to put into practice the principles of critical thinking discussed here. Use and enjoy the Internet, but think critically and proceed with caution.

Develop principles or rules for evaluating evidence. Develop a written list of principles to serve as guidelines for evaluating evidence and claims. Continually evaluate and modify this list on the basis of your experience.

Become a seeker of wisdom, not a vessel of information. Many people believe that the main goal of their education is to learn as much as they can by gathering more and more information. We believe that the primary goal is to learn how to sift through mountains of facts and ideas to find the few *nuggets of wisdom* that are the most useful for understanding the world and for making decisions. This book is full of facts and numbers, but they are useful only to the extent that they lead to an understanding of key ideas, concepts, connections, and scientific laws and theories. The major goals of the study of environmental science are to find out how nature works and sustains itself *(environmental wisdom)* and to use *principles of environmental wisdom* to help make human societies and economies more sustainable, more just, and more beneficial and enjoyable for all. As writer Sandra Carey observed, "Never mistake knowledge for wisdom. One helps you make a living; the other helps you make a life."

To help you practice critical thinking, we have supplied questions throughout this book, found within each chapter in brief boxes labeled *Thinking About,* in the captions of many figures, and at the end of each chapter. There are no right or wrong answers to many of these questions. A good way to improve your critical thinking skills is to compare your answers with those of your classmates and to discuss how you arrived at your answers.

Use the Learning Tools We Offer in This Book

We have included a number of tools throughout this textbook that are intended to help you improve your learning skills and apply them. First, consider the *Key Concepts* list at the beginning of each chapter section. You can use these to preview a chapter and to review the material after you've read it.

Next, note that we use three different special notations throughout the text. Each chapter opens with a **Core Case Study**, and each time we tie material within the chapter back to this core case, we note it in bold, colored type as we did in this sentence. You will also see two icons appearing regularly in the text margins. When you see the *sustainability* icon, you will know that you have just read something that relates directly to the overarching theme of this text, summarized by our six **principles of sustainability** which are introduced in Figures 1.2, p. 6, and 1.5, p. 9, and summarized in Supplement 7 (pp. S50–S51). The *Good News* icon appears near each of many examples of successes that people have had in dealing with the environmental challenges we face.

GOOD NEWS

We also include several brief *Connections* boxes to show you some of the often surprising connections between environmental problems or processes and some of the products and services we use every day or some of the activities we partake in. These, along with the *Thinking About* boxes scattered throughout the text (both designated by the *Consider This* heading), are intended to get you to think carefully about the activities and choices we take for granted, and about how they might affect the environment.

At the end of each chapter, we list what we consider to be the *three big ideas* that you should take away from the chapter. Following that list in each chapter is a *Tying It All Together* box. This feature quickly reviews the Core Case Study and how key chapter material relates to it, and it explains how the principles of sustainability can be applied to deal with challenges discussed in the Core Case Study and throughout the chapter.

Finally, we have included a *Chapter Review* section at the end of each chapter, with questions listed for each chapter section. These questions cover all of the key material and key terms in each chapter. In each chapter, they are followed by *Critical Thinking* questions that help you to apply chapter material to the real world and to your own life; a *Doing Environmental Science* exercise to help you experience the work of scientists; a *Global Environment Watch* exercise, in which you can use the exciting GREENR online global environmental database; and a *Data Analysis* or *Ecological Footprint Analysis* exercise to help you learn how to interpret and use scientific research data.

Know Your Own Learning Style

People have different ways of learning and it can be helpful to know your own learning style. *Visual learners* learn best from reading and viewing illustrations and diagrams. *Auditory learners* learn best by listening and discussing. They might benefit from reading aloud while studying and using a tape recorder in lectures for study and review. *Logical learners* learn best by using concepts and logic to uncover and understand a subject rather than relying mostly on memory.

This book and the related MindTap contain plenty of tools for all types of learners. Visual learners can benefit from the animations and videos in MindTap that support many of the concepts presented. In addition, features such as an easy-to-use note-taking feature and flash cards help you learn important terms and concepts. This is a highly visual book with many photographs and diagrams carefully selected to illustrate important ideas, concepts, and processes. Auditory learners can make use of our *Read-Speaker app* in MindTap, which can read the chapter aloud in various voices and speeds. Additionally, podcasts featuring interviews of National Geographic Explorers and grantees add context to many environmental issues. For logical learners, the book is organized by key concepts that are revisited throughout any chapter and related carefully to other concepts, major principles, and case studies and other examples. We urge you to become aware of your own learning style and make the most of these various tools.

This Book Presents a Positive, Realistic Environmental Vision of the Future

Our goal is to present a positive vision of our environmental future based on realistic optimism. To do so, we strive not only to present the facts about environmental issues, but also to give a balanced presentation of different viewpoints. We consider the advantages and disadvantages of various technologies and proposed solutions to environmental problems. We argue that environmental solutions usually require *trade-offs* among opposing parties, and that the best solutions are *win-win* solutions. And we present the good news as well as the bad news about efforts to deal with environmental problems.

One cannot study a subject as important and complex as environmental science without forming conclusions, opinions, and beliefs. However, we argue that any such results should be based on use of critical thinking to evaluate conflicting positions and to understand the trade-offs involved in most environmental solutions. To that end, we emphasize critical thinking throughout this textbook, and we encourage you to develop a practice of thinking critically about everything you read and hear, both in school, and throughout your life.

Help Us Improve This Book

Researching and writing a book that covers and connects the numerous major concepts from the wide variety of environmental science disciplines is a challenging and exciting task. Almost every day, we learn about some new connection in nature. However, in a book this complex, there are bound to be some errors—some typographical mistakes that slip through and some statements that you might question, based on your knowledge and research. We invite you to contact us to correct any errors you find, point out any bias you see, and suggest ways to improve this book. Please e-mail your suggestions to Tyler Miller at mtg89@hotmail.com or Scott Spoolman at spoolman@tds.net.

Now start your journey into this fascinating and important study of how the earth's life-support system works and how we can leave our planet in a condition at least as good as what we now enjoy. Have fun.

Online Learning Solutions and Resources for Students

You have a large variety of electronic and other supplemental materials available to you to help you take your learning experience beyond this textbook:

MindTap Environmental Science. MindTap is a new approach to highly personalized online learning. Beyond an e-book, homework solution, digital supplement, or premium website, MindTap is a digital learning platform that works alongside your campus Learning Management System (LMS) to deliver course curriculum across the range of electronic devices in your life. MindTap is built on an "app" model allowing enhanced digital collaboration and delivery of engaging content across a spectrum of Cengage and non-Cengage resources.

Aplia for Environmental Science. Aplia™ is an online interactive learning solution that helps you improve comprehension—and your grade—by integrating a variety of mediums and tools such as videos, tutorials, practice tests, and interactive e-books. Aplia provides automatically graded assignments with detailed, immediate feedback on every question, and innovative teaching materials. More than 2 million students like you have used Aplia at over 1,800 institutions. Aplia should be purchased only when assigned by your instructor as part of your course.

Global Environment Watch. Updated several times a day, the Global Environment Watch is a focused portal into GREENR—the Global Reference on the Environment, Energy, and Natural Resources—an ideal one-stop site for classroom discussion and research projects. This resource center keeps courses up to date with the most current news on the environment. Users get access to information from trusted academic journals, news outlets, and magazines, as well as statistics, an interactive world map, videos, primary sources, case studies, podcasts, and much more. Log in or purchase access at www.cengagebrain.com to complete the exercises found at the end of each chapter. Links to GREENR for in-text activities are also provided via MindTap.

Virtual Field Trips in Environmental Issues. *Virtual Field Trips in Environmental Issues* brings the field to you, with dynamic panoramas, videos, photographs, maps, and quizzes covering important topics within environmental science. A case study approach covers the issues of *keystone species, climate change's role in extinctions, invasive species, the evolution of a species due to its environment,* and *an ecosystem approach to sustaining biodiversity.* Engage and interact with these real issues to help you think critically about the world.

ENVIRONMENTAL SCIENCE

FIFTEENTH EDITION

1

ENVIRONMENTAL PROBLEMS, THEIR CAUSES, AND SUSTAINABILITY

No civilization has survived
the ongoing destruction of
its natural support system.
Nor will ours.

1.3 Why do we have environmental problems?

1.4 What is an environmentally sustainable society?

Forests such as this one in California's
Sequoia National Park help to sustain
all life and economies.

Robert Harding World Imagery/Alamy

Sustainability is the capacity of the earth's natural systems and human cultural systems to survive, flourish, and adapt to changing environmental conditions into the very long-term future. It is the overarching theme of this textbook.

Since the mid-1980s, there has been a boom in environmental awareness on college campuses and in public and private schools around the world. In the United States, hundreds of colleges and universities have now taken the lead in a quest to become more sustainable and to educate their students about sustainability.

For example, at Oberlin College in Ohio, a group of students worked with faculty members and architects to design a more sustainable environmental studies building (Figure 1.1) powered by solar panels, which produce 30% more electricity than the building uses. Closed-loop underground geothermal wells provide heating and cooling. In its solar greenhouse, a series of open tanks populated by plants and other organisms purifies the building's wastewater. The building collects rainwater for irrigating the surrounding grasses, gardens, and meadow, which contain a diversity of plant and animal species.

Berea College in Kentucky boasts an innovative environmental science curriculum including a Sustainable Appalachian Communities course. The school also features its Ecovillage, a 50-unit experimental residence complex that uses passive solar heating, solar panels, and filtered rainwater.

At the University of California, Santa Cruz, in 2012, students reused, recycled, or composted more than 70% of their solid waste with a goal of reaching 100% by 2020. And in Ashville, North Carolina, Warren Wilson College gets more than a third of its food from regional farms, including its own large on-campus organic garden.

In addition to making campuses greener, colleges are increasingly offering environmental sustainability courses and programs. At Pfeiffer University, many students have accompanied Professor Luke Dollar, a National Geographic Emerging Explorer, on trips to Madagascar to take part in his research on that country's endangered species and ecosystems. At the University of Wisconsin–Madison, the Nelson Institute for Environmental Studies seeks to integrate sustainability content throughout the academic departments, as well as to serve communities outside of the university.

These and many other institutions are educating students who will provide leadership in helping us to make our societies and economies more sustainable during the next few decades. Maybe you will join the ranks of such environmental leaders.

Robb Williamson/NREL

FIGURE 1.1 The Adam Joseph Lewis Center for Environmental Studies at Oberlin College in Oberlin, Ohio.

1.1 WHAT ARE SOME PRINCIPLES OF SUSTAINABILITY?

CONCEPT 1.1A Life on the earth has been sustained for billions of years by solar energy, biodiversity, and chemical cycling.

CONCEPT 1.1B Our lives and economies depend on energy from the sun and on natural resources and ecosystem services *(natural capital)* provided by the earth.

CONCEPT 1.1C We could shift toward living more sustainably by applying full-cost pricing, searching for win-win solutions, and committing to preserving the earth's life-support system for future generations.

Environmental Science Is a Study of Our Interactions with the World

The **environment** is everything around us. It includes the living and the nonliving things (air, water, and energy) with which we interact in a complex web of relationships that connect us to one another and to the world we live in. Despite our many scientific and technological advances, we are utterly dependent on the earth for clean air and water, food, shelter, energy, fertile soil, and all other components of the planet's *life-support system.*

This textbook is an introduction to **environmental science**, an *interdisciplinary* study of how humans interact with the living and nonliving parts of their environment. It integrates information and ideas from the *natural sciences* such as biology, chemistry, and geology; the *social sciences* such as geography, economics, and political science; and the *humanities* such as ethics. The three goals of environmental science are **(1)** to learn how life on the earth has survived and thrived, **(2)** to understand how we interact with the environment, and **(3)** to find ways to deal with environmental problems and live more sustainably.

A key component of environmental science is **ecology**, the biological science that studies how living things interact with one another and with their environment. These living things are called **organisms**. Each organism belongs to a **species**, a group of organisms that has a unique set of characteristics that distinguish it from other groups of organisms.

A major focus of ecology is the study of ecosystems. An **ecosystem** is a set of organisms within a defined area of land or volume of water that interact with one another and with their environment of nonliving matter and energy. For example, a forest ecosystem consists of plants (especially trees; see chapter-opening photo), animals, and various other organisms that decompose organic materials, all interacting with one another, with solar energy, and with the chemicals in the forest's air, water, and soil.

We should not confuse environmental science and ecology with **environmentalism**, a social movement dedicated to trying to sustain the earth's life-support systems for all forms of life. Environmentalism is practiced more in the political and ethical arenas than in the realm of science. Environmentalism and environmental science are both being practiced vigorously on many college and university campuses (**Core Case Study**).

Three Scientific Principles of Sustainability

How has the incredible variety of life on the earth been sustained for at least 3.8 billion years in the face of catastrophic changes in environmental conditions? Such changes included gigantic meteorites impacting the earth, ice ages lasting for hundreds of millions of years, and long warming periods during which melting ice raised sea levels by hundreds of feet.

The latest version of our species has been around for only about 200,000 years—less than the blink of an eye, relative to the 3.8 billion years that life has existed on the planet (see the Geologic and Biological Time Scale in Supplement 6, p. S49). Yet, there is mounting scientific evidence that, as we have expanded into and dominated almost all of the earth's ecosystems during that short time, and especially since 1900, we have seriously degraded these natural systems that support all species, including our own, and our economies.

Our science-based research leads us to believe that three major natural factors have played the key roles in the long-term sustainability of life on this planet, as summarized below and in Figure 1.2 (**Concept 1.1A**). We use these three **scientific principles of sustainability**, or *lessons from nature*, throughout the book to suggest how we might move toward a more sustainable future.

- **Dependence on solar energy:** The sun's input of energy, called **solar energy**, warms the planet and provides energy that plants use to produce **nutrients**, the chemicals necessary for their own life processes and for those of most other animals, including humans. The sun also powers *indirect forms of solar energy* such as wind and flowing water, which we use to produce electricity.

- **Biodiversity:** The variety of genes, organisms, species, and ecosystems in which organisms exist and interact are referred to as **biodiversity** (short for *biological diversity*). The interactions among species, especially the feeding relationships, provide vital ecosystem services and keep any population from growing too large. Biodiversity also provides countless ways for life to adapt to changing environmental conditions, even catastrophic changes that wipe out large numbers of species.

- **Chemical cycling:** The circulation of chemicals necessary for life from the environment (mostly from soil and water) through organisms and back to the environment is called **chemical cycling**, or **nutrient**

Solar Energy

FIGURE 1.2 Three **scientific principles of sustainability** based on how nature has sustained a huge variety of life on the earth for 3.8 billion years, despite drastic changes in environmental conditions (**Concept 1.1A**).

Chemical Cycling

Biodiversity

© Cengage Learning

cycling. The earth receives a continuous supply of energy from the sun, but it receives no new supplies of life-supporting chemicals. Thus through their complex interactions with their living and nonliving environment, organisms must continually recycle the chemicals they need in order to survive. This means that there is little waste in nature, other than in the human world, because the wastes and decayed bodies of any organism become nutrients or raw materials for other organisms. In nature,

waste = useful resources

Ecology and environmental science reveal that *interdependence, not independence, is what sustains life* and allows it to adapt to a continually changing set of environmental conditions. Many environmental scientists argue that understanding this interdependence is the key to learning how to live more sustainably.

Sustainability Has Certain Key Components

Sustainability, the central integrating theme of this book, has several critical components that we use as subthemes. One such component is **natural capital**—the natural resources and ecosystem services that keep us and other species alive and support human economies (Figure 1.3).

Natural resources are materials and energy in nature that are essential or useful to humans. They are often classified as *inexhaustible resources* (such as energy from the sun and wind), *renewable resources* (such as air, water, topsoil, plants, and animals) or *nonrenewable* or *depletable resources* (such as copper, oil, and coal). **Ecosystem services** are processes provided by healthy ecosystems that support life and human economies at no monetary cost to us. Examples include purification of air and water, renewal of topsoil, nutrient cycling, pollination, and pest control.

Natural Capital

Natural Capital = Natural Resources + Ecosystem Services

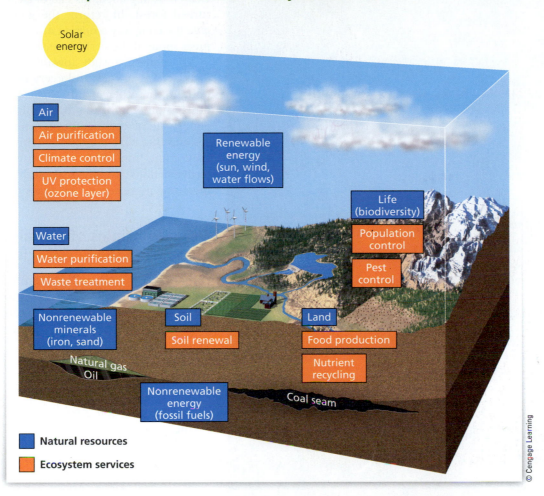

FIGURE 1.3 Natural capital consists of natural resources (blue) and ecosystem services (orange) that support and sustain the earth's life and human economies (**Concept 1.1B**).

One essential ecosystem service is chemical, or nutrient, cycling—the basis for one of the three **scientific principles of sustainability** (Figure 1.2). Chemical cycling helps to turn wastes into resources. An important component of nutrient cycling is *topsoil*—a vital natural resource that provides us and most other land-dwelling species with food. Without nutrient cycling in topsoil, life as we know it could not exist on the earth's land.

Natural capital is also supported by energy from the sun—the focus of another of the **scientific principles of sustainability** (Figure 1.2). Thus, our lives and economies depend on energy from the sun, and on natural resources and ecosystem services *(natural capital)* provided by the earth (**Concept 1.1B**).

A second component of sustainability—and another subtheme of this text—is to recognize that many human activities can *degrade natural capital* by using normally renewable resources such as trees and topsoil faster than

nature can restore them and by overloading the earth's normally renewable air and water systems with pollution and wastes. For example, in some parts of the world, we are replacing diverse and naturally sustainable forests (Figure 1.4) with crop plantations that can be sustained only with large inputs of water, fertilizer, and pesticides. We are also adding harmful chemicals and wastes to some rivers, lakes, and oceans faster than these bodies of water can cleanse themselves through natural processes. In addition, we are disrupting the nutrient cycles that support life because many of the plastics and other synthetic materials that we have created cannot be broken down and used as nutrients by other organisms.

This leads us to a third component of sustainability: *solutions*. While environmental scientists search for scientific solutions to problems such as the degradation of forests and other forms of natural capital, social scientists are looking for economic and political solutions. For example,

The clearing of vast areas of forest is an example of natural capital degradation.

John Lee/Aurora Photos

FIGURE 1.4 Small remaining area of once diverse Amazon rain forest surrounded by vast soybean fields in the Brazilian state of Mato Grosso.

a scientific solution to the problems of depletion of forests is to stop burning or cutting down biologically diverse, mature forests (Figure 1.4). A scientific solution to the problem of pollution of rivers is to prevent the excessive dumping of harmful chemicals and wastes into streams and to allow them to recover naturally. However, to implement such solutions, governments often have to enact and enforce environmental laws and regulations.

The search for solutions often involves conflicts. For example, when a scientist argues for protecting a long-undisturbed forest to help preserve its important biodiversity, the timber company that had planned to harvest the trees in that forest might protest. Dealing with such conflicts often involves making *trade-offs*, or compromises—

another component of sustainability. For example, the timber company might be persuaded to plant and harvest trees in an area that it had already cleared or degraded, instead of clearing the undisturbed forest. In return, the government might give the company a *subsidy*, or financial support, to meet some of the costs for planting the trees.

In making a shift toward sustainability, the daily actions of each and every individual are important. In other words, *individuals matter*—another subtheme of this book. History shows that almost all of the significant changes in human systems that have improved environmental quality have come from the bottom up, through the collective actions of individuals and from individuals inventing more sustainable ways of doing things.

Other Principles of Sustainability Come from the Social Sciences

Our study of environmental problems, proposed solutions, and trade-offs has led us to propose three **social science principles of sustainability** (Figure 1.5), derived from studies of economics, political science, and ethics:

- **Full-cost pricing** (from economics): Many economists urge us to find ways to include the harmful environmental and health costs of producing and using goods and services in their market prices—a practice called **full-cost pricing**. This would give consumers better information about the environmental impacts of their lifestyles, and it would allow them to make more informed choices about the goods and services they use.

- **Win-win solutions** (from political science): We can learn to work together in dealing with environmental problems by recognizing our interdependent connections with others and with our life-support system. This means shifting from a *win-lose* approach based on competition and dominance of other humans and of ecosystems to *win-win* solutions that are based on compromise in light of our interdependence and that benefit both people and the environment.

- **A responsibility to future generations** (from ethics): We should leave the planet's life-support systems in at least as good a condition as that which we now enjoy, if not better, for future generations.

Other researchers have proposed additional sustainability principles, but we believe that our six **principles of sustainability** (Figure 1.2, Figure 1.5, and Supplement 7, pp. S50–S51) can serve as key guidelines for helping us live more sustainably.

Resources Are Inexhaustible, Renewable, or Nonrenewable

A **resource** is anything that we can obtain from the environment to meet our needs and wants. Some resources, such as surface water, trees, and edible wild plants, are directly available for use. Other resources, such as petroleum, minerals, wind, and underground water, become useful to us only with some effort and technological ingenuity.

Resources can be classified as inexhaustible, renewable, or nonrenewable (exhaustible) (Figure 1.6). Solar energy is called an **inexhaustible resource** because its continuous supply is expected to last for at least 6 billion years until the sun dies. It also provides us with inexhaustible wind and flowing water that we use to produce electricity. A **renewable resource** is one that can be replenished by natural processes within hours to centuries, as long as we do not use it up faster than natural processes can renew it. Examples include forests, grasslands, fishes, fertile topsoil, clean air, and freshwater. The highest rate at which we can use a renewable resource indefinitely without reducing its available supply is called its **sustainable yield**.

FIGURE 1.5 Three **social science principles of sustainability** can help us make a transition to a more environmentally and economically sustainable future.

Inexhaustible
Solar energy
Wind energy
Geothermal energy

Renewable
Trees
Topsoil
Freshwater

Nonrenewable (Exhaustible)
Fossil fuels (oil, natural gas, coal)
Iron and copper

FIGURE 1.6 We depend on a combination of inexhaustible, renewable, and exhaustible (nonrenewable) natural resources.

Nonrenewable or exhaustible resources exist in a fixed quantity, or *stock*, in the earth's crust. On a time scale of millions to billions of years, geologic processes can renew such resources. However, on the much shorter human time scale, we can deplete these resources much faster than nature can form them. Such exhaustible stocks include *energy resources* such as oil and coal, *metallic mineral resources* such as copper and aluminum, and *nonmetallic mineral resources* such as salt and sand. As we deplete such resources, human ingenuity can often find substitutes. However, sometimes there is no acceptable or affordable substitute for a widely used nonrenewable resource.

Countries Differ in Resource Use and Environmental Impact

The United Nations (UN) classifies the world's countries as economically more developed or less developed, based primarily on their average income per person. **More-developed countries**—industrialized nations with high average income per person—have 17% of the world's population and include the United States, Japan, Canada, Australia, Germany, and most other European countries.

All other nations, in which 83% of the world's people live, are classified as **less-developed countries**, most of them in Africa, Asia, and Latin America. Some are *middle-income, moderately developed countries* such as China, India, Brazil, Thailand, and Mexico. Others are *low-income, least-developed countries* including Nigeria, Bangladesh, Congo, and Haiti. For a map of less-developed and more-developed countries see Figure 3, p. S18, in Supplement 4.

1.2 **HOW ARE OUR ECOLOGICAL FOOTPRINTS AFFECTING THE EARTH?**

CONCEPT 1.2 As our ecological footprints grow, we are depleting and degrading more of the earth's natural capital.

We Are Living Unsustainably

According to a large and growing body of scientific evidence, we are living unsustainably by wasting, depleting, and degrading the earth's natural capital (Figure 1.3)—a process known as **environmental degradation**, summarized in Figure 1.7. Scientists also refer to this as **natural capital degradation**.

In many parts of the world, renewable forests are shrinking (Figure 1.4), deserts are expanding, and topsoil is eroding. In addition, the lower atmosphere is warming, floating ice and some glaciers are melting at unexpected rates, sea levels are rising, ocean acidity is increasing, and floods, droughts, severe weather, and forest fires are more frequent in some areas. In a number of regions, rivers are running dry, harvests of many species of fish are dropping sharply, and coral reefs are dying. Species are becoming extinct at least 100 times faster than in pre-human times, and extinction rates are projected to increase by at least another 100-fold during this century.

In 2005, the UN released its *Millennium Ecosystem Assessment*, a 4-year study by 1,360 experts from 95 countries.

FIGURE 1.7 Natural capital degradation: Examples of the degradation of normally renewable natural resources and natural services (Figure 1.3) in parts of the world, mostly as a result of growing populations and rising rates of resource use per person.

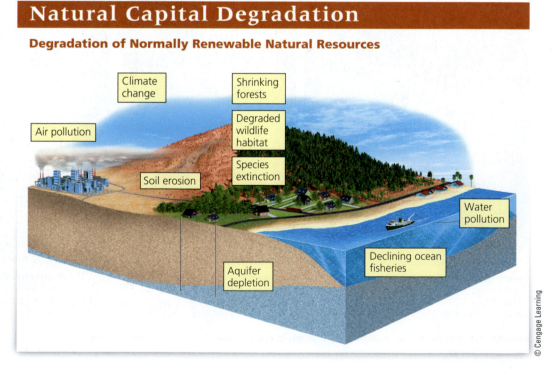

Natural Capital Degradation

Degradation of Normally Renewable Natural Resources

Climate change
Shrinking forests
Air pollution
Degraded wildlife habitat
Soil erosion
Species extinction
Water pollution
Aquifer depletion
Declining ocean fisheries

© Cengage Learning

According to this study, human activities have degraded or overused about 60% of the earth's ecosystem services (Figure 1.3, orange boxes), mostly since 1950. The report's summary statement warned that "human activity is putting such a strain on the natural functions of Earth that the ability of the planet's ecosystems to sustain future generations can no longer be taken for granted." The report also concluded that we have scientific, economic, and political solutions to these problems that we could implement within a few decades, as you will learn in reading this book. **GOOD NEWS**

Pollution Comes from a Number of Sources

One major environmental problem is **pollution**, which is contamination of the environment by any chemical or other agent such as noise or heat to a level that is harmful to the health, survival, or activities of humans or other organisms. Polluting substances, or *pollutants*, can enter the environment naturally, such as from volcanic eruptions, or through human activities, such as the burning of coal and gasoline, and the dumping of chemicals into rivers, lakes, and oceans. At a high enough concentration in the air, in water, or in our bodies, almost any chemical can cause harm and be classified as a pollutant.

The pollutants we produce come from two types of sources. **Point sources** are single, identifiable sources. Examples are the smokestack of a coal-burning power or industrial plant (Figure 1.8), the drainpipe of a factory, and the exhaust pipe of an automobile. **Nonpoint sources** are

dispersed and often difficult to identify. Examples are pesticides and particles of topsoil blown from the land into the air and the runoff of fertilizers, pesticides, and trash from the land into streams and lakes (Figure 1.9). It is much easier and less costly to identify and control or prevent pollution from point sources than from widely dispersed nonpoint sources.

We have tried to deal with pollution in two very different ways. One approach is **pollution cleanup**, which involves cleaning up or diluting pollutants after we have produced them. For example, we can use tall smokestacks (Figure 1.8) to dilute and reduce the local effects of air pollutants. However, while tall smokestacks can reduce local air pollution they can increase air pollution in downwind areas.

The other approach is **pollution prevention**, efforts focused on greatly reducing or eliminating the production of pollutants. For example, we can enact pollution control laws that ban, or set low levels for, the emission of various pollutants into the atmosphere or into bodies of water.

We Are Degrading Commonly Shared Renewable Resources: The Tragedy of the Commons

Some renewable resources, known as *open-access renewable resources*, are not owned by anyone and can be used by almost anyone. Examples are the atmosphere, the open ocean and its fishes, and the earth's life-support system. Other examples of less open, but often *shared resources*, are

FIGURE 1.8 *Point-source air pollution* from smokestacks in a coal-burning industrial plant.

FIGURE 1.9 The trash in this river came from a large area of land and is an example of *nonpoint-source water pollution*.

grasslands, forests, and streams. Many of these renewable resources have been environmentally degraded. In 1968, biologist Garrett Hardin (1915–2003) called such degradation the *tragedy of the commons*.

Degradation of such shared or open-access renewable resources occurs because each user reasons, "The little bit that I use or pollute is not enough to matter, and anyway, it's a renewable resource." When the number of users is small, this logic works. Eventually, however, the cumulative effect of large numbers of people trying to exploit a widely available or shared resource can degrade it and eventually exhaust or ruin it. Then no one can benefit from it. That is the tragedy.

Our Ecological Footprints Are Growing

When people use renewable resources, it can result in natural capital degradation (Figure 1.7), pollution, and wastes. We can think of this harmful environmental impact as an **ecological footprint**—the amount of land and water needed to supply a population or an area with renewable resources and to absorb and recycle the wastes and pollution produced by such resource use. The **per capita**

ecological footprint is the average ecological footprint of an individual in a given country or area. Figure 1.10 shows the human ecological impact in different parts of the world. (See Figure 6 in Supplement 4, p. S21, for the human ecological footprint in North America.)

If the total ecological footprint for a city, a country, or the world is larger than its *biological capacity* to replenish its renewable resources and absorb the resulting wastes and pollution, it is said to have an *ecological deficit*. In other words, its people are living unsustainably by depleting natural capital instead of living off the renewable supply of resources and ecosystem services provided by such capital. Globally we are running up a huge ecological deficit, as shown by the map in Figure 4, p. S19, in Supplement 4. According to the 2013 *Living Planet Report* from the World Wide Fund for Nature (WWF), we would need 1.5 planet Earths to indefinitely sustain the world's 2012 rate of total resource use.

A relatively new school of thought, led by architect William McDonough and scientist Michael Braungart, takes a slightly different approach to the notion of human environmental impacts and how to reduce them. Called *upcycling*, this approach acknowledges our growing eco-

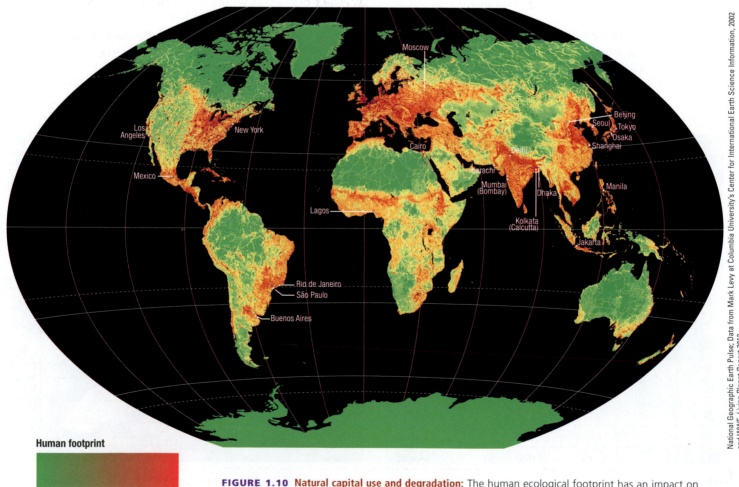

National Geographic Earth Pulse; Data from Mark Levy at Columbia University's Center for International Earth Science Information, 2002 and WWF, *Living Planet Report 2012.*

Human footprint

Least impact   Most impact

FIGURE 1.10 Natural capital use and degradation: The human ecological footprint has an impact on about 83% of the earth's total land surface.

logical footprints and the need to reduce them, but it stresses that in pursuing our activities and in creating products and services, we need to consider how we can improve some aspect of the environment while we serve our needs and wants. In other words, the goals of sustainability will be served by our working to *create and expand our beneficial environmental impact*. This is in keeping with the win-win **principle of sustainability**, and it is being demonstrated on many college and university campuses (**Core Case Study**).

Throughout this book we discuss ways to use existing and emerging technologies and economic tools to reduce the size of our ecological footprints, while increasing our beneficial environmental impact.

IPAT Is Another Environmental Impact Model

In the early 1970s, scientists Paul Ehrlich and John Holdren developed a simple model showing how *population size* (P), *affluence* (A), or wealth, as measured by rates of resource consumption per person, and the beneficial and harmful environmental effects of *technologies* (T) help to determine the *environmental impact* (I) of human activities. We can summarize this model by the simple equation:

Impact (I) = Population (P) × Affluence (A) × Technology (T)

Table 1.1 shows the relative importance of these three factors in selected high-, middle-, and low-income countries. See Figure 3, p. S18, in Supplement 4 for a map comparing the world's high-, middle-, and low-income countries. While the ecological footprint model emphasizes the use of renewable resources, the IPAT model includes the environmental impact of using both renewable and nonrenewable resources.

Some forms of technology such as polluting factories, gas-guzzling motor vehicles, and coal-burning power plants increase environmental impact by raising the harmful T factor. But other technologies reduce environmental impact by decreasing the T factor. Examples are pollution control and prevention technologies, fuel-efficient cars, and wind turbines and solar cells that generate electricity with a low environmental impact. These and newer technologies to come can help us reduce our ecological footprints and expand our beneficial environmental impact.

1.3 WHY DO WE HAVE ENVIRONMENTAL PROBLEMS?

CONCEPT 1.3A Major causes of environmental problems are population growth, unsustainable resource use, poverty, avoidance of full-cost pricing, and increasing isolation from nature.

CONCEPT 1.3B Our environmental worldviews play a key role in determining whether we live unsustainably or more sustainably.

TABLE 1.1 Environmental Impact of Selected High-, Middle-, and Low-Income Countries

Country	Population Size	Population Growth Rate	Resource Use Per Person	Use of Harmful Technology	Use of Beneficial Technology	Overall Environmental Impact
High-Income Countries						
United States	316 million	Moderate (0.5%)	Very high	Moderate	High	High
Japan	128 million	Negative (−0.2%)	High	Moderate	High	Moderate
Germany	82 million	Negative (−0.2%)	High	Moderate	High	Moderate
Middle-Income Countries						
China	1.35 billion	Moderate (0.5%)	Low	High	Moderate	High
India	1.26 billion	High (1.5%)	Low	High	Low	High
Brazil	194 million	Moderate (1%)	Low	High	Moderate	Moderate
Low-Income Countries						
Nigeria	402 million	High (2.6%)	Very low	High	Low	Moderate
Bangladesh	228 million	High (1.6%)	Very low	High	Low	Moderate
Congo	194 million	High (2.8%)	Very low	High	Low	Moderate

Compiled by the authors using data from Population Reference Bureau, Global Footprint Network, World Wide Fund for Nature, and Earth Policy Institute.

Causes of Environmental Problems

Population growth

Unsustainable resource use

Poverty

Excluding environmental costs from market prices

Increasing isolation from nature

© Cengage Learning

FIGURE 1.11 Environmental and social scientists have identified five basic causes of the environmental problems we face (**Concept 1.3A**). *Question:* For each of these causes, what are two environmental problems that result?

Left: Jeremy Richards/Shutterstock. Left center: steve estvanik/Shutterstock.com. Center: Lucian Coman/Shutterstock.com. Right center: El Greco/Shutterstock.com. Right: CREATISTA/Shutterstock.

Experts Have Identified Several Causes of Environmental Problems

According to a number of environmental and social scientists, the major causes of the environmental problems we face are **(1)** population growth, **(2)** wasteful and unsustainable resource use, **(3)** poverty, **(4)** failure to include the harmful environmental and health costs of goods and services in their market prices, and **(5)** increasing isolation from nature (**Concept 1.3A**) (Figure 1.11). We discuss each of these causes in detail in later chapters of this book. Let us begin with a brief overview of them.

The Human Population Is Growing at a Rapid Rate

Exponential growth occurs when a quantity such as the human population increases at a fixed percentage per unit of time, such as 0.5% or 2% per year. Exponential growth starts off slowly. But after only a few doublings, it grows to enormous numbers because each doubling is twice the total of all earlier growth.

For an example of the awesome power of exponential growth, consider a simple form of bacterial reproduction in which one bacterium splits into two every 20 minutes. Starting with one bacterium, after 20 minutes, there would be two; after an hour, there would be eight; ten hours later, there would be more than 1,000, and after just 36 hours (assuming that nothing interfered with their reproduction), there would be enough bacteria to form a layer 0.3 meters (1 foot) deep over the entire earth's surface.

The human population has grown exponentially (Figure 1.12), and in 2013, the rate of growth was 1.2%. As a result, there are now about 7.1 billion people with about 85 million more people added each year. There could be 9.7 billion of us by 2050. This projected addition of 2.6 billion more people within your lifetime is more than eight times the current U.S. population and almost twice that of China, the world's most populous nation.

CONSIDER THIS ...

CONNECTIONS Exponential Growth and Doubling Time: The Rule of 70

The doubling time of the human population or of any growing quantity can be calculated by using the rule of 70: doubling time (years) = 70/annual growth rate (%). The world's population is growing at about 1.2% per year. At this rate how long will it take to double its size?

No one knows how many people the earth can support indefinitely, or at what level of average resource consumption per person, without seriously degrading the planet's life-support system. However, our large and expanding ecological footprints (Figure 1.10 and Figure 6, p. S21, in Supplement 4) and the resulting widespread natural capital degradation are disturbing warning signs. Some scientists argue that we could control such severe degradation by slowing population growth with the goal of having it level off at around 8 billion by 2050. We examine the possible ways to do this in Chapter 6.

Affluence Has Harmful and Beneficial Environmental Effects

The lifestyles of the world's expanding population of consumers are built on growing affluence as more people achieve higher incomes. This results in higher levels of total and per capita resource consumption along with more environmental degradation, waste, and pollution.

These results can be dramatic. The WWF has estimated that the United States is responsible for almost half of the global ecological footprint. The average American consumes about 30 times the amount of resources that the average Indian consumes and 100 times the amount con-

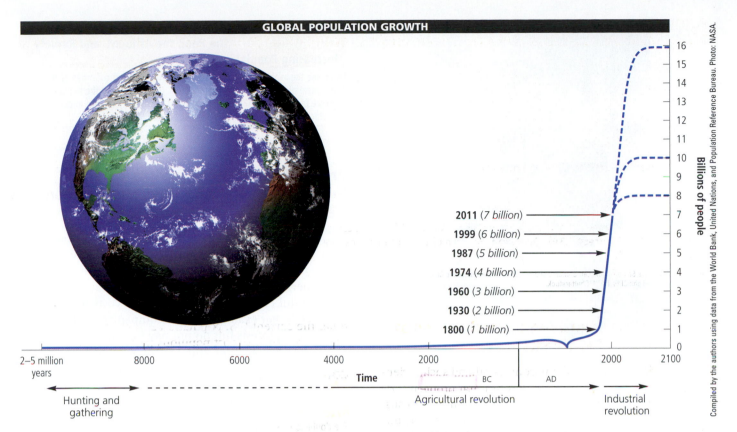

Compiled by the authors using data from the World Bank, United Nations, and Population Reference Bureau. Photo: NASA.

FIGURE 1.12 *Exponential growth:* The J-shaped curve represents past exponential world population growth, with projections to 2100 showing possible population stabilization as the J-shaped curve of growth changes to an S-shaped curve. (This figure is not to scale.)

sumed by the average person in the world's poorest countries. As a result, the WWF projected that we would need 5 planet Earths to indefinitely sustain the rate of resource use of the average American.

The problem is that providing such resources contributes to air pollution and water pollution from factories and motor vehicles and land degradation from the mining of raw materials used to make the products we consume. Another downside to wealth is that it allows affluent consumers to obtain their resources from almost anywhere in the world without seeing the harmful effects of their high-consumption lifestyles.

On the other hand, affluence can allow for more widespread and better education, which can lead people to become more concerned about environmental quality. Affluence can also make more money available for developing technologies to reduce pollution, environmental degradation, and resource waste along with other ways to increase our beneficial environmental impact.

As a result, in the United States and most other affluent countries, the air is clearer, drinking water is purer, and most rivers and lakes are cleaner than they were in the 1970s. In addition, the food supply is more abundant and safer, the incidence of life-threatening infectious diseases has been greatly reduced, life spans are longer, and some endangered species are being rescued

GOOD NEWS

from extinction hastened by human activities. These improvements were largely financed by affluence.

Poverty Can Have Harmful Environmental and Health Effects

Poverty is a condition in which people are unable to fulfill their basic needs for adequate food, water, shelter, health care, and education. According to the World Bank, about 900 million people—almost three times the U.S. population—live in *extreme poverty,* struggling to live on the equivalent of less than $1.25 a day, which is less than what many people spend for a bottle of water or a cup of coffee. About one of every three, or 2.6 billion, of the world's people struggles to live on less than $2.25 a day. Could you do this?

Poverty can cause a number of harmful environmental and health effects. The daily lives of the world's poorest people are focused on getting enough food, water, and cooking and heating fuel to survive. Desperate for short-term survival, these individuals do not have the luxury of worrying about long-term environmental quality or sustainability. Thus, collectively, they can degrade forests, topsoil, and grasslands, and deplete fisheries and wildlife populations in order to stay alive.

However, poverty does not necessarily lead to environmental degradation. Some of the world's poor people have

FIGURE 1.13 One of every three children younger than age 5 in less-developed countries, such as this starving child in Bangladesh, suffers from severe malnutrition caused by a lack of calories and protein.

learned how to increase their beneficial environmental impact by planting and nurturing trees and conserving the soils that they depend on, as a part of their long-term survival strategy.

CONSIDER THIS. . .

CONNECTIONS Poverty and Population Growth

To many poor people, having more children is a matter of survival. Their children help them gather firewood, haul water, and tend crops and livestock. The children also help to care for their aging parents, most of whom do not have social security, health care, and retirement funds. This daily struggle for survival is largely why populations in some of the poorest countries continue to grow at high rates.

Environmental degradation can have severe health effects on poor people. One such problem is life-threatening *malnutrition*, a lack of protein and other nutrients needed for good health (Figure 1.13), resulting from environmental degradation that interferes with food production.

Another such effect is illness caused by limited access to adequate sanitation facilities and clean drinking water. About one of every eight of the world's people get water for drinking, washing, and cooking from sources polluted by human and animal feces. According to the World Health Organization, air pollution also threatens millions of people in less-developed countries. Outdoor air pollution causes more than 1 million deaths per year in China, and worldwide, about 4.3 million people are killed every year by indoor air pollution, mostly smoke from open fires or poorly vented stoves used for heating and cooking.

CONSIDER THIS. . .

THINKING ABOUT The Poor, the Affluent, and Rapidly Increasing Population Growth

Some see the rapid population growth in less-developed countries as the primary cause of our environmental problems. Others say that the high rate of resource use per person in more-developed countries is a more important factor. Which factor do you think is more important? Why?

Prices of Goods and Services Rarely Include Their Harmful Environmental and Health Costs

Another basic cause of environmental problems has to do with how goods and services are priced in the marketplace. Companies using resources to provide goods for consumers generally are not required to pay for most of the harmful environmental and health costs of supplying such goods. For example, timber companies pay the cost of clear-cutting forests but do not pay for the resulting environmental degradation and loss of wildlife habitat. The primary goal of a company is to maximize profits for its owners or stockholders, so it is not inclined to add these costs to its prices voluntarily. Because the prices of goods and services do not include most of their harmful environmental and health costs, consumers and decision makers have no effective way to evaluate these harmful effects.

Another problem can arise when governments (taxpayers) give companies *subsidies* such as tax breaks and payments to assist them with using resources to run their businesses. This helps to create jobs and stimulate economies, but environmentally harmful subsidies encourage the depletion and degradation of natural capital. (See the online Guest Essay by Norman Myers on this topic.)

We could live more sustainably and increase our beneficial environmental impact by finding ways to include in market prices the harmful environmental and health costs of the goods and services that we use. Such full-cost pricing is the basis for one of the three **social science principles of sustainability**. Two ways to do this over the next two decades would be to shift from environmentally harmful government subsidies to environmentally beneficial subsidies, and to tax pollution and waste heavily while reducing taxes on income and wealth. We discuss such *subsidy shifts* and *tax shifts* in Chapter 17.

We Are Increasingly Isolated from Nature

Today, more than half of the world's people (and three out of four people in the more-developed countries) live in urban areas, and this shift from rural to urban living is continuing at a rapid pace. Artificial urban environments and the increasing use of cell phones, computers, and other electronic devices are isolating more and more people, especially children, from the natural world. Some ar-

REBECCA HALE/National Geographic Creative

Background photo: djgis/Shutterstock.com

Juan Martinez—Working to Reconnect People with Nature

National Geographic Emerging Explorer Juan Martinez learned first-hand about the value of connecting with nature, and now he spends his time working to pass along his experience and knowledge of that value to others, particularly to disadvantaged youths.

Martinez grew up in a poor area of Los Angeles, California, where as a boy he was in danger of becoming absorbed by a gang culture. One of his teachers recognized Martinez's potential and gave him a chance to pass a class that he was failing by joining the school's Eco Club. Martinez took that opportunity and when the club planned a field trip to see the Grand Teton Mountains of Wyoming, he jumped at the chance. As a result, he says, "I still can't find words to describe the first moment I saw those mountains rising up from the valley. Watching bison, seeing a sky full of stars, and hiking through that scenery was overwhelming."

The experience transformed Martinez's life. Today, he spearheads the Natural Leaders Network of the Children & Nature Network, an organization creating links between environmental organizations, corporations, government, education, and individuals to reconnect children with nature. His work as an environmental leader has inspired many others to do similar work.

Martinez has received a great deal of recognition for his efforts, including invitations to White House forums on environmental education. His greatest reward, however, is in seeing how his efforts help others. He notes: "Some kids on my trips have been in foster care their whole lives, feeling very disconnected from other people and nature. Suddenly they're out in the backcountry relying on each other. Nature can be a real facilitator for skills that are so crucial in life—communicating, working together, and realizing you can do things you never thought you could."

gue that this has led to a phenomenon known as *nature deficit disorder* (see About the Cover on page ii.)

Thus, it is not surprising that many people do not know the full story of where their food, water, and other goods come from. Similarly, many people are unaware of the amounts of wastes and pollutants they produce, where these wastes and pollutants go, and how they affect the environment. Some analysts ask: How will we live more sustainably by shrinking our ecological footprints and expanding our beneficial environmental impact, if we do not appreciate the beauty and importance of nature and understand that we are utterly dependent on the earth's natural systems and the natural capital they provide? Some environmental leaders are focusing on this problem (Individuals Matter 1.1).

People Have Different Views about Environmental Problems and Their Solutions

People differ over the nature and seriousness of the world's environmental problems and over what we should do to help solve them, and these disagreements arise mostly because of differing environmental worldviews. Your **environmental worldview** is your set of assumptions and values reflecting how you think the world works and what you think your role in the world should be. **Environmental ethics**, the study of varying beliefs about what is right and wrong with how we treat the environment, provides useful tools for examining worldviews. For example, here are some important *ethical questions* relating to the environment:

- Why should we care about the environment?

- Are we the most important species on the planet or are we just another one of the earth's millions of different forms of life?

- Do we have an obligation to see that our activities do not cause the extinction of other species? If so, should we try to protect all species or only some? How do we decide which to protect?

- Do we have an ethical obligation to pass the natural world on to future generations in a condition that is as good as or better than what we inherited?

- Should every person be entitled to equal protection from environmental hazards regardless of race, gender, age, national origin, income, social class, or any other factor? (This is the central ethical and political issue for what is known as the *environmental justice* movement; see Chapter 17 and the online Guest Essay by Robert D. Bullard for more on this topic.)

- Should we seek to live more sustainably, and if so, how?

People with widely differing environmental worldviews can take the same data, be logically consistent with it, and arrive at quite different answers to such questions because they start with different assumptions and moral, ethical, or religious beliefs. Environmental worldviews are discussed in detail in Chapter 17, but here is a brief introduction.

There are three major categories of environmental worldviews: human-centered, life-centered, and earth-centered. A **human-centered environmental worldview** sees the natural world primarily as a support system for human life. One such worldview, the *planetary management worldview,* holds that humans are separate from and in charge of nature and that we can manage the earth mostly for our benefit, into the distant future. Another is the *stewardship worldview,* which holds that we can and should manage the earth for our benefit, but that we have an ethical responsibility to be caring and responsible managers, or *stewards,* of the earth.

According to the **life-centered environmental worldview**, all species have value as participating members of the biosphere, regardless of their potential or actual use to humans. Most people with a life-centered worldview believe we have an ethical responsibility to avoid hastening the extinction of species through our activities.

The **earth-centered environmental worldview** holds that we are part of, and dependent on, nature and that the earth's life-support system exists for all species, not just for us. According to this view, our economic success and the long-term survival of our cultures and our species depend on learning how life on the earth has sustained itself for billions of years and integrating such lessons from nature into the ways we think and act.

The Rise of Environmental Conservation and Protection in the United States

European colonists arriving in the early 1600s viewed North America as a continent with seemingly inexhaustible resources and as a wilderness to be conquered and managed for human use. As the colonists spread across the continent, they cleared forests for cropland and settlements, plowed up grasslands to plant crops, and mined gold, lead, and other minerals.

© Cengage Learning

FIGURE 1.14 As leader of the preservationist movement, John Muir (1838–1914) called for setting aside some of the country's public lands as protected wilderness, an idea that was not enacted into law until 1964. Muir also proposed creating a national park system on public lands and was largely responsible for establishing Yosemite National Park in 1890. In 1892 he founded the Sierra Club, which is to this day a political force working on behalf of the environment.

In 1864, George Perkins Marsh, a scientist and member of Congress from Vermont, questioned the idea that the country's resources were inexhaustible. He also used scientific studies and case studies to show how the rise and fall of past civilizations were linked to the use and misuse of their soils, water supplies, and other resources. Thus he was one of the founders of the U.S. conservation movement.

Early in the 20th century, this movement split into two factions with differing views over how public lands should be used. The *preservationist school,* led by naturalist John Muir (Figure 1.14), wanted wilderness areas on some public lands to be left untouched. The *conservationist school,* led by Teddy Roosevelt (Figure 1.15) and Gifford Pinchot, the first chief of the U.S. Forest Service, believed all public lands should be managed wisely and scientifically, primarily to provide resources for people.

U.S. conservation efforts continued to straddle these two schools of thought, and Aldo Leopold (Figure 1.16), wildlife manager, professor, writer, and conservationist, personified this dual approach. Trained in the conservation school, he eventually shifted toward the preservation school. He became a pioneer in forestry, soil conservation, wildlife ecology, and wilderness preservation. In 1935, he helped to found the U.S. Wilderness Society. Largely through his writings, especially his 1949 book *A Sand County Almanac,* he laid the groundwork for the field of environmental ethics. He contended that the role of the human species should be to protect nature, not to conquer it.

Later in the 20th century, it became necessary to broaden the concept of resource conservation to include

FIGURE 1.15 Effective protection of forests and wildlife on federal lands did not begin until Theodore (Teddy) Roosevelt (1858–1919) became president. His term of office, 1901–1909, has been called the country's *Golden Age of Conservation*. He established 36 national wildlife reserves and more than tripled the size of the national forest reserves.

preservation of the *quality* of the planet's air, water, soil, and wildlife. One prominent pioneer in that effort was biologist Rachel Carson (Figure 1.17), who in 1962 wrote *Silent Spring,* which documented the pollution of air, water, and wildlife from the use of harmful pesticides such as DDT. This influential book heightened public awareness of

FIGURE 1.16 Aldo Leopold (1887–1948) became a leading conservationist and his book, *A Sand County Almanac,* is considered an environmental classic that helped to inspire the modern conservation and environmental movements. With more than 350 books and articles published, he has been called "probably the most quoted voice in the history of conservation."

FIGURE 1.17 Rachel Carson (1907–1964) alerted us to the harmful effects of the widespread use of pesticides. Many environmental historians mark Carson's wake-up call as the beginning of the modern environmental movement in the United States.

pollution problems, which led to the regulation of dangerous pesticides.

Between 1940 and 1970, the United States underwent rapid economic growth and industrialization. The by-products were increasing pollution of air and water and growing mounds of solid and hazardous wastes. Air pollution got so bad in many industrial cities that drivers had to use their car headlights during the daytime, and many thousands died each year from the harmful effects of air pollution. A stretch of the Cuyahoga River running through Cleveland, Ohio, was so polluted with oil and other flammable pollutants that it caught fire several times. And there was a devastating oil spill off the California coast in 1969. In addition, well-known wildlife species such as the American bald eagle, the grizzly bear, the whooping crane, and the peregrine falcon became endangered.

Growing publicity over these problems led the American public to demand government action. When the first *Earth Day* was held on April 20, 1970, some 20 million people in more than 2,000 U.S. communities and college and university campuses (**Core Case Study**) took to the streets to demand improvements in environmental quality. This led to the establishment of the Environmental Protection Agency (EPA) in 1970 and to passage of most of the U.S. environmental laws now in place during the 1970s—known as the *decade of the environment.*

In the 1980s there was a backlash against environmental laws and regulations, led by some corporate leaders and members of Congress who argued that environmental laws were hindering economic growth. Since 1990, leaders and supporters of the environmental movement have had to spend much of their time and resources fighting efforts to discredit the movement and to weaken or eliminate most environmental laws passed during the 1970s.

Since 1970, many grassroots environmental organizations have sprung up to help deal with environmental threats. Interest in environmental issues grew on many college and university campuses, resulting in the expansion of environmental studies courses and programs (**Core Case Study**). In addition, there came a growing awareness of critical, complex, and largely invisible environmental issues, including losses in biodiversity, aquifer depletion, ocean pollution and acidification, and atmospheric warming and the threat of climate disruption.

In the 1970s, the United States led the world in environmental awareness, wildlife conservation, and environmental protection. Some analysts point to this fact and call for new U.S. leadership in dealing with these larger and less visible long-term threats and in finding paths toward more environmentally sustainable societies and economies.

1.4 WHAT IS AN ENVIRONMENTALLY SUSTAINABLE SOCIETY?

CONCEPT 1.4 Living sustainably means living off the earth's natural income without depleting or degrading the natural capital that supplies it.

Environmentally Sustainable Societies Protect Natural Capital and Live Off Its Income

According to most environmental scientists, our ultimate goal should be to achieve an **environmentally sustainable society**—one that meets the current and future basic resource needs of its people in a just and equitable manner without compromising the ability of future generations to meet their basic resource needs. This is in keeping with the future generations **principle of sustainability**.

Imagine that you win $1 million in a lottery. Suppose you invest this money (your capital) and earn 10% interest per year. If you live on just the interest, or the income made by your capital, you will have a sustainable annual income of $100,000 that you can spend each year indefinitely without depleting your capital. However, if you consistently spend more than your income, you will deplete your capital. Even if you spend just $110,000 per year while still allowing the interest to accumulate, your money will be gone within 18 years.

The lesson here is an old one: *Protect your capital and live on the income it provides.* Deplete or waste your capital and you will move from a sustainable to an unsustainable lifestyle.

The same lesson applies to our use of the earth's natural capital (Figure 1.3)—the global trust fund of free natural resources and ecosystem services that nature has provided for us, for future generations, and for the earth's other spe-

cies. *Living sustainably* means living on **natural income**, the renewable resources such as plants, animals, soil, clean air, and clean water, provided by the earth's natural capital. By working to preserve and replenish the earth's natural capital, which supplies this income, we can find the best ways to reduce our ecological footprints while expanding our beneficial environmental impact (**Concept 1.4**).

A More Sustainable Future Is Possible

Making a shift toward a more sustainable future will involve some tough challenges. Ecologists note that, given enough time, nature can recover from many of our environmentally harmful impacts. However, natural recovery can take hundreds to thousands of years, while harmful human impacts are expanding exponentially within a time period of 10 to 100 years. Thus, in learning to live more sustainably, time is our most scarce resource.

Here are two pieces of good news: *First,* research by social scientists suggests that it takes only 5–10% of the population of a community, a country, or the world to bring about major social and environmental change. *Second,* such research also shows that such change can occur in a much shorter time than most people think. Anthropologist Margaret Mead summarized our potential for social change: "Never doubt that a small group of thoughtful, committed citizens can change the world. Indeed, it is the only thing that ever has." This is now being demonstrated on many college campuses around the world (**Core Case Study**).

One of our goals in writing this book has been to provide a realistic vision of how we can change the world, for the benefit of us all, as well as for the environment. We base this vision not on immobilizing fear, gloom, and doom, but on energizing and realistic hopes.

BIG IDEAS

- A more sustainable future will require that we rely more on energy from the sun and other renewable energy sources, protect biodiversity through the preservation of natural capital, and avoid disrupting the earth's vitally important chemical cycles.

- A major goal for becoming more sustainable is full-cost pricing—the inclusion of harmful environmental and health costs in the market prices of goods and services.

- We will benefit ourselves and future generations if we commit ourselves to finding win-win solutions to our problems and to leaving the planet's life-support system in a condition as good as or better than what we now enjoy.

The Greening of American Campuses and Sustainability

We opened this chapter with a **Core Case Study** about college students around the United States who are working to address many of the serious environmental problems we face and to improve environmental quality on their campuses. Thousands of students on campuses all over the world are doing the same. Many of these students are learning that a key to most solutions is to apply the three **scientific principles of sustainability** (Figure 1.2) and the three **social science principles of sustainability** (Figure 1.5) to the design of our economic and social systems, and to our individual lifestyles.

We can use such strategies to try to slow the rapidly expanding losses of biodiversity, to sharply reduce our production of wastes and pollution, to switch to more sustainable

Pecold/Shutterstock.com

sources of energy, and to promote more sustainable forms of agriculture and other uses of land and water. We can also use these principles to sharply reduce poverty and slow human population growth.

You and all of your fellow students have the good fortune to be members of the 21st century's *transition generation* that will play a major role in deciding whether humanity creates a more sustainable future or continues on a path toward further environmental degradation and disruption. This means confronting the urgent challenges presented by the major environmental problems discussed in this book. It is an incredibly exciting and challenging time to be alive as we struggle to develop a more sustainable relationship with this planet that is our only home.

Chapter Review

Core Case Study

1. Define **sustainability**, and summarize the story of how many college campuses are working to become more environmentally sustainable.

Section 1.1

2. What are the three key concepts for this section? Define **environment**. Distinguish among **environmental science**, **ecology**, and **environmentalism**. Distinguish between an **organism** and a **species**. What is an **ecosystem**? What are three **scientific principles of sustainability** derived from how the natural world works? What is **solar energy** and why is it important to life on the earth? What is **biodiversity** and why is it important to life on the earth? Define **nutrients**. Define **chemical cycling** (or **nutrient cycling**) and explain why it is important to life on the earth.

3. Define **natural capital**. Define **natural resources** and **ecosystem services**, and give two examples of each. Give three examples of how we are degrading natural capital. Explain how finding solutions to environmental problems involves making trade-offs. Explain why individuals matter in dealing with the environmental problems we face. What are three

social science principles of sustainability? What is **full-cost pricing** and why is it important?

4. What is a **resource**? Distinguish between an **inexhaustible resource** and a **renewable resource** and give an example of each. What is the **sustainable yield** of a renewable resource? Define and give two examples of a **nonrenewable** or **exhaustible resource**. Distinguish between **more-developed countries** and **less-developed countries** and give one example each of a high-income, middle-income, and low-income country.

Section 1.2

5. What is the key concept for this section? Define and give three examples of **environmental degradation (natural capital degradation)**. About what percentage of the earth's natural or ecosystem services have been degraded by human activities? Define **pollution**. Distinguish between **point sources** and **nonpoint sources** of pollution and give an example of each. Distinguish between **pollution cleanup** and **pollution prevention** and give an example of each. What is the tragedy of the commons?

6. What is an **ecological footprint**? What is a **per capita ecological footprint**? Use the ecological footprint concept to explain how we are living unsustainably. What is meant by upcycling? What is the IPAT model for estimating our environmental impact?

Section 1.3

7. What are the two key concepts for this section? Identify five basic causes of the environmental problems that we face. What is **exponential growth**? What is the rule of 70? What is the current size of the human population? About how many people are added each year? How big is the population projected to be in 2050? How do Americans, Indians, and the average people in the poorest countries compare in terms of average resource consumption per person? Summarize the potential environmental harms and environmental benefits of affluence.

8. What is **poverty** and what are three of its harmful environmental and health effects? About what percentage of the world's people struggle to live on the equivalent of $1.25 a day? About what percentage have to live on $2.25 a day? How are poverty and population growth connected? List three major health problems faced by many poor people.

9. Explain how exclusion of the harmful environmental and health costs of production from the prices of goods and services is related to the environmental problems we face. What is the connection between government subsidies, resource use, and environmental degradation? What are two ways to include the harmful environmental and health costs of the goods and services in their market prices? Explain how a lack of knowledge about nature and the importance of natural capital, along with our increasing isolation from nature, can intensify the environmental problems we face. Describe how Juan Martinez is working to reconnect people with nature. What is an **environmental worldview**? What is **environmental ethics**? Distinguish among the **human-centered**, **life-centered**, and **earth-centered worldviews**. Describe the rise of environmental conservation and protection in the United States.

Section 1.4

10. What is the key concept for this section? What is an **environmentally sustainable society**? What is **natural income** and what does it mean to live off of natural income? What are two pieces of good news about making the transition to a more sustainable society? Based on the three **scientific principles of sustainability** and the three **social science principles of sustainability**, what are three important ways to make a transition to sustainability as summarized in this chapter's *three big ideas*? How can we use these six **principles of sustainability** to move toward a more sustainable future?

Note: Key terms are in bold type. Knowing the meanings of these terms will help you in the course you are taking.

Critical Thinking

1. What do you think are the three most environmentally unsustainable components of your lifestyle? List two ways in which you could apply each of the three **scientific principles of sustainability** (Figure 1.2) and each of the three **social science principles of sustainability** (Figure 1.5) to making your lifestyle more environmentally sustainable.

2. What are three ways in which college campuses (**Core Case Study**) are employing upcycling by expanding their beneficial environmental impacts? What are three ways in which you could do this in your daily life?

3. For each of the following actions, state one or more of the three **scientific principles of sustainability** that are involved: **(a)** recycling aluminum cans; **(b)** using a rake instead of a leaf blower; **(c)** walking or bicycling to class instead of driving; **(d)** taking your own reusable bags to a store to carry your purchases home; and **(e)** volunteering to help restore a prairie.

4. Explain why you agree or disagree with the following propositions:
   a. Stabilizing population is not desirable because, without more consumers, economic growth would stop.
   b. The world will never run out of resources because we can use technology to find substitutes and to help us reduce resource waste.
   c. We can shrink our ecological footprints while creating beneficial environmental impacts.

5. When you read that the average American consumes 30 times more resources than the average citizen of India, are you skeptical, indifferent, concerned, or angry about this fact? Can you think of something that you and others could do to address this problem?

6. When you read that at least 19,000 children of ages 5 and younger die each day (13 per minute) from preventable malnutrition and infectious disease, what is your response? Can you think of something that you and others could do to address this problem?

7. Explain why you agree or disagree with each of the following statements: **(a)** humans are superior to other forms of life; **(b)** humans are in charge of the

earth; **(c)** the value of other forms of life depends only on whether they are useful to humans; **(d)** all forms of life have a right to exist; **(e)** all economic growth is good; **(f)** nature has an almost unlimited storehouse of resources for human use; **(g)** technology can solve our environmental problems; **(h)** I don't have any obligation to future generations; and **(i)** I don't have any obligation to other forms of life.

8. What are the basic beliefs within your environmental worldview? Record your answer. Then, at the end of this course, return to your answer to see if your environmental worldview has changed. Are the beliefs included in your environmental worldview consistent with the answers you gave to Question 7 above? Are your actions that affect the environment consistent with your environmental worldview? Explain.

Doing Environmental Science

Estimate your own ecological footprint by using one of the many estimator tools available on the Internet. Is your ecological footprint larger or smaller than you thought it would be, according to this estimate? Why do you think this is so? List three ways in which you could reduce your ecological footprint. Try one of them for a week, and write a report on this change.

Global Environment Watch Exercise

Use the world maps in Figure 1, p. S14, in Supplement 4 and Figure 1.10 to choose one more-developed country and one less-developed country to compare their ecological footprints (found under Quick Facts on the country portal). Click on the ecological footprint number to view a graph of both the ecological footprint and biocapacity of each country. Using those graphs, determine whether these countries are living sustainably or not. What would be some reasons for these trends?

Ecological Footprint Analysis

If the *ecological footprint per person* of a country or the world is larger than its *biological capacity per person* to replenish its renewable resources and absorb the resulting waste products and pollution, the country or the world is said to have an *ecological deficit*. If the reverse is true, the country or the world has an *ecological credit* or *reserve*. Use the data to the right to calculate the ecological deficit or credit for the countries listed. (As an example, this value has been calculated and filled in for World.)

1. Which three countries have the largest ecological deficits? For each of these countries, why do you think it has a deficit?

2. Rank the four countries with ecological credits in order from highest to lowest credit. For each of the four, why do you think it has an ecological credit?

3. Rank all of the countries in order from the largest to the smallest per capita ecological footprint.

Place	Per Capita Ecological Footprint (hectares per person)	Per Capita Biological Capacity (hectares per person)	Ecological Credit (+) or Deficit (−) (hectares per person)
World	2.7	1.8	−0.9
United States	7.2	3.9	
Canada	6.4	14.9	
Mexico	3.3	1.4	
Brazil	2.9	9.6	
South Africa	2.6	1.2	
Saudi Arabia	4.0	0.7	
Israel	4.0	0.3	
Germany	4.6	2.0	
Russia	4.4	6.6	
India	0.9	0.5	
China	2.1	0.9	
Australia	6.7	14.6	

Compiled by the authors using data from World Wide Fund for Nature *Living Planet Report 2012*.

2
SCIENCE, MATTER, ENERGY, AND SYSTEMS

Science is built up of facts, as a house is built of stones; but an accumulation of facts is no more a science than a heap of stones is a house.

HENRI POINCARÉ

2.3 What is energy and what happens when it undergoes change?

2.4 What are systems and how do they respond to change?

Researchers measuring a 3,200-year-old giant sequoia in California's Sequoia National Park.

Michael Nichols/National Geographic Creative

How Do Scientists Learn about Nature? Experimenting with a Forest

Suppose a logging

company plans to cut down all of the trees on a hillside in back of your house. You are very concerned and want to know about the possible harmful environmental effects of this action.

One way to learn about such effects is to conduct a *controlled experiment,* just as environmental scientists do. They begin by identifying key *variables,* such as water loss and soil nutrient content, that might change after the trees are cut down. Then, they set up two groups. One is the *experimental group,* in which a chosen variable is changed in a known way. The other is the *control group,* in which the chosen variable is not changed. They then compare the results from the two groups.

In 1963, botanist F. Herbert Bormann, forest ecologist Gene Likens, and their colleagues began carrying out such a controlled experiment. Their goal was to compare the loss of water and soil nutrients from an area

of uncut forest (the *control site*) with one that had been stripped of its trees (the *experimental site*).

They built V-shaped concrete dams across the creeks at the bottoms of several forested valleys in the Hubbard Brook Experimental Forest in New Hampshire (Figure 2.1). The dams were designed so that all surface water leaving each forested valley had to flow across a dam, where scientists could measure its volume and dissolved nutrient content.

First, the researchers measured the amounts of water and dissolved soil nutrients flowing from an undisturbed forested area in one of the valleys (the control site) (Figure 2.1, left). These measurements showed that an undisturbed mature forest is very efficient at storing water and retaining chemical nutrients in its soils.

Next, they set up an experimental forest area in a nearby valley (Figure 2.1, right). One winter, they cut down all the trees and shrubs in that valley, left them where they fell, and sprayed

the area with herbicides to prevent the regrowth of vegetation. Then, for 3 years, they compared outflow of water and nutrients in this experimental site with those in the control site.

The scientists found that, with no plants to help absorb and retain water, the amount of water flowing out of the deforested valley increased by 30–40%. As this excess water ran rapidly over the ground, it eroded soil and carried dissolved nutrients out of the topsoil in the deforested site. Overall, the loss of key soil nutrients from the experimental forest was six to eight times that in the nearby uncut control forest.

In this chapter, you will learn more about how scientists study nature and about the matter and energy that make up the world within and around us. You will also learn about three *scientific laws,* or rules of nature, that govern the changes that matter and energy undergo. And you will learn the important difference between a scientific hypothesis and a scientific theory.

© Cengage Learning

FIGURE 2.1 This controlled field experiment measured the loss of water and soil nutrients from a forest due to deforestation. The forested valley (left) was the control site; the cutover valley (right) was the experimental site.

CONCEPT 2.1 Scientists collect data and develop hypotheses, theories, and laws about how nature works.

Scientists Use Observations, Experiments, and Models to Answer Questions about How Nature Works

Science is a broad field of study focused on discovering how nature works and using that knowledge to describe what is likely to happen in nature. It is based on the assumption that events in the physical world follow orderly cause-and-effect patterns that can be understood through careful observation, measurements, and experimentation. Figure 2.2 summarizes the scientific process.

In this chapter's **Core Case Study**, experimenters Bormann and Likens carried out this scientific process to find out how clearing forested land can affect its ability to store water and retain soil nutrients. They designed an experiment to collect **data**, or information, to answer their question (Figure 2.1). They then proposed a **scientific hypothesis**—a possible and testable explanation of the data. Bormann and Likens came up with the following hypothesis to explain their data: When a forest is cleared of its vegetation and exposed to rain and melting snow, it retains less water and loses large quantities of soil nutrients. They tested this hypothesis for the soil nutrient nitrogen and then repeated their controlled experiment for phosphorus.

The experimenters wrote scientific articles describing their research and these were evaluated by other scientists in their fields. These reviews and further research by other scientists supported their results and hypothesis. Another way to test projections is to develop a **model**, an approximate representation or simulation of a system. Data from the research carried out by Bormann and Likens and from other scientists' research was fed into such models, which also supported their research.

A well-tested and widely accepted scientific hypothesis or a group of related hypotheses is called a **scientific theory**. The research conducted by Bormann and Likens and other scientists led to the scientific theory that trees and other plants hold soil in place and help it to retain water and nutrients needed to support the plants.

Scientists Are Curious and Skeptical and Demand Evidence

Good scientists are extremely curious about how nature works and are keen observers of what is happening in nature (see Individuals Matter 2.1). Scientists tend to be highly skeptical of new data and hypotheses. They say, "Show me your evidence and explain the reasoning be-

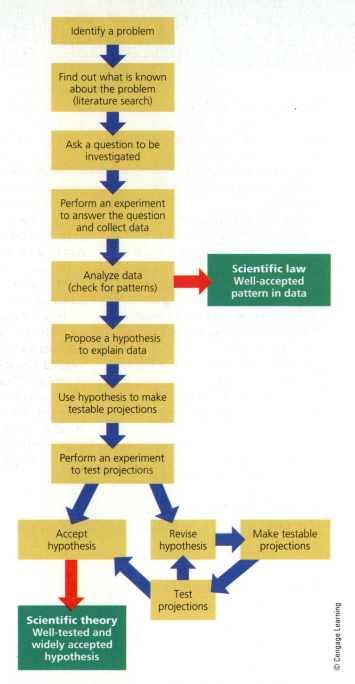

© Cengage Learning

FIGURE 2.2 The general process that scientists use for discovering and testing ideas about how the natural world works.

hind the scientific ideas or hypotheses that you propose to explain your data."

An important part of the scientific process is **peer review**. It involves scientists openly publishing details of the methods they used, the results of their experiments, and the reasoning behind their hypotheses for other scientists working in the same field (their *peers*) to evaluate. Scientific knowledge advances in this self-correcting way, with scientists continually questioning and checking the data and hypotheses of their peers.

Jane Goodall: Chimpanzee Researcher and Protector

Jane Goodall is an ethologist and conservationist with a PhD from Cambridge University. She is also a National Geographic Explorer-in-Residence Emeritus. At age 26, she began a more-than-50-year career of studying chimpanzee social and family life in the Gombe Stream Game Reserve in what is now the African country of Tanzania.

One of her major scientific discoveries was that chimpanzees make and use tools. She observed some chimpanzees modifying twigs or blades of grass and then poking them into termite mounds. When the termites latched on to these primitive tools, the chimpanzees pulled them out and ate the termites. Goodall and several other scientists have also observed that chimpanzees, including captive chimpanzees, can learn simple sign language, do simple arithmetic, play computer games, develop relationships, and worry about and protect one another.

In 1977, she established the Jane Goodall Institute, an organization that works to preserve great ape populations and their habitats. In 1991 Goodall started *Roots & Shoots,* an environmental education program for youth that is active in more than 130 countries. She has received many awards and prizes for her scientific contributions and conservation efforts and has written 27 books for adults and children and has been involved with more than 15 films about the lives and importance of chimpanzees.

Goodall spends nearly 300 days a year traveling and educating people throughout the world about chimpanzees and the need to protect the environment. She says, "I can't slow down. . . . If we're not raising new generations to be better stewards of the environment, what's the point?"

Critical Thinking and Creativity Are Important in Science

Scientists use logical reasoning and critical thinking skills (p. xxiv) to learn about the natural world. Thinking critically involves four important steps:

1. Be skeptical about everything you read or hear.
2. Look at the evidence and evaluate it and any related information, along with inputs and opinions from a variety of reliable sources.
3. Be open to many viewpoints and evaluate each one before coming to a conclusion.
4. Identify and evaluate your personal assumptions, biases, and beliefs, being careful to distinguish between facts and opinions.

Logic and critical thinking are very important tools in science, but imagination, creativity, and intuition are just as vital. According to physicist Albert Einstein, "There is no completely logical way to a new scientific idea."

Scientific Theories and Laws Are the Most Important and Certain Results of Science

We should never take a scientific theory lightly. It has been tested widely, is supported by extensive evidence, and is accepted as being a useful explanation of some phenomenon by most scientists in a particular field or related fields of study. So when you hear someone say, "Oh, that's just a theory," you will know that he or she does not have a clear understanding of what a scientific theory is.

Another important and reliable outcome of science is a **scientific law**, or **law of nature**—a well-tested and widely accepted description of what we find happening repeatedly and in the same way in nature. An example is the *law of gravity.* After making many thousands of observations and measurements of objects falling from different heights, scientists developed the following scientific law: all objects fall to the earth's surface at predictable speeds.

We can break a society's law, for example, by driving faster than the speed limit. But *we cannot break a scientific law* such as the law of gravity.

The Results of Science Can Be Reliable, Unreliable, or Tentative

Reliable science consists of data, hypotheses, models, theories, and laws that are widely accepted by all or most of the scientists who are considered experts in the field under study. Scientific hypotheses and results that are presented as reliable without having undergone the rigors of peer review, or that have been discarded as a result of peer review or additional research, are considered to be **unreliable science**.

Preliminary scientific results that have not been widely tested and accepted by peer review or tested and reproduced by other scientists are not yet considered to be reliable, and can be thought of as **tentative science**. Some of these results and hypotheses will be validated and classified as reliable and some will be discredited and classified as unreliable. This is how scientific knowledge advances.

Science Has Some Limitations

Environmental science and science in general have four important limitations. *First,* scientists cannot prove or disprove anything absolutely, because there is always some degree of uncertainty in scientific measurements, observations, and models. Instead, scientists try to establish that a particular scientific theory has a very high *probability* or *certainty* (typically 90% to 95%) of being useful for understanding some aspect of the natural world.

Many scientists do not use the word *proof* because it implies "absolute proof" to many. For example, most scientists will rarely say something like, "We have proven that cigarettes cause lung cancer." Rather, they might say, "Overwhelming evidence from thousands of studies indicates that people who smoke regularly for many years have a greatly increased chance of developing lung cancer."

CONSIDER THIS. . .

THINKING ABOUT Scientific Proof

Does the fact that science can never prove anything absolutely mean that its results are not valid or useful? Explain.

A *second* limitation of science is that scientists are human and thus are not totally free of bias about their own results and hypotheses. However, the high standards of evidence required through peer review can help to uncover or greatly reduce personal bias and expose occasional cheating by scientists who falsify their results.

A *third* limitation is that many systems in the natural world involve a huge number of variables with complex interactions. This makes it too difficult, costly, and time consuming to test one variable at a time in controlled experiments such as the one described in this chapter's **Core Case Study**. To deal with this problem, scientists develop *mathematical models* that can take into account the interac-

tions of many variables, and they run such models on high-speed computers.

A *fourth* limitation of science involves the use of statistical tools. For example, there is no way to measure accurately how many metric tons of soil are eroded annually worldwide. Instead, scientists use statistical sampling and mathematical methods to estimate such numbers.

Despite these limitations, science is the most useful way that we have of learning about how nature works and projecting how it might behave in the future. But we still know too little about how the earth works, about its present state of environmental health, and about the likely current and future environmental impacts of our activities.

2.2 WHAT IS MATTER AND WHAT HAPPENS WHEN IT UNDERGOES CHANGE?

CONCEPT 2.2A Matter consists of elements and compounds, which in turn are made up of atoms, ions, or molecules.

CONCEPT 2.2B Whenever matter undergoes a physical or chemical change, no atoms are created or destroyed (the law of conservation of matter).

Matter Consists of Elements and Compounds

Matter is anything that has mass and takes up space. It can exist in three *physical states*—solid, liquid, and gas—and two *chemical forms*—elements and compounds. (See Supplement 3 for an expanded discussion of basic chemistry.)

An **element** such as gold or mercury (Figure 2.3) is a type of matter that has a unique set of properties and that cannot be broken down into simpler substances by chemical means. Chemists refer to each element with a unique one- or two-letter symbol (such as C for carbon and Au for gold) and have arranged the known elements on the basis of their chemical behavior, placing them on a chart known as the **periodic table of elements** (see Figure 1, p. S5, Supplement 3).

Andraž Cerar/Shutterstock.com

Hurst Photo/ Shutterstock.com

FIGURE 2.3 Mercury (left) and gold (right) are chemical elements. Each has a unique set of properties and cannot be broken down into simpler substances.

TABLE 2-1 Chemical Elements Used in This Book

Element	Symbol	Element	Symbol
arsenic	As	lead	Pb
bromine	Br	lithium	Li
calcium	Ca	mercury	Hg
carbon	C	nitrogen	N
copper	Cu	phosphorus	P
chlorine	Cl	sodium	Na
fluorine	F	sulfur	S
gold	Au	uranium	U

© Cengage Learning

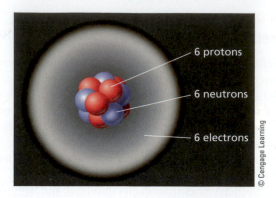

FIGURE 2.4 A greatly simplified model of a carbon-12 atom. It consists of a nucleus containing six protons, each with a positive electrical charge, and six neutrons with no electrical charge. Six negatively charged electrons are found outside its nucleus.

The periodic table of the elements contains 118 elements, not all of which occur naturally. Table 2.1 lists the elements and their symbols that you need to know to understand the material in this book.

Most matter consists of **compounds**, combinations of two or more different elements held together in fixed proportions. For example, water is a compound made of the elements hydrogen and oxygen that have chemically combined with one another.

Atoms, Molecules, and Ions Are the Building Blocks of Matter

The basic building block of matter is an **atom**, the smallest unit of matter into which an element can be divided and still have its distinctive chemical properties. The idea that all elements are made up of atoms is called the **atomic theory** and it is the most widely accepted scientific theory in chemistry.

Atoms are incredibly small. For example, more than 3 million hydrogen atoms could sit side by side on the period at the end of this sentence. If you could view atoms with a supermicroscope, you would find that each different type of atom contains a certain number of three types of *subatomic particles:* **neutrons**, with no electrical charge; **protons**, each with a positive electrical charge (+); and **electrons**, each with a negative electrical charge (−).

Each atom has an extremely small center called the **nucleus**, which contains one or more protons and, in most cases, one or more neutrons. Outside of the nucleus we find one or more electrons in rapid motion (Figure 2.4).

Each element has a unique **atomic number** equal to the number of protons in the nucleus of its atom. Carbon (C), with 6 protons in its nucleus, has an atomic number of 6, whereas uranium (U), a much larger atom, has 92 protons in its nucleus and thus an atomic number of 92.

Because electrons have so little mass compared to protons and neutrons, *most of an atom's mass is concentrated in its nucleus*. The mass of an atom is described by its **mass number**, the total number of neutrons and protons in its nucleus. For example, a carbon atom with 6 protons and 6 neutrons in its nucleus has a mass number of 12 (Figure 2.4).

Each atom of a particular element has the same number of protons in its nucleus. But the nuclei of atoms of a particular element can vary in the number of neutrons they contain, and, therefore, in their mass numbers. The forms of an element having the same atomic number but different mass numbers are called **isotopes** of that element. Scientists identify isotopes by attaching their mass numbers to the name or symbol of the element. For example, the three most common isotopes of carbon are carbon-12 (with six protons and six neutrons, Figure 2.4), carbon-13 (with six protons and seven neutrons), and carbon-14 (with six protons and eight neutrons). Carbon-12 makes up about 98.9% of all naturally occurring carbon.

TABLE 2-2 Compounds Used in This Book

Compound	Formula	Compound	Formula
sodium chloride	NaCl	methane	CH_4
sodium hydroxide	NaOH	glucose	$C_6H_{12}O_6$
carbon monoxide	CO	water	H_2O
carbon dioxide	CO_2	hydrogen sulfide	H_2S
nitric oxide	NO	sulfur dioxide	SO_2
nitrogen dioxide	NO_2	sulfuric acid	H_2SO_4
nitrous oxide	N_2O	ammonia	NH_3
nitric acid	HNO_3	calcium carbonate	$CaCO_3$

© Cengage Learning

TABLE 2-3 Chemical Ions Used in This Book

Positive Ion	Symbol	Components
hydrogen ion	H^+	One hydrogen atom, one positive charge
sodium ion	Na^+	One sodium atom, one positive charge
calcium ion	Ca^{2+}	One calcium atom, two positive charges
aluminum ion	Al^{3+}	One aluminum atom, three positive charges
ammonium ion	NH_4^+	One nitrogen atom, four hydrogen atoms, one positive charge

Negative Ion	Symbol	Components
chloride ion	Cl^-	One chlorine atom, one negative charge
hydroxide ion	OH^-	One oxygen atom, one hydrogen atom, one negative charge
nitrate ion	NO_3^-	One nitrogen atom, three oxygen atoms, one negative charge
carbonate ion	CO_3^{2-}	One carbon atom, three oxygen atoms, two negative charges
sulfate ion	SO_4^{2-}	One sulfur atom, four oxygen atoms, two negative charges
phosphate ion	PO_4^{3-}	One phosphorus atom, four oxygen atoms, three negative charges

A second building block of matter is a **molecule**, a combination of two or more atoms of the same or different elements held together by forces known as *chemical bonds.* Molecules are the basic building blocks of many compounds. For example, water is a compound with a molecule that consists of two atoms of hydrogen and one atom of oxygen held together by chemical bonds. (For more examples, see Table 2.2 and Figure 4, p. S7, in Supplement 3.)

A third building block of some types of matter is an **ion**—an atom or a group of atoms with one or more net positive or negative electrical charges as a result of losing or gaining one or more electrons. Chemists use a superscript after the symbol of an ion to indicate the number of positive or negative electrical charges, as shown in Table 2.3. An important example of an ion is the nitrate ion, a nutrient essential for plant growth. In this chapter's (**Core Case Study**), Bormann and Likens measured the loss of nitrate ions from the deforested area (Figure 2.1, right) in their controlled experiment (Figure 2.5).

Ions are also important for measuring a substance's **acidity** in a water solution, which is based on the comparative amounts of hydrogen ions (H^+) and hydroxide ions (OH^-) contained in a particular volume of the solution. Scientists use **pH** as a measure of acidity. Pure water (not tap water or rainwater) has an equal number of H^+ and OH^- ions. It is called a *neutral solution* and has a pH of 7. An *acidic solution* has more hydrogen ions than hydroxide ions and has a pH less than 7. A *basic solution* has more hydroxide ions than hydrogen ions and has a pH greater

than 7. For more details see Figure 5, p. S8, in Supplement 3.

Chemists use a **chemical formula** to show the number of each type of atom or ion in a compound. This shorthand contains the symbol for each element present (Table 2.1) and uses subscripts to show the number of atoms or ions of each element in the compound's basic structural unit. Examples of compounds and their formulas encountered in this book are sodium chloride (NaCl) and water (H_2O, read as "H-two-O"). These and other compounds important to the study of environmental science in this textbook are listed in Table 2.2.

Organic Compounds Are the Chemicals of Life

Plastics, table sugar, vitamins, aspirin, penicillin, and most of the chemicals in your body are called **organic compounds**, because they contain at least two carbon atoms combined with atoms of one or more other elements. An exception is methane (CH_4), the simplest organic compound, with only one carbon atom.

The millions of known organic (carbon-based) compounds include *hydrocarbons*—compounds of carbon and hydrogen atoms—such as methane (CH_4), the main component of natural gas. They also include *simple carbohydrates (simple sugars)* that contain carbon, hydrogen, and oxygen

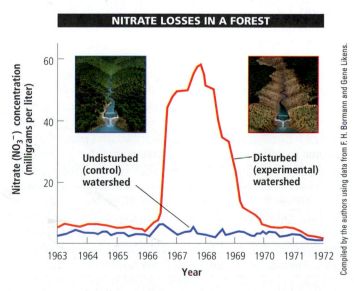

FIGURE 2.5 Loss of nitrate ions (NO_3^-) from a deforested watershed in the Hubbard Brook Experimental Forest (**Core Case Study**, Figure 2.1, right).

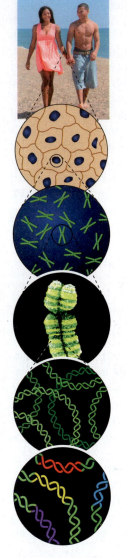

A human body contains trillions of cells, each with an identical set of genes.

Each human cell (except for red blood cells) contains a nucleus.

Each cell nucleus has an identical set of chromosomes, which are found in pairs.

A specific pair of chromosomes contains one chromosome from each parent.

Each chromosome contains a long DNA molecule in the form of a coiled double helix.

Genes are segments of DNA on chromosomes that contain instructions to make proteins—the building blocks of life.

© Cengage Learning

FIGURE 2.6 The relationships among cells, nuclei, chromosomes, DNA, and genes.

Photo: Flashon Studio/Shutterstock.com

atoms. An example is glucose ($C_6H_{12}O_6$), which most plants and animals break down in their cells to obtain energy.

Several types of larger and more complex organic compounds, essential to life, are called *polymers.* They form when a number of simple organic molecules *(monomers)* are linked together by chemical bonds—somewhat like rail cars linked in a freight train. Three major types of organic polymers are *complex carbohydrates* such as cellulose and starch (see Figure 7, p. S9, in Supplement 3), *proteins,* which play many vital roles in the body (see Figure 8, p. S9, in Supplement 3), and *nucleic acids* such as RNA and DNA, formed by monomers called *nucleotides,* and critical for reproduction (see Figures 9 and 10, p. S10, in Supplement 3). *Lipids,* which include fats and waxes, are not made of monomers but are a fourth type of macromolecule essential for life (see Figure 11, p. S11, in Supplement 3).

Matter Comes to Life through Cells, Genes, and Chromosomes

All organisms are composed of one or more **cells**—the fundamental structural and functional units of life. The idea that all living things are composed of cells is called the *cell theory* and it is the most widely accepted scientific theory in biology.

Within some DNA molecules (see Figure 10, p. S10, in Supplement 3) are certain sequences of nucleotides called **genes**. Each of these distinct pieces of DNA contains instructions, or codes, called *genetic information,* for making specific proteins. Each of these coded units of genetic information leads to a specific **trait**, or characteristic, passed on from parents to offspring during reproduction in an animal or plant.

In turn, thousands of genes make up a single **chromosome**, a double-helix DNA molecule wrapped around one or more proteins. Genetic information coded in your chromosomal DNA is what makes you different from an oak leaf, an alligator, or a mosquito, and from your parents. The relationships of genetic material to cells are depicted in Figure 2.6.

Matter Undergoes Physical, Chemical, and Nuclear Changes

When a sample of matter undergoes a **physical change**, there is no change in its *chemical composition*. A piece of aluminum foil cut into small pieces is still aluminum foil. When solid water (ice) melts and when liquid water boils, the resulting liquid water and water vapor are still made up of H_2O molecules.

When a **chemical change**, or **chemical reaction**, takes place, there is a change in the chemical composition of the substances involved. Chemists use a *chemical equation* to show how chemicals are rearranged in a chemical reaction. For example, coal is made up almost entirely of the element carbon (C). When coal is burned completely in a power plant, the solid carbon in the coal combines with oxygen gas (O_2) from the atmosphere to form the gaseous compound carbon dioxide (CO_2). Chemists use the following shorthand chemical equation to represent this chemical reaction:

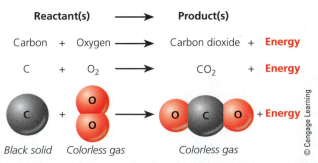

In addition to physical and chemical changes, matter can undergo three types of **nuclear change**, or change in the nuclei of its atoms. **Radioactive decay** occurs when the nuclei of unstable isotopes spontaneously emit fast-moving chunks of matter (alpha particles or beta particles), high-energy radiation (gamma rays), or both at a fixed rate. **Nuclear fission** occurs when the nuclei of certain isotopes with large mass numbers (such as uranium-235) are split apart into lighter nuclei and release energy when struck by a neutron. Each fission releases neutrons, which can cause more nuclei to fission. This cascade of fissions can result in a chain reaction that releases an enormous amount of energy in a short time. **Nuclear fusion** occurs when two nuclei of lighter atoms, such as hydrogen, are forced together at extremely high temperatures until they fuse to form a heavier nucleus and release a tremendous amount of energy. (See Figure 14, p. S13, in Supplement 3 for diagrams showing radioactive decay, nuclear fission, and nuclear fusion.)

We Cannot Create or Destroy Atoms: The Law of Conservation of Matter

We can change elements and compounds from one physical or chemical form to another, but we cannot create or destroy any of the atoms involved in any physical or chemical change. All we can do is rearrange the atoms, ions, or molecules into different spatial patterns (physical changes) or chemical combinations (chemical changes). These facts, based on many thousands of measurements, describe a scientific law known as the **law of conservation of matter**: Whenever matter undergoes a physical or chemical change, no atoms are created or destroyed (**Concept 2.2B**).

Chemical equations can be confusing in that they can seem to show atoms appearing or disappearing in chemical reactions. Chemists use a process of *balancing the equation* to account for all atoms involved in any reaction. See Supplement 3, p. S12, to learn how to balance chemical equations.

2.3 WHAT IS ENERGY AND WHAT HAPPENS WHEN IT UNDERGOES CHANGE?

CONCEPT 2.3A Whenever energy is converted from one form to another in a physical or chemical change, no energy is created or destroyed (first law of thermodynamics).

CONCEPT 2.3B Whenever energy is converted from one form to another in a physical or chemical change, we end up with lower-quality or less-usable energy than we started with (second law of thermodynamics).

Energy Comes in Many Forms

Suppose you find this book on the floor and you pick it up and put it on your desktop. To do this you have to use a certain amount of muscular force, or *work*, to move the book from one place to another. In scientific terms, work is done when any object is moved a certain distance (work = force × distance). **Energy** is the capacity to do work.

Wind and electricity are two forms of kinetic energy.

Yegor Korzh/Shutterstock.com

FIGURE 2.7 Kinetic energy, created by the gaseous molecules in a mass of moving air, turns the blades of these wind turbines. The turbines then convert this kinetic energy to electrical energy, which is another form of kinetic energy.

There are two major types of energy: *moving energy* (called kinetic energy) and *stored energy* (called potential energy). Matter in motion has **kinetic energy**, or energy associated with motion. Examples are flowing water, a car speeding down the highway, electricity (electrons flowing through a wire or other conducting material), and wind (a mass of moving air that we can use to produce electricity, as shown in Figure 2.7).

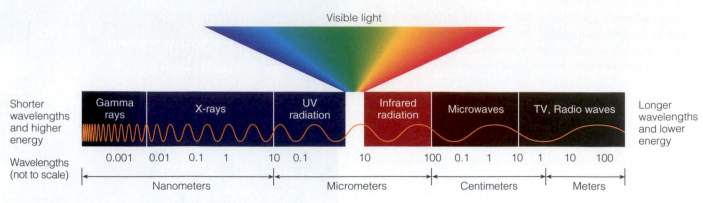

ANIMATED FIGURE 2.8 The *electromagnetic spectrum* consists of a range of electromagnetic waves, which differ in wavelength (the distance between successive peaks or troughs) and energy content.

© Cengage Learning

Another form of kinetic energy is **heat**, or **thermal energy**, the total kinetic energy of all moving atoms, ions, or molecules in an object, a body of water, or a volume of gas such as the atmosphere. If the atoms, ions, or molecules in a sample of matter move faster, the matter will become warmer. When two objects at different temperatures come in contact with one another, heat flows from the warmer object to the cooler object. You learned this the first time you touched a hot stove.

In another form of kinetic energy called **electromagnetic radiation**, energy travels in the form of a *wave* as a result of changes in electrical and magnetic fields. There are many different forms of electromagnetic radiation (Figure 2.8), each having a different *wavelength* (the distance between successive peaks or troughs in the wave), and *energy content*. Those with short wavelengths have more energy than do those with longer wavelengths.

The other major type of energy is **potential energy**, which is stored and potentially available for use. Examples of this type of energy include a rock held in your hand, the water in a reservoir behind a dam, and the chemical energy stored in the carbon atoms of coal or in the molecules of any food that you eat.

We can change potential energy to kinetic energy. If you hold this book in your hand, it has potential energy. However, if you drop it on your foot, the book's potential energy changes to kinetic energy. When a car engine burns gasoline, the potential energy stored in the chemical bonds of the gasoline molecules changes into kinetic energy that propels the car, and into heat that flows into the environment. When water in a reservoir flows through channels in a dam (Figure 2.9), its potential energy becomes kinetic energy used to spin turbines in the dam to produce electricity—another form of kinetic energy.

About 99% of the energy that keeps us warm and supports the plants that we and other organisms eat comes from the sun at no cost to us, in keeping with the solar energy **principle of sustainability** (see Figure 1.2, p. 6 and Supplement 7, p. S50). Without this es-

sentially inexhaustible solar energy, the earth would be frozen and life as we know it would not exist.

Commercial energy—energy that is sold in the marketplace—makes up the remaining 1% of the energy we use to supplement the earth's direct input of solar energy. About 87% of the commercial energy used in the world and 87% of that used in the United States comes from the burning of nonrenewable *fossil fuels*—oil, coal, and natural gas. They are called fossil fuels because they were formed over hundreds of thousands to millions of years as layers of the decaying remains of ancient plants and animals were exposed to intense heat and pressure within the earth's crust.

Some Types of Energy Are More Useful Than Others

Energy quality is a measure of the capacity of a type of energy to do useful work. **High-quality energy** is concentrated energy that has a high capacity to do useful work. Examples are very high-temperature heat, concentrated sunlight, high-speed wind, and the energy released when we burn wood, gasoline, natural gas, or coal.

By contrast, **low-quality energy** is energy that is so dispersed that it has little capacity to do useful work. For example, the enormous number of moving molecules in the atmosphere or in an ocean together have such low-quality energy, and such a low temperature, that we cannot use them to move things or to heat things to high temperatures.

Energy Changes Are Governed by Two Scientific Laws

After observing and measuring energy being changed from one form to another in millions of physical and chemical changes, scientists have summarized their results in the **first law of thermodynamics**, also known as the **law of**

Water behind a dam has potential energy that becomes kinetic energy when the water flows out of the reservoir.

FIGURE 2.9 The water stored in this reservoir behind a dam has potential energy, which becomes kinetic energy when the water flows through channels built into the dam where it spins a turbine and produces electricity—another form of kinetic energy.

conservation of energy. According to this scientific law, whenever energy is converted from one form to another in a physical or chemical change, no energy is created or destroyed (**Concept 2.3A**).

This scientific law tells us that no matter how hard we try or how clever we are, we cannot get more energy out of a physical or chemical change than we put in. This is one of nature's basic rules that we cannot violate.

Because the first law of thermodynamics states that energy cannot be created or destroyed, but only converted from one form to another, you may be tempted to think we will never have to worry about running out of energy. Yet if you fill a car's tank with gasoline and drive around

or run your cell phone battery down, something has been lost. What is it? The answer is *energy quality,* the amount of energy available for performing useful work.

Thousands of experiments have shown that whenever energy is converted from one form to another in a physical or chemical change, we end up with lower-quality or less usable energy than we started with (**Concept 2.3B**). This is a statement of the **second law of thermodynamics**. The resulting low-quality energy usually takes the form of heat that flows into the environment. In the environment, the random motion of air or water molecules further disperses this heat, decreasing its temperature to the point where its energy quality is too low to do much useful work.

In other words, *when energy is changed from one form to another, it always goes from a more useful to a less useful form.* No one has ever witnessed a violation of this fundamental scientific law. This means that we cannot recycle or reuse high-quality energy to perform useful work. Once the concentrated, high-quality energy in a serving of food, a tank of gasoline, or a chunk of uranium nuclear fuel is released, it is degraded to low-quality heat and dispersed into the environment.

2.4 WHAT ARE SYSTEMS AND HOW DO THEY RESPOND TO CHANGE?

CONCEPT 2.4 Systems have inputs, flows, and outputs of matter and energy, and feedback can affect their behavior.

There Are Nonliving and Living Systems

A **system** is a set of components that function and interact in some regular way. A cell, the human body, a forest, a river, an economy, and the earth are all systems. *Nonliving systems* such as a car or a TV do not change their size or how they perform in response to changes in environmental conditions. In contrast, *living systems* can change their size and components and how they behave in response to changing environmental conditions. For example, our bodies produce a new layer of skin cells about every 28 days.

Most living systems have the following key components: **inputs** of matter and energy from the environment, **flows** or **throughputs** of matter and energy within the system, and **outputs** of matter and energy to the environment (Figure 2.10) (**Concept 2.4**). A living system can become unsustainable if the throughputs of matter and energy resources exceed the abilities of the system's environment to provide the required resource inputs and to absorb or dilute the system's outputs of matter and energy.

Systems Respond to Change through Feedback Loops

Most systems are affected in one way or another by **feedback**, any process that increases (positive feedback) or decreases (negative feedback) a change to a system

(**Concept 2.4**). Such a process, called a **feedback loop**, occurs when an output of matter, energy, or information is fed back into the system as an input and leads to changes in that system. A **positive feedback loop** causes a system to change further in the same direction (Figure 2.11).

For example, in the Hubbard Brook experiments (**Core Case Study**), researchers found that when vegetation was removed from a stream valley, flowing water from precipi-

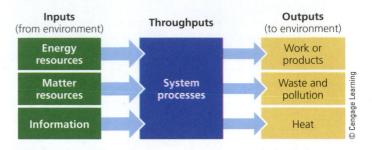

FIGURE 2.10 A greatly simplified model of a system.

© Cengage Learning

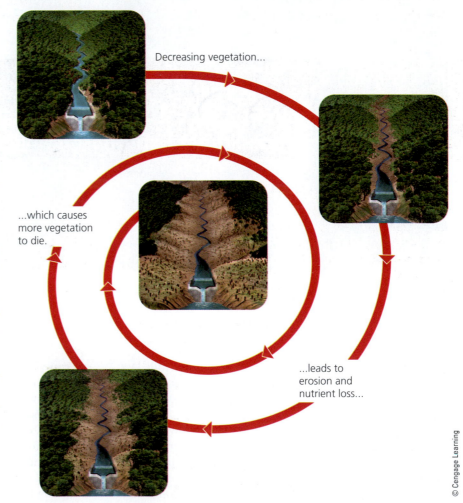

Decreasing vegetation...

...which causes more vegetation to die.

...leads to erosion and nutrient loss...

FIGURE 2.11 A *positive feedback loop.* Decreasing vegetation in a valley causes increasing erosion and nutrient losses that in turn cause more vegetation to die, resulting in more erosion and nutrient losses. **Question:** Can you think of another positive feedback loop in nature?

© Cengage Learning

tation caused erosion and losses of nutrients, which caused more vegetation to die. With even less vegetation to hold soil in place, flowing water caused even more erosion and nutrient loss, which caused even more plants to die.

When a natural system gets locked into such a positive feedback loop, it can reach an **ecological tipping point** beyond which the system can be drastically changed, experiencing severe degradation or collapse. There are many types of ecological tipping points that we discuss throughout this book.

A **negative**, or **corrective, feedback loop** causes a system to change in the opposite direction from which it is moving. An example of a negative feedback loop is the recycling of aluminum. An aluminum can is an output of mining and manufacturing systems that require large inputs of energy and matter and that produce pollution and solid waste. When we recycle the can, that output becomes an input that lessens the need for the mining and manufacturing processes, as well as their energy and matter inputs and harmful environmental effects. Such a negative feedback loop is an application of the chemical cycling **principle of sustainability**.

TYING IT ALL TOGETHER

The Hubbard Brook Forest Experiment and Sustainability

In the controlled experiment discussed in this chapter's **Core Case Study**, the clearing of a mature forest degraded some of its natural capital (see Figure 1.3, p. 7, and photo at right). Specifically, the loss of trees and vegetation altered the ability of the forest to retain and recycle water and other critical plant nutrients—a crucial ecological function based on one of the three **scientific principles of sustainability.**

This clearing of vegetation also violated the other two **scientific principles of sustainability**. For example, the cleared forest lost most of its plants that had used solar energy to produce food for the forest's animals, which supplied nutrients to the soil when they died. Thus the forest lost many of its key nutrients that would normally have been recycled, and it lost much of its life-sustaining biodiversity.

Many of the results of environmental science are based on this sort of experimentation. Throughout this textbook, we explore other examples of scientists' work, as well as ways to use their results to understand how our actions affect the environment and how we can try to solve some of our environmental problems.

steve estvanik/Shutterstock.com

Chapter Review

Core Case Study

1. Describe the controlled scientific experiment carried out in the Hubbard Brook Experimental Forest.

Section 2.1

2. What is the key concept for this section? What is **science**? List the steps involved in a scientific process. What is **data**? What is a **model**? Distinguish among a **scientific hypothesis**, a **scientific theory**, and a **scientific law (law of nature)**. What is **peer review** and why is it important?

3. Explain why scientific theories and laws are the most important and most certain results of science and why people often use the term *theory* incorrectly.

4. Distinguish among **reliable science**, **unreliable science**, and **tentative science**. What are four limitations of science?

Section 2.2

5. What are the two key concepts for this section? What is **matter**? Distinguish between an **element** and a **compound** and give an example of each. What is the **periodic table of elements**? Define **atoms**, **molecules**, and **ions** and give an example of each. What is the **atomic theory**? Distinguish among **protons**, **neutrons**, and **electrons**. What is the **nucleus** of an atom? Distinguish between the **atomic number** and the **mass number** of an element. What is an **isotope**? What is **acidity**? What is **pH**? Define **chemical formula** and give an example.

6. Define and give two examples of an **organic compound**. What are three types of organic polymers that are important to life? What is a **cell**? Define **gene**, **trait**, and **chromosome**.

7. Define and distinguish between a **physical change** and a **chemical change (chemical reaction)** in matter and give an example of each. What is a **nuclear change**? Define and explain the differences among **natural radioactive decay**, **nuclear fission**, and **nuclear fusion**. What is the **law of conservation of matter**?

Section 2.3

8. What are the two key concepts for this section? What is **energy**? Define and distinguish between **kinetic energy** and **potential energy** and give an example of each. What is **heat (thermal energy)**? Define and give two examples of **electromagnetic radiation**. What is **energy quality**? Distinguish between **high-quality energy** and **low-quality energy** and give an example of each. What is the **first law of thermodynamics (law of conservation of energy)** and why is it important? What is the **second law of thermodynamics** and why is it important? Explain why the second law means that we can never recycle or reuse high-quality energy.

Section 2.4

9. What is the key concept for this section? Define and give an example of a **system**. Distinguish among the **inputs, flows (throughputs)**, and **outputs** of a system. What is **feedback**? What is a **feedback loop**? Distinguish between a **positive feedback loop** and a **negative (corrective) feedback loop** in a system, and give an example of each. What is an **ecological tipping point**?

10. What are this chapter's *three big ideas*? Explain how the Hubbard Brook Experimental Forest controlled experiments illustrated the three **scientific principles of sustainability**.

Note: Key terms are in bold type.

Critical Thinking

1. What ecological lesson can we learn from the controlled experiment on the clearing of forests described in the **Core Case Study** that opened this chapter?

2. Suppose you observe that all of the fish in a pond have disappeared. Explain how you might use the scientific process described in the **Core Case Study** and in Figure 2.2 to determine the cause of this fish kill.

3. Respond to the following statements:
   a. Scientists have not absolutely proven that anyone has ever died from smoking cigarettes.
   b. The *natural greenhouse effect*—that certain gases such as water vapor and carbon dioxide help to warm the lower atmosphere—is not a reliable idea because it is just a scientific theory.

4. A tree grows and increases its mass. Explain why this is not a violation of the law of conservation of matter.

5. If there is no "away" where organisms can get rid of their wastes due to the law of conservation of matter, why is the world not filled with waste matter?

6. Suppose someone wants you to invest money in an automobile engine, claiming that it will produce more energy than is found in the fuel used to run it. What would be your response? Explain.

7. Use the second law of thermodynamics to explain why we can use oil only once as a fuel, or in other words, why we cannot recycle its high-quality energy.

8. Imagine that for one day (a) you have the power to revoke the law of conservation of matter, and (b) you have the power to violate the first law of thermodynamics. For each of these scenarios, list three ways in which you would use your new power. Explain your choices.

Doing Environmental Science

Find (a) a newspaper or magazine article or a report on the Web that attempts to discredit a scientific hypothesis because it has not been proven, or (b) a report of a new scientific hypothesis that has the potential to be controversial. Analyze the piece by doing the following: (1) determine its source (author or organization); (2) detect an alternative hypothesis, if any, that is offered by the author; (3) determine the primary objective of the author (for example, to debunk the original hypothesis, to state an alternative hypothesis, or to raise new questions); (4) summarize the evidence given by the authors for their position; and (5) compare the authors' evidence with the evidence for the original hypothesis. Write a report summarizing your analysis and compare it with those of your classmates.

Global Environment Watch Exercise

Starting on the GREENR home page, under *Browse Issues and Topics,* choose *Forests and Deforestation* from the *Resource Management* category. Browse the articles listed there and find one that involves a controlled experiment or some other form of scientific research in a forest. Determine what the hypothesis was that the researchers were testing. Briefly summarize their research methods and any conclusions that were reached. Was the research similar in any way to that described in the **Core Case Study**? Explain.

Data Analysis

Consider the graph to the right that compares the losses of calcium from the experimental and control sites in the Hubbard Brook Experimental Forest (**Core Case Study**). Note that this figure is very similar to Figure 2.5, which compares loss of nitrates from the two sites. After studying this graph, answer these questions.

1. In what year did the calcium loss from the experimental site begin a sharp increase? In what year did it peak? In what year did it again level off?

2. In what year were the calcium losses from the two sites closest together? In the span of time between 1963 and 1972, did they ever get that close again?

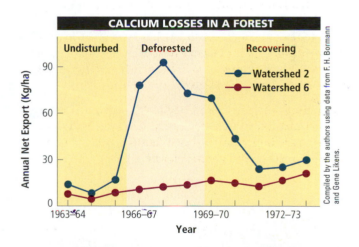

3. Does this graph support the hypothesis that cutting the trees from a forested area causes the area to lose nutrients more quickly than leaving the trees in place? Explain.

CENGAGE brain.com To access course materials, including Aplia homework, please visit www.cengagebrain.com.

WWW.CENGAGEBRAIN.COM   **39**

3
ECOSYSTEMS: WHAT ARE THEY AND HOW DO THEY WORK?

To halt the decline of an ecosystem, it is necessary to think like an ecosystem.

DOUGLAS WHEELER

Scientists studying life in the treetops of a tropical forest.

Bill Hatcher/National Geographic Creative

Tropical Rain Forests Are Disappearing

Tropical rain forests are found near the earth's equator and contain an incredible variety of life. These lush forests are warm year round and have high humidity and heavy rainfall almost daily. Although they cover only about 2% of the earth's land surface, studies indicate that they contain up to half of the world's known terrestrial plant and animal species. For these reasons, they make an excellent natural laboratory for the study of *ecosystems*—communities of organisms interacting with one another and with the physical environment of matter and energy in which they live (see chapter-opening photo).

So far, at least half of these forests have been destroyed or disturbed by humans cutting down trees, growing crops, grazing cattle, and building settlements (Figure 3.1), and the degradation of these centers of biodiversity is increasing. Ecologists warn that without strong protective measures, most of these forests will probably be gone or severely degraded by the end of this century.

So why should we care that tropical rain forests are disappearing? Scientists give three reasons. *First,* clearing these forests will reduce the earth's vital biodiversity by destroying or degrading the habitats of many of the unique plant and animal species that live in them, which could lead to the early extinction of these species and other species that depend on them.

Second, the destruction of these forests is helping to accelerate atmospheric warming, and thus projected climate change (as discussed in more detail in Chapter 15). The reason is that eliminating large areas of trees faster than they can grow back decreases the forests' ability to remove from the atmosphere the human-generated emissions of carbon dioxide (CO_2), a gas that contributes to atmospheric warming and projected climate change.

Third, large-scale rain forest losses can change regional weather patterns in ways that can prevent the return of diverse tropical rain forests in cleared or severely degraded areas. When this irreversible *ecological tipping point* is reached, tropical rain forests in such areas become much less diverse tropical grasslands.

Agronomist and National Geographic Explorer Vitor O. Becker has studied tropical forests in Latin America for decades. Like many ecologists, he studies ecosystems to learn how their varieties of organisms interact with their living *(biotic)* environment of other organisms and with their nonliving *(abiotic)* environment of soil, water, other forms of matter, and energy, mostly from the sun. In effect, ecologists study *connections in nature*. In this chapter, we look more closely at how tropical rain forests and other ecosystems work, how human activities are affecting them, and how we can try to preserve them.

17 Jun 1975

FIGURE 3.1 Natural capital degradation: Satellite image of the loss of topical rain forest, cleared for farming, cattle grazing, and settlements, near the Bolivian city of Santa Cruz between June 1975 (left) and May 2003 (right).

6 May 2003

HOW DOES THE EARTH'S LIFE-SUPPORT SYSTEM WORK?

CONCEPT 3.1A The four major components of the earth's life-support system are the atmosphere (air), the hydrosphere (water), the geosphere (rock, soil, and sediment), and the biosphere (living things).

CONCEPT 3.1B Life is sustained by the flow of energy from the sun through the biosphere, the cycling of nutrients within the biosphere, and gravity.

Earth's Life-Support System Has Four Major Components

The earth's life-support system consists of four main spherical systems (Figure 3.2) that interact with one another—the atmosphere (air), the hydrosphere (water), the geosphere (rock, soil, and sediment), and the biosphere (living things) (**Concept 3.1A**).

The **atmosphere** is a thin spherical envelope of gases surrounding the earth's surface. Its inner layer, the **troposphere**, extends about 17 kilometers (11 miles) above sea level at the tropics and about 7 kilometers (4 miles) above the earth's north and south poles. It contains the air we

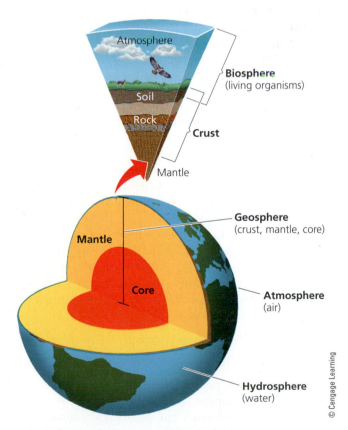

FIGURE 3.2 Natural capital: The earth consists of a land sphere *(geosphere)*, an air sphere *(atmosphere)*, a water sphere *(hydrosphere)*, and a life sphere *(biosphere)* (**Concept 3.1A**).

© Cengage Learning

breathe, consisting mostly of nitrogen (78% of the total volume) and oxygen (21%). Most of the remaining 1% of the air consists of water vapor, carbon dioxide, and methane. The next layer, reaching from 17 to 50 kilometers (11–31 miles) above the earth's surface, is called the **stratosphere**. Its lower portion holds enough ozone (O_3) gas to filter out about 95% of the sun's harmful *ultraviolet (UV) radiation*. This global sunscreen allows life to exist on the surface of the planet.

The **hydrosphere** is made up of all of the water on or near the earth's surface. It is found as *water vapor* in the atmosphere, as *liquid water* on the surface and underground, and as *ice*—polar ice, icebergs, glaciers, and ice in frozen soil-layers called *permafrost*. The oceans, which cover about 71% of the globe, contain about 97% of the earth's water and support about half of the world's species.

The **geosphere** consists of the earth's intensely hot *core*, a thick *mantle* composed mostly of rock, and a thin outer *crust*. The crust's upper portion contains soil chemicals that organisms need in order to live, grow, and reproduce (nutrients), as well as nonrenewable *fossil fuels*—coal, oil, and natural gas—and minerals that we use.

The **biosphere** consists of the parts of the atmosphere, hydrosphere, and geosphere where life is found. If the earth were an apple, the biosphere would be no thicker than the apple's skin. An important goal of environmental science is to understand the key interactions that occur within this thin layer of air, water, soil, and organisms and how we interact with the biosphere.

Three Factors Sustain the Earth's Life

Life on the earth depends on three interconnected factors (**Concept 3.1B**):

1. The *one-way flow of high-quality energy* from the sun, through living things in their feeding interactions, into the environment as low-quality energy (mostly heat dispersed into air or water at a low temperature), and eventually to outer space as heat (Figure 3.3). This is the basis for the solar energy **principle of sustainability**. As solar energy interacts with carbon dioxide (CO_2), water vapor, and several other gases in the troposphere, it warms the troposphere— a process known as the **greenhouse effect**. Without this natural process, the earth would be too cold to support most of the forms of life we find here today.

2. The *cycling of nutrients* (the atoms, ions, and molecules of elements and compounds needed for survival by living organisms) through parts of the biosphere. Because the earth does not get significant inputs of matter from space, its essentially fixed supply of nutrients must be continually recycled to support life, in keeping with the chemical cycling **principle of sustainability**.

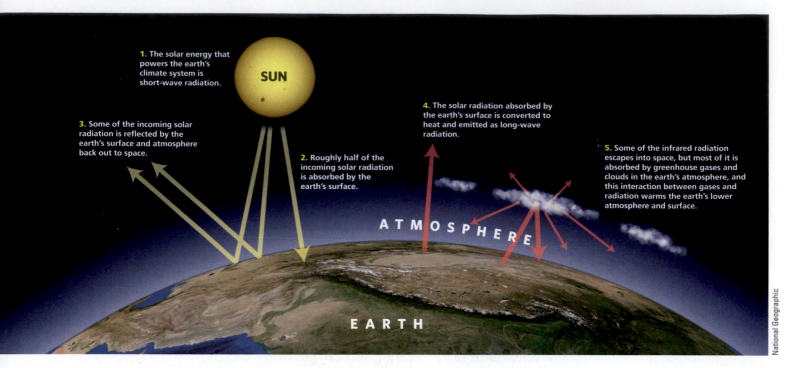

ANIMATED FIGURE 3.3 *Greenhouse Earth.* High-quality solar energy flows from the sun to the earth. It is degraded to lower quality energy (mostly heat) as it interacts with the earth's air, water, soil, and life forms, and eventually returns to space. Certain gases in the earth's atmosphere retain enough of the sun's incoming energy as heat to warm the planet in what is known as the *greenhouse effect.*

Text inside figure:

1. The solar energy that powers the earth's climate system is short-wave radiation.

2. Roughly half of the incoming solar radiation is absorbed by the earth's surface.

3. Some of the incoming solar radiation is reflected by the earth's surface and atmosphere back out to space.

4. The solar radiation absorbed by the earth's surface is converted to heat and emitted as long-wave radiation.

5. Some of the infrared radiation escapes into space, but most of it is absorbed by greenhouse gases and clouds in the earth's atmosphere, and this interaction between gases and radiation warms the earth's lower atmosphere and surface.

SUN

ATMOSPHERE

EARTH

National Geographic

3. *Gravity,* which allows the planet to hold onto its atmosphere and helps to enable the movement and cycling of chemicals through air, water, soil, and organisms.

3.2 WHAT ARE THE MAJOR COMPONENTS OF AN ECOSYSTEM?

CONCEPT 3.2 Some organisms produce the nutrients they need, others get the nutrients they need by consuming other organisms, and some recycle nutrients back to producers by decomposing the wastes and remains of other organisms.

Ecosystems Have Several Important Components

Ecology is the science that focuses on how organisms interact with one another and with their nonliving environment of matter and energy. Scientists classify matter into levels of organization ranging from atoms to galaxies. Ecologists study interactions within and among five of these levels—**organisms**, **populations**, **communities**, **ecosystems**, and the **biosphere**, which are illustrated and defined in Figure 3.4.

The biosphere and its ecosystems are made up of living *(biotic)* and nonliving *(abiotic)* components (Figure 3.5). Examples of nonliving components are water, air, nutrients, rocks, heat, and solar energy. Living components include plants, animals, microbes, and all other organisms.

Ecologists assign each organism in an ecosystem to a *feeding level,* or **trophic level**, depending on its source of nutrients. We can broadly classify living organisms as *producers* and *consumers.*

Producers, sometimes called **autotrophs** (self-feeders), make the nutrients they need from compounds and energy obtained from their environment (**Concept 3.2**). In a process called **photosynthesis**, plants capture solar energy that falls on their leaves and use it in combination with carbon dioxide and water to form organic molecules, including energy-rich carbohydrates (such as glucose, $C_6H_{12}O_6$), which store the chemical energy that plants need. We can summarize the overall chemical reaction for photosynthesis as follows:

carbon dioxide + water + **solar energy** → glucose + oxygen

$$6\ CO_2 + 6\ H_2O + \textbf{solar energy} \rightarrow C_6H_{12}O_6 + 6\ O_2$$

(See Supplement 3, p. S12, for information on how to balance chemical equations such as this one.)

About 2.8 billion years ago, producer organisms called *cyanobacteria,* most of them floating on the surface of the ocean, began to carry out the process of photosynthesis. At that time the atmosphere contained essentially no oxygen.

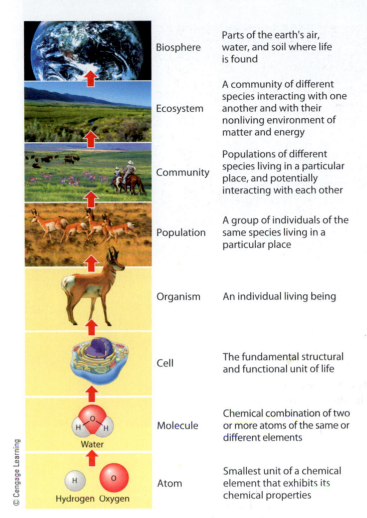

Biosphere	Parts of the earth's air, water, and soil where life is found
Ecosystem	A community of different species interacting with one another and with their nonliving environment of matter and energy
Community	Populations of different species living in a particular place, and potentially interacting with each other
Population	A group of individuals of the same species living in a particular place
Organism	An individual living being
Cell	The fundamental structural and functional unit of life
Molecule	Chemical combination of two or more atoms of the same or different elements
Atom	Smallest unit of a chemical element that exhibits its chemical properties

ANIMATED FIGURE 3.4 Some of the levels of the organization of matter in nature. Ecology focuses on the top five of these levels.

It took several hundred million years before the percentage of oxygen in the air, generated by photosynthesis, reached its current level of about 21%—high enough to keep animals like us alive.

Today, most producers on land are trees and other green plants. In freshwater and ocean ecosystems, algae and aquatic plants growing near shorelines are the major producers. In open water, the dominant producers are *phytoplankton*—mostly microscopic organisms that float or drift in the water.

The other organisms in an ecosystem are **consumers**, or **heterotrophs** ("other-feeders"), which cannot produce the nutrients they need through photosynthesis or other processes (**Concept 3.2**). They get their nutrients by feeding on other organisms (producers or other consumers) or their wastes and remains.

There are several types of consumers. **Primary consumers**, or **herbivores** (plant eaters), are animals that eat mostly green plants. Examples are caterpillars, giraffes, and zooplankton (tiny sea animals that feed on phytoplankton). **Carnivores** (meat eaters) are animals that feed on the flesh of other animals. Some carnivores such as

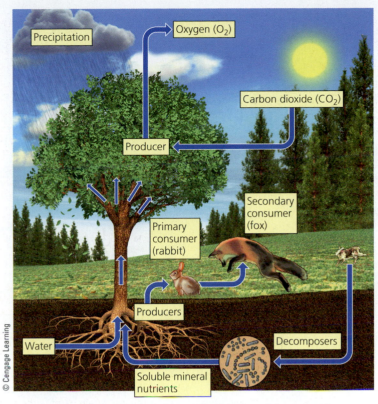

ANIMATED FIGURE 3.5 Key living (biotic) and nonliving (abiotic) components of an ecosystem in a field.

spiders, lions (Figure 3.6), and most small fishes are **secondary consumers** that feed on the flesh of herbivores. Other carnivores such as tigers, hawks, and killer whales (orcas) are **tertiary** (or higher-level) **consumers** that feed on the flesh of herbivores and other carnivores. Some of these relationships are shown in Figure 3.5. **Omnivores** such as pigs, rats, and humans eat both plants and animals.

CONSIDER THIS...

THINKING ABOUT What You Eat

When you ate your most recent meal, were you an herbivore, a carnivore, or an omnivore?

Decomposers are consumers that, in the process of obtaining their nutrients, release nutrients from the wastes or remains of plants and animals and return those nutrients to the soil, water, and air for reuse by producers (**Concept 3.2**). Most decomposers are bacteria and fungi (Science Focus 3.1). Other consumers, called **detritus feeders**, or **detritivores**, feed on the wastes or dead bodies (detritus) of other organisms. Examples are earthworms, hyenas, and vultures.

Hordes of detritivores and decomposers can transform a fallen tree trunk into wood particles and, finally, into simple inorganic molecules that plants can absorb as nutrients

FIGURE 3.6 This lioness (a carnivore) is feeding on a freshly killed zebra (an herbivore) in Kenya, Africa.

Natasha Chamberlain/Photos.com

SCIENCE**FOCUS**3.1

MANY OF THE WORLD'S MOST IMPORTANT ORGANISMS ARE INVISIBLE TO US

They are everywhere. Trillions can be found inside your body, on your body, in a handful of soil, and in a cup of ocean water.

These mostly invisible rulers of the earth are *microbes,* or *microorganisms,* catchall terms for many thousands of species of bacteria, protozoa, fungi, and floating phytoplankton. They play key roles in the earth's life-support system. Bacteria in our intestinal tracts break down the food we eat, and microbes in our noses help to prevent harmful bacteria from reaching our lungs. Other microbes help to purify the water we drink by breaking down plant and animal wastes in the water. Bacteria and fungi in the soil decompose organic wastes into nutrients that can be taken up by plants that are then eaten by humans and many other plant eaters. Without these tiny creatures, we would go hungry and be up to our necks in waste matter.

Some microorganisms, particularly phytoplankton in the ocean, provide much of the planet's oxygen, and help to regulate the atmosphere's average temperature by removing some of the carbon dioxide produced when we burn coal, natural gas, and gasoline. Other microbes help us to control diseases that affect plants and to limit populations of insects that attack our food crops. In short, microbes are a vital part of the earth's natural capital.

Critical Thinking

What are two advantages that microbes have over humans for thriving in the world?

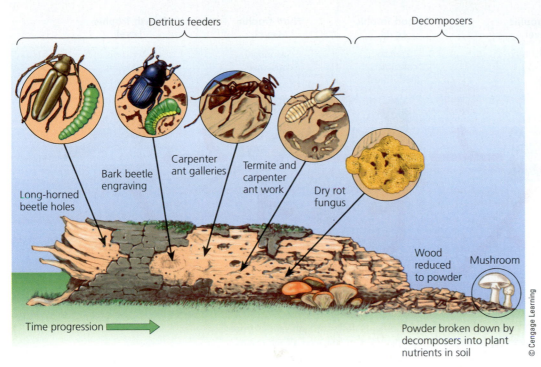

FIGURE 3.7 Various detritivores and decomposers (mostly fungi and bacteria) can "feed on" or digest parts of a log and eventually convert its complex organic chemicals into simpler inorganic nutrients that can be taken up by producers.

Detritus feeders

Decomposers

Bark beetle
engraving

Carpenter
ant galleries

Long-horned
beetle holes

Termite and
carpenter
ant work

Dry rot
fungus

Wood
reduced
to powder

Mushroom

Time progression

Powder broken down by
decomposers into plant
nutrients in soil

© Cengage Learning

(Figure 3.7). Thus, in natural ecosystems, the wastes and dead bodies of organisms serve as resources for other organisms in keeping with the chemical cycling **principle of sustainability**. Without decomposers and detritivores, the planet would be overwhelmed with plant litter, animal wastes, dead animal bodies, and garbage.

Producers, consumers, and decomposers use the chemical energy stored in glucose and other organic compounds to fuel their life processes. In most cells, this energy is released by **aerobic respiration**, which uses oxygen to convert glucose (or other organic nutrient molecules) back into carbon dioxide and water. We can summarize the overall reaction for the process of aerobic respiration as follows:

$$\text{glucose} + \text{oxygen} \rightarrow \text{carbon dioxide} + \text{water} + \textbf{energy}$$
$$C_6H_{12}O_6 + 6\,O_2 \rightarrow 6\,CO_2 + 6\,H_2O + \textbf{energy}$$

Note that the net chemical change for aerobic respiration is the opposite of that for photosynthesis. Plants and animals complement one other with each using the other's waste and decomposed remains as food.

To summarize, ecosystems and the biosphere are sustained through a combination of *one-way energy flow* from the sun through these systems and the *nutrient cycling* of key materials within them (**Concept 3.1B**)—in keeping with two of the **scientific principles of sustainability** (Figure 3.8).

Heat

Chemical nutrients
(carbon dioxide,
oxygen, nitrogen,
minerals)

Solar
energy

Heat

Heat

Heat

Decomposers
(bacteria, fungi)

Producers
(plants)

© Cengage Learning

Consumers
(plant eaters,
meat eaters)

Heat

Heat

ANIMATED FIGURE 3.8 Natural capital: The main components of an ecosystem are energy, chemicals, and organisms. Nutrient cycling and the flow of energy—first from the sun, then through organisms, and finally into the environment as low-quality heat—link these components.

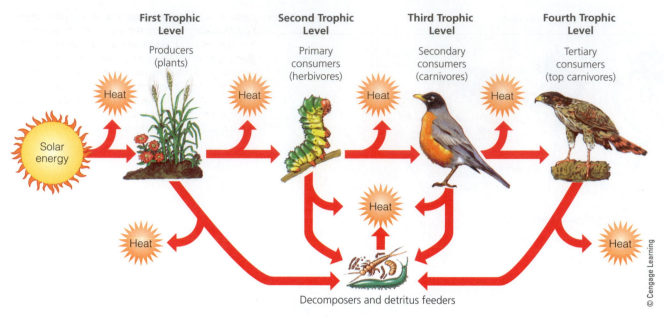

First Trophic Level
Producers (plants)

Second Trophic Level
Primary consumers (herbivores)

Third Trophic Level
Secondary consumers (carnivores)

Fourth Trophic Level
Tertiary consumers (top carnivores)

Heat

Solar energy

Decomposers and detritus feeders

© Cengage Learning

ANIMATED FIGURE 3.9 In a food chain, chemical energy in nutrients flows through various trophic levels. **Question:** Think about what you ate for breakfast. At what level or levels on a food chain were you eating?

3.3 WHAT HAPPENS TO ENERGY IN AN ECOSYSTEM?

CONCEPT 3.3 As energy flows through ecosystems in food chains and food webs, the amount of high-quality chemical energy available to organisms at each successive feeding level decreases.

Energy Flows through Ecosystems in Food Chains and Food Webs

The chemical energy stored as nutrients in the bodies and wastes of organisms flows through ecosystems from one trophic (feeding) level to another. A sequence of organisms, each of which serves as a source of nutrients or energy for the next, is called a **food chain** as shown in Figure 3.9. Every use and transfer of energy by organisms involves a loss of some high-quality energy to the environment as low-quality energy in the form of heat, in accordance with the second law of thermodynamics.

In natural ecosystems, most consumers feed on more than one type of organism, and most organisms are eaten or decomposed by more than one type of consumer. Because of this, organisms in most ecosystems form a complex network of interconnected food chains called a **food web** (Figure 3.10). Food chains and food webs show how producers, consumers, and decomposers are connected to one another as energy flows through trophic levels in an ecosystem. The **pyramid of energy flow** in Figure 3.11 illustrates this energy loss for a simple food chain, assuming a 90% energy loss with each transfer.

Some Ecosystems Produce Plant Matter Faster Than Others Do

Gross primary productivity (GPP) is the *rate* at which an ecosystem's producers (usually plants) convert solar energy into chemical energy stored in compounds found in their tissues. To stay alive, grow, and reproduce, producers must use some of their stored chemical energy for their own respiration. **Net primary productivity (NPP)** is the *rate* at which producers use photosynthesis to produce and store chemical energy *minus* the *rate* at which they use some of this stored chemical energy through aerobic respiration. NPP measures how fast producers can make the chemical energy that is stored in their tissues and that is potentially available to other organisms (consumers) in an ecosystem.

Terrestrial ecosystems and aquatic life zones differ in their NPP as illustrated in Figure 3.12. Despite its low NPP, the open ocean produces more of the earth's biomass per year than any other ecosystem or life zone. This occurs because of the enormous volume of the global ocean, which covers 71% of the earth's surface and contains huge numbers of phytoplankton and other producers.

Tropical rain forests have a very high net primary productivity because they have a large number and variety of producer trees and other plants. When such forests are cleared (**Core Case Study**) or burned to make way for crops or for grazing cattle, they suffer a sharp drop in net primary productivity and a loss of many of their diverse array of plant and animal species.

Only the plant matter represented by NPP is available as nutrients for consumers, and they use only a portion of it. Thus, *the planet's NPP ultimately limits the number of con-*

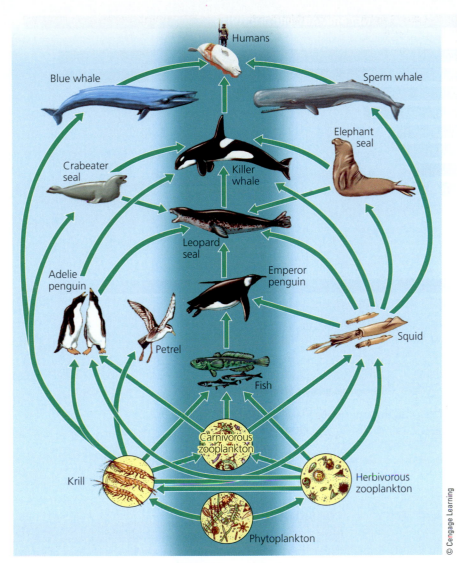

ANIMATED FIGURE 3.10 This is a greatly simplified *food web* found in the southern hemisphere. The shaded middle area shows a simple food chain that is part of these complex interacting feeding relationships. Many more participants in the web, including an array of decomposer and detritus feeder organisms, are not shown here. ***Question:*** Can you imagine a food web of which you are a part? Try drawing a simple diagram of it.

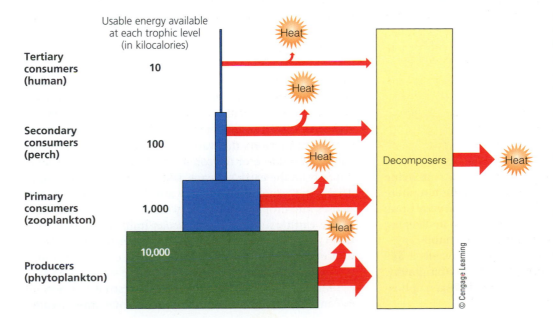

ANIMATED FIGURE 3.11
Generalized *pyramid of energy flow* showing the decrease in usable chemical energy available at each succeeding trophic level in a food chain or food web. This model assumes that with each transfer from one trophic level to another, there is a 90% loss of usable energy to the environment in the form of low-quality heat. Calories and joules are used to measure energy. 1 kilocalorie = 1,000 calories = 4,184 joules. ***Question:*** Why is a vegetarian diet more energy efficient than a meat-based diet?

Concept 3.3   **49**

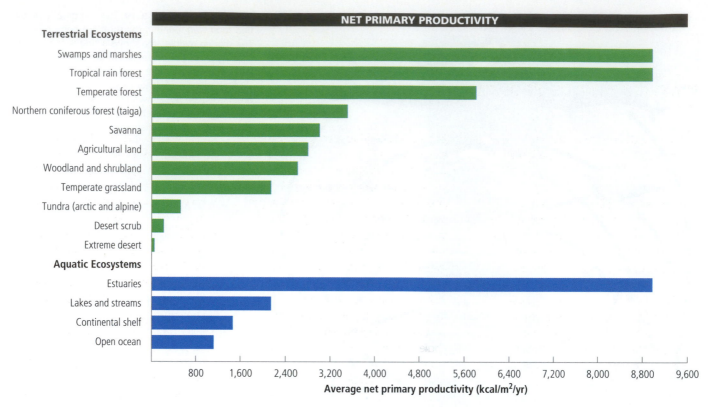

Terrestrial Ecosystems

Swamps and marshes
Tropical rain forest
Temperate forest
Northern coniferous forest (taiga)
Savanna
Agricultural land
Woodland and shrubland
Temperate grassland
Tundra (arctic and alpine)
Desert scrub
Extreme desert

Aquatic Ecosystems

Estuaries
Lakes and streams
Continental shelf
Open ocean

800 1,600 2,400 3,200 4,000 4,800 5,600 6,400 7,200 8,000 8,800 9,600

Average net primary productivity (kcal/m²/yr)

FIGURE 3.12 Estimated annual average *net primary productivity* in major life zones and ecosystems expressed as kilocalories of energy produced per square meter per year (kcal/m²/yr). *Question:* What are the three most productive and the three least productive systems?

Compiled by the authors using data from R. H. Whittaker, *Communities and Ecosystems,* 2nd ed., New York: Macmillan, 1975.

sumers *(including humans) that can survive on the earth.* This is an important lesson from nature.

3.4 WHAT HAPPENS TO MATTER IN AN ECOSYSTEM?

CONCEPT 3.4 Matter, in the form of nutrients, cycles within and among ecosystems and the biosphere, and human activities are altering these chemical cycles.

Nutrients Cycle within and among Ecosystems

The elements and compounds that make up nutrients move continually through air, water, soil, rock, and living organisms within ecosystems, as well as in the biosphere in cycles called **nutrient cycles**, or *biogeochemical cycles* (literally, life-earth-chemical cycles). This is in keeping with the chemical cycling **principle of sustainability**. These cycles, which are driven directly or indirectly by incoming solar energy and by the earth's gravity, include the hydrologic (water), carbon, nitrogen, phosphorus, and sulfur cycles.

CONSIDER THIS. . .

CONNECTIONS Nutrient Cycles and Life

Nutrient cycles connect past, present, and future forms of life. Some of the carbon atoms in your skin may once have been part of an oak leaf, a dinosaur's skin, or a layer of limestone rock. Your grandmother, George Washington, or a hunter–gatherer who lived 25,000 years ago may have inhaled some of the nitrogen molecules you just inhaled.

The Water Cycle

Water (H_2O) is an amazing substance (Science Focus 3.2) that is necessary for life on the earth. The **hydrologic cycle**, or **water cycle**, collects, purifies, and distributes the earth's fixed supply of water, as shown in Figure 3.13.

The sun powers the water cycle. Incoming solar energy causes *evaporation,* or the conversion of water from liquid to vapor from the earth's oceans, lakes, rivers, soil, and plants. This water vapor rises into the atmosphere, where it condenses into droplets, and gravity then draws the water back to the earth's surface as *precipitation* (rain, snow, sleet, and dew). Over land, about 90% of the water that reaches the atmosphere evaporates from the surfaces of plants through a process called *transpiration* and from the soil.

Most precipitation falling on terrestrial ecosystems becomes **surface runoff**. This water flows into streams,

WATER'S UNIQUE PROPERTIES

Water (H₂O) is a remarkable substance with a unique combination of properties:

- *Forces of attraction,* called *hydrogen bonds* (see Figure 6, p. S8, in Supplement 3), *hold water molecules together* and are the major factor determining water's distinctive properties.
- *Water exists as a liquid over a wide temperature range because of the hydrogen bonds between its molecules.* If liquid water had a much narrower range of temperatures between freezing and boiling, the oceans might have frozen solid or boiled away long ago.
- *Liquid water changes temperature slowly because it can store a large amount of heat without a large change in its own temperature.* This high heat storage capacity helps to protect living organisms from temperature changes, moderates the earth's climate, and makes water an excellent coolant for car engines and power plants.
- *It takes a large amount of energy to evaporate water because of its hydrogen bonds.* Water absorbs large amounts of heat as it changes into water vapor and releases this heat as the vapor condenses back to liquid water. This helps to distribute heat throughout the world and to determine regional and local climates. It also makes evaporation a cooling process—explaining why you feel cooler when perspiration evaporates from your skin.
- *Liquid water can dissolve a variety of compounds* (see Figure 3, p. S6, in Supplement 3). It carries dissolved nutrients into the tissues of living organisms, flushes waste products out of those tissues, serves as an all-purpose cleanser, and helps to remove and dilute the water-soluble wastes of civilization. This property also means that water-soluble wastes can easily pollute water.
- *Water filters out wavelengths of the sun's ultraviolet radiation that would harm some aquatic organisms.* This helps to sustain aquatic ecosystems on which many land animals, including humans, depend.
- *Unlike most liquids, water expands when it freezes.* This means that ice floats on water because it has a lower density (mass per unit of volume) than liquid water has. Otherwise, lakes and streams in cold climates would freeze solid, losing most of their aquatic life. Because water expands upon freezing, it can break pipes, crack a car's engine block (if it doesn't contain antifreeze), break up pavement, and fracture rocks. Thus water plays a major role in shaping landscapes and forming soil.

Critical Thinking

Pick two of the special properties listed above and, for each property, explain how life on the earth would be different if it did not exist.

rivers, lakes, wetlands, and oceans, from which it can evaporate to repeat the cycle. Some precipitation seeps into the upper layers of soils where it is used by plants, and some evaporates from the soils back into the atmosphere. Some precipitation also sinks through soil into underground layers of rock, sand, and gravel called **aquifers**, where it is stored as **groundwater.** Some precipitation is converted to ice that is stored in *glaciers,* usually for long periods of time.

Because water is good at dissolving many different compounds, it can easily be polluted. Throughout the hydrologic cycle, many natural processes purify water by drawing pollutants out of it. Thus, *the hydrologic cycle can be viewed as a cycle of natural renewal of water quality*—an important and free ecosystem service.

Only about 0.024% of the earth's vast water supply is available to humans and other species as liquid freshwater in accessible groundwater deposits and in lakes, rivers, and streams. The rest is too salty for us to use, is stored as ice, or is too deep underground to extract at affordable prices.

Humans alter the water cycle in three major ways (see the red arrows and boxes in Figure 3.13). *First,* we withdraw freshwater from rivers, lakes, and aquifers, sometimes at rates faster than natural processes can replace it. As a result, some aquifers are being depleted and some rivers no longer flow to the ocean. *Second,* we clear vegetation from land for agriculture, mining, road building, and other activities, and cover much of the land with buildings, concrete, and asphalt. This increases runoff and reduces infiltration that would normally recharge groundwater supplies. *Third,* we drain and fill wetlands for farming and urban development. Left undisturbed, wetlands provide the ecosystem service of flood control, acting like sponges to absorb and hold overflows of water from drenching rains or rapidly melting snow.

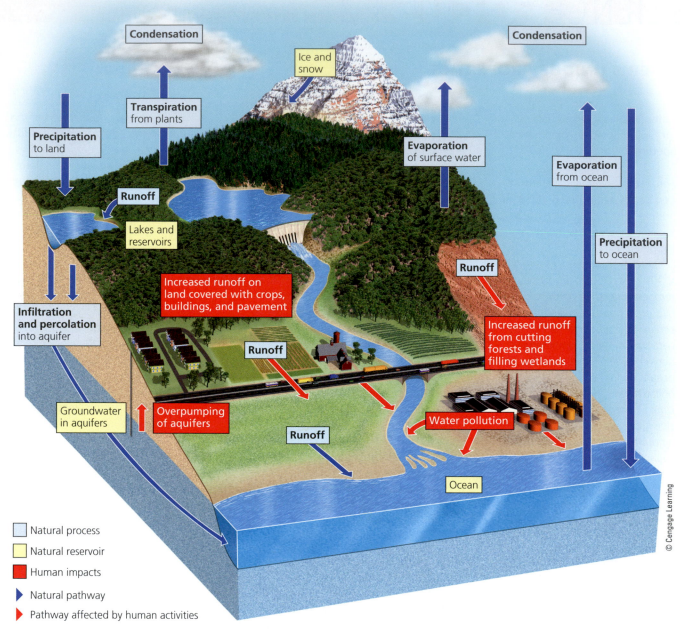

Condensation

Ice and snow

Condensation

Transpiration
from plants

Precipitation
to land

Evaporation
of surface water

Evaporation
from ocean

Runoff

Lakes and
reservoirs

Precipitation
to ocean

Runoff

Increased runoff on
land covered with crops,
buildings, and pavement

Infiltration
and percolation
into aquifer

Runoff

Increased runoff
from cutting
forests and
filling wetlands

Groundwater
in aquifers

Overpumping
of aquifers

Runoff

Water pollution

Runoff

Ocean

© Cengage Learning

☐ Natural process

☐ Natural reservoir

☐ Human impacts

▶ Natural pathway

▶ Pathway affected by human activities

ANIMATED FIGURE 3.13 Natural capital: Simplified model of the *water cycle*, or *hydrologic cycle*, in which water circulates in various physical forms within the biosphere. The red arrows and boxes identify major effects of human activities on this cycle. ***Question:*** What are three ways in which your lifestyle directly or indirectly affects the hydrologic cycle?

CONSIDER THIS. . .

CONNECTIONS Clearing a Rain Forest Can Affect Local Weather and Climate

Clearing vegetation can alter weather patterns by reducing transpiration, especially in dense tropical rain forests (**Core Case Study**). Because so many plants in such a forest transpire water into the atmosphere, vegetation is the primary source of local rainfall. Cutting down large areas of forest raises ground temperatures (by reducing shade) and can reduce local rainfall so much that the forest cannot grow back. If this occurs over a large area for a long enough time, the climate of the affected area can change, and much less diverse tropical grasslands can replace biologically diverse rain forests.

The Carbon Cycle

Carbon is the basic building block of the carbohydrates, fats, proteins, DNA, and other organic compounds necessary for life. Various compounds of carbon circulate through the biosphere, the atmosphere, and parts of the hydrosphere, in the **carbon cycle** shown in Figure 3.14.

A key component of the carbon cycle is carbon dioxide (CO_2) gas, which makes up about 0.040% of the volume of the earth's atmosphere and is also dissolved in water.

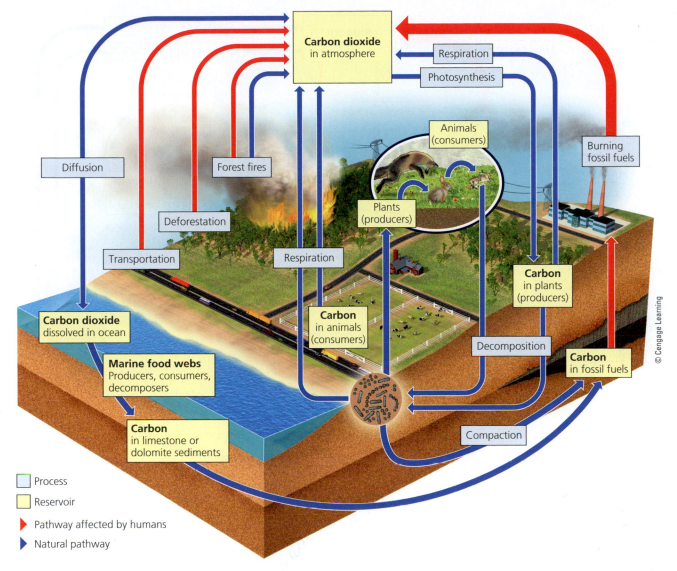

ANIMATED FIGURE 3.14 Natural capital: Simplified model showing the circulation of various chemical forms of carbon in the global *carbon cycle*, with major harmful impacts of human activities shown by the red arrows. (Yellow box sizes do not show relative reservoir sizes.) *Question:* What are three ways in which you directly or indirectly affect the carbon cycle?

Carbon dioxide affects the temperature of the earth's atmosphere through the greenhouse effect (Figure 3.3) and thus plays a major role in determining the earth's climate.

Carbon is cycled through the biosphere by a combination of *photosynthesis* by producers, which remove CO_2 from the air and water, and *aerobic respiration* by producers, consumers, and decomposers, which adds CO_2 in the atmosphere (Figure 3.8). Typically, CO_2 remains in the atmosphere for 100 years or more. In water, decomposers release carbon that can be stored as insoluble carbonates in bottom sediment for very long periods of time. Every living thing on the earth is part of the natural carbon cycle.

Over millions of years, the carbon in deeply buried deposits of dead plant matter and algae were converted by high pressure from the weight of overlying sediments and heat released during the decomposition of dead matter into carbon-containing *fossil fuels* such as coal, oil, and natural gas (Figure 3.14). In only a few hundred years, we have extracted and burned huge quantities of fossil fuels that took millions of years to form and added large quantities of CO_2 to the atmosphere, thus altering the carbon cycle (see the red arrows in Figure 3.14). Another way in which we alter the cycle is by clearing carbon-absorbing vegetation from many forests, especially tropical forests, faster than it can grow back (**Core Case Study**). These alterations are contributing to environmental problems affecting the atmosphere and oceans—topics we discuss further in Chapters 9 and 15.

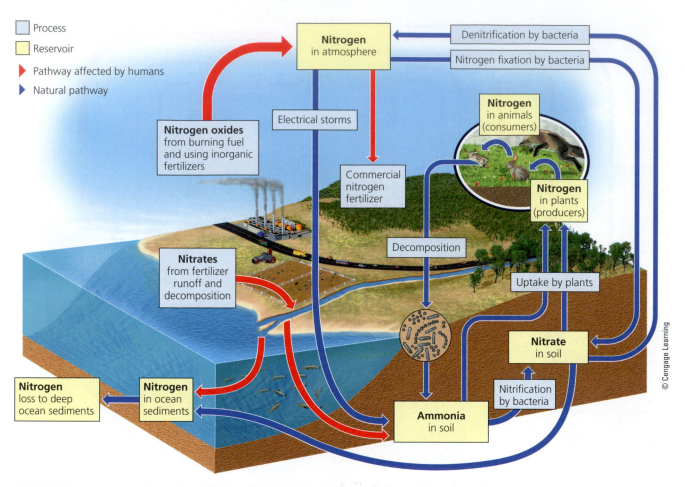

ANIMATED FIGURE 3.15 **Natural capital:** Simplified model showing the circulation of various chemical forms of nitrogen in the *nitrogen cycle*, with major harmful human impacts shown by the red arrows. (Yellow box sizes do not show relative reservoir sizes.) *Question:* What are two ways in which the carbon cycle and the nitrogen cycle are linked?

Legend:
- Process
- Reservoir
- Pathway affected by humans
- Natural pathway

Diagram labels:
- Denitrification by bacteria
- Nitrogen fixation by bacteria
- Nitrogen in atmosphere
- Electrical storms
- Nitrogen oxides from burning fuel and using inorganic fertilizers
- Nitrogen in animals (consumers)
- Commercial nitrogen fertilizer
- Nitrogen in plants (producers)
- Nitrates from fertilizer runoff and decomposition
- Decomposition
- Uptake by plants
- Nitrate in soil
- Nitrogen loss to deep ocean sediments
- Nitrogen in ocean sediments
- Ammonia in soil
- Nitrification by bacteria

© Cengage Learning

The Nitrogen Cycle: Bacteria in Action

Nitrogen gas (N_2) makes up 78% of the volume of the atmosphere. Nitrogen cannot be absorbed and used directly as a nutrient by plants or animals, but as a component of compounds such as ammonia (NH_3) and ammonium ions (NH_4^+), it becomes a plant nutrient.

These compounds are created within the **nitrogen cycle** (Figure 3.15) by reactions involving lightning and by specialized bacteria found in the top layer of soil. Other specialized bacteria convert most of the NH_3 and NH_4^+ in the soil to *nitrate ions* (NO_3^-), which are easily taken up by the roots of plants. The plants then use these forms of nitrogen to produce various proteins, nucleic acids, and vitamins. Animals that eat plants eventually consume these nitrogen-containing compounds, as do detritus feeders and decomposers. Different specialized bacteria in waterlogged soil and in the bottom sediments of lakes, oceans, swamps, and bogs convert these various nitrogen compounds back into nitrogen gas (N_2), which is released to the atmosphere to begin the nitrogen cycle again.

We intervene in the nitrogen cycle in several ways (see the red arrows in Figure 3.15). We add large amounts of nitric oxide (NO) as a product of combustion when we burn gasoline and other fuels. In the atmosphere, NO can be converted to nitrogen dioxide gas (NO_2) and nitric acid vapor (HNO_3), which can return to the earth's surface as damaging *acid deposition,* commonly called *acid rain.*

We also remove large amounts of nitrogen (N_2) from the atmosphere to make ammonia (NH_3) and ammonium ions (NH_4^+), which are used to make fertilizers. In addition, we alter the nitrogen cycle in aquatic ecosystems by adding excess nitrates (NO_3^-) to bodies of water through agricultural runoff of fertilizers and animal manure and through discharges from municipal sewage treatment systems. This can cause excessive growth of algae that can disrupt aquatic systems.

According to the 2005 Millennium Ecosystem Assessment, since 1950, human activities have more than doubled the annual release of nitrogen from the land into the rest of the environment, mostly from the greatly increased use of inorganic fertilizers to grow crops. The amount re-

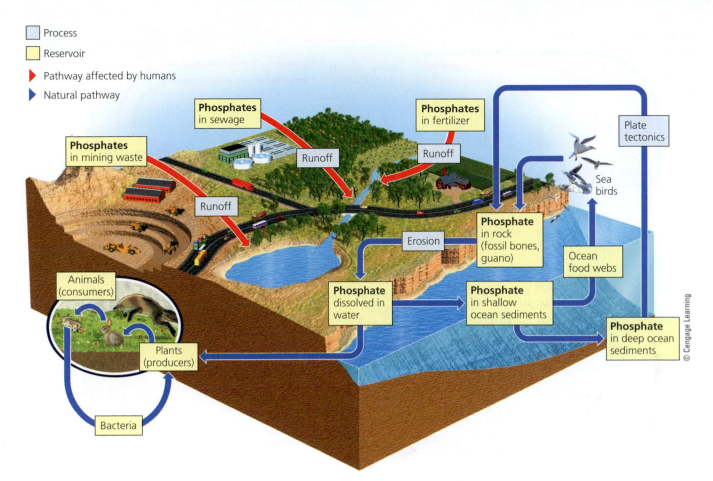

Process
Reservoir
Pathway affected by humans
Natural pathway

Phosphates in sewage

Phosphates in mining waste

Phosphates in fertilizer

Runoff

Runoff

Runoff

Plate tectonics

Sea birds

Phosphate in rock (fossil bones, guano)

Erosion

Ocean food webs

Animals (consumers)

Phosphate dissolved in water

Phosphate in shallow ocean sediments

Phosphate in deep ocean sediments

Plants (producers)

Bacteria

© Cengage Learning

FIGURE 3.16 Natural capital: Simplified model showing the circulation of various chemical forms of phosphorus (mostly phosphates) in the *phosphorus cycle*, with major harmful human impacts shown by the red arrows. (Yellow box sizes do not show relative reservoir sizes.) **Questions:** What are two ways in which the phosphorus cycle and the nitrogen cycle are linked? What are two ways in which the phosphorus cycle and the carbon cycle are linked?

leased is projected to double again by 2050 (see Figure 12, p. S44, in Supplement 5) and this would seriously alter the nitrogen cycle.

The Phosphorus Cycle

Compounds of phosphorus (P) circulate through water, the earth's crust, and living organisms in the **phosphorus cycle**, depicted in Figure 3.16. Most of these compounds contain *phosphate* ions (PO_4^{3-}), which serve as an important plant nutrient. Phosphorus does not cycle through the atmosphere and its cycle is slow compared to the cycling of water, carbon, and nitrogen.

As water runs over exposed rocks, it slowly erodes away inorganic compounds that contain phosphate ions and carries these ions into the soil where they can be absorbed by the roots of plants and by other producers. Phosphate compounds are then transferred by food webs from producers to consumers and eventually to detritus feeders and decomposers.

Phosphate can be lost from the cycle for long periods of time when it is washed into the ocean where it is typically deposited as marine sediment and can remain trapped for millions of years. Over time, geological processes can uplift and expose these seafloor deposits, from which phosphate can be eroded and freed up to reenter the phosphorus cycle.

Because most soils contain little phosphate, the lack of it often limits plant growth on land unless phosphorus (as phosphate salts mined from the earth) is applied to the soil as a fertilizer. Lack of phosphorus also limits the growth of producer populations in many freshwater streams and lakes because phosphate salts are only slightly soluble in water and thus do not release many phosphate ions to producers in aquatic systems.

Human activities, including the removal of large amounts of phosphate from the earth to make fertilizer, are affecting the phosphorus cycle (see red arrows in Figure 3.16). Also, by clearing tropical forests (**Core Case Study**), we reduce phosphate levels in tropical soils. Topsoil

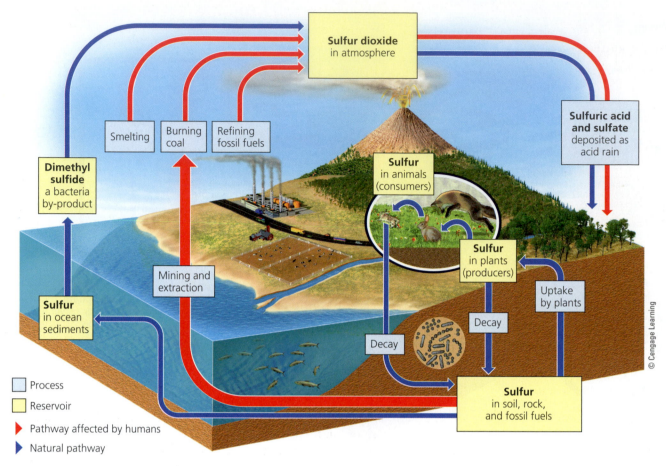

Process
Reservoir
Pathway affected by humans
Natural pathway

FIGURE 3.17 **Natural capital:** Simplified model showing the circulation of various chemical forms of sulfur, with major harmful human impacts shown by the red arrows. (Yellow box sizes do not show relative reservoir sizes.) **Questions:** What are two ways in which the sulfur cycle and the nitrogen cycle are linked? What are two ways in which the sulfur cycle and the carbon cycle are linked?

that is eroded from, and fertilizer that is washed from, fertilized crop fields, lawns, and golf courses carry large quantities of phosphate ions into streams, lakes, and oceans. There they stimulate the growth of producers such as algae and various aquatic plants, which can upset chemical cycling and other processes in bodies of water.

The Sulfur Cycle

Sulfur circulates through the biosphere in the **sulfur cycle**, shown in Figure 3.17. Much of the earth's sulfur is stored underground in rocks and minerals and in the form of sulfate (SO_4^{2-}) salts buried deep under ocean sediments.

Sulfur also enters the atmosphere from several natural sources. Hydrogen sulfide (H_2S)—a colorless, highly poisonous gas with a rotten-egg smell—is released from active volcanoes and from organic matter broken down by anaerobic decomposers in flooded swamps, bogs, and tidal flats. Sulfur dioxide (SO_2), a colorless and suffocating gas, also comes from volcanoes.

Particles of sulfate (SO_4^{2-}) salts, such as ammonium sulfate, enter the atmosphere from sea spray, dust storms, and forest fires. Plant roots absorb sulfate ions and incorporate the sulfur as an essential component of many proteins.

In the oxygen-deficient environments of flooded soils, freshwater wetlands, and tidal flats, specialized bacteria convert sulfate ions to sulfide ions (S^{2-}). The sulfide ions can then react with metal ions to form insoluble metallic sulfides, which are deposited as rock or metal ores (often extracted by mining and converted to various metals), and the cycle continues.

Human activities have affected the sulfur cycle primarily by releasing large amounts of sulfur dioxide (SO_2) into the atmosphere (as shown by the red arrows in Figure 3.17). We release sulfur to the atmosphere in three ways. *First,* we burn sulfur-containing coal and oil to produce electric power. *Second,* we refine sulfur-containing oil (petroleum) to make gasoline, heating oil, and other useful products. *Third,* we extract metals such as copper, lead, and zinc from sulfur-containing compounds in rocks that

individuals matter 3.1

Thomas E. Lovejoy—Forest Researcher and Biodiversity Educator

For several decades, conservation biologist and National Geographic Explorer Thomas E. Lovejoy has played a major role in educating scientists and the public about the need to understand and protect tropical forests. He has carried out research in the Amazon forests of Brazil since 1965, especially with regard to estimating the minimum area necessary for sustaining biodiversity in national parks and biological reserves in tropical forests. In 1980, he coined the term *biological diversity.*

Lovejoy served as the principal adviser for the popular and widely acclaimed public television series *Nature* in its early days and has written numerous articles and books on issues related to conservation of biodiversity. In addition to teaching environmental science and policy at George Mason University, he has held several prominent posts, including director of the World Wildlife Fund's conservation program, president of the Society for Conservation Biology, and executive director of the U.N. Environment Programme (UNEP). In 2012, he was awarded the Blue Planet Prize for his efforts to understand and sustain the earth's biodiversity.

are mined for these metals. In the atmosphere, SO_2 is converted to droplets of sulfuric acid (H_2SO_4) and particles of sulfate (SO_4^{2-}) salts, which return to the earth as acid deposition, which in turn can damage ecosystems.

3.5 HOW DO SCIENTISTS STUDY ECOSYSTEMS?

CONCEPT 3.5 Scientists use both field research and laboratory research, as well as mathematical and other models, to learn about ecosystems.

Some Scientists Study Nature Directly

Scientists use both field and laboratory research and mathematical and other models to learn about ecosystems (**Concept 3.5**). *Field research*, sometimes called "muddy-boots biology," involves going into forests and other natural settings to study the structure of ecosystems and what happens in them (see Chapter 2 opening photo). Most of what we know about ecosystems has come from such research (Individuals Matter 3.1). **GREEN CAREER: Ecologist**

Scientists use a variety of methods to study tropical forests (**Core Case Study**). In a few cases, ecologists have erected tall construction cranes that provide them access to the canopies of tropical forests. This, along with climbing trees (see Chapter 2 opening photo) and erecting rope walkways between treetops, has helped them to identify

and observe the rich diversity of species living or feeding in these treetop habitats.

Conservationist, pilot, and National Geographic Explorer Michael Fay has flown over vast areas of African forests. The government of Gabon used the 116,000 photographs that he took to create a system of 13 protected national parks. Recently he spent a year hiking through what remains of California's redwood forests and collecting ecological data to help understand and protect this one-of-a-kind ecosystem.

Sometimes ecologists carry out controlled experiments by isolating and changing a variable in part of an area and comparing the results with nearby unchanged areas. We learned about a classic example of this in the Core Case Study of Chapter 2 (p. 26).

Scientists also use aircraft and satellites equipped with sophisticated cameras and other *remote sensing* devices to scan and collect data on the earth's surface (see Figure 2, p. S16, in Supplement 4). Then they use *geographic information system (GIS)* software to capture, store, analyze, and display such information. For example, a GIS can convert digital satellite images into global, regional, and local maps showing variations in vegetation, gross primary productivity, air pollution emissions, and many other variables.

Some researchers attach tiny radio transmitters to animals and use global positioning systems (GPS) to learn about the animals by tracking where and how far they go. This technology is very important for studying endangered species (which we discuss in Chapter 8). **GREEN CAREERS: GIS analyst; remote sensing analyst**

Concept 3.5   **57**

TESTING PLANETARY BOUNDARIES: FROM HOLOCENE TO ANTHROPOCENE

For most of the past 10,000–12,000 years we have been living in an era called the **Holocene**—a period of relatively stable climate and other environmental conditions following a long glacial period. This general stability has allowed the human population to grow, develop agriculture, and take over a large and growing share of the earth's land and other resources.

According to a number of scientists, since the industrial revolution began around 1750, we have entered an era called the **Anthropocene**. In this new era, humans have become major agents of change in the functioning of the earth's life-support system as their ecological footprints have spread over the earth (see Figure 1-11, p. 14).

In 2009, an international group of 28 scientists, led by Johan Rockström of the Stockholm Resilience Centre, estimated nine major *planetary boundaries* or *ecological tipping points*, three of which they say we have already

exceeded (Figure 3.A). According to these scientists, if we exceed too many of these boundaries, we could trigger abrupt and long-lasting or irreversible environmental changes that could seriously degrade the earth's life-support system.

Some scientists argue that there is an urgent need for more research to fill in the missing data on these

planetary boundaries and on how our exceeding them could affect the health of humans, other species, and the earth's life-support systems. Such information combined with action could help us to change the nature of the Anthropocene era by shrinking our ecological footprints while creating and expanding beneficial environmental impacts.

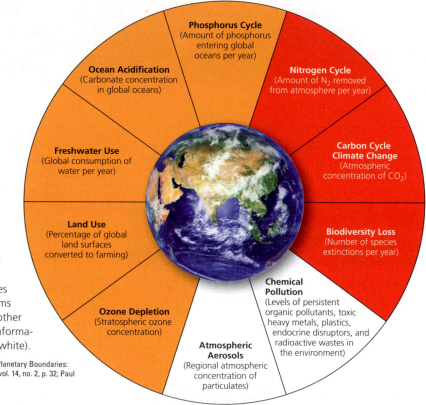

FIGURE 3.A Planetary boundaries for nine major components of the earth's life-support system. A team of scientists estimated that human activities have exceeded the boundary limits for three systems (shown in red) and are close to the limits for five other systems (shown in orange). There is not enough information to evaluate the other two systems (shown in white).

Compiled by the authors using data from Johan Rockström et al., 2009, "Planetary Boundaries: Exploring the Safe Operating Space for Humanity," *Ecology and Society*, vol. 14, no. 2, p. 32; Paul Crutzen; and James Hansen.

Photo: Sailorr/Shutterstock.com

CONSIDER THIS. . .

CONNECTIONS Drone Technology and Environmental Science

Researchers are increasingly using small drones to track whale migrations, monitor the size of the Arctic ice pack, and evaluate soil erosion, deforestation, water usage, and pest outbreaks in various areas of the world.

Some Ecologists Use Laboratory Experiments or Modeling

Since the 1960s, ecologists have increasingly supplemented field research by using *laboratory research*—setting up, observing, and making measurements of model ecosystems and populations under laboratory conditions. They have created such simplified systems in containers such as culture tubes, bottles, aquariums, and greenhouses, and in indoor and outdoor chambers where they can control temperature, light, CO_2, humidity, and other variables.

Such systems make it easier for scientists to carry out controlled experiments, which are often quicker and less costly than similar experiments in the field. But scientists must consider how well their scientific observations and measurements in a simplified, controlled system under laboratory conditions reflect what actually takes place under the more complex and often changing conditions found in nature.

Since the late 1960s, ecologists have developed mathematical models that simulate ecosystems. By running such models on high-speed supercomputers, they try to understand large and very complex systems, such as lakes, oceans, forests, and the earth's climate system, that cannot be adequately studied and modeled in field or laboratory research. **GREEN CAREER: Ecosystem modeler**

We Need to Learn More about the Health of the World's Ecosystems

According to the 2005 Millennium Ecosystem Assessment, scientists do not have enough basic ecological data to fully evaluate the status of the world's ecosystems, to see how they are changing, and to develop effective strategies for preventing or slowing their degradation. Ecologists have called for a massive program to develop baseline data to meet these needs. A high priority for some scientists is to clearly identify certain *planetary boundaries* or *ecological tipping points* (see p. 37 and Science Focus 3.3) to help us avoid reaching or passing them.

TYING IT ALL TOGETHER
Tropical Rain Forests and Sustainability

This chapter began with a discussion of the importance of the world's incredibly diverse tropical rain forests (**Core Case Study**). These ecosystems showcase the functioning of the three **scientific principles of sustainability**, which apply as well to the world's other ecosystems.

First, producers within rain forests rely on *solar energy* to produce a vast amount of biomass through photosynthesis. *Second*, species living in the forests take part in and depend on the *cycling of nutrients* and the flow of energy within the forests and throughout the biosphere. *Third*, tropical rain forests contain a huge and vital part of the earth's *biodiversity*, and interactions among species living in these forests help to sustain these complex ecosystems.

We also reported recent research on the possible long-lasting, harmful effects of our exceeding any key planetary boundaries. In many of the chapters to follow, we will further examine such risks, and we will consider ways in which we can apply the six **principles of sustainability** (see Supplement 7, pp. S50–S51) to try to stay within the key planetary boundaries, to live more sustainably, and to create and expand beneficial environmental impacts.

Anneka/Shutterstock.com

Chapter Review

Core Case Study

1. What are three harmful effects of the clearing and degradation of tropical rain forests?

Section 3.1

2. What are the two key concepts for this section? Define and distinguish among the **atmosphere**, **troposphere**, **stratosphere**, **hydrosphere**, **geosphere**, and **biosphere**. What three interconnected factors sustain life on the earth? Describe the flow of energy to and from the earth. What is the **greenhouse effect** and why is it important?

Section 3.2

3. What is the key concept for this section? Define **ecology**. Define **organism**, **population**, **community**, and **ecosystem**, and give an example of each.

4. Distinguish between the living and nonliving components in ecosystems and give two examples of each.

5. What is a **trophic level**? Distinguish among **producers (autotrophs)**, **consumers (heterotrophs)**, **decomposers** and **detritus feeders (detritivores)**, and give an example of each. Summarize the processes of **photosynthesis**. Distinguish among **primary consumers (herbivores)**, **carnivores**, **secondary consumers**, **tertiary consumers**, and **omnivores**, and give an example of each.

6. Explain the importance of microbes. What is **aerobic respiration**? What two processes sustain ecosystems and the biosphere and how are they linked?

Section 3.3

7. What is the key concept for this section? Define and distinguish between a **food chain** and a **food web**.

Explain what happens to energy as it flows through food chains and food webs. What is a **pyramid of energy flow**?

8. Distinguish between **gross primary productivity (GPP)** and **net primary productivity (NPP)**, and explain their importance. What are the two most productive land ecosystems and the two most productive aquatic ecosystems?

Section 3.4

9. What is the key concept for this section? What happens to matter in an ecosystem? What is a **nutrient cycle**? Explain how nutrient cycles connect past, present, and future life. Summarize the unique properties of water. Describe the **hydrologic cycle**, or **water cycle**. What three major processes are involved in the water cycle? What is **surface runoff**? Define **groundwater**. What is an **aquifer**? What percentage of the earth's water supply is available to humans and other species as liquid freshwater? Explain how human activities are affecting the water cycle. Explain how clearing a rain forest can affect local weather and climate. Describe the **carbon**, **nitrogen**, **phosphorus**, and **sulfur cycles**, and explain how human activities are affecting each cycle.

Section 3.5

10. What is the key concept for this section? Describe three ways in which scientists study ecosystems. Explain why we need much more basic data about the structure and condition of the world's ecosystems. Distinguish between the **Holocene** and **Anthropocene** eras. List nine planetary boundaries that scientists have identified. Which three of these boundaries have already been exceeded, according to these scientists? What are this chapter's *three big ideas*? How are the three **scientific principles of sustainability** showcased in tropical rain forests?

Note: Key terms are in bold type.

Critical Thinking

1. How would you explain the importance of tropical rain forests (**Core Case Study**) to people who think that such forests have no connection to their lives?

2. Explain **(a)** why the flow of energy through the biosphere depends on the cycling of nutrients, and **(b)** why the cycling of nutrients depends on gravity.

3. Explain why microbes are so important. What are two ways in which they benefit your health or lifestyle? Write a brief description of what you think would happen to you if microbes were eliminated from the earth.

4. Make a list of the foods you ate for lunch or dinner today. Trace each type of food back to a particular producer species. Describe the sequence of feeding levels that led to your feeding.

5. Use the second law of thermodynamics (see Chapter 2, p. 35) to explain why many poor people in less-developed countries live on a mostly vegetarian diet.

6. How might your life and the lives of any children or grandchildren you might have be affected if human activities as a whole continue to intensify the water cycle?

7. What would happen to an ecosystem if (a) all of its decomposers and detritus feeders were eliminated, (b) all of its producers were eliminated, and (c) all of its insects were eliminated? Could a balanced ecosystem exist with only producers and decomposers and no consumers such as humans and other animals? Explain.

8. For each of the earth's nine major planetary boundaries (Figure 3.A), describe how our exceeding that boundary might affect (a) you, (b) any child you might have, and (c) any grandchild you might have.

Doing Environmental Science

Visit a nearby terrestrial ecosystem or aquatic life zone and try to identify major producers, primary and secondary consumers, detritus feeders, and decomposers. Take notes and describe at least one example of each of these types of organisms. Make a simple sketch showing how these organisms might be related to each other or to other organisms in a food chain or food web. Think of two ways in which this food web or chain could be disrupted. Write a report summarizing your research and conclusions.

Global Environment Watch Exercise

Search for *Nitrogen Cycle* and look for information on how humans are affecting the nitrogen cycle. Specifically look for impacts on the atmosphere and on human health from emissions of nitrogen oxides, and look for the harmful ecological effects of the runoff of nitrate fertilizers into rivers and lakes. Make a list of these impacts and use this information to review your daily activities. Find three things that you do regularly that contribute to these impacts.

Data Analysis

Recall that net primary productivity (NPP) is the *rate* at which producers can make the chemical energy that is stored in their tissues and that is potentially available to other organisms (consumers) in an ecosystem. In Figure 3.15, it is expressed as units of energy (kilocalories, or *kcal*) produced in a given area (square meters, or m^2) over a period of time (a year). Look again at Figure 3.12 and consider the differences in NPP among various ecosystems. Then answer the following questions:

1. What is the approximate NPP of a tropical rain forest in kcal/m^2/yr? Which terrestrial ecosystem produces about one-third of that rate? Which aquatic ecosystem has about the same NPP as a tropical rain forest?

2. Early in the 20th century, large areas of temperate forestland in the United States were cleared to make way for agricultural land. For each unit of this forest area that was cleared and replaced by farmland, about how much NPP was lost?

3. Why do you think deserts and grasslands have dramatically lower NPP than swamps and marshes?

4. About how many times more NPP do estuaries produce, compared to lakes and streams? Why do you think this is so?

CENGAGE **brain**.com To access course materials, including Aplia homework, please visit www.cengagebrain.com.

WWW.CENGAGEBRAIN.COM **61**

4 BIODIVERSITY AND EVOLUTION

There is grandeur to this view of life. . .
that, whilst this planet has gone cycling
on. . . endless forms most beautiful and
most wonderful have been, and are
being, evolved.

CHARLES DARWIN

4.3 How does the earth's life change over time?

4.4 What factors affect biodiversity?

Endangered San Lucas marsupial frog in
Ecuador.

Pete Oxford/Minden Pictures

Why Are Amphibians Vanishing?

Amphibians—frogs, toads, and salamanders—were among the earliest vertebrates (animals with backbones) to emerge from the earth's waters and live on the land. They have adjusted to and survived environmental changes more effectively than many other species. However, the amphibian world is changing rapidly.

An amphibian lives part of its life in water and part on land. Now, many of the 6,700 or more amphibian species are having difficulty adapting to rapid changes that have taken place in their water and land habitats during the past few decades. Such changes have resulted primarily from human activities such as use of pesticides and other chemicals that can become water pollutants.

Since 1980, populations of hundreds of amphibian species throughout the world have declined or vanished (Figure 4.1). According to the International Union for Conservation of Nature (IUCN), about 41% of all known amphibian species are threatened with *extinction,* the process in which they could cease to exist (see chapter-opening photo).

No single cause has been found to explain the declines of many amphibian species. However, scientists have identified a number of factors that affect amphibians at various points in their life cycles. For example, frog eggs have no shells to protect frog embryos from water pollutants, and adult frogs are often exposed to insecticides contained in many of the insects they eat. We explore these and other factors later in this chapter.

Why should we care if some amphibian species become extinct? Scientists give three reasons. *First,* amphibians are sensitive biological *indicators* of changes in environmental conditions such as habitat loss, air and water pollution, ultraviolet (UV) radiation, and climate change. The growing threats to the survival of an increasing number of amphibian species indicate that environmental conditions are deteriorating in many parts of the world.

Second, adult amphibians play important ecological roles in biological communities. For example, amphibians eat more insects (including mosquitoes) than do birds. In some habitats, the extinction of certain amphibian species could lead to the extinction of other species of amphibians, as well as some aquatic insects, reptiles, birds, fish, and mammals that feed primarily on amphibians or their larvae.

Third, amphibians are a genetic storehouse of pharmaceutical products waiting to be discovered. For example, compounds in secretions from the skin of certain amphibians have been isolated and used as painkillers and antibiotics, and in treatments for burns and heart disease.

Many scientists believe that the threats to amphibians present a warning about a number of environmental threats to the earth's biodiversity. In this chapter, we will learn about biodiversity, how it has arisen, why it is important, and how it is threatened. We will also look at some possible solutions to these problems.

FIGURE 4.1 Specimens of some of the nearly 200 amphibian species that have gone extinct since the 1970s.

4.1 WHAT IS BIODIVERSITY AND WHY IS IT IMPORTANT?

CONCEPT 4.1 The biodiversity found in genes, species, ecosystems, and ecosystem processes is vital to sustaining life on the earth.

Biodiversity Is a Crucial Part of the Earth's Natural Capital

Biological diversity, or **biodiversity**, is the variety of the earth's species, the genes they contain, the ecosystems in which they live, and the ecosystem processes such as energy flow and nutrient cycling that sustain all life (Figure 4.2). Acting together, these four interacting compo-

nents of the earth's biodiversity provide us with the ecosystem services (Figure 1.3, orange items, p. 7) that sustain our lives and economies and that tend to make ecosystems more resilient and adaptable.

Estimates of the number of species on the earth range from 7 million to 100 million, with a best guess of 7–10 million species. So far, biologists have identified about 2 million species—most of them being insects (Science Focus 4.1).

Species diversity, the number and variety of the species present in any biological community, is the most obvious component of biodiversity. Scientists think that ecosystems with high levels of species diversity tend to be more stable. Another important component is *genetic diversity,* the variety of genes found in a population or in a species (Figure 4.3), which enable the earth's species to survive and adapt to dramatic environmental changes.

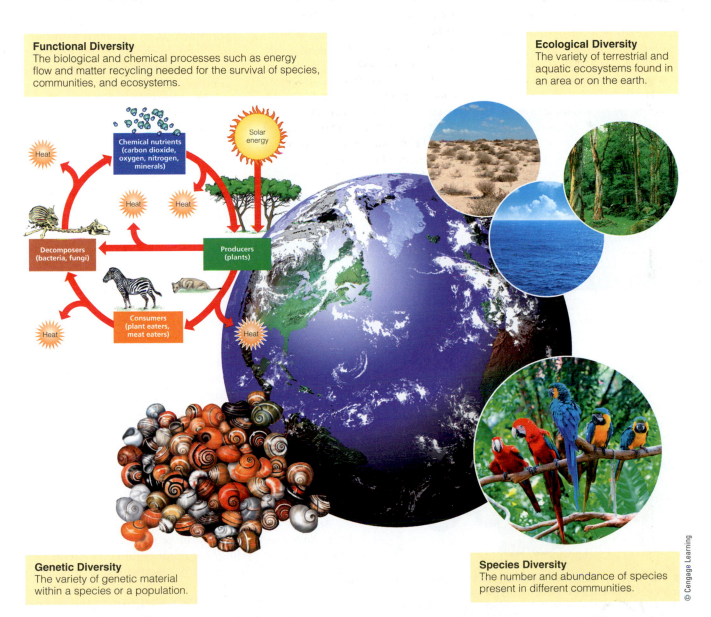

Functional Diversity
The biological and chemical processes such as energy flow and matter recycling needed for the survival of species, communities, and ecosystems.

Ecological Diversity
The variety of terrestrial and aquatic ecosystems found in an area or on the earth.

Chemical nutrients (carbon dioxide, oxygen, nitrogen, minerals)

Solar energy

Heat

Producers (plants)

Decomposers (bacteria, fungi)

Consumers (plant eaters, meat eaters)

Heat

Genetic Diversity
The variety of genetic material within a species or a population.

Species Diversity
The number and abundance of species present in different communities.

© Cengage Learning

FIGURE 4.2 Natural capital: The major components of the earth's *biodiversity*—one of the planet's most important renewable resources and a key component of its natural capital (see Figure 1.3, p. 7)
Question: Do you think we should protect the earth's biodiversity from harmful human activities? Explain.

INSECTS PLAY A VITAL ROLE IN OUR WORLD

We classify many insect species as *pests* because they compete with us for food, spread human diseases such as malaria, bite or sting us, and invade our lawns, gardens, and houses. Some people fear insects and many think the only good bug is a dead bug. They fail to recognize the vital roles insects play in helping to sustain life on the earth.

For example, *pollination* is a vital ecosystem service that allows flowering plants to reproduce sexually when pollen grains are transferred from the flower of one plant to a receptive part of the flower of another plant of the same species. Many of the earth's plant flowering species depend on insects to pollinate their flowers (Figure 4.A, left). Also, insects that eat other insects—such as the praying mantis (Figure 4.A, right)—help to control the populations of at least half the insect species we call pests. This free source of pest control is another vital ecosystem service.

Certain environmental changes, some of them caused by human activities, are threatening insect populations and their ecosystem services around the world. *Entomologists*—scientists who study insects—are expanding their research in areas related to such environmental threats. For example, entomologist Diana Cox-Foster of Pennsylvania State University is studying the decline of honeybees, which are extremely important pollinators. This decline threatens to disrupt whole ecosystems that depend on bees for pollination, as well as much of the human food supply. We discuss this serious environmental problem more fully in Chapter 8.

Critical Thinking

Can you think of three insect species not discussed above that benefit your life?

Darlyne A. Murawski/National Geographic Creative

Dr. Morley Read/Shutterstock.com

FIGURE 4.A *Importance of insects:* Bees (left) and numerous other insects pollinate flowering plants that serve as food for many plant eaters, including humans. This praying mantis, which is eating a moth (right), and many other insect species help to control the populations of most of the insect species we classify as pests.

© Cengage Learning

FIGURE 4.3 *Genetic diversity* among individuals in this population of a species of Caribbean snail is reflected in the variations in shell color and banding patterns. Genetic diversity can also include other variations such as slight differences in chemical makeup, sensitivity to various chemicals, and behavior.

Ecosystem diversity—the earth's variety of deserts, grasslands, forests, mountains, oceans, lakes, rivers, and wetlands—is another major component of biodiversity. Biologists have classified the terrestrial (land) ecosystems into **biomes**—large regions such as forests, deserts, and grasslands with distinct climates and certain species (especially vegetation) typically occurring within them. Figure 4.4 shows different major biomes along the 39th parallel spanning the United States. We discuss biomes in more detail in Chapter 7.

Another important component of biodiversity is *functional diversity*—the variety of processes such as energy flow and matter cycling that occur within ecosystems (see Figure 3.8, p. 47) as species interact with one another in food chains and food webs.

The earth's biodiversity is a vital part of the natural capital (see Figure 1.3, p. 7) on which we depend. With the help of technology, we use the earth's biodiversity as

individuals matter 4.1

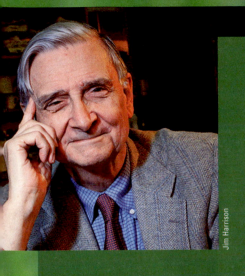
Jim Harrison

Edward O. Wilson: A Champion of Biodiversity

As a boy growing up in the southeastern United States, Edward O. Wilson became interested in insects at age 9. He has said, "Every kid has a bug period. I never grew out of mine."

Before entering college, Wilson had decided he would specialize in the study of ants. He became one of the world's experts on ants and then steadily widened his focus, eventually to include the entire biosphere. One of Wilson's landmark works is *The Diversity of Life*, published in 1992, in which he presented the principles and practical issues of biodiversity more completely than anyone had to that point. He is now deeply involved in writing and lecturing about the need for global conservation efforts and is promoting the goal of completing a global survey of biodiversity.

Wilson has won more than 100 national and international awards and has written 28 books, two of which won the Pulitzer Prize for General Nonfiction. About the importance of biodiversity, he writes: "How can we save Earth's life forms from extinction if we don't even know what most of them are? . . . I like to call Earth a little known planet."

Background photo: Christian Musat/Shutterstock.com

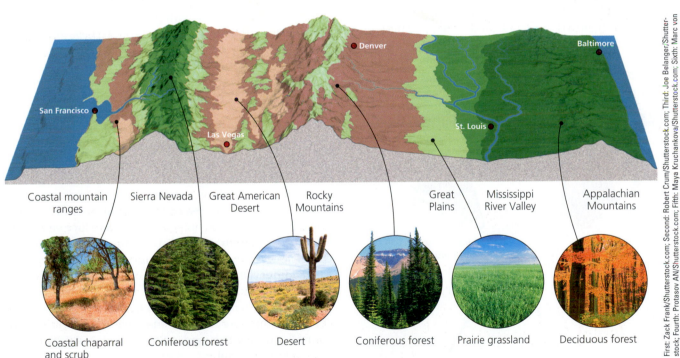

FIGURE 4.4 The variety of biomes found along the 39th parallel across the United States.

First: Zack Frank/Shutterstock.com; Second: Robert Crum/Shutterstock.com; Third: Joe Belanger/Shutterstock; Fourth: Protasov AN/Shutterstock.com; Fifth: Maya Kruchankova/Shutterstock.com; Sixth: Marc von Hacht/Shutterstock.com.

a source of food, medicine, building materials, and energy. Biodiversity also provides critical ecosystem services such as air and water purification, renewal of topsoil, decomposition of wastes, and pollination. In addition, the earth's variety of species and ecosystems serve as raw materials for the evolution of new species and ecosystem services, in response to changing environmental conditions. We owe much of what we know about biodiversity to researchers such as Edward O. Wilson (Individuals Matter 4.1).

4.2 WHAT ROLES DO SPECIES PLAY IN ECOSYSTEMS?

CONCEPT 4.2A Each species plays a specific ecological role called its *niche*.

CONCEPT 4.2B Any given species may play one or more of four important roles—native, nonnative, indicator, or keystone—in a particular ecosystem.

Each Species Plays a Role in Its Ecosystem

An important principle of ecology is that *each species has a specific role to play in the ecosystems where it is found* (**Concept 4.2A**). Scientists describe the role that a species plays in its ecosystem as its **ecological niche**, or simply **niche**. It is a species' way of life in a community and includes everything that affects its survival and reproduction, such as how much water and sunlight it needs, how much space it requires, what it feeds on, what feeds on it, and the temperatures and other conditions it can tolerate. A species' niche should not be confused with its **habitat**, which is the place, or type of ecosystem, in which it lives.

Scientists use the niches of species to classify them mostly as *generalists* or *specialists*. **Generalist species** such as raccoons have broad niches (Figure 4.5, right curve). They can live in many different places, eat a variety of foods, and often tolerate a wide range of environmental conditions. Other generalist species are flies, cockroaches, rats, white-tailed deer, and humans.

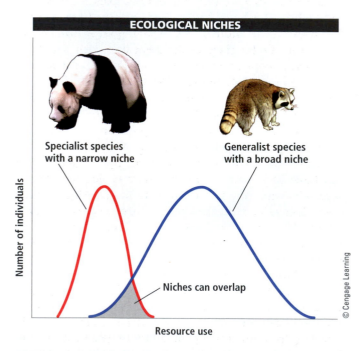

ECOLOGICAL NICHES

Specialist species with a narrow niche

Generalist species with a broad niche

Niches can overlap

Number of individuals

Resource use

© Cengage Learning

FIGURE 4.5 Specialist species such as the giant panda have a narrow niche (left curve) and generalist species such as the raccoon have a broad niche (right curve).

In contrast, **specialist species** such as the giant panda occupy narrow niches (Figure 4.5, left curve). They may be able to live in only one type of habitat, use only one or a few types of food, or tolerate a narrow range of environmental conditions. For example, some shorebirds are specialized to feed on crustaceans, insects, or other organisms found on sandy beaches and their adjoining coastal wetlands (Figure 4.6).

Because of their narrow niches, specialists are more prone to extinction when environmental conditions change. For example, China's *giant panda* (Figure 4.5, left) is highly endangered because of a combination of habitat loss, low birth rate, and its specialized diet consisting mostly of bamboo.

Is it better to be a generalist or a specialist? It depends. When environmental conditions are fairly constant, as in a tropical rain forest, specialists have an advantage because they have fewer competitors. But under rapidly changing environmental conditions, the more adaptable generalist usually is better off.

Species Can Play Four Major Roles within Ecosystems

Niches can be classified further in terms of specific roles that certain species play within ecosystems. Ecologists describe *native, nonnative, indicator,* and *keystone* roles. Any given species may play one or more of these four roles in a particular ecosystem (**Concept 4.2B**).

Native species are those species that normally live and thrive in a particular ecosystem. Other species that migrate into, or are deliberately or accidentally introduced into an ecosystem are called **nonnative species**, also referred to as *invasive, alien,* and *exotic species*.

People often think of nonnative species as threatening. In fact, most domesticated species, including many food crops, flowers, chickens, cattle, and fish, benefit people in the areas where they were introduced. However, some nonnative species can compete with and reduce an ecosystem's native species, causing unintended and unexpected consequences. In 1957, for example, Brazil imported wild African honeybees to help increase honey production, but the bees displaced some native honeybee populations, which led to a reduced honey supply. Also, these bees are aggressive and unpredictable and have killed thousands of domesticated animals and an estimated 1,000 people in the western hemisphere, many of whom were allergic to bee stings.

Nonnative species can spread rapidly if they find a new location with favorable conditions. In their new niches, some of these species may not face the predators and diseases they faced in their native niches, or they may be able to out-compete some native species in their new locations. We examine this environmental threat in greater detail in Chapter 8.

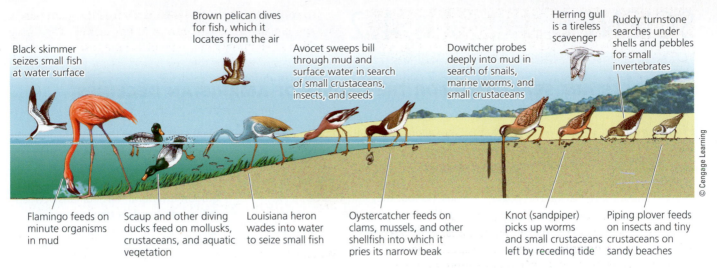

Black skimmer seizes small fish at water surface

Brown pelican dives for fish, which it locates from the air

Avocet sweeps bill through mud and surface water in search of small crustaceans, insects, and seeds

Dowitcher probes deeply into mud in search of snails, marine worms, and small crustaceans

Herring gull is a tireless scavenger

Ruddy turnstone searches under shells and pebbles for small invertebrates

© Cengage Learning

Flamingo feeds on minute organisms in mud

Scaup and other diving ducks feed on mollusks, crustaceans, and aquatic vegetation

Louisiana heron wades into water to seize small fish

Oystercatcher feeds on clams, mussels, and other shellfish into which it pries its narrow beak

Knot (sandpiper) picks up worms and small crustaceans left by receding tide

Piping plover feeds on insects and tiny crustaceans on sandy beaches

FIGURE 4.6 Various bird species in a coastal wetland occupy specialized feeding niches. This specialization reduces competition and allows for sharing of limited resources.

Indicator Species Serve as Biological Smoke Alarms

Species that provide early warnings of environmental change in a community or an ecosystem are called **indicator species**. For example, in this chapter's Core Case Study, we learned that some amphibians are classified as indicator species. One reason for this is that a 2005 study found an apparent correlation between climate change caused by atmospheric warming and the extinction of about two-thirds of the 110 known species of harlequin frogs in tropical forests of Central and South America.

The decline of amphibian populations likely results from a number of factors. Scientists have fanned out to several fronts to explore some of the possible causes (Science Focus 4.2).

Keystone Species Play Critical Roles in Their Ecosystems

Keystone species are species whose roles have a large effect on the types and abundance of other species in an ecosystem. Many of these species occur in small and dwindling numbers, so some keystone species are more vulnerable to extinction than other species are.

Keystone species can play several critical roles in helping to sustain ecosystems. One such role is the pollination of flowering plant species by butterflies, honeybees (Figure 4.A), hummingbirds, bats, and other species. In addition, top predator keystone species feed on and help to regulate the populations of other species. Examples are wolves, leopards, lions, the American alligator (see the following Case Study), and some shark species (see second Case Study that follows).

The loss of a keystone species can lead to population crashes and extinctions of other species in a community

that depends on them for certain ecosystem services. This is why it so important for scientists to identify keystone species and for us to protect them.

The American Alligator—A Keystone Species That Almost Went Extinct

The American alligator (Figure 4.7) is a keystone species because it plays a number of important roles in the ecosystems where it is found in the southeastern United States.

In the 1930s, hunters began killing large numbers of these animals for their exotic meat and their soft belly skin, used to make expensive shoes, belts, and pocketbooks. Other people hunted alligators for sport or out of dislike for the large reptile. By the 1960s, hunters and poachers had wiped out 90% of the alligators in the state of Louisiana, and the alligator population in the Florida Everglades was also near extinction.

Those who did not care much for the alligator might not have been aware of its important ecological role—its *niche*—in subtropical wetland ecosystems. Alligators dig deep depressions, or gator holes. These depressions hold freshwater during dry spells, serve as refuges for aquatic life, and supply freshwater and food for fishes, insects, snakes, turtles, birds, and other animals.

The large nesting mounds that alligators build provide nesting and feeding sites for some herons and egrets, and red-bellied turtles lay their eggs in old gator nests. In addition, alligators eat large numbers of gar, a predatory fish, which helps to maintain populations of game fish such as bass and bream that the gar eat.

As alligators create gator holes and nesting mounds, they help to keep shore and open water areas free of in-

SCIENTISTS ARE SEARCHING FOR THE CAUSES OF AMPHIBIAN DECLINES

Herpetologists, the scientists who study frogs and other amphibians, have identified a number of factors—both natural and human-caused—that threaten these species at various points in their life cycles. One of the natural causes is *parasites* such as flatworms that feed on certain amphibian eggs. Scientists think this has caused birth defects such as missing limbs or extra limbs in some amphibians.

Some herpetologists hypothesize that *viral and fungal diseases,* especially the chytrid fungus (Figure 4.B) that attacks the skin of frogs, are reducing the frogs' ability to ingest water through their skin. This leads to death from dehydration. Such diseases can spread fairly easily, because adults of many amphibian species congregate in large numbers to breed.

Another major threat to amphibians is *habitat loss and fragmentation.* This is mostly a human-caused problem resulting from the clearing of forests and the draining and filling of freshwater wetlands for farming and urban development. Another threat is *prolonged drought,* which can dry up breeding pools that frogs and other amphibians depend on for reproduction and survival through their early stages of life (Figure 4.C).

Another human-influenced problem is *higher levels of UV radiation,* which can harm embryos of amphibians in shallow ponds as well as adults basking in the sun for warmth. Historically, such radiation has been screened by ozone in the stratosphere, but during the past few decades, ozone-depleting chemicals released into the troposphere from human sources have drifted upward into the stratosphere and destroyed some of the protective ozone found there.

Pollution is another human-caused threat to amphibians. Frogs and other species are increasingly exposed to pesticides in ponds and in the bodies of insects that they eat. This can make them more vulnerable to bacterial, viral, and fungal diseases and to some parasites. Amphibian expert and National Geographic Explorer Tyrone Hayes, a professor of biology at University of California, Berkeley, is conducting research on how some pesticides can harm frogs and other animals by disrupting their endocrine systems.

Overhunting is another problem, especially in areas of Asia and Europe, where frogs are hunted for their leg meat. Yet another threat is the invasion of amphibian habitats by *nonnative predators and competitors,* such as certain fish species. Some of this immigration is natural, but humans accidentally or deliberately transport many species to amphibian habitats.

Most herpetologists believe that a combination of these factors, which vary from place to place, probably is responsible for most of the decline and disappearances among amphibian species.

Critical Thinking

Of the factors listed above, which three do you think could be most effectively controlled by human efforts?

FIGURE 4.B Frogs killed by the chytrid fungus at a high-elevation lake in California.

Joel Sartore/National Geographic Creative

Michael & Patricia Fogden/Minden Pictures/Getty Images

FIGURE 4.C This golden toad lived in Costa Rica's high-altitude Monteverde Cloud Forest Reserve. The species became extinct in 1989, apparently because its habitat dried up.

The American alligator plays a key role in its habitat.

FIGURE 4.7 *Keystone species:* The American alligator plays an important ecological role in its marsh and swamp habitats in the southeastern United States.

vading vegetation. Without this free ecosystem service, freshwater ponds and coastal wetlands where alligators live would be filled in with shrubs and trees, and dozens of species would disappear from these ecosystems. For these reasons, most ecologists classify the American alligator as a keystone species vital to the sustainability of the ecosystems in which it is found.

In 1967, the U.S. government placed the American alligator on the endangered species list. By 1977, because it was protected, its populations had made a strong enough comeback to be removed from the endangered species list. Today, there are well over a million alligators in Florida. The state now allows property owners to kill alligators that stray onto their land.

To conservation biologists, the comeback of the American alligator is an important success story in wildlife conservation. Recently, however, large and rapidly reproducing Burmese and African pythons released deliberately or accidently by humans have invaded the Florida Everglades, as discussed in Chapter 8. These nonnative invaders are capable of feeding on young alligators and thus could threaten the long-term survival of the American alligators of the Everglades.

CASE STUDY

Why Should We Protect Sharks?

As *keystone species,* certain shark species play crucial roles in helping to keep their ecosystems functioning. Sharks that feed at or near the tops of their food webs remove injured and sick animals from the ocean. Without this free ecosystem service, the oceans would be teeming with dead and dying fish and marine mammals.

FIGURE 4.8 The threatened whale shark (left), which feeds on plankton, is the largest fish in the ocean and is quite friendly to humans. The scalloped hammerhead shark (right) is endangered.

Left: Colin Parker/National Geographic My Shot/National Geographic Creative; Below: Westend61/SuperStock

Media reports on shark attacks greatly exaggerate the dangers presented by sharks. Every year, members of a few species, including the great white, bull, tiger, oceanic white tip, and hammerhead sharks, injure 60–75 people and typically kill 6–10 people worldwide. For every shark that injures or kills a person, people kill about 1.2 million sharks. As many as 73 million sharks are caught each year for their valuable fins and then thrown back alive into the water, fins removed, to bleed to death or drown because they can no longer swim. Sharks are also killed for their livers, meat, hides, and jaws, and because we fear them.

Harvested shark fins are used widely in Asia as an ingredient in expensive soup (up to $100 a bowl) and as a supposed pharmaceutical cure-all. According to the wildlife conservation group WildAid, there is no reliable evidence that the fins provide flavor or have any nutritional or medicinal value. The group also warns that consumption of shark fins and shark meat can threaten human health because these foods often contain very high levels of mercury and other toxins.

According to a 2014 IUCN study, 25% of the world's open-ocean shark species are threatened with extinction (Figure 4.8), primarily due to overfishing. Sharks are especially vulnerable to population declines because they grow slowly, mature late, and have only a few offspring per generation. Today, they are among the earth's most vulnerable and least protected animals.

With research support from the National Geographic Society, biologist Samuel H. Gruber has been studying lemon sharks in the Florida Keys and the Bahamas, primarily focusing on the Florida East Coast from Palm Beach to Cape Canaveral. His goal is to help us understand and reduce the slaughter of these sharks, which he calls "fantastic and amazing creatures." Protecting sharks is one way for us to live more sustainably by increasing our beneficial environmental impacts.

4.3 HOW DOES THE EARTH'S LIFE CHANGE OVER TIME?

CONCEPT 4.3A The scientific theory of evolution through natural selection explains how life on the earth changes over time due to changes in the genes of populations.

CONCEPT 4.3B Populations evolve when genes mutate and give some individuals genetic traits that enhance their abilities to survive and to produce offspring with these traits (natural selection).

Biological Evolution by Natural Selection Explains How Life Changes over Time

Most of what we know about the long history of life on the earth comes from **fossils**: mineralized or petrified replicas of skeletons, bones, teeth, shells, leaves, and seeds, or impressions of such items found in rocks. Scientists have found such fossil evidence in successive layers of sedimentary rock such as limestone and sandstone, and they have also studied ancient pollen from samples of glacial ice.

The body of evidence gathered through the use of these methods is called the *fossil record*. This record is uneven and incomplete because many forms of life left no fossils and some fossils have decomposed. Scientists estimate that the fossils found so far represent probably only 1% of all species that have ever lived.

How did we end up with such an amazing diversity of species? The scientific answer is **biological evolution** (or simply **evolution**): the process whereby the earth's life changes over time through changes in the genes of populations of organisms in succeeding generations. According to this scientific theory, species have evolved from earlier, ancestral species through **natural selection**. In this process, individuals with certain genetic traits are more likely to survive and reproduce under a particular set of environmental conditions, and to pass these traits on to their offspring, than are individuals without these traits (**Concept 4.3B**).

A large body of evidence supports this idea. As a result, *biological evolution through natural selection* has become a widely accepted scientific theory that generally explains how life has changed over the past 3.8 billion years and why life is so diverse today (**Concept 4.3A**). However, there are still many unanswered questions about some of the details of evolution by natural selection, and research continues in this area.

Biological Evolution by Natural Selection Depends on Genetic Diversity

The idea that organisms change over time and are descended from a single common ancestor has been around in one form or another since the time of the early Greek philosophers. But no one had developed an explanation of how this could happen until 1858 when naturalists Charles Darwin (1809–1882) and Alfred Russel Wallace (1823–1913) independently proposed the concept of *natural selection* as a mechanism for biological evolution. Darwin meticulously gathered evidence for this idea and published it in his 1859 book, *On the Origin of Species by Means of Natural Selection*.

Biological evolution by natural selection involves changes in a population's genetic makeup through successive generations. Recall that a *population* is a group of individuals of the same species living in a particular space. Note that *populations—not individuals—evolve by becoming genetically different.*

The first step in this process is the development of *genetic variability*, or variety in the genetic makeup of individuals in a population. This occurs through **mutations**: changes in the DNA molecules of a gene in any cell that can be inherited by offspring. During an organism's lifetime, the DNA in its cells (see Figure 10, p. S10, Supplement 3) is copied each time one of its cells divides and whenever it reproduces—millions of times, in all. Most mutations result from random changes in the DNA's coded genetic instructions that occur in some tiny fraction of these millions of times. Some mutations also occur from exposure to external agents such as radioactivity, ultraviolet radiation from the sun, and certain natural and human-made chemicals (called *mutagens*).

Mutations can occur in any cell, but only those that take place in the genes of reproductive cells are passed on to offspring. Sometimes a mutation can result in a new genetic trait, called a *heritable trait*, which can be passed from one generation to the next. In this way, populations develop genetic differences among their individuals. Some mutations are harmful to offspring and some are beneficial.

The next step in biological evolution is *natural selection*, in which environmental conditions favor some individuals over others. The favored individuals possess heritable traits that give them some advantage over other individuals in a given population. Such a trait is called an **adaptation**, or **adaptive trait**—any heritable trait that improves the ability of an individual organism to survive and to reproduce at a higher rate than other individuals in a population are able to do under prevailing environmental conditions. Thus, the scientific concept of natural selection explains how populations adapt to changes in environmental conditions by changing their overall genetic makeup.

An example of natural selection at work is *genetic resistance*—the ability of one or more organisms in a population to tolerate a chemical such as a pesticide or antibiotic designed to kill it and to reproduce more rapidly than the members of the population that do not have such genetic traits. Genetic resistance can develop fairly quickly in populations of organisms such as many species of bacteria and insects that can rapidly produce large numbers of offspring. For example, certain disease-causing bacteria have developed genetic resistance to widely used antibacterial drugs, or *antibiotics* (Figure 4.9).

Our own species is another example of evolution by natural selection. We have evolved certain traits that have allowed us to dominate most of the earth's land area and many of its aquatic systems. Evolutionary biologists attribute our success to three major adaptations: *strong opposable thumbs* that allowed us to grip and use tools better than the few other animals that have thumbs could do; an *ability to walk upright*, which gave us agility and freed up our hands for many uses; and a *complex brain*, which allowed us to develop many skills, including the ability to use speech and to read and write in order to transmit complex ideas.

To summarize the process of biological evolution by natural selection: *Genes mutate, individuals are selected, and the resulting populations are better adapted to survive and reproduce under existing environmental conditions* (**Concept 4.3B**).

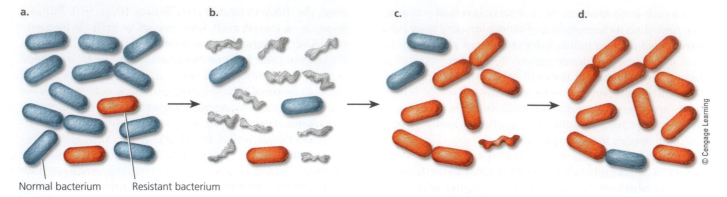

a.   b.   c.   d.

Normal bacterium   Resistant bacterium

FIGURE 4.9 *Evolution by natural selection:* **(a)** A population of bacteria is exposed to an antibiotic, which **(b)** kills all individuals except those possessing a trait that makes them resistant to the drug. **(c)** The resistant bacteria multiply and eventually **(d)** replace all or most of the nonresistant bacteria.

Adaptation through Natural Selection Has Limits

In the not-too-distant future, will adaptations to new environmental conditions through natural selection allow our skin to become more resistant to the harmful effects of UV radiation, our lungs to cope with air pollutants, and our livers to better detoxify pollutants in our bodies?

According to scientists in this field, the answer is *no* because of two limitations on adaptation through natural selection. *First,* a change in environmental conditions can lead to such an adaptation only for genetic traits already present in a population's gene pool or for traits resulting from mutations, which occur randomly.

Second, even if a beneficial heritable trait is present in a population, the population's ability to adapt may be limited by its reproductive capacity. Populations of genetically diverse species that reproduce quickly—such as weeds, mosquitoes, rats, bacteria, and cockroaches—often adapt to a change in environmental conditions in a short time (days to years). By contrast, species that cannot produce large numbers of offspring rapidly—such as elephants, tigers, sharks, and humans—take a much longer time (typically thousands or even millions of years) to adapt through natural selection.

Some Common Myths about Evolution through Natural Selection

There are a number of misconceptions about biological evolution through natural selection. Here are four common ones:

- *Survival of the fittest means survival of the strongest.* To biologists, *fitness* is a measure of reproductive success, not strength. Thus, the fittest individuals are those that leave the most descendants, not those that are physically the strongest.

- *Evolution explains the origin of life.* It does not. However, it does explain how species have evolved after life came into being.

- *Humans evolved from apes or monkeys.* Actually, humans, apes, and monkeys evolved along different paths from a common ancestor that lived 5–8 million years ago.

- *Evolution by natural selection involves a grand plan of nature in which species are to become more perfectly adapted.* There is no evidence of any such plan. Instead, evidence indicates that the forces of natural selection and random mutations can push evolution along any number of paths.

4.4 WHAT FACTORS AFFECT BIODIVERSITY?

CONCEPT 4.4A As environmental conditions change, the balance between the formation of new species and the extinction of existing species determines the earth's biodiversity.

CONCEPT 4.4B Human activities are decreasing biodiversity by causing the extinction of many species and by destroying or degrading habitats needed for the development of new species through natural selection.

How Do New Species Arise?

Under certain circumstances, natural selection can lead to an entirely new species. In this process, called **speciation**, one species splits into two or more different species. For sexually reproducing organisms, a new species forms when one population of a species has evolved to the point where its members can no longer breed and produce fertile offspring with members of another population that did not change or that evolved differently.

The most common way in which speciation occurs, especially among sexually reproducing species, is when a barrier or distant migration separates two or more populations of a species and prevents the flow of genes between them. This happens in two phases: first geographic isolation, and then reproductive isolation.

Geographic isolation occurs when different groups of the same population of a species become physically isolated from one another for a long period of time. For example, part of a population may migrate in search of food and then begin living as a separate population in another area with different environmental conditions. Populations can also be separated by a physical barrier (such as a flooding stream or a new road), a volcanic eruption, geological processes (Science Focus 4.3), or winds or flowing water that carry a few individuals to a distant area. These separated populations can develop quite different characteristics. For example, populations of poison dart frogs (**Core Case Study**) living on different islands or in different parts of a region can have dramatic differences in coloration, as shown in Figure 4.10.

In **reproductive isolation**, mutation and change by natural selection operate independently in the gene pools of geographically isolated populations. If this process continues for a long enough time, members of the geographically and reproductively isolated populations of sexually reproducing species can become so different in genetic makeup that they cannot produce live, fertile offspring if they are rejoined and attempt to interbreed. As a result, one species has become two, and speciation has occurred (Figure 4.11).

Scientists have found fascinating ways to study speciation and evolution. For example, molecular biologist and Emerging National Geographic Explorer Beth Shapiro analyzes DNA samples from the fossils of ancient animals and plants and uses statistical models to discover the course of evolution of populations over time and territory. Such research can pinpoint changes in the genetic diversity of populations, as well as extinctions and speciation resulting from factors such as new predators and changes in habitat and climate. This information could be used to help devise strategies for protecting species today.

Humans are having an increasingly larger effect on the earth's biodiversity by manipulating and combining the genes of species. We have used **artificial selection** to change the genetic characteristics of populations with similar genes. In this process, we select one or more desirable genetic traits in the population of a plant or animal such as a type of wheat, fruit, or dog. Then we use *selective breeding,* or *crossbreeding,* to generate populations of the species containing large numbers of individuals with the desired traits.

Note that artificial selection involves crossbreeding between genetic varieties of the same species or between species that are genetically close to one another, and thus it is not a form of speciation. Most of the grains, fruits, and vegetables we eat are produced by artificial selection. It has also given us food crops with higher yields, cows that give more milk, trees that grow faster, and many different types of dogs and cats. But traditional crossbreeding is a slow process and it can be used only on species that are genetically similar to one another.

Now scientists are using **genetic engineering** to speed up our ability to manipulate genes. In this process, scientists alter an organism's genetic material by add-

Brandon Alms/Shutterstock.com

FIGURE 4.10 These poison dart frogs vary in coloration, partly because they were exposed to different environmental conditions.

GEOLOGICAL PROCESSES AFFECT BIODIVERSITY

The earth's surface has changed dramatically over its long history. Scientists have discovered that huge flows of molten rock within the earth's interior have broken its surface into a number of gigantic solid plates, called *tectonic plates.* For hundreds of millions of years, these plates have drifted slowly on the planet's mantle (Figure 4.D).

Rock and fossil evidence indicates that 200–250 million years ago, all of the earth's present-day continents were connected in a supercontinent called Pangaea (Figure 4.D, left). About 135 million years ago, Pangaea began splitting apart as the earth's tectonic plates moved, eventually resulting in the present-day locations of the continents (Figure 4.D, right).

The fact that tectonic plates drift has had two important effects on the evolution and distribution of life on the earth. *First,* the locations of continents and oceanic basins have greatly influenced the earth's climate and thus have helped to determine where plants and animals can live. *Second,* the breakup, movement, and joining of continents has allowed species to move, adapt to new environments, and form new species through speciation.

Adjoining tectonic plates that are grinding along slowly next to one another sometimes shift quickly. Such sudden movement of tectonic plates can cause *earthquakes,* which can also affect biological evolution by causing fissures in the earth's crust that, on rare occasions, can separate and isolate

populations of species. Over long periods, this can lead to the formation of new species as each isolated population changes genetically in response to new environmental conditions. *Volcanic eruptions* that occur along the boundaries of tectonic plates can also affect extinction and speciation by destroying habitats and reducing, isolating, or wiping out populations of species. We discuss these processes in greater detail in Chapter 12.

Critical Thinking

The earth's tectonic plates, including the one you are riding on, typically move at about the rate at which your fingernails grow. If they stopped moving, how might this affect the future biodiversity of the planet?

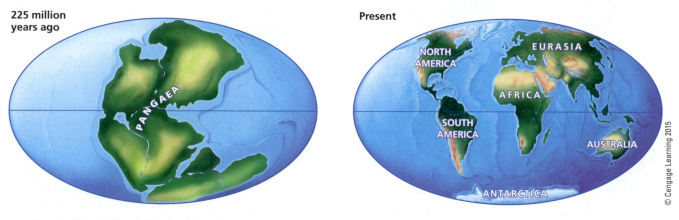

FIGURE 4.D Over millions of years, the earth's continents have moved very slowly on several gigantic tectonic plates. **Question:** How might an area of land splitting apart cause the extinction of a species?

ing, deleting, or changing segments of its DNA to produce desirable traits or to eliminate undesirable ones. For example, scientists have used genetic engineering to develop modified crop plants, new drugs, pest-resistant plants, and animals that grow rapidly (Figure 4.12).

Genetic engineering also enables scientists to transfer genes between different species that would not interbreed

in nature. For example, we can put genes from a cold-water fish species into a tomato plant to give it properties that help it to resist cold weather.

A new and rapidly growing form of genetic engineering is **synthetic biology**. It is a technology that enables scientists to make new sequences of DNA and to use such genetic information to design and create new cells, tissues,

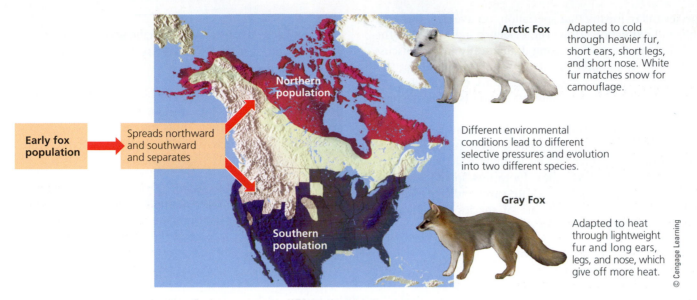

Arctic Fox Adapted to cold through heavier fur, short ears, short legs, and short nose. White fur matches snow for camouflage.

Early fox population → Spreads northward and southward and separates

Northern population

Different environmental conditions lead to different selective pressures and evolution into two different species.

Gray Fox Adapted to heat through lightweight fur and long ears, legs, and nose, which give off more heat.

Southern population

© Cengage Learning

FIGURE 4.11 *Geographic isolation* can lead to reproductive isolation, divergence of gene pools, and speciation.

MCT/Landov

FIGURE 4.12 The larger Atlantic salmon was genetically engineered with a growth hormone from a Chinook salmon. The smaller fish is an unmodified Atlantic salmon of the same age. *Questions:* Would you have any concerns about eating the genetically engineered salmon? Why or why not?

organisms, and devices, and to redesign existing natural biological systems.

All Species Eventually Become Extinct

Another factor affecting the number and types of species on the earth is **extinction**, the process in which an entire species ceases to exist (also referred to as *biological extinction*). When environmental conditions change dramatically or rapidly, a population of a species faces three possible futures: *adapt* to the new conditions through natural selection, *migrate* (if possible) to another area with more favorable conditions, or *become extinct*.

Species that are found in only one area, called **endemic species**, are especially vulnerable to extinction.

They exist on islands and in other unique areas, especially in tropical rain forests where most species have highly specialized roles. For these reasons, they are unlikely to be able to migrate or adapt in the face of rapidly changing environmental conditions. Many of these endangered species are amphibians (**Core Case Study**), such as the now-extinct golden toad (Figure 4.B).

Throughout most of the earth's long history, species have disappeared at a low rate, called the **background extinction rate**—the rate that existed before the human population began growing exponentially. In contrast, a **mass extinction** is a significant rise in extinction rates above the background rate. In such a catastrophic, widespread, and often global event, large groups of species (25–95% of all species) are wiped out, primarily because of major, widespread environmental changes. Fossil and geological evidence indicate that there have been at least three and probably five mass extinctions (at intervals of 20–60 million years) during the past 500 million years (see Supplement 6, p. S49).

A mass extinction provides an opportunity for the evolution of new species that can fill unoccupied ecological niches or newly created ones. Scientific evidence indicates that each occurrence of mass extinction has been followed by an increase in species diversity over several million years as new species have arisen to occupy new habitats or to exploit newly available resources.

As environmental conditions change, the balance between speciation and extinction determines the earth's biodiversity (**Concept 4.4A**). The existence of millions of species today means that speciation, on average, has kept ahead of extinction. However, evidence indicates that the global extinction rate is rising dramatically (as we discuss more fully in Chapter 8). Many scientists argue that this and other evidence indicate that we are experiencing the beginning of a new sixth mass extinction (see Supplement 6, p. S49).

There is also considerable evidence that much of the current rise in the extinction rate and the resulting loss of biodiversity are primarily due to human activities (**Concept 4.4B**), as our ecological footprints spread over the planet (see Figure 1.10, p. 12). Research indicates that the largest cause of the rising rate of species extinctions is the loss, fragmentation, and degradation of habitats. We examine this issue further in Chapters 8 and 9.

TYING IT ALL TOGETHER
Amphibians and Sustainability

This chapter's Core Case Study on amphibians reports on the increasing losses of amphibian species and explains why these species are important, ecologically. In this chapter, we studied the importance of biodiversity—the

Robert King/Shutterstock.com

numbers and varieties of species found in different parts of the world, along with genetic, ecosystem, and functional diversity.

We examined the variety of niches and roles played by species in ecosystems. For example, we saw that some species, including many amphibians, are indicator species that clue us in to the presence of threats to biodiversity, to ecosystems, and to the biosphere. Others such as the American alligator and some shark species are keystone species that play vital roles in sustaining the ecosystems where they live.

We also studied the process whereby all species came to be, according to the scientific theory of biological evolution through natural selection and how new species can arise. We learned that the earth's biodiversity is the result of a balance between the formation of new species (speciation) and extinctions of existing species in response to changing environmental conditions.

The ecosystems where amphibians and other species live are functioning examples of the three **scientific principles of sustainability** in action. They depend on solar energy, the cycling of nutrients, and biodiversity. Disruptions in any of these forms of natural capital can result in degradation of these species' populations and their ecosystems.

Chapter Review

Core Case Study

1. Describe the threats to many of the world's amphibian species and explain why we should avoid hastening the extinction of any amphibian species through our activities.

Section 4.1

2. What is the key concept for this section? Define **biodiversity (biological diversity)** and list and describe its four major components. Why is biodiversity important? Summarize the importance of insects. Define and give three examples of **biomes**. Summarize the scientific contributions of Edward O. Wilson.

Section 4.2

3. What are the two key concepts for this section? Define and distinguish between a **niche**, or **ecological niche**, and a **habitat**. Distinguish between **generalist species** and **specialist species** and give an example of each.

4. Define and distinguish among **native**, **nonnative**, **indicator**, and **keystone species** and give an example of each. What major ecological roles do many amphibian species play (**Core Case Study**)? List six factors that contribute to the threats of extinction for frogs and other amphibians. Describe the role of the American alligator as a keystone species. Explain why we should protect sharks.

Section 4.3

5. What are the two key concepts for this section? What are **fossils** and how do scientists use them? Define **biological evolution (evolution)** and

natural selection and explain how they are related. What is the scientific theory of biological evolution through natural selection?

6. What is a **mutation** and what role do mutations play in evolution through natural selection? What is an **adaptation**, or **adaptive trait**? Explain how harmful bacteria can become genetically resistant to antibiotics. What three genetic adaptations have helped humans to become such a powerful species?

7. What are two limitations on evolution through natural selection? What are four common myths about evolution through natural selection, and for each of them, why is it a myth?

Section 4.4

8. What are the two key concepts for this section? Define **speciation**. Distinguish between **geographic isolation** and **reproductive isolation**, and explain how they can lead to the formation of a new species. Explain how geological processes can affect biodiversity. Define and distinguish between **artificial selection** and **genetic engineering** and give an example of each. Define **synthetic biology**.

9. What is **extinction**? What is an **endemic species** and why can such a species be vulnerable to extinction? Define and distinguish between the **background extinction rate** and a **mass extinction**. What is one of the leading causes of the rising rate of extinction?

10. What are this chapter's *three big ideas*? How are ecosystems where amphibians and other species live functioning examples of the three **scientific principles of sustainability**?

Note: Key terms are in bold type.

Critical Thinking

1. How might we and other species be affected if most or all amphibians (**Core Case Study**) were to go extinct?

2. Is the human species a keystone species? Explain. If humans were to become extinct, what are three species that might also become extinct and what are three species whose populations would probably grow?

3. If you were forced to choose between saving the giant panda from extinction and saving a shark species, which would you choose? Explain.

4. How would you respond to someone who tells you that:
   a. he or she does not believe in biological evolution because it is "just a theory"?
   b. we should not worry about air pollution because natural selection will enable humans to develop lungs that can detoxify pollutants?

5. How would you respond to someone who says that because extinction is a natural process, we should not worry about the loss of biodiversity when species become extinct largely as a result of our activities?

6. What role does each of the following processes play in helping to implement the three **scientific principles of sustainability**: (a) natural selection, (b) speciation, and (c) extinction?

7. List three aspects of your lifestyle that could be contributing to some of the losses of the earth's biodiversity. For each of these, what are some ways to avoid making this contribution?

8. Congratulations! You are in charge of the future evolution of life on the earth. What are the three things that you would consider to be the most important to do?

Doing Environmental Science

Study an ecosystem of your choice, such as a meadow, a patch of forest, a garden, or an area of wetland. (If you cannot do this physically, do so virtually by reading about an ecosystem online or in a library.) Determine and list five major plant species and five major animal species in your ecosystem. Write hypotheses about (a) which of these species, if any, are indicator species and (b) which of them, if any, are keystone species. Explain how you arrived at your hypotheses. Then design an experiment to test each of your hypotheses, assuming you would have unlimited means to carry them out.

Global Environment Watch Exercise

Search for *Amphibians* to find out more about the current state of these species with regard to threats to their existence (**Core Case Study**). What actions are being taken by various nations and organizations to protect amphibians? Write a short summary report on your research.

Data Analysis

The following table is a sample of a very large body of data reported by J. P. Collins, M. L. Crump, and T. E. Lovejoy III in their book *Extinction in Our Times–Global Amphibian Decline*. It compares various areas of the world in terms of the number of amphibian species found and the number of amphibian species that were endemic, or unique to each area. Scientists like to know these percentages because endemic species tend to be more vulnerable to extinction than do non-endemic species. Study the table below and then answer the following questions.

1. Fill in the fourth column to calculate the percentage of amphibian species that are endemic to each area.

2. Which two areas have the highest numbers of endemic species? Name the two areas with the highest percentages of endemic species.

3. Which two areas have the lowest numbers of endemic species? Which two areas have the lowest percentages of endemic species?

4. Which two areas have the highest percentages of non-endemic species?

Area	Number of Species	Number of Endemic Species	Percentage Endemic
Pacific/Cascades/Sierra Nevada Mountains of North America	52	43	
Southern Appalachian Mountains of the United States	101	37	
Southern Coastal Plain of the United States	68	27	
Southern Sierra Madre of Mexico	118	74	
Highlands of Western Central America	126	70	
Highlands of Costa Rica and Western Panama	133	68	
Tropical Southern Andes Mountains of Bolivia and Peru	132	101	
Upper Amazon Basin of Southern Peru	102	22	

© Cengage Learning

CENGAGE **brain**.com To access course materials, including Aplia homework, please visit www.cengagebrain.com.

WWW.CENGAGEBRAIN.COM **81**

5

SPECIES INTERACTIONS, ECOLOGICAL SUCCESSION, AND POPULATION CONTROL

In looking at nature, never
forget that every single
organic being around us
may be said to be striving
to increase its numbers.

CHARLES DARWIN, 1859

5.3 What limits the growth of populations?

A clownfish gains protection by living
among sea anemones and helps protect

The Southern Sea Otter: A Species in Recovery

Southern sea otters

(Figure 5.1, left) live in giant kelp forests (Figure 5.1, right) in shallow waters along parts of the Pacific coast of North America. Most of the remaining members of this endangered species are found off the western coast of the United States between the cities of Santa Cruz and Santa Barbara, California.

Southern sea otters are fast and agile swimmers that dive to the ocean bottom looking for shellfish and other prey, including sea urchins. On the surface they swim on their backs and feed on their prey, using their bellies as a table (Figure 5.1, left). Each day, a sea otter consumes 20–35% of its weight in clams, mussels, crabs, sea urchins, abalone, and about 40 other species of bottom-dwelling organisms.

It is estimated that between 13,000 and 20,000 southern sea otters once lived in California's coastal waters. By the early 1900s, they were hunted almost to extinction in this region by fur traders who killed them for their thick, luxurious fur. Commercial fisherman also killed otters because they viewed them as competitors in the hunt for valuable abalone and other shellfish.

Since that time this population has generally grown from a low of about 50 in 1938 to an estimated 2,941 in 2013. Their partial recovery got a boost in 1977 when the U.S. Fish and Wildlife Service declared the species endangered in most of its range, with a total population of only 1,850 individuals. Despite such progress, the population has a long way to go to justify removing it from the endangered species list.

Why should we care about the southern sea otters of California? One reason is *ethical:* many people believe it is wrong to allow human activities to cause the extinction of a species. Another reason is that people love to look at these appealing and highly intelligent animals as they play in the water. As a result, they help to generate millions of dollars a year in tourism revenues. A third reason—and a key reason in our study of environmental science—is that biologists classify them as a *keystone species* (see p. 69). Scientists hypothesize that in the absence of southern sea otters, sea urchins and other kelp-eating species would probably destroy the Pacific coast kelp forests and much of the rich biodiversity they support.

Biodiversity is an important part of the earth's natural capital and is the focus of one of the three **scientific principles of sustainability**. In this chapter, we will look at how species interact and help control one another's population sizes and how communities, ecosystems, and populations of species respond to changes in environmental conditions.

FIGURE 5.1 An endangered southern sea otter in Monterey Bay, California (USA) uses a stone to crack the shells of the clams that it feeds on (left). It lives in a bed of seaweed called giant kelp (right).

Left: Kirsten Wahlquist | Dreamstime.com. Right: Paul Whitted/Shutterstock.com.

5.1 HOW DO SPECIES INTERACT?

CONCEPT 5.1 Five types of interactions among species—interspecific competition, predation, parasitism, mutualism, and commensalism—affect the resource use and population sizes of species.

Most Species Compete with One Another for Certain Resources

Ecologists have identified five basic types of interactions among species as they share limited resources such as food, shelter, and space. These types of interactions are called *interspecific competition, predation, parasitism, mutualism,* and *commensalism,* and they all have significant effects on the population sizes of the species in an ecosystem and on how they use resources (**Concept 5.1**).

The most common interaction among species is *competition,* which occurs when members of one or more species interact to use the same limited resources such as food, water, light, and space. Competition within a species is called *intraspecific competition,* but **interspecific competition**, or competition among different species, plays a larger role in most ecosystems. Most interspecific competition involves one species becoming more efficient than others in obtaining the resources it needs.

When two species compete with one another for the same resources, their niches *overlap* (see Figure 4.5, p. 68). The greater this overlap, the more they compete for key resources. If one species can take over the largest share of one or more key resources, each of the other competing species must move to another area (if possible), adapt by shifting its feeding habits or behavior through natural selection to reduce or alter its niche, suffer a sharp population decline, or become extinct in that area.

Humans compete with many other species for space, food, and other resources. As our ecological footprints grow and spread (see Figure 1.10, p. 12), we are taking over or degrading the habitats of many other species and depriving them of resources they need in order to survive.

Over a time scale long enough for natural selection to occur, populations of some species develop adaptations that allow them to reduce or avoid competition with other species for resources. One way this happens is through **resource partitioning**, which occurs when species competing for similar scarce resources evolve specialized traits that allow them to share resources by using parts of them, using them at different times, or using them in different ways. Figure 5.2 shows resource partitioning by some insect-eating bird species whose adaptations allow them to reduce competition by feeding in different portions of certain spruce trees and by feeding on different insect species. Similarly, Figure 4.6 (p. 69) shows how the evolution of specialized feeding niches has reduced competition for resources among bird species in a coastal wetland.

Consumer Species Feed on Other Species

In **predation**, a member of one species (the **predator**, or hunter) feeds directly on all or part of a living organism (the **prey**, or hunted) as part of a food web. Together, the two different species—such as a lion, the predator, and a zebra, its prey (see Figure 3.6, p. 46)—are engaged in a

Blackburnian Warbler   Black-throated Green Warbler   Cape May Warbler   Bay-breasted Warbler   Yellow-rumped Warbler

After R. H. MacArthur, "Population Ecology of Some Warblers in Northeastern Coniferous Forests," *Ecology* 36:533–536, 1958.

FIGURE 5.2 *Sharing the wealth:* Resource partitioning among five species of insect-eating warblers in the spruce forests of the U.S. state of Maine. Each species spends at least half its feeding time in its associated yellow-highlighted areas of these spruce trees.

THREATS TO KELP FORESTS

A kelp forest is composed of large concentrations of a seaweed called *giant kelp.* Anchored to the ocean floor, its long blades grow toward the sunlit surface waters (Figure 5.1, right). Under good conditions, the blades can grow 0.6 meter (2 feet) in a day and the plant can grow as tall as a ten-story building. The blades are very flexible and can survive all but the most violent storms and waves.

Kelp forests are one of the most biologically diverse ecosystems found in marine waters, supporting large numbers of marine plants and animals. These forests help reduce shore erosion by blunting the force of incoming waves and trapping some of the outgoing sand.

Sea urchins (Figure 5.A) prey on kelp plants. Large populations of these predators can rapidly devastate a kelp forest because they eat the bases of young kelp plants. Scientific studies by

biologists, including James Estes of the University of California at Santa Cruz, indicate that the southern sea otter (**Core Case Study**) is a keystone species that helps to sustain kelp forests by controlling populations of sea urchins.

A second threat to kelp forests is polluted water running off the land and into the coastal waters where kelp forests grow. The pollutants in this runoff include pesticides and herbicides that can kill kelp plants and other species and upset the food webs in these aquatic forests. Another runoff pollutant is fertilizer. Its plant nutrients (mostly nitrates) can cause excessive growth of algae and other plants, which block some of the sunlight needed to support the growth of giant kelp.

Kokhanchikov/Shutterstock.com

FIGURE 5.A The purple sea urchin inhabits the coastal waters of the U.S. state of California and feeds on kelp.

Some scientists warn that the warming of the world's oceans now presents a growing threat to kelp forests, which require fairly cool water. If coastal waters get warmer during this century, as projected by climate models, many or most of the California's coastal kelp forests could disappear.

Critical Thinking

List three ways in which we could reduce the degradation of giant kelp forest ecosystems.

---

predator–prey relationship, a species interaction that has a strong effect on population sizes and other factors in many ecosystems.

In a giant kelp forest ecosystem, sea urchins prey on kelp, a type of seaweed (Science Focus 5.1). However, as a keystone species, southern sea otters (**Core Case Study**) prey on the sea urchins and help to keep them from destroying the kelp forests.

Predators have a variety of methods that help them capture prey. *Herbivores* can walk, swim, or fly to the plants they feed on. *Carnivores* such as the cheetah catch prey by running fast. Others, such as the American bald eagle, can fly and have keen eyesight. Still others work together to capture their prey, as female African lions often do in preying on zebras (see Figure 3.6, p. 46), wildebeest, antelopes, and other fast-running large animals of the open savannah grasslands.

Other predators use *camouflage* to hide in plain sight and ambush their prey. For example, praying mantises (see Figure 4.A, right, p. 66) sit on flowers or plants of a

color similar to their own and ambush visiting insects. White ermines (a type of weasel), snowy owls, and arctic foxes hunt their prey in snow-covered areas. People camouflage themselves to hunt wild game and use camouflaged traps to capture wild animals. Some predators use *chemical warfare* to attack their prey. For example, some spiders and poisonous snakes use venom to paralyze their prey and to deter their predators.

Prey species have evolved many ways to avoid predators, including abilities to run, swim, or fly fast, and some have highly developed senses of sight, sound, or smell that alert them to the presence of predators. Other adaptations include protective shells (as on armadillos and turtles), thick bark (on giant sequoia trees), spines (on porcupines), and thorns (on cacti and rose bushes).

Other prey species use the camouflage of certain shapes or colors. Some insect species have shapes that look like twigs (Figure 5.3a), or bird droppings on leaves. A leaf insect can be almost invisible against its background (Figure 5.3b), as can an arctic hare in its white winter fur.

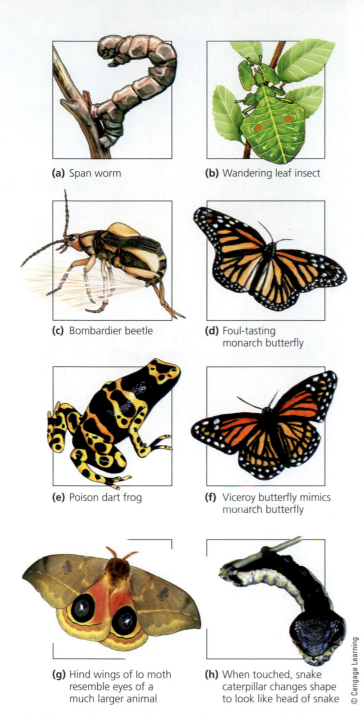

(a) Span worm

(b) Wandering leaf insect

(c) Bombardier beetle

(d) Foul-tasting monarch butterfly

(e) Poison dart frog

(f) Viceroy butterfly mimics monarch butterfly

(g) Hind wings of Io moth resemble eyes of a much larger animal

(h) When touched, snake caterpillar changes shape to look like head of snake

© Cengage Learning

FIGURE 5.3 These prey species have developed specialized ways to avoid their predators: (a, b) *camouflage*, (c, d, e) *chemical warfare*, (d, e, f) *warning coloration*, (f) *mimicry*, (g) *deceptive looks*, and (h) *deceptive behavior*.

Chemical warfare is another common strategy for prey species. Some discourage predators by containing or emitting chemicals that are *poisonous* (oleander plants), *irritating* (stinging nettles and bombardier beetles, Figure 5.3c), *foul smelling* (skunks and stinkbugs), or *bad tasting* (buttercups and monarch butterflies, Figure 5.3d). When attacked, some species of squid and octopus emit clouds of black ink, allowing them to escape by confusing their predators.

Many bad-tasting, bad-smelling, toxic, or stinging prey species have evolved *warning coloration*, brightly colored advertising that helps experienced predators to recognize and avoid them. They flash a warning: "Eating me is risky." Examples are the brilliantly colored, foul-tasting monarch butterflies (Figure 5.3d) and poisonous frogs (Figure 5.3e and Figure 4.10, p. 75). When a bird such as a blue jay eats a monarch butterfly, it usually vomits and learns to avoid monarchs.

CONSIDER THIS. . .

CONNECTIONS Coloration and Dangerous Species

Biologist Edward O. Wilson gives us two rules for evaluating the possible dangers posed by various brightly-colored animal species. *First*, if they are small and strikingly beautiful, they are probably poisonous. *Second*, if they are strikingly beautiful and easy to catch, they are probably deadly.

Some butterfly species gain protection by looking and acting like other, more dangerous species, a protective device known as *mimicry*. For example, the nonpoisonous viceroy butterfly (Figure 5.3f) mimics the monarch butterfly. Other prey species use *behavioral strategies* to avoid predation. Some attempt to scare off predators by puffing up (blowfish), spreading their wings (peacocks), or mimicking a predator (Figure 5.3h). Some moths have wings that look like the eyes of much larger animals (Figure 5.3g). Other prey species gain some protection by living in large groups such as schools of fish and herds of antelope.

Interactions between Predator and Prey Species Can Drive Each Other's Evolution

At the individual level, members of the predator species benefit from their predation and members of the prey species are harmed. At the population level, predation plays a role in natural selection. Animal predators, for example, tend to kill the sick, weak, aged, and least fit members of a prey population because they are the easiest to catch. Individuals with better defenses against predation thus tend to survive longer and leave more offspring with adaptations that can help them avoid predation. Over time, as a prey species develops traits that make it more difficult to catch, its predators face selection pressures that favor traits increasing their ability to catch their prey. Then the prey species must get better at eluding the more effective predators.

When populations of two different species interact in such a way over a long period of time, changes in the gene pool of populations one species can lead to changes in the gene pool of the other. Such changes can help both competing species to become more competitive or to avoid or reduce competition. Biologists call this natural selection process **coevolution**.

For example, bats prey on certain species of moths, and they hunt at night using *echolocation*, emitting pulses of

FIGURE 5.4 *Parasitism:* This blood-sucking, parasitic sea lamprey has attached itself to an adult lake trout from one of the Great Lakes (USA, Canada).

Great Lakes Fishery Commission

high-frequency sound that bounce off objects and capturing the returning echoes that tell them where their prey is located. Over time, certain moth species have evolved ears that are especially sensitive to the sound frequencies that bats use to find them. When they hear these frequencies, they try to escape by dropping to the ground or flying evasively. Some bat species have evolved ways to counter this defense by changing the frequency of their sound pulses. In turn, some moths have evolved their own high-frequency clicks to jam the bats' echolocation systems. Some bat species have then adapted by turning off their echolocation systems and using the moths' clicks to locate their prey.

Thus, the complex predator–prey relationship plays an important role in controlling population growth of predator and prey species, and it helps these species to thrive and to contribute to important ecosystem services. However, these relationships and ecosystem services can be disrupted when nonnative predator species are introduced, either accidentally or deliberately, into ecosystems. We discuss the effects of such disruption in Chapter 8.

Some Species Feed Off Other Species by Living On or Inside Them

Parasitism occurs when one species (the *parasite*) feeds on another organism (the *host*), usually by living on or inside the host. In this relationship, the parasite benefits and the host is often harmed.

A parasite usually is much smaller than its host and rarely kills it. However, most parasites remain closely associated with their hosts, draw nourishment from them, and may gradually weaken them.

Some parasites such as tapeworms live inside their hosts. Others such as mistletoe plants and blood-sucking sea lampreys (Figure 5.4) attach themselves to the outsides of their hosts. Some parasites, including fleas and ticks, move from one host to another while others, including tapeworms, spend their adult lives within a single host.

Villiers Steyn/Dreamstime.com

FIGURE 5.5 *Mutualism:* Oxpeckers feed on parasitic ticks that infest animals such as this impala and warn of approaching predators.

Parasites harm their hosts but help to keep the populations of their hosts in check.

In Some Interactions, Both Species Benefit

In **mutualism**, two species behave in ways that benefit both by providing each with food, shelter, or some other resource. One example is pollination of flowering plants by species such as honeybees, hummingbirds, and butterflies that feed on the nectar of flowers (Figure 4.A, p. 66).

Figure 5.5 shows an example of a mutualistic relationship that combines *nutrition* and *protection*. It involves birds that ride on the backs or heads of large animals such as elephants, rhinoceroses, and impalas. The birds remove and eat parasites and pests (such as ticks and flies) from

FIGURE 5.6 *Commensalism:* this pitcher plant is attached to a branch of a tree without penetrating or harming the tree. This carnivorous plant feeds on insects that become trapped inside it.

the animals' bodies and often make noises warning the larger animals when predators are approaching.

Another example of mutualism involves clownfish, which usually live within sea anemones (see chapter-opening photo), whose tentacles sting and paralyze most fish that touch them. The clownfish, which are not harmed by the tentacles, gain protection from predators and feed on the waste matter left from the anemones' meals. The sea anemones benefit because the clownfish protect them from some of their predators and parasites.

In *gut inhabitant mutualism,* armies of bacteria in the digestive systems of animals help to break down (digest) the animals' food. In turn, the bacteria receive a sheltered habitat and food from their hosts. Hundreds of millions of

bacteria in your gut secrete enzymes that help you digest the food you eat.

It is tempting to think of mutualism as an example of cooperation between species. In reality, the species in a mutualistic interaction benefit one another unintentionally and are each in it for their own survival.

In Some Interactions, One Species Benefits and the Other Is Not Harmed

Commensalism is an interaction that benefits one species but has little, if any, beneficial or harmful effect on the other. One example involves plants called *epiphytes* (air plants), which attach themselves to the trunks or branches of trees (Figure 5.6) in tropical and subtropical forests. Epiphytes benefit by having a solid base on which to grow in an elevated location that gives them better access to sunlight, water from the humid air and rain, and nutrients falling from the tree's upper leaves and limbs. Their presence apparently does not harm the tree. Similarly, birds benefit by nesting in trees, generally without harming them.

5.2 HOW DO COMMUNITIES AND ECOSYSTEMS RESPOND TO CHANGING ENVIRONMENTAL CONDITIONS?

CONCEPT 5.2 The species composition of a community or ecosystem can change in response to changing environmental conditions through a process called *ecological succession.*

Communities and Ecosystems Change over Time: Ecological Succession

The types and numbers of species in biological communities and ecosystems change in response to changing environmental conditions such as a fires, volcanic eruptions, climate change, and the clearing of forests to plant crops. The normally gradual change in species composition in a given area is called **ecological succession** (**Concept 5.2**).

Ecologists recognize two major types of ecological succession, depending on the conditions present at the beginning of the process. **Primary ecological succession** involves the gradual establishment of communities of different species in lifeless areas where there is no soil in a terrestrial ecosystem or no bottom sediment in an aquatic ecosystem. Examples include bare rock exposed by a retreating glacier (Figure 5.7), newly cooled lava, an abandoned highway or parking lot, and a newly created shallow pond or reservoir. Primary succession usually takes hundreds to thousands of years because of the need

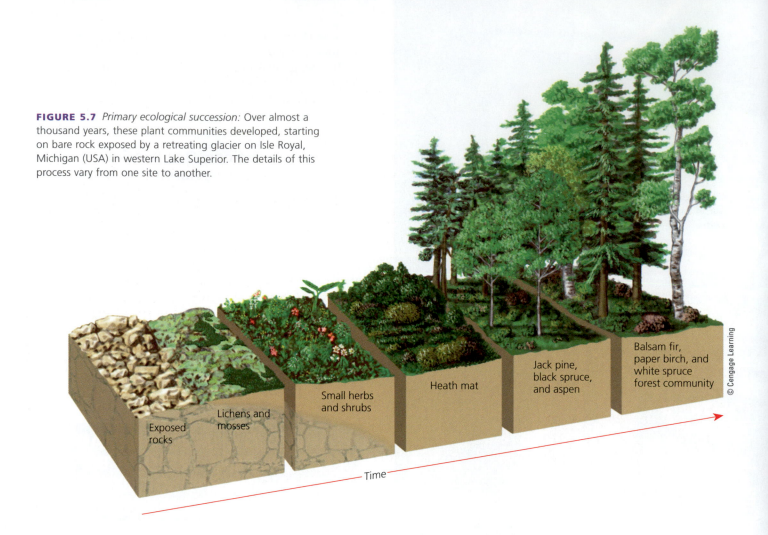

FIGURE 5.7 *Primary ecological succession:* Over almost a thousand years, these plant communities developed, starting on bare rock exposed by a retreating glacier on Isle Royal, Michigan (USA) in western Lake Superior. The details of this process vary from one site to another.

Exposed rocks

Lichens and mosses

Small herbs and shrubs

Heath mat

Jack pine, black spruce, and aspen

Balsam fir, paper birch, and white spruce forest community

© Cengage Learning

Time

to build up fertile soil or aquatic sediments to provide the nutrients needed to establish a plant community.

The other, more common type of ecological succession is called **secondary ecological succession**, in which a series of communities or ecosystems with different species develop in places containing soil or bottom sediment. This type of succession begins in an area where an ecosystem has been disturbed, removed, or destroyed, but some soil or bottom sediment remains. Candidates for secondary succession include abandoned farmland (Figure 5.8), burned or cut forests, heavily polluted streams, and flooded land. Because some soil or sediment is present, new vegetation can begin to grow, usually within a few weeks. It begins with the germination of seeds already in the soil and seeds imported by wind or in the droppings of birds and other animals.

Ecological succession is an important ecosystem service that tends to enrich the biodiversity of communities and ecosystems by increasing species diversity and interactions among species. Such interactions in turn enhance sustainability by promoting population control and by increasing the complexity of food webs, which enhances energy flow and nutrient cycling. As part of the earth's natural capital,

primary and secondary ecological succession are examples of *natural ecological restoration.*

Living Systems Are Sustained through Constant Change

All living systems, from a cell to the biosphere, are constantly changing in response to changing environmental conditions. However, living systems contain complex processes that interact to provide some degree of stability, or sustainability. This *stability,* or capacity to withstand external stress and disturbance, is maintained only by constant change in response to changing environmental conditions. For example, in a mature tropical rain forest, some trees die and others take their places. However, unless the forest is cut, burned, or otherwise destroyed, you would still recognize it as a tropical rain forest 50 or 100 years from now.

It is useful to distinguish between two aspects of stability or sustainability in ecosystems. One is **inertia**, or **persistence**: the ability of an ecosystem to survive moderate disturbances. A second factor is **resilience**: the ability of

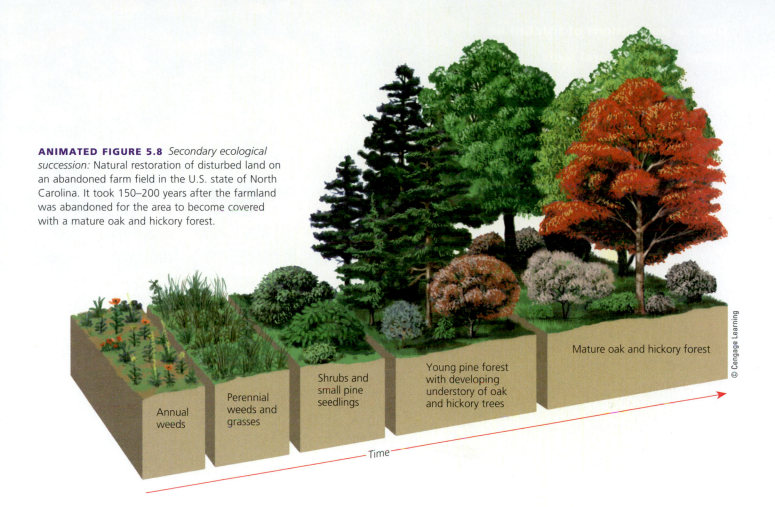

ANIMATED FIGURE 5.8 *Secondary ecological succession:* Natural restoration of disturbed land on an abandoned farm field in the U.S. state of North Carolina. It took 150–200 years after the farmland was abandoned for the area to become covered with a mature oak and hickory forest.

Annual weeds

Perennial weeds and grasses

Shrubs and small pine seedlings

Young pine forest with developing understory of oak and hickory trees

Mature oak and hickory forest

Time

© Cengage Learning

an ecosystem to be restored through secondary ecological succession after a more severe disturbance.

Evidence suggests that some ecosystems have one of these properties but not the other. For example, tropical rain forests have high species diversity and high inertia and thus are resistant to lower levels of change or damage. But once a large tract of tropical rain forest is cleared or severely damaged, the resilience of the resulting degraded forest ecosystem may be so low that it reaches an ecological tipping point after which it might not be restored by secondary ecological succession. One reason for this is that most of the nutrients in a typical rain forest are stored in its vegetation, not in the topsoil, as in most other terrestrial ecosystems. Once the nutrient-rich vegetation is gone, daily rains can remove most of the remaining soil nutrients and thus prevent the return of a tropical rain forest on a large cleared area.

By contrast, grasslands are much less diverse than most forests, and consequently they have low inertia and can burn easily. However, because most of their plant matter is stored in underground roots, these ecosystems have high resilience and can recover quickly after a fire, as their root systems produce new grasses. Grassland can be destroyed only if its roots are plowed up and something else is planted in its place, or if it is severely overgrazed by livestock or other herbivores.

5.3 WHAT LIMITS THE GROWTH OF POPULATIONS?

CONCEPT 5.3 No population can grow indefinitely because of limitations on resources and because of competition among species for those resources.

Populations Can Grow, Shrink, or Remain Stable

A **population** is a group of interbreeding individuals of the same species (Figure 5.9). Most populations live together in *clumps* such as packs of wolves, schools of fish, and flocks of birds. This allows them to cluster where resources are available. Living in groups can also provide

Diverse populations of fish can be found living on coral reefs.

FIGURE 5.9 A population, or *school*, of Anthias fish on coral in Australia's Great Barrier Reef.

iStockphoto.com/Rich Carey

some protection from predators, and living in packs gives some predator species a better chance of getting a meal.

Four variables—*births, deaths, immigration,* and *emigration*—govern changes in population size. A population increases through birth and immigration (arrival of individuals from outside the population) and decreases through death and emigration (departure of individuals from the population):

Population change = (Births + Immigration) − (Deaths + Emigration)

A population's **age structure**—its distribution of individuals among various age groups—can have a strong effect on how rapidly it grows or declines. Age groups are usually described in terms of organisms not mature enough to reproduce (the *prereproductive stage*), those capable of reproduction (the *reproductive stage*), and those too old to reproduce (the *postreproductive stage*).

The size of a population will likely increase if it is made up mostly of individuals in their reproductive stage, or soon to enter this stage. In contrast, the size of a popula-

tion dominated by individuals in their postreproductive stage will tend to decrease over time.

Some Factors Can Limit Population Size

Each population in an ecosystem has a **range of tolerance**—a range of variations in its physical and chemical environment under which it can survive (Figure 5.10). Individuals within a population may also have slightly different tolerance ranges for temperature or other physical or chemical factors because of small differences in their genetic makeup, health, and age. For example, a trout population may do best within a narrow band of temperatures (*optimum level* or *range*), but a few individuals can survive above and below that band. However, if the water becomes too hot or too cold, none of the trout can survive.

A number of physical or chemical factors can help to determine the number of organisms in a population. Sometimes one or more factors, known as **limiting**

factors, are more important than other factors in regulating population growth. On land, precipitation often is the limiting factor. For example, low precipitation levels in desert ecosystems limit desert plant growth. Important limiting physical factors for populations in *aquatic systems* include water temperature (Figure 5.10), water depth and clarity (allowing for more or less sunlight), nutrient availability, acidity, salinity, and the level of oxygen gas in the water *(dissolved oxygen content)*.

Too much of a physical or chemical factor can also be limiting. For example, too much water or fertilizer can kill land plants. If acidity levels are too high in an aquatic environment, some of its organisms can be harmed.

An additional factor that can limit the sizes of some populations is **population density**, the number of individuals in a population found within a defined area or volume. Some limiting factors, called *density-dependent factors*, become more important as a population's density increases. For example, in a dense population, parasites and diseases can spread more easily, resulting in higher death rates. On the other hand, a higher population density can help sexually reproducing individuals to find mates more easily in order to produce offspring. Other factors such as drought and climate change are considered *density-independent*, because they can affect population sizes regardless of density.

No Population Can Grow Indefinitely: J-Curves and S-Curves

Some species have an incredible ability to increase their numbers and grow exponentially (see p. 14). Plotting these numbers against time yields a J-shaped curve of ex-

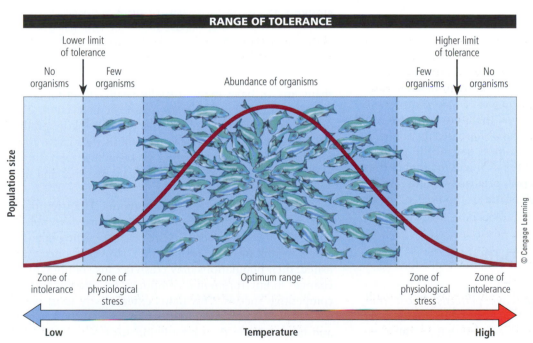

FIGURE 5.10 Range of tolerance for a population of trout to changes in water temperature.

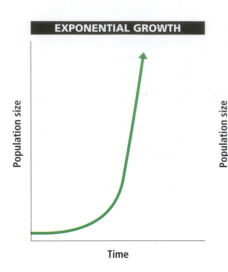

EXPONENTIAL GROWTH

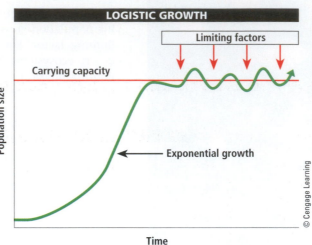

LOGISTIC GROWTH

Limiting factors

Carrying capacity

Exponential growth

Time

Time

FIGURE 5.11 Populations of species can undergo *exponential growth* represented by a J-shaped curve (left) when resource supplies are plentiful. As resource supplies become limited, a population undergoes *logistic growth*, represented by an S-shaped curve (right), when the size of the population approaches the carrying capacity of its habitat.

ponential growth (Figure 5.11, left). Members of such populations typically reproduce at an early age, have many offspring each time they reproduce, and reproduce many times, with short intervals between successive generations.

However, *there are always limits to population growth in nature*. Research reveals that a rapidly growing population of any species eventually reaches some size limit imposed by limiting factors such as sunlight, water, temperature, space, or nutrients, or by exposure to predators or infectious diseases (**Concept 5.2**). The sum of all such factors in any habitat is called **environmental resistance**. These limiting factors largely determine any area's **carrying capacity**: the maximum population of a given species that a particular habitat can sustain indefinitely. As a population approaches the carrying capacity of its habitat, the J-shaped curve of its exponential growth (Figure 5.11, left) is converted to an S-shaped curve of *logistic growth*, or growth that fluctuates around a certain level (Figure 5.11, right).

Some populations do not make a smooth transition from exponential growth to logistic growth. Instead, they use up their resource supplies and temporarily *overshoot*, or exceed, the carrying capacity of their environment. In such cases, the population suffers a sharp decline, called a *dieback*, or **population crash**, unless part of the population can switch to new resources or move to an area that has more resources. Such a crash occurred when reindeer were introduced onto a small island in the Bering Sea in the early 1900s (Figure 5.12).

Different Species Have Different Reproductive Patterns

Species vary in their reproductive patterns. Those that have a capacity for a high rate of population increase *(r)* are called *r-selected species*. They tend to have short life

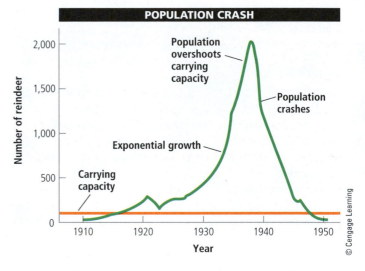

POPULATION CRASH

Population overshoots carrying capacity

Population crashes

Exponential growth

Carrying capacity

FIGURE 5.12 Exponential growth, overshoot, and population crash of a population of reindeer introduced onto the small Bering Sea island of St. Paul in 1910.

spans and to have many, usually small, offspring and to give them little or no parental care or protection. They overcome typically massive losses of offspring by producing so many offspring that a few will likely survive to reproduce many more offspring to keep this reproductive pattern going. Examples include algae, bacteria, and most insects.

Such species tend to be *opportunists*. They reproduce and disperse rapidly when conditions are favorable or when a disturbance such as a fire or clear-cutting opens up a new habitat or niche for invasion. However, once established, their populations may crash because of unfavorable changes in environmental conditions or invasion by more competitive species. This helps explain why most opportunist species go through irregular and unstable boom-and-bust cycles in their population sizes.

WHY DO CALIFORNIA'S SOUTHERN SEA OTTERS FACE AN UNCERTAIN FUTURE?

The population size of southern sea otters (**Core Case Study**) has fluctuated in response to changes in environmental conditions. One such change has been a rise in populations of the orcas (killer whales) that feed on them. Scientists hypothesize that orcas began feeding more on southern sea otters when populations of their normal prey, sea lions and seals, began declining. Also, between 2010 and 2012, the number of southern sea otters killed or injured by sharks increased for reasons that scientists are trying to understand.

Another factor may be parasites known to breed in the intestines of cats. Scientists hypothesize that some southern sea otters might be dying because coastal area cat owners flush feces-laden cat litter down their toilets or dump it in storm drains that empty into coastal waters. The feces contain parasites that then infect the otters.

Otters are also threatened by blooms of toxic algae that are fed by urea, a key ingredient in fertilizer that washes into coastal waters. Other pollutants released by human activities are PCBs and other fat-soluble toxic chemicals that can kill otters by accumulating to high levels in the tissues of the shellfish on which otters feed. Because southern sea otters feed at high

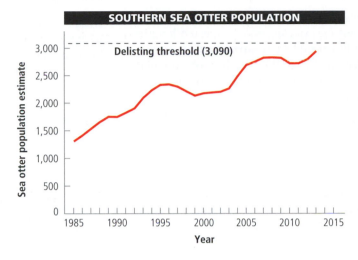

SOUTHERN SEA OTTER POPULATION

FIGURE 5.B
Changes in the population size of southern sea otters off the coast of the U.S. state of California, 1983–2013.

Compiled by the authors using data from U.S. Geological Survey.

trophic levels and live close to the shore, they are vulnerable to these and other pollutants in coastal waters.

The factors listed here, mostly resulting from human activities, plus a fairly low reproductive rate and a rising mortality rate, have hindered the ability of the endangered southern sea otter to rebuild its population (Figure 5.B). In 2012, the National Geographic Society funded a project led by Nicole Thometz to learn more about why juvenile sea otters, in particular, were suffering a high mortality rate. The aim of this study was to track changes in the physiological development of the otters and to understand how physiological

variables affected their foraging ability and success. Such information could be used to help biologists to refine recovery plans for the southern sea otter. According to the U.S. Geological Survey, the California southern sea otter population would have to reach at least 3,090 animals for 3 years in a row before it could be considered for removal from the endangered species list.

Critical Thinking

How would you design a controlled experiment to test the hypothesis that cat litter flushed down toilets might be killing southern sea otters?

At the other extreme are **K-selected species**. They tend to reproduce later in life and have a small number of offspring with fairly long life spans. Typically, the offspring of *K*-selected mammal species develop inside their mothers (where they are safe), and are born fairly large. After birth, they mature slowly and are cared for and protected by one or both parents, and in some cases by living in herds or groups, until they reach reproductive age and begin the cycle again. Most organisms have reproductive patterns between the extremes of *r*-selected and *K*-selected species.

Such species are called *K*-selected species because they tend to do well in competitive conditions when their population size is near the carrying capacity *(K)* of their environment. Most large mammals (such as elephants, whales, and humans), birds of prey, and large and long-lived plants (such as the saguaro cactus, and most tropical rain forest trees) are *K*-selected species. Many of these species—especially those with low reproductive rates, such as elephants, sharks, giant redwood trees, and California's southern sea otters (Science Focus 5.2)—are vulnerable to extinction.

CONSIDER THIS. . .

THINKING ABOUT *r*-Selected and *K*-Selected Species

If the earth experiences significant warming during this century as projected, is this likely to favor *r*-selected or *K*-selected species? Explain.

CONSIDER THIS. . .

THINKING ABOUT Survivorship Curves

Which type of survivorship curve applies to the human species?

Species Vary in Their Typical Life Spans

Individuals of species with different reproductive strategies tend to have different *life expectancies,* or expected lengths of life. This can be illustrated by a **survivorship curve**, which shows the percentages of the members of a population surviving at different ages. There are three generalized types of survivorship curves: late loss, early loss, and constant loss (Figure 5.13). A *late loss* population (such as elephants and rhinoceroses) typically has high survivorship to a certain age, then high mortality. A *constant loss* population (such as many songbirds) shows a fairly constant death rate at all ages. For an *early loss* population (such as annual plants and many bony fish species), survivorship is low early in life. These generalized survivorship curves only approximate the realities of nature.

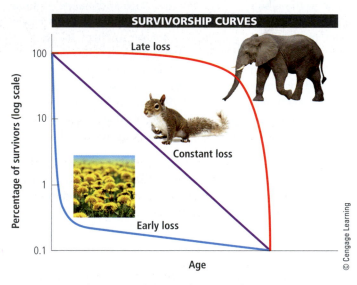

FIGURE 5.13 Three general survivorship curves for populations of different species, obtained by showing the percentages of the members of a population surviving at different ages.

Top: gualtiero boffi/Shutterstock.com. Center: IrinaK/Shutterstock.com. Bottom: ultimathule/Shutterstock.com.

Humans Are Not Exempt from Nature's Population Controls

Humans are not exempt from population crashes. In 1845, Ireland experienced such a crash after a fungus destroyed its potato crop. About 1 million people died from hunger or diseases related to malnutrition, and millions more migrated to other countries, sharply reducing the Irish population.

During the 14th century, the *plague* spread through densely populated European cities and killed at least 25 million people, amounting to one-third of the European population. The bacterium causing this disease normally lives in rodents. It was transferred to humans by fleas that fed on infected rodents and then bit humans. The disease spread like wildfire through crowded cities, where sanitary conditions were poor and rats were abundant. Today several antibiotics can be used to treat bubonic plague.

So far, technological, social, and other cultural changes have expanded the earth's carrying capacity for the human species. We have used large amounts of energy and matter resources to occupy formerly uninhabitable areas, to expand agriculture, and to control the populations of other species that compete with us for resources. Some say we can keep expanding our ecological footprint in this way indefinitely, mostly because of our technological ingenuity. Others say that sooner or later, we will reach the limits that nature eventually imposes on any population that exceeds or degrades its resource base. We discuss these issues in Chapter 6.

BIG IDEAS

- Certain interactions among species affect their use of resources and their population sizes.

- The species composition and population sizes of a community or ecosystem can change in response to changing environmental conditions through a process called *ecological succession.*

- There are always limits to population growth in nature.

Southern Sea Otters and Sustainability

On the western coast of North America, sea otters are part of a complex ecosystem made up of large underwater kelp forests, bottom-dwelling creatures, whales, and other species that depend on one another for survival. The sea otters act as keystone species mostly by feeding on sea urchins and keeping them from destroying the kelp. However, the kelp forests are potentially in danger because sea otter populations have been vulnerable to a combination of overhunting for their pelts, predation by other species, and exposure to harmful chemicals and parasites introduced into ocean waters by human activities.

In this chapter, we focused on how biodiversity promotes sustainability, provides a variety of species to restore damaged ecosystems through ecological succession, and limits the sizes of populations. Populations of most plants and animals depend, directly or indirectly, on solar energy, and all populations play roles in the cycling of nutrients in the ecosystems where they live. In addition, the biodiversity found in the variety of species in different terrestrial and aquatic ecosystems provides alternative paths for energy flow and nutrient

cycling, better opportunities for natural selection as environmental conditions change, and natural population control mechanisms. When we disrupt these paths, we violate all three **scientific principles of sustainability**.

fred goldstein/shutterstock.com

Chapter Review

Core Case Study

1. Explain how southern sea otters act as a keystone species in their environment. Explain why we should care about protecting this species from extinction, which could result primarily from human activities.

Section 5.1

2. What is the key concept for this section? Define and give an example of **interspecific competition**. How is it different from intraspecific competition? Define and give an example of **resource partitioning** and explain how it can increase species diversity. Define **predation** and distinguish between a **predator** species and a **prey** species and give an example of each. What is a **predator–prey relationship** and why is it important?

3. Explain why we should preserve kelp forests. Describe three ways in which predators can increase their chances of feeding on their prey and three ways

in which prey species can avoid their predators. Define and give an example of **coevolution**.

4. Define **parasitism**, **mutualism**, and **commensalism** and give an example of each. Explain how each of these species interactions, along with predation, can affect the population sizes of species in ecosystems.

Section 5.2

5. What is the key concept for this section? What is **ecological succession**? Distinguish between **primary ecological succession** and **secondary ecological succession** and give an example of each.

6. Explain how living systems achieve some degree of sustainability by undergoing constant change in response to changing environmental conditions. In terms of the stability of ecosystems, distinguish between **inertia (persistence)** and **resilience** and give an example of each.

Section 5.3

7. What is the key concept for this section? Define **population**. Why do most populations live in clumps?

List four variables that govern changes in population size. Write an equation showing how these variables interact. What is a population's **age structure** and what are the three major age groups called? Define **range of tolerance**. Define **limiting factor** and give an example. Define **population density** and explain how some limiting factors can become more important as a population's density increases.

8. Distinguish between the exponential and logistic growth of a population and describe the nature of their growth curves. Define **environmental resistance**. What is the **carrying capacity** of an environment? Define and give an example of a **population crash**.

9. Describe two different reproductive strategies for species. Distinguish between *r*-**selected species** and *K*-**selected species** and give an example of each. Define **survivorship curve** and describe three types of curves. Why is the recovery of the southern sea otters slow and what factors are threatening this recovery? Explain why humans are not exempt from nature's population controls.

10. What are this chapter's *three big ideas*? Explain how the interactions among plant and animal species in any ecosystem are related to the **scientific principles of sustainability**.

Note: Key terms are in bold type.

Critical Thinking

1. What difference would it make if the southern sea otter (**Core Case Study**) became extinct primarily because of human activities? What are three things we could do to help prevent the extinction of this species?

2. Use the second law of thermodynamics (Chapter 2, p. 35) and the concept of food chains and food webs to explain why predators are generally less abundant than their prey.

3. How would you reply to someone who argues that we should not worry about the effects that human activities have on natural systems because ecological succession will repair whatever damage we do?

4. How would you reply to someone who contends that efforts to preserve species and ecosystems are not worthwhile because nature is largely unpredictable?

5. Explain why most species with a high capacity for population growth (such as bacteria, flies, and cockroaches) tend to have small individuals, while those with a low capacity for population growth (such as humans, elephants, and whales) tend to have large individuals.

6. Which reproductive strategy do most species of insect pests and harmful bacteria use? Why does this make it difficult for us to control their populations?

7. List two factors that may limit human population growth in the future. Do you think that we are close to reaching those limits? Explain.

8. If the human species were to suffer a population crash, what are three species that might move in to occupy part of our ecological niche?

Doing Environmental Science

Visit a nearby land area, such as a partially cleared or burned forest, a grassland, or an abandoned crop field, and record signs of secondary ecological succession. Take notes on your observations and formulate a hypothesis about what sort of disturbance led to this succession. Include your thoughts about whether this disturbance was natural or caused by humans. Study the area carefully to see whether you can find patches that are at different stages of succession and record your thoughts about what sorts of disturbances have caused these differences. You might want to research the topic of ecological succession in such an area.

Global Environment Watch Exercise

Search for *kelp forests* (also sometimes called *kelp beds*), and use the results to find sources of information about how a warmer ocean, as a result of climate change, might affect California's coastal kelp forests on which the southern sea otters depend (**Core Case Study**). Write a report on what you found. Try to include information on current effects of warmer water on the kelp beds as well as projections about future effects. Also, summarize any information you might find on possible ways to prevent harm to these kelp forests.

Data Analysis

The graph below shows changes in the size of an Emperor penguin population in terms of numbers of breeding pairs on the island of Terre Adelie in the Antarctic. Scientists used this data along with data on the penguins' shrinking ice habitat to project a general decline in the island's Emperor penguin population, to the point where they will be endangered in 2100. Use the graph to answer the questions on the right.

1. Assuming that the penguin population fluctuates around the carrying capacity, what was the approximate carrying capacity of the island for the penguin population from 1960 to 1975? What was the approximate carrying capacity of the island for the penguin population from 1980 to 2010?

2. What was the overall percentage decline in the penguin population from 1975 to 2010?

3. What is the projected overall percentage decline in the penguin population between 2010 and 2100?

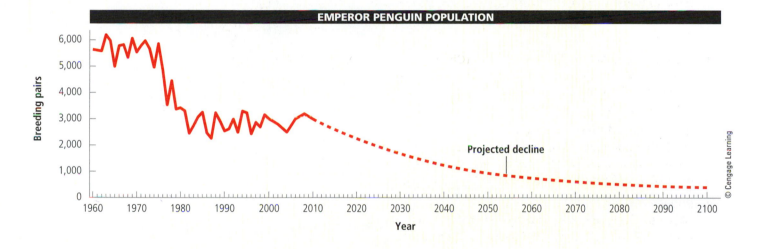

CENGAGE**brain**.com To access course materials, including Aplia homework, please visit www.cengagebrain.com.

WWW.CENGAGEBRAIN.COM   **99**

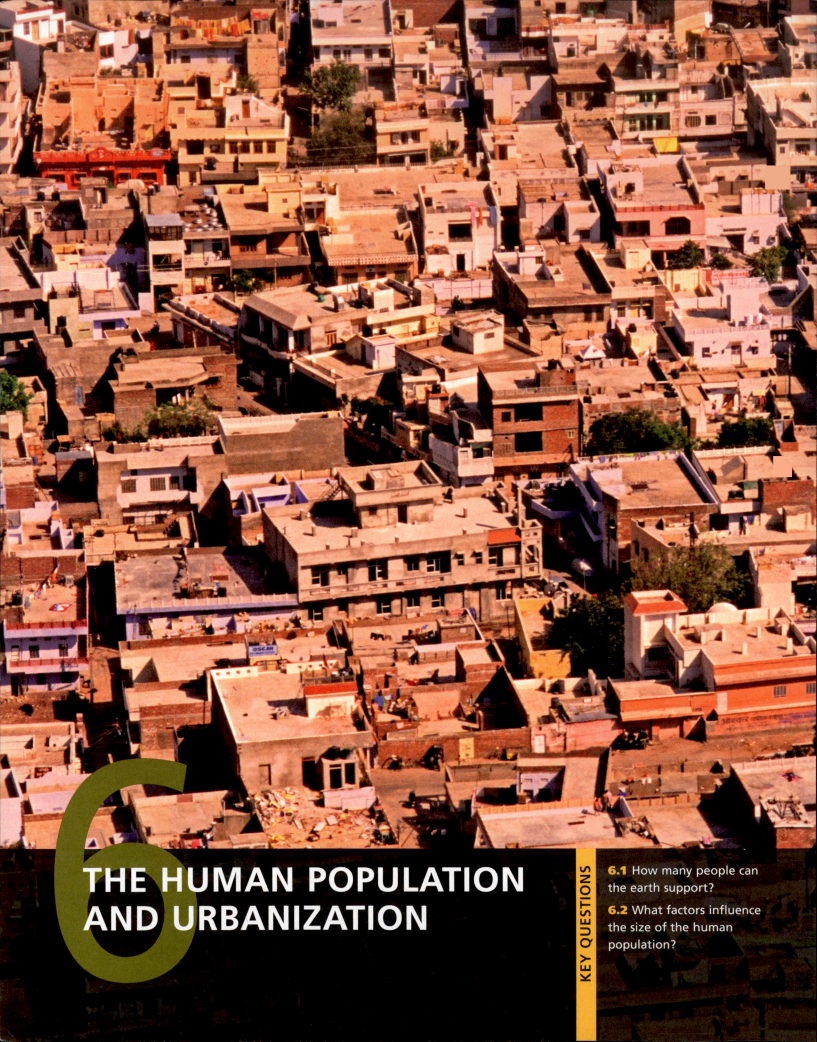

6 THE HUMAN POPULATION AND URBANIZATION

Either we limit our population growth or the natural world will do it for us.

SIR DAVID ATTENBOROUGH

Population and urban pressures in Jaipur, India.

Eastland Photo/Alamy

Portland, Oregon: On a Quest for Urban Sustainability

There are about

7.1 billion people on the earth. Each year we add about 85 million people—an average of more than 232,000 people each day. There might be 8.1 billion of us by 2025, 9.7 billion by 2050, and perhaps 11 billion by 2100.

The subject of population is not just about the number of people. It is also about how the world's people are distributed among rural and urban areas. More than half (52%) of the world's people live in urban areas and by 2050, two of every three people are likely to be urban dwellers—most of them in rapidly growing cities in less-developed countries (see chapter-opening photo). Most of the world's urban areas have huge ecological footprints that extend far beyond their boundaries, and they are not self-sustaining systems. Four of the biggest challenges of this century are to find ways to slow population growth, to reduce our ecological footprints (see Figure 1.10, p. 12), to increase our beneficial environmental impact, and

to make rapidly growing urban areas more sustainable and livable.

Portland, Oregon, a city that lies on the banks of the Willamette River (Figure 6.1), is leading the way in meeting at least three of those challenges. It has consistently ranked at or near the top in several lists of the most sustainable and livable U.S. cities for over four decades.

Since the 1970s, Portland has used strong land-use policies to control its growth, reduce dependence on automobiles, and preserve green space. Portland encourages the development of mixed-use neighborhoods with stores, light industries, professional offices, high-density housing, and access to mass transit, which allows most people to meet most of their daily needs without using a car. The city has excellent light-rail and bus lines, and has further reduced car use by developing an extensive network of bike lanes and walkways. This decreased reliance on the automobile has saved its residents more than $1 billion a year in transportation costs. It

has also contributed to public health by cutting air pollution and encouraging higher levels of physical activity.

Portland implemented a recycling system in 1987 and, by 2013, was recycling and composting 67% of its municipal solid waste—one of the highest rates in the country. In 1993, Portland became the first U.S. city to develop a plan to reduce its greenhouse gas emissions. By 2012, it had reduced its per capita greenhouse gas emissions to 26% below 1990 levels—more than any other major U.S. city had done. In 2009, Portland implemented a Climate Action Plan with the goal of cutting its greenhouse gas emissions to 40% below 1990 levels by 2030 and to 80% below those levels by 2050.

In this chapter, we examine population growth trends and the environmental impacts of our growing population, and we look at ways to slow population growth and to make the world's rapidly growing urban areas more sustainable and livable.

FIGURE 6.1 Portland, Oregon is one of the most environmentally friendly and sustainable cities in the United States.

6.1 HOW MANY PEOPLE CAN THE EARTH SUPPORT?

CONCEPT 6.1 The continuing rapid growth of the human population and its impact on natural capital raises questions about how long the human population can keep growing.

Human Population Growth Shows Certain Trends

For most of history, the human population grew slowly (see Figure 1.12, p. 15, left part of curve). But for the past 200 years, the human population has grown rapidly, resulting in the characteristic J-curve of exponential growth (Figure 1.12, right part of curve).

Three major factors account for this rapid rise of the human population: early and modern agriculture allowed us to feed more people; technologies helped us expand into almost all of the planet's climate zones and habitats (see Figure 1.10, p. 12); and death rates dropped sharply because of improved sanitation and health care and the development of antibiotics and vaccines to help control infectious diseases.

Demographers, or population experts, recognize three important trends related to the current size and impact of the human population. *First*, the rate of population growth has slowed since 1960 (Figure 6.2), but the world's population is still growing at a rate of about 1.2%. This may not seem like much but in 2013, this growth added about 85 million people to the population—an average of about 162 people every minute.

Second, human population growth is unevenly distributed. About 96% of the 85 million new arrivals on the planet in 2013 were added to the world's less-developed countries, where the population is growing 14 times faster than the population of the more-developed countries. At least 95% of the 2.6 billion people projected to be added to the world's population between 2013 and 2050 will live in less-developed countries, most of which are not equipped to deal with the pressures of such rapid growth.

Third, people have moved in large numbers from rural areas to urban areas. About 52% of world's people now live in urban areas and this trend is increasing.

Scientists and other analysts have long pondered the question: How long can the human population continue to grow while sidestepping many of the factors that sooner or later limit the growth of any population? These experts disagree over how many people the earth can support indefinitely (Science Focus 6.1). So far, advances in food production and health care have staved off widespread population declines. But there is extensive and growing evidence that we are depleting and degrading much of the earth's irreplaceable natural capital (Figure 6.3) and that we are approaching or exceeding various planetary boundaries (see Figure 3.A, p. 58).

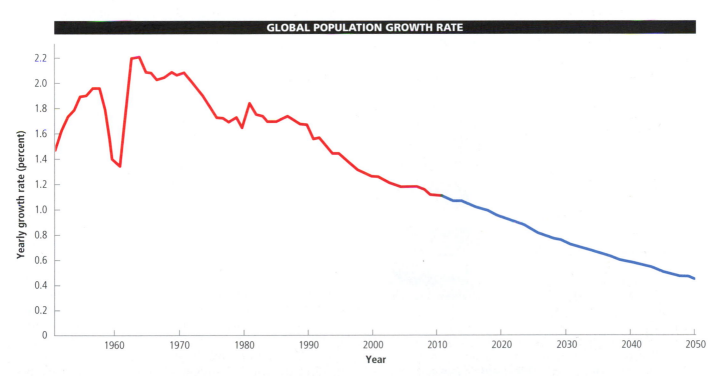

FIGURE 6.2 Global human population growth rate, 1950–2013, with projection to 2050 (in blue).
Question: Why do you think that while the annual growth *rate* of world population has generally dropped since the 1960s, the population has continued to grow? (Hint: see Figure 1-12, p. 15.)

Compiled by the authors using data from United Nations Population Division, U.S. Census Bureau, and Population Reference Bureau.

HOW LONG CAN THE HUMAN POPULATION KEEP GROWING?

Are there physical limits to human population growth and economic growth on a finite planet? Some say yes. Others say no.

The debate over possible limits to the growth of human populations and economies has been going on for more than 200 years. Meanwhile, natural capital degradation (Figure 6.2) has occurred widely and grown more intense. The earth's life-support system has been resilient enough to withstand such widespread disturbances.

However, at some point, we could reach one or more planetary boundaries or ecological tipping points (see Figure 3.A, p. 58). Exceeding such boundaries could lead to damaging long-term changes and the possibility of a sharp decline in the human population due to increasing death rates.

To some analysts, the key problem is the large and rapidly growing number of people in less-developed countries (see Table 1.1, p. 13). To others, the key factor is *overconsumption* in afflu-

ent, more-developed countries because of their high rates of resource use per person. Thus, they debate over which is more important for shrinking the human ecological footprint: slowing population growth or reducing resource consumption. Some call for doing both.

Another view of population growth is that, so far, technological advances have allowed us to overcome the environmental limits that all populations of other species face and that this has had the effect of increasing the earth's

Natural Capital Degradation

Altering Nature to Meet Our Needs

Reducing biodiversity

Increasing use of net primary productivity

Increasing genetic resistance in pest species and disease-causing bacteria

Eliminating many natural predators

Introducing harmful species into natural communities

Using some renewable resources faster than they can be replenished

Disrupting natural chemical cycling and energy flow

Relying mostly on polluting and climate-changing fossil fuels

FIGURE 6.3 Human activities have altered the natural systems and ecosystem services that sustain our lives and economies in at least eight major ways to meet the increasing needs and wants of our growing population (**Concept 6.1**). *Questions:* In your daily living, do you think you contribute directly or indirectly to any of these harmful environmental impacts? Which ones? Explain.

Top: Dirk Ercken/Shutterstock.com. Center: Fulcanelli/Shutterstock.com. Bottom: Werner Stoffberg/Shutterstock.com.

6.2 WHAT FACTORS INFLUENCE THE SIZE OF THE HUMAN POPULATION?

CONCEPT 6.2A Population size increases through births and immigration, and decreases through deaths and emigration.

CONCEPT 6.2B The key factor that determines the size of a human population is the average number of children born to the women in that population (*total fertility rate*).

The Human Population Can Grow, Decline, or Remain Fairly Stable

The basics of global population change are quite simple. If there are more births than deaths during a given period of time, the human population increases, and when the opposite is true, it decreases.

Human populations grow or decline in particular countries, cities, or other areas through the interplay of three factors: *births (fertility), deaths (mortality),* and *migration*. We can calculate the **population change** of an area by subtracting the number of people leaving a population (through death and emigration) from the number entering it (through birth and immigration) during a specified period of time (usually 1 year) (**Concept 6.2A**).

$$\text{Population change} = (\text{Births} + \text{Immigration}) - (\text{Deaths} + \text{Emigration})$$

carrying capacity for our species. Proponents of this view point out that average life expectancy in most of the world has been steadily rising, despite warnings that we are seriously degrading our life-support system.

Some of these analysts argue that because of our technological ingenuity, there are few, if any, limits to human population growth and resource use per person. They believe that we can continue ever-increasing economic growth and avoid serious damage to our life-support systems by making technological advances in areas such as food production and medicine, and by finding substitutes for resources that we are depleting. As a result, they see no need to slow population growth or resource consumption.

Proponents of slowing and eventually stopping population growth point out that in addition to degrading our life-support system, we are failing to provide the basic necessities for about 1.4 billion people—one of every five on the planet—who struggle to survive on the equivalent of about $1.25 per day. This raises a serious question: How will we meet the basic needs of the additional 2.6 billion people projected to be added between 2013 and 2050?

No one knows how close we are to environmental limits that some analysts say eventually will reduce the size of the human population primarily by sharply increasing the human death rate. These analysts call for us to confront this vital scientific, political, economic, and ethical issue.

Critical Thinking

Do you think there are environmental limits to human population growth? If so, how close do you think we are to such limits? Very close, moderately close, or far away? Explain.

When births plus immigration exceed deaths plus emigration, a population grows; when the reverse is true, a population declines. See Figure 10, p. S26, in Supplement 4 for a map comparing generalized rates of population growth among countries and regions in 2013. See Figure 13 on p. S44 in Supplement 5 for more detailed population-related data for high-, middle-, and low-income countries.

Women Are Having Fewer Babies but the World's Population Is Still Growing

A key factor affecting human population growth and size is the **total fertility rate (TFR)**: the average number of children born to the women in a population during their reproductive years (**Concept 6.2B**). See the Case Study that follows and see Figure 12, p. S29, in Supplement 4 for a map showing how TFRs vary globally.

Between 1955 and 2013, the global TFR dropped from 5 to 2.5. Those who support slowing the world's population growth view this as good news. However, to eventually halt population growth, the global TFR would have to drop to 2.1—the rate necessary for replacing both parents after taking infant mortality into account.

CASE STUDY

The U.S. Population—Third Largest and Growing

The United States has the world's third largest population. Between 1900 and 2013, its population grew fourfold from 76 million to 316 million, despite oscillations in the country's TFR (Figure 6.4) and population growth rate. During the period of high birth rates between 1946 and 1964, known as the *baby boom,* 79 million people were added to the U.S. population. At the peak of the baby boom in 1957, the average TFR was 3.7 children per woman. In most years since 1972, it has been at or below 2.1 children per woman, compared to a global TFR of 2.5.

The drop in the TFR has slowed the rate of population growth in the United States, but the country's population is still growing. According to the U.S. Census Bureau, about 2.7 million people were added to the U.S. population in 2013. About 1.6 million (59% of the total) were added because there were that many more births than deaths, and about 1.1 million (41% of the total) were legal immigrants. Since 1820 the United States has admitted almost twice as many legal immigrants and refugees as all other countries combined. In 2013, the country also had about 11.5 million illegal immigrants.

In addition to the fourfold increase in population growth since 1900, some amazing changes in lifestyles took place in the United States during the 20th century (Figure 6.5), which led to Americans living longer. Along with this came dramatic increases in per capita resource use and much larger total and per capita ecological footprints.

The U.S. Census Bureau projected that between 2013 and 2050, the U.S. population would likely grow from 316 million to 400 million—an increase of 84 million people. The United States has the world's largest total and per capita ecological footprints, mostly because of its very high

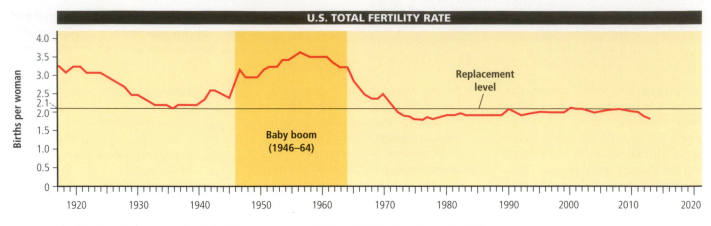

U.S. TOTAL FERTILITY RATE

FIGURE 6.4 Total fertility rates for the United States between 1917 and 2013. **Question:** The U.S. fertility rate has declined and remained at or below replacement levels since 1972. So why is the population of the United States still growing?

Compiled by the authors using data from the Population Reference Bureau and U.S. Census Bureau.

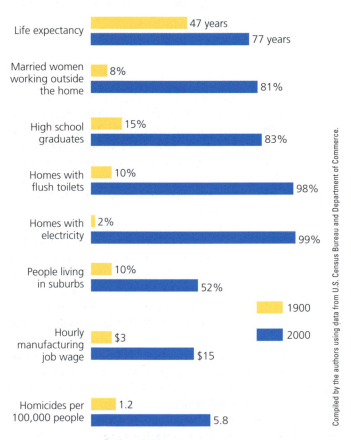

FIGURE 6.5 Some major changes took place in the United States between 1900 and 2000. **Question:** Which two of these changes do you think had the biggest impacts on the U.S. ecological footprint?

Compiled by the authors using data from U.S. Census Bureau and Department of Commerce.

rate of resource use per person, multiplied by the large size of its population, and the resulting wastes and pollution. (See a map of the U.S. ecological footprint in Figure 6, p. S21, Supplement 4.) This explains why some analysts consider the United States to be the world's most overpopulated country.

Several Factors Affect Birth and Fertility Rates

Many factors affect a country's average birth rate and TFR. One is the *importance of children as a part of the labor force,* especially in less-developed countries. Many poor couples in those countries struggling to survive on less than $2.25 a day have a large number of children to help haul daily drinking water, gather wood for heating and cooking, and grow or find food. Many children in such countries have to work for wages to help their families survive (Figure 6.6).

Another economic factor is the *cost of raising and educating children.* Birth and fertility rates tend to be lower in more-developed countries, where raising children is much more costly because they do not enter the labor force until they are in their late teens or twenties. In the United States, the cost of raising a child born in 2013 to age 18 will range from $169,000 to $390,000 depending on household income.

The *availability of, or lack of, private and public pension systems* can influence the number of children some couples have, especially the poor in less-developed countries. Pensions reduce a couple's need to have several children to

FIGURE 6.6 Young girl breaking granite into gravel in the Kerala State of India.

Priit Vesilind/National Geographic Creative

replace those that die at an early age and to help support them in old age.

Urbanization also plays a role. People living in urban areas usually have better access to family planning services and tend to have fewer children than do those living in the rural areas of poorer countries.

Another important factor is the *educational and employment opportunities available for women.* Total fertility rates tend to be low when women have access to education and paid employment outside the home. In less-developed countries, a woman with no education typically has two more children than does a woman with a high school education.

Average age at marriage (or, more precisely, the average age at which a woman has her first child) also plays a role. Women normally have fewer children when their average age at marriage is 25 or older.

Birth rates and TFRs are also affected by the *availability of legal abortions.* According to the World Health Organization and the Guttmacher Institute, each year, more than 208 million women become pregnant and at least 40 million of them get abortions—about 20 million of them legal and the other 20 million illegal (and often unsafe). The *availability of reliable birth control methods* also allows women to control the number and spacing of the children they have.

Religious beliefs, traditions, and cultural norms also play a role. In some countries, these factors favor large families, as many people strongly oppose abortion and some forms of birth control.

Several Factors Affect Death Rates

The rapid growth of the world's population over the past 100 years is not primarily the result of a rise in the birth rate. Instead, it is largely the result of declining death rates, especially in less-developed countries. More people in some of these countries are living longer and fewer infants are dying because of larger food supplies, improvements in food distribution, better nutrition, medical advances such as immunizations and antibiotics, improved sanitation, and safer water supplies.

A useful indicator of the overall health of people in a country or region is **life expectancy**: the average number of years a person born in the year of the estimate can be expected to live. Between 1955 and 2013, the average global life expectancy increased from 48 years to 70 years. In 2013, Japan had the world's longest life expectancy of 83 years. Between 1900 and 2013, the average U.S. life expectancy rose from 47 years to 79 years. In the world's poorest countries, life expectancy in 2013 was 55 years or less. Research indicates that poverty, which reduces the average life span by 7–10 years, is the single most important factor affecting life expectancy.

Another important indicator of overall health in a population is its **infant mortality rate**, the number of babies out of every 1,000 born who die before their first birthday. It is viewed as one of the best measures of a society's quality of life because it reflects a country's general level of nutrition and health care. (See Figure 13, p. S29, in Supplement 4 for a map comparing generalized infant mortality rates among the world's countries.) A high infant mortality rate usually indicates insufficient food (*undernutrition*), poor nutrition (*malnutrition,* see Figure 1.13, p. 16), and a high incidence of infectious disease. Infant mortality also affects the TFR. In areas with low infant

GOOD NEWS

mortality rates, women tend to have fewer children because fewer of their children die at an early age.

Infant mortality rates in most countries have declined dramatically since 1965. Even so, every year more than 4 million infants (most of them in less-developed countries) die of *preventable* causes during their first year of life. This average of nearly 11,000 mostly unnecessary infant deaths per day is equivalent to 55 jet airliners, each loaded with 200 infants, crashing *every day* with no survivors—an ongoing tragedy rarely reported in the media.

Between 1900 and 2013, the U.S. infant mortality rate dropped from 165 to 5.9. This sharp decline was a major factor in the marked increase in U.S. average life expectancy during this period. However, 40 other nations (most in Europe) had lower infant mortality rates than the United States in 2013.

Migration Affects an Area's Population Size

A third factor in population change is **migration**: the movement of people into *(immigration)* and out of *(emigration)* specific geographic areas. Most people migrate to another area within their country or to another country to seek jobs and economic improvement. But many are driven to migrate by religious persecution, ethnic conflicts, political oppression, or war. There are also *environmental refugees*—people who have to leave their homes and sometimes their countries because of water or food shortages, soil erosion, or some other form of environmental degradation or depletion.

6.3 HOW DOES A POPULATION'S AGE STRUCTURE AFFECT ITS GROWTH OR DECLINE?

CONCEPT 6.3 The numbers of males and females in young, middle, and older age groups determine how fast a population grows or declines.

A Population's Age Structure Helps Us to Make Projections

An important factor determining whether the population of a country increases or decreases is its **age structure**: the numbers or percentages of males and females in young, middle, and older age groups in that population (**Concept 6.3**).

Population experts construct a population *age-structure diagram* by plotting the percentages or numbers of males and females in the total population in each of three age categories: *prereproductive* (ages 0–14), consisting of individuals normally too young to have children; *reproductive* (ages 15–44), consisting of those normally able to have children; and *postreproductive* (ages 45 and older), with individuals normally too old to have children. Figure 6.7 presents generalized age-structure diagrams for countries with rapid, slow, zero, and negative population growth rates.

A country with a large percentage of its people younger than age 15 (represented by a wide base in Figure 6.7, far left) will experience rapid population growth unless death rates rise sharply. Because of this *demographic momentum,*

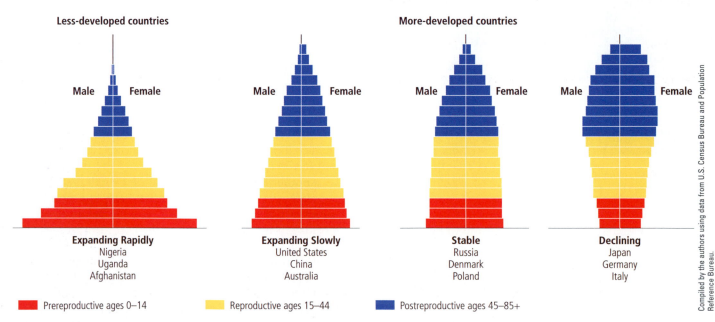

ANIMATED FIGURE 6.7 Generalized population age-structure diagrams for countries with rapid (1.5–3%), slow (0.3–1.4%), stable (0–0.2%), and negative (declining) population growth rates.
Question: Which of these diagrams best represents the country where you live?

Compiled by the authors using data from U.S. Census Bureau and Population Reference Bureau.

the number of births in such a country will rise for several decades even if women have an average of only one or two children each, due to the large number of girls entering their prime reproductive years. Most future human population growth will take place in less-developed countries because of their typically youthful age structure and rapid population growth rates.

The global population of seniors—people who are 65 and older—is projected to triple between 2013 and 2050, when one of every six people will be a senior. This graying of the world's population is due largely to declining birth rates and medical advances that have extended life spans. In 2013, the three nations with the largest senior populations (by percentage of their total population) were Japan (25%), Germany (21%), and Italy (21%). In such countries, the number of working adults is shrinking in proportion to the number of seniors, which in turn is slowing the growth of tax revenues in these countries. This raises questions about how such societies will support their growing populations of seniors.

CASE STUDY

The American Baby Boom

Changes in the distribution of a country's age groups have long-lasting economic and social impacts. For example, consider the American baby boom, which added 79 million people to the U.S. population between 1946 and 1964. Over time, this group looks like a bulge moving up through the country's age structure, as shown in Figure 6.8.

For decades, members of the baby-boom generation have strongly influenced the U.S. economy because they make up about 36% of all adult Americans. Baby boomers

created the youth market in their teens and twenties and are now creating the late-middle-age and senior markets. In addition to having this economic impact, the baby-boom generation plays an increasingly important role in deciding who gets elected to public office and what laws are passed or weakened.

Since 2011, when the first baby boomers began turning 65, the number of Americans older than age 65 has grown at the rate of about 10,000 a day and will do so through 2030. This process has been called the *graying of America*. As the number of working adults declines in proportion to the number of seniors, there may be political pressure from baby boomers to increase tax revenues to help support the growing senior population. This could lead to economic and political conflicts between older and younger Americans.

Aging Populations Can Decline Rapidly

As the percentage of people age 65 or older increases, more countries will begin experiencing population declines. If population decline is gradual, its harmful effects usually can be managed. However, some countries are experiencing fairly rapid declines and feeling such effects more severely.

Japan has the world's highest percentage of elderly people (above age 65) and the world's lowest percentage of young people (below age 15). In 2013, Japan's population was 127 million. By 2050, its population is projected to be 97 million, a 24% drop. As its population declines, there will be fewer adults working and paying taxes to support an increasing elderly population. Because Japan discourages immigration, this could threaten its economic future.

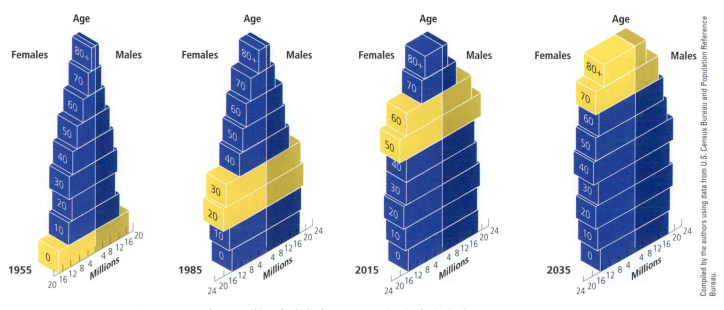

ANIMATED FIGURE 6.8 Age structure charts tracking the baby-boom generation in the United States, 1955, 1985, 2015, and 2035 (projected).

Compiled by the authors using data from U.S. Census Bureau and Population Reference Bureau.

Some Problems with Rapid Population Decline

Can threaten economic growth

Labor shortages

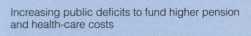

Less government revenues with fewer workers

Less entrepreneurship and new business formation

Less likelihood for new technology development

Increasing public deficits to fund higher pension and health-care costs

Pensions may be cut and retirement age increased

© Cengage Learning

FIGURE 6.9 Rapid population decline can cause several problems. *Question:* Which two of these problems do you think are the most urgent?

Top: Slavoljub Pantelic/Shutterstock.com. Center: Iofoto/Shutterstock.com. Bottom: sunabesyou/Shutterstock.

Figure 6.9 lists some of the problems associated with rapid population decline. Countries with rapidly declining populations, in addition to Japan, include Germany, Italy, Bulgaria, Hungary, Serbia, Greece, and Portugal. Other countries facing population declines in the not-too-distant future are Thailand (with a TFR drop from 7.0 in the 1970s to 1.6 in 2013) and South Korea (with a TFR of 1.3). Population declines are very difficult to reverse.

6.4 WHAT ARE SOME WAYS TO SLOW HUMAN POPULATION GROWTH?

CONCEPT 6.4 We can slow human population growth by reducing poverty through economic development, elevating the status of women, and encouraging family planning.

Promoting Economic Development Can Stabilize a Population

There is controversy over whether we should slow population growth (Science Focus 6.1). Many analysts have argued that, because population growth can be linked to environmental degradation, we need to slow population growth in order to reduce such degradation. These experts have suggested several ways to do this, one of which is to reduce poverty through economic development.

Demographers, examining the birth and death rates of western European countries that became industrialized during the 19th century, have developed a hypothesis on population change known as the **demographic transition**: As countries become industrialized and economically developed, their per capita incomes rise, poverty declines, and their populations tend to grow more slowly. According to the hypothesis, this transition takes place in four stages, as shown in Figure 6.10.

Some analysts believe that most of the world's less-developed countries will make a demographic transition over the next few decades, primarily because newer technologies will help them to develop economically and to reduce poverty by raising their per capita incomes. Other analysts fear that rapid population growth, extreme poverty, and increasing environmental degradation and resource depletion could leave some low-income, less-developed countries stuck in stage 2 of the demographic transition. This highlights the need to reduce poverty as a key to improving human health and stabilizing the population.

Empowerment of Women Tends to Slow Population Growth

A number of studies show that women tend to have fewer children if they are educated, have the ability to control their own fertility, earn an income of their own, and live in societies that do not suppress their rights. Although women make up roughly half of the world's population, in most societies, they have fewer rights and educational and economic opportunities than men have.

Women do almost all of the world's domestic work and child care for little or no pay and provide more unpaid health care (within their families) than do all of the world's organized health-care services combined. In rural areas of Africa, Latin America, and Asia, women do 60–80% of the work associated with growing food, hauling water, and gathering, and hauling wood (Figure 6.11) and animal dung for use as fuel. As one Brazilian woman observed, "For poor women, the only holiday is when you are asleep."

While women account for 66% of all hours worked, they receive only 10% of the world's income and own just 2% of the world's land. They also make up 70% of the world's poor and 66% of its 800 million illiterate adults. Poor women who cannot read often have an average of

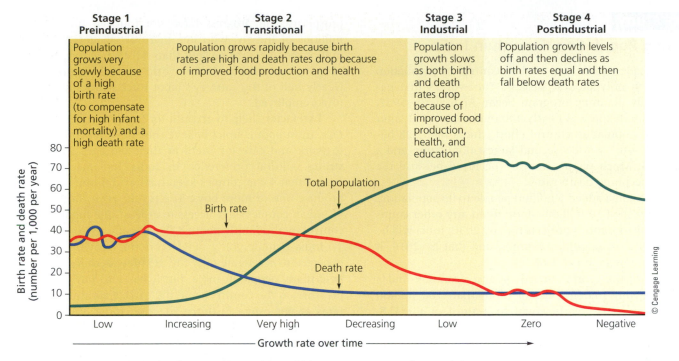

Stage 1 Preindustrial	Stage 2 Transitional	Stage 3 Industrial	Stage 4 Postindustrial
Population grows very slowly because of a high birth rate (to compensate for high infant mortality) and a high death rate	Population grows rapidly because birth rates are high and death rates drop because of improved food production and health	Population growth slows as both birth and death rates drop because of improved food production, health, and education	Population growth levels off and then declines as birth rates equal and then fall below death rates

Birth rate and death rate (number per 1,000 per year)

80 70 60 50 40 30 20 10 0

Total population

Birth rate

Death rate

Low   Increasing   Very high   Decreasing   Low   Zero   Negative

Growth rate over time

© Cengage Learning

ANIMATED FIGURE 6.10 The *demographic transition,* which a country can experience as it becomes industrialized and more economically developed, can take place in four stages. **Question:** At what stage is the country where you live?

Iv Nikolny/Shutterstock.com

FIGURE 6.11
This woman in Nepal was bringing home firewood. Typically, she spends 2 hours a day, two or three times a week, gathering and hauling wood.

five to seven children, compared to two or fewer children in societies where almost all women can read. This points to the need to see that all children get at least an elementary school education. Coupling this with a free school lunch program for the poorest children could encourage poor families to send their children to school while improving their children's ability to study and learn.

A growing number of women in less-developed countries are taking charge of their lives and reproductive behavior. As it expands, such bottom-up change driven by individual women will play an important role in stabilizing populations, improving human health, reducing poverty and environmental degradation, and allowing more access to basic human rights.

Some Argue for Promoting Family Planning

Family planning involves the provision of education and clinical services that can help couples to choose how many children to have and when to have them. Such programs vary from culture to culture, but most of them provide information on birth spacing, birth control, and health care for pregnant women and infants.

Family planning enables women to limit the size of their families if they wish to do so, and to plan their pregnancies. According to studies by the UN Population Division and other population agencies, family planning has been a major factor in reducing the number of unintended pregnancies and births, the number of safe and unsafe abortions, the number of mothers and fetuses dying during pregnancy, rates of infant mortality, rates of HIV/AIDS infection, and population growth rates. It also has financial benefits. Studies have shown that each dollar spent on family planning in countries such as Thailand, Egypt, and Bangladesh saves $10–$16 in health, education, and social service costs by preventing unwanted births.

Slowing Population Growth in India

For six decades, India has tried to control its population growth with only modest success. The world's first national family planning program began in India in 1952, when its population was nearly 400 million. In 2013, after 61 years of population control efforts, India had 1.28 billion people—the world's second largest population and a TFR of 2.4. Much of this increase occurred because the country's declining death rates.

In 1952, India added 5 million people to its population. In 2013, it added 19 million—more than any other country. The United Nations projects that by 2030, India will be the world's most populous country, and that by 2050, it will have a population of 1.65 billion.

India has the world's fourth largest economy and a rapidly growing middle class of more than 100 million people—a number nearly equal to a third of the U.S. population. However, the country faces serious poverty, malnutrition, and environmental problems that could worsen as its population continues to grow rapidly. About one-fourth of all people in India's cities live in slums, and prosperity and progress have not touched hundreds of millions of Indians who live in rural villages. With 400 million people earning less than $1.25 per day, India is home to one-third of the world's poor (Figure 6.12). Nearly half of the country's labor force is unemployed or underemployed.

Two factors help to account for larger families in India. *First,* most poor couples believe they need several children to work and care for them in their old age. *Second,* the strong cultural preference in India for male children means that some couples keep having children until they produce one or more boys. The result: even though 90% of Indian couples have access to at least one modern birth control method, only about 47% actually use one.

India is undergoing rapid economic growth, which is expected to accelerate over the next few decades. This will help many people in India, but it will also put increasing pressure on the country's and the earth's natural capital as rates of per capita resource use rise. India already faces serious soil erosion, overgrazing, water pollution, and air pollution problems. On the other hand, economic growth may help India to slow its population growth by accelerating its demographic transition.

FIGURE 6.12 Homeless people in Kolkata, India in 2011.

Slowing Population Growth in China

China is the world's most populous country, with 1.36 billion people in 2013 (Figure 6.13). In 2011, the U.S. Census Bureau projected that if current trends continue, China's population is expected to decline to about 1.3 billion by 2050 and to as low as 750 million by the end of this century.

In the 1960s, China's large population was growing so rapidly that there was a serious threat of mass starvation. To avoid this, government officials took measures that eventually led to the establishment of the world's most extensive, intrusive, and strict family planning and birth control program.

The goal has been to sharply reduce population growth and avoid mass starvation and social upheaval by promoting one-child families. The government provides contraceptives, sterilizations, and abortions for married couples. In addition, married couples pledging to have no more than one child receive a number of benefits, including better housing, more food, free health care, salary bonuses, and preferential job opportunities for their child. Couples who break their pledge lose such benefits.

Since this government-controlled program began, China has made impressive efforts to feed its people and bring its population growth under control. Between 1972 and 2013, the country cut its birth rate in half and reduced its TFR from 5.7 to 1.5 (compared to 1.9 in 2013 in the United States and 2.4 in India). China's population is growing more slowly than the U.S. population, even with legal immigration included. Although China has avoided mass starvation, its strict population control program has been accused of violating human rights.

Since 1980, China has undergone rapid industrialization and economic growth. According to the Earth Policy Institute, between 1990 and 2010, this process reduced the number of people living in extreme poverty by almost 500 million. It has also helped at least 300 million Chinese—a number almost equal to the entire U.S. population—to become middle-class consumers. Over time, China's rapidly growing middle class will consume more resources per person, expanding China's ecological footprint within its own borders and in other parts of the world that provide it with resources. This will put a strain on China's and the earth's natural capital. Like India, China faces serious soil erosion, overgrazing, water pollution, and air pollution problems.

Because of its one-child policy, during the past two decades, the average age of China's population has been increasing at one of the fastest rates ever recorded. In

FIGURE 6.13 Thousands of people crowd a street in China, where almost one-fifth of all people on the planet live.

2012, there were 194 million Chinese people aged 60 and over—the largest number of people in this age group of all the world's countries. While China's population is not yet declining, the UN estimates that by 2030, the country is likely to have too few young workers to support its rapidly aging population. This graying of the Chinese population could lead to a declining work force, limited funds for supporting continued economic development, and fewer children and grandchildren to care for the growing number of elderly people. These concerns and other factors may slow China's economic growth and lead to some relaxation of the country's one-child population control policy.

6.5 WHAT ARE THE MAJOR URBAN RESOURCE AND ENVIRONMENTAL PROBLEMS?

CONCEPT 6.5 Most cities are unsustainable because of high levels of resource use, waste, pollution, and poverty.

Population Experts See Three Important Urban Trends

About 52% of the world's people, 81% of all Americans (see Case Study that follows), and 53% of China's population live in urban areas. Every day there are about 200,000 more urban dwellers. See Figure 10, p. S26 in Supplement 4 for a map of global population density.

Urban areas grow in two ways—by *natural increase* (more births than deaths) and by *immigration,* mostly from rural areas. Rural people are *pulled* to urban areas in search of jobs, food, housing, educational opportunities, better health care, and entertainment. Some are also *pushed* from rural to urban areas by factors such as famine, losses of land for growing food, deteriorating environmental conditions, war, and religious, racial, and political conflicts.

Three major trends in urban population dynamics are important for understanding the problems and challenges of urban growth:

1. *The percentage of the global population that lives in urban areas has grown sharply and this trend is projected to continue.* Between 1850 and 2013, the percentage of the world's people living in urban areas increased from 2% to 52% and is likely to reach 67% by 2050, with most new urban dwellers living in less-developed countries.

2. *The numbers and sizes of urban areas are mushrooming.* Today there are 26 *megacities*—cities with 10 million or more people—19 of them in less-developed countries (Figure 6.14). Nine of these urban areas are *hypercities* with more than 20 million people. The largest hypercity is Tokyo, Japan with 37.1 million—more than the entire population of Canada. By 2025, the number of megacities is expected to reach 37 with 21

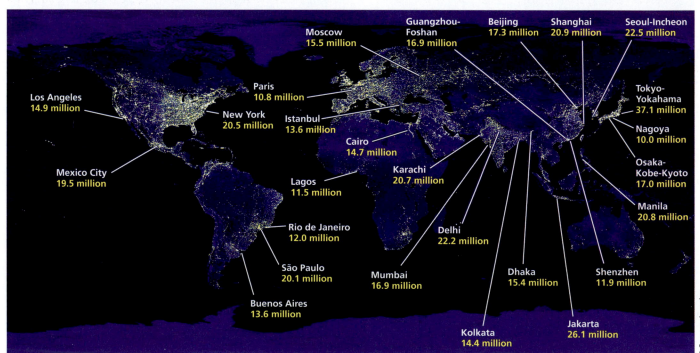

FIGURE 6.14 *Global outlook:* Megacities, or major urban areas with 10 million or more people, in 2012. **Question:** In order, what were the world's five most populous urban areas in 2012?

Compiled by the authors using data from National Geophysics Data Center, Demographia, National Oceanic and Atmospheric Administration, and United Nations Population Division.

of them in Asia. Some of the world's megacities and hypercities are merging into vast urban *megaregions,* each with more than 100 million people. The largest megaregion is the Hong Kong-Shenhzen-Guangzhou region in China with about 120 million people.

3. *Poverty is becoming increasingly urbanized, mostly in less-developed countries.* The United Nations estimates that at least 1 billion people live in the slums and shantytowns of most of the major cities in less-developed countries (see chapter-opening photo).

CONSIDER THIS. . .

THINKING ABOUT Urban Trends

If you could reverse one of the three urban trends discussed here, which one would it be? Explain.

CASE STUDY

Urbanization in the United States

Between 1800 and 2013, the percentage of the U.S. population living in urban areas increased from 5% to 81%. This population shift has occurred in three phases.

First, *people migrated from rural areas to large central cities.* In 2013, about 71% of all Americans lived in urban areas with at least 50,000 people, and about 54% lived in urban areas with 1 million or more residents (Figure 6.15, shaded areas).

Second, *many people migrated from large central cities to smaller cities and suburbs.* Currently, about half of all urban Americans live in the suburbs, nearly a third in central cities, and the rest in rural housing developments beyond suburbs.

Third, *many people migrated from the North and East to the South and West.* Since 1980, about 80% of the U.S. population increase, occurred in the South and West.

Since 1920, and especially since 1970, many of the worst urban environmental problems in the United States have been reduced significantly (Figure 6.5). Most people have better working and housing conditions and air and water quality have improved. Better sanitation, clean public water supplies, and expanded medical care have slashed death rates and incidences of sickness from infectious diseases. Also, the concentration of most of the population in urban areas has helped to protect some of the country's biodiversity by reducing the destruction and degradation of wildlife habitat.

However, a number of U.S. central cities—especially older ones—have deteriorating services and aging *infrastructures* (streets, bridges, dams, power lines, schools, water supply pipes, and sewers). Funds for repairing and upgrading urban infrastructure have declined in many urban areas as the flight of people and businesses to the suburbs and beyond has led to lower central city property tax revenues. However, this trend, too, has been reversed in some cities, including Portland, Oregon (**Core Case Study**).

Urban Sprawl Gobbles Up the Countryside

In the United States and some other countries, **urban sprawl**—the growth of low-density development on the edges of cities and towns—is eliminating agricultural and wild lands around many cities (Figure 6.16). It results in a dispersed jumble of housing developments, shopping malls, parking lots, and office complexes that are loosely connected by multilane highways and freeways.

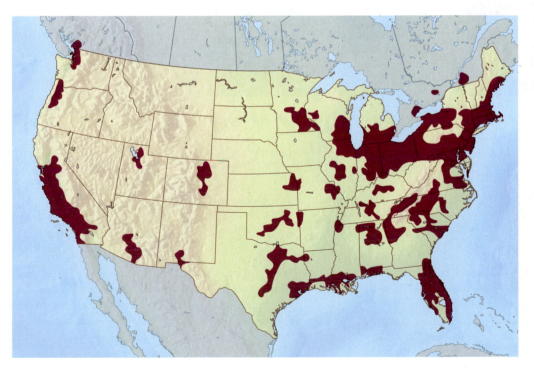

FIGURE 6.15 Urbanized areas (shaded) in the United States where cities, suburbs, and towns dominate the land area. **Question:** Why do you think many of the largest urban areas are located near water?

Compiled by the authors using data from National Geophysical Data Center/National Oceanic and Atmospheric Administration and U.S. Census Bureau.

1973

2009

Courtesy of U.S. Geological Survey

ANIMATED FIGURE 6.16 *Urban sprawl* in and around the U.S. city of Las Vegas, Nevada, between 1973 and 2009. ***Question:*** What might be a limiting factor on population growth in Las Vegas?

Natural Capital Degradation

Urban Sprawl

Land and Biodiversity	Water	Energy, Air, and Climate	Economic Effects
Loss of cropland	Increased use and pollution of surface water and groundwater	Increased energy use and waste	Decline of downtown business districts
Loss and fragmentation of forests, grasslands, wetlands, and wildlife habitat	Increased runoff and flooding	Increased emissions of carbon dioxide and other air pollutants	More unemployment in central cities

© Cengage Learning

FIGURE 6.17 Some of the undesirable impacts of urban sprawl, or car-dependent development. ***Question:*** Which five of these effects do you think are the most harmful?

Left: Condor 36/Shutterstock.com.
Left center: spirit of america/Shutterstock.com.
Right center: ssuaphotos/Shutterstcok.com.
Right: ronfromyork/Shutterstock.com.

Urban sprawl is largely the product of ample affordable land, automobiles, federal and state funding of highways, and inadequate urban planning. Many people prefer living in suburbs and *exurbs*—housing developments scattered over vast areas that lie beyond suburbs. Compared to central cities, these areas provide lower-density living and access to larger lot sizes and single-family homes. Often these areas also have newer public schools and lower crime rates. On the other hand, urban sprawl has caused or contributed to a number of environmental problems, as summarized in Figure 6.17.

Urbanization Has Advantages

Urbanization has many benefits. From an *economic standpoint,* cities are centers of economic development, innovation, education, technological advances, social and cultural diversity, and job markets. Urban residents in many parts of the world tend to live longer than do rural residents and to have lower infant mortality and fertility rates. They typically also have better access to medical care, family planning, education, and social services than do their rural counterparts.

Urban areas also have some environmental advantages. Recycling is more economically feasible because of the high concentrations of recyclable materials in urban areas. Concentrating people in cities helps to preserve biodiversity by reducing the stress on wildlife habitats. Heating and cooling multistory apartment and office buildings in central cities takes less energy per person than does heating and cooling single-family homes and smaller office buildings, more common in the suburbs. Central-city dwellers also tend to drive less and rely more on mass transportation, car-pooling, walking, and bicycling.

Urbanization Has Disadvantages

Most urban areas are unsustainable systems. Urban populations occupy only about 3% of the earth's land area, but they consume about 75% of its resources and produce about 75% of the world's pollution and wastes. Because of this high input of food, water, and other resources, and the resulting high waste output (Figure 6.18), most of the world's cities have huge ecological footprints that extend far beyond their boundaries, and

they typically are not self-sustaining systems (**Concept 6.5**), for a number of reasons.

Most Cities Lack Vegetation. In urban areas, most trees, shrubs, grasses, and other plants are cleared to make way for buildings, roads, parking lots, and housing developments. Thus, most cities do not benefit from the free ecosystem services provided by vegetation, including air purification, generation of oxygen, removal of atmospheric CO_2, control of soil erosion, and the provision of habitat for wildlife.

Many Cities Have Water Problems. Often, as cities grow and their water demands increase, expensive reservoirs and canals must be built and deeper wells must be drilled. This can deprive rural and wild areas of surface water and it can deplete groundwater supplies. Also, projected climate change is expected to melt some mountaintop glaciers, and cities that depend on this ice for much of their annual water supplies will face severe water shortages.

Flooding also tends to be greater in cities that are built on floodplains near rivers or along low-lying coastlines. In most cities, buildings and paved surfaces cause precipitation to run off quickly and overload storm drains. Urban development has often destroyed or degraded large areas of wetlands that have served as natural sponges to help absorb excess storm water. Many of the world's largest coastal cities (Figure 6.14) will very likely face a new flooding threat at some time in this century as sea levels rise because of projected climate change.

Cities Tend to Concentrate Pollution and Health Problems. Because of their high population densities and rates of resource consumption, cities produce most of the world's air pollution, water pollution, and solid and hazardous wastes. Pollutant levels are generally higher because the pollution is produced in a confined area and

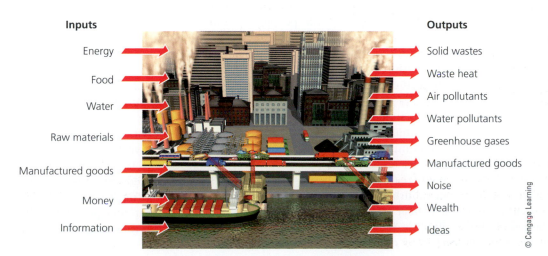

Inputs		Outputs
Energy		Solid wastes
Food		Waste heat
Water		Air pollutants
Raw materials		Water pollutants
Manufactured goods		Greenhouse gases
Money		Manufactured goods
Information		Noise
		Wealth
		Ideas

© Cengage Learning

FIGURE 6.18 Natural capital degradation: The typical city depends on nonurban areas for huge inputs of matter and energy resources, while it generates and concentrates large outputs of pollution, waste matter, and heat. *Question:* How would you apply the three **scientific principles of sustainability** to lessen some of these impacts?

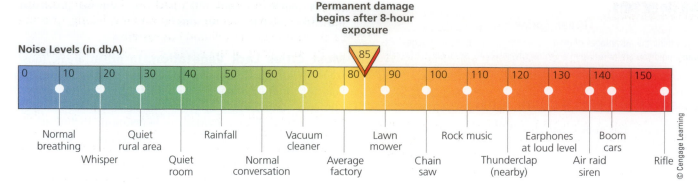

FIGURE 6.19 *Noise levels* (in decibel-A [dbA] sound pressure units) of some common sounds. **Question:** How often are your ears subjected to noise levels of 85 or more dbA?

cannot be dispersed and diluted as readily as pollution produced in rural areas can. In addition, high population densities in urban areas can promote the spread of infectious diseases, especially if adequate drinking water and sewage systems are not in place.

Cities Have Excessive Noise. Most urban dwellers are subjected to **noise pollution**: any unwanted, disturbing, or harmful sound that damages, impairs, or interferes with hearing, causes stress, hampers concentration and work efficiency, or causes accidents. Noise levels are measured in decibel-A (dbA) sound pressure units that vary with different human activities (Figure 6.19).

Sound pressure becomes damaging at about 85 dbA and painful at around 120 dbA. At 180 dbA, sound can kill. Prolonged exposure to sound levels above 85 dbA can cause permanent hearing damage. Just one-and-a-half minutes of exposure to 110 decibels or more can cause such damage. You are being exposed to a sound level high enough to cause permanent hearing damage if a noise requires you to raise your voice to be heard, if a noise causes your ears to ring, or if a noise makes nearby speech seem muffled. Prolonged exposure to lower noise levels and occasional loud sounds may not damage your hearing, but these sound levels can be very stressful.

Noise pollution can be reduced by modifying noisy activities, shielding noisy activities or processes, shielding workers or other persons from the noise, moving noisy operations or machines away, and using *antinoise* (various technologies that cancel or muffle one noise with another).

Cities Affect Local Climates. On average, cities tend to be warmer, rainier, foggier, and cloudier than suburbs and nearby rural areas. In cities, the enormous amount of heat generated by cars, factories, furnaces, lights, air conditioners, and heat-absorbing dark roofs and streets creates an *urban heat island* that is surrounded by cooler suburban and rural areas. As cities grow and merge, their heat islands merge, which can reduce the natural dilution and cleansing of polluted air. The urban heat island effect can

also greatly increase dependence on air conditioning. This in turn leads to higher energy consumption, greenhouse gas emissions, and other forms of air pollution.

CONSIDER THIS. . .

THINKING ABOUT Disadvantages of Urbanization

Which two of these disadvantages of urbanization do you think are the most serious? Explain.

Life Is a Desperate Struggle for the Urban Poor in Less-Developed Countries

Poverty is a way of life for many urban dwellers in less-developed countries. According to a UN study, the number of urban residents living in poverty—now about 1 billion—could reach 1.4 billion by 2020.

Some of these people live in crowded *slums*—areas dominated by dilapidated tenements, or rooming houses where several people might live in a single room (see chapter-opening photo). Other poor people live in *squatter settlements* and *shantytowns* on the outskirts of cities. They build shacks from corrugated metal, plastic sheets, scrap wood, and other scavenged building materials, or they live in rusted shipping containers and junked cars.

Poor people living in shantytowns and squatter settlements, or on the streets, usually lack clean water supplies, sewers, electricity, and roads, and are subject to severe air and water pollution and hazardous wastes from nearby factories. Many of these settlements are in locations especially prone to landslides, flooding, or earthquakes. Some city governments regularly bulldoze squatter shacks and send police to drive illegal settlers out. The people usually move back in within a few days or weeks, or develop another shantytown elsewhere.

Some governments have addressed these problems by legally recognizing slums and granting legal titles to the land. They base this on evidence that poor people usually improve their living conditions once they know they have a permanent place to live.

Mexico City

With 19.5 million people, Mexico City is one of the world's megacities (Figure 6.14) and will soon become a hypercity with more than 20 million people. More than one-third of its residents live in slums called *barrios* or in squatter settlements that lack running water and electricity. At least 3 million people in the barrios have no sewage facilities, so human waste from these slums is deposited in gutters, vacant lots, and open ditches every day, attracting rats and flies. When the winds pick up dried excrement, a *fecal snow* blankets parts of the city. This bacteria-laden fallout leads to widespread salmonella and hepatitis infections, especially among children.

In 1992, the United Nations named Mexico City "the most polluted city on the planet." Since then Mexico City has made progress in reducing the severity of some of its air pollution problems. In 2013, the Institute for Transportation and Development awarded Mexico City its Sustainable Transportation Award for expanding its bus rapid-transit system, rebuilding its public parks, reducing crime, and expanding its bike sharing program and its bike lanes. The percentage of days each year in which air pollution standards are violated has fallen from 50% to 20% and ozone and other air pollutants are now at about the same level as those of Los Angeles, California.

GOOD NEWS

The city government has moved refineries and factories out of the city, banned cars in its central zone, and required air pollution controls on all cars made after 1991. It has also phased out the use of leaded gasoline, expanded public transportation, and replaced some old buses, taxis, and delivery trucks with vehicles that produce fewer emissions.

Mexico City still has a long way to go as its human population increases along with its number of motor vehicles. However, this story shows what can be done to improve environmental quality once a community decides to act.

6.6 HOW DOES TRANSPORTATION AFFECT URBAN ENVIRONMENTAL IMPACTS?

CONCEPT 6.6 In some countries, many people live in widely dispersed urban areas and depend mostly on motor vehicles for their transportation, which greatly expands their ecological footprints.

Cities Can Grow Outward or Upward

If a city cannot spread outward, it must grow vertically—upward and downward (below ground)—so that it occupies a small land area with a high population density. Most people living in *compact cities* such as Hong Kong, China, and Tokyo, Japan, get around by walking, biking, or using mass transit such as rail and bus systems.

In other parts of the world, a combination of plentiful land and networks of highways have produced *dispersed cities* whose residents depend on motor vehicles for most travel (**Concept 6.6**). Such car-centered cities are found in the United States, Canada, Australia, and some other countries where ample land is available for cities to expand outward. The resulting urban sprawl can have a number of undesirable effects (Figure 6.17).

The United States is a prime example of a car-centered nation. With 4.4% of the world's people, the country has about 25% of the world's 1 billion motor vehicles, according to the U.S. Department of Transportation. In its dispersed urban areas, passenger vehicles are used for 86% of all transportation and 76% of urban residents drive alone to work every day (up from 64% in 1980).

Use of Motor Vehicles Has Advantages and Disadvantages

Motor vehicles provide mobility and offer a convenient and comfortable way to get from one place to another. For many people, driving is personally satisfying. Also, much of the world's economy is built on producing motor vehicles and supplying fuel, roads, services, and repairs for them.

Despite their important benefits, motor vehicles have many harmful effects on people and the environment. Globally, automobile accidents kill about 1.3 million people per year—an average of more than 3,500 deaths per day—and injure another 50 million people. They also kill about 50 million wild animals and family pets every year.

Each year, motor vehicle accidents in the United States kill about 32,000 people and injure another 2 million, at least 300,000 of them severely. Car accidents have killed more Americans than have all the wars in the country's history.

Motor vehicles are the world's largest source of outdoor air pollution, which causes 30,000–60,000 premature deaths per year in the United States, according to the Environmental Protection Agency. They are also the fastest-growing source of climate-changing CO_2 emissions. At least a third of the world's urban land and half of that in the United States is devoted to roads, parking lots, gasoline stations, and other automobile-related uses.

Another problem is congestion. If current trends continue, U.S. motorists will spend an average of 2 years of their lives in traffic jams, as streets and freeways will more often resemble parking lots. Traffic congestion in some cities in less-developed countries is much worse. Building more roads is not likely the answer because more roads usually encourage more people to use motor vehicles.

FIGURE 6.20 Widespread bicycle use and a light-rail system, in operation since 1986, have helped to reduce car use in Portland, Oregon (**Core Case Study**).

Ken Hawkins/Alamy

Reducing Automobile Use Is Not Easy, but It Can Be Done

Some environmental scientists and economists suggest that we can reduce the harmful effects of automobile use by making drivers pay directly for most of the environmental and health costs caused by their automobile use—a *user-pays* approach.

One way to phase in such *full-cost pricing*, in keeping with one of the **principles of sustainability**, would be to charge a tax or fee on gasoline to cover the estimated harmful costs of driving. According to a study by the International Center for Technology Assessment, such a tax would amount to about $3.18 per liter ($12 per gallon) of gasoline in the United States. Gradually phasing in such a tax, as has been done in many European nations, could spur the use of more energy-efficient motor vehicles and mass transit. It would also reduce pollution and environmental degradation and help to slow projected climate change and ocean acidification.

Proponents of higher gasoline taxes urge governments to do two major things. *First,* fund programs to educate people about the hidden costs they are paying for their automobile use. *Second,* use gasoline tax revenues to help finance mass transit systems, bike lanes, and sidewalks as alternatives to cars, and also to reduce taxes on income, wages, and wealth to offset the increased taxes on gasoline. Such a *tax shift* would help to make higher gasoline taxes more politically and economically acceptable.

Taxing gasoline heavily would be difficult in the United States, for three reasons. *First,* it faces strong opposition from people who feel they are already overtaxed, many of whom are largely unaware of the hidden costs they are paying for gasoline. The other opposition group is made up of the powerful transportation-related industries such as carmakers, oil and tire companies, road builders, and many real estate developers. *Second,* the dispersed nature of most U.S. urban areas makes people dependent on cars, and thus higher taxes would be an economic burden for them. In 2012, according to the U.S. Department of Transportation, 10% of U.S. workers carpooled (down from 20% in 1980), 5% used mass transit (down from 6% in 1980), and 3% biked or walked to work (down from 6% in 1980). *Third,* fast, efficient, reliable, and affordable mass transit options, bike lanes, and sidewalks are not widely available in the United States, primarily because most of the revenue from gasoline taxes is used for building and improving highways for motor vehicles.

Another way to reduce automobile use and urban congestion is to raise parking fees and charge tolls on roads, tunnels, and bridges leading into cities—especially during peak traffic times. Densely populated Singapore is rarely congested because it auctions the rights to buy a car, and drivers are charged a fee every time they enter the city. Several European cities have also imposed stiff fees for motor vehicle use in their central cities, while others have banned the parking of cars on city streets and established networks of bike lanes.

More than 300 European cities have *car-sharing* networks that provide short-term rental of cars. Portland, Oregon (**Core Case Study**), was the first U.S. city to develop a car-sharing system. Network members reserve a car in advance or contact the network and are directed to the closest car. In Berlin, Germany, car sharing has cut car ownership by 75%. According to the Worldwatch Institute, car sharing in Europe has reduced the average driver's CO_2 emissions by 40–50%. Car-sharing networks have sprouted in several U.S. cities and on some college campuses, and some large car-rental companies have begun renting cars by the hour.

Bicycling accounts for about a third of all urban trips in the Netherlands and in Copenhagen, Denmark, compared to less than 1% of in the United States. Portland, Oregon (**Core Case Study**), has the nation's highest percentage of bicycle commuters and a goal of reaching 25% by 2030. It also has widely used bus transit and light rail systems (Figure 6.20).

Each of the several alternatives to motor vehicles has advantages and disadvantages. Figures 6.21 through 6.24 summarize the pros and cons of using, respectively, bicycles, bus rapid-transit systems, mass-transit rail systems (within urban areas), and high-speed rail systems (between urban areas).

Trade-Offs

Bicycles

Advantages	Disadvantages
Are quiet and nonpolluting	Provide little protection in an accident
Take few resources to manufacture	Provide no protection from bad weather
Burn no fossil fuels	Are impractical for long trips
Require little parking space	Bike lanes and secure bike storage not yet widespread

© Cengage Learning

FIGURE 6.21 Bicycle use has advantages and disadvantages. **Question:** Which single advantage and which single disadvantage do you think are the most important?

Tyler Olson/Shutterstock.com

Trade-Offs

Buses

Advantages	Disadvantages
Reduce car use and air pollution	Can lose money because they require affordable fares
Can be rerouted as needed	Can get caught in traffic and add to noise and pollution
Cheaper than heavy-rail system	Commit riders to transportation schedules

© Cengage Learning

FIGURE 6.22 Bus rapid-transit (BRT) systems and conventional bus systems in urban areas have advantages and disadvantages. BRT systems make bus use more convenient by including fast express routes, allowing riders to pay at machines at each bus stop before they board a bus, and having 3 or 4 doors for quicker boarding. **Question:** Which single advantage and which single disadvantage do you think are the most important?

Isaak/Shutterstock.com

Trade-Offs

Mass Transit Rail

Advantages	Disadvantages
Uses less energy and produces less air pollution than cars do	Expensive to build and maintain
Uses less land than roads and parking lots use	Cost-effective only in densely populated areas
Causes fewer injuries and deaths than cars	Commits riders to transportation schedules

© Cengage Learning

FIGURE 6.23 Mass-transit rail systems in urban areas have advantages and disadvantages. They include *heavy-rail* systems (subways, elevated railways, and metro trains) and *light-rail* systems (streetcars, trolley cars, and tramways). **Question:** Which single advantage and which single disadvantage do you think are the most important?

Steve Rosset/Shutterstock.com

Trade-Offs

Rapid Rail

Advantages	Disadvantages
Much more energy efficient per rider than cars and planes are	Costly to run and maintain
Produces less air pollution than cars and planes	Causes noise and vibration for nearby residents
Can reduce need for air travel, cars, roads, and parking areas	Adds some risk of collision at car crossings

© Cengage Learning

FIGURE 6.24 Rapid-rail systems between urban areas have advantages and disadvantages. Western Europe, Japan, and China have a number of high-speed bullet trains that travel between cities at up to 306 kilometers (190 miles) per hour. **Question:** Which single advantage and which single disadvantage do you think are the most important?

Alfonso d'Agostino/Shutterstock.com

6.7 HOW CAN CITIES BECOME MORE SUSTAINABLE AND LIVABLE?

CONCEPT 6.7 An *eco-city* allows people to choose walking, biking, or mass transit for most transportation needs; to recycle or reuse most of their wastes; to grow much of their food; and to protect biodiversity by preserving surrounding land.

Smart Growth Can Promote Environmental Sustainability

Smart growth is a set of policies and tools that allow and encourage more environmentally sustainable urban development with less dependence on cars. It uses zoning laws and other tools to channel growth in order to reduce its ecological footprint.

Some critics contend that by limiting urban expansion, smart growth can lead to higher land and housing prices. Supporters counter that it controls and directs sprawl, protects ecologically sensitive and important lands and waterways, and results in neighborhoods that are enjoyable places to live. Figure 6.25 lists some of the widely used smart growth tools.

The Eco-City Concept: Cities for People, Not Cars

Many environmental scientists and urban planners call for us to make new and existing urban areas more sustainable and enjoyable places to live through good ecological design—an important way to increase our beneficial environmental impact.

An eco-city is a people-oriented city, not a car-oriented city. Its residents are able to walk, bike, or use low-polluting mass transit for most of their travel. Its buildings, vehicles, and appliances meet high energy-efficiency standards. Trees and plants adapted to the local climate and soils are planted throughout the city to provide shade, beauty, and wildlife habitats, and to reduce air pollution, noise, and soil erosion.

In an eco-city, abandoned lots and industrial sites are cleaned up and used. Nearby forests, grasslands, wetlands, and farms are preserved. Much of the food that people eat comes from nearby organic farms, solar greenhouses, community gardens, and small gardens on rooftops, in yards, and in window boxes. Parks are easily available to everyone. People who design and live in eco-cities take seriously the advice that U.S. urban planner Lewis Mumford gave more than 3 decades ago: "Forget the damned motor car and build cities for lovers and friends."

Solutions

Smart Growth Tools

Limits and Regulations

Limit building permits

Draw urban growth boundaries

Create greenbelts around cities

Zoning
Promote mixed use of housing and small businesses

Concentrate development along mass transportation routes

Planning
Ecological land-use planning

Environmental impact analysis

Integrated regional planning

Protection

Preserve open space

Buy new open space

Prohibit certain types of development

Taxes

Tax land, not buildings

Tax land on value of actual use instead of on highest value as developed land

Tax Breaks

For owners agreeing not to allow certain types of development

For cleaning up and developing abandoned urban sites

Revitalization and New Growth

Revitalize existing towns and cities

Build well-planned new towns and villages within cities

© Cengage Learning

FIGURE 6.25 Smart growth tools can be used to prevent or control urban growth and sprawl. ***Questions:*** Which five of these tools do you think would be best for preventing or controlling urban sprawl? Which, if any, of these tools are used in your community?

Top: Tungphoto/Shutterstock.com. Bottom: istockphoto.com/schmidt-z.

The eco-city model is not a futuristic dream, but a growing reality in a number of cities, including Portland, Oregon (**Core Case Study**), that are striving to become more environmentally sustainable and livable. Other examples are Curitiba, Brazil (see the following Case Study); Bogotá, Colombia; Waitakere City, New Zealand; Stockholm, Sweden; Helsinki, Finland; Copenhagen, Denmark; Melbourne, Australia; Vancouver, Canada; Leicester, England; Neerlands, the Netherlands; and in the United States, Davis, California; Olympia, Washington; and Chattanooga, Tennessee.

The Eco-City Concept in Curitiba, Brazil

An example of an eco-city is Curitiba ("koor-i-TEE-ba"), a city of 3.2 million people, known as the "ecological capital" of Brazil. In 1969, planners in this city decided to focus on an inexpensive and efficient mass-transit system rather than on the car.

Curitiba's superb bus rapid-transit (BRT) system moves large numbers of passengers efficiently, including 72% of the city's commuters. Each of the system's five major "spokes," connecting the city center to outlying districts (see map in Figure 6.26), has two express lanes used only by buses. Double- and triple-length bus sections are coupled together as needed to carry up to 300 passengers. Boarding is speeded up by the use of extra-wide bus doors and covered boarding platforms where passengers can pay before getting on the bus (Figure 6.26).

Only high-rise apartment buildings are allowed near major bus routes, and the bottom two floors of each building must be devoted to stores—a practice that reduces the need for residents to travel. Cars are banned from 49 blocks in the center of the downtown area, which has a network of pedestrian walkways connected to bus sta-tions, parks, and bicycle paths running throughout most of the city. Consequently, Curitiba uses less energy per person and has lower emissions of greenhouse gases and other air pollutants and less traffic congestion than do most comparably sized cities.

Along the six streams that run within Curitiba's borders, the city removed most buildings and lined the streams with a series of interconnected parks. Volunteers have planted more than 1.5 million trees throughout the city, and no one can cut down a tree without a permit, which also requires that two trees must be planted for each one that is cut down.

Curitiba recycles roughly 70% of its paper and 60% of its metal, glass, and plastic. Recovered materials are sold mostly to the city's 500 or more major industries, which must meet strict pollution standards.

Curitiba's poor residents receive free medical and dental care, child care, and job training, and 40 feeding centers are available for street children. People who live in areas not served by garbage trucks can collect garbage and exchange filled garbage bags for surplus food, bus tokens, and school supplies. The city uses old buses as roving classrooms to train its poor in basic job skills. Other retired buses have become health clinics, soup kitchens, and day-care centers that are free for low-income parents.

FIGURE 6.26 Solutions: Curitiba's bus rapid-transit system has greatly reduced car use in this Brazilian city.

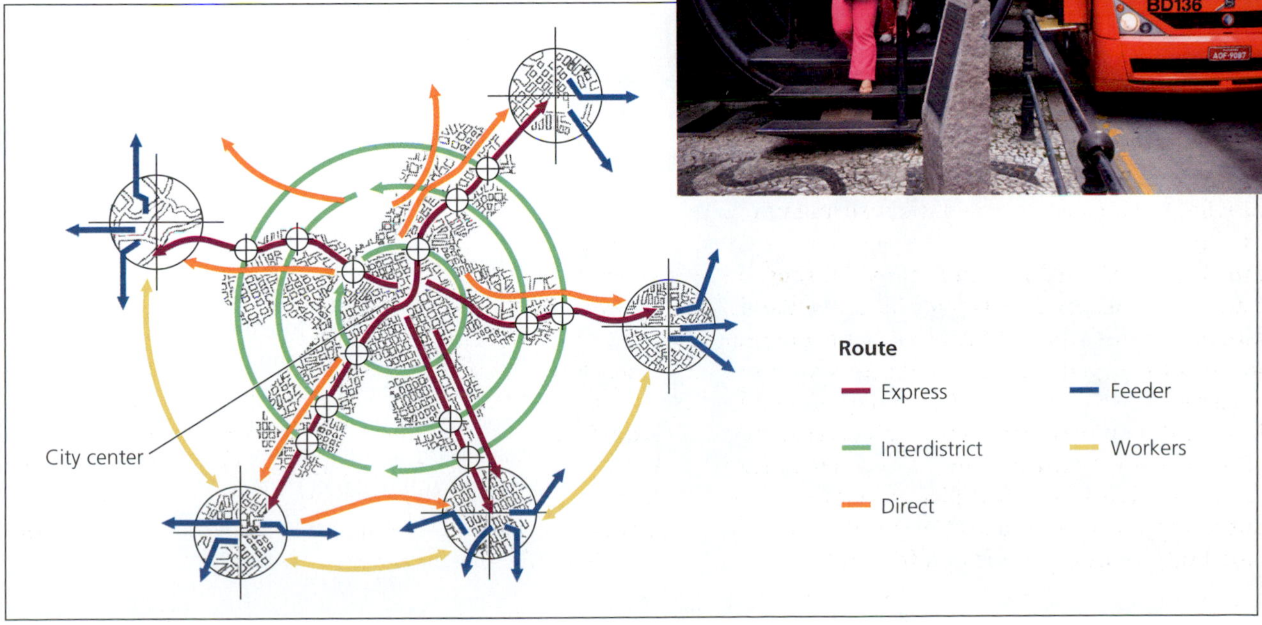

Route

- ▬ Express
- ▬ Interdistrict
- ▬ Direct
- ▬ Feeder
- ▬ Workers

City center

Carlos Cazalis/Corbis

© Cengage Learning

About 95% of Curitiba's citizens can read and write and 83% of its adults have at least a high school education. All school children study ecology. Polls show that 99% of the city's inhabitants would not want to live anywhere else.

Curitiba does face challenges, as do all cities, mostly due to a fivefold increase in its population since 1965. Its once-clear streams are often overloaded with pollutants. The bus system is nearing capacity, and car ownership is on the rise. The city is considering building a light-rail system to relieve some of the pressure.

This internationally acclaimed model of urban planning and sustainability is the brainchild of architect and former college professor Jaime Lerner, who has served as the city's mayor three times since 1969.

TYING IT ALL TOGETHER
Portland, Oregon, Population Growth, and Sustainability

In this chapter, we looked at the growth of the human population and urban areas, their environmental impacts, ways to slow population growth, and how we can make urban areas more sustainable and livable as Portland, Oregon (**Core Case Study**) and Curitiba, Brazil are doing.

The three **scientific principles of sustainability**—reliance on solar energy, chemical cycling, and biodiversity—can guide us in dealing with the problems brought on by population growth and by urban growth. By employing solar and other renewable energy technologies more widely, we can cut pollution and emissions of climate-changing gases that are increasing as the population, resource use per person, and urban areas grow. By reusing and recycling more materials, we could cut our waste and reduce our ecological footprints. And in focusing on preserving biodiversity, we could help to sustain the life-support system on which we and all other species depend, thereby increasing our beneficial environmental impact.

Making this transition toward sustainability is also in keeping with the three **social science principles of sustainability**. Full-cost pricing requires that the harmful environmental costs of urbanization be included in the market prices of goods and services. To achieve this, people would have to work together to find win-win solutions to population and urban problems. And by implementing these solutions, we would be applying the ethical principle that calls for our leaving the world in a more sustainable condition for future generations.

Jeremy Richards/Shutterstock

Chapter Review

Core Case Study

1. Explain how Portland, Oregon has attempted to become a more sustainable city.

Section 6.1

2. What is the key concept for this section? List three factors that account for the rapid increase in the world's human population over the past 200 years. Summarize the three major population growth trends recognized by demographers. About how many people are added to the world's population each year? List eight major ways in which we have altered the earth's ecosystem services to meet our needs. Summarize the debate over whether and how long the human population can keep growing.

Section 6.2

3. What are the two key concepts for this section? List three variables that affect the growth and decline of human populations. How can we calculate the **population change** of an area? Define the **total fertility rate (TFR)**. How has the global TFR changed since 1955? Summarize the story of population growth in the United States. About how much of the annual U.S. population growth is due to legal immigration? List six changes in lifestyles that have taken place in the United States during the 20th century, leading to a rise in per capita resource use.

4. List nine factors that affect birth rates and fertility rates. Define **life expectancy** and **infant mortality rate** and explain how they affect the population size of a country. What is **migration**? What factors can promote migration?

Section 6.3

5. What is the key concept for this section? What is the **age structure** of a population? Explain how age structure affects population growth and economic growth. Describe the American baby boom and some of its economic and social effects. What are some problems related to rapid population decline due to an aging population?

Section 6.4

6. What is the key concept for this section? What is the **demographic transition** and what are its four stages? Explain how the reduction of poverty and empowerment of women can help countries to slow their population growth. What is **family planning** and how can it help to stabilize populations? Describe India's efforts to control its population growth. Describe China's population control program and compare it with that of India.

Section 6.5

7. What is the key concept for this section? What percentage of the world's people lives in urban areas? List two ways in which urban areas grow. List three trends in global urban growth. Describe the three phases of urban growth in the United States. What is **urban sprawl**? List five factors that have promoted urban sprawl in the United States. List five undesirable effects of urban sprawl.

8. What are the major advantages and disadvantages of urbanization? Define **noise pollution**. Explain why most urban areas are unsustainable systems. Describe the major aspects of poverty in urban areas. Summarize Mexico City's major urban and environmental problems and what government officials are doing about them.

Section 6.6

9. What is the key concept for this section? Distinguish between compact and dispersed cities, and give an example of each. What are the major advantages and disadvantages of using motor vehicles? List four ways to reduce dependence on motor vehicles. List the major advantages and disadvantages of relying more on **(a)** bicycles, **(b)** bus rapid-transit systems, **(c)** mass-transit rail systems within urban areas, and **(d)** rapid-rail systems between urban areas.

Section 6.7

10. Define **smart growth** and explain its benefits. Describe the eco-city model. Give five examples of how Curitiba, Brazil, has attempted to become an eco-city. What are this chapter's *three big ideas*? Explain how Portland, Oregon, and other cities are applying the six **principles of sustainability** to become more sustainable urban areas.

Note: Key terms are in bold type.

Critical Thinking

1. Portland, Oregon (**Core Case Study**) has made significant progress in becoming a more environmentally sustainable and desirable place to live. If you live in an urban area, what steps, if any, has your community taken toward becoming more environmentally sustainable? What further steps could be taken?

2. Do you think that the global population of 7.1 billion is too large? Explain. If your answer was *yes,* what do you think should be done to slow human population growth? If your answer was *no,* do you believe that there is a population size that would be too big? Explain. Do you think that the population of the country where you live is too large? Explain.

3. If you could say hello to a new person every second without taking a break and working around the clock, how many years would it take you to greet the 85 million people who were added to the world's population in 2013? (Hint: start by dividing 85 million seconds by 60 to find the number of minutes, and go from there to find the number of years.) How many years would it take for you to greet 7.1 billion people?

4. Identify a major local, national, or global environmental problem, and describe the role that population growth plays in this problem.

5. Some people think that our most important environmental goal should be to sharply reduce the rate of population growth in less-developed countries, where at least 92% of the world's population growth is expected to take place between now and 2050. Others argue that the most serious environmental problems stem from high levels of resource consumption per person in more-developed countries, which have much larger ecological footprints per person than do less-developed countries. What is your view on this issue? Explain.

6. If you own a car or hope to own one, what conditions, if any, would encourage you to rely less on your car and to travel to school or work by bicycle, on foot, by mass transit, or by carpool?

7. Do you think the United States (or the country in which you live) should develop a comprehensive and integrated mass-transit system over the next 20 years, including an efficient rapid-rail network for travel within and between its major cities? Explain. If so, how would you pay for such a system?

8. Consider the characteristics of an eco-city listed on p. 122. How close to this eco-city model is the city in which you live or the city nearest to where you live? Pick what you think are the five most important characteristics of an eco-city and, for each of these characteristics, describe a way in which your city could attain it.

Doing Environmental Science

The campus where you go to school is something like an urban community. Choose five eco-city characteristics (p. 122) and apply them to your campus. For each characteristic:

1. Create a scale of 1 to 10 in order to rate the campus on how well it does in having that characteristic. (For example, how well does it do in giving students options for getting around, other than by using a car? A rating of 1 could be *not at all,* while a rating of 10 could be *excellent.*)

2. Do some research and rate your campus for each characteristic.

3. Write an explanation of your research process and why you chose each rating.

Write a proposed plan for how the campus could improve its ratings.

Global Environment Watch Exercise

Find three different projections for the size of the global population in 2050 (**Core Case Study**). Explain how the projections were made. To do this, try to find out the assumptions behind each of the projections with regard to total fertility rates, crude death rates, infant mortality rates, life expectancies, and other factors. Based on your reading, choose the projection that you believe to be the closest to reality, and explain why you chose this projection.

Data Analysis

The chart below shows selected population data for two different countries, A and B. Study the chart and answer the questions that follow.

	Country A	Country B
Population (millions)	144	82
Crude birth rate (number of live births per 1,000 people per year)	43	8
Crude death rate (number of deaths per 1,000 people per year)	18	10
Infant mortality rate (number of babies per 1,000 born who die in first year of life)	100	3.8
Total fertility rate (average number of children born to women during their childbearing years)	5.9	1.3
% of population under 15 years old	45	14
% of population older than 65 years	3	19
Average life expectancy at birth	47	79
% urban	44	75

© Cengage Learning

1. Calculate the rates of natural increase (due to births and deaths, not counting immigration) for the populations of country A and country B. Based on these calculations and the data in the table, for each of the countries, suggest whether it is a more-developed country or a less-developed country and explain the reasons for your answers.

2. Describe where each of the two countries might be in the stages of demographic transition (Figure 6.10). Discuss factors that could hinder either country from progressing to later stages in the demographic transition.

3. Explain how the percentages of people under age 15 in each country could affect its per capita and total ecological footprints.

CENGAGE**brain**.com To access course materials, including Aplia homework, please visit www.cengagebrain.com.

WWW.CENGAGEBRAIN.COM **127**

7 CLIMATE AND BIODIVERSITY

When we try to pick out anything by itself, we find it hitched to everything else in the universe.

JOHN MUIR

7.3 What are the major types of marine aquatic systems and how are human activities affecting them?

7.4 What are the major types of freshwater systems and how are human activities affecting them?

Coral reef in Egypt's Red Sea.
Vlad61/Shutterstock.com

Why Should We Care about Coral Reefs?

Coral reefs form in clear, warm coastal waters in tropical areas. These stunningly beautiful natural wonders (see chapter-opening photo) are among the world's oldest, most diverse, and most productive ecosystems.

Coral reefs are formed by massive colonies of tiny animals called *polyps* (close relatives of jellyfish). They slowly build reefs by secreting a protective crust of limestone (calcium carbonate) around their soft bodies. When the polyps die, their empty crusts remain behind as part of a platform for more reef growth. The resulting elaborate network of crevices, ledges, and holes serves as calcium carbonate "condominiums" for a variety of marine animals.

Coral reefs are the result of a mutually beneficial relationship between polyps and tiny single-celled algae called *zooxanthellae*

("zoh-ZAN-thel-ee") that live in the tissues of the polyps. In this example of mutualism (see Chapter 5, p. 88), the algae provide the polyps with food and oxygen through photosynthesis, and help the corals produce calcium carbonate. Algae also give the reefs their stunning coloration. The polyps, in turn, provide the algae with a well-protected home and some of their nutrients.

Although shallow and deep-water coral reefs occupy only about 0.2% of the ocean floor, they provide important ecosystem and economic services. They act as natural barriers that help to protect 15% of the world's coastlines from flooding and erosion caused by battering waves and storms. They also provide habitats, food, and spawning grounds for one-quarter to one-third of the organisms that live in the ocean, and they produce about one-tenth of the global fish catch. Through tourism

and fishing they provide goods and services worth about $40 billion a year.

Coral reefs are vulnerable to damage because they grow slowly and are disrupted easily. Runoff of soil and other materials from the land can cloud the water and block the sunlight that the algae in shallow reefs need for photosynthesis. Also, the water in which shallow reefs live must have a temperature of 18–30°C (64–86°F) and cannot be too acidic. This explains why the two major long-term threats to coral reefs are projected *climate change,* which could raise the water temperature above tolerable limits in most reef areas, and *ocean acidification,* which could make it harder for polyps to build reefs and could even dissolve some of their calcium carbonate formations.

One result of stresses such as pollution and rising ocean water temperatures is *coral bleaching* (Figure 7.1). Such stressors can cause the colorful algae, upon which corals depend for food, to die off. Without food, the coral polyps die, leaving behind a white skeleton of calcium carbonate. Studies by the Global Coral Reef Monitoring Network and other scientist groups estimate that since the 1950s, some 45% to 53% of the world's shallow coral reefs have been destroyed or degraded and that another 25% to 33% could be lost within 20 to 40 years. These centers of biodiversity are by far the most threatened marine ecosystems.

In this chapter, we explore the factors that determine climate, the nature of terrestrial and aquatic ecosystems, and the effects of human activities on these forms of natural capital.

FIGURE 7.1 This bleached coral has lost most of its algae because of changes in the environment such as warming of the waters and deposition of sediments.

istockphoto.com/RainervonBrandis

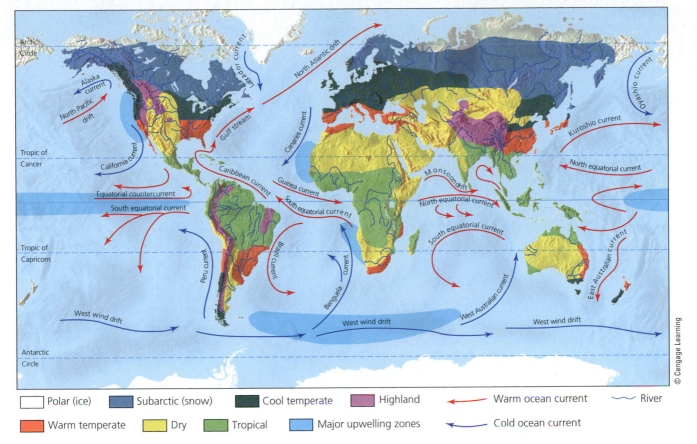

□ Polar (ice)	■ Subarctic (snow)	■ Cool temperate	■ Highland	→ Warm ocean current	⌇ River
■ Warm temperate	■ Dry	■ Tropical	■ Major upwelling zones	→ Cold ocean current	

ANIMATED FIGURE 7.2 Natural capital: This generalized map of the earth's current climate zones also shows the major ocean currents and upwelling areas (where currents bring nutrients from the ocean bottom to the surface). ***Question:*** Based on this map, what is the general type of climate where you live?

7.1 WHAT FACTORS INFLUENCE CLIMATE?

CONCEPT 7.1 Key factors that influence an area's climate are incoming solar energy, the earth's rotation, global patterns of air and water movement, gases in the atmosphere, and the earth's surface features.

The Earth Has Many Different Climates

It is important to understand the difference between weather and climate. **Weather** is a set of physical conditions of the lower atmosphere, including temperature, precipitation, humidity, wind speed, cloud cover, and other factors, in a given area over a period of hours or days. The most important factors in the weather of any area are atmospheric temperature and precipitation.

Weather differs from **climate**, which is the general pattern of atmospheric conditions in a given area over periods ranging from at least three decades to thousands of years. Weather often fluctuates daily, from one season to

another, and from one year to the next. However, climate tends to change slowly because it is the average of weather conditions and their fluctuations in a given area over at least 30 years.

Based on such data, scientists describe the various regions of the earth according to their climates. Figure 7.2 shows these major climate zones along with the major **ocean currents**—mass movements of surface and deep ocean water. These currents help to determine regional climates.

Climate varies among the earth's different regions primarily because, over long periods of time, patterns of global air circulation and ocean currents distribute heat and precipitation unevenly between the tropics and other parts of the world (Figure 7.3). Three major factors affect the circulation of air in the lower atmosphere:

1. *Uneven heating of the earth's surface by the sun.* Air is heated much more at the equator, where the sun's rays strike directly, than at the poles, where sunlight strikes at an angle and spreads out over a much greater area (Figure 7.3, left). These differences in the input of solar energy to the atmosphere help explain why tropical

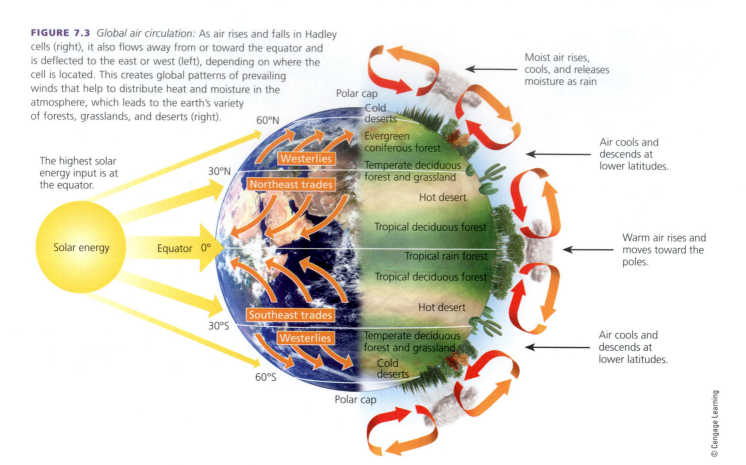

FIGURE 7.3 *Global air circulation:* As air rises and falls in Hadley cells (right), it also flows away from or toward the equator and is deflected to the east or west (left), depending on where the cell is located. This creates global patterns of prevailing winds that help to distribute heat and moisture in the atmosphere, which leads to the earth's variety of forests, grasslands, and deserts (right).

The highest solar energy input is at the equator.

Solar energy

Equator 0°

60°N

30°N

Westerlies

Northeast trades

Southeast trades

30°S

Westerlies

60°S

Polar cap

Cold deserts

Evergreen coniferous forest

Temperate deciduous forest and grassland

Hot desert

Tropical deciduous forest

Tropical rain forest

Tropical deciduous forest

Hot desert

Temperate deciduous forest and grassland

Cold deserts

Polar cap

Moist air rises, cools, and releases moisture as rain

Air cools and descends at lower latitudes.

Warm air rises and moves toward the poles.

Air cools and descends at lower latitudes.

© Cengage Learning

regions near the equator are hot, why polar regions are cold, and why temperate regions in between generally have both warm and cool temperatures (Figure 7.2). The intense input of solar radiation in tropical regions leads to greatly increased evaporation of moisture from forests, grasslands, and bodies of water. As a result, tropical regions normally receive more precipitation than do other areas of the earth. The amount of solar energy reaching the earth also varies slightly (typically by less than 0.1%) over the 11-year solar activity cycle that includes sunspots and other solar events.

2. *Rotation of the earth on its axis.* As the earth rotates around its axis, the equator spins faster than the regions to its north and south. This means that air masses moving to the north or south are deflected to the east, because they are also moving easterly (the direction of the earth's rotation) faster than the land below them—a process known as the *Coriolis effect.* For the same reason, air moving from the north or south toward the equator will curl in a westerly direction (Figure 7.3, left) because it is moving east more slowly than the land below it as it moves toward the equator. Also, the atmosphere is divided into six huge regions, three on either side of the equator (Figure 7.3, right), called *Hadley cells,* in which warm air rises and cools, then falls and heats up again in great rolling patterns. Those cells that bring the surface air toward the

equator cause westerly winds (hooking west because of the Coriolis effect, Figure 7.3, left) and those in which surface air moves toward the poles cause easterly winds. These easterly and westerly wind patterns are called *prevailing winds*—major surface winds that blow almost continuously and help to distribute heat and moisture over the earth's surface and to drive surface ocean currents.

3. *Properties of air, water, and land.* Heat from the sun evaporates ocean water and transfers heat from the oceans to the atmosphere, especially near the hot equator. This evaporation of water creates the giant cyclical Hadley cells (Figure 7.3, right) that circulate air, heat, and moisture both vertically and from place to place in the atmosphere, as shown in Figure 7.4.

Driven by prevailing winds and the earth's rotation, the earth's major ocean currents (Figure 7.2) help to redistribute heat from the sun, thereby influencing climate and vegetation, especially near coastal areas. This solar heat, along with differences in water *density* (mass per unit volume), creates warm and cold ocean currents. Prevailing winds and the Coriolis effect drive these currents, and continental coastlines change their directions. As a result, they flow in roughly circular patterns between the continents, clockwise in the northern hemisphere and counterclockwise in the southern hemisphere.

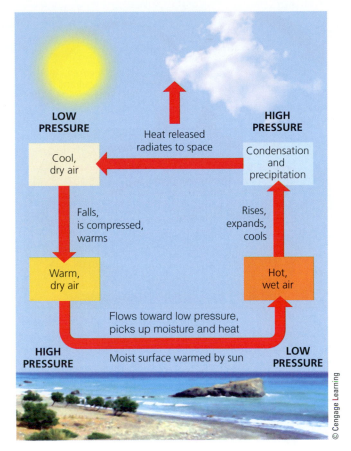

FIGURE 7.4 Energy is transferred by *convection* in the atmosphere—the process by which warm, wet air rises, then cools and releases heat and moisture as precipitation (right side and top, center). Then the cooler, denser, and drier air sinks, warms up, and absorbs moisture as it flows across the earth's surface (left side and bottom) to begin the cycle again.

Water also moves vertically in the oceans as denser water sinks while less dense water rises. This creates a connected loop of deep and shallow ocean currents (which are separate from those shown in Figure 7.2). This loop acts somewhat like a giant conveyer belt that moves heat from the surface to the deep sea and transfers warm and cold water between the tropics and the poles (Figure 7.5).

The oceans and the atmosphere are strongly linked in two ways: ocean currents are affected by winds in the atmosphere, and heat from the oceans affects atmospheric circulation. One example of the interactions between the oceans and the atmosphere is the *El Niño–Southern Oscillation,* or *ENSO* (Figure 7.6). This large-scale weather phenomenon occurs every few years when prevailing winds in the tropical Pacific Ocean weaken and change direction. The resulting above-average warming of Pacific waters alters the weather over at least two-thirds of the earth for 1 or 2 years by, for example, leading to one or two milder winters in some areas.

The earth's air circulation patterns, prevailing winds, and configuration of continents and oceans are all factors

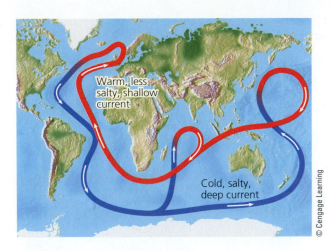

FIGURE 7.5 A connected loop of deep and shallow ocean currents transports warm and cool water to various parts of the earth.

in the formation of the six Hadley cells. Together, all of these factors lead to an irregular distribution of climates and of the resulting deserts, grasslands, and forests, as shown in Figure 7.3, right (**Concept 7.1**).

Greenhouse Gases Warm the Lower Atmosphere

As energy flows from the sun to the earth, some of it is reflected by the earth's surface back into the atmosphere. Molecules of certain gases in the atmosphere, including water vapor (H_2O), carbon dioxide (CO_2), methane (CH_4), and nitrous oxide (N_2O), absorb some of this solar energy and release a portion of it as infrared radiation (heat) that warms the lower atmosphere and the earth's surface. These gases, called **greenhouse gases**, play a role in determining the lower atmosphere's average temperatures and thus the earth's climates. This natural warming of the lower atmosphere is called the **greenhouse effect** (see Figure 3.3, p. 44). Without this natural warming effect, the earth would be a very cold and mostly lifeless planet.

Human activities such as the production and burning of fossil fuels, clearing of forests, and growing of crops release large quantities of the greenhouse gases carbon dioxide and methane into the atmosphere. A considerable body of scientific evidence, combined with climate model projections, indicates that we are emitting greenhouse gases into the atmosphere faster than they can be removed by the earth's carbon and nitrogen cycles (see Figures 3.14, p. 53, and 3.15, p. 54). These emissions are very likely to enhance the earth's natural greenhouse effect and change the earth's climate during this century. If this occurs, it will alter temperature and precipitation patterns, raise average sea levels, and shift areas where we can grow crops and where many types of plants and animals (including humans) can live, as discussed more fully in Chapter 15.

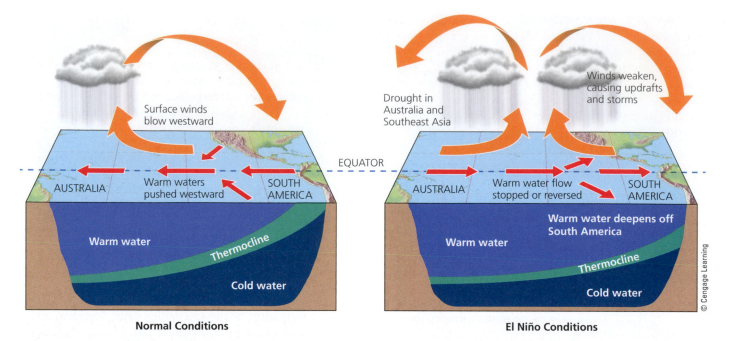

FIGURE 7.6 Normal prevailing, or trade winds blowing east to west cause shore upwellings of cold, nutrient-rich bottom water in the tropical Pacific Ocean near the coast of Peru (left). Every few years, a shift in trade winds, known as the *El Niño–Southern Oscillation (ENSO),* disrupts this pattern for 1–2 years.

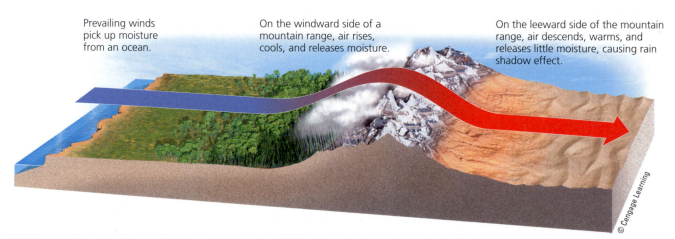

FIGURE 7.7 The *rain shadow effect* is a reduction of rainfall and loss of moisture from the landscape on the leeward side of a mountain. Warm, moist air in onshore winds loses most of its moisture as rain and snow that fall on the windward slopes of a mountain range. This leads to semiarid and arid conditions on the leeward side of the mountain range and on the land beyond.

The Earth's Surface Features Affect Local Climates

Various topographic features of the earth's surface can create local and regional climatic conditions that differ from the general climate in some regions. For example, mountains interrupt the flow of prevailing surface winds and the movement of storms. When moist air blowing inland from an ocean reaches a mountain range, it is forced upward. As it rises, it cools and expands, and loses most of its moisture as rain and snow that fall on the windward slope of the mountain.

As the drier air mass passes over the mountaintops, it flows down the leeward slopes (facing away from the wind), and warms up. This warmer air can hold more moisture, but it typically does not release much of this moisture and instead tends to dry out plants and soil below. This process is called the **rain shadow effect** (Figure 7.7), and over many decades, it results in *semiarid* or *arid* conditions on the leeward side of a high mountain range. Sometimes this effect leads to the formation of deserts such as Death Valley, a part of the Mojave Desert, which lies within the U.S. states of California, Nevada, Utah, and Arizona.

Cities also create distinct microclimates. Bricks, concrete, asphalt, and other building materials absorb and hold heat, and buildings block wind flow. Motor vehicles and the heating and cooling systems of buildings release large quantities of heat and pollutants. As a result, cities on average tend to have more haze and smog, higher temperatures, and lower wind speeds than the surrounding countryside, and these factors make them *heat islands.*

7.2 WHAT ARE THE WORLD'S MAJOR TERRESTRIAL ECOSYSTEMS AND HOW ARE HUMAN ACTIVITIES AFFECTING THEM?

CONCEPT 7.2A Differences in long-term average annual precipitation and temperature lead to the formation of tropical, temperate, and cold deserts, grasslands, and forests, and largely determine their locations.

CONCEPT 7.2B Human activities are disrupting ecosystem and economic services provided by many of the earth's deserts, grasslands, forests, and mountains.

Climate Helps to Determine Where Terrestrial Organisms Can Live

Differences in climate (Figure 7.2) help to explain why one area of the earth's land surface is a desert, another a grassland, and another a forest. (See Figure 2, p. S16, Supplement 4.) Different combinations of varying average annual precipitation and temperatures, along with global air circulation patterns and ocean currents, lead to the formation of tropical (hot), temperate (moderate), and polar (cold) deserts, grasslands, and forests, as summarized in Figure 7.8 (**Concept 7.2A**).

Climate and vegetation vary according to *latitude* and also according to *elevation,* or height above sea level. If you climb a tall mountain, from its base to its summit, you can observe changes in plant life similar to those you would encounter in traveling from the equator to the earth's northern polar region (Figure 7.9).

Figure 7.10 shows how scientists have divided the world into several major **biomes**—large terrestrial regions, each characterized by a particular type of climate and a certain combination of dominant plant life. The variety of terrestrial biomes and aquatic systems is one of the four components of the earth's biodiversity (see Figure 4.2, p. 65)—a vital part of the earth's natural capital. Fig-

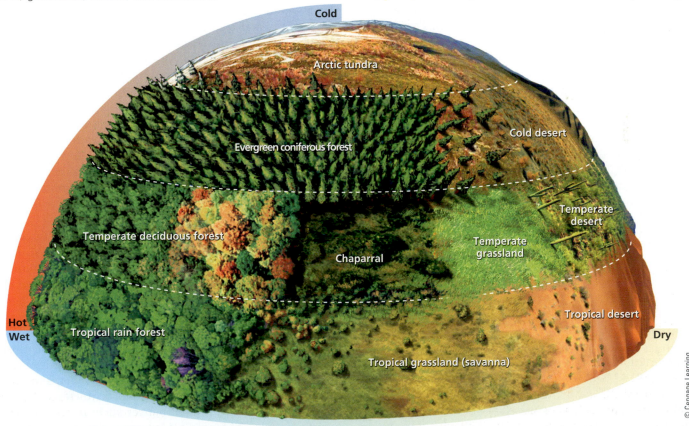

FIGURE 7.8 Natural capital: Average precipitation and average temperature, acting together as limiting factors over a long time, help to determine the type of desert, grassland, or forest in any particular area, and thus the types of plants, animals, and decomposers found in that area (assuming it has not been disturbed by human activities).

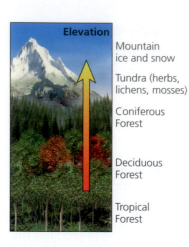

Elevation

Mountain
ice and snow

Tundra (herbs,
lichens, mosses)

Coniferous
Forest

Deciduous
Forest

Tropical
Forest

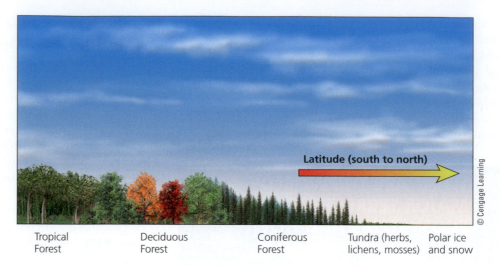

Latitude (south to north)

© Cengage Learning

Tropical
Forest

Deciduous
Forest

Coniferous
Forest

Tundra (herbs,
lichens, mosses)

Polar ice
and snow

FIGURE 7.9 Biomes and climate both change with elevation (left), as well as with latitude (right).

ure 4.4 (p. 67) shows how major biomes along the 39th parallel in the United States are related to different climates. The map in Figure 5, p. S20, in Supplement 4 shows the major biomes in North America.

On maps such as the one in Figure 7.10, biomes are shown with sharp boundaries, and each biome is covered with one general type of vegetation. In reality, biomes are not uniform. They consist of a *mosaic of patches,* each with somewhat different biological communities but with similarities typical of the biome. These patches occur primarily because of the irregular distribution of the resources needed by plants and animals and because human activities have removed or altered the natural vegetation in many areas.

There are also differences along the transition zone (called the *ecotone*) between two different ecosystems or biomes. This zone contains habitats that are common to both ecosystems along with other habitats that are unique to the transition zone. This results in the **edge effect**, or the tendency for a transition zone between two different ecosystems to have greater species diversity and a higher density of organisms than are found in either of the individual ecosystems.

CONSIDER THIS. . .

There Are Three Major Types of Deserts

In a *desert,* annual precipitation is low and often scattered unevenly throughout the year. During the day, the baking sun warms the ground and evaporates water from plant leaves and from the soil. But at night, most of the heat stored in the ground radiates quickly into the atmosphere. This explains why in a desert, you may roast during the day but shiver at night.

A combination of low rainfall and varying average temperatures creates a variety of desert types—tropical, temperate, and cold (Figures 7.8 and 7.10 and **Concept 7.2A**). *Tropical deserts* (Figure 7.11, top photo) such as the Sahara and the Namib of Africa are hot and dry most of the year (Figure 7.11, top graph). They have few plants and a hard, windblown surface strewn with rocks and sand.

In *temperate deserts* (Figure 7.11, center photo) daytime temperatures are high in summer and low in winter and there is more precipitation than in tropical deserts (Figure 7.11, center graph). The sparse vegetation consists mostly of widely dispersed, drought-resistant shrubs and cacti or other succulents adapted to the dry conditions and temperature variations.

In *cold deserts* such as the Gobi Desert in Mongolia, vegetation is sparse (Figure 7.11, bottom photo). Winters are cold, summers are warm or hot, and precipitation is low (Figure 7.11, bottom graph). In all types of deserts, plants and animals have evolved adaptations that help them to stay cool and to get enough water to survive (Science Focus 7.1).

Desert ecosystems are fragile because they have slow plant growth, low species diversity, slow nutrient cycling (due to low bacterial activity in the soils), and very little water. It can take decades to centuries for their soils to recover from disturbances such as off-road vehicle traffic, which can also destroy the habitats for a variety of animal species that live underground. The lack of vegetation, especially in tropical and polar deserts, also makes them vulnerable to heavy wind erosion from sandstorms.

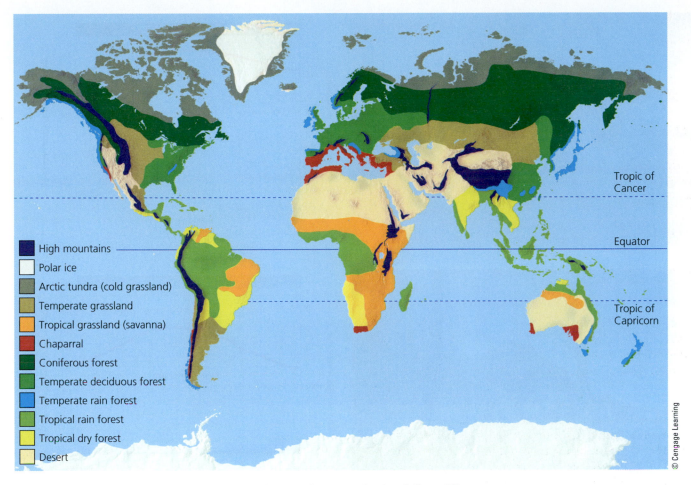

High mountains
Polar ice
Arctic tundra (cold grassland)
Temperate grassland
Tropical grassland (savanna)
Chaparral
Coniferous forest
Temperate deciduous forest
Temperate rain forest
Tropical rain forest
Tropical dry forest
Desert

Tropic of Cancer

Equator

Tropic of Capricorn

© Cengage Learning

ANIMATED FIGURE 7.10 **Natural capital:** The earth's major *biomes* result primarily from differences in climate.

There Are Three Major Types of Grasslands

Grasslands occur primarily in the interiors of continents in areas that are too moist for deserts to form and too dry for forests to grow (Figures 7.8 and 7.10). Grasslands persist because of a combination of seasonal drought, grazing by large herbivores, and occasional fires—all of which keep shrubs and trees from growing in large numbers.

The three main types of grassland—tropical, temperate, and cold (arctic tundra)—result from combinations of low average precipitation and varying average temperatures (**Concept 7.2A**). One type of tropical grassland, called a *savanna* (Figure 7.12, top photo), contains widely scattered clumps of trees. This biome usually has warm temperatures year-round and alternating dry and wet seasons (Figure 7.12, top graph).

Tropical savannas in East Africa are home to *grazing* (primarily grass-eating) and *browsing* (twig- and leaf-nibbling) hoofed animals, including wildebeests, gazelles, zebras, giraffes, and antelopes, as well as their predators such as lions, hyenas, and humans. Herds of these grazing and browsing animals migrate to find water and food in response to seasonal and year-to-year variations in rainfall

(Figure 7.12, blue areas in top graph) and food availability. Savanna plants, like those in deserts, are adapted to survive drought and extreme heat; many have deep roots that can tap into groundwater.

CONSIDER THIS. . .

CONNECTIONS Grassland Niches and Feeding Habits

As an example of differing niches, some large herbivores have evolved specialized eating habits that minimize competition among species for the vegetation found on the savanna. For example, giraffes eat leaves and shoots from the tops of trees, elephants eat leaves and branches farther down, wildebeests prefer short grasses, and zebras graze on longer grasses and stems.

In a *temperate grassland*, winters can be bitterly cold, summers are hot and dry, and annual precipitation is fairly sparse and falls unevenly throughout the year (Figure 7.12, center graph). Because the aboveground parts of most of the grasses die and decompose each year, organic matter accumulates to produce deep, fertile topsoil. This topsoil is held in place by a thick network of the grasses' intertwined roots (unless the topsoil is plowed up, which exposes it to high winds found in these biomes). This bi-

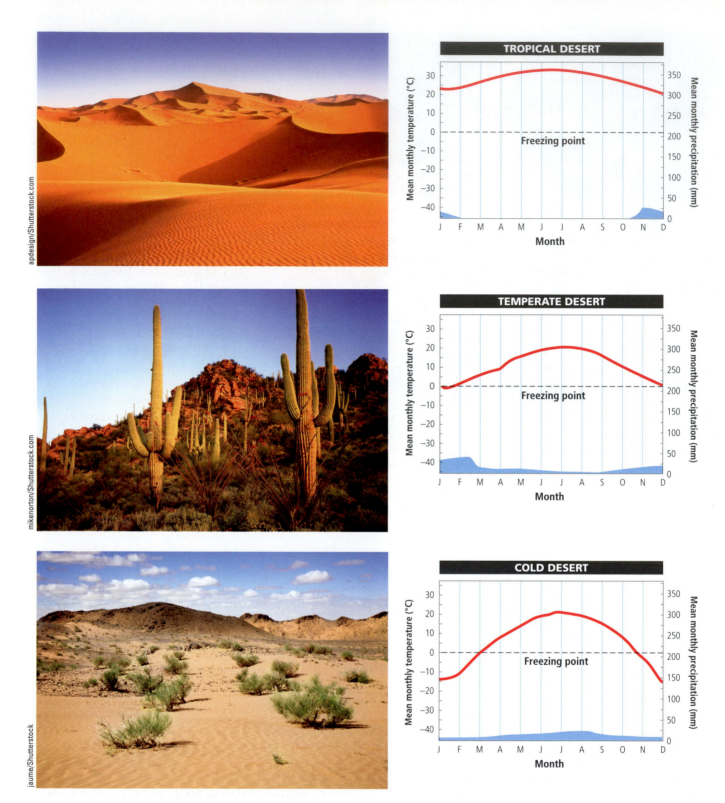

FIGURE 7.11 These climate graphs track the typical variations in annual temperature (red) and precipitation (blue) in tropical, temperate, and cold deserts. Top photo: a *tropical desert* in Morocco. Center photo: a *temperate desert* in southeastern California, with saguaro cactus, a prominent species in this ecosystem. Bottom photo: a *cold desert*, Mongolia's Gobi Desert. ***Question:*** Which month of the year has the highest temperature and which month has the lowest rainfall for each of the three types of deserts?

© Cengage Learning

STAYING ALIVE IN THE DESERT

Adaptations for survival in the desert have two themes: *beat the heat* and *every drop of water counts.*

Desert plants have evolved a number of strategies based on such adaptations. During long hot and dry spells, plants such as mesquite and creosote drop their leaves to survive in a dormant state. *Succulent* (fleshy) *plants* such as the saguaro ("sah-WAH-ro") cactus (Figure 7.11, middle photo) have no leaves that can lose water to the atmosphere through *transpiration.* They also store water and synthesize food in their expandable, fleshy tissue and they reduce water loss by opening their pores only at night to take up carbon dioxide (CO_2). The spines of these and many other desert plants guard them from being eaten by herbivores seeking the precious water they hold.

Some desert plants use deep roots to tap into groundwater. Others such as prickly pear and saguaro cacti use widely spread shallow roots to collect water after brief showers and store it in their spongy tissues.

Some desert plants conserve water by having wax-coated leaves that reduce water loss. Others such as annual wildflowers and grasses store much of their biomass in seeds that remain inactive, sometimes for years, until they receive enough water to germinate. Shortly after a rain, these seeds germinate, grow, and carpet some deserts with dazzling arrays of colorful flowers that last for up to a few weeks.

Most desert animals are small. Some beat the heat by hiding in cool burrows or rocky crevices by day and coming out at night or in the early morning. Others become dormant during periods of extreme heat or drought. Some larger animals such as camels can drink massive quantities of water when it is available and store it in their fat for use as needed. Also, the camel's thick fur actually helps it to keep cool because the air spaces in the fur insulate the camel's skin against the outside heat. And camels do not sweat, which reduces their water loss through evaporation. Kangaroo rats never drink water. They get the water they need by breaking down fats in seeds that they consume.

Insects and reptiles such as rattlesnakes have thick outer coverings to minimize water loss through evaporation, and their wastes are dry feces and a dried concentrate of urine. Many spiders and insects get their water from dew or from the food they eat.

Critical Thinking

What are three steps you would take to survive in the open desert if you had to?

---

ome's grasses are adapted to periodic droughts and to fires that burn the plant parts above the ground but do not harm the roots, from which new grass can grow. Many of the world's natural temperate grasslands have been converted to farmland, because their fertile soils are useful for growing crops (Figure 7.13) and grazing cattle.

Cold grasslands, or *arctic tundra,* lie south of the arctic polar ice cap (Figures 7.8 and 7.10). During most of the year, these treeless plains are bitterly cold (Figure 7.12, bottom graph), swept by frigid winds, and covered with ice and snow. Winters are long with few hours of daylight, and the scant precipitation falls primarily as snow.

Under the snow, this biome is carpeted with a thick, spongy mat of low-growing plants. Trees and tall plants cannot survive in the cold and windy tundra because they would lose too much of their heat. Most of the annual growth of the tundra's plants occurs during the 7- to 8-week summer, when there is daylight almost around the clock.

One outcome of the extreme cold is the formation of **permafrost**, underground soil in which captured water stays frozen for more than two consecutive years. During the brief summer, the permafrost layer keeps melted snow and ice from draining into the ground. As a consequence, many shallow lakes, marshes, bogs, ponds, and other seasonal wetlands form when snow and frozen surface soil melt on the waterlogged tundra. Hordes of mosquitoes, black flies, and other insects thrive in these shallow surface pools. They serve as food for large colonies of migratory birds (especially waterfowl) that migrate from the south to nest and breed in the tundra's summer bogs and ponds.

Animals in this biome survive the intense winter cold through adaptations such as thick coats of fur (arctic wolf, arctic fox, and musk oxen) or feathers (snowy owl) and living underground (arctic lemming). In the summer, caribou (often called reindeer) and other types of deer migrate to the tundra to graze on its vegetation.

Tundra is a fragile biome. Tundra soils usually are nutrient poor. Because of the short growing season, tundra soil and vegetation recover very slowly from damage or disturbance. Human activities in the arctic tundra—primarily

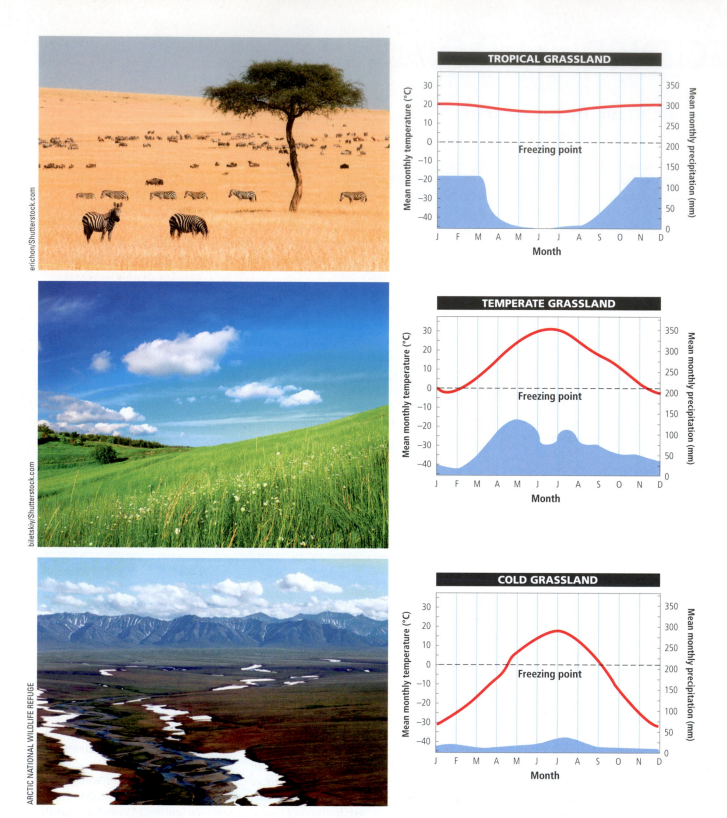

FIGURE 7.12 These climate graphs track the typical variations in annual temperature (red) and precipitation (blue) in tropical, temperate, and cold (arctic tundra) grasslands. Top photo: *savanna (tropical grassland)* in Kenya, Africa, with zebras grazing. Center photo: *prairie (temperate grassland)* in the U.S. state of Illinois. Bottom photo: *arctic tundra (cold grassland)* in Alaska's Arctic National Wildlife Refuge in summer. **Question:** Which month of the year has the highest temperature and which month has the lowest rainfall for each of the three types of grassland?

FIGURE 7.13 Natural capital degradation: This intensively cultivated cropland is an example of the replacement of biologically diverse temperate grasslands (such as in the center photo of Figure 7.12) with a monoculture crop.

National Archives/EPA Documerica

on and around oil drilling sites, pipelines, mines, and military bases—leave scars that persist for centuries.

There Are Three Major Types of Forests

Forests are lands that are dominated by trees. The three main types of forest—*tropical, temperate,* and *cold* (northern coniferous, or boreal)—result from combinations of varying precipitation levels and varying average temperatures (**Concept 7.2A**) (Figures 7.8 and 7.10).

Tropical rain forests (Figure 7.14, top photo) are found near the equator (Figure 7.8), where hot, moisture-laden air rises and dumps its moisture (Figure 7.3). These lush forests have year-round, uniformly warm temperatures, high humidity, and almost daily heavy rainfall (Figure 7.14, top graph). This fairly constant warm, wet climate is ideal for a wide variety of plants and animals.

Tropical rain forests are dominated by *broadleaf evergreen plants,* which keep most of their leaves year-round. The tops of the trees form a dense *canopy* (Figure 7.14, top

photo) that blocks most light from reaching the forest floor. Many of the plants that do live at the ground level have enormous leaves to capture what little sunlight filters down to them.

Some trees are draped with vines (called *lianas*) that reach for the treetops to gain access to sunlight. In the canopy, the vines grow from one tree to another, providing walkways for many species living there. When a large tree is cut down, its network of lianas can pull down other trees.

Tropical rain forests have a very high net primary productivity (see Figure 3.12, p. 50). They are teeming with life and possess incredible biological diversity. Although tropical rain forests cover only about 2% of the earth's land surface, ecologists estimate that they contain at least 50% of the known terrestrial plant and animal species. For example, a single tree in these forests may support several thousand different insect species. Plants from tropical rain forests are a source of a variety of chemicals, many of which have been used as blueprints for making most of the world's prescription drugs.

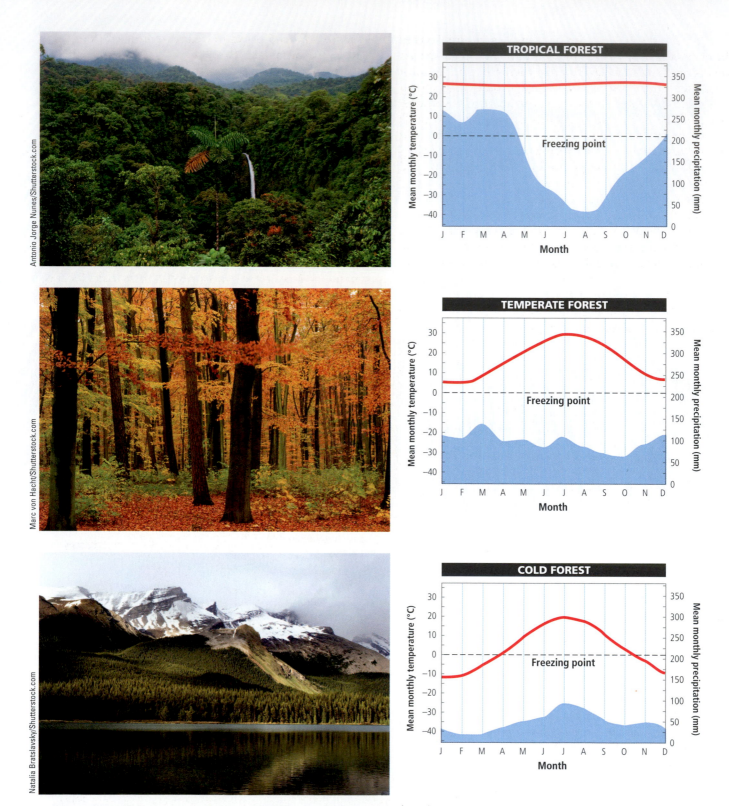

FIGURE 7.14 These climate graphs track the typical variations in annual temperature (red) and precipitation (blue) in tropical, temperate, and cold (northern coniferous, or boreal) forests. Top photo: the closed canopy of a *tropical rain forest* in Costa Rica. Middle photo: a *temperate deciduous forest* near Hamburg, Germany in autumn. Bottom photo: a *northern coniferous forest* in Canada's Jasper National Park. **Question:** Which month of the year has the highest temperature and which month has the lowest rainfall for each of the three types of forest?

© Cengage Learning

Rain forest species occupy a variety of specialized niches in distinct layers, which contribute to their high species diversity. Vegetation layers are structured, for the most part, according to the plants' needs for sunlight, as shown in Figure 7.15. Much of the animal life, particularly insects, bats, and birds, lives in the sunny canopy layer, with its abundant shelter and supplies of leaves, flowers, and fruits. To study life in the treetops, ecologists climb trees (see Chapter 2 opening photo) and build platforms and boardwalks in the upper canopy.

Dropped leaves, fallen trees, and dead animals decompose quickly in tropical rain forests because of the warm, moist conditions and the hordes of decomposers. About 90% of the nutrients released by this rapid decomposition are quickly taken up and stored by trees, vines, and other plants. Nutrients that are not taken up are soon leached from the thin topsoil by the almost daily rainfall. As a result, very little plant litter builds up on the ground. The resulting lack of fertile soil helps to explain why rain for-

ests are not good places to clear and grow crops or graze cattle on a sustainable basis.

At least half of all tropical rain forests have been destroyed or disturbed by human activities such as farming, and the pace of this destruction and degradation is increasing (see Chapter 3 Core Case Study, p. 42). Ecologists warn that without strong protective measures, most of these forests, along with their rich biodiversity and other highly valuable ecosystem services, could be gone by the end of this century.

The second major type of forest is the *temperate deciduous forest* (Figure 7.14, middle photo). Such forests typically see warm summers, cold winters, and abundant precipitation—rain in summer and snow in winter months (Figure 7.14, middle graph). They are dominated by a few species of *broadleaf deciduous trees* such as oak, hickory, maple, aspen, and birch. Animal species living in these forests include predators such as wolves, foxes, and wildcats. They feed on herbivores such as white-tailed deer,

FIGURE 7.15 Specialized plant and animal niches are *stratified*, or arranged roughly in layers, in a tropical rain forest. Filling such specialized niches enables many species to avoid or minimize competition for resources and results in the coexistence of a great variety of species.

squirrels, rabbits, and mice. Warblers, robins, and other bird species live in these forests during the spring and summer, mating and raising their young.

In these forests, most of the trees' leaves, after developing their vibrant colors in the fall (Figure 7.14, middle photo), drop off the trees. This allows the trees to survive the cold winters by becoming dormant. Each spring, they sprout new leaves and spend their summers growing and producing until the cold weather returns.

Because they have cooler temperatures and fewer decomposers than tropical forests have, these forests also have a slower rate of decomposition. As a result, they accumulate a thick layer of slowly decaying leaf litter, which becomes a storehouse of nutrients.

On a global basis, temperate forests have been degraded by various human activities, especially logging and urban expansion, more than any other terrestrial biome. However, within 100 to 200 years, forests of this type that have been cleared can return through secondary ecological succession (see Figure 5.8, p. 91).

Another type of temperate forest, the *coastal coniferous forests* or *temperate rain forests* are found in scattered coastal temperate areas with ample rainfall and moisture from dense ocean fogs. Thick stands of these forests with large conifers such as Sitka spruce, Douglas fir, giant sequoia (see Chapter 2 opening photo), and redwoods once dominated undisturbed areas of these biomes along the coast of North America, from Canada to Northern California in the United States.

Cold, or *northern coniferous forests* (Figure 7.14, bottom photo), also called *boreal forests* or *taigas* ("TIE-guhs"), are found south of arctic tundra. In their subarctic, cold and moist climate, winters are long and extremely cold, with winter sunlight available only 6–8 hours per day in the northernmost taigas. Summers are short, with cool to warm temperatures (Figure 7.14, bottom graph), and the sun shines as long as 19 hours a day during mid-summer.

Most boreal forests are dominated by a few species of *coniferous* (cone-bearing) *evergreen trees* or *conifers* such as spruce, fir, cedar, hemlock, and pine that keep most of their leaves (or needles) year-round. Most of these species have small, needle-shaped, wax-coated leaves that can withstand the intense cold and drought of winter, when snow blankets the ground. Plant diversity is low because few species can survive the winters when soil moisture is frozen.

Beneath the stands of trees in these forests is a deep layer of partially decomposed conifer needles. Decomposition is slow because of low temperatures, the waxy coating on the needles, and high soil acidity. The decomposing conifer needles make the thin, nutrient-poor topsoil acidic, which prevents most other plants (except certain shrubs) from growing on the forest floor.

Year-round wildlife includes bears, wolves, moose, lynx, and many burrowing rodent species. Caribou spend the winter in taiga and the summer in the arctic tundra (Figure 7.12, bottom). During the brief summer, warblers and other insect-eating birds feed on flies, mosquitoes, and caterpillars.

Mountains Play Important Ecological Roles

Some of the world's most spectacular environments are high on *mountains* (Figure 7.16), steep or high-elevation lands that cover about one-fourth of the earth's land surface. Mountains are places where dramatic changes in altitude, slope, climate, soil, and vegetation take place over very short distances (Figure 7.9, left).

About 1.2 billion people (17% of the world's population) live in mountain ranges or in their foothills, and 4 billion people (56% of the world's population) depend on mountain systems for all or some of their water. Because of the steep slopes, mountain soils are easily eroded when the vegetation holding them in place is removed by natural disturbances such as landslides and avalanches, or by human activities such as timber cutting and agriculture. Many mountains are *islands of biodiversity* surrounded by a sea of lower-elevation landscapes transformed by human activities.

Mountains play an important ecological role. They contain the majority of the world's forests, which are habitats for much of the planet's terrestrial biodiversity. They often are habitats for *endemic species*—those that are found nowhere else on earth. They also serve as sanctuaries for animals that are capable of migrating to higher altitudes and surviving in such environments. Every year, more of these animals are driven from lowland areas to mountain habitats by human activities and by a warming climate.

Finally, mountains play a critical role in the hydrologic cycle (see Figure 3.13, p. 52) by serving as major storehouses of water. During winter, precipitation is stored as ice and snow. In the warmer weather of spring and summer, much of this snow and ice melts, releasing water to streams for use by wildlife and by humans for drinking and for irrigating crops. As the atmosphere has warmed over the last 40 years, some mountaintop snow packs and glaciers have been melting earlier in the spring each year. This is leading to lower food production in certain areas, because much of the water needed throughout the summer to irrigate crops gets released too early in the season and too quickly.

Scientific measurements and climate models indicate that a large number of the world's mountaintop glaciers may disappear during this century if the atmosphere keeps getting warmer as projected. This could force many people to move from their homelands in search of new water supplies and places to grow their crops. Despite the ecological, economic, and cultural importance of mountain ecosystems, methods for protecting these areas have largely eluded governments and environmental organizations.

Mountains play important ecological roles.

FIGURE 7.16 Mountains, such as this one near Aspen, Colorado, provide vital ecosystem services.

Charles Kogod/National Geographic Stock

Conservationist, mountain explorer, and Emerging National Geographic Explorer Gregg Treinish is trying to change this situation. He founded Adventurers and Scientists for Conservation (ASC). This nonprofit organization connects outdoor adventurers who volunteer to collect data during their travels with researchers who are focused on identifying the effects of climate change on mountain ecosystems, as well as the conservation needs for threatened wildlife species, diminishing habitats, and rapidly changing ecosystems. Treinish, who was also National Geographic Adventurer of the Year in 2008–09, leads his own research expeditions to many of the world's rugged mountain regions.

Humans Have Disturbed Much of the Earth's Land

According to the 2005 Millennium Ecosystem Assessment and later updates of such research, about 60% of the world's major terrestrial ecosystems are being degraded or used unsustainably, as the human ecological footprint gets bigger and spreads across the globe (see Figure 1.10, p. 12). Figure 7.17 summarizes some of the most harmful human impacts on the world's deserts, grasslands, forests, and mountains (**Concept 7.2B**).

How long can we keep eating away at these terrestrial forms of natural capital without threatening our econo-

Major Human Impacts on Terrestrial Ecosystems

Deserts	Grasslands	Forests	Mountains
Large desert cities	Conversion to cropland	Clearing for agriculture, livestock grazing, timber, and urban development	Agriculture
Destruction of soil and underground habitat by off-road vehicles	Release of CO_2 to atmosphere from burning grassland	Conversion of diverse forests to tree plantations	Timber and mineral extraction
Depletion of groundwater	Overgrazing by livestock	Damage from off-road vehicles	Hydroelectric dams and reservoirs
Land disturbance and pollution from mineral extraction	Oil production and off-road vehicles in arctic tundra	Pollution of forest streams	Air pollution blowing in from urban areas and power plants
			Soil damage from off-road vehicles

© Cengage Learning

FIGURE 7.17 Human activities have had major impacts on the world's deserts, grasslands, forests, and mountains, as summarized here. *Question:* For each of these biomes, which two of the impacts listed do you think are the most harmful?

Left: somchaij/Shutterstock.com. Left center: Orientaly/Shutterstock.com. Right center: Eppic | Dreamstime.com. Right: Vasik Olga/Shutterstock.com.

mies and the long-term survival of our own and many other species? No one knows. But there are increasing signs that we need to come to grips with this vital issue.

7.3 WHAT ARE THE MAJOR TYPES OF MARINE AQUATIC SYSTEMS AND HOW ARE HUMAN ACTIVITIES AFFECTING THEM?

CONCEPT 7.3 Oceans dominate the planet and provide vital ecosystem and economic services that are being disrupted by human activities.

Water Covers Most of the Planet

When viewed from outer space, the earth appears as a mostly blue planet with about 71% of its surface covered with ocean water. Although the *global ocean* is a single and continuous body of saltwater, geographers divide it into five large areas—the Arctic, Atlantic, Pacific, Indian, and Southern Oceans—separated by the continents. Together, the oceans hold almost 98% of the earth's water. Each of us is connected to, and utterly dependent on, the earth's global ocean through the water cycle (see Figure 3.13, p. 52).

The aquatic equivalents of biomes are called **aquatic life zones**—saltwater and freshwater portions of the biosphere that can support life. The distribution of many aquatic organisms is determined largely by the water's *salinity*—the amounts of various salts such as sodium chloride dissolved in a given volume of water. As a result, aquatic life zones are classified into two major types: **saltwater** or **marine life zones** (oceans and their bays, estuaries, coastal wetlands, shorelines, coral reefs, and mangrove forests) and **freshwater life zones** (lakes, rivers, streams, and inland wetlands).

In most aquatic systems, the key factors determining the types and numbers of organisms found at various depths are *water temperature, dissolved oxygen content, avail-*

ability of food, and *availability of light and nutrients required for photosynthesis,* such as carbon (as dissolved CO_2 gas), nitrogen (as NO_3^-), and phosphorus (mostly as PO_4^{3-}).

Oceans Provide Vital Ecosystem and Economic Services

Oceans provide enormously valuable ecosystem and economic services (Figure 7.18) that help to keep us and other species alive and that support our economies. They are also enormous reservoirs of biodiversity. Marine life is found in three major *life zones:* the coastal zone, the open sea, and the ocean bottom (Figure 7.19).

The **coastal zone** is the warm, nutrient-rich, shallow water that extends from the high-tide mark on land to the gently sloping, shallow edge of the *continental shelf* (the submerged part of the continents). It makes up less than 10% of the world's ocean area, but it contains 90% of all marine species and is the site of most large commercial marine fisheries. This zone's aquatic systems include estuaries, coastal marshes, mangrove forests, and coral reefs.

An **estuary** is where a river meets the sea (Figure 7.20). It is a partially enclosed body of water where seawater mixes with the river's freshwater, as well as nutrients and pollutants in runoff from the land.

Estuaries are associated with **coastal wetlands**—coastal land areas covered with water all or part of the year. These wetlands, which are some of the earth's most productive ecosystems (see Figure 3.12, p. 50) include *coastal marshes* (Figure 7.21) and *mangrove forests* (Figure 7.22). *Sea-grass beds* (Figure 7.23), another component of coastal marine biodiversity, are underwater ecosystems in shallow coastal waters that host as many as 60 species of grasses and other plants, located along most continental coastlines.

These coastal aquatic systems provide important ecosystem and economic services. They help to maintain water quality in tropical coastal zones by filtering toxic pollutants, excess plant nutrients, and sediments, and by absorbing other pollutants. They provide food, habitats, and nursery sites for a variety of aquatic and terrestrial species. They also reduce storm damage and coastal erosion by absorbing waves and storing excess water produced by storms and tsunamis.

CASE STUDY

Revisiting Coral Reefs—Amazing Centers of Biodiversity

As we noted in the **Core Case Study**, coral reefs (see chapter-opening photo) are some of the world's oldest and most diverse and productive ecosystems. These centers of aquatic biodiversity are the marine equivalents of tropical rain forests, with complex interactions among their diverse

Natural Capital

Marine Ecosystems

Ecosystem Services	Economic Services
Oxygen supplied through photosynthesis	Food
Water purification	Energy from waves and tides
Climate moderation	Pharmaceuticals
CO_2 absorption	Harbors and transportation routes
Nutrient cycling	
Reduced storm damage (mangroves, barrier islands, coastal wetlands)	Recreation and tourism
	Employment
Biodiversity: species and habitats	Minerals

© Cengage Learning

FIGURE 7.18 Marine systems provide a number of important ecosystem and economic services (**Concept 7.3**). *Questions:* Which two ecosystem services and which two economic services do you think are the most important? Why?

Top: Willyam Bradberry/Shutterstock.com. Bottom: James A. Harris/Shutterstock.com.

populations of species. Many scientists are carrying out research on the world's shallow and deep coral reefs. For example, with a National Geographic research grant, ecologist Rhian Waller has studied life on deep-sea coral reefs to gather information about how they function and about how to protect them from damaging human activities.

Worldwide, coral reefs are being damaged and destroyed at an alarming rate by a variety of human activities. The newest growing threat is **ocean acidification**—the rising levels of acidity in ocean waters. This is occurring because the oceans absorb about 25% of the CO_2 emitted into the atmosphere by human activities, especially the burning of carbon-containing fossil fuels. The CO_2 reacts with ocean water to form a weak acid (carbonic acid, H_2CO_3). This reaction decreases the levels of carbonate ions (CO_3^{2-}) necessary for the formation of coral reefs and the shells and skeletons of marine organisms such as crabs, oysters, mussels, and some phytoplankton. This is making it harder for these species to thrive and reproduce, and at some point, this rising acidity could slowly dissolve corals and the shells and skeletons of these marine species.

Ocean acidification and other forms of degradation could have devastating effects on the biodiversity of coral reefs, on the food webs dependent on them, and on the ecosystem services they provide. Coral reef degradation and destruction will also have a severe impact on the ap-

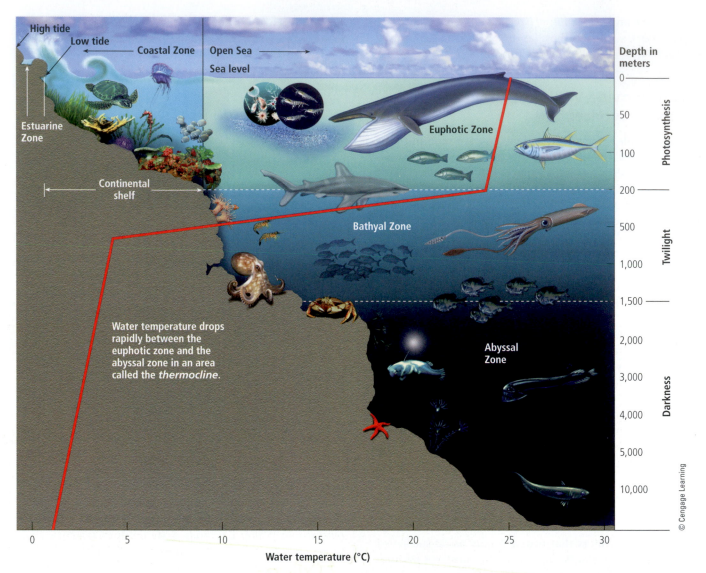

High tide
Low tide
Coastal Zone
Open Sea
Sea level

Depth in meters

Estuarine Zone

Continental shelf

Euphotic Zone

Bathyal Zone

Water temperature drops rapidly between the euphotic zone and the abyssal zone in an area called the *thermocline*.

Abyssal Zone

Photosynthesis

Twilight

Darkness

0
50
100
200
500
1,000
1,500
2,000
3,000
4,000
5,000
10,000

© Cengage Learning

0   5   10   15   20   25   30
Water temperature (°C)

FIGURE 7.19 Major life zones and vertical zones (not drawn to scale) in an ocean. Actual depths of zones may vary. Available light determines the euphotic, bathyal, and abyssal zones. Temperature zones also vary with depth, shown here by the red line. **Question:** How is an ocean like a rain forest? (Hint: See Figure 7.15.)

FIGURE 7.20 Satellite photo of an *estuary*. A sediment plume (cloudiness caused by runoff) forms at the mouth of Madagascar's Betsiboka River as it flows through the estuary and into the Mozambique Channel.

NASA

FIGURE 7.21 Coastal marsh near Cape Cod, Massachusetts.

Michael Melford/National Geographic Stock

FIGURE 7.22 Mangrove forest on the coast of Thailand.

Manit Larpluechai | Dreamstime.com

Concept 7.3   **149**

FIGURE 7.23 Sea-grass beds, such as this one near the coast of San Clemente Island, California, support a variety of marine species.

James Forte/National Geographic Stock

proximately 500 million people who depend on them for their food or for income from fishing and tourism.

According to a 2013 study by more than 500 of the world's leading experts on ocean acidification, since about 1800 on average, there has been a 26% rise in the acidity of ocean water—15% since the 1990s—with the largest increase occurring in deep cold waters near the poles (especially in the Arctic Sea) and along the West Coast waters of the United States. These scientists project that by 2100,

the oceans may be, on average, as much as 170% more acidic than they were in 1800, unless we sharply reduce our CO_2 emissions.

In 2013, Richard Vevers and teams of marine scientists launched the Catlin Seaview Survey and the Underwater Earth Project. These researchers are using sophisticated underwater cameras to create 3-D digital images of the world's major coral reefs. These two global data projects aim to provide a baseline of the health of the world's coral reefs and to identify areas that need emergency protection to keep the reefs from dying. They are also aimed at greatly increasing public support for protection of coral reefs and other marine ecosystems.

The Open Sea and the Ocean Floor Host a Variety of Species

The sharp increase in water depth at the edge of the continental shelf separates the coastal zone from the vast volume of the ocean called the **open sea**. This aquatic life zone is divided into three *vertical zones* (Figure 7.19), or layers, primarily based on the degree of penetration of sunlight. Temperatures also change with depth (Figure 7.19, red line) and we can use them to define zones of varying species diversity in these layers.

The *euphotic zone* is the brightly lit upper zone, where drifting phytoplankton carry out about 40% of the world's photosynthetic activity. Large, fast-swimming predatory fishes such as swordfish, sharks, and bluefin tuna populate the euphotic zone.

The *bathyal zone* is the dimly lit middle zone, which receives little sunlight and therefore does not contain photosynthesizing producers. Zooplankton and smaller fishes, many of which migrate to feed on the surface at night, are found in this zone.

The deepest zone, called the *abyssal zone*, is dark and very cold. There is no sunlight to support photosynthesis, and this zone has little dissolved oxygen. Nevertheless, the deep ocean floor is teeming with life because it contains enough nutrients to support a large number of species. Most organisms in the deep waters and on the ocean floor get their food from showers of dead and decaying organisms—called *marine snow*—drifting down from upper, more lighted levels of the ocean.

Net primary productivity (NPP) is quite low in the open sea (Figure 3.12, p. 50), except in upwelling areas, where currents bring up nutrients from the ocean bottom. However, because the open sea covers so much of the earth's surface, it makes the largest contribution to the earth's overall NPP.

Human Activities Are Disrupting and Degrading Marine Ecosystems

Certain human activities are disrupting and degrading many of the ecosystem and economic services provided by marine aquatic systems, especially coastal marshes, shore-

Natural Capital Degradation

Major Human Impacts on Marine Ecosystems and Coral Reefs

Marine Ecosystems

Half of coastal wetlands lost to agriculture and urban development

Over one-fifth of mangrove forests lost to agriculture, aquaculture, and development

Beaches eroding due to development and rising sea levels

Ocean-bottom habitats degraded by dredging and trawler fishing

At least 20% of coral reefs severely damaged and 25–33% more threatened

Coral Reefs

Ocean warming

Rising ocean acidity

Rising sea levels

Soil erosion

Algae growth from fertilizer runoff

Bleaching

Increased UV exposure

Damage from anchors and from fishing and diving

© Cengage Learning

FIGURE 7.24 Human activities are having major harmful impacts on all marine ecosystems (left) and particularly on coral reefs (right) (**Concept 7.3**). *Questions:* Which two of the threats to marine ecosystems do you think are the most serious? Why? Which two of the threats to coral reefs do you think are the most serious? Why?

Top left: Jorg Hackemann/Shutterstock.com. Top right: Rich Carey/Shutterstock.com. Bottom left: Piotr Marcinski/Shutterstock.com. Bottom right: Rostislav Ageev/Shutterstock.com.

lines, mangrove forests, and coral reefs, as summarized in Figure 7.24 (**Concept 7.3**). We examine these harmful effects and possible ways to lessen them in Chapter 9.

CONSIDER THIS...

THINKING ABOUT Coral Reef Destruction

How might the loss of most of the world's remaining tropical coral reefs (**Core Case Study**) affect your life and the lives of any children or grandchildren you might have? What are two things you could do to help reduce this loss?

WHAT ARE THE MAJOR TYPES OF FRESHWATER SYSTEMS AND HOW ARE HUMAN ACTIVITIES AFFECTING THEM?

CONCEPT 7.4 Freshwater lakes, rivers, and wetlands provide important ecosystem and economic services that are being disrupted by human activities.

Water Stands in Some Freshwater Systems and Flows in Others

Precipitation that does not sink into the ground or evaporate becomes **surface water**—freshwater that flows or is stored in bodies of water on the earth's surface. *Freshwater aquatic life zones* include *standing* bodies of freshwater such as lakes, ponds, and inland wetlands, and *flowing* systems such as streams and rivers. Surface water that flows into such bodies of water is called **runoff**. A **watershed**, or **drainage basin**, is the land area that delivers runoff, sediment, and dissolved substances to a stream, lake, or wetland. Although freshwater systems cover less than 2.2% of the earth's surface, they provide a number of important ecosystem and economic services (Figure 7.25).

Lakes are large natural bodies of standing freshwater formed when precipitation, runoff, streams, rivers, and

Natural Capital

Freshwater Systems

Ecosystem Services	Economic Services
Climate moderation	Food
Nutrient cycling	Drinking water
Waste treatment	Irrigation water
Flood control	Hydroelectricity
Groundwater recharge	Transportation corridors
Habitats for many species	Recreation
Genetic resources and biodiversity	Employment
Scientific information	

© Cengage Learning

FIGURE 7.25 Freshwater systems provide many important ecosystem and economic services (**Concept 7.4**). *Questions:* Which two ecosystem services and which two economic services do you think are the most important? Why?

Top: Galyna Andrushko/Shutterstock.com. Bottom: Kletr/Shutterstock.com.

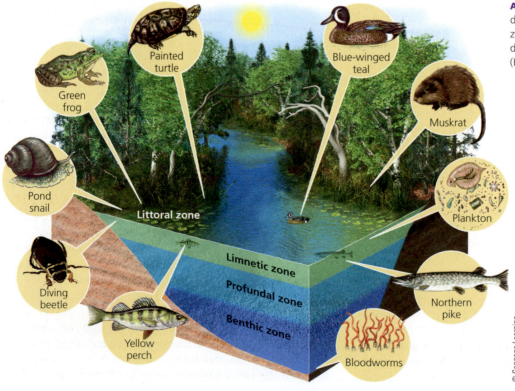

ANIMATED FIGURE 7.26 A typical deep temperate-zone lake has distinct zones of life. *Question:* How are deep lakes like tropical rain forests? (Hint: See Figure 7.15.)

Painted turtle
Blue-winged teal
Green frog
Muskrat
Pond snail
Plankton
Littoral zone
Diving beetle
Limnetic zone
Profundal zone
Benthic zone
Northern pike
Yellow perch
Bloodworms

© Cengage Learning

FIGURE 7.27 Trillium Lake in the U.S. state of Oregon with a view of Mount Hood.

tusharkoley/Shutterstock.com

groundwater seepage fill depressions in the earth's surface. Causes of such depressions include glaciation (as in Lake Louise in Alberta, Canada), displacement of the earth's crust (Lake Nyasa in East Africa), and volcanic activity. A lake's watershed supplies it with water from rainfall, melting snow, and streams.

Freshwater lakes vary tremendously in size, depth, and nutrient content. Deep lakes normally consist of four distinct zones that are defined by their depth and distance from shore (Figure 7.26).

Ecologists classify lakes according to their nutrient content and primary productivity. Lakes that have a small supply of plant nutrients are called **oligotrophic lakes**. This type of lake (Figure 7.27) is often deep and can have steep banks. Glaciers and mountain streams supply water to many of these lakes, which usually have crystal-clear water and small populations of phytoplankton and fish

species, such as smallmouth bass and trout. Because of their low levels of nutrients, these lakes have a low net primary productivity.

Over time, sediments, organic material, and inorganic nutrients wash into most oligotrophic lakes, and plants grow and decompose to form bottom sediments. A lake with a large supply of nutrients is called a **eutrophic lake** (Figure 7.28). Such lakes typically are shallow and have murky brown or green water. Because of their high levels of nutrients, these lakes have a high net primary productivity.

Human inputs of nutrients through the atmosphere and from urban and agricultural areas within a lake's watershed can accelerate the eutrophication of the lake. This process, called **cultural eutrophication**, often puts excessive nutrients into lakes. Most lakes fall somewhere between the two extremes of nutrient enrichment.

This lake has been overfertilized with plant nutrients flowing off the surrounding land.

FIGURE 7.28 This eutrophic lake has received large flows of plant nutrients. As a result, its surface is covered with mats of algae.

Nicholas Rjabow | Dreamstime.com

Freshwater Streams and Rivers Carry Large Volumes of Water

In drainage basins, water accumulates in small streams that join to form rivers, which, collectively, carry huge amounts of water from highlands to lakes and oceans. Typically, a stream flows through three zones (Figure 7.29): the *source zone,* which contains *headwater* streams found in highlands and mountains; the *transition zone,* which contains wider, lower-elevation streams; and the *floodplain zone,* which contains rivers that empty into larger rivers or into the ocean.

As streams flow downhill, they shape the land through which they pass. Over millions of years, the friction of moving water has leveled mountains and cut deep canyons, and sand, gravel, and soil carried by streams and rivers have been deposited as sediment in low-lying areas.

At its mouth, a river may divide into many channels as it flows through its **delta**—an area at the mouth of a river built up by deposited sediment and often containing estuaries (Figure 7.20) and coastal wetlands (Figure 7.21). These important forms of natural capital absorb and slow the velocity of floodwaters from coastal storms, hurri-

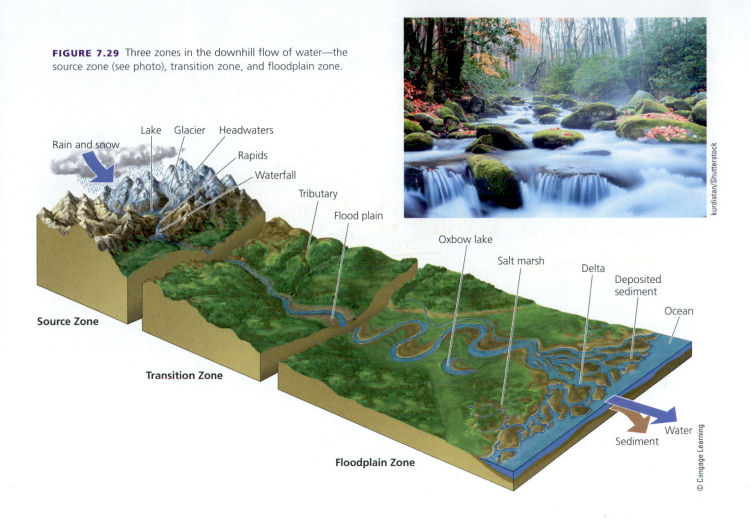

FIGURE 7.29 Three zones in the downhill flow of water—the source zone (see photo), transition zone, and floodplain zone.

Rain and snow
Lake Glacier Headwaters
Rapids
Waterfall
Tributary
Flood plain
Source Zone
Oxbow lake
Salt marsh
Delta
Deposited sediment
Ocean
Transition Zone
Water
Sediment
Floodplain Zone

kurdistan/Shutterstock

© Cengage Learning

canes, and tsunamis and provide habitats for a wide variety of marine life.

Freshwater Inland Wetlands Are Vital Sponges

Inland wetlands are lands located away from coastal areas that are covered with freshwater all or part of the time—excluding lakes, reservoirs, and streams. They include *marshes, swamps,* and *prairie potholes* (depressions carved out by ancient glaciers). Other examples are *floodplains,* which receive excess water from streams or rivers during heavy rains and floods.

Some wetlands are covered with water year-round and others remain under water for only a short time each year. The latter include prairie potholes, floodplain wetlands, and arctic tundra (see Figure 7.12, bottom).

Inland wetlands provide a number of free ecosystem and economic services, which include:

- filtering and degrading toxic wastes and pollutants;
- reducing flooding and erosion by absorbing storm water and releasing it slowly, and by absorbing overflows from streams and lakes;

- helping to sustain stream flows during dry periods;
- helping to recharge groundwater aquifers;
- helping to maintain biodiversity by providing habitats for a variety of species;
- supplying valuable products such as fishes and shellfish, blueberries, cranberries, and wild rice; and
- providing recreation for birdwatchers, nature photographers, boaters, anglers, and waterfowl hunters.

Human Activities Are Disrupting and Degrading Freshwater Systems

Human activities are disrupting and degrading many of the ecosystem and economic services provided by freshwater rivers, lakes, and wetlands (**Concept 7.4**) in four major ways. *First,* dams and canals restrict the flows of about 40% of the world's 237 largest rivers. This alters or destroys terrestrial and aquatic wildlife habitats along these rivers and in their coastal deltas and estuaries by reducing water flow and the flow of sediments to river deltas. *Second,* flood control levees and dikes built along rivers dis-

Alexandra Cousteau: Environmental Storyteller and National Geographic Emerging Explorer

Alexandra Cousteau is proud of her heritage as granddaughter of Captain Jacques-Yves Cousteau and daughter of Philippe Cousteau. Her father and grandfather were legendary underwater explorers who brought the mysteries and wonders of the oceans into living rooms around the world with their films and books.

The focus of Alexandra's work is advocating the importance of conservation and sustainable management of water in order to preserve a healthy planet. She seeks to make water one of the defining issues of this century, stating: "We live on a water planet, which means we're all downstream from one another. Where water comes from, where it goes, and its quality is intricately connected to our quality of life."

She is utilizing tools not even imagined by her grandfather—those of social networking and other modes of mobile communication. She believes that environmental advocates can use such new media tools and technology to inform people about how their actions affect our water. For example, she imagines a day in the future when knowing the quality and quantity of our water is as easy as checking the weather on our smart phones.

Alexandra's nonprofit Blue Legacy International harnesses technology to tell the stories of our water planet and provides film and digital resources to allow others to explore and understand water issues.

Courtesy of Alexandra Cousteau

Background photo: Lisa Heeter | Dreamstime.com

connect the rivers from their floodplains, destroy aquatic habitats, and alter or degrade the functions of adjoining wetlands.

Third, cities and farms add pollutants and excess plant nutrients to nearby streams, rivers, and lakes. For example, runoff of nutrients into a lake (cultural eutrophication, Figure 7.28) causes explosions in the populations of algae and cyanobacteria, which deplete the lake's dissolved oxygen. Fishes and other species may then die off, which can mean a major loss in biodiversity.

Fourth, many inland wetlands have been drained or filled to grow crops or have been covered with concrete, asphalt, and buildings. More than half of the inland wetlands estimated to have existed in the continental United States during the 1600s no longer exist. About 80% of these lost wetlands were drained to grow crops. This loss of natural capital has been an important factor in increasing flood damage in parts of the United States. Many other countries have suffered similar losses. For example, 80% of all inland wetlands in Germany and France have been destroyed.

A large number of scientists and other individuals are devoting their lives to understanding aquatic systems and learning how we can use them more sustainably (Individuals Matter 7.1).

BIG IDEAS

- Differences in climate, based primarily on long-term differences in average temperature and precipitation, largely determine the types and locations of the earth's deserts, grasslands, and forests.

- Saltwater and freshwater aquatic systems cover almost three-fourths of the earth's surface, and oceans dominate the planet.

- The earth's terrestrial and aquatic systems provide important ecosystem and economic services that are being degraded and disrupted by human activities.

Coral Reefs and Sustainability

This chapter's Core Case Study pointed out the ecological and economic importance of the world's incredibly diverse coral reefs. They are living examples of the three **scientific principles of sustainability** in action. They thrive on solar energy, play key roles in the cycling of carbon and other chemicals, and sustain a great deal of aquatic biodiversity.

We also discussed the influence of climate on terrestrial biodiversity in the formation of biomes—deserts, grasslands, and forests—as well as the life forms that live in those systems. These relationships are also in keeping with the three **scientific principles of sustainability**. The earth's dynamic climate system helps to distribute heat from solar energy and to recycle the earth's nutrients. This in turn helps to generate and support the biodiversity found in the earth's various biomes.

Finally, we looked at how human activities are degrading the vital ecosystem and economic services that the earth's terrestrial and aquatic systems provide. Scientists call for much more research on the components and workings of the world's terrestrial and aquatic systems, on how they are interconnected, and on which systems are in the greatest danger of being disrupted by human activities.

JonMilnes/Shutterstock.com

Chapter Review

Core Case Study

1. What are **coral reefs** and why should we care about them? What is coral bleaching? What are the major threats to coral reefs?

Section 7.1

2. What is the key concept for this section? Distinguish between **weather** and **climate**. Define **ocean currents**. Describe three major factors that determine how air circulates in the lower atmosphere. Explain how varying combinations of temperature and precipitation, along with global air circulation and ocean currents, lead to the formation of various types of forests, grasslands, and deserts.

3. Define and give three examples of a **greenhouse gas**. What is the **greenhouse effect** and why is it important to the earth's life and climate? What is the **rain shadow effect** and how can it lead to the formation of deserts? Why do cities tend to have more haze and smog, higher temperatures, and lower wind speeds than the surrounding countryside?

Section 7.2

4. What are the two key concepts for this section? What is a **biome**? Explain why there are three major types of each of the major biomes (deserts, grasslands, and forests). Explain why biomes are not uniform. Describe how climate and vegetation vary with latitude and elevation. What is the **edge effect**?

5. Explain how the three major types of deserts differ in their climate and vegetation. Why are desert ecosystems fragile? How do desert plants and animals survive? Explain how the three major types of grasslands differ in their climate and vegetation. What is a savanna? Why have many of the world's temperate grasslands disappeared? What is **permafrost**? Explain how the three major types of forests differ in their climate and vegetation. Why is biodiversity so high in tropical rain forests? Explain why most soils in tropical rain forests hold few plant nutrients. Why does a thick layer of decaying litter typically cover the floors of temperate deciduous forests? What are coastal coniferous or temperate rain forests? How do most species of coniferous evergreen trees survive the cold winters in boreal forests? What important ecological roles do mountains play?

6. About what percentage of the world's major terrestrial ecosystems are being degraded or used unsustainably? Summarize the ways in which human activities have affected the world's deserts, grasslands, forests, and mountains.

Section 7.3

7. What is the key concept for this section? What percentage of the earth's surface is covered with ocean water? What is an **aquatic life zone**? Distinguish between a **saltwater (marine) life zone** and a **freshwater life zone**, and give two examples of each. List five factors that determine the types and numbers of organisms found in the layers of aquatic life zones.

8. What major ecosystem and economic services are provided by marine systems? What are the three major life zones in an ocean? Define and distinguish between the **coastal zone** and the **open sea**. Distinguish between an **estuary** and a **coastal wetland**. Explain the ecological and economic importance of coastal marshes, mangrove forests, and sea-grass beds. What is **ocean acidification** and why is it a threat to coral reefs? Describe the three major zones in the open sea. List five human activities that pose major threats to marine systems and eight human activities that threaten coral reefs.

Section 7.4

9. What is the key concept for this section? Define **surface water**, **runoff**, and **watershed (drainage basin)**. What major ecosystem and economic services do freshwater systems provide? What is a **lake**? What four zones are found in deep lakes? Distinguish between **oligotrophic** and **eutrophic lakes**. What is **cultural eutrophication**? Describe the three zones that a stream passes through as it flows from highlands to lower elevations. What is a **delta**? Give three examples of **inland wetlands** and describe the ecological and economic importance of such wetlands. List four ways in which human activities are disrupting and degrading many of the ecosystem and economic services provided by freshwater rivers, lakes, and wetlands. How is Alexandra Cousteau attempting to educate people about the importance of aquatic systems?

10. What are this chapter's *three big ideas*? Explain how terrestrial and aquatic systems are living examples of the **scientific principles of sustainability** in action.

Note: Key terms are in bold type.

Critical Thinking

1. What are three steps that governments and private interests could take to protect the world's remaining coral reefs (**Core Case Study**)?

2. Why do most animals in a tropical rain forest live in its trees?

3. How might the distribution of the world's forests, grasslands, and deserts shown in Figure 7.10 differ if the prevailing winds shown in Figure 7.3 did not exist?

4. Which biomes are best suited for **(a)** raising crops and **(b)** grazing livestock? Use the three **scientific principles of sustainability** to come up with three guidelines for growing crops and grazing livestock more sustainably in these biomes.

5. What type of biome do you live in? (If you live in a developed area, what type of biome was the area before it was developed?) List three ways in which your lifestyle could be contributing to the degradation of this biome. What are three lifestyle changes that you could make in order to reduce your contribution?

6. You are a defense attorney arguing in court for sparing a tropical rain forest from being cut down. Give your three best arguments for the defense of this ecosystem. Do the same for sparing a threatened coral reef (**Core Case Study**). If you had to choose between protecting a tropical rain forest and a coral reef, which one would you select? Explain.

7. Why is ocean acidification considered to be a very serious problem? If acidity levels in the ocean rise sharply during your lifetime, how might this affect you? Can you think of ways in which you might be contributing to this problem? What could you do to reduce your impact?

8. Suppose you have a friend who owns property that includes a freshwater wetland and the friend tells you she is planning to fill the wetland to make more room for her lawn and garden. What would you say to this friend?

Doing Environmental Science

Find a natural ecosystem near where you live or go to school, either a terrestrial ecosystem such as a forest, or an aquatic system such as a lake or wetland. Study and write a description of the system, including its dominant vegetation and any animal life that you are aware of. Also, note how any human disturbances have changed the system. Compare your notes with those of your classmates.

Global Environment Watch Exercise

Search for *Coral reefs* and use the topic portal to find information on **(a)** trends in the global rate of coral reef destruction; **(b)** what areas of the world are seeing rising rates of coral reef destruction and what areas are seeing falling rates; and **(c)** what is being done to protect coral reefs in various areas. Write a report on your findings.

Data Analysis

In this chapter, you learned how long-term variations in average temperatures and average precipitation play a major role in determining the types of deserts, forests, and grasslands found in different parts of the world. Below are typical annual climate graphs for a tropical grassland (savanna) in Africa and a temperate grassland in the Midwestern United States.

1. In what month (or months) does the most precipitation fall in each of these areas?

2. What are the driest months in each of these areas?

3. What is the coldest month in the tropical grassland?

4. What is the warmest month in the temperate grassland?

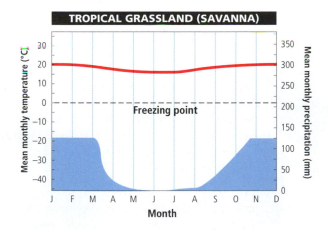

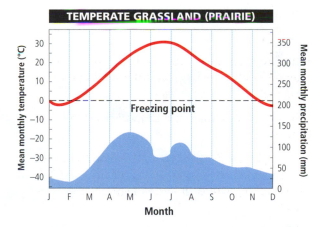

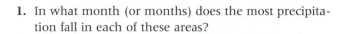

CENGAGEbrain To access course materials, including Aplia homework, please visit www.cengagebrain.com.

WWW.CENGAGEBRAIN.COM   **159**

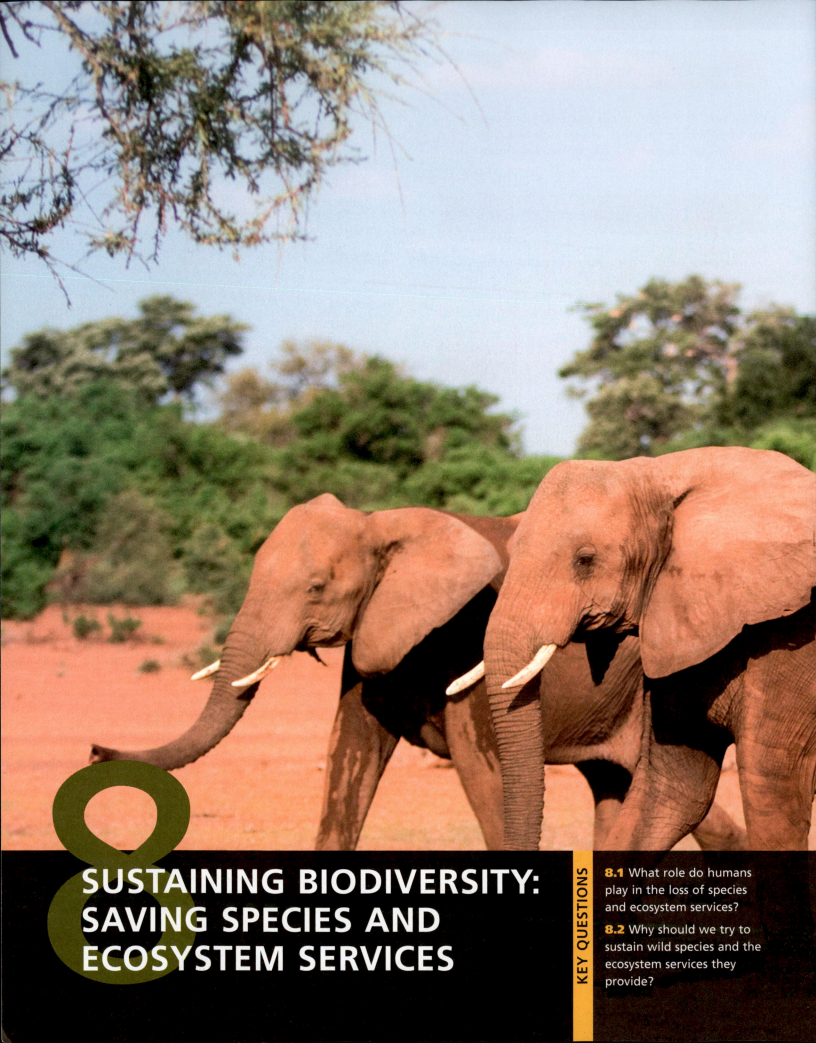

8

SUSTAINING BIODIVERSITY: SAVING SPECIES AND ECOSYSTEM SERVICES

KEY QUESTIONS

8.1 What role do humans play in the loss of species and ecosystem services?

8.2 Why should we try to sustain wild species and the ecosystem services they provide?

The last word in ignorance is the person who says of an animal or plant: "What good is it?" . . . If the land mechanism as a whole is good, then every part of it is good, whether we understand it or not.

ALDO LEOPOLD

8.3 How do humans accelerate species extinction and degradation of ecosystem services?

8.4 How can we sustain wild species and the ecosystem services they provide?

African elephants are being killed illegally in large numbers for their valuable ivory tusks.

Villiers Steyn/Shutterstock

Where Have All the Honeybees Gone?

In meadows, forests, farm fields, and gardens around the world, industrious honeybees (Figure 8.1) flit from one flowering plant to another, collecting nectar and pollen that they take back to their hives. They feed their young the protein-rich pollen, and the adults live on the honey made from the collected nectar and stored in the hive.

Bees provide us with one of nature's most important ecosystem services: *pollination,* the transfer of pollen within and among flowering plants that enables them to produce seeds and fruit. Bees pollinate many flower species and some of our most important food crops, including many vegetables, fruits, and tree nuts such as almonds. European honeybees pollinate about 71 of the 100 fruit and vegetable crop species that provide 90% of the world's food and a third of the U.S. food supply.

Many U.S. growers rent European honeybees from commercial beekeepers that truck about 2.7 million hives to farms across the country to pollinate different crops. Whereas nature relies on the earth's free pollination service provided by a diversity of bees and other wild pollinators, farmers who practice industrialized agriculture on vast croplands rely mostly on a single bee species.

The problem is that since the 1980s, European honeybees have been in decline. Since 2006, massive numbers of these bees in the United States and in parts of Europe have been disappearing from their colonies during the winter and not returning as expected in the spring—a phenomenon called **colony collapse disorder (CCD)**. Every winter since 2006, according to the U.S. National Academy of Sciences, about 30% to 40% of the European honeybee colonies in the United States have suffered from CCD, well above the historical loss rates of 10% to 15%.

Some producers believe we need the industrialized honeybee pollination system in order to grow enough food.

Others see such heavy dependence on a single bee species as a potentially dangerous violation of the earth's biodiversity **principle of sustainability**. They argue that this dependence could put food supplies at risk if the European honeybees continue to decline. Also, if this decline leads to smaller harvests, food prices will rise.

This is a classic case of how the decline of a species can threaten vital ecosystem and economic services. Scientists project that during this century, human activities, especially those that contribute to habitat loss and climate change, are likely to play a key role in the extinction of one-fourth to one-half of the world's plant and animal species. Many scientists view this threat as one of the most serious and long-lasting environmental and economic problems we face. In this chapter, we discuss the causes of this problem and possible ways to deal with it.

FIGURE 8.1 European honeybee sipping nectar from a flower.

Darlyne A. Murawski/National Geographic Creative

8.1 WHAT ROLE DO HUMANS PLAY IN THE LOSS OF SPECIES AND ECOSYSTEM SERVICES?

CONCEPT 8.1 Species are becoming extinct 100 to 1,000 times faster than they were before the human population grew exponentially, and by the end of this century, the extinction rate is projected to be 10,000 times higher than the historical rate.

Extinctions Are Natural but Sometimes They Increase Sharply

When a species can no longer be found anywhere on the earth, it has suffered **biological extinction**. The extinction of many species in a relatively short period of geologic time is called a **mass extinction**. Geologic, fossil, and other records indicate that the earth has experienced five mass extinctions, during which 50–90% of all species appear to have become extinct (see Supplement 6, p. S49). The third mass extinction, about 250 million years ago, was the worst, with the disappearance of about 90% of the world's species, including 96% of all marine species. Scientists hypothesize that one of its causes was massive volcanic eruptions over a long time period that released huge amounts of the greenhouse gases CO_2 and methane (CH_4) into the atmosphere. This caused major and prolonged warming of the earth's atmosphere and oceans, along with ocean acidification.

Scientific evidence indicates that after each mass extinction, the earth's overall biodiversity eventually returned to equal or higher levels, but each recovery required several million years. Such long-time recovery is an example of the biodiversity **principle of sustainability** in action.

The causes of past mass extinctions are poorly understood but probably involved global changes in environmental conditions. Examples are sustained and significant global warming or cooling, large changes in sea levels and ocean water acidity, and catastrophes such as multiple large-scale volcanic eruptions and large asteroids hitting the planet.

Some Human Activities Hasten Extinctions and Threaten Ecosystem Services

Scientists who study extinction base much of their work on an estimated **background extinction rate**—the rate that existed before modern humans evolved some 200,000 years ago—which scientists estimate was about 1 species per year for every 1 million wild species living on the earth.

Scientific evidence indicates that extinction rates have risen in some areas as human populations have spread over most of the globe, destroying and degrading habitats, consuming huge quantities of resources, and creating large and growing ecological footprints (see Figure 1.10, p. 12). In the words of biodiversity expert Edward O. Wilson (see Individuals Matter 4.1, p. 67), "The natural world is everywhere disappearing before our eyes—cut to pieces, mowed down, plowed under, gobbled up, replaced by human artifacts."

Scientists from around the world who conducted the 2005 Millennium Ecosystem Assessment estimated that the current annual rate of species extinction is at least 100 to 1,000 times the estimated background extinction rate (Science Focus 8.1) (**Concept 8.1**). We have identified about 2 million species so far, but scientists estimate that there are many millions more. Assuming that there are 10 million species on earth, at the background extinction rate of 1 species per million per year, about 10 species would disappear naturally each year. However, at today's estimated rate of 100 to 1,000 times the background rate, we are losing between 1,000 and 10,000 species per year, or between 2 and 27 species every day, on average.

Biodiversity researchers project that during this century, the extinction rate is likely to rise to at least 10,000 times the background rate—mostly because of habitat loss and degradation, climate change, ocean acidification, and other environmentally harmful effects of human activities (**Concept 8.1**). At this rate, if there were 10 million species on the earth, then about 100,000 species would be expected to disappear each year—an average of about 274 species per day or about 11 every hour. By the end of this century, most of the big carnivorous cats, including cheetahs, tigers, and lions, will probably exist only in zoos and small wildlife sanctuaries. And most elephants (see chapter-opening photo), rhinoceroses, gorillas, chimpanzees, and orangutans will likely disappear from the wild.

So why is this a big deal? According to biodiversity researchers Edward O. Wilson and Stuart Pimm, at this extinction rate, at least 25% and as many as 50% of the world's roughly 2 million identified animal and plant species could vanish from the wild by the end of this century, along with many of the millions of unidentified species. This would amount to a sixth mass extinction caused primarily by human activities and taking place within one century (see Supplement 6, p. S49).

Such a large-scale loss of species would also impair some of the earth's vital ecosystem services, including air and water purification, natural pest control, and pollination (Core Case Study). According to the 2005 Millennium Ecosystem Assessment, 15 of 24 of the earth's major ecosystem services are in decline. If such estimates are only half correct, we can see why many biologists warn that such a massive loss of biodiversity and ecosystem services within the span of a single human lifetime is one of the most important and long-lasting environmental and economic problems we face. By saving as many species as possible from extinction—especially keystone species (see p. 69)—we could increase our beneficial environmental impact and help to sustain and enrich our own lives and economies.

SCIENCE**FOCUS**8.1

ESTIMATING EXTINCTION RATES

Scientists who try to catalog extinctions, estimate past extinction rates, and project future extinction rates face three problems. *First,* because the natural extinction of a species typically takes a very long time, it is difficult to document. *Second,* we have identified only about 2 million of the world's estimated 7–10 million and perhaps as many as 100 million species. *Third,* scientists know little about the ecological roles of most of the species that have been identified.

One approach to estimating future extinction rates is to study records documenting past rates at which easily observable mammals and birds have become extinct. Most of these extinctions have occurred since humans began to dominate the planet about 10,000 years ago, when we began the shift from hunting and gathering food in the wild to growing our food and

living in towns and cities. This information can be compared with fossil records of extinctions that occurred before that time.

Another approach is to observe how reductions in habitat area affect extinction rates. The *species–area relationship,* studied by Edward O. Wilson (see Individuals Matter 4.1, p. 67) and Robert MacArthur, suggests that, on average, a 90% loss of land habitat in a given area can cause the extinction of about 50% of the species living in that area. Thus we can base extinction rate estimates on the rates of habitat destruction and degradation, which is increasing around the world.

Scientists also use mathematical models to estimate the risk of a particular species becoming endangered or extinct within a certain period of time. These models include factors such as

trends in population size, past and projected changes in habitat availability, interactions with other species, and genetic factors.

Researchers are continually striving to get more and better data and to improve the models they use in order to make better estimates of extinction rates and to project the effects of such extinctions on vital ecosystem services such as pollination (**Core Case Study**). These scientists contend that our need for better data and models should not delay our acting now to keep from hastening extinctions and the accompanying losses of ecosystem services through human activities.

Critical Thinking

Does the fact that extinction rates can only be estimated make them unreliable? Why or why not?

In fact, Wilson, Pimm, and other extinction experts consider a projected extinction rate of 10,000 times the background extinction rate to be low, for several reasons. *First,* both the rate of extinction and the resulting threats to ecosystem services are likely to increase sharply during the next 50–100 years because of the harmful environmental impacts of the rapidly growing human population and its growing per capita use of resources (see Table 1.1, p. 13).

Second, the current and projected extinction rates in the world's *biodiversity hotspots*—areas that are highly endangered centers of biodiversity—are much higher than the global average. Degradation of these hotspot areas affects many more species than do human activities in already degraded areas.

Third, we are eliminating, degrading, fragmenting, and simplifying many biologically diverse environments—including tropical forests, coral reefs, wetlands, and estuaries—that serve as potential sites for the emergence of new species. Thus, in addition to greatly increasing the rate of extinction, we may be limiting the long-term recovery of biodiversity by eliminating these places where

new species can evolve. In other words, we are also creating a *speciation crisis.* (See the online Guest Essay by Norman Myers on this topic.)

Biologists Philip Levin, Donald Levin, and others warn that, while our activities are likely to reduce the speciation rates for some species, they might increase the speciation rates for other rapidly reproducing species such as weeds and rats, as well as cockroaches and many other species of insects. Rapidly expanding populations of such species could crowd and compete with various other species, further accelerating their extinction and also threatening key ecosystem services.

Endangered and Threatened Species Are Ecological Smoke Alarms

Biologists classify species that are heading toward biological extinction as either *endangered* or *threatened.* An **endangered species** has so few individual survivors that the species could soon become extinct. A **threatened species**, such as the cheetah or the loggerhead sea turtle,

a. Mexican gray wolf: About 42 in the forests of Arizona and New Mexico

b. California condor: 226 in the southwestern United States (up from 9 in 1986)

c. Whooping crane: 437 in North America

d. Sumatran tiger: No more than 500 on the Indonesian island of Sumatra

FIGURE 8.2 Endangered natural capital: These four critically endangered species are threatened with extinction, largely because of human activities. The number below each photo indicates the estimated total number of individuals of that species remaining in the wild.

Characteristic	Examples
Low reproductive rate (K-strategist)	Blue whale, giant panda, rhinoceros
Specialized niche	Blue whale, giant panda, Everglades kite
Narrow distribution	Elephant seal, desert pupfish
Feeds at high trophic level	Bengal tiger, bald eagle, grizzly bear
Fixed migratory patterns	Blue whale, whooping crane, sea turtle
Rare	African violet, some orchids
Commercially valuable	Snow leopard, tiger, elephant, rhinoceros, rare plants and birds
Require large territories	California condor, grizzly bear, Florida panther

© 2016 Cengage Learning

FIGURE 8.3 Certain characteristics can put a species in greater danger of becoming extinct.

still has enough remaining individuals to survive in the short term, but because of declining numbers, it is likely to become endangered in the near future.

Figure 8.2 shows four of the 21,286 species listed in 2013 by the International Union for Conservation of Nature (IUCN) as critically endangered, endangered, or threatened. The real number of species in trouble is very likely much higher. Some species have characteristics that increase their chances of becoming extinct (Figure 8.3). As biodiversity expert Edward O. Wilson puts it, "The first animal species to go are the big, the slow, the tasty, and those with valuable parts such as tusks and skins."

8.2 WHY SHOULD WE TRY TO SUSTAIN WILD SPECIES AND THE ECOSYSTEM SERVICES THEY PROVIDE?

CONCEPT 8.2 We should avoid speeding up the extinction of wild species because of the ecosystem and economic services they provide, because it can take millions of years for nature to recover from large-scale extinctions, and because many people believe that species have a right to exist regardless of their usefulness to us.

Species Are a Vital Part of the Earth's Natural Capital

According to the World Wildlife Fund (WWF), only 50,000–60,000 orangutans (Figure 8.4) remain in the wild, most of them in the tropical forests of Borneo, Asia's

FIGURE 8.4 **Natural capital degradation:** These endangered orangutans depend on a rapidly disappearing tropical forest habitat in Borneo. **_Question:_** What difference will it make if human activities hasten the extinction of the orangutan?

Seatraveler/Dreamstime.com

largest island. These highly intelligent animals are disappearing at an estimated rate of 1,000–2,000 per year, partly because much of their tropical forest habitat is being cleared for plantations of oil palms that supply palm oil used in the production of cosmetics and biodiesel fuel. Their numbers are also declining because they bear young only about once every 8 years. Another reason for their decline is smuggling. An illegally smuggled, live orangutan sells for a street price of up to $10,000. Without urgent protective action, the endangered orangutan may disappear in the wild within the next two decades.

Cheryl Knott—a biological anthropologist, National Geographic Emerging Explorer, and Associate Professor of Anthropology at Boston University—has been studying endangered orangutans in a national park in Borneo. Her research goals are to understand how fluctuations in food availability in the orangutan's environment have shaped their behavioral and physiological adaptations and how this can help us better understand human evolutionary history.

Does it matter that orangutans might soon become extinct largely due to human activities, or that some unknown insect or plant species may meet the same fate? New species eventually evolve to take the places of those species lost through mass extinctions, so why should we care if we greatly speed up the extinction rate over the next 50–100 years? According to biologists, there are three major reasons why we should work to prevent our activities from causing or hastening the extinction of other species.

First, the world's species provide vital _ecosystem services_ (see Figure 1.3, p. 7) that help to keep us alive and support our economies (**Concept 8.2**). For example, we depend on honeybees (**Core Case Study**) and other insects for pollination of many food crops and on certain species of birds, amphibians (see Chapter 4, Core Case Study, p. 64), and spiders for natural control of insect pests. By eliminating a species or sharply reducing its population, especially a species that plays a keystone role (see p. 69), we can speed up the extinction of other species. This can upset an ecosystem and degrade its important ecosystem services.

CONSIDER THIS. . .

CONNECTIONS Species and Ecosystem Services

Plant and animal species provide us with services in ways that you might not expect. For example, those that live in streams help to purify the flowing water. Trees and other forest plants produce oxygen, without which we could not survive. And earthworms aerate topsoil that we use for growing our food. These are vital ecosystem services, provided free of charge.

A _second_ major reason for preventing extinctions caused or hastened by human activities is that many species also contribute to _economic services_ on which we depend (**Concept 8.2**). Various plant species provide economic value as food crops, fuelwood, lumber, and paper from trees, and useful scientific knowledge. To biologist Edward O. Wilson, carelessly and rapidly eliminating species that make up an essential part of the world's biodiversity is like burning millions of books that we have never read.

For example, *bioprospectors* are scientists who search tropical forests and other ecosystems to find plants and animals that we can use to make medicinal drugs. According to a United Nations University report, 62% of all cancer drugs were derived from the discoveries of bioprospectors. Despite their economic and medicinal potential, less than 0.5% of the world's known plant species have been examined for their medicinal properties. **GREEN CAREER: Bioprospecting**

Another economic benefit that can be gained through preserving species and their habitats is the revenues from *ecotourism*, which generates more than $1 million per minute in tourist expenditures, worldwide. Conservation biologist Michael Soulé estimates that a male lion living to age 7 generates about $515,000 through ecotourism in Kenya, but only about $10,000 if it is killed for its skin.

A *third* major reason for not hastening extinctions through our activities is that a sharp reduction in biodiversity will hinder speciation. Analysis of past mass extinctions indicates it will take 5 million to 10 million years for natural speciation to rebuild the biodiversity that is likely to be lost during this century as a result of human activities.

Fourth, many people believe that wild species have a right to exist, regardless of their usefulness to us (**Concept 8.2**). According to this worldview, we have an ethical responsibility to protect the earth's species from becoming extinct as a result of our activities, and to prevent the degradation of the world's ecosystem services that are vital to our life-support system and economies.

8.3 HOW DO HUMANS ACCELERATE SPECIES EXTINCTION AND DEGRADATION OF ECOSYSTEM SERVICES?

CONCEPT 8.3 The greatest threats to species and ecosystem services are (in order) loss or degradation of habitat, harmful invasive species, human population growth, pollution, climate change, and overexploitation.

Loss of Habitat Is the Single Greatest Threat to Species: Remember HIPPCO

Biodiversity researchers summarize the most important direct causes of extinction and threats to ecosystem services using the acronym **HIPPCO**: **H**abitat destruction, degradation, and fragmentation; **I**nvasive (nonnative) species; **P**opulation growth and increasing use of resources; **P**ollution; **C**limate change; and **O**verexploitation (**Concept 8.3**).

According to biodiversity researchers, the greatest threat to wild species is habitat loss (Figure 8.5), degradation, and

fragmentation. Specifically, deforestation in tropical areas (see Figure 3.1, p. 42) is the greatest threat to species and to the ecosystem services they provide, followed by the destruction and degradation of coastal wetlands and coral reefs (see Chapter 7, Core Case Study, p. 130), the plowing of grasslands for planting of crops (see Figure 7.13, p. 141), and the pollution of streams, lakes, and oceans.

Island species—many of them found nowhere else on earth—are especially vulnerable to extinction when their habitats are destroyed, degraded, or fragmented, because they have nowhere else to go. This is why the collection of islands that make up the U.S. state of Hawaii is America's "extinction capital"—with 63% of its species at risk.

Habitat fragmentation occurs when a large, intact area of habitat such as a forest or natural grassland is divided, typically by roads, logging operations, crop fields, and urban development, into smaller, isolated patches or *habitat islands*. This process can reduce tree cover in forests and block animal migration routes. It can also divide populations of a species into increasingly isolated small groups that are more vulnerable to predators, competitor species, disease, and catastrophic events such as storms and fires. In addition, habitat fragmentation creates barriers that limit the abilities of some species to disperse and colonize areas, to locate adequate food supplies, and to find mates.

We Have Moved Disruptive Species into Some Ecosystems

After habitat loss and degradation, the next biggest cause of animal and plant extinctions and loss of the ecosystem services they provide is the deliberate or accidental introduction of harmful species into ecosystems (**Concept 8.3**).

Many introductions of nonnative species have been beneficial to us. According to a study by ecologist David Pimentel, nonnative species such as corn, wheat, rice, and other food crops, as well as some species of cattle, poultry, and other livestock, provide more than 98% of the U.S. food supply. Similarly, nonnative tree species are grown in about 85% of the world's tree plantations. Some deliberately introduced species have also helped to control pests. In the 1600s English settlers brought highly beneficial European honeybees (**Core Case Study**) to North America to provide honey and we now use them to pollinate most major crops.

The problem is that, in their new habitats, some introduced species do not face the natural predators, competitors, parasites, viruses, bacteria, or fungi that had helped to control their numbers in their original habitats. This can allow such nonnative species to crowd out populations of many native species, disrupt ecosystem services, cause human health problems, and lead to economic losses.

Figure 8.6 shows some of the 7,100 or more invasive species that, after being deliberately or accidentally introduced into the United States, have caused ecological and economic harm. According to the U.S. Fish and Wildlife

GOOD NEWS

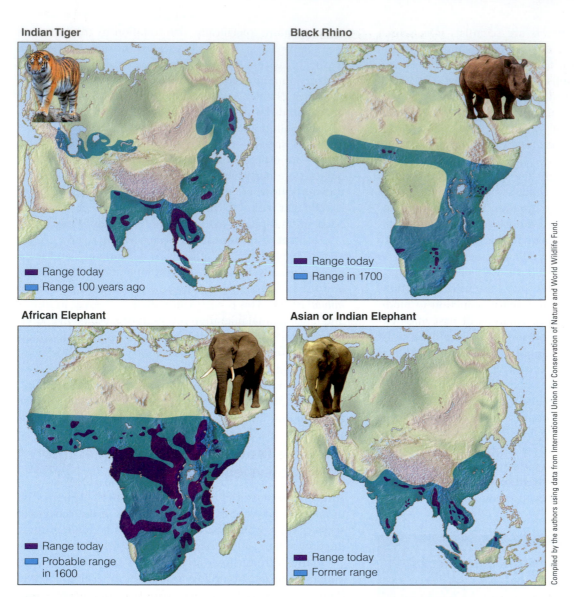

Indian Tiger

Range today
Range 100 years ago

Black Rhino

Range today
Range in 1700

African Elephant

Range today
Probable range in 1600

Asian or Indian Elephant

Range today
Former range

Compiled by the authors using data from International Union for Conservation of Nature and World Wildlife Fund.

ANIMATED FIGURE 8.5 **Natural capital degradation:** These maps reveal the reductions in the ranges of four wildlife species, mostly as the result of severe habitat loss and fragmentation and illegal hunting for some of their valuable body parts. **Question:** Would you support expanding these ranges even though this would reduce the land available for human habitation and farming? Explain.

Service (USFWS), about 40% of the species listed as endangered in the United States and 95% of those in the U.S. state of Hawaii are on the list because of threats from invasive species.

In the 1930s, the extremely aggressive Argentina fire ant (Figure 8.6) was accidentally introduced into the United States probably on shiploads of lumber or coffee imported from South America. These ants have no natural predators in the southern United States where they have spread rapidly by land and water. (They can float.) They have also stowed away on exported goods in shipping containers and have invaded other countries, including China, Taiwan, Malaysia, and Australia.

When these ants invade an area, they can wipe out as much as 90% of native ant populations, which we depend

on for ecosystem services such as enrichment of topsoil, dispersal of plant seeds, and control of pest species such as flies, bedbugs, and cockroaches. If you walk on a fire ant mound, as many as 100,000 ants might swarm out of their nest to attack you with painful, burning stings. They have killed deer fawns, ground-nesting birds, baby sea turtles, newborn calves, pets, and at least 80 people who were allergic to their venom.

Widespread pesticide spraying in the 1950s and 1960s temporarily reduced fire ant populations. But this chemical warfare actually hastened the advance of the rapidly multiplying fire ants by reducing populations of many native ant species. Even worse, it promoted the development of genetic resistance to pesticides in the fire ants through natural selection. Pest management scientist

Deliberately Introduced Species

Purple loosestrife

African honeybee ("Killer bee")

Kudzu

Nutria

European wild boar (Feral pig)

Accidentally Introduced Species

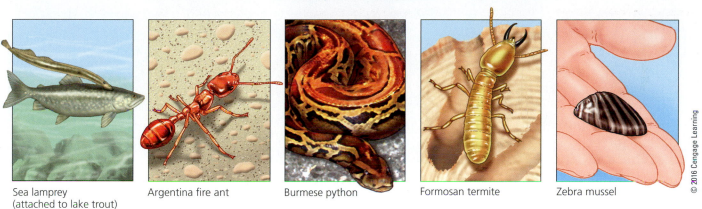

Sea lamprey (attached to lake trout)

Argentina fire ant

Burmese python

Formosan termite

Zebra mussel

© 2016 Cengage Learning

FIGURE 8.6 These are some of the estimated 7,100 harmful invasive species that have been deliberately or accidentally introduced into the United States.

Scott Ludwig has reported some success in using tiny parasitic flies to reduce fire ant populations. But more research is needed to see how well this approach will work.

Other troublesome invasive species include Burmese pythons, African pythons, and several species of boa constrictors, all of which have ended up in the Florida Everglades. About a million of these snakes, imported from Africa and Asia, have been sold as pets. Some buyers, after learning that these reptiles do not make good pets, have dumped them into the wetlands of the Everglades.

The Burmese python (Figure 8.7) can live 20–25 years, growing as long as 5 meters (16 feet). It can weigh as much 77 kilograms (170 pounds) and be as big around as a telephone pole. Pythons are hard to find and kill and they reproduce rapidly. They have huge appetites and feed at night, eating a variety of birds and mammals (such as rabbits, foxes, raccoons, and white-tailed deer) and occasionally other reptiles, including the American alligator—a keystone species in the Everglades ecosystem (see Chapter 4, Case Study, p. 64). They seize their prey with their

sharp teeth, wrap themselves around the prey, and squeeze them to death before feeding on them. They have also been known to eat pet cats and dogs, small farm animals, and geese. Research indicates that predation by these snakes is altering the complex food webs and ecosystem services of the Everglades.

Researchers say that the Burmese python population in Florida's wetlands cannot be controlled. In 2014, researchers Shannon Pittman and Kristen Hart found that trapping and moving the snakes from one area to another is not a viable control strategy because the snakes have the ability to return to the areas where they were captured. Some researchers warn that the Burmese python could spread to other swampy wetlands in the southern half of the United States.

Some invasive species, such as *kudzu* (Figure 8.6), have been deliberately introduced into ecosystems. In the 1930s, this plant was imported and planted in the southeastern United States to help control soil erosion. Kudzu does control erosion, but it grows so rapidly that it can engulf hillsides, gardens, trees, stream banks, cars, and anything else

Python populations are rapidly expanding in the Florida Everglades.

AP images/Michael R. Rochford/University of Florida

FIGURE 8.7 University of Florida researchers hold a 4.6-meter-long (15-foot-long), 74-kilogram (162-pound) Burmese python captured in Everglades National Park shortly after it had eaten a 1.8-meter (6-foot) long American alligator.

in its path. This plant—sometimes called "the vine that ate the South"—has spread throughout much of the southeastern United States and could spread to the north if the climate gets warmer as scientists project.

Bioinvaders also affect aquatic systems and are blamed for about two-thirds of all fish extinctions in the United States since 1990. The Great Lakes of North America have been invaded by at least 180 nonnative species and the number keeps rising. One of the biggest threats is the fish-killing sea lamprey (see Figure 5.4, p. 88), which has depleted some Great Lakes populations of important sport fish species such as lake trout.

Another invader is a thumbnail-sized mollusk called the *zebra mussel* (Figure 8.6), which reproduces rapidly and has no known natural enemies in the Great Lakes. It has displaced other mussel species and depleted the food supply for some native aquatic species. The mussels have also caused massive economic damages by clogging irrigation pipes, shutting down cooling water intake pipes for power plants and city water supplies, jamming ship rudders, and growing in large masses on boat hulls, piers, and other exposed aquatic surfaces.

Prevention Is the Best Way to Reduce Threats from Invasive Species

Once a harmful nonnative species becomes established in an ecosystem, removing it is almost impossible—somewhat like trying to collect smoke after it has come out of a chimney. Americans are paying more than $160 billion a year to eradicate or control an increasing number of invasive species, without much success. Clearly, the best way to limit the harmful impacts of nonnative species is to prevent them from being introduced into ecosystems.

Scientists suggest several ways to do this, including:

- Funding an intensive research program to identify the major characteristics of successful invaders, the types of ecosystems that are vulnerable to invaders, and the natural predators, parasites, bacteria, and viruses that could be used to control populations of established invaders.

- Greatly increasing ground surveys and satellite observations to track invasive plant and animal species, and continuing the development of better models for predicting how they will spread and what harmful effects they might have.

- Identifying major harmful invader species and establishing international treaties banning their transfer from one country to another, as is now done for endangered species, while stepping up inspection of imported goods to enforce such bans.

- Educating the public about the effects of releasing exotic plants and pets into the environment near where they live.

Figure 8.8 shows some of the things you can do to help prevent or slow the spread of harmful invasive species.

Population Growth, High Rates of Resource Use, Pollution, and Climate Change All Contribute to Species Extinctions

Past and projected *human population growth* and rising rates of *resource use per person* have greatly expanded the human ecological footprint (see Figure 1.10, p. 12). People have eliminated, degraded, and fragmented vast areas of wildlife habitat as they have spread out all over the planet, using resources at increasing rates wherever they go (Figure 8.5), and this has caused the extinction of many species (**Concept 8.3**).

Pollution also threatens some species with extinction, as has been shown by the unintended effects of certain pesticides. According to the USFWS, each year, pesticides kill about one-fifth of the European honeybee colonies that pollinate almost a third of all U.S. food crops (**Core Case Study** and Science Focus 8.2). The USFWS estimates that pesticides also kill more than 67 million birds and 6–14 million fish each year, and they threaten about 20% of the country's endangered and threatened species.

During the 1950s and 1960s, populations of fish-eating birds such as ospreys, brown pelicans, and bald eagles plummeted. The concentration of a chemical derived from the pesticide DDT was magnified as it moved up through their food web (Figure 8.9). The chemical made these top predator birds' eggshells so fragile that they could not reproduce successfully. Also hard hit in those years were predatory birds such as the prairie falcon, sparrow hawk, and peregrine falcon, which help to control populations of rabbits, ground squirrels, and other crop eaters. Since the U.S. ban on DDT in 1972, most of these bird species **GOOD NEWS** have made a comeback.

According to a study by Conservation International, projected *climate change* could help to drive a quarter to half of all land animals and plants to extinction by the end of this century. For example, scientific studies indicate that the polar bear is threatened because of higher temperatures and melting sea ice in its polar habitat. The shrinkage of floating ice is making it harder for polar bears to find prey such as seals, which use the ice as a platform for diving to find food and to give birth to their pups (Figure 8.10). According to the IUCN, the world's total polar bear population is likely to decline by 30–35% by 2050. By the end of this century, polar bears might be found only in zoos.

The Illegal Killing, Capturing, and Selling of Wild Species Threatens Biodiversity

Some protected species are illegally killed (poached) for their valuable parts or are sold live to collectors. Globally, this illegal trade in wildlife brings in an average of at least

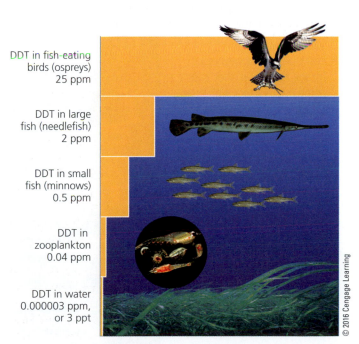

DDT in fish-eating birds (ospreys) 25 ppm

DDT in large fish (needlefish) 2 ppm

DDT in small fish (minnows) 0.5 ppm

DDT in zooplankton 0.04 ppm

DDT in water 0.000003 ppm, or 3 ppt

© 2016 Cengage Learning

FIGURE 8.9 *Bioaccumulation and biomagnification:* DDT is a fat-soluble chemical that can accumulate in the fatty tissues of animals. In a food chain or food web, the accumulated DDT is biologically magnified in the bodies of animals at each higher trophic level, as it was in the case of a food chain in the U.S. state of New York, illustrated here. **Question:** How does this story demonstrate the value of pollution prevention?

What Can You Do?

Controlling Invasive Species

- Do not buy wild plants and animals or remove them from natural areas.

- Do not release wild pets in natural areas.

- Do not dump aquarium contents or unused fishing bait into waterways or storm drains.

- When camping, use only local firewood.

- Brush or clean pet dogs, hiking boots, mountain bikes, canoes, boats, motors, fishing tackle, and other gear before entering or leaving wild areas.

© 2016 Cengage Learning

FIGURE 8.8 **Individuals matter:** Some ways to prevent or slow the spread of harmful invasive species. **Questions:** Which two of these actions do you think are the most important ones to take? Why? Which of these actions do you plan to take?

HONEYBEE LOSSES: A SEARCH FOR CAUSES

Each winter since 2006, 30–35% of the U.S. population of European honeybees has disappeared because of colony collapse disorder (**Core Case Study**). This problem also occurs in parts of Europe, China, and India.

Scientific research has found a number of possible of reasons for this decline, including:

- The *varroa mite*, a parasitic insect, can weaken and kill honeybee adults by feeding on their blood and larvae. It has killed millions of bees since it appeared in the U.S. in 1987—probably from infected bees imported from South America.
- *Harmful interactions between viruses and fungi* found in European honeybees and in almost all bee colonies suffering from colony collapse disorder can kill bees by weakening their immune systems.
- *Pesticides.* As honeybees forage for nectar, they can come into contact with insecticides that they sometimes carry back to their hives. Some research indicates that several types of widely used *neonicotinoids,* which are derived from nicotine that tobacco plants use as a natural insecticide, may play a role in CCD. These pesticides are deliberately incorporated into food plants so they end up in the nectar and pollen that the bees take back to their colonies. Some studies suggest that exposure to trace amounts of these chemicals can disrupt the nervous systems of bees, decrease their ability to find their

way back to their hives, weaken their immune systems, and make them more susceptible to mites, viruses, and fungi. The explosive growth in the use of neonicotinoids since 2005 has roughly tracked the rise in CCD since 2006. Pesticide manufacturers deny that the levels of neonicotinoids and other pesticides that bees are exposed to are high enough to cause harm if the chemicals are used as directed.

- *Stress and poor nutrition,* for some of the bees that are transported long distances around the United States for the industrial pollination business (Figure 8.A), can weaken the bees' immune systems and make them more vulnerable to death due to mites, viruses, fungi, and pesticides.

- *Lack of natural foraging areas.* Widespread conversion of grasslands to crop fields has dramatically reduced the diversity of natural plants on which bees naturally thrive.

The growing consensus among bee researchers is that the likely cause of colony collapse disorder is a combination of the often-interacting factors listed here. Some scientists see the decline of honeybees as an indicator of the environmental degradation of our life-support system.

Critical Thinking

Can you think of some ways in which commercial beekeepers could lessen one or more of the threats described here? Explain.

FIGURE 8.A European honeybee hive boxes in an acacia orchard. Each year, commercial beekeepers rent and deliver several million hives by truck to farmers throughout the United States.

Cristi111 | Dreamstime.com

FIGURE 8.10 On floating ice in the Arctic sea, this polar bear has killed a bearded seal, one of its major sources of food. **Question:** Do you think it matters that the polar bear may become extinct in the wild during this century primarily because of human activities? Explain.

$1.8 million an hour and at least two-thirds of all live animals smuggled around the world die in transit. Few of the smugglers are caught or punished.

To poachers, a highly endangered, live eastern mountain gorilla (of which there are about 720 left in the wild) is worth $150,000, and the pelt of an endangered giant panda (1,600 to 3,000 left in the wild in China) can bring $100,000. A single poached black rhinoceros horn (Figure 8.11) can be worth as much as $500,000 by the time it reaches the black market in Asia. Four of the five rhino species are critically endangered, mostly because so many have been illegally killed for their horns.

The illegal killing of elephants, especially African elephants (see chapter-opening photo) for their valuable ivory tusks has increased in recent years, despite an international ban on the trade of ivory. The two ivory tusks on an adult male elephant can be worth $375,000 on the black market. In 2013, according to WildAid, elephants were being killed for their ivory at a rate of about 30,000 per year, or one about every 17 minutes, on average. China has the largest market for illegal ivory, followed by the United States.

In 2014, researchers were testing the use of small aerial drones to find illegal poachers during the night by detecting the presence of their weapons. The drones could deter poachers and expose their locations to wildlife rangers by using bright strobe lights when they detected the poachers.

In 1900, there were an estimated 100,000 tigers in the wilds of Asia in a rapidly shrinking range (Figure 8.5, upper left). Today there are about 3,200, half of them in India. This is mostly due to a 90% loss of habitat, caused by rapid human population growth, and poaching, much of which is motivated by poverty. The Indian, or Bengal, tiger is at risk because a coat made from its fur can sell for as much as $100,000 in Tokyo. The bones and penis of a single tiger can fetch as much as $70,000 in China, the world's biggest market for such illegal items. According to the World Wildlife Fund, without emergency action to curtail poaching and preserve tiger habitat, few if any tigers, including the Sumatran tiger (Figure 8.2d), will be left in the wild by 2022.

In India, conservation biologist and National Geographic Emerging Explorer Krithi Karanth is studying conflicts between the rapidly growing number of humans and the declining populations of wildlife such as tigers and Asian elephants. As habitat for these animals shrinks, they damage farmers' crops while trying to feed themselves. Karanth has visited more than 8,000 villages in India, and also involves 500 "citizen scientists" to help her interview villagers and collect data. Her goals are to document the disappearance of wildlife and the conflicts between humans and wildlife, and to find effective ways to reduce such conflicts.

FIGURE 8.11 A poacher in South Africa killed this critically endangered northern white rhinoceros for its two horns. This species is now extinct in the wild. **Question:** What would you say if you could talk to the poacher who killed this animal?

Photoshot Holdings Ltd/Alamy

CONSIDER THIS. . .

THINKING ABOUT Tigers

Would it matter to you if all of the world's wild tigers were to disappear? Explain. List two steps you could take to help protect the world's remaining wild tigers from extinction.

Around the globe, the legal and illegal trade in wild species for use as pets is a huge and very profitable business. Many owners of wild pets do not know that, for every live animal captured and sold in the pet market, many others are killed or die in transit. According to the IUCN, more than 60 bird species, mostly parrots, are endangered or threatened because of the wild bird trade (see the Case Study that follows).

Buyers of wild animals might also be unaware that some imported exotic animals carry diseases such as Hantavirus, Ebola virus, Asian bird flu, herpes B virus (carried by most adult macaques), and salmonella (from pets such as hamsters, turtles, and iguanas). These diseases can spread quite easily from pets to their owners and then to other people.

Other wild species whose populations are depleted because of the pet trade include many amphibians (see Chapter 4, Core Case Study, p. 64), various reptiles, and tropical fishes taken mostly from the coral reefs of Indonesia and the Philippines. Some divers catch tropical fish by using plastic squeeze bottles of poisonous cyanide to stun them. For each fish caught alive, many more die. In addition, the cyanide solution kills the polyps, the tiny animals that create the reef.

Some exotic plants, especially orchids and cacti (see Figure 7.11, center, p. 138), are endangered because they are gathered, often illegally, and sold to collectors to decorate houses, offices, and landscapes. A mature crested saguaro cactus can earn a cactus rustler as much as $15,000, and an orchid collector might pay $5,000 for a single rare orchid.

According to the USFWS, collectors of exotic birds might pay $10,000 or more for an endangered hyacinth macaw smuggled out of Brazil. However, during its lifetime, a single hyacinth macaw left in the wild could attract an estimated $165,000 in ecotourism revenues.

A Rising Demand for Bushmeat Threatens Some African Species

For centuries, indigenous people in much of West and Central Africa have sustainably hunted wildlife for *bushmeat* as a source of food. But in the last three decades, bushmeat hunting in some areas has skyrocketed as hunters have tried to provide food for rapidly growing populations or to make a living by supplying restaurants in major cities with exotic meats from gorillas (Figure 8.12) and other species. Logging roads in once-inaccessible forests have made such hunting much easier.

Bushmeat hunting has driven at least one species— Miss Waldron's red colobus monkey—to complete extinction. It is also a factor in the reduction of some populations

FIGURE 8.12 *Bushmeat* such as this severed head of an endangered lowland gorilla in the Congo is consumed as a source of protein by local people in parts of West and Central Africa and is sold in national and international marketplaces and served in some restaurants where wealthy patrons regard gorilla meat as a source of status and power. **Question:** How, if at all, is this different from killing a cow for food?

Photoshot Holdings Ltd/Alamy

of orangutans (Figure 8.4), gorillas, chimpanzees, elephants, and hippopotamuses. Another problem is that butchering and eating some forms of bushmeat has helped to spread fatal diseases such as HIV/AIDS and the Ebola virus from animals to humans.

The U.S. Agency for International Development (USAID) is trying to reduce unsustainable hunting for bushmeat in some areas of Africa by introducing alternative sources of food, including farmed fish. They are also showing villagers how to breed large rodents such as cane rats as a source of protein.

A Disturbing Message from the Birds

Approximately 70% of the world's more than 10,000 known bird species are declining in numbers, and much of this decline is related to human activities, summarized by HIPPCO. According to the IUCN 2013 Red List of Endangered Species, roughly one of every eight (13%) of all bird species is threatened with extinction mostly by habitat loss, degradation, and fragmentation (the H in HIPPCO)—primarily in tropical forests.

According to a 2011 study, *State of the Birds,* almost one-third of the more than 800 bird species in the United States are endangered (Figure 8.2b and c), threatened, or in decline, mostly because of habitat loss and degradation, invasive species, and climate change. Sharp declines in bird populations have occurred among songbird species that migrate long distances. These birds nest deep in North American woods in the summer and spend their winters in Central or South America or on the Caribbean Islands. The primary causes of these population declines

appear to be habitat loss and fragmentation of the birds' breeding habitats in North America and Central and South America. In addition, the populations of 40% of the world's water birds are in decline because of the global loss of wetlands.

After habitat loss, the intentional or accidental introduction of nonnative species such as bird-eating rats is the second greatest danger, affecting about 28% of the world's threatened birds. Other such invasive species (the I in HIPPCO) include snakes (such as the brown tree snake) and mongooses. In the United States, feral cats and pet cats kill at least 1.4 billion birds each year, according to a 2012 study by Peter Mara of the Smithsonian Conservation Biology Institute.

Population growth, the first P in HIPPCO, also threatens some bird species, as more people spread out over the landscape each year, increasing their use of timber, food, and other resources, which results in destruction or disturbance of bird habitats. Pollution, the second P in HIPPCO, is another major threat to birds. Countless birds are exposed to oil spills, insecticides, herbicides, and toxic lead from shotgun pellets that fall into wetlands and from lead sinkers left by anglers. (However, the use of lead in shotgun pellets and sinkers is being phased out.)

Another rapidly growing threat to birds is climate change, the C in HIPPCO. A study done for the WWF found that the effects of climate change, such as heat waves and flooding, are causing declines of some bird populations in every part of the globe. Such losses are expected to increase sharply during this century.

Overexploitation (the O in HIPPCO) is also a major threat to bird populations. Fifty-two of the world's 388 parrot species are threatened, partly because so many parrots are captured (often illegally) for sale as pets, usually to

Çağan Hakkı Şekercioğlu: Protector of Birds and National Geographic Emerging Explorer

Çağan Şekercioğlu, assistant professor in the University of Utah Department of Biology, is a bird expert, a tropical biologist, an accomplished wildlife photographer, and a National Geographic Emerging Explorer. He has seen over 64% of the planet's known bird species in 75 countries, developed a global database on bird ecology, and become an expert on the causes and consequences of bird extinctions around the world. One example of an endangered species he has studied is the great green macaw of South and Central America (shown here in the background photo).

In 2007 Şekercioğlu founded KuzeyDoğa, an award-winning ecological research and community-based conservation organization to help conserve and protect the wildlife of northeastern Turkey. He also developed Turkey's first protected wildlife corridor, which would stretch across the eastern half of the country, according to his plan. In 2011, he was named as Turkey's Scientist of the Year.

Based on his extensive research Şekercioğlu estimates that the percentage of the world's known bird species that are endangered could approximately double from 13% in 2013 to 25% by the end of this century. He says, "My ultimate goal is to prevent extinctions and consequent collapses of critical ecosystem processes while making sure that human communities benefit from conservation as much as the wildlife they help conserve. . . . I don't see conservation as people versus nature, I see it as a collaboration."

Rebecca Hale/National Geographic Stock

Background photo: Courtesy of Dr. Şekercioğlu

---

buyers in Europe and the United States. At least 23 species of seabirds face extinction, largely due to the harmful effects of industrialized fishing. For example, many diving birds drown after becoming hooked on baited lines or trapped in huge nets that are set out by fishing boats.

Biodiversity scientists view this decline of bird species with alarm. One reason is that birds are excellent *indicator species* because they live in every climate and biome, respond quickly to environmental changes in their habitats, and are relatively easy to track and count. To these scientists, the decline of many bird species indicates widespread environmental degradation.

A second reason for alarm is that birds perform critically important economic and ecosystem services throughout the world. For example, many birds play specialized roles in pollination and seed dispersal, especially in tropical areas, so extinctions of these bird species could lead to extinctions of plants that depend on the birds for pollination. Then, some specialized animals that feed mostly on these plants might also become extinct. Such a *cascade of extinctions,* in turn, could affect our own food supplies and well-being. Biodiversity scientists (Individuals Matter 8.1) urge us to listen more carefully to what birds are telling us about the state of the environment, for the birds' sake, as well as for ours.

8.4 HOW CAN WE SUSTAIN WILD SPECIES AND THE ECOSYSTEM SERVICES THEY PROVIDE?

CONCEPT 8.4 We can reduce species extinction and sustain ecosystem services by establishing and enforcing national environmental laws and international treaties and by creating and protecting wildlife sanctuaries.

International Treaties and National Laws Can Help to Protect Species

Some countries have strong laws that help to sustain wild species and ecosystem services (see the Case Study that follows). Several international treaties and conventions also help to protect endangered and threatened wild species (**Concept 8.4**).

One of the most far reaching of international agreements is the 1975 *Convention on International Trade in Endangered Species (CITES).* This treaty, signed by 178 countries, bans the hunting, capturing, and selling of threatened or endangered species. It lists 926 species that are in danger of extinction and that cannot be commercially traded as live specimens or for their parts or products. It restricts the international trade of roughly 5,000 animal species and 29,000 plant species that are at risk of becoming threatened.

CITES has helped to reduce the international trade of many threatened animals, including elephants (see chapter-opening photo), crocodiles, cheetahs, and chimpanzees. But the effects of this treaty are limited because enforcement varies from country to country, and convicted violators often pay only small fines. Also, member countries can exempt themselves from protecting any listed species, and much of the highly profitable illegal trade in wildlife and wildlife products goes on in countries that have not signed the treaty.

Another important treaty is the *Convention on Biological Diversity (CBD)*, ratified or accepted by 193 countries (but as of 2014, not by the United States). It legally commits participating governments to reducing the global rate of biodiversity loss and to equitably sharing the benefits from use of the world's genetic resources. This includes efforts to prevent or control the spread of ecologically harmful invasive species.

This convention is a landmark in international law because it focuses on ecosystems rather than on individual species, and it links biodiversity protection to issues such as the traditional rights of indigenous peoples. However, because some key countries, including the United States, have not ratified it, implementation has been slow. Also, the law contains no severe penalties or other enforcement mechanisms.

CASE STUDY

The U.S. Endangered Species Act

The *Endangered Species Act of 1973* (ESA; amended in 1982, 1985, and 1988) was designed to identify and protect endangered species in the United States and abroad (**Concept 8.4**). This act is probably the most successful and far-reaching environmental law adopted by any nation, which has made it controversial.

Under the ESA, the National Marine Fisheries Service (NMFS) is responsible for identifying and listing endangered and threatened ocean species, while the U.S. Fish and Wildlife Service (USFWS) is to identify and list all other endangered and threatened species. Any decision by either agency to list or delist a species must be based on biological factors alone, without consideration of economic or political factors. However, the two agencies can use economic factors in deciding whether and how to protect endangered habitat and in developing recovery plans for listed species. The ESA also forbids federal agencies (except the Defense Department) to carry out, fund, or authorize projects that would jeopardize any endangered or threatened species or destroy or modify its critical habitat.

For offenses committed on private lands, fines as high as $100,000 and 1 year in prison can be imposed to ensure protection of the habitats of endangered species. Although this provision has rarely been used, it has been controversial because at least 90% of the listed species live totally or partially on private land. Since 1982, the ESA has been amended to give private landowners various economic incentives to help save endangered species living on their lands. The ESA also makes it illegal for Americans to sell or buy any product made from an endangered or threatened species or to hunt, kill, collect, or injure such species in the United States.

Between 1973 and April 2014, the number of U.S. species on the official endangered and threatened species lists increased from 92 to 1,524. According to a study by the Nature Conservancy, about 33% of the country's species are at risk of extinction, and 15% of all species are at high risk—far more than the current number listed.

All seven species of sea turtles are classified as critically endangered or threatened under the ESA. Much of the beach habitat on which they lay their eggs has been taken over by human activities, and poachers often take their eggs. Many turtles have choked to death on plastic bags that they mistake for jellyfish. Others have drowned or starved to death after becoming entangled in fishing lines and nets (Figure 8.13) or in lobster or crab traps.

Behavioral ecologist and National Geographic Explorer Katsufumi Sato, of the University of Tokyo's Atmosphere and Ocean Research Institute, uses tiny electronic data loggers to record images and detailed movements of animals (still pictures and videos), including endangered sea turtles. These devices, attached to turtles' shells, can record photos, videos, and data about the migrations of the turtles. The data loggers are retrieved to obtain and analyze data about the turtles, including their migration routes and dive profiles. These data can help scientists to learn more about the plight of the turtles and how we can help to protect them.

According to a 2013 study by the Center for Biological Diversity (CBD), 90% of the ESA-protected species are recovering at the rate projected in their recovery plans and 99% of the listed species have been saved from extinction. And between 2003 and 2014, the cumulative area designated as critical habitats increased almost tenfold. Successful recoveries plans include those for the American alligator (see Chapter 4 Case Study, p. 69), the gray wolf, the peregrine falcon, the bald eagle, the humpback whale, and the brown pelican. **GOOD NEWS**

The ESA also requires that all commercial shipments of wildlife and wildlife products enter or leave the country through one of 17 designated airports and ocean ports. The 120 full-time USFWS inspectors can inspect only a small fraction of the more than 200 million wild animals brought legally into the United States annually. Each year, tens of millions of wild animals are also brought in illegally, but few illegal shipments of endangered or threatened animals or plants are confiscated.

Since 1995, there have been numerous efforts to weaken the ESA and to reduce its already meager annual budget. Opponents of the act contend that it puts the rights and welfare of endangered plants and animals above those of people. Some critics would do away with this act.

Fishing nets entangle and kill many sea turtles.

Mark Foley/State Archives of Florida, Florida Memory

FIGURE 8.13 This endangered green sea turtle died after being caught in a fishing net.

They call it an expensive failure because only 26 species have recovered enough to be removed from the endangered species list, and because 10 became extinct while on the list, even though eight of those species were very likely extinct when they were put on the list.

Most biologists insist that the act has not been a failure and indeed, is one of the world's most successful environmental laws, for several reasons. *First*, species are listed only when they face serious danger of extinction. ESA supporters argue that this is similar to a hospital emergency room set up to take only the most desperate cases, often with little hope for recovery. Such a facility could not be expected to save all or even most of its patients.

Second, according to federal data, the conditions of more than half of the listed species are stable or improving, 90% are recovering at rates specified by their recovery plans, and 99% of the protected species are still surviving. A hospital emergency room having similar results would be considered to be an astounding success story.

Third, it takes many decades for a species to reach the point where it is in danger of extinction. Thus, it takes many decades to bring a species back to the point where it can be removed from the endangered list.

Fourth, the 2013 budget for protecting endangered species amounted to an average expenditure of about 61 cents per U.S. citizen, and the budget has not increased significantly since then. To ESA supporters, it is amazing that the federal agencies responsible for enforcing the act have managed to stabilize or improve the conditions of 99% of the listed species on such a small budget.

A national poll conducted by the CBD found that two out of three Americans want the ESA strengthened or left alone. However, in 2014, some members of Congress were attempting to gut the law by supporting legislation that would **(1)** require state and congressional approval for adding any new species to the protected list, which would essentially make it impossible to list a new species, **(2)** automatically delist any species after 5 years—a way to phase out this popular law without voting to repeal it, and **(3)** allow state governors to decide whether and how their states would obey ESA regulations.

Some ESA supporters agree that the act can be improved. They cite a U.S. National Academy of Sciences study that recommended three major changes in the way the law is being implemented, in order to make it more scientifically sound and effective:

- Greatly increase the meager funding for implementing the act.

- Put greater emphasis on developing recovery plans more quickly.

- When a species is first listed, establish the core of its habitat as critical for its survival and give that area the maximum protection.

Other suggestions include providing more technical and financial assistance to landowners who want to help protect endangered species on their property. At least 80% of the habitat for about half of the protected species is on privately owned land.

We Can Establish Wildlife Refuges and Other Protected Areas

In 1903, President Theodore Roosevelt established the first U.S. federal wildlife refuge at Pelican Island, Florida, to help protect the brown pelican and other birds from extinction (Figure 8.14). This approach worked well. In 2009, the brown pelican was removed from the U.S. Endangered Species list. By 2013, there were more than 560 refuges in the National Wildlife Refuge System. Each year, more than 47 million Americans visit these refuges to hunt, fish, hike, and watch birds and other wildlife.

More than three-fourths of the refuges serve as wetland sanctuaries that are vital for protecting migratory waterfowl. At least one-fourth of all U.S. endangered and threatened species have habitats in the refuge system, and some refuges have been set aside specifically for certain endangered species (**Concept 8.4**). Such areas have helped Florida's key deer, the brown pelican, and the trumpeter swan to recover.

There is also bad news about the U.S. refuge system. According to a General Accounting Office study, activities that are harmful to wildlife, such as mining, oil drilling, and use of off-road vehicles, take place in nearly 60% of the nation's wildlife refuges. Biodiversity researchers urge the U.S. government to set aside more refuges and to increase the long-underfunded budget for the refuge system.

Elsewhere in the world, reserves and refuges have also been successful, and public awareness has played a big role in their success. Dereck and Beverly Joubert are National Geographic Explorers and award-winning filmmakers who, for more than 30 years, have been studying, filming, and writing about threatened lions, leopards, and cheetahs, and other big-cat predators in Africa, hoping to heighten public awareness of the plight of these animals. Their efforts have contributed to the establishment of protected reserves for big cats and other African wildlife in Botswana, Tanzania, and Kenya.

National Geographic is funding several other efforts to preserve wild species, including that of Maia Raymundo, who is studying a critically endangered species of fruit bat in the Philippines, threatened by hunting and high rates of deforestation. Conservation biologists are alarmed about the steep decline of many bat species. Some populations of these fruit bats have declined by as much as 98% in large areas of their range. Raymundo's goal is to identify and protect critical habitat areas for the endangered bats.

Seed Banks, Botanical Gardens, and Wildlife Farms Can Help to Protect Species

We can use *seed banks* to preserve genetic information and endangered plant species, storing their seeds in refrigerated, low-humidity environments. More than 1,000 seed banks around the world collectively hold about 3 million samples.

FIGURE 8.14 The Pelican Island National Wildlife Refuge in Florida was America's first National Wildlife Refuge.

Some species cannot be preserved in seed banks, which vary in quality, are expensive to operate, and are vulnerable to destruction by fire or other mishaps. However, the Svalbard Global Seed Vault, an underground facility on a remote island in the Arctic, will eventually contain 100 million of the world's seeds. It is not vulnerable to fires, storms, or war.

The world's 1,600 *botanical gardens* and *arboreta* contain living plants that represent almost one-third of the world's known plant species. But they contain only about 3% of the world's rare and threatened plant species and have too little space and funding to preserve most of those species.

We can take pressure off some endangered or threatened species by raising individuals of these species on *farms* for commercial sale. In Florida, for example, alligators are raised on farms for their meat and hides. Butterfly farms established to raise and protect endangered species flourish in Papua New Guinea, where many butterfly species are threatened by development activities. These farms are also used to educate visitors about the need to protect butterfly species.

Zoos and Aquariums Protect Some Species

Zoos, aquariums, game parks, and animal research centers are being used to preserve some individuals of critically endangered animal species, with the long-term goal of reintroducing the species into protected wild habitats.

Two techniques for preserving endangered terrestrial species are *egg pulling* and *captive breeding*. Egg pulling involves collecting wild eggs laid by critically endangered bird species and then hatching them in zoos or research centers. In captive breeding, some or all of the wild individuals of a critically endangered species are collected for breeding in captivity, with the aim of reintroducing the offspring into the wild. Captive breeding has been used to save the peregrine falcon and the California condor (Figure 8.2b).

Other techniques for increasing the populations of captive species include artificial insemination, embryo transfer (surgical implantation of eggs of one species into a surrogate mother of another species), use of incubators, and cross fostering (in which the young of a rare species are raised by parents of a similar species). Scientists also match individuals for mating by using DNA analysis along with computer databases that hold information on family lineages of endangered zoo animals—a computer dating service for zoo animals.

The ultimate goal of captive breeding programs is to build populations to a level where they can be reintroduced into the wild. Successes include the black-footed ferret, the golden lion tamarin (a highly endangered monkey species), the Arabian oryx, and the California condor (Figure 8.2b). However, most reintroductions fail because of a lack of suitable habitat, an inability of the individuals bred in captivity to survive in the wild, renewed overhunting or poaching, or pollution and other hazards in the environment.

One problem for captive breeding programs is that a captive population of an endangered species must typically number 100–500 individuals in order to avoid extinction resulting from accidents, diseases, or the loss of genetic diversity through inbreeding. Recent genetic research indicates that 10,000 or more individuals are needed for an endangered species to maintain its capacity for biological evolution. Zoos and research centers do not have the funding or space to house such large populations.

Public aquariums that exhibit unusual and attractive species of fish and marine animals such as seals and dolphins help to educate the public about the need to protect such species. However, mostly because of limited funds, public aquariums have not served as effective gene banks for endangered marine species, especially marine mammals that need large volumes of water.

Efforts to Protect Species Raise Difficult Questions

Efforts to prevent the extinction of wild species and the accompanying losses of ecosystem services require the use of financial and human resources that are limited. This raises some challenging questions:

- Should we focus on protecting species or should we focus more on protecting ecosystems and the ecosystem services they provide?

- How do we allocate limited resources between these two priorities?

- How do we decide which species should get the most attention in our efforts to protect as many species as possible? For example, should we focus on protecting the most threatened species or on protecting keystone species?

- Protecting species that are appealing to humans, such as panda bears and orangutans (Figure 8.4), can increase public awareness of the need for wildlife conservation. Is this more important than focusing on the ecological importance of species when deciding which ones to protect?

- How do we determine which habitat areas are the most critical to protect?

- How do we allocate limited resources among such biodiversity hotspots?

Conservation biologists struggle with these questions all the time. Regardless of the answers, each of us can help in the efforts to protect species from extinction due largely to human activities. Figure 8.15 lists some guidelines that you can follow to play your part and to help protect species and to increase your beneficial environmental impact.

`GOOD NEWS`

What Can You Do?

Protecting Species

- Do not buy furs, ivory products, or other items made from endangered or threatened animal species.

- Do not buy wood or wood products from tropical or old-growth forests.

- Do not buy pet animals or plants taken from the wild.

- Tell friends and relatives what you're doing about this problem.

© 2016 Cengage Learning

FIGURE 8.15 Individuals matter: You can help to prevent the extinction of species. *Questions:* Which two of these actions do you believe are the most important ones to take? Why?

BIG IDEAS

- We are hastening the extinction of wild species and degrading the ecosystem services they provide by destroying and degrading natural habitats, introducing harmful invasive species, and increasing human population growth, pollution, climate change, and overexploitation.

- We should avoid causing or hastening the extinction of wild species because of the ecosystem and economic services they provide and because their existence should not depend primarily on their usefulness to us.

- We can work to prevent the extinction of species and to protect overall biodiversity and ecosystem services by establishing and enforcing environmental laws and treaties and by creating and protecting wildlife sanctuaries.

Honeybees and Sustainability

In this chapter, we learned about the human activities that are hastening the extinction of many species and about how we might curtail those activities. We learned that there is considerable evidence that as many as half of the world's wild species could go extinct during this century, largely as a result of human activities that threaten many species and some of the vital ecosystem services they provide. For example, populations of honeybees, vital for pollinating crops that supply much of our food, have been declining for a variety of reasons (**Core Case Study**), many of them related to human activities. One of the key reasons for such problems is that most people are simply

Malwina Szweda/Shutterstock.com

unaware of the highly valuable ecosystem and economic services provided by the earth's species.

In keeping with two of the three **scientific principles of sustainability**, by acting to prevent the extinction of species as a result of human activities, we would be helping to preserve not only the earth's biodiversity, but also the vital ecosystem services that sustain us, including chemical cycling. And it is not only for these species that we ought to act, but also for the long-term health and well-being of our own species in keeping with two of the three **social science principles of sustainability** (see Supplement 7, p. S51).

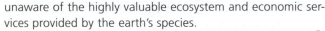

Chapter Review

Core Case Study

1. What economic and ecological services do honeybees provide? How are human activities contributing to the decline of many populations of European honeybees? What is **colony collapse disorder (CCD)**?

Section 8.1

2. What is the key concept for this section? Define and distinguish between **biological extinction** and **mass extinction**. What is the **background extinction rate**, and how do estimated current and projected extinction rates compare with it? What percentage of the earth's land and what percentage of the earth's oceans have been disturbed by human activities? Explain how scientists estimate extinction rates and describe the challenges they face in doing so. Give three reasons why many extinction experts believe that projected extinction rates are probably on the low side. What percentage of the world's species are likely to go extinct, largely as a result of human activities, during this century? Distinguish between **endangered species** and **threatened species** and give an example of each. List four characteristics that make some species especially vulnerable to extinction.

Section 8.2

3. What is the key concept for this section? What are three reasons for trying to avoid hastening the extinction of wild species? Describe two economic and two ecological benefits of species diversity. Explain how saving other species and the ecosystem services they provide can help us to save our own species and our cultures and economies.

Section 8.3

4. What is the key concept for this section? What is **HIPPCO**? What is the greatest threat to wild species? What is **habitat fragmentation**? Describe the major effects of habitat loss and fragmentation. Why are island species especially vulnerable to extinction? What are habitat islands?

5. Give two examples of the benefits that have been gained by the introduction of nonnative species. Give two examples of the harmful effects of nonnative species that have been introduced deliberately and two examples of the same for accidentally introduced species. Explain why prevention is the best way to reduce threats from invasive species and list four proposed ways to implement this strategy.

6. Summarize the roles of population growth, overconsumption, pollution, and climate change in the extinction of wild species. Explain how concentrations of pesticides such as DDT can accumulate to high levels in food webs. List possible causes of the decline of European honeybee populations in the United States.

Give three examples of species that are threatened by poaching. Why are wild tigers likely to disappear within a few decades? What is the connection between infectious diseases in humans and the pet trade? Describe the threat to some forms of wildlife from the increased hunting for bushmeat.

7. List the major threats to the world's bird populations and give two reasons for protecting bird species from extinction. Summarize environmental scientist Çağan Şekercioğlu's contributions to our understanding of the ecological importance of birds and threats to their extinction.

Section 8.4

8. What is the key concept for this section? Name two international treaties that are used to help protect species. What is the U.S. Endangered Species Act? How successful has it been, and why is it controversial?

9. Summarize the roles and limitations of wildlife refuges, gene banks, botanical gardens, wildlife farms, zoos, and aquariums in protecting some species. Describe the role of captive breeding in efforts to prevent species extinction and give an example of success in returning a nearly extinct species to the wild. What are three important questions related to protecting wild species from extinction by human activities?

10. What are this chapter's *three big ideas*? Why is it in keeping with two of the three **scientific principles of sustainability** to protect honeybees and other wild species from extinction along with protecting the ecosystem services provided by species? What two **social science principles of sustainability** are involved in protecting species from extinction due to human activities?

Note: Key terms are in bold type.

Critical Thinking

1. What are three aspects of your lifestyle that might directly or indirectly contribute to declines in European honeybee populations and the endangerment of other pollinator species (**Core Case Study**)?

2. Give your response to the following statement: "Eventually, all species become extinct. So it does not really matter that the world's remaining tiger species or a tropical forest plant are endangered mostly because of human activities." Be honest about your reaction, and give arguments to support your position.

3. Do you accept the ethical position that each species has the inherent right to survive without human interference, regardless of whether it serves any useful purpose for humans? Explain. Would you extend this right to the *Anopheles* mosquito, which transmits malaria, and to harmful infectious bacteria? Explain. If your answer is no, where would you draw the line?

4. Wildlife ecologist and environmental philosopher Aldo Leopold wrote this with respect to preventing the extinction of wild species: "To keep every cog and wheel is the first precaution of intelligent tinkering." Explain how this statement relates to the material in this chapter.

5. What would you do if fire ants invaded your yard and house? Explain your reasoning behind your course of action. How might your actions affect other species or the ecosystem you are dealing with?

6. How do you think your daily habits might contribute directly or indirectly to the extinction of some bird species? What are three things that you think should be done to reduce the rate of extinction of bird species?

7. Which of the following statements best describes your feelings toward wildlife?
   a. As long as it stays in its space, wildlife is okay.
   b. As long as I do not need its space, wildlife is okay.
   c. I have the right to use wildlife habitat to meet my own needs.
   d. When you have seen one redwood tree, elephant, or some other form of wildlife, you have seen them all, so preserve a few of each species in a zoo or wildlife park and do not worry about protecting the rest.
   e. Wildlife should be protected in its current ranges.

8. How might your lifestyle change if human activities were to contribute to the extinction of 25–50% of the world's identified species during this century? How might this affect the lives of any children or grandchildren you eventually might have? List two aspects of your lifestyle that contribute to this threat to the earth's natural capital.

Doing Environmental Science

Identify examples of habitat destruction or degradation in the area in which you live or go to school. Try to determine and record any harmful effects that these activities have had on the populations of one wild plant and one animal species. (Name each of these species and describe how they have been affected.) Do some research on the Internet and/or in a school library on *wildlife management plans*, and then develop a management plan for restoring the habitats and species you have studied. Try to determine whether trade-offs are necessary with regard to the human activities you have observed, and account for these trade-offs in your management plan. Compare your plan with those of your classmates.

Global Environment Watch Exercise

Search for *Extinction*, and scroll to statistics on the portal's page. Click on "Known Causes of Animal Extinction since 1600." You will find four general categories of causes.

Thinking about history from 1600 through today, how do you think humans have changed their impact on species in each of these categories? Has the impact increased or decreased over this time period? Give specific examples of changes in this timeframe to support your answers.

Data Analysis

Examine the following data released by the World Resources Institute and answer these questions:

1. Complete the table by filling in the last column. For example, to calculate this value for Costa Rica, divide the number of threatened breeding bird species by the total number of known breeding bird species and multiply the answer by 100 to get the percentage.

2. Arrange the countries from largest to smallest according to total land area. Does there appear to be any correlation between the size of country and the percentage of threatened breeding bird species? Explain your reasoning.

Country	Total Land Area in Square Kilometers (Square Miles)	Protected Area as Percent of Total Land Area (2003)	Total Number of Known Breeding Bird Species (1992–2002)	Number of Threatened Breeding Bird Species (2002)	Threatened Breeding Bird Species as Percent of Total Number of Known Breeding Bird Species
Afghanistan	647,668 (250,000)	0.3	181	11	
Cambodia	181,088 (69,900)	23.7	183	19	
China	9,599,445 (3,705,386)	7.8	218	74	
Costa Rica	51,114 (19,730)	23.4	279	13	
Haiti	27,756 (10,714)	0.3	62	14	
India	3,288,570 (1,269,388)	5.2	458	72	
Rwanda	26,344 (10,169)	7.7	200	9	
United States	9,633,915 (3,718,691)	15.8	508	55	

Compiled by the authors using data from World Resources Institute, *Earth Trends, Biodiversity and Protected Areas, Country Profiles.*

CENGAGEbrain.com To access course materials, including Aplia homework, please visit www.cengagebrain.com.

WWW.CENGAGEBRAIN.COM   **183**

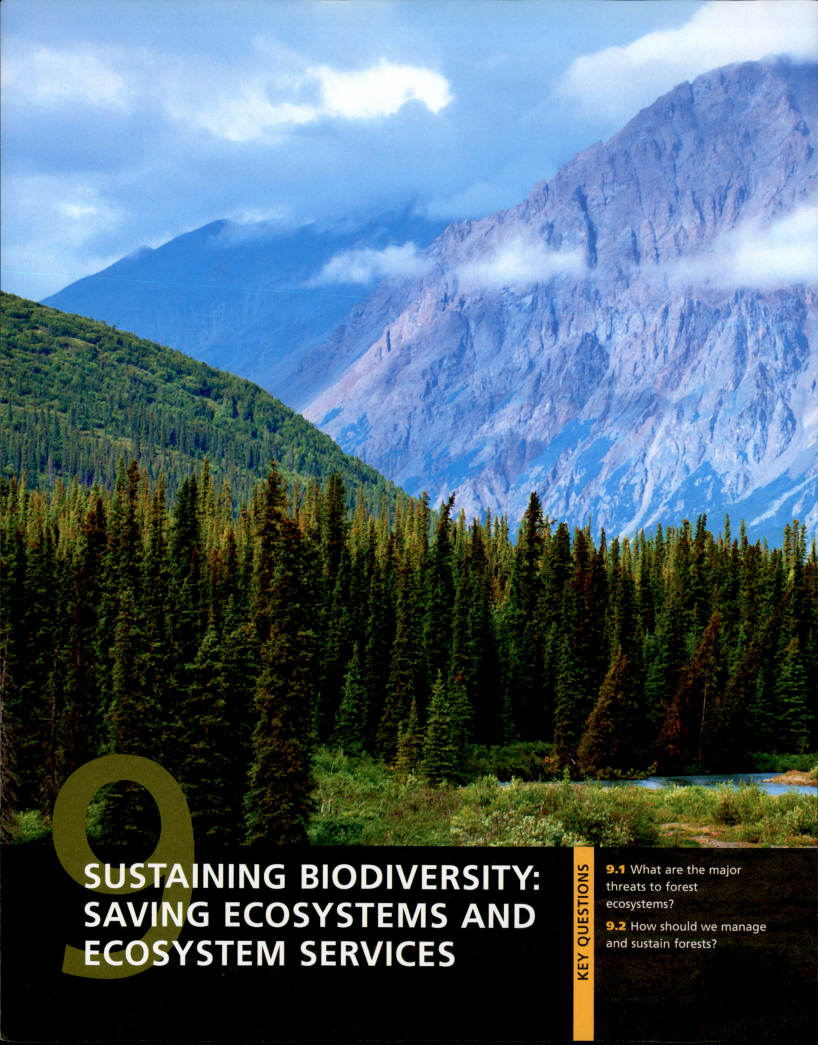

9

SUSTAINING BIODIVERSITY: SAVING ECOSYSTEMS AND ECOSYSTEM SERVICES

There is no solution, I assure you, to save Earth's biodiversity other than preservation of natural environments in reserves large enough to maintain wild populations sustainably.

EDWARD O. WILSON

9.3 How should we manage and sustain grasslands?

9.4 How should we manage and sustain parks and nature reserves?

9.5 What is the ecosystem approach to sustaining terrestrial biodiversity and ecosystem services?

9.6 How can we help to sustain aquatic biodiversity?

Denali National Park, Alaska (USA).
Jerryway | Dreamstime.com

Tropical forests once completely covered Central America's Costa Rica, which is smaller in area than the U.S. state of West Virginia and about one-tenth the size of France. Between 1963 and 1983, politically powerful ranching families cleared much of the country's forests to graze cattle.

Despite such widespread forest loss, tiny Costa Rica is a superpower of biodiversity, with an estimated 500,000 plant and animal species. A single park in Costa Rica is home to more bird species than are found in all of North America. This oasis of biodiversity is also home to an amazing variety of other exotic wildlife, including, monkeys, jaguars, snakes, spiders, and frogs.

This biodiversity results mostly from two factors: first, the country's tropical geographic location, lying between two oceans and having both coastal and mountainous regions that provide a variety of microclimates and habitats for wildlife; and second, the government's strong conservation efforts.

In the mid-1970s, Costa Rica established a system of nature reserves and national parks (Figure 9.1) that, by 2012, included more than 25% of its land—6% of it reserved for indigenous peoples. Costa Rica has increased its beneficial environmental impact by devoting a larger proportion of its land than any other country has to biodiversity conservation, in keeping with the biodiversity **principle of sustainability**.

To reduce *deforestation,* or the widespread removal of forests, the government has eliminated subsidies for converting forestland to rangeland. Instead, it pays landowners to maintain or restore tree cover. The strategy has worked: Costa Rica has gone from having one of the world's highest deforestation rates to having one of the lowest. **GOOD NEWS**

Ecologists warn that human population growth, economic development, and poverty are exerting increasing pressure on the earth's ecosystems and on the ecosystem services they provide that help to sustain biodiversity. In 2010, a report by two United Nations environmental bodies warned that unless radical and creative action is taken to conserve the earth's biodiversity, many local and regional ecosystems that help to support human lives and livelihoods are at risk of collapsing.

This chapter is devoted to helping us understand the threats to the earth's terrestrial and aquatic biodiversity, and to seeking ways to help sustain these vital ecosystems and the ecosystem services they provide.

FIGURE 9.1 La Fortuna Falls is located in a tropical rain forest in Costa Rica's Arenal Volcano National Park.

gary yim/Shutterstock.com

WHAT ARE THE MAJOR THREATS TO FOREST ECOSYSTEMS?

CONCEPT 9.1A Forest ecosystems provide ecosystem services far greater in value than the value of raw materials obtained from forests.

CONCEPT 9.1B Unsustainable cutting and burning of forests and projected climate change are the chief threats to forest ecosystems.

Forests Vary in Age and Makeup

Scientists have divided forests into major types, based on their age and structure. An **old-growth forest**, or **primary forest**, is an uncut or regenerated forest that has not been seriously disturbed by human activities or natural disasters for 200 years or more (Figure 9.2). Old-growth forests are reservoirs of biodiversity because they provide ecological niches for a multitude of wildlife species (see Figure 7.15, p. 143).

A **second-growth forest** is a stand of trees resulting from secondary ecological succession (see Figure 5.8, p. 91). These forests develop after the trees in an area have been removed by human activities, such as clear-cutting for timber or conversion to cropland, or by natural forces such as fires and hurricanes.

A **tree plantation**, also called a **tree farm** or **commercial forest** (Figure 9.3), is a managed forest containing only one or two species of trees that are all of the same age. They are often grown on land that was cleared of an old-growth or second-growth forest and are usually harvested by clear-cutting as soon as they become commercially valuable. The land is then replanted and clear-cut

again in a regular cycle. When managed carefully, such plantations can produce wood at a rapid rate and could supply most of the wood used for industrial purposes, including papermaking. This would help to protect the world's remaining old-growth and second-growth forests, as long as they are not cleared to make room for tree plantations.

The downside of tree plantations is that, with only one or two tree species, they are much less biologically diverse and less sustainable than old-growth and second-growth forests because they violate nature's biodiversity **principle of sustainability**. In addition, repeated cycles of cutting and replanting can eventually deplete the nutrients in the forest's topsoil and hinder the regrowth of any type of forest on such land.

Forests Provide Important Economic and Ecosystem Services

By sustaining forests, we can increase our beneficial environmental impact, because forests provide highly valuable economic and ecosystem services (Figure 9.4 and **Concept 9.1A**). For example, through photosynthesis, forests remove CO_2 from the atmosphere and store it in organic compounds, playing a role in the global carbon cycle (see Figure 3.14, p. 53). By performing this ecosystem service, forests help to stabilize average atmospheric temperatures and climate conditions.

Forests provide habitats for about two-thirds of the earth's terrestrial species and are home to more than 300 million people. About 1 billion people living in extreme poverty depend on forests for their survival.

Forests also provide us with important health benefits. For example, traditional medicines, used by 80% of the world's people, are derived mostly from plant species that are native to forests. Some of the chemicals found in tropi-

FIGURE 9.2 Old-growth forest in Poland.

FIGURE 9.3 Olive tree plantation.

PUTTING A PRICE TAG ON NATURE'S ECOSYSTEM SERVICES

Currently, forests and other ecosystems are valued mostly for their economic services (Figure 9.4, right). Ecologists and ecological economists call for us to include in such valuations the estimated monetary value of the ecosystem services provided by forests (Figure 9.A). This is a way to implement the full-cost pricing **principle of sustainability**.

In 1997, a team of ecologists, economists, and geographers, led by ecological economist Robert Costanza, estimated the monetary worth of 17 ecosystem services provided by the earth's ecosystems (including, for example, pollination and regulation of atmospheric temperatures) to be at least $33.2 trillion per year. The researchers also estimated that the amount of money we would need to put into a savings account to earn that amount of interest income would be at least $500 trillion—an average of more than $70,400 for every person on the earth in 2014.

According to Costanza's study, the world's forests alone provide us with ecosystem services worth at least $4.7 trillion per year—hundreds of times more than their economic value in terms of lumber, paper, and other

wood products (Figure 9.A). The researchers pointed out that their estimates were very conservative.

In 2002, Costanza and other researchers estimated that preserving ecosystems in a global network of nature reserves occupying 15% of the earth's land surface and 30% of the ocean would provide $4.4 trillion to $5.2 trillion worth of ecosystem services—about 100 times the economic value of converting those systems to human uses.

If the estimates from these studies are close to correct, we can draw three important conclusions: **(1)** the earth's

ecosystem services are essential for all humans and their economies; **(2)** the economic value of these services is huge; and **(3)** these ecosystem services will be an ongoing source of ecological income, as long as they are used sustainably.

Critical Thinking

Some analysts believe that we should not try to put economic values on the world's irreplaceable ecosystem services because their value is infinite. Do you agree with this view? Explain. What is the alternative?

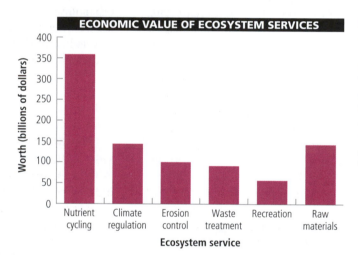

FIGURE 9.A Estimates of the annual global economic values of some ecosystem services provided by forests, compared to the value of the raw materials they produce (in billions of dollars).

Compiled by the authors using data from Robert Costanza.

cal forest plants are used as blueprints for making most of the world's prescription drugs.

Scientists and economists have found various ways to estimate the economic value of major ecosystem services provided by the world's forests and other ecosystems (Science Focus 9.1).

There Are Several Ways to Harvest Trees

Because of the immense economic value of forests, the harvesting of wood is one of the world's major industries. The first step in harvesting trees is to build roads for access

and timber removal. Even carefully designed logging roads can have a number of harmful effects (Figure 9.5)—namely, topsoil erosion, sediment runoff into waterways, habitat fragmentation, and loss of biodiversity. Logging roads also expose forests to invasion by disease-causing organisms and nonnative pests, and to disturbances from human activities such as farming and ranching.

Loggers can use a variety of methods to harvest trees. With *selective cutting*, intermediate-aged or mature trees in a forest are cut singly or in small groups (Figure 9.6a). However, loggers often remove all the trees from an area in what is called a *clear-cut* (Figure 9.6b and Figure 9.7).

Natural Capital

Forests

Ecosystem Services	Economic Services
Support energy flow and chemical cycling	Fuelwood
Reduce soil erosion	Lumber
Absorb and release water	Pulp to make paper
Purify water and air	Mining
Influence local and regional climate	Livestock grazing
Store atmospheric carbon	Recreation
Provide numerous wildlife habitats	Jobs

Photo: Val Thoermer/Shutterstock.com; © 2016 Cengage Learning

FIGURE 9.4 Forests provide many important ecosystem and economic services (**Concept 9.1**). **Question:** Which two ecosystem services and which two economic services do you think are the most important?

Clear-cutting is the most efficient and sometimes least-costly way to cut trees, and it provides benefits for landowners and timber companies. But it can also harm an ecosystem by causing increased erosion, sediment pollution of nearby waterways, and losses in biodiversity.

A variation of clear-cutting that allows a more sustainable timber yield without widespread destruction is *strip cutting* (Figure 9.6c). It involves clear-cutting a strip of trees along the contour of the land within a corridor narrow enough to allow natural forest regeneration within a few years. After regeneration, loggers cut another strip next to the first, and so on.

Fire Can Threaten or Benefit Forest Ecosystems

Two types of fires can affect forest ecosystems. *Surface fires* (Figure 9.8, left) usually burn only undergrowth and leaf litter on the forest floor. They kill seedlings and small trees, but they spare most mature trees and allow most wild animals to escape.

Occasional surface fires have a number of ecological benefits. They:

- burn away flammable material such as dry brush and help to prevent more destructive fires;
- free valuable plant nutrients tied up in slowly decomposing litter and undergrowth;
- release seeds from the cones of tree species such as lodgepole pines and stimulate the germination of other seeds such as those of the giant sequoia; and
- help to control destructive insects and tree diseases.

Another type of fire, called a *crown fire* (Figure 9.8, right), is an extremely hot fire that leaps from treetop to treetop, burning whole trees. Crown fires usually occur in forests that have not experienced surface fires for several decades, a situation that allows dead wood, leaves, and other flammable ground litter to accumulate. These rapidly burning fires can destroy most vegetation, kill wildlife, increase topsoil erosion, and burn or damage human structures in their paths.

Almost Half of the World's Forests Have Been Cut Down

Deforestation is the temporary or permanent removal of large expanses of forest for agriculture, settlements, or other uses. Surveys by the World Resources Institute (WRI) indicate that during the past 8,000 years, human activities have reduced the earth's old-growth forest cover by about 47%, with most of this loss occurring in the last 60 years.

According to the WRI, if current deforestation rates continue, about 40% of the world's remaining intact for-

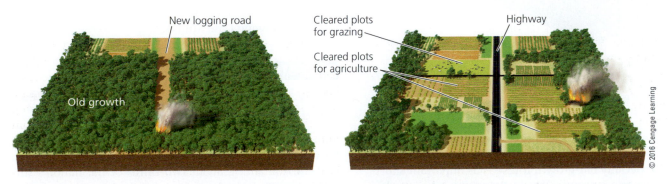

FIGURE 9.5 **Natural capital degradation:** Building roads into previously inaccessible forests is the first step in harvesting timber, but it also paves the way to fragmentation, destruction, and degradation of forest ecosystems.

© 2016 Cengage Learning

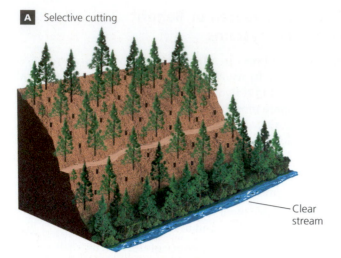

A Selective cutting

Clear stream

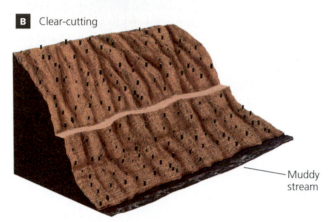

B Clear-cutting

Muddy stream

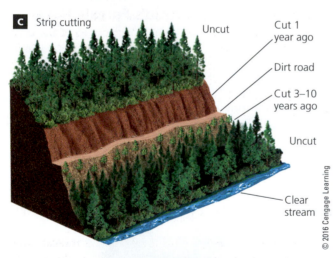

C Strip cutting

Uncut

Cut 1 year ago

Dirt road

Cut 3–10 years ago

Uncut

Clear stream

© 2016 Cengage Learning

FIGURE 9.6 Three major ways to harvest trees. **Question:** If you were cutting trees in a forest you owned, which method would you choose and why?

ests will have been logged or converted to other uses within two decades if not sooner. Clearing large areas of forests, especially old-growth forests, has important short-term economic benefits (Figure 9.4, right column), but it also has a number of harmful environmental effects (Figure 9.9).

Eppic/Dreamstime.com

FIGURE 9.7 Clear-cut forest.

In 2011, the Food and Agriculture Organization of the United Nations (FAO) reported that the net total forest cover in several countries, including the United States, changed very little or even increased between 2000 and 2010. Some of the increases resulted from natural reforestation by secondary ecological succession on cleared forest areas and abandoned croplands (see Figure 5.8, p. 91). Other increases in forest cover were due to the spread of commercial tree plantations and to a global program, sponsored by the United Nations Environment Programme (UNEP), to plant billions of trees throughout much of the world—many of them in tree plantations. China now leads the world in new forest cover, mostly due to its plantations of fast-growing trees. Other countries that have increased their forest cover are Costa Rica (**Core Case Study**), the Philippines, Russia, and the United States (see the following Case Study).

GOOD NEWS

CASE STUDY

Many Cleared Forests in the United States Have Grown Back

Forests cover about 30% of the U.S. land area, providing habitats for more than 80% of the country's wildlife species and containing about two-thirds of the nation's surface water. Today, forests in the United States (including tree plantations) cover more area than they did in 1920. The primary reason is that many of the old-

GOOD NEWS

FIGURE 9.8 Surface fires (left) usually burn only undergrowth and leaf litter on a forest floor. They can help to prevent more destructive crown fires (right) by removing flammable ground material.

Left: David J. Moorhead, University of Georgia, Bugwood.org; Right: Xneo/Dreamstime.com

growth forests that were cleared or partially cleared between 1620 and 1920 have grown back naturally through secondary ecological succession (Figure 9.10).

There are now fairly diverse second-growth (and in some cases third-growth) forests in every region of the United States except in much of the West. Environmental writer Bill McKibben has cited this forest regrowth in the United States—especially in the East—as "the great environmental success story of the United States, and in some ways, the whole world."

Protected forests make up about 40% of the country's total forest area, mostly in the *National Forest System,* which consists of 155 national forests managed by the U.S. Forest Service (USFS). On the other hand, since the mid-1960s, a large area of the nation's remaining old-growth and fairly diverse second-growth forests has been cut down and replaced with biologically simplified tree plantations.

Tropical Forests Are Disappearing Rapidly

Tropical forests (see Figure 7.14, top, p. 142) cover about 6% of the earth's land area—roughly the area of the continental United States. Climatic and biological data indicate that mature tropical forests once covered at least twice as much area as they do today. Most of this loss of half of the world's tropical forests has taken place since 1950 (see Chapter 3, Core Case Study, p. 42). Between 2000 and 2013, the world lost the equivalent of more than 50 soccer fields of tropical forest every minute. In 2014, the WRI and more than 40 other organizations, including Google and UNEP, created Global Forest Watch. This free online monitoring and mapping system provides near-real time reliable data about what is happening to the world's forests.

Satellite scans and ground-level surveys indicate that large areas of tropical forests are being cut rapidly in parts of Africa, Southeast Asia, and South America—especially in Brazil's vast Amazon Basin, which has more than 40% of the world's remaining tropical forests. Currently, these forests absorb and store about one-third of the world's terrestrial carbon emissions as part of the carbon cycle, so by reducing these forests, we reduce their carbon absorption and contribute to climate change. Also, the burning and clearing of tropical forests adds carbon to the atmosphere, accounting for 10–15% of global greenhouse gas emissions.

Natural Capital Degradation

Deforestation

- Water pollution and soil degradation from erosion
- Acceleration of flooding
- Local extinction of specialist species
- Habitat loss for native and migrating species
- Release of CO_2 and loss of CO_2 absorption

FIGURE 9.9 Deforestation has some harmful environmental effects that can reduce biodiversity and degrade the ecosystem services provided by forests (Figure 9.4, left).

© 2016 Cengage Learning

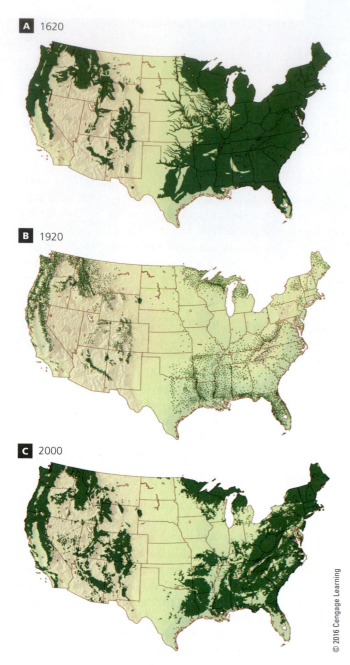

© 2016 Cengage Learning

FIGURE 9.10 In 1620, **(a)** when European settlers were moving to North America, forests covered more than half of the current land area of the continental United States. By 1920, **(b)** most of these forests had been decimated. In 2000, **(c)** secondary and commercial forests covered about a third of U.S. land in the lower 48 states.

Water evaporating from trees and vegetation in tropical rain forests plays a large role in determining the amount of rainfall there. Removing large areas of trees can lead to a drier climate that dehydrates the topsoil by exposing it to sunlight, allowing it to be blown away. This makes it difficult for a forest to grow back in the area, and such forest areas are often replaced by tropical grassland or savanna. Scientists project that if current burning and deforestation rates continue, 20–30% of the Amazon Basin will be turned into savanna within the next 50 years, and most of it could become savanna by 2080.

Studies indicate that at least half of the world's known species of terrestrial plants, animals, and insects live in tropical forests. Because of their specialized niches, many of these species are highly vulnerable to extinction when their forest habitats are destroyed or degraded. The FAO warns that at the current global rate of tropical deforestation, as much as 50% of the world's remaining old-growth tropical forests will be gone or severely degraded by the end of this century (**Concept 9.1B**)

Tropical deforestation results from a number of underlying and direct causes. Underlying causes, such as pressures from population growth and poverty, push subsistence farmers and the landless poor into tropical forests, where they cut or burn trees for firewood or try to grow enough food to survive. Government subsidies can accelerate other direct causes such as large-scale logging and ranching by reducing the costs of these enterprises.

The major direct causes of deforestation vary in different tropical areas. Tropical forests in the Amazon and other South American countries are cleared or burned primarily for cattle grazing and large soybean plantations (see Figure 1.4, p. 8). In Indonesia, Malaysia, and other areas of Southeast Asia, tropical forests are being replaced by large plantations of oil palm, which produce palm oil used to make several products, including biodiesel fuel and cosmetics. In Africa, the primary direct cause of deforestation is people clearing plots for small-scale farming and harvesting wood for fuel.

The degradation of a tropical forest usually begins when a road is cut deep into the forest interior for logging and settlement (Figure 9.5). Loggers then use selective cutting (Figure 9.6a) to remove the largest and best trees. When these big trees fall, many other trees often fall with them because of their shallow roots and the network of vines connecting the trees in the forest's canopy.

International corporations do much of this logging and then they often sell the land to ranchers who burn the remaining timber to clear the land for cattle grazing. Within a few years, their cattle typically overgraze the land and the ranchers move on, selling the degraded land to large-scale farm operators who plow it up to plant large crops such as soybeans (see Figure 1.4, p. 8), or to settlers for small-scale farming. After a few years of crop growing and erosion from rain, the nutrient-poor topsoil is depleted of nutrients. Then the farmers and settlers move on to newly cleared land to repeat this environmentally destructive process.

CONSIDER THIS...

THINKING ABOUT Tropical Forests

Why should you care if most of the world's remaining tropical forests are burned or cleared or converted to savanna within your lifetime? What are three ways in which this might affect your life or the lives of any children and grandchildren that you eventually might have?

CONCEPT 9.2 We can sustain forests by emphasizing the economic value of their ecosystem services, removing government subsidies that hasten their destruction, protecting old-growth forests, harvesting trees no faster than they are replenished, and planting trees.

We Can Manage Forests More Sustainably

Biodiversity researchers and a growing number of foresters have called for more sustainable forest management (Figure 9.11) (**Concept 9.2**). Certification of sustainably grown timber and of sustainably produced forest products can help consumers to play their part in reaching this goal. The nonprofit Forest Stewardship Council (FSC) oversees the certification of forestry operations that meet certain standards. To gain such certification, operators must make sure that: any cutting of trees does not exceed long-term forest regeneration in a given area; roads and harvesting systems do not cause unreasonable ecological damage; forest topsoil does not become unreasonably damaged; and downed wood and standing dead trees be left to provide wildlife habitat.

The FSC reported that, by 2012, about 5% of the world's forest area in 80 countries had been certified according to FSC standards. The FSC also certifies 5,400 manufacturers and distributors of wood products. The

Solutions

More Sustainable Forestry

- Include ecosystem services of forests in estimates of their economic value

- Identify and protect highly diverse forest areas

- Stop logging in old-growth forests

- Stop clear-cutting on steep slopes

- Reduce road-building in forests and rely more on selective and strip cutting

- Leave most standing dead trees and larger fallen trees for wildlife habitat and nutrient cycling

- Put tree plantations only on deforested and degraded land

- Certify timber grown by sustainable methods

© 2016 Cengage Learning

FIGURE 9.11 Ways to grow and harvest trees more sustainably (**Concept 9.2**). *Questions:* Which three of these methods of more sustainable forestry do you think are the best methods? Why?

paper used in this book was produced with the use of sustainably grown timber, as certified by the FSC, and contains recycled paper fibers.

Loggers could use more sustainable selective cutting (Figure 9.6a) and strip cutting (Figure 9.6c) to harvest tropical trees for lumber instead of clear-cutting the forests (Figure 9.6b). They could also be more careful when cutting and removing individual trees by taking care to cut canopy vines (lianas) before felling a tree to reduce damage to neighboring trees.

Many economists are urging governments to begin making a shift to more sustainable forest management by phasing out government subsidies and tax breaks that encourage forest degradation and deforestation and replacing them with forest-sustaining economic rewards. This would likely lead to higher prices on unsustainably produced timber and wood products, in keeping with the full-cost pricing **principle of sustainability**. Costa Rica (**Core Case Study**) is taking a lead in using this approach. Governments can also encourage tree-planting programs to help restore degraded forests. **GREEN CAREER: Sustainable forestry**

We Can Improve the Management of Forest Fires

In the United States, the Smokey Bear educational campaign undertaken by the Forest Service and the National Advertising Council has prevented many forest fires, saved many lives, and prevented billions of dollars in losses of trees, wildlife, and human structures. At the same time, it has convinced much of the public that all forest fires are bad and should be prevented or put out. Ecologists warn that trying to prevent all forest fires can make matters worse by increasing the likelihood of destructive crown fires (Figure 9.8, right) due to the accumulation of highly flammable underbrush in some forests.

Ecologists and forest fire experts have proposed several strategies for reducing fire-related harm to forests and to people who use or live in the forests, including:

- Using carefully planned and controlled *prescribed burns* to remove flammable small trees and underbrush in the highest-risk forest areas.

- Allowing some fires on public lands to burn underbrush and smaller trees, as long as the fires do not threaten human structures or human lives.

- Protecting houses and other buildings in fire-prone areas by thinning trees and other vegetation in a zone around them and eliminating the use of highly flammable construction materials such as wood shingles.

- Using solar-powered micro-drones, equipped with infrared sensors, to detect forest fires and monitor progress in fighting them.

FIGURE 9.12 Solutions: The pressure to cut trees to make paper could be greatly reduced by planting and harvesting a fast-growing plant known as kenaf.

We Can Reduce the Demand for Harvested Trees

According to the Worldwatch Institute and to forestry analysts, *up to 60% of the wood consumed in the United States is wasted unnecessarily.* This results from inefficient use of construction materials, excessive packaging, overuse of junk mail, inadequate paper recycling, and the failure to reuse or find substitutes for wooden shipping containers.

One reason for cutting trees is to provide pulp for making paper, but we can make paper by using fiber from sources other than trees. China uses rice straw and other agricultural residues to make much of its paper. Most of the small amount of tree-free paper produced in the United States is made from the fibers of a rapidly growing woody annual plant called *kenaf* (pronounced "kuh-NAHF," Figure 9.12). Kenaf and other nontree fibers such as hemp yield more paper pulp per area of land than tree farms do and require less use of pesticides and herbicides.

Another way to reduce the demand for tree cutting is to sharply reduce the use of throwaway paper products made from trees. We can instead choose reusable plates, cups, cloth napkins and handkerchiefs, and cloth bags.

More than 2 billion people in less-developed countries use fuelwood (see Figure 6.11, p. 111) and charcoal made from wood for heating and cooking. Most of these countries are suffering from fuelwood shortages because people are cutting trees for fuelwood and forest products 10–20 times faster than new trees are being planted (Figure 9.13). In Guatemala, which is threatened by severe deforestation, scientists estimate that fuelwood harvesting accounts for more than 55% of the annual deforestation.

One way to reduce the severity of the fuelwood crisis in less-developed countries is to establish small plantations of fast-growing trees and shrubs around farms and in community woodlots. Another way, as is being done in Guatemala, is to produce biomass briquettes as a substitute for fuelwood. The National Geographic Society is funding a project led by the Legacy Foundation in cooperation with Fundación Progresar that aims to establish local enterprises that will create biomass briquette production businesses, training businesses, and a network of producers to spread the word about the viability of biomass as a fuelwood alternative. This could bolster the Guatemalan economy and support women's empowerment, while reducing deforestation.

There Are Several Ways to Reduce Tropical Deforestation

Analysts have suggested various ways to protect tropical forests and use them more sustainably (Figure 9.14).

At the international level, *debt-for-nature swaps* can make it financially attractive for countries to protect their tropical forests. In such swaps, participating countries act as custodians of protected forest reserves in return for foreign aid or debt relief. In a similar strategy, called *conservation concessions*, governments or private conservation organizations pay nations for agreeing to preserve their natural resources.

National governments can also take important steps to reduce deforestation (**Core Case Study**). Between 2005 and 2013, Brazil cut its deforestation rate by 80% by cracking down on illegal logging and setting aside a large conservation reserve in the Amazon Basin. Governments can also subsidize sustainable forestry and the planting of trees.

GOOD NEWS

Consumers can reduce the demand for unsustainable and illegal logging in tropical forests by buying only wood and wood products that have been certified as sustainably produced by the FSC and other organizations, including the Rainforest Alliance and the Sustainability Action Network. The late Wangari Maathai, a Nobel Peace Prize winner, promoted tree planting in her native country of Kenya and throughout the world in what became the Green Belt Movement. Her efforts inspired the UNEP to implement a global effort to plant at least 1 billion trees a year beginning in 2006. By 2012, the year Maathai died, about 12.6 billion trees had been planted in 193 countries.

FIGURE 9.13 Natural capital degradation: Haiti's deforested brown landscape (left) contrasts sharply with the heavily forested green landscape of its neighboring country, the Dominican Republic.

Solutions

Sustaining Tropical Forests

Prevention

Protect the most diverse and endangered areas

Educate settlers about sustainable agriculture and forestry

Subsidize only sustainable forest use

Protect forests through debt-for-nature swaps and conservation concessions

Certify sustainably grown timber

Reduce poverty and slow population growth

Restoration

Encourage regrowth through secondary succession

Rehabilitate degraded areas

Concentrate farming and ranching in already-cleared areas

FIGURE 9.14 Some effective ways to protect tropical forests and to use them more sustainably (**Concept 9.2**). *Questions:* Which three of these solutions do you think are the best ones? Why?

FIGURE 9.15 Natural capital degradation: To the left of the fence is overgrazed rangeland. The land to the right of the fence is lightly grazed.

9.3 HOW SHOULD WE MANAGE AND SUSTAIN GRASSLANDS?

CONCEPT 9.3 We can sustain the productivity of grasslands by controlling the numbers and distribution of grazing livestock and by restoring degraded grasslands.

Some Rangelands Are Overgrazed

Grasslands provide many important ecosystem services, including soil formation, erosion control, chemical cycling, storage of atmospheric carbon dioxide in biomass, and maintenance of biodiversity.

After forests, grasslands are the ecosystems most widely used and altered by human activities. **Rangelands** are unfenced grasslands in temperate and tropical climates that supply *forage,* or vegetation for grazing (grass-eating) and browsing (shrub-eating) animals. Cattle, sheep, and goats graze on about 42% of the world's grassland. The 2005 UN Millennium Ecosystem Assessment—a 4-year study by 1,360 experts from 95 countries—estimated that this could increase to 70% by 2050. Livestock also graze in **pastures**, which are managed grasslands or fenced meadows often planted with domesticated grasses or other forage crops such as alfalfa and clover.

Blades of rangeland grass grow from the base, not at the tip as broadleaf plants do. Thus, as long as only the upper portion of the blade is eaten and its lower portion remains, rangeland grass is a renewable resource that can be grazed again and again. Moderate levels of grazing are healthy for grasslands, because removal of mature vegetation stimulates rapid regrowth and encourages greater plant diversity.

Overgrazing occurs when too many animals graze for too long, damaging the grasses and their roots, and exceeding the carrying capacity of a rangeland area (Figure 9.15, left). Overgrazing reduces grass cover, exposes the topsoil to erosion by water and wind, and compacts the soil, which lessens its capacity to hold water. Overgrazing also encourages the invasion of rangeland by species such as sagebrush, mesquite, cactus, and cheatgrass, which cattle will not eat. The FAO has estimated that overgrazing by livestock has reduced productivity on as much as 20% of the world's rangeland.

We Can Manage Rangelands More Sustainably

The most widely used way to manage rangelands more sustainably is to control the number of grazing animals and the duration of their grazing in a given area so the carrying capacity of the area is not exceeded (**Concept 9.3**). One method for doing this is called *rotational grazing,* in which cattle are confined by portable fencing to one area for a few days and then moved to a new location.

Cattle like to graze around natural water sources, especially along streams or rivers lined by strips of vegetation known as *riparian zones,* and around ponds. Overgrazing can destroy the vegetation in such areas (Figure 9.16, left). Ranchers can protect overgrazed land through rotational grazing and by fencing off damaged areas, which eventually leads to their natural restoration by ecological succession (Figure 9.16, right).

A more expensive and less widely used method of rangeland management is to suppress the growth of unwanted invader plants by the use of herbicides, mechanical removal, or controlled burning. A cheaper way to discourage unwanted vegetation in some areas is through controlled, short-term trampling by large numbers of livestock such as sheep, goats, and cattle that destroy the invasive plants' root systems.

FIGURE 9.16 Natural capital restoration: In the mid-1980s, cattle had degraded the vegetation and soil on this stream bank along the San Pedro River in the U.S. state of Arizona (left). Within 10 years, the area was restored through secondary ecological succession (right) after grazing and off-road vehicle use were banned (**Concept 9.3**).

Left: U.S. Bureau of Land Management; Right: U.S. Bureau of Land Management

9.4 HOW SHOULD WE MANAGE AND SUSTAIN PARKS AND NATURE RESERVES?

CONCEPT 9.4 Sustaining biodiversity will require more effective protection of existing parks and nature reserves, as well as the protection of much more of the earth's remaining undisturbed land area.

National Parks Face Many Environmental Threats

According to the International Union for the Conservation of Nature (IUCN), there are now more than 6,600 major national parks located in more than 120 countries (see chapter-opening photo). However, most of these parks are too small to sustain many large animal species. And many parks suffer from invasions by harmful nonnative species that compete with and reduce the populations of native species. Some national parks are so popular that large numbers of visitors are degrading the natural features that make them attractive (see the Case Study that follows).

Parks in less-developed countries have the greatest biodiversity of all the world's parks, but only about 1% of these parklands are protected. Local people in many of these countries enter the parks illegally in search of wood, game animals, and other natural products that they need for their daily survival. Loggers and miners operate illegally in many of these parks, as do wildlife poachers who kill animals to obtain and sell items such as rhino horns (see Figure 8.11, p. 174), elephant tusks, and furs. Park services in most of the less-developed countries have too little money and too few personnel to fight these invasions, either by force or through education.

Stresses on U.S. Public Parks

The U.S. National Park System, established in 1912, includes 59 major national parks, sometimes called the country's crown jewels (see chapter-opening photo), along with 339 monuments and historic sites. States, counties, and cities also operate public parks.

Popularity is one of the biggest problems for many parks. Between 1960 and 2012, the number of recreational visitors to U.S. national parks more than tripled, reaching about 282 million. In some U.S. parks and other public lands, noisy and polluting dirt bikes, dune buggies, jet skis, snowmobiles, and other off-road vehicles destroy or damage fragile vegetation, disturb wildlife, and degrade the aesthetic experience for many visitors.

A number of parks also suffer damage from the migration or deliberate introduction of nonnative species. For example, European wild boars, imported into the state of North Carolina in 1912 for hunting, now threaten vegetation in parts of the Great Smoky Mountains National Park. Nonnative mountain goats in Washington State's Olympic National Park trample and destroy the root systems of native vegetation and accelerate soil erosion.

At the same time, native species—some of them threatened or endangered—are killed in, or illegally removed from, almost half of all U.S. national parks. This is what happened to the gray wolf in Yellowstone National Park until it was successfully reintroduced there after a 50-year absence (Science Focus 9.2).

Many U.S. national parks have become threatened islands of biodiversity surrounded by commercial develop-

REINTRODUCING THE GRAY WOLF TO YELLOWSTONE NATIONAL PARK

FIGURE 9.B After becoming almost extinct in much of the western United States, the *gray wolf* was listed and protected as an endangered species in 1974.

Volodymyr Burdiak/Shutterstock.com

Around 1800, at least 350,000 gray wolves (Figure 9.B) roamed over about three-quarters of America's lower 48 states, especially in the West. They survived mostly by preying on abundant bison, elk, caribou, and deer. However, between 1850 and 1900, most of them were shot, trapped, or poisoned by ranchers, hunters, and government employees, and they were driven to near extinction in the lower 48 states.

Ecologists recognize the important role that this keystone predator species once played. In the Yellowstone National Park region, the wolves culled herds of bison, elk, moose, and mule deer, and kept down coyote populations. By leaving some of their kills partially uneaten, they provided meat for scavengers such as ravens, bald eagles, ermines, grizzly bears, and foxes.

When the number of gray wolves declined, herds of plant-browsing elk, moose, and mule deer expanded and devastated the willow and aspen trees growing near streams and rivers. This led to increased soil erosion and to declining populations of other wildlife species such as beaver, which eat willow and aspen. This in turn affected species that depended on wetlands created by dam-building beavers.

When Congress passed the U.S. Endangered Species Act in 1973, only a few hundred gray wolves remained outside of Alaska, primarily in Minnesota and Michigan. In 1974, the gray wolf was listed as an endangered species in the lower 48 states.

In 1987, the U.S. Fish and Wildlife Service (USFWS) proposed reintroducing gray wolves into the Yellowstone National Park to try to help stabilize the ecosystem. The proposal brought angry protests from area ranchers who feared the wolves would leave the park and attack their cattle and sheep, and from hunters who feared the wolves would kill too many big-game animals. Mining and logging companies objected, fearing that the government

ment. Nearby human activities that threaten wildlife and recreational values in many national parks include mining, logging, livestock grazing, use of coal-fired power plants, water diversion, and urban development. According to the National Park Service, air pollution, mostly from coal-fired power plants and dense vehicle traffic, degrades scenic views in many U.S. national parks more than 90% of the time.

Nature Reserves Occupy Only a Small Part of the Earth's Land

Most ecologists and conservation biologists argue that the best way to preserve biodiversity is to create a worldwide network of protected areas. In 2013, less than 13% of the earth's land area (not including Antarctica) was protected either strictly or partially in more than 177,000 wildlife refuges, nature reserves, parks, and wilderness areas. However, no more than 6% of the earth's land is strictly protected from potentially harmful human activities. Figure 9.17 is a map of the earth's remaining wildlands.

Conservation biologists call for fully protecting at least 20% of the earth's land area in a global system of biodiversity reserves that would include multiple examples of all the earth's biomes (**Concept 9.4**). In 2012, the IUCN estimated that making this investment to protect a vital part of our life-support system would cost roughly $23 billion a year—more than 4 times what is spent on such protection today.

Most developers and resource extractors oppose protecting even 13% of the earth's land, arguing that these areas might contain valuable resources that would provide

would halt their operations on wolf-populated federal lands.

In 1995 and 1996, federal wildlife officials caught gray wolves in Canada and northwest Montana and relocated 41 of them in Yellowstone National Park. Scientists estimate that the long-term carrying capacity of the park is 110 to 150 gray wolves. By 2007, the park had 171 wolves, but that number had dropped to 82 by the end of 2012, for various reasons having to do mostly with the supply of prey for the wolves. Such population fluctuations are natural, according to the National Park Service.

Wildlife ecologist Robert Crabtree and other scientists have been studying the effects of this reintroduction. They have put radio collars on most of the park's wolves to gather data and track their movements. They have also studied changes in vegetation and in the populations of various plant and animal species.

Studies indicate that the return of this keystone predator has contributed to a decline in populations of elk, the wolves' primary food source. The leftovers of elk killed by wolves have also been an important food source for scavengers such as bald eagles and ravens. The wolves' presence, with a projected decline in elk numbers, was supposed to promote the regrowth of young aspen trees that elk feed on and had depleted. However, a 2010 study led by U.S. Geological Survey scientist Matthew Kauffman indicated that the aspen were not recovering despite a 60% decline in elk numbers. Declining populations of elk were also supposed to allow for the return of willow trees along streams. Research indicates that willows have only partly recovered.

The wolves have cut in half the Yellowstone population of coyotes—the top predators in the absence of wolves. This has reduced coyote attacks on cattle from area ranches and has led to larger populations of small animals such as ground squirrels, mice, and gophers, which are hunted by coyotes, eagles, and hawks. Overall, this experiment has had some important ecological benefits for the Yellowstone ecosystem. However, some scientists hypothesize that the long-term absence of wolves led to a number of changes in plant and animal numbers and diversity that are difficult to reverse.

The wolf reintroduction has produced economic benefits for the region. One of the main attractions of the park for many visitors is the hope of spotting wolves chasing their prey across its vast meadows.

Between 1974 and 2012, the wolf population in the lower 48 states grew from around 100 to roughly 5,400 individuals. In 2013, the USFWS proposed removing the gray wolf from the U.S. endangered species list in the lower 48 states and turning over wolf protection responsibilities to state wildlife agencies. More than 750,000 Americans sent comments to the USFWS stating their opposition to this plan. Conservation groups plan to fight the proposal in court, arguing that it could lead to greatly increased killing of the ecologically important gray wolf.

Critical Thinking

Do you think gray wolves should remain on the U.S. endangered species list? Explain.

short-term economic benefits. In contrast, ecologists and conservation biologists view protected areas as islands of biodiversity and ecosystem services that help to sustain all life and economies indefinitely and that serve as centers of future evolution. In other words, they serve as an "ecological insurance policy" for us and other species. (See the online Guest Essay on this topic by Norman Myers.)

Conservation biologists call for using the *buffer zone concept* whenever possible to design and manage nature reserves. This means strictly protecting an inner core of a reserve, usually by establishing two buffer zones in which local people can extract resources sustainably without harming the inner core (see the Case Study that follows). By 2012, the United Nations had used this concept to create a global network of 621 *biosphere reserves* in 117 countries.

Identifying and Protecting Biodiversity in Costa Rica

For several decades, Costa Rica (**Core Case Study**) has been using government and private research agencies to identify the plants and animals that make it one of the world's most biologically diverse countries. The government has consolidated the country's parks and reserves into several large conservation areas, or *megareserves,* designed with the goal of sustaining about 80% of the country's biodiversity (Figure 9.18).

Each reserve contains a protected inner core surrounded by two buffer zones that local and indigenous people can use for sustainable logging, crop farming, cattle grazing, hunting, fishing, and ecotourism. Instead of shut-

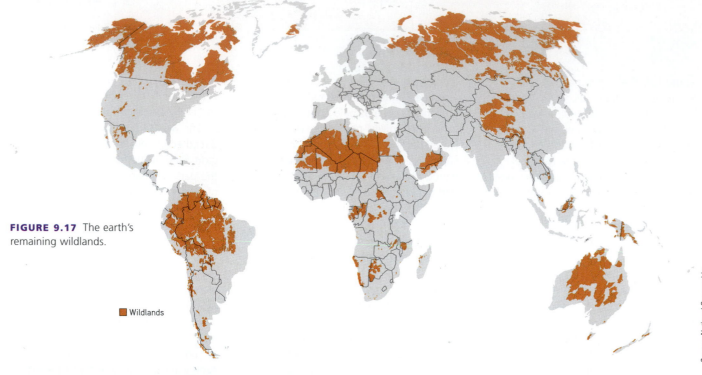

FIGURE 9.17 The earth's remaining wildlands.

■ Wildlands

Source: National Geographic

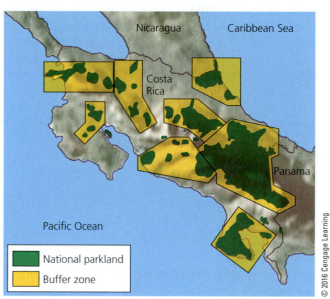

National parkland

Buffer zone

© 2016 Cengage Learning

FIGURE 9.18 Solutions: Costa Rica has created several *mega-reserves.* Green areas are protected natural parklands and yellow areas are the surrounding buffer zones.

ting people out of the protected areas, this approach enlists local people as partners in protecting a reserve from unsustainable uses such as illegal logging and poaching. It is an application of the biodiversity and win-win **principles of sustainability**.

In addition to its ecological benefits, this strategy has paid off financially. Today, Costa Rica's largest source of income is its $1-billion-a-year tourism industry, almost two-thirds of which involves ecotourism.

Protecting Wilderness Is an Important Way to Preserve Biodiversity

One way to protect undeveloped lands from human exploitation is to set them aside as **wilderness**—land officially designated as an area where natural communities have not been seriously disturbed by humans and where harmful human activities are limited by law (**Concept 9.4**). Theodore Roosevelt (see Figure 1.15, p. 19), the first U.S. president to set aside protected areas, summarized his thoughts on what to do with wilderness: "Leave it as it is. You cannot improve it."

Some critics oppose protecting large areas for their scenic and recreational value, arguing that such areas could contribute greatly to economic development. Conservation biologists contend that the most important reasons for protecting wilderness areas is not for their scenic, recreational, or economic value, but for the long-term needs of all species—to *preserve biodiversity* as a vital part of the earth's natural capital and to *protect wilderness areas as centers for evolution* in response to mostly unpredictable changes in environmental conditions.

In 1964, the U.S. Congress passed the Wilderness Act, which allowed the government to protect undeveloped tracts of U.S. public land from development as part of the National Wilderness Preservation System (Figure 9.19). The area of protected wilderness grew by nearly twelve-fold between 1964 and 2012. Even so, only about 5% of all U.S. land is protected as wilderness—more than 54% of it in Alaska. Only about 2.7% of the land area of the lower 48 states is protected as wilderness, most of it in the West, and most of these areas are threatened habitat islands in a sea of development.

FIGURE 9.19 Diablo Lake in the wilderness area of North Cascades National Park in the U.S. state of Washington.

tusharkoley/Shutterstock

9.5 WHAT IS THE ECOSYSTEM APPROACH TO SUSTAINING TERRESTRIAL BIODIVERSITY AND ECOSYSTEM SERVICES?

CONCEPT 9.5 We can help to sustain terrestrial biodiversity and increase our beneficial environmental impact by identifying and protecting biodiversity hotspots and employing restoration ecology and reconciliation ecology.

The Ecosystems Approach: A Strategy for Sustaining Terrestrial Biodiversity

Most wildlife biologists and conservationists believe that the best way to keep from hastening the extinction of wild species through human activities is to protect threatened habitats and ecosystem services. This *ecosystems approach* would generally employ the following five-point plan:

1. Map the world's terrestrial ecosystems and create an inventory of the species contained in each of them, along with the ecosystem services they provide.

2. Identify terrestrial ecosystems that are resilient and can recover if not overwhelmed by harmful human activities, along with ecosystems that are fragile and need protection.

3. Protect the most endangered terrestrial ecosystems and species, with emphasis on protecting plant biodiversity and ecosystem services.

4. Seek to restore as many degraded ecosystems as possible.

5. Make development *biodiversity-friendly* by providing significant financial incentives (such as tax breaks and subsidies) and technical help to private landowners who agree to help protect endangered ecosystems.

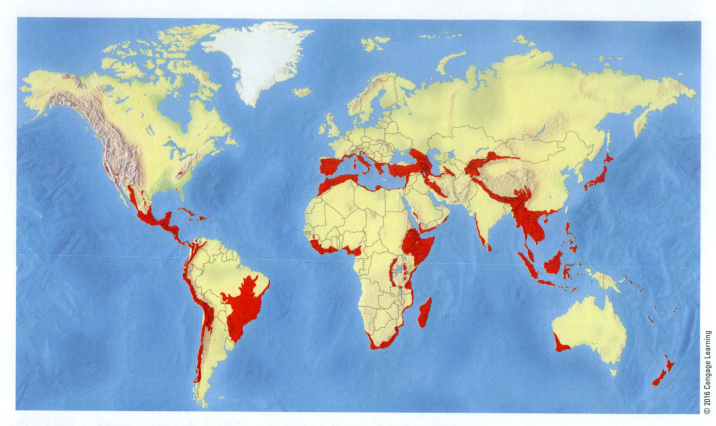

ANIMATED FIGURE 9.20 *Endangered natural capital:* Biologists have identified these 34 biodiversity hotspots. Compare this map with the global map of the human ecological footprint, shown in Figure 1.10, p. 12. **Question:** Why do you think so many hotspots are located near coastal areas? (Compiled by the authors using data from the Center for Applied Biodiversity Science at Conservation International.)

Protecting Global Biodiversity Hotspots Is an Urgent Priority

To protect as much of the earth's remaining biodiversity as possible, some biodiversity scientists urge the adoption of an *emergency action* strategy to identify and quickly protect **biodiversity hotspots**—areas especially rich in highly endangered species that are found nowhere else (**Concept 9.5**). These areas have suffered serious ecological disruption, mostly because of rapid human population growth and the resulting pressure on natural resources and ecosystem services. Environmental scientist Norman Myers first proposed this idea in 1988 (see his online Guest Essay on this topic).

Figure 9.20 shows 34 terrestrial biodiversity hotspots identified by biologists. (For a map of hotspots in the United States, see Figure 7, p. S22, in Supplement 4.) These areas cover little more than 2% of the earth's land surface, but they contain an estimated 50% of the world's flowering plant species and 42% of all terrestrial vertebrates (mammals, birds, reptiles, and amphibians), according to the IUCN. Yet, only about 5% of the total area of these hotspots is truly protected with government funding and law enforcement.

Biodiversity hotspots are also home to a large majority of the world's endangered or critically endangered species,

as well as to 1.2 billion people—one-sixth of the world's population. In 2012, the IUCN began publishing its Red List of Ecosystems that are vulnerable, endangered, or critically endangered as a companion to its Red List of Threatened Species.

Protecting Ecosystem Services Is Another Urgent Priority

Another way to help sustain the earth's biodiversity is to identify and protect areas where vital ecosystem services (see the orange boxed labels in Figure 1.3, p. 7) are being impaired enough to reduce biodiversity and harm local residents.

This approach has received more attention since the release of the 2005 UN Millennium Ecosystem Assessment. It identified key ecosystem services that provide numerous ecological and economic benefits, including those provided by forests (Figure 9.4). The study pointed out that humans are degrading or overusing about 60% of the ecosystem services provided by various ecosystems around the world, and it outlined ways to help sustain these vital services. One such way is to set aside and protect reserves and wilderness areas, especially highly endangered biodiversity hotspots (Figure 9.20).

Proponents of this strategy would also identify highly stressed *life raft ecosystems.* These would be areas where poverty levels are high and where a large part of the economy depends on various ecosystem services that are being degraded severely enough to threaten humans and other species. In such areas, residents, public officials, and conservation scientists would work together to develop strategies to help protect human communities along with the natural biodiversity and ecosystem services that support all life and economies. Thus, instead of pitting people against nature, this approach applies the win-win **principle of sustainability**.

We Can Rehabilitate and Partially Restore Ecosystems That We Have Damaged

Almost every natural place on the earth has been affected or degraded to some degree by human activities. We can partially reverse much of this harm through **ecological restoration**: the process of repairing damage caused by humans to various ecosystems. Examples include replanting forests (see the Case Study that follows), reintroducing native species (Science Focus 9.2), removing harmful invasive species, freeing river flows by removing dams, and restoring grasslands, coral reefs, wetlands, and stream banks (Figure 9.16, right). This is an important way to expand our beneficial environmental impact.

By studying how natural ecosystems recover, scientists are learning how to employ and enhance ecological succession processes by using a variety of approaches, including the following four:

- *Restoration:* returning a degraded habitat or ecosystem to a condition as similar as possible to its natural state in cases where this is feasible.

- *Rehabilitation:* turning a degraded ecosystem into a functional or useful ecosystem without trying to restore it to its original condition. Examples include removing pollutants from abandoned mining or industrial sites and replanting trees to reduce soil erosion in clear-cut forests.

- *Replacement:* replacing a degraded ecosystem with another type of ecosystem. For example, a degraded forest could be replaced by a productive pasture or tree plantation.

- *Creating artificial ecosystems:* for example, artificial wetlands have been created in some areas to help reduce flooding and to treat sewage.

Researchers have suggested a science-based, four-step strategy for carrying out most forms of ecological restoration and rehabilitation.

1. Identify the causes of the degradation (such as pollution, farming, overgrazing, mining, or invasive species).

2. Stop the degradation by eliminating or sharply reducing these factors.

3. If necessary, reintroduce key species to help restore natural ecological processes, as was done with gray wolves in the Yellowstone ecosystem (Science Focus 9.2).

4. Protect the area from further degradation to allow natural recovery (Figure 9.16, right).

By following this general plan, conservationist and National Geographic Explorer Sean Gerrity is working with his 30-person team in the state of Montana to create the American Prairie Reserve, the largest terrestrial wildlife reserve in the continental United States. Their goal is to restore the wildlife and ecosystem services common to this unique area of North America's grasslands for more than 11,000 years, dating back to the most recent ice age.

CASE STUDY

Ecological Restoration of a Tropical Dry Forest in Costa Rica

Costa Rica (**Core Case Study**) is the site of one of the world's largest ecological restoration projects. In the lowlands of its Guanacaste National Park, a tropical dry forest was burned, degraded, and fragmented for large-scale conversion to cattle ranches and farms. Now it is being restored and reconnected to a rain forest on nearby mountain slopes. The goal is to eliminate damaging nonnative grasses and reestablish a tropical dry-forest ecosystem during the next 100–300 years.

Daniel Janzen, professor of conservation biology at the University of Pennsylvania and a leader in the field of restoration ecology, used his own MacArthur Foundation grant money to purchase the Guanacaste forestland for designation as a national park. He also raised more than $10 million for restoring the park.

Janzen recognizes that ecological restoration and protection of the park will fail unless the people in the surrounding area believe they will benefit from such efforts. His vision is to see that the nearly 40,000 people who live near the park play an essential role in the restoration of the forest, a concept he calls *biocultural restoration.*

In the park, local farmers are paid to remove nonnative species and to plant tree seeds and seedlings started in Janzen's lab. Local grade school, high school, and university students and citizens' groups study the park's ecology during field trips. The park's location near the Pan American Highway makes it an ideal area for ecotourism, which stimulates the local economy.

This project also serves as a training ground in tropical forest restoration for scientists from all over the world. Research scientists working on the project give guest classroom lectures and lead field trips. Janzen believes that education, awareness, and involvement—not guards and fences—are the best ways to protect largely intact ecosystems from unsustainable use so they can be restored. This is an application of the biodiversity and win-win **principles of sustainability**.

We Can Share Areas We Dominate with Other Species

Ecologist Michael L. Rosenzweig suggests that we develop a form of conservation biology called **reconciliation ecology**. This scientific approach focuses on establishing and maintaining new habitats to conserve species diversity in places where people live, work, or play. In other words, the focus is on increasing our beneficial environmental impact by learning how to share some of the spaces we dominate with other species.

For example, people can learn how protection of local wildlife and ecosystems can provide economic resources for their communities by encouraging sustainable forms of ecotourism. In the Central American country of Belize, for instance, conservation biologist Robert Horwich helped establish a local sanctuary for the black howler monkey. He convinced local farmers to set aside strips of forest to serve as habitats and corridors through which these monkeys can travel. The reserve, run by a local women's cooperative, has attracted ecotourists and biologists. Local residents receive income for housing and guiding these visitors.

Without proper controls, ecotourism can lead to degradation of popular sites if they are overrun by visitors or are degraded by the construction of nearby hotels and other tourist facilities. However, when managed properly, ecotourism can be a useful form of reconciliation ecology.

Reconciliation ecology is also a way to protect vital ecosystem services. For example, some people are learning how to protect insect pollinators, such as butterflies and honeybees (see Chapter 8, Core Case Study, p. 162), which are vulnerable to pesticides and habitat loss. Neighborhoods and municipal governments are doing this by reducing or eliminating the use of pesticides on their lawns, fields, golf courses, and parks. Neighbors also work together to plant gardens of flowering plants as a source of food for bees and other pollinators. According to some honeybee experts, people trying to help bees in this way must avoid using glyphosate herbicides (such as Roundup®) and plants that contain neonicotinoid insecticides.

People have also worked together to protect bluebirds within human-dominated habitats where most of the bluebirds' nesting trees have been cut down and bluebird populations have declined. Special boxes were designed for nesting bluebirds, and with their widespread use, bluebird numbers are rising again.

These and many other examples of people working together on projects to restore degraded ecosystems all involve applications of the biodiversity and win-win **principles of sustainability**. Figure 9.21 lists some ways in which you can help to sustain the earth's terrestrial biodiversity.

9.6 HOW CAN WE HELP TO SUSTAIN AQUATIC BIODIVERSITY?

CONCEPT 9.6 We can help to sustain aquatic biodiversity and increase our beneficial environmental impact by establishing protected sanctuaries, managing coastal development, reducing water pollution, and preventing overfishing.

Human Activities Are Destroying and Degrading Aquatic Biodiversity

Human activities have destroyed or degraded a large portion of the world's coastal wetlands, coral reefs, mangroves, and ocean bottom, and disrupted many of the world's freshwater ecosystems. Scientists reported in 2006 that coastal habitats are disappearing at rates 2–10 times higher than the rate of tropical forest loss. During this century, rising sea levels, primarily caused by projected climate change, are likely to destroy many coral reefs (see Chapter 7, Core Case Study, p. 130, and Figure 7.24, p. 151) and flood some low-lying islands along with their protective coastal mangrove forests (see Figure 7.22, p. 149).

Sea-bottom habitats are faring no better, being threatened by dredging operations and trawler fishing boats. Like giant submerged bulldozers, trawlers drag huge nets weighted down with chains and steel plates over the ocean floor to harvest a few species of bottom fish and shellfish (Figure 9.22). Each year, thousands of trawlers scrape and

---

What Can You Do?

Sustaining Terrestrial Biodiversity

- Plant trees and take care of them

- Recycle paper and buy recycled paper products

- Buy sustainably produced wood and wood products and wood substitutes such as recycled plastic furniture and decking

- Help restore a degraded forest or grassland

- Landscape your yard with a diversity of native plants

© 2016 Cengage Learning

FIGURE 9.21 Individuals matter: Some ways to help to sustain terrestrial biodiversity. **Questions:** Which two of these actions do you think are the most important ones to take? Why? Which of these things do you already do?

Courtesy of Peter J. Auster/National Undersea Research Center

Courtesy of Peter J. Auster/National Undersea Research Center

FIGURE 9.22 Natural capital degradation: An area of ocean bottom before (left) and after (right) a trawler net scraped it. **Question:** What land activities are comparable to this?

disturb an area of ocean floor many times larger than the annual global total area of forests that are clear-cut. According to marine scientist Elliot Norse, "Bottom trawling is probably the largest human-caused disturbance to the biosphere."

A 2011 WRI study estimated that 75% of the world's shallow coral reefs are threatened by warmer waters, due mostly to heat absorbed from a warmer atmosphere, as well as by over-fishing, pollution, and ocean acidification, which decreases the carbonate ions in ocean water that corals need to build their calcium carbonate skeletons (Science Focus 9.3). Today, coral reefs, on average, are exposed to the warmest and most acidic ocean waters of the past 400,000 years. According to the WRI, by 2050, some 90% of the world's shallow tropical coral reefs will be threatened by coral bleaching caused by warmer ocean water and by ocean acidification.

Habitat disruption is also a problem in fresh-water aquatic zones. The main causes of disruption are dam building and excessive water withdrawal from rivers for irrigation and urban water supplies. These activities destroy aquatic habitats, degrade water flows, and disrupt fresh-water biodiversity.

Another problem that threatens aquatic bio-diversity is the deliberate or accidental introduction of hundreds of harmful invasive species (Figure 9.23) into coastal waters, wetlands, and lakes throughout the world. According to the

Cigdem Cooper/Shutterstock

FIGURE 9.23 The *common lionfish* has invaded the eastern coastal waters of North America, where it has few if any predators. One scientist described it as "an almost perfectly designed invasive species."

OCEAN ACIDIFICATION: THE OTHER CO₂ PROBLEM

By burning an increasingly large amount of carbon-containing fossil fuels, especially since 1950, we have added carbon dioxide (CO_2) to the atmosphere faster than it can be removed by the carbon cycle (see Figure 3.14, p. 53). Globally, CO_2 emissions have risen by about 45% between 1992 and 2013. Extensive research indicates that if we continue to increase CO_2 levels in the atmosphere, we will play a role in disrupting the earth's climate during this century.

Another problem related to CO_2 emissions is *ocean acidification,* a change in ocean chemistry. The oceans have absorbed about one-fourth of the excess CO_2 that human activities have added to the atmosphere. When this absorbed CO_2 combines with ocean water, it forms carbonic acid (H_2CO_3), a weak acid also found in carbonated drinks. As a result, the level of hydrogen ions (H^+) in the water rises, making the water less basic, while the level of carbonate ions (CO_3^{2-}) in the water drops because these ions react with hydrogen ions (H^+) to form bicarbonate ions (HCO_3^-).

The problem is that many aquatic species, including phytoplankton, corals, sea snails, crabs, and oysters, use carbonate ions to produce calcium carbonate ($CaCO_3$), the main component of their shells and bones. As a result, in less basic waters, shell-building species and coral reefs grow more slowly (Figure 9.C), and when the hydrogen ion concentration of the surrounding seawater gets high enough, their calcium carbonate begins to dissolve. These effects will occur first in colder ocean waters, because they have a greater capacity for absorption of CO_2 from the atmosphere than warmer waters have.

Since we began burning fossil fuels in large quantities during the Industrial Revolution of the 18th and 19th centuries, there has been a 30% rise in the average acidity (actually a 30% decrease in average basicity) of surface ocean water, according to a 2010 UNEP summary of research on this problem. A 2013 report prepared by 540 scientists from 37 countries projected that, by 2100, we could see a 170% drop in the average basicity of surface ocean water because of increasing acidity levels. The report also warned that this would reduce the ability of the oceans to help in regulating the rate of climate change by absorbing CO_2, and that it could jeopardize the stability of marine ecosystems and lead to a loss of some $130 billion a year for the fishing industry.

According to most marine scientists, the only way to slow these changes is through a quick and sharp reduction in the use of fossil fuels around the world, which would lessen the massive inputs of CO_2 into the air and from there into the ocean. We can also slow the rise of acidity levels in ocean waters by protecting and restoring mangrove forests, sea grasses, and coastal wetlands, because these aquatic systems take up and store some of the atmospheric CO_2 that is at the heart of this problem.

Critical Thinking

How might widespread losses of some forms of marine aquatic life due to ocean acidification affect life on land? How might it affect your life? (Hint: Think food webs.)

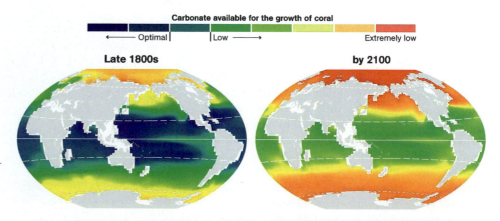

FIGURE 9.C Calcium carbonate levels in ocean waters, calculated from historical data (left), and projected for 2100 (right). Colors shifting from blue to red indicate waters becoming less basic.

Sources: Andrew G. Dickson, Scripps Institution of Oceanography, U.C. San Diego, and Sarah Cooley, Woods Hole Oceanographic Institution. Used by permission from National Geographic.

National Geographic Maps/National Geographic Creative

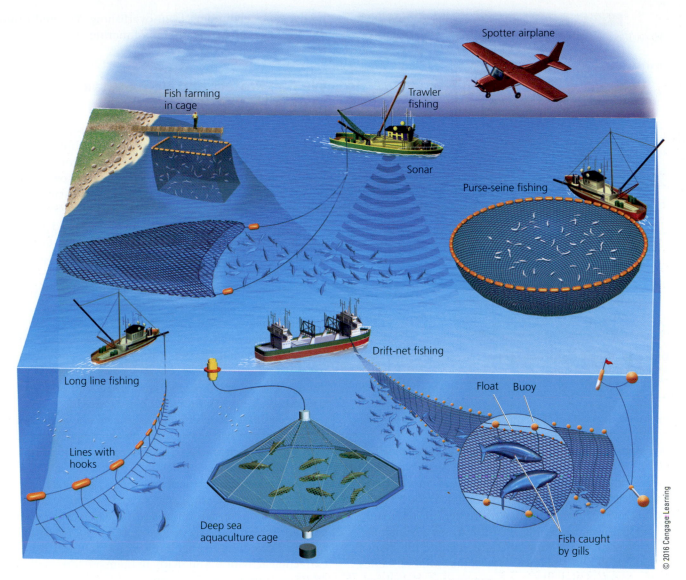

FIGURE 9.24 Major commercial fishing methods used to harvest various marine species (along with methods used to raise fish through aquaculture).

USFWS, bioinvaders are blamed for about two-thirds of all fish extinctions in the United States since 1900 and have caused huge economic losses.

CONSIDER THIS. . .

Researchers have found that invasive lionfish (Figure 9.23) eat at least 50 species of prey fish, including parrotfish, that normally consume enough algae around coral reefs to keep the algae from overgrowing and killing the corals. Scientists warn that, where lionfish are now the dominant species, such as in the Bahamas, where the tourist trade is largely dependent on healthy reefs, unchecked algal growth could overwhelm and destroy some reefs.

According to the IUCN, 34% of the world's known marine fish species and 71% of the world's freshwater fish species face premature extinction within your lifetime.

Overfishing: Gone Fishing, Fish Gone

Fish and fish products provide about 20% of the world's protein consumed by billions of people. A **fishery** is a concentration of a particular wild aquatic species suitable for commercial harvesting in a given ocean area or inland body of water. Today, fish are hunted throughout the world's oceans by a global fleet of about 4.4 million fishing boats.

Industrial fishing fleets use global satellite positioning equipment, sonar fish-finding devices, huge nets and long fishing lines, spotter planes, and refrigerated factory ships that can process and freeze their enormous catches. These highly efficient fleets help to supply the growing demand for seafood, but critics say that they are vacuuming the seas, reducing marine biodiversity, and degrading important marine ecosystem services. Figure 9.24 shows the

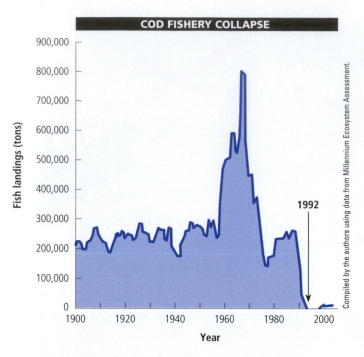

Compiled by the authors using data from Millennium Ecosystem Assessment.

FIGURE 9.25 Natural capital degradation: The collapse of New-foundland's Atlantic cod fishery.

major methods used for the commercial harvesting of various marine fishes and shellfish.

A **fishprint** is defined as the area of ocean needed to sustain the fish consumption of an average person, a nation, or the world. It is based on the concept of an ecological footprint (see Figure 1.10, p. 12). According to the *Fishprint of Nations 2006*, all fishing nations together are harvesting well over half again as many wild fish as the fish populations can sustain in the long run. According to the Woods Hole National Fisheries Service, 57% of the world's fisheries are fully exploited and 30% are overexploited or depleted. Such overharvesting has led to the collapse of some of the world's major fisheries (Figure 9.25).

One result of the increasingly efficient global hunt for fish is that larger individuals of commercially valuable wild species—including cod, marlin, swordfish, and tuna—are becoming scarce. Between 1950 and 2006, according to a study led by marine ecologist Boris Worm, 90% or more of these and other large, predatory, open-ocean fishes disappeared. Another effect of overfishing is that when larger predatory species dwindle, rapidly reproducing invasive species such as jellyfish can more easily take over and disrupt ocean food webs.

Also, as commercially valuable large species are overfished, the fishing industry has begun working its way down marine food webs by shifting to smaller marine species such as anchovies, herring, sardines, and shrimp-like krill—known as forage fish. About 90% of this catch is converted to fishmeal and fish oil, most of which is fed to farmed fish. Scientists warn that this will reduce the food supply for larger species, which will likely then have a harder time rebounding from overfishing. The end result will likely be further disruption of marine ecosystems and their ecosystem services.

We Can Protect and Help Sustain Marine Biodiversity

Protecting marine biodiversity is difficult for several reasons. *First,* the human ecological footprint and fishprint are expanding so rapidly that it is difficult to monitor their impacts. *Second,* much of the damage to the oceans and other bodies of water is not visible to most people. *Third,* many people incorrectly view the seas as an inexhaustible resource that can absorb an almost infinite amount of waste and pollution and still produce all the seafood we want. *Fourth,* most of the world's ocean area lies outside the legal jurisdiction of any country. Thus, much of it is an open-access resource, subject to overexploitation—a classic example of the tragedy of the commons (see Chapter 1, pp. 11–12).

Nevertheless, there are several ways to protect and sustain marine biodiversity, thereby increasing our beneficial environmental impact (**Concept 9.6**). For example, we can *protect endangered and threatened aquatic species,* as discussed in Chapter 8, and we can restore and sustain streams, wetlands, and other aquatic systems.

We can also *establish protected marine sanctuaries.* Since 1986, the IUCN has helped to establish a global system of *marine protected areas (MPAs)*—areas of ocean partially protected from human activities. According to the U.S. National Ocean Service, there are more than 5,800 MPAs worldwide (more than 1,600 in U.S. waters), covering about 2.8% of the world's ocean surface, and their numbers are growing. However, most MPAs allow dredging, trawler fishing, and other ecologically harmful resource extraction activities. And many of them are too small to be effective in protecting larger species.

Many scientists and policymakers call for protecting and sustaining entire marine ecosystems within a global network of fully protected *marine reserves,* some of which already exist. These areas are declared off-limits to destructive human activities such as commercial fishing, dredging, mining, and waste disposal in order to enable their ecosystems to recover and flourish.

Marine reserves work and they work quickly. **GOOD NEWS** Scientific studies show that in fully protected marine reserves, on average, commercially valuable fish populations double, fish size grows by almost a third, fish reproduction triples, and species diversity increases by almost one-fourth. Furthermore, these improvements can happen within 2–4 years after strict protection begins.

Despite the importance of such protection, only about 1.2% of the world's oceans are fully protected, compared to about 5% of the world's land. In other words, 98.8% of the world's oceans are not effectively protected from harmful human activities. Also, many of the existing reserves are fully protected only on paper because of short-

individuals matter 9.1

Sylvia Earle—Advocate for the Oceans

Tyrone Turner/National Geographic

Sylvia Earle is one of the world's most respected oceanographers and is a National Geographic Society Explorer-in-Residence. She has taken a leading role in helping us to understand the world's oceans and to protect them. *Time* magazine named her the first *Hero for the Planet* and the U.S. Library of Congress calls her "a living legend."

Earle has led more than 100 ocean research expeditions and has spent more than 7,000 hours underwater, either diving or descending in research submarines to study ocean life. She has focused her research on the ecology and conservation of marine ecosystems, with an emphasis on developing deep-sea exploration technology.

She is the author of more than 175 publications and has been a participant in numerous radio and television productions. During her long career, Earle has also served as the Chief Scientist of the U.S. National Oceanic and Atmospheric Administration (NOAA) and she has founded three companies devoted to developing submarines and other devices for deep-sea exploration and research. She has received more than 100 major international and national honors, including a place in the National Women's Hall of Fame.

These days, Earle is leading a campaign called *Mission Blue* to finance research and to ignite public support for a global network of marine protected areas, which she dubs "hope spots." Her goal is to help save and restore the oceans, which she calls "the blue heart of the planet." She says, "There is still time, but not a lot, to turn things around."

Background photo: Igor Kovalchuk/Shutterstock.com

ages of funding and a need for more trained staff to manage and monitor them.

Many marine scientists want to set aside 10% to 30% of the world's oceans as fully protected marine reserves—a very important way to increase our beneficial environmental impact. One such researcher is marine ecologist and National Geographic Explorer-in-Residence Enric Sala, who is searching for and studying pristine marine ecosystems worldwide. His scientific publications are used as guidelines for the creation of marine reserves. Another leading marine scientist is National Geographic Explorer-in-Residence Sylvia Earle (Individuals Matter 9.1).

CONSIDER THIS. . .

THINKING ABOUT Marine Reserves

Do you support setting aside at least 30% of the world's oceans as fully protected marine reserves? Explain. How would this affect your life? How would you fund this protection?

Taking an Ecosystem Approach to Sustaining Aquatic Biodiversity

Edward O. Wilson (see Individuals Matter 4.1, p. 67) and other biodiversity experts have promoted an ecosystem approach to sustaining terrestrial biodiversity. Wilson has proposed the following strategies for applying this approach to aquatic biodiversity:

- Complete the mapping of the world's aquatic biodiversity, identifying and locating as many plant and animal species as possible.

- Identify and preserve the world's aquatic biodiversity hotspots and areas where deteriorating ecosystem services threaten people and other forms of life.

- Create large and fully protected marine reserves to allow damaged marine ecosystems to recover and to allow fish stocks to be replenished.

- Protect and restore the world's lakes and river systems, which are the most threatened ecosystems of all.

- Initiate worldwide ecological restoration projects in systems such as coral reefs and inland and coastal wetlands.

- Find ways to raise the incomes of people who live in or near protected lands and waters so that they can become partners in the protection and sustainable use of ecosystems.

There is growing evidence that many of the harmful effects of human activities on both terrestrial and aquatic biodiversity and ecosystem services could be reversed over the next two decades. Doing this will require implementing an ecosystem approach to sustaining both terrestrial and aquatic ecosystems. According to Edward O. Wilson, such a conservation strategy would cost about $30 billion per year—an amount that could be provided by a tax of one penny per cup of coffee consumed in the world each year.

This strategy for protecting the earth's vital biodiversity and increasing our beneficial environmental impact will not be implemented without bottom-up political pressure on elected officials from individual citizens and groups. It will also require cooperation among scientists, engineers, and key people in government and the private sector. And it will be important for individuals to "vote with their wallets" by trying to buy only products and services that do not have harmful impacts on terrestrial and aquatic biodiversity.

BIG IDEAS

- The economic values of the ecosystem services provided by the world's ecosystems are far greater than the value of raw materials obtained from those systems.

- We can sustain terrestrial biodiversity and ecosystem services and increase our beneficial environmental impact by protecting severely threatened areas, protecting remaining undisturbed areas, restoring damaged ecosystems, and sharing with other species much of the land we dominate.

- We can sustain aquatic biodiversity and increase our beneficial environmental impact by establishing protected marine sanctuaries, managing coastal development, reducing water pollution, and preventing overfishing.

TYING IT ALL TOGETHER

Sustaining Costa Rica's Biodiversity

In this chapter, we looked at how human activities are destroying or degrading much of the earth's terrestrial and aquatic biodiversity. We discussed the importance of preserving what remains of richly diverse and highly endangered biodiversity hotspots and of sustaining the earth's ecosystem services. We also saw how we can reduce this destruction and degradation by using the earth's resources more sustainably and by employing restoration ecology and reconciliation ecology. The **Core Case Study** introduced much of this by reporting on what Costa Rica is doing to protect and restore its precious biodiversity.

Preserving terrestrial and aquatic biodiversity involves applying the three **scientific principles of sustainability**. First, it means respecting biodiversity and understanding the value of sustaining it. Also, if we rely less on fossil fuels and more on direct solar energy and its indirect forms, such as wind and flowing water, we will generate less pollution and interfere less with chemical cycling and other forms of natural capital that sustain biodiversity and our own lives and economies.

We can also apply the three **social science principles of sustainability** to help preserve biodiversity. By placing

economic value on ecosystem services, we would more clearly acknowledge their importance. By working together to find win-win solutions to problems of environmental degradation, we would benefit both the earth and its people. The search for such solutions could be guided by an ethical responsibility to sustain biodiversity and ecosystem services for current and future generations.

Eduardo Rivero/Shutterstock.com

Chapter Review

Core Case Study

1. Summarize the story of Costa Rica's efforts to preserve its rich biodiversity.

Section 9.1

2. What are the two key concepts for this section? Distinguish among **old-growth (primary) forests**, **second-growth forests**, and **tree plantations** (**tree farms** or **commercial forests**). What major ecological and economic benefits do forests provide? Describe the efforts of scientists and economists to put a price tag on the major ecosystem services provided by forests and other ecosystems.

3. Explain how building roads into previously inaccessible forests can harm the forests. Distinguish among selective cutting, clear-cutting, and strip cutting in the harvesting of trees. What are two types of forest fires? What are some ecological benefits of occasional surface fires?

4. What is **deforestation** and what parts of the world are experiencing the greatest forest losses? List some major harmful environmental effects of deforestation. Summarize the story of reforestation in the United States. Explain how increased reliance on tree plantations can reduce overall forest biodiversity and degrade forest topsoil. Summarize the trends in tropical deforestation. What are four major causes of tropical deforestation? Explain how widespread tropical deforestation can convert a tropical forest to tropical grassland (savanna).

Section 9.2

5. What is the key concept for this section? What is certified sustainably grown timber? List four ways to manage forests more sustainably. What are four ways to reduce the harm caused by forest fires to forests and to people? What are three ways to reduce the need to harvest trees? Describe the global fuelwood crisis. What are five ways to protect tropical forests and use them more sustainably?

Section 9.3

6. What is the key concept for this section? Distinguish between **rangelands** and **pastures**. What is **overgrazing** and what are its harmful environmental effects? What are three ways to reduce overgrazing and use rangelands more sustainably?

Section 9.4

7. What is the key concept for this section? What are the major environmental threats to national parks in the world and in the United States? Describe some of the ecological effects of reintroducing the gray wolf to Yellowstone National Park. What percentage of the world's land has been set aside and protected as nature reserves, and what percentage should be protected, according to conservation biologists? What is the buffer zone concept? How has Costa Rica applied this approach? What is **wilderness** and why is it important, according to conservation biologists? Summarize the history of wilderness protection in the United States.

Section 9.5

8. What is the key concept for this section? Summarize the five-point strategy recommended by biologists for protecting terrestrial ecosystems. What is a **biodiversity hotspot** and why is it important to protect such areas? Explain the importance of protecting ecosystem services and list three ways to do this. Define **ecological restoration**. What are four approaches to restoration? Summarize the science-based, four-step strategy for carrying out ecological restoration and rehabilitation. Describe the ecological restoration of Guanacaste National Park in Costa Rica. Define and give three examples of **reconciliation ecology**.

Section 9.6

9. What is the key concept for this section? Summarize the threats to aquatic biodiversity resulting from human activities. What is ocean acidification and why is it a major threat? Define **fishery** and summarize the threats to marine fisheries. Briefly describe the major industrial fish harvesting methods. What is a **fishprint**? Why is it difficult to protect marine biodiversity? What are three ways in which we could try to protect marine biodiversity? What percentage of the world's oceans is strictly protected from harmful human activities in marine reserves? Summarize the contributions of Sylvia Earle to the protection of aquatic biodiversity. How can the ecosystem approach be applied to protecting aquatic biodiversity?

10. What are this chapter's *three big ideas*? Explain the relationship between preserving biodiversity as it is done in Costa Rica and the six **principles of sustainability**.

Note: Key terms are in bold type.

Critical Thinking

1. Why do you think Costa Rica (**Core Case Study**) has set aside a much larger percentage of its land for biodiversity conservation than the United States has? Should the United States reserve more of its land for this purpose? Explain.

2. If we fail to protect a much larger percentage of the world's remaining old-growth forests and tropical rain forests, what are three harmful effects that this failure is likely to have on any children and grandchildren you eventually might have?

3. In the early 1990s, Miguel Sanchez, a subsistence farmer in Costa Rica, was offered $600,000 by a hotel developer for a piece of land that he and his family had been using sustainably for many years. An area under rapid development surrounded the land, which contained an old-growth rain forest and a black sand beach. Sanchez refused the offer. Explain how Sanchez's decision was an application of one of the **social science principles of sustainability**. What would you have done if you were Sanchez? Explain.

4. Should more-developed countries provide at least half of the money needed to help preserve the remaining tropical forests in less-developed countries?

Explain. Do you think that the long-term economic and ecological benefits of doing this would outweigh the short-term economic costs? Explain.

5. Are you in favor of establishing more wilderness areas in the United States (or in the country where you live)? Explain. What might be some drawbacks of doing this?

6. You are a defense attorney arguing in court for preserving an old-growth forest that developers want to clear for a suburban development. Give your three strongest arguments for preserving this ecosystem. How would you counter the argument that preserving the forest would harm the economy by causing a loss of jobs in the timber industry?

7. What do you think are the three greatest threats to aquatic biodiversity and aquatic ecosystem services? For each of them, explain your thinking. Imagine that you are a national official in charge of setting policy for preserving aquatic biodiversity and outline a plan for dealing specifically with these threats.

8. Some scientists consider ocean acidification to be one of the most serious environmental and economic threats that the world faces. How do you think you might be contributing to ocean acidification in your daily life? What are three things you could do to help reduce the threat of ocean acidification?

Doing Environmental Science

Pick an area near where you live or go to school that hosts a variety of plants and animals. It could be a yard, an abandoned lot, a park, a forest, or some part of your campus. Visit this area at least three times and make a survey of the plants and animals that you find there, including any trees, shrubs, groundcover plants, insects, reptiles, amphibians, birds, and mammals. Also, take a small sample of the topsoil and find out what organisms

are living there. (Be careful to get permission from whoever owns or manages the land before doing any digging.) Using guidebooks and other resources to help identify different species, record your findings and categorize them into the general types of organisms listed above. Then do some research to find out about the ecosystem services that some or all of these organisms provide. Try to find and record five of these services. Finally, do some research to find a range of values that economists have assigned to these ecosystem services at the global level. Write a report summarizing your findings.

Global Environment Watch Exercise

Go to the *Forests and Deforestation* portal and next to the Statistics heading click "View All." On this page, click on "Share of Tropical Deforestation, 2000–2005." Choose one

of these countries and research the deforestation in this country further (tip: use the World Map feature). Write a report on your findings and include possible solutions for this deforestation problem. Solutions may include those legislated by governments, as well as those being tried by private individuals or companies.

Ecological Footprint Analysis

A fishprint provides a measure of a country's fish harvest in terms of area. The unit of area used in fishprint analysis is the global hectare (gha), a unit weighted to reflect the relative ecological productivity of the area fished. When compared with the fishing area's *sustainable biocapacity* (its ability to provide a stable supply of fish year after year, expressed in terms of yield per area), its fishprint indicates whether the country's annual fishing harvest is sustainable. The fishprint and biocapacity are calculated using the following formulas:

Fishprint in (gha) = metric tons of fish harvested per year/productivity in metric tons per hectare × weighting factor

Biocapacity in (gha) = sustained yield of fish in metric tons per year/productivity in metric tons per hectare × weighting factor

The following graph shows the earth's total fishprint and biocapacity between 1950 and 2000. Study it and answer the following questions.

1. Based on the graph,
   a. In what year did the global fishprint begin to exceed the biological capacity of the world's oceans?
   b. By how much did the global fishprint exceed the biological capacity of the world's oceans in 2000?

2. Assume a country harvests 18 million metric tons of fish annually from an ocean area with an average productivity of 1.3 metric tons per hectare and a weighting factor of 2.68. What is the annual fishprint of that country?

3. If biologists determine that this country's sustained yield of fish is 17 million metric tons per year,
   a. What is the country's sustainable biological capacity?
   b. Is the county's annual fishing harvest sustainable?
   c. To what extent, as a percentage, is the country undershooting or overshooting its biological capacity?

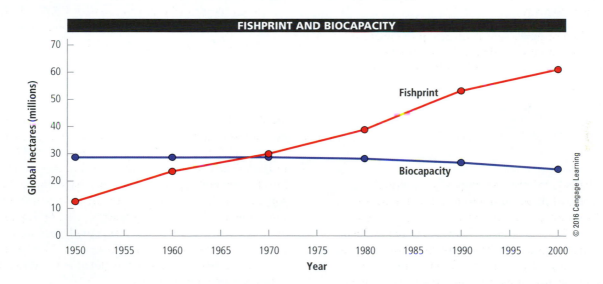

FISHPRINT AND BIOCAPACITY

© 2016 Cengage Learning

CENGAGEbrain.com To access course materials, including Aplia homework, please visit www.cengagebrain.com.

WWW.CENGAGEBRAIN.COM   **213**

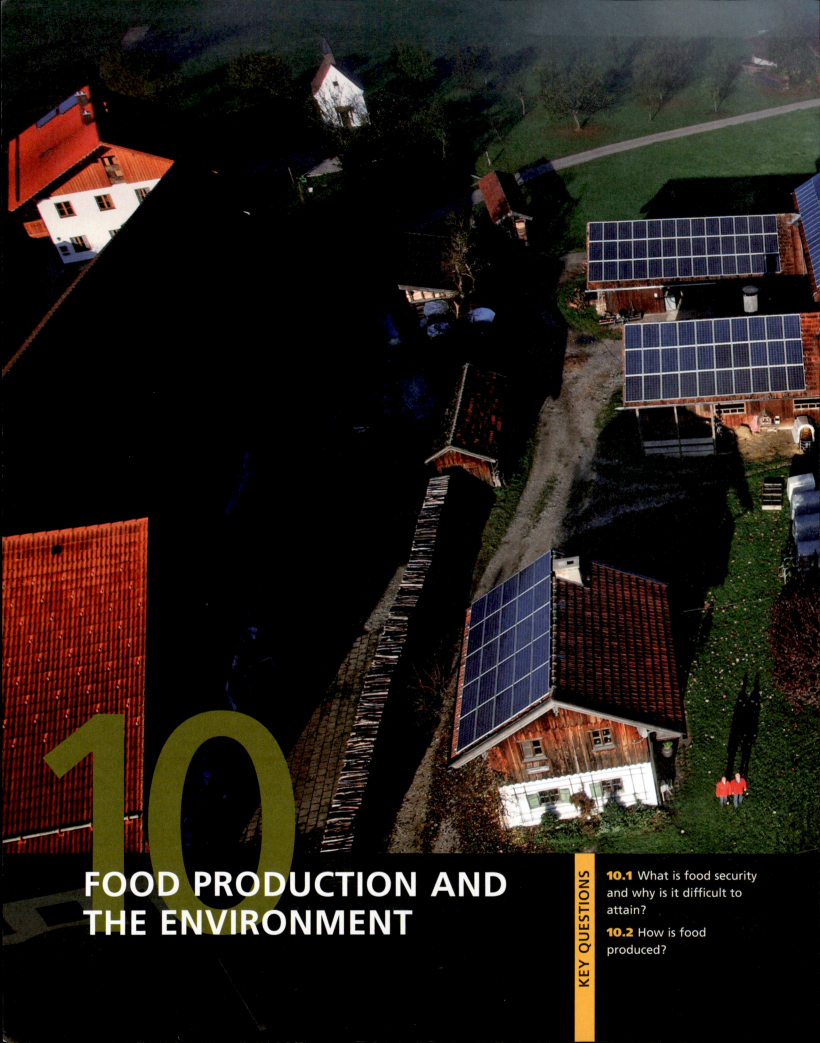

10

FOOD PRODUCTION AND THE ENVIRONMENT

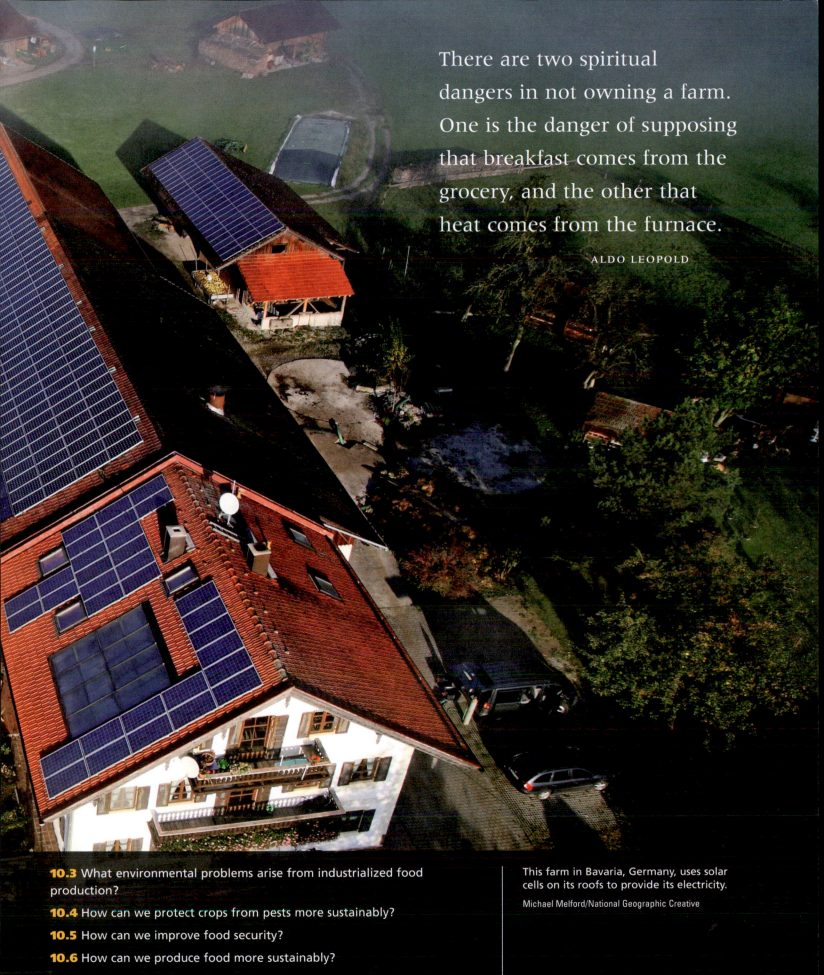

There are two spiritual dangers in not owning a farm. One is the danger of supposing that breakfast comes from the grocery, and the other that heat comes from the furnace.

ALDO LEOPOLD

10.3 What environmental problems arise from industrialized food production?

10.4 How can we protect crops from pests more sustainably?

10.5 How can we improve food security?

10.6 How can we produce food more sustainably?

This farm in Bavaria, Germany, uses solar cells on its roofs to provide its electricity.

Michael Melford/National Geographic Creative

Growing Power—An Urban Food Oasis

A *food desert* is an urban area where people have little or no easy access to nutritious food without traveling long distances. In the United States, an estimated 23.5 million people, including 6.5 million children, live in such areas, according to the U.S. Centers for Disease Control and Prevention. People living in these urban neighborhoods tend to rely on convenience stores and fast food restaurants that mainly offer high-calorie, highly processed foods that can lead to higher risks of obesity, diabetes, and heart disease.

Will Allen (Figure 10.1), one of six children of a sharecropper, grew up on a farm in Maryland, but left the farming life for college and a professional basketball career, followed by a successful corporate marketing career. In 1993, Allen had decided to return to his roots. He bought the last working farm within the city limits of Milwaukee, Wisconsin, and, in time, created a food oasis in a food desert.

On this small urban plot, Allen developed Growing Power, Inc., an ecologically based farm and a showcase for forms of agriculture that apply all three **scientific principles of sustainability**. It is powered partly by solar electricity and solar hot water systems and makes use of several greenhouses to capture solar energy for growing food throughout the year. The farm produces an amazing diversity of crops—150 varieties of organic vegetables along with organic herbs and sunflowers. It also produces organically raised chickens, turkeys, goats, fish such as tilapia and perch, and honeybees. And the farm's nutrients are recycled in creative ways. For

Courtesy of Growing Power, Inc.

FIGURE 10.1 In 1996, Will Allen founded Growing Power—an urban farm in Milwaukee, Wisconsin.

example, wastes from the farmed fish are used as nutrients to raise some of the crops.

The farm's products are sold locally at Growing Power farm stands throughout the region and to various restaurants. Allen also worked with the city of Milwaukee to establish the Farm-to-City Market Basket program through which people can sign up for weekly deliveries of organic produce at modest prices.

In addition, Growing Power runs an education program in which schoolchildren visit the farm and learn about where their food comes from. Allen also runs a training program for about 1,000 people every year who want to learn organic farming methods. The farm has also partnered with the city of Milwaukee to create 150

new green jobs for unemployed and low-income workers, building greenhouses, and growing food organically. Growing Power has expanded, locating another urban farm in a neighborhood of Chicago, Illinois, and setting up satellite training sites in five other states.

For his creative and energetic efforts, Allen has won several prestigious awards. However, he is most proud of the fact that his urban farm helps to feed more than 10,000 people every year and puts people to work raising good food.

In this chapter we look at different ways to produce food, the environmental effects of food production, and how to produce food more sustainably.

10.1 WHAT IS FOOD SECURITY AND WHY IS IT DIFFICULT TO ATTAIN?

CONCEPT 10.1A Many people in less-developed countries have health problems from not getting enough food, while many people in more-developed countries suffer health problems from eating too much.

CONCEPT 10.1B The greatest obstacles to providing enough food for everyone are poverty, war, bad weather, climate change, and the harmful environmental effects of industrialized food production.

Poverty Is the Root Cause of Food Insecurity

Food security is the condition under which all or most of the people in a population have daily access to enough nutritious food to live active and healthy lives. More than 1 billion people work in agriculture to grow crops on about 38% of the earth's ice-free land. They produce more than enough food to meet the basic nutritional needs of every person on the earth. Despite this food surplus, one of every six people in the world is not getting enough to eat. These people face **food insecurity**—living with chronic hunger and poor nutrition, which threatens their ability to lead healthy and productive lives (**Concept 10.1A**).

Most agricultural experts agree that *the root cause of food insecurity is poverty*, which prevents poor people from growing or buying enough food to meet their needs. This is not surprising given that about half of the world's people are trying to survive on the equivalent of $2.25 a day and one out of six people struggle to get by on $1.25 a day.

Other obstacles to food security are war, corruption, political upheaval, bad weather (such as prolonged drought, flooding, and heat waves), and climate change (**Concept 10.1B**). Another reason for food insecurity is that about 33% of the calories in the world's grain production are used to produce meat that many poor people cannot afford, 5% are used to produce biofuels for cars, and another 33% are wasted.

Each day, there are about 232,000 more people at the world's dinner tables and many of them will have little or no food on their plates. By 2050, there will likely be at least 2.6 billion more people to feed. This projected population increase will be more than twice China's current population and more than eight times the current U.S. population. Most of these people will live in the major cities of less-developed countries. A critical question is, how will we feed the projected 9.7 billion people in 2050 without causing serious harm to the environment? We explore possible answers to this question throughout this chapter.

Many People Suffer from Chronic Hunger and Malnutrition

To maintain good health and resist disease, individuals need fairly large amounts of *macronutrients* (such as carbohydrates, proteins, and fats) and smaller amounts of *micronutrients*—vitamins, such as A, B, C, and E, and minerals such as iron, iodine, and calcium.

People who cannot grow or buy enough food to meet their basic energy needs suffer from **chronic undernutrition**, or **hunger** (**Concept 10.1A**). Most of the world's hungry people live in low-income, less-developed countries and typically can afford only a low-protein, high-carbohydrate, vegetarian diet consisting mainly of grains such as wheat, rice, and corn. In more-developed countries, people living in food deserts (**Core Case Study**) have a similar problem, except that their diet is heavy on cheap food loaded with fats, sugar, and salt. In both cases, people often suffer from **chronic malnutrition**, a condition in which they do not get enough protein and other key nutrients. This can weaken them, make them more vulnerable to disease, and hinder the normal physical and mental development of children.

According to the United Nations Food and Agriculture Organization (FAO), there are about 1 billion chronically undernourished and malnourished people (see Figure 1.13, p. 16)—one of every seven people on the planet. The FAO estimates that each year, nearly 6 million children younger than age 5 die from chronic hunger and malnutrition and from increased susceptibility to normally nonfatal infectious diseases (such as measles and diarrhea) because of their weakened condition. That is an average of 11 children per minute dying from these causes.

Many People Do Not Get Enough Vitamins and Minerals

About 2 billion people, most of them in less-developed countries, suffer from a deficiency of one or more vitamins and minerals, usually *vitamin A, iron,* and *iodine* (**Concept 10.1A**). According to the World Health Organization (WHO), at least 250,000 children younger than age 6, most of them in less-developed countries, go blind every year from a lack of vitamin A. Within a year, more than half of them die.

Having too little *iron* (Fe)—a component of the hemoglobin that transports oxygen in the blood—causes *anemia*, which causes fatigue, makes infection more likely, and increases a woman's chances of dying from hemorrhage in childbirth. According to the WHO, one of every five people in the world—mostly women and children in less-developed countries—suffers from iron deficiency.

The chemical element *iodine* (I) is essential for proper functioning of the thyroid gland, which produces hormones that control the body's rate of metabolism. Chronic lack of iodine can cause stunted growth, mental retardation, and goiter—a severely swollen thyroid gland that can lead to

FIGURE 10.2 This woman in Niger, West Africa suffers from goiter, an enlargement of the thyroid gland, caused by a lack of iodine in her diet.

deafness (Figure 10.2). According to the United Nations (UN), some 600 million people (almost twice the current U.S. population) suffer from goiter, most of them in less-developed countries. Also, 26 million children suffer irreversible brain damage every year from lack of iodine. The FAO and the WHO estimate that eliminating this serious health problem by adding traces of iodine to salt would cost the equivalent of only 2–3 cents per year for every person in the world.

Many People Have Health Problems from Eating Too Much

Overnutrition occurs when food energy intake exceeds energy use and causes excess body fat. Too many calories, too little exercise, or both can cause overnutrition. People who are underfed and underweight and those who are overfed and overweight share similar health problems: *lower life expectancy, greater susceptibility to disease and illness,* and *lower productivity and life quality* (**Concept 10.1A**).

We live in a world where, according to the WHO, about 1 billion people face health problems because they do not get enough nutritious food to eat and another 1.6 billion have health problems because they eat too much. The WHO estimates that obesity contributes to the deaths of 2.8 million people per year.

In the United States, according to the U.S. Centers for Disease Control and Prevention (CDC), about 69% of adults over age 20 and 33% of all children are overweight or obese. A 2013 study by Columbia University and the Robert Wood Johnson Foundation found that obesity plays an important role in nearly one in five deaths in the United States from heart disease, stroke, type 2 diabetes, and some forms of cancer.

10.2 HOW IS FOOD PRODUCED?

CONCEPT 10.2 We have used high-input industrialized agriculture and lower-input traditional agriculture to greatly increase food supplies.

Food Production Has Increased Dramatically

Three systems supply most of the world's food. *Croplands* that produce grains—primarily rice, wheat, and corn—provide about 77% of the world's food. The rest is provided by *rangelands, pastures,* and *feedlots* that produce meat and meat products and *fisheries* and *aquaculture* (fish farming) that supply fish and shellfish. Since 1960, there has been a staggering increase in global food production from all three of the major food production systems (**Concept 10.2**).

These three systems depend on a small number of plant and animal species. About two out of three of the world's people survive primarily by eating three grain crops—*rice, wheat,* and *corn*—because they cannot afford to eat meat. Only a few species of mammals and fish provide most of the world's meat and seafood. (See Figure 14 in Supplement 5, p. S46, for a closer look at the loss of food diversity in the United States.)

Such food specialization puts us in a vulnerable position, should any of the small number of crop strains, livestock breeds, and fish and shellfish species we depend on become depleted as a result of factors such as disease, environmental degradation, and climate change. This violates the biodiversity **principle of sustainability**, which calls for depending on a variety of food sources as an ecological insurance policy for dealing with changes in environmental conditions that have occurred throughout human history.

Industrialized Crop Production Relies on High-Input Monocultures

The two major types of agriculture are industrialized agriculture and subsistence agriculture. **Industrialized agriculture**, or **high-input agriculture**, uses heavy equipment along with large amounts of financial capital, fossil fuels, water, commercial inorganic fertilizers, and pesticides to produce single crops, or *monocultures*. The major goal of industrialized agriculture is to steadily increase each crop's *yield*—the amount of food produced per unit of land. Industrialized agriculture is practiced on 25% of all cropland, mostly in more-developed countries, and produces about 80% of the world's food (**Concept 10.2**).

Plantation agriculture is a form of industrialized agriculture used primarily in tropical less-developed countries. It involves growing *cash crops* such as bananas, coffee, vegetables, soybeans (mostly to feed livestock; see Figure 1.4, p. 8), sugarcane (to produce sugar and ethanol fuel), and palm oil (to produce cooking oil and biodiesel fuel).

These crops are grown on large monoculture plantations, mostly for export to more-developed countries.

Traditional Agriculture Often Relies on Low-Input Polyculture

Traditional, low-input agriculture provides about 20% of the world's food crops on about 75% of its cultivated land, mostly in less-developed countries. **Traditional subsistence agriculture** supplements energy from the sun with the labor of humans and draft animals to produce enough crops for a farm family's survival, with little left over to sell or store as a reserve for hard times. In **traditional intensive agriculture**, farmers try to obtain higher crop yields by increasing their inputs of human and draft-animal labor, animal manure for fertilizer, and water.

Some traditional farmers focus on cultivating a single crop, but many grow several crops on the same plot simultaneously, a practice known as **polyculture**. This method relies on solar energy and natural fertilizers such as animal manure. The various crops mature at different times, providing food throughout the year and keeping the topsoil covered to reduce erosion from wind and water. Polyculture also lessens the need for fertilizer and water because root systems at different depths in the soil capture nutrients and moisture efficiently.

Polyculture is an application of the biodiversity **principle of sustainability**, because crop diversity helps protect and replenish the soil and reduces the chance of losing most or all of the year's food supply to pests, bad weather, and other misfortunes. Research shows that, on average, low-input polyculture produces higher yields than does high-input industrialized monoculture, while using less energy and fewer resources and providing more food security **GOOD NEWS** for small landowners. For example, ecologists Peter Reich and David Tilman found that carefully controlled polyculture plots with 16 different species of plants consistently out-produced plots with 9, 4, or only 1 type of plant species.

Such research explains why some analysts argue for greatly expanding the use of polyculture to produce food more sustainably. The Growing Power farm (**Core Case Study**) practices polyculture by growing a variety of crops in inexpensive greenhouses—an application of the solar energy and biodiversity **principles of sustainability**.

Organic Agriculture Is on the Rise

A fast-growing sector of the U.S. and world economies is **organic agriculture**, in which crops are grown without the use of synthetic pesticides, synthetic inorganic fertilizers, and genetically engineered varieties, and animals must be raised on 100% organic feed without the use of antibiotics or growth hormones. Growing Power (**Core Case Study**) has become a well-known model for such food production. Figure 10.3 compares organic agriculture with industrialized agriculture.

Industrialized Agriculture

Uses synthetic inorganic fertilizers and sewage sludge to supply plant nutrients

Makes use of synthetic chemical pesticides

Uses conventional and genetically modified seeds

Depends on nonrenewable fossil fuels (mostly oil and natural gas)

Produces significant air and water pollution and greenhouse gases

Is globally export-oriented

Uses antibiotics and growth hormones to produce meat and meat products

Organic Agriculture

Emphasizes prevention of soil erosion and the use of organic fertilizers such as animal manure and compost, but no sewage sludge, to supply plant nutrients

Employs crop rotation and biological pest control

Uses no genetically modified seeds

Reduces fossil fuel use and increases use of renewable energy such as solar and wind power for generating electricity

Produces less air and water pollution and greenhouse gases

Is regionally and locally oriented

Uses no antibiotics or growth hormones to produce meat and meat products

© 2016 Cengage Learning

FIGURE 10.3 Some major differences between industrialized agriculture and organic agriculture.

In the United States, by law, a label of *100 percent organic* (or *USDA Certified Organic*) means that a product is produced only by organic methods and contains all organic ingredients. About 13,000 of the 2.2 million farms in the United States are USDA certified organic. Products labeled *organic* must contain at least 95% organic ingredients and those labeled *made with organic ingredients* must contain at least 70% organic ingredients. The word *natural* is used on food labels primarily as an advertising ploy.

A Closer Look at Industrialized Crop Production

Farmers have two ways to produce more food: farming more land or getting higher yields from existing cropland. Since 1950, about 88% of the increase in global grain production has come from using high-input industrialized agriculture to increase crop yields. This process, called the **green revolution**, involves three steps. *First*, develop and plant monocultures of selectively bred or genetically engineered varieties of key grain crops such as rice, wheat, and corn. *Second*, produce high yields by using large inputs of water, synthetic inorganic fertilizers, and pesticides. *Third*, increase the number of crops grown per year on a plot of land through *multiple cropping*.

Between 1950 and 1970, in what was called the *first green revolution*, this high-input approach dramatically raised crop yields in most of the world's more-developed countries, especially the United States (see the Case Study that follows). In the *second green revolution*, which began in 1967, fast-growing varieties of rice and wheat, specially bred for tropical and subtropical climates, were introduced into middle-income, less-developed countries, including India, China, and Brazil.

Largely because of the two green revolutions, between 1950 and 2012, world grain production (Figure 10.4, left) increased by 72% and per capita grain production (Figure 10.4, right) grew by 28%. The world's three largest grain-producing countries—China, India, and the United States—produce almost half of the world's grains.

People directly consume about 46% of the world's grain production. Most of the rest is fed to livestock and thus is consumed by people who can afford to eat meat and meat products and to purchase biofuels such as ethanol and biodiesel. In 2012, for example, about 40% of the U.S. corn crop was used to produce biofuels (mostly energy-inefficient ethanol).

CASE STUDY

Industrialized Food Production in the United States

In the United States, industrialized farming has evolved into *agribusiness*, as a small number of giant multinational corporations increasingly control the growing, processing, distribution, and sale of food in U.S. and global markets. Since 1960, U.S. industrialized agriculture has more than doubled the yields of key crops such as wheat, corn, and soybeans without the need for cultivating more land. Such yield increases have kept large areas of U.S. forests, grasslands, and wetlands from being converted to farmland.

However, if we view productivity not as the yield per area of land but as the number of people fed per area of land, then China and India are more productive than the United States, according to research by the ETS Group. This has occurred because about 38% of the grain produced in the United States is used to feed livestock for meat production and to produce biofuels for cars.

Low-income people in less-developed countries typically spend 50% to 70% of their income on food, according to 2012 data from the FAO. The world's 1.4 billion poorest people, struggling to live on the equivalent of less

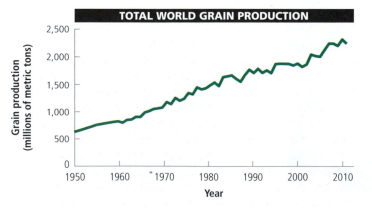
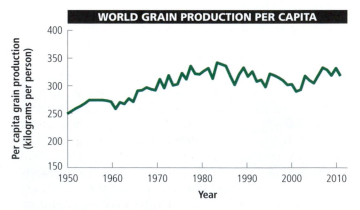

FIGURE 10.4 Growth in worldwide grain production (left) of wheat, corn, and rice, and per capita grain production (right) between 1950 and 2012. Since 1978, per capita grain production has generally leveled off. **Question:** Why do you think per capita grain production has grown less consistently than total grain production?

Compiled by the authors using data from U.S. Department of Agriculture, Worldwatch Institute, UN Food and Agriculture Organization, and Earth Policy Institute.

than $2.25 a day, typically spend about 70% of their meager income on food. By contrast, because of the efficiency of U.S. agriculture, Americans spend the lowest percentage of disposable income in the world—an average of 7%—on food.

However, because of a number of *hidden costs* related to their food production and consumption, most American consumers are not aware that their actual food costs are much higher than the market prices they pay. Such hidden costs include taxes to pay for farm *subsidies,* or government payments intended to help farmers stay in business and increase their yields. (Most subsidies go to producers of corn, wheat, soybeans, and rice.) Hidden costs also include the costs of pollution, environmental degradation, and higher health insurance bills related to the harmful environmental and health effects of industrialized agriculture (discussed in the next section). As food expert Michael Pollan puts it, "There is no such thing as cheap food. The real cost of food is paid somewhere. . . . if it isn't paid at the cash register, it's charged to the environment or to the public purse in the form of subsidies. And it's charged to your health."

Crossbreeding and Genetic Engineering Produce New Varieties of Crops and Livestock

For centuries, farmers and scientists have used *crossbreeding* through *artificial selection* to develop genetically improved varieties of crops and livestock animals. Such selective breeding in this first *gene revolution* has yielded amazing results. For example, ancient ears of corn were about the size of your little finger, and wild tomatoes were once the size of grapes, but most of the large varieties used now were selectively bred.

Traditional crossbreeding is a slow process, typically taking 15 years or more to produce a commercially valuable new crop variety, and it can combine traits only from species that are genetically similar. Typically, resulting varieties remain useful for only 5–10 years before pests and diseases reduce their yields. Important advances are still being made with this method.

Today, scientists are creating a *second gene revolution* by using *genetic engineering* to develop genetically modified strains of crops and livestock animals. Engineers use a process called *gene splicing* to alter an organism's genetic material through adding, deleting, or changing segments of its DNA. The goal of this process is to add desirable traits or to eliminate undesirable ones by enabling scientists to transfer genes between different species that would not normally interbreed in nature. The resulting organisms are called *genetically modified organisms (GMOs)*.

Compared to traditional crossbreeding, developing a new crop variety through gene splicing takes about half as long, usually costs less, and allows for the insertion of genes from almost any other organism into crop or animal cells. According to the U.S. Department of Agriculture (USDA), at least 80% of the food products on U.S. supermarket shelves contain some form of genetically engineered food or ingredients, and that proportion is growing.

Globally, 61 countries require the labeling of food with genetically modified ingredients. In the United States, federal laws do not require such labeling, even though polls indicate that 90% of the public wants to have such information included on food labels. This is largely because some seed and food companies have lobbied the U.S. Congress and state legislatures to oppose GMO food labeling. Foods labeled *100 percent organic* or *USDA Certified Organic* can make no use of genetically modified seeds or ingredients.

In 2013, the citizens of Washington state approved a measure to require the labeling of food with genetically modified ingredients. In 2014, the food industry was fighting such labeling by proposing voluntary national labeling with the provision that no state could mandate such labeling. The industry was also pushing to allow the use of the word *natural* on food with genetically modified ingredients.

Bioengineers hope to develop new GMO varieties of crops that are resistant to heat, cold, drought, insect pests, parasites, viral diseases, herbicides, and salty or acidic soil. They also hope to develop crop plants that can grow faster and survive with little or no irrigation and with less use of fertilizer and pesticides. Some scientists contend that such innovations hold great promise for helping to improve global food security. Others warn that genetic engineering is not free of drawbacks, which we examine later in this chapter.

Meat Consumption Has Grown Steadily

Meat and animal products such as eggs and milk are sources of high-quality protein and represent the world's second major food-producing system. Between 1961 and 2012, according to the FAO, production of meat and meat products—mostly beef, pork, poultry, mutton, milk, cheese, and eggs—increased more than twofold and the average consumption per person of meat and meat products more than doubled. Global meat consumption is likely to more than double again by 2050 as incomes rise and millions of people in rapidly developing countries move up the food chain and consume more meat and meat products every year. For example, meat consumption in China increased ninefold between 1978 and 2013.

About half of the world's meat comes from livestock grazing on grass in unfenced rangelands and enclosed pastures. The other half is produced through an industrialized factory farm system. It involves raising large numbers of animals bred to gain weight quickly, mostly in crowded *feedlots* (Figure 10.5) or in very crowded pens and cages in huge buildings called *concentrated animal feeding operations (CAFOs)* (Figure 10.6).

FIGURE 10.5 *Industrialized beef production:* On this cattle feedlot in west Texas, thousands of cattle are fattened on grain for a few months before being slaughtered.

FIGURE 10.6 Concentrated chicken feeding operation in Iowa (USA). Such operations can house up to 100,000 chickens.

In these facilities, the animals are fed grain, soybeans, fishmeal, or fish oil, and some of this feed is doctored with growth hormones and antibiotics to accelerate livestock growth. Because of the crowding, animal wastes and run-off from feedlots and CAFOs can have serious impacts on the air and water, which we examine later in this chapter.

Fish Production and Shellfish Production Have Risen Dramatically

The world's third major food-producing system consists of fisheries and aquaculture. A **fishery** is a concentration of particular aquatic species suitable for commercial harvesting in a given ocean area or inland body of water. Industrial fishing fleets harvest most of the world's marine catch of wild fish by using a variety of methods (see Figure 9.24, p. 207). Fish and shellfish are also produced through **aquaculture** or **fish farming**—the practice of raising fish in freshwater ponds, lakes, reservoirs, and rice paddies, and in underwater cages in coastal and deeper ocean waters.

Between 1950 and 2012, global seafood production of wild and farmed fish increased ninefold (**Concept 10.2**). In 2012, about 58% of the world's fish and shellfish were caught mostly by industrial fishing fleets and 42% were produced through aquaculture—the world's fastest-growing type of food production (Figure 10.7). In 2011, farmed fish production topped beef production, with China accounting for 62% of the world total. According to a 2012 report on world fisheries by the FAO, about 87% of the world's commercial ocean fisheries are being harvested at full capacity (57%) or overfished (30%).

Most of the world's aquaculture involves raising species that feed on algae or other plants—mainly carp in China and India, catfish in the United States, tilapia in several

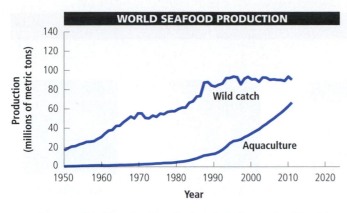

WORLD SEAFOOD PRODUCTION

FIGURE 10.7 World seafood production, including both wild catch (marine and inland) and aquaculture, grew between 1950 and 2012, with the wild catch generally leveling off since 1996 and aquaculture production rising sharply since 1990. **Question:** Why do you think the wild catch has leveled off?

Compiled by the authors using data from UN Food and Agriculture Organization, Worldwatch Institute, and Earth Policy Institute.

countries, and shellfish in a number of coastal countries. However, the farming of meat-eating species such as shrimp and salmon is growing rapidly, especially in more-developed countries. Such species are often fed fishmeal and fish oil produced from other fish and their wastes, although by 2014, this practice had declined slightly.

Industrialized Food Production Requires Huge Inputs of Energy

The industrialization of food production has been made possible by using fossil fuels—mostly oil and natural gas—to run farm machinery and fishing vessels, to pump irrigation water for crops, and to produce synthetic pesticides (mostly from petrochemicals produced when oil is refined) and synthetic inorganic fertilizers. Fossil fuels are also used to process food and transport it long distances within and between countries. Agriculture uses about 20% of all the energy used in the United States, where food items travel an average of 2,400 kilometers (1,300 miles) from farm to plate.

When we consider the energy used to grow, store, process, package, transport, refrigerate, and cook all plant and animal food, it takes about 10 units of fossil fuel energy to put 1 unit of food energy on the table in the United States. Also, according to a study led by ecological economist Peter Tyedmers, the world's fishing fleets use about 12.5 units of energy to put 1 unit of food energy from seafood on the table. In other words, today's systems for producing, processing, transporting, and preparing food are highly dependent on fossil fuels, and together, they result in a large *net energy loss.*

On the other hand, the amount of energy per calorie used to produce crops in the United States has declined by about 50% since the 1970s. One factor in this decline is

that the amount of energy used to produce synthetic nitrogen fertilizer has dropped sharply. Another is the rising use of conservation tillage, which sharply reduces energy use and the harmful environmental effects of plowing, as is explained further in the next section of this chapter.

10.3 WHAT ENVIRONMENTAL PROBLEMS ARISE FROM INDUSTRIALIZED FOOD PRODUCTION?

CONCEPT 10.3 Future food production may be limited by soil erosion and degradation, desertification, irrigation water shortages, air and water pollution, climate change, and loss of biodiversity.

Producing Food Has Major Environmental Impacts

Industrialized agriculture has allowed farmers to use less land to produce more food, and this has helped to protect biodiversity in many areas by reducing the destruction of forests and grasslands for farming. However, many analysts point out that industrialized agriculture has greater overall harmful environmental impacts (Figure 10.8) than any other human activity and that these environmental effects may limit future food production (**Concept 10.3**).

According to a 2010 study by 27 experts assembled by the United Nations Environment Programme (UNEP), agriculture uses massive amounts of the world's resources. It accounts for about 70% of the freshwater removed from aquifers and surface waters, worldwide. It also uses about 38% of the world's ice-free land, emits about 25% of the world's greenhouse gas emissions, and produces about 60% of all water pollution. As a result, many analysts view today's industrialized agriculture as environmentally and economically unsustainable, in the long run. However, proponents of industrialized agriculture argue that its benefits outweigh its harmful effects.

Topsoil Erosion Is a Serious Problem in Parts of the World

Soil is literally the foundation of life on land (Science Focus 10.1). Specifically, the fertile top layer of many soils, called **topsoil**, is a vital component of natural capital, because it stores the water and nutrients needed by plants. Irreplaceable topsoil is one of the earth's most important renewable resources, because our food production largely depends on it. Topsoil nutrients are recycled endlessly as long as they are not removed faster than natural processes replenish them. Thus, topsoil renewal is one of the earth's most important ecosystem services.

Food Production

Biodiversity Loss	Soil	Water	Air Pollution	Human Health
Conversion of grasslands, forests, and wetlands to crops or rangeland	Erosion	Aquifer depletion	Emissions of greenhouse gases CO_2 from fossil fuel use, N_2O from inorganic fertilizer use, and methane (CH_4) from cattle	Nitrates in drinking water (blue baby)
Fish kills from pesticide runoff	Loss of fertility	Increased runoff, sediment pollution, and flooding from cleared land		Pesticide residues in water, food, and air
Killing of wild predators to protect livestock	Salinization	Pollution from pesticides	Other air pollutants from fossil fuel use and pesticide sprays	Livestock wastes in drinking and swimming water
Loss of agrobiodiversity replaced by monoculture strains	Waterlogging	Algal blooms and fish kills caused by runoff of fertilizers and farm wastes		Bacterial contamination of meat
	Desertification			

© 2016 Cengage Learning

FIGURE 10.8 Food production has a number of harmful environmental effects (**Concept 10.3**).
Question: Which item in each of these categories do you think is the most harmful?

Left: Orientaly/Shutterstock.com. Left center: pacopi/Shutterstock.com. Center: Tim McCabe/USDA Natural Resources Conservation Service.
Right center: Mikhail Malyshev/Shutterstock.com. Right: B Brown/Shutterstock.com.

FIGURE 10.9 Natural capital degradation: Flowing water from rainfall is the leading cause of topsoil erosion as seen on this farm in the U.S. state of Tennessee (left). Severe water erosion can become gully erosion, which has damaged this cropland in western Iowa (right).

One major problem related to agriculture is **soil erosion**—the movement of soil components, especially surface litter and topsoil, from one place to another by the actions of wind and water. Some topsoil erosion is natural, but much of it is caused by human activities.

Flowing water, the largest cause of erosion, carries away particles of topsoil that have been loosened by rainfall (Figure 10.9, left). Severe erosion of this type leads to the formation of gullies (Figure 10.9, right). Wind also loosens and blows topsoil particles away, especially in areas with a dry climate and relatively flat and exposed land (Figure 10.10).

In undisturbed, vegetated ecosystems, the roots of plants help to anchor the topsoil and to prevent some erosion.

Lynn Betts/USDA Natural Resources Conservation Service

FIGURE 10.10 Wind is an important cause of topsoil erosion in dry areas that are not covered by vegetation such as this bare crop field in the U.S. state of Iowa.

However, topsoil can erode when we remove soil-holding grasses, trees, and other vegetation through activities such as farming (see Figure 7.13, p. 141), deforestation (see Figure 9.7, p. 190), and overgrazing (see Figure 9.15, p. 196). A joint survey by the UNEP and the World Resources Institute indicated that topsoil is eroding faster than it forms on about one-third of the world's cropland (Figure 10.11). (See the online Guest Essay on soil erosion by David Pimentel.)

Erosion of topsoil has three major harmful effects. One is *loss of soil fertility* through depletion of plant nutrients in topsoil. A second effect is *water pollution* in surface waters where eroded topsoil ends up as sediment, which can kill fish and shellfish and clog irrigation ditches, boat channels, reservoirs, and lakes. Additional water pollution occurs when the eroded sediment contains pesticide residues that can be ingested by aquatic organisms and in some cases biomagnified within food webs (see Figure 8.9, p. 171). Third, erosion releases carbon that was stored in the soil by vegetation into the air and water, thus altering the carbon cycle and adding to atmospheric levels of carbon dioxide (CO_2).

The rise of industrialized agriculture has exposed irreplaceable topsoil to erosion by water and wind and reduced the plant nutrient content of topsoil in many areas. This erosion of soil nutrients from topsoil, and from synthetic chemical fertilizers added to the soil, sends the nutrients on a one-way trip to crops and then to nearby bodies of surface water, which often become overloaded with plant nutrients. The continuing disruption of the nitrogen and phosphorus cycles (see Figures 3.15 and 3.16, pp. 54 and 55), due to the loss of topsoil and depletion of its key nutrients, is another factor that could eventually make industrialized agriculture unsustainable (**Concept 10.3**).

Soil pollution is also a problem in parts of the world. Some of the chemicals emitted into the atmosphere by industrial and power plants and by motor vehicles can pollute soil and water used to irrigate soil. Pesticides can also contaminate soil.

In 2014, China's environment ministry reported on a 4-year survey finding that 19% of China's arable (farmable) land is contaminated, especially with toxic metals such as cadmium, arsenic, and nickel, and that about 2.5% of the country's cropland is too contaminated to grow food safely. China, with 19% of the world's people and only 7% of the world's arable land, cannot afford to lose 2.5% of its cropland.

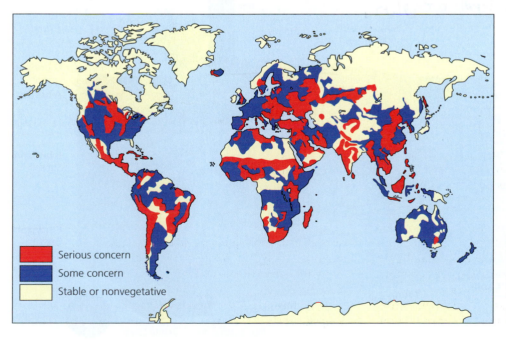

Serious concern

Some concern

Stable or nonvegetative

FIGURE 10.11 Natural capital degradation: Topsoil erosion is a serious problem in some parts of the world. **Question:** Can you see any geographical pattern associated with this problem?

Compiled by the authors using data from the UN Environment Programme and the World Resources Institute.

SOIL IS THE FOUNDATION OF LIFE ON LAND

Soil is a complex mixture of eroded rock, mineral nutrients, decaying organic matter, water, air, and billions of living organisms, most of them microscopic decomposers. Soil formation begins when bedrock is slowly broken down into fragments and particles by physical, chemical, and biological processes, called *weathering.* Figure 10.A shows profiles of different-aged soils.

Soil, on which all terrestrial life depends, is one of the most important components of the earth's natural capital. It supplies most of the nutrients needed for plant growth and purifies and stores water. Organisms living in the soil remove carbon dioxide from the atmosphere and store it as organic carbon compounds, thereby helping to control the earth's climate as part of the carbon cycle (see Figure 3.14, p. 53). You might think of topsoil as something found only in farm fields, but it underlies all forests and grasslands, as well as croplands.

Most soils that have developed over long periods of time, called *mature soils,* contain horizontal layers, or *horizons* (Figure 10.A), each with a distinct texture and composition that vary with different types of soils. Most mature soils have at least three of the four possible horizons.

The roots of most plants and the majority of a soil's organic matter are concentrated in the soil's two upper layers, the *O horizon* of leaf litter and the *A horizon* of topsoil. In healthy soils, these two layers teem with bacteria, fungi, earthworms, and small insects, all interacting in complex food webs. Bacteria and other decomposer microorganisms, found by the billions in every handful of topsoil, break down some of the soil's complex organic compounds into a mixture of the partially decomposed bodies of dead plants and animals, called *humus,* and inorganic materials such as clay, silt, and sand. Soil moisture carrying these dissolved nutrients is drawn up by the roots of plants and transported through their stems and into their leaves as a key component of chemical cycling, the basis for one of the **principles of sustainability**.

The *B horizon (subsoil)* and the *C horizon (parent material)* contain most of a soil's inorganic matter, mostly broken-down rock consisting of varying mixtures of sand, silt, clay, and gravel. Much of it is transported by water from the A horizon. The C horizon lies on a base of parent material, which is often *bedrock.*

The spaces, or *pores,* between the solid organic and inorganic particles in the upper and lower soil layers contain varying amounts of air (mostly nitrogen and oxygen gas) and water. Plant roots use the oxygen for cellular respiration. As long as the O and A horizons are anchored by vegetation, the soil layers as a whole act as a sponge, storing water and nutrients, and releasing them in a nourishing trickle.

Although topsoil is a renewable resource, it is renewed very slowly, which means it can be depleted. Just 1 centimeter (0.4 inch) of topsoil can take hundreds of years to form, but it can be washed or blown away in a matter of weeks or months when we plow grassland or clear a forest and leave its topsoil unprotected.

Critical Thinking

How does soil contribute to each of the four components of biodiversity described in Figure 4.2 (p. 65)?

ANIMATED FIGURE 10.A
Generalized soil profile and formation of soil.
Questions: What role do you think the tree in this figure plays in soil formation? How might the soil formation process change if the tree were removed?

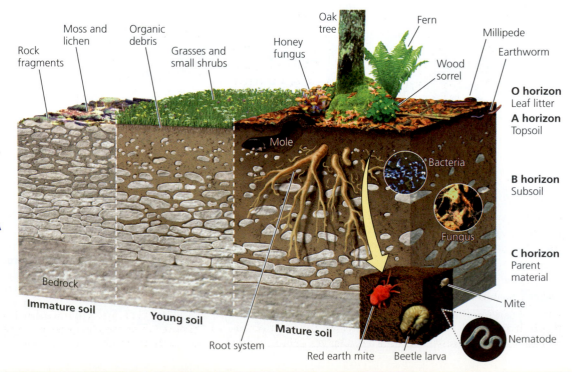

Rock fragments — Moss and lichen — Organic debris — Grasses and small shrubs — Honey fungus — Oak tree — Fern — Millipede — Earthworm — Wood sorrel — Mole — Bacteria — Fungus — Root system — Red earth mite — Beetle larva — Mite — Nematode

Bedrock

Immature soil   **Young soil**   **Mature soil**

O horizon Leaf litter
A horizon Topsoil
B horizon Subsoil
C horizon Parent material

FIGURE 10.12 Severe soil erosion due to over-grazing by cattle has led to desertification.

Dick Ercken/Shutterstock.com

Drought and Human Activities Are Degrading Drylands

A major threat to food security in some of the world's water-short drylands, which are home for some 2 billion people, is **desertification**—the process in which the productive potential of topsoil falls by 10% or more because of a combination of prolonged drought and human activities that expose topsoil to erosion.

Desertification can be *moderate* (with a 10–25% drop in productivity), *severe* (with a drop of 25–50%), or *very severe* (with a drop of more than 50%, usually resulting in large gullies and sand dunes; see Figure 10.12). Only in extreme cases does desertification lead to what we call desert.

Over thousands of years, the earth's deserts have expanded and contracted, primarily because of climate change. However, human use of the land, especially for agricultural purposes, has increased desertification in some parts of the world mostly because of deforestation, excessive plowing, and overgrazing that leave the topsoil bare and unprotected against erosion by flowing water and wind.

Excessive Irrigation Has Serious Consequences

A major reason for the success of farmers in boosting productivity on farms is the use of irrigation, which accounts for about 70% of the water that humanity uses. Currently, the 20% of the world's cropland that is irrigated produces about 40% of the world's food.

However, irrigation has a downside. Most irrigation water is a dilute solution of various salts, such as sodium chloride, that are picked up as the water flows over or through soil and rocks. Irrigation water that is not absorbed into the topsoil evaporates and leaves behind a thin

crust of dissolved mineral salts in the topsoil. Repeated applications of irrigation water in dry climates lead to the gradual accumulation of salts in the upper soil layers—a soil degradation process called **soil salinization**. It stunts crop growth, lowers crop yields, and can eventually kill plants and ruin the land.

The FAO estimates that severe soil salinization has reduced yields on at least 10% of the world's irrigated cropland, and that by 2020, 30% of the world's arable land will be salty. Salinization affects almost one-fourth of irrigated cropland in the United States, especially in western states (Figure 10.13).

USDA Natural Resources Conservation Service

FIGURE 10.13 Natural capital degradation: White alkaline salts have displaced crops that once grew on this heavily irrigated land in the U.S. state of Colorado.

Concept 10.3   **227**

Another problem with irrigation is **waterlogging**, in which water accumulates underground and gradually raises the water table, especially when farmers apply large amounts of irrigation water in an effort to reduce salinization by leaching salts deeper into the soil. Waterlogging lowers the productivity of crop plants and kills them after prolonged exposure, because it deprives plants of the oxygen they need to survive. At least 10% of the world's irrigated land suffers from this worsening problem, according to the FAO.

Agriculture Contributes to Air Pollution and Climate Change

Agricultural activities, including the clearing and burning of forests to raise crops or livestock, create a great deal of air pollution. They also account for more than a quarter of all human-generated emissions of carbon dioxide (CO_2), which is helping to warm the atmosphere and lead to climate change that is projected to play an important role in making some areas unsuitable for growing crops during this century.

According to the 2006 FAO study, *Livestock's Long Shadow,* industrialized livestock production generates about 18% of the world's greenhouse gases—more than all of the world's cars, trucks, buses, and planes emit. In particular, cattle and dairy cows release methane (CH_4)—a greenhouse gas with about 25 times the warming potential of CO_2 per molecule—mostly through belching. Along with the methane generated by liquid animal manure stored in feedlot waste lagoons, this accounts for about 18% of the global annual emissions of methane. And nitrous oxide (N_2O), with about 300 times the atmospheric warming capacity of CO_2 per molecule, is released in huge quantities by synthetic inorganic fertilizers, as well as by livestock manure.

Food and Biofuel Production Systems Have Caused Major Losses of Biodiversity

Natural biodiversity and some ecosystem services are threatened when tropical and other forests are cleared (see Figure 9.9, p. 191) and when grasslands are plowed up and replaced with croplands used to produce food and biofuels (**Concept 10.3**).

For example, one of the fastest-growing threats to the world's biodiversity is the cutting or burning of large areas of tropical forest in Brazil's Amazon Basin and the clearing of areas of its *cerrado,* a huge tropical grassland region south of the Amazon Basin. This land is being burned or cleared for cattle ranches, large plantations of soybeans grown for cattle feed (see Figure 1.4, p. 8), and sugarcane used for making ethanol fuel. In Indonesia, tropical forests are burned to make way for plantations of oil palm trees increasingly used to produce biodiesel fuel.

A related problem is the increasing loss of **agrobiodiversity**—the genetic variety of animal and plant species used on farms to produce food. Scientists estimate that since 1900, we have lost 75% of the genetic diversity of agricultural crops that existed then. For example, India once planted 30,000 varieties of rice. Now more than 75% of its rice production comes from only ten varieties and soon, almost all of its production might come from just one or two varieties. In the United States, about 97% of the food plant varieties available to farmers in the 1940s no longer exist, except perhaps in small amounts in seed banks and in the backyards of a few gardeners (see Figure 14, p. S46, in Supplement 5).

In losing agrobiodiversity, ecologists warn that we are rapidly shrinking the world's genetic "library" of plant varieties, which are critical for increasing food yields. This failure to preserve agrobiodiversity is a serious violation of the biodiversity **principle of sustainability** that could reduce the sustainability of food production.

Individual plants and seeds from endangered varieties of crops and wild plant species important to the world's food supply are stored in about 1,400 refrigerated seed banks, as well as in agricultural research centers and botanical gardens scattered around the world. However, power failures, fires, storms, wars, and unintentional disposal of seeds can cause irreversible losses of these stored plants and seeds. The world's most secure seed bank is the underground Doomsday Seed Vault, which was carved into the Arctic permafrost on a frozen Norwegian arctic island (Figure 10.14). It is being stocked with duplicates of much of the world's seed collections.

However, the seeds of many plants cannot be stored successfully in seed banks. Because stored seeds do not remain alive indefinitely, they must be planted and germinated periodically, and new seeds must be collected for storage. Unless this is done, seed banks become *seed morgues.*

There Is Controversy over Genetically Engineered Foods

While genetic engineering could help to improve food security for some, controversy has arisen over the use of this technology. Its producers and investors see genetically modified (GM) food production as a potentially sustainable way to solve world hunger problems. However, some critics consider it potentially dangerous "Frankenfood" that would allow a small number of seed companies to patent genetically modified crops and control most of the world's food production, and thus food prices. Figure 10.15 summarizes the major projected benefits and drawbacks of this new technology.

Some critics recognize the potential benefits of GM crops (Figure 10.15, left) but they point out that most of the GM crops developed so far have provided very few of these benefits. In 2011, an international team of scientists and

FIGURE 10.14 Svalbard Global Seed Vault.

Jim Richardson/National Geographic Creative

Trade-Offs

Genetically Modified Crops and Foods

Projected Advantages	Projected Disadvantages
May need less fertilizer, pesticides, and water	Have unpredictable genetic and ecological effects
Can be resistant to insects, disease, frost, and drought	May put toxins in food
Can grow faster	Can promote pesticide-resistant insects, herbicide-resistant weeds, and plant diseases
May tolerate higher levels of herbicides	Could disrupt seed market and reduce biodiversity

FIGURE 10.15 Use of genetically modified crops and foods has advantages and disadvantages. ***Questions:*** Which two advantages and which two disadvantages do you think are the most important? Why?

Top: Lenar Musin/Shutterstock.com. Bottom: oksix/Shutterstock.com.

There Are Limits to Expansion of the Green Revolutions

So far, several factors have limited the success of the green revolutions and may limit them even more in the future (**Concept 10.3**). For example, without large inputs of water and synthetic inorganic fertilizers and pesticides, most green revolution and genetically engineered crop varieties produce yields that are no higher (and are sometimes lower) than those from traditional strains. These high inputs also cost too much for most subsistence farmers in less-developed countries.

Scientists point out that where such inputs do increase yields, there comes a point where yields stop growing because of the inability of crop plants to take up nutrients from additional fertilizer and irrigation water. This helps to explain the slowdown in the rate of growth in global grain yields since 1990.

Can we expand the green revolutions by irrigating more cropland? Since 1978, the amount of irrigated land per person has been declining, and it is projected to fall much more by 2050. One reason for this is population growth, which is projected to add 2.6 billion more people between 2013 and 2050. Other factors are limited availability of irrigation water, soil salinization, and the fact that most of the world's farmers do not have enough money to irrigate their crops. In addition, projected climate change during this century is likely to melt some of the mountain glaciers that provide irrigation and drinking water for many millions of people in China, India, and South America.

analysts published the *Global Citizens' Report on the State of GMOs*. It calls into serious question industry claims that genetic engineering will increase crop yields, lessen the need for pesticides, and yield drought-tolerant crops. The report also summarized findings indicating that GM crops with built-in toxins, such as Bt toxins, widely used to fend off insects in corn production, could threaten human health by triggering an inflammatory response leading to diseases such as diabetes and heart disease. In addition, herbicide-resistant genetically engineered crops have led to increased herbicide use and to herbicide-resistant superweeds, some of which can rapidly grow more than 2 meters (7 feet) tall.

The Ecological Society of America and various critics of genetically engineered crops call for more controlled field experiments and long-term testing, to better understand the ecological and health risks, and stricter regulation of this rapidly growing technology.

Can we increase the food supply by cultivating more land? We have already cleared or converted about 38% of the world's ice-free land surface for use as croplands and pastures. By clearing tropical forests and irrigating arid land, we could more than double the area of the world's cropland. The problem is that such massive clearing of forests would speed up climate change and increase topsoil erosion and biodiversity losses. Also, much of this land has poor soil fertility, steep slopes, or both, and cultivating such land would be expensive and probably not ecologically sustainable.

In addition, during this century, fertile croplands in coastal areas, including many of the major rice-growing floodplains and river deltas in Asia, are likely to be flooded by rising sea levels resulting from projected climate change. Food production could also drop sharply in some major food-producing areas because of longer and more intense droughts and heat waves, also resulting from projected climate change.

Industrialized Meat Production Has Harmful Environmental Effects

Proponents of industrialized meat production point out that it has increased meat supplies, reduced overgrazing, and kept food prices down. But feedlots (Figure 10.5) and concentrated animal feeding operations (Figure 10.6) use large amounts of water to grow feed for livestock and to wash away their wastes. In his book *The Food Revolution*, John Robbins estimated that "you'd save more water by not eating a pound of California beef than you would by not showering for a year."

Analysts also point out that meat produced by industrialized agriculture is artificially cheap because most of its harmful environmental and health costs are not included in the market prices of meat and meat products, a violation of the full-cost pricing **principle of sustainability**. Figure 10.16 summarizes the advantages and disadvantages of industrialized meat production.

In 2008, the FAO reported that overgrazing and erosion by livestock had degraded about 20% of the world's grasslands and pastures. The same report estimated that rangeland grazing and industrialized livestock production caused about 55% of all topsoil erosion and sediment pollution, and fully one-third of the water pollution resulting from the runoff of nitrogen and phosphorus from excessive inputs of synthetic fertilizers.

CONSIDER THIS. . .

CONNECTIONS Meat Production and Ocean Dead Zones

Huge amounts of synthetic inorganic fertilizers are used in the Midwestern United States to produce corn for animal feed and ethanol fuel for cars. Much of this fertilizer runs off cropland and eventually goes into the Mississippi River. The added nitrate and phosphate nutrients over-fertilize coastal waters in the Gulf of Mexico, where the river flows into the ocean. Each year, this creates a "dead zone" often larger than the U.S. state of Massachusetts. This oxygen-depleted zone threatens one-fifth of the nation's seafood yield. In other words, growing corn in the Midwest, largely to feed cattle and fuel cars, degrades aquatic biodiversity and seafood production in the Gulf of Mexico.

Trade-Offs

Animal Feedlots and CAFOs

Advantages	Disadvantages
Increased meat production	Animals unnaturally confined and crowded
Higher profits	Large inputs of grain, fishmeal, water, and fossil fuels
Less land use	Greenhouse gas (CO_2 and CH_4) emissions
Reduced overgrazing	Concentration of animal wastes that can pollute water
Reduced soil erosion	Use of antibiotics can increase genetic resistance to microbes in humans
Protection of biodiversity	

FIGURE 10.16 Use of animal feedlots and concentrated animal feeding operations has advantages and disadvantages. **Questions:** Which single advantage and which single disadvantage do you think are the most important? Why?

Top: Mikhail Malyshev/Shutterstock.com. Bottom: Maria Dryfhout/Shutterstock.com.

Industrialized meat production uses large amounts of energy (mostly from oil), which helps to make it one of the chief sources of air and water pollution and greenhouse gas emissions. The Environmental Working Group has estimated that production of meat and meat products generates 10–20 times more greenhouse gases per unit of weight than does production of common vegetables and grains.

Another growing problem is the use of antibiotics in industrialized livestock production facilities. In 2011, the U.S. Food and Drug Administration (FDA) estimated that about 80% of all antibiotics sold in the United States (and 50% of those in the world) are added to animal feed. This is done to try to prevent the spread of diseases in crowded feedlots and CAFOs and to promote the growth of the animals before they are slaughtered. According to FDA data and several studies, this plays a role in the rise of genetic resistance among many disease-causing bacteria (see Figure 4.9, p. 74). Such resistance can reduce the effectiveness of some antibiotics used to treat humans for bacterial infections, and it can promote the development of new, more genetically resistant infectious disease organisms.

Finally, according to the USDA, animal waste produced by the American meat industry amounts to about 130 times the amount of waste produced by the country's human population. Globally, only about half of all manure is returned to the land as nutrient-rich fertilizer—a violation of the chemical cycling **principle of sustainability**. Much of the other half ends up polluting aquatic systems, producing foul odors, and emitting large quantities of climate-changing greenhouse gases into the atmosphere.

Aquaculture Can Harm Aquatic Ecosystems

Figure 10.17 lists the major benefits and drawbacks of aquaculture, which in 2012 accounted for about 42% of all seafood produced for human consumption. Some analysts warn that the harmful environmental effects of aquaculture could limit its future production potential (**Concept 10.3**).

One major environmental problem associated with aquaculture is that about a third of the wild fish caught from the oceans are used to make the fishmeal and fish oil that are to fed to farmed fish. This is contributing to the depletion of many populations of wild fish that are crucial to marine food webs—a serious threat to marine biodiversity and ecosystem services.

Another problem is that some fishmeal and fish oil fed to farm-raised fish is contaminated with long-lived toxins such as PCBs and dioxins that are picked up from the ocean floor. Aquaculture producers contend that the concentrations of these chemicals are not high enough to threaten human health, but some scientists disagree. Fish farms, especially those that raise carnivorous fish such as

salmon and tuna, also produce large amounts of wastes, including pesticides and antibiotics used on fish farms. Yet another problem is that farmed fish can escape their pens and mix with wild fish, possibly disrupting the gene pools of wild populations.

Major seed companies are now pushing to use patented, genetically modified soybeans as the primary feed for farm-raised fish and shellfish. This could increase water pollution because fish that are fed soy tend to produce more waste than those that are not fed soy. It would also give a small number of seed companies control over much of the world's seafood production, along with seafood prices. And it could encourage more deforestation and loss of biodiversity wherever soy plantations replace tropical forests (see Figure 1.4, p. 8).

10.4 HOW CAN WE PROTECT CROPS FROM PESTS MORE SUSTAINABLY?

CONCEPT 10.4 We can sharply cut pesticide use without decreasing crop yields by using a mix of cultivation techniques, biological pest controls, and small amounts of selected chemical pesticides as a last resort (integrated pest management).

Nature Controls the Populations of Most Pests

A **pest** is any species that interferes with human welfare by competing with us for food, invading our homes, lawns, or gardens, destroying building materials, spreading disease, invading ecosystems, or simply being a nuisance. Worldwide, only about 100 species of plants (weeds), animals (mostly insects), fungi, and microbes cause most of the damage to the crops we grow.

In natural ecosystems and in many polyculture crop fields, *natural enemies* (predators, parasites, and disease organisms) control the populations of most potential pest species. This free ecosystem service is an important part of the earth's natural capital. For example, biologists estimate that the world's 30,000 known species of spiders kill far more crop-eating insects every year than humans do by using insecticides. Most spiders, including the wolf spider (Figure 10.18), do not harm humans.

When we clear forests and grasslands, plant monoculture crops, and douse fields with chemicals that kill pests, we upset many of these natural population checks and balances that are in keeping with the biodiversity **principle of sustainability**. Then we must devise ways to protect our monoculture crops, tree plantations, lawns, and golf courses from insects, weeds, and other pests that nature has helped to control at no charge.

Trade-Offs

Aquaculture

Advantages	Disadvantages
High efficiency	Use of fish oil and fishmeal on fish farms depletes wild fisheries
High yield	Large waste output
Reduces over-harvesting of fisheries	Loss of mangrove forests and estuaries
Jobs and profits	Dense populations vulnerable to disease

© 2016 Cengage Learning

FIGURE 10.17 Use of aquaculture has advantages and disadvantages. *Questions:* Which single advantage and which single disadvantage do you think are the most important? Why?

Top: Vladislav Gajic/Shutterstock.com. Bottom: Nordling/Shutterstock.com.

FIGURE 10.18 **Natural capital:** This ferocious-looking wolf spider with a grasshopper in its mouth is one of many important insect predators that can be killed by some pesticides.

Cathy Keifer/Shutterstock.com

We Use Pesticides to Help Control Pest Populations

We have developed a variety of synthetic **pesticides**— chemicals used to kill or control populations of organisms that we consider undesirable. Common types of pesticides include *insecticides* (insect killers), *herbicides* (weed killers), *fungicides* (fungus killers), and *rodenticides* (rat and mouse killers).

We did not invent the use of chemicals to repel or kill other species. For nearly 225 million years, plants have been producing chemicals to ward off, deceive, or poison the insects and herbivores that feed on them. Scientists have used such chemicals to create *biopesticides* to kill some pests. This battle produces a never-ending, ever-changing coevolutionary process: insects and herbivores overcome various plant defenses through natural selection and new plant defenses are favored by natural selection.

Since 1950, synthetic pesticide use has increased more than 50-fold and most of today's pesticides are 10–100 times more toxic to pests than those used in the 1950s. Some synthetic pesticides, called *broad-spectrum agents,* are toxic to beneficial species as well as to pests. Examples are organochlorine compounds (such as DDT), organophosphates (such as malathion and parathion), carbamates, pyrethroids, and neonicotinoids. Others, called *selective,* or *narrow-spectrum, agents,* are each effective against a narrowly defined group of organisms. Examples are chitins that inhibit the molting process of insects, juvenile hormones that disrupt the life cycle of insects, and fungicides.

Pesticides vary in their *persistence*, the length of time they remain deadly in the environment. Some, such as DDT and related compounds, remain in the environment for years and can be biologically magnified in food chains and webs (see Figure 8.9, p. 171). Others, such as organophosphates, are active for days or weeks and are not biologically magnified but can be highly toxic to humans.

About one-fourth of the pesticides used in the United States are aimed at ridding houses, gardens, lawns, parks, playing fields, swimming pools, and golf courses of insects and other species that we view as pests. According to the U.S. Environmental Protection Agency (EPA), the amount of synthetic pesticides used on the average U.S. homeowner's lawn is ten times the amount (per unit of land area) typically used on U.S. croplands.

Synthetic Pesticides Provide Several Benefits

Use of synthetic pesticides has its advantages and disadvantages. Proponents contend that the benefits of pesticides (Figure 10.19, left) outweigh their harmful effects (Figure 10.19, right). They point to the following benefits:

- *They have saved human lives.* Since 1945, DDT and other insecticides probably have prevented the premature deaths of at least 7 million people (some say as many as 500 million) from insect-transmitted diseases such as malaria (carried by the *Anopheles* mosquito), bubonic plague (carried by rat fleas), and typhus (carried by body lice and fleas).

- *They can increase food supplies* by reducing food losses due to pests.

- *They can help farmers to increase their profits.* Officials of pesticide companies estimate that, under certain

Trade-Offs

Conventional Chemical Pesticides

Advantages		Disadvantages
Expand food supplies		Promote genetic resistance
Raise profits		Can kill pests' natural enemies and harm wildlife and people
Work fast		Can pollute air, water, and land
Are safe if used properly		Are expensive for farmers

© 2016 Cengage Learning

FIGURE 10.19 Use of synthetic pesticides has advantages and disadvantages. **Questions:** Which single advantage and which single disadvantage do you think are the most important? Why?

Photo: B Brown/Shutterstock.com

conditions, a dollar spent on pesticides can lead to an increase in crop yields worth as much as $4.

- *They work fast.* Pesticides control most pests quickly, have a long shelf life, and are easily shipped and applied.

- *When used properly, the health risks of some pesticides are very low, relative to their benefits,* according to some scientific studies.

- *Newer pesticides are safer to use and more effective than many older ones.* Greater use is being made of chemicals derived originally from plants (biopesticides), which are generally safer to use and less damaging to the environment than are many older pesticides. Genetic engineering is also being used to develop pest-resistant crop strains and genetically altered crops that produce biopesticides.

Synthetic Pesticides Have Several Drawbacks

Opponents of widespread use of synthetic pesticides contend that the harmful effects of these chemicals outweigh their benefits. They cite several problems.

- *They accelerate the development of genetic resistance to pesticides in pest organisms.* Insects breed rapidly, and within 5–10 years (much sooner in tropical areas), they can develop immunity to widely used pesticides through natural selection. Since 1945, about 1,000 species of insects and rodents (mostly rats) and 550 types of weeds and plant diseases have developed genetic resistance to one or more pesticides. Since 1996, the widespread use of glyphosate herbicide has led to at least 15 species of "superweeds" that are genetically resistant to it.

- *They can put farmers on a financial treadmill.* Because of genetic resistance, farmers can find themselves having to pay more and more for a chemical pest control program that can become less and less effective.

- *Some insecticides kill natural predators and parasites that help to control the pest populations.* About 100 of the 300 most destructive insect pests in the United States were minor pests until widespread use of insecticides wiped out many of their natural predators. (See the Case Study that follows.)

- *Pesticides are usually applied inefficiently and often pollute the environment.* According to the USDA, about 98–99.9% of the insecticides and more than 95% of the herbicides applied by aerial spraying or ground spraying do not reach the target pests. They end up in the air, surface water, groundwater, bottom sediments, food, and nontarget organisms, including humans.

- *Some pesticides harm wildlife.* According to the USDA and the U.S. Fish and Wildlife Service, each year, some of

What Can You Do?

Reducing Exposure to Pesticides

- Grow some of your food using organic methods
- Buy certified organic food
- Wash and scrub all fresh fruits and vegetables
- Eat less meat, no meat, or certified organically produced meat
- Before cooking, trim the fat from meat

FIGURE 10.20 **Individuals matter:** You can reduce your exposure to pesticides. **Questions:** Which three of these steps do you think are the most important ones to take? Why?

the pesticides applied to cropland poison honeybee colonies on which we depend for pollination of many food crops (see Chapter 8, Core Case Study, p. 162, and Science Focus 8.2, p. 172). According to a study by the Center for Biological Diversity, pesticides menace about a third of all endangered and threatened species in the United States.

- *Some pesticides threaten human health.* The WHO and UNEP have estimated that pesticides annually poison at least 3 million agricultural workers in less-developed countries and at least 300,000 workers in the United States. They also cause 20,000–40,000 deaths per year, worldwide. According to studies by the National Academy of Sciences, pesticide residues in food cause an estimated 4,000–20,000 cases of cancer per year in the United States.

The pesticide industry disputes these claims, arguing that if used as directed, pesticides do not remain in the environment at levels high enough to cause serious environmental or health problems. Figure 10.20 lists some ways to reduce your exposure to synthetic pesticides.

CASE STUDY

Ecological Surprises: The Law of Unintended Consequences

Malaria once infected nine of every ten people in North Borneo, now known as the eastern Malaysian state of Sabah. In 1955, the WHO sprayed the island with dieldrin (a DDT relative) to kill malaria-carrying mosquitoes. The program was so successful that the dreaded disease was nearly eliminated.

Then unexpected things began to happen. The dieldrin also killed other insects, including flies and cockroaches living in houses, which made the islanders happy. Next, small insect-eating lizards living in the houses died after gorging themselves on dieldrin-contaminated insects. Then cats began dying after feeding on the lizards. In the absence of cats, rats flourished in and around the villages. When the residents became threatened by sylvatic plague carried by rat fleas, the WHO parachuted healthy cats onto the island to help control the rats. Operation Cat Drop worked.

But then the villagers' roofs began to fall in. The dieldrin had killed wasps and other insects that fed on a type of caterpillar that had either avoided or was not affected by the insecticide. With most of its predators eliminated, the caterpillar population exploded, munching its way through its favorite food: the leaves used in thatch roofs.

Ultimately, this story ended well. Both malaria and the unexpected effects of the spraying program were brought under control. Nevertheless, this chain of unintended and unforeseen events reminds us that whenever we intervene in nature and affect organisms that interact with one another, we need to ask, "Now what will happen?"

Pesticide Use Has Not Consistently Reduced U.S. Crop Losses to Pests

Largely because of genetic resistance and the loss of many natural predators, synthetic pesticides have not always succeeded in reducing U.S. crop losses. When David Pimentel, an expert on insect ecology, evaluated data from more than 300 agricultural scientists and economists, he reached three major conclusions. *First,* between 1942 and 1997, estimated crop losses from insects almost doubled from 7% to 13%, despite a 10-fold increase in the use of synthetic insecticides. *Second,* according to the International Food Policy Research Institute, the estimated environmental, health, and social costs of pesticide use in the United States are $5–$10 in damages for every dollar spent on pesticides. *Third,* experience indicates that alternative pest management practices could cut the use of synthetic pesticides by half on 40 major U.S. crops without reducing crop yields (**Concept 10.4**).

The pesticide industry disputes these findings. However, numerous studies and experience support them. For example, Sweden has cut its pesticide use in half with almost no decrease in crop yields. And the soup company Campbell's® uses no pesticides on the tomatoes it grows in Mexico, and yields have not dropped.

Laws and Treaties Can Help to Protect Us from the Harmful Effects of Pesticides

More than 20,000 different pesticide products are used in the United States. Three federal agencies, the EPA, the USDA, and the Food and Drug Administration (FDA), regulate the use of these pesticides under the Federal Insecticide, Fungicide, and Rodenticide Act (FIFRA), first passed in 1947 and amended in 1972.

Under FIFRA, the EPA was supposed to assess the health risks of the active ingredients in synthetic pesticide products already in use. However, since 1972, less than 10% of the active ingredients in pesticide products have been tested for chronic health effects. And serious evaluation of the health effects of the 1,200 inactive ingredients is only partially done. The EPA says that the U.S. Congress has not provided them with enough funds to carry out this complex and lengthy evaluation process.

In 1996, Congress passed the Food Quality Protection Act, mostly because of growing scientific evidence and citizen pressure concerning the effects of small amounts of pesticides on children. This act requires the EPA to reduce the allowed levels of pesticide residues in food by a factor of 10 when there is inadequate information on the potentially harmful effects on children.

Between 1972 and 2013, the EPA used FIFRA to ban or severely restrict the use of 64 active pesticide ingredients, including DDT and most other chlorinated hydrocarbon insecticides. However, according to studies by the National Academy of Sciences, federal laws regulating pesticide use generally are inadequate and poorly enforced by the three agencies. One study found that as much as 98% of the potential risk of developing cancer from pesticide residues on food grown in the United States would be eliminated if EPA standards were as strict for pesticides developed before 1972 as they are for newer pesticides.

CONSIDER THIS. . .

CONNECTIONS Pesticides and Food Choices

According to the Environmental Working Group (EWG), a research organization that helped to push through the Food Quality Protection Act, you could reduce your pesticide intake by up to 90% by eating only 100% USDA Certified Organic versions of 12 types of fruits and vegetables that tend to have the highest pesticide residues. In 2014, these foods, which the EWG calls the "dirty dozen," were apples, strawberries, grapes, celery, peaches, spinach, sweet bell peppers, imported nectarines, cucumbers, cherry tomatoes, imported snap peas, and potatoes.

In what environmental scientists call a *circle of poison,* or the *boomerang effect,* residues of synthetic pesticides that have been banned in one country but exported to other countries can return to the exporting countries on imported food. Winds can also carry persistent pesticides from one country to another.

In 1998, more than 50 countries developed an international treaty that requires exporting countries to have informed consent from importing counties for exports of 22 synthetic pesticides and 5 industrial chemicals. In 2000, more than 100 countries developed an international agreement to ban or phase out the use of 12 especially hazardous persistent organic pollutants (POPs)—9 of them persistent hydrocarbon pesticides such as DDT and other chemically similar pesticides. By 2011, the initial list of 12 chemicals had been expanded to 21. In 2004, the POPs treaty went into effect and by 2013, had been signed or ratified by 172 countries, not including the United States.

There Are Alternatives to Synthetic Pesticides

Many scientists urge us to greatly increase the use of biological, ecological, and other alternative methods for controlling pests and diseases that affect crops and human health (**Concept 10.4**). Here are some of these alternatives:

- *Fool the pest.* A variety of *cultivation practices* can be used to fake out pests. Examples include rotating the types of crops planted in a field each year and adjusting planting times so that major insect pests either starve or get eaten by their natural predators.

- *Provide homes for pest enemies.* Farmers can increase the use of polyculture, which uses plant diversity to reduce losses to pests by providing habitats for the predators of pest species.

- *Implant genetic resistance.* Use genetic engineering to speed up the development of pest- and disease-resistant crop strains. But controversy persists over whether the projected advantages of using GM plants outweigh their projected disadvantages (Figure 10.15).

- *Bring in natural enemies.* Use *biological control* by importing natural predators (Figures 10.18 and 10.21), parasites, and disease-causing bacteria and viruses to help regulate pest populations. This approach is nontoxic to other species and is usually less costly than applying pesticides. However, some biological control agents are difficult to mass-produce and are often slower acting and more difficult to apply than synthetic pesticides are. Sometimes the agents can multiply and become pests themselves.

- *Use insect perfumes.* Trace amounts of *sex attractants* (called *pheromones*) can be used to lure pests into traps or to attract their natural predators into crop fields. Each of these chemicals attracts only one species. They have little chance of causing genetic resistance and are not harmful to nontarget species. However, they are costly and time-consuming to produce.

- *Bring in the hormones.* Hormones are chemicals produced by animals to control their developmental processes at different stages of life. Scientists have learned how to identify and use hormones that disrupt an insect's normal life cycle, thereby preventing it from reaching maturity and reproducing. Use of insect hormones has some of the same advantages and disadvantages as use of sex attractants has. Also, they take weeks to kill an insect, are often ineffective with large infestations of insects, and sometimes break down before they can act.

- *Reduce the use of synthetic herbicides to control weeds.* Organic farmers control weeds by methods such as crop rotation, mechanical cultivation, hand weeding, and the use of cover crops and mulches.

FIGURE 10.21 Natural capital: In this example of biological pest control, a wasp is parasitizing a gypsy moth caterpillar.

Integrated Pest Management Is a Component of More Sustainable Agriculture

Many pest control experts and farmers believe the best way to control crop pests is through **integrated pest management (IPM)**, a carefully designed program in which each crop and its pests are evaluated as parts of an ecosystem, and farmers use a carefully coordinated combination of cultivation, biological, and chemical tools and techniques (**Concept 10.4**).

The overall aim of IPM is to reduce crop damage to an economically tolerable level. Each year, crops are rotated in an effort to disrupt pest infestations. When farmers detect an economically damaging level of pests in any field, they first use biological methods (natural predators, parasites, and disease organisms) and cultivation controls (such as altering planting time and using large machines to vacuum up harmful bugs). They apply small amounts of synthetic insecticides—preferably biopesticides—only when insect or weed populations reach a threshold where the potential cost of pest damage to crops outweighs the cost of applying the pesticide.

IPM has a good track record. In Sweden and Denmark, farmers have used it to cut their synthetic pesticide use by more than half. In Cuba, where organic farming is used almost exclusively, farmers make extensive use of IPM. In Brazil, IPM has reduced pesticide use on soybeans by as much as 90%. In Japan, many farmers save money by using ducks for pest control in rice paddies. The ducks' droppings provide nutrients for the rice plants.

According to the U.S. National Academy of Sciences, these and other experiences show that a well-designed IPM program can reduce synthetic pesticide use and pest control costs by 50–65%, without reducing crop yields and food quality. IPM can also reduce inputs of fertilizer and irrigation water, and slow the development of genetic resistance, because pests are attacked less often and with lower doses of pesticides. IPM is an important form of *pollution prevention* that reduces risks to wildlife and human health, applies the biodiversity **principle of sustainability**, and expands our beneficial environmental impact.

Despite its promise, IPM—like any other form of pest control—has some drawbacks. It requires expert knowledge about each pest situation and takes more time than does using conventional pesticides. Methods developed for a crop in one area might not apply to areas with even slightly different growing conditions. Initial costs may be higher, although long-term costs typically are lower than those of using conventional pesticides. Widespread use of IPM has been hindered in the United States and other countries by government subsidies that support use of synthetic chemical pesticides, as well as by opposition from pesticide manufacturers, and a shortage of IPM experts. **GREEN CAREER: Integrated pest management**

A growing number of scientists are urging the USDA to use a three-point strategy to promote IPM in the United States. *First*, add a small sales tax on synthetic pesticides and use the revenue to fund IPM research and education. *Second*, set up a federally supported IPM demonstration project on at least one farm in every county in the United States. *Third*, train USDA field personnel and county farm agents in IPM so they can help farmers use this alternative. Several UN agencies and the World Bank have joined together to establish an IPM facility. Its goal is to promote the use of IPM by disseminating information and establishing networks among researchers, farmers, and agricultural extension agents involved in IPM.

10.5 HOW CAN WE IMPROVE FOOD SECURITY?

CONCEPT 10.5 We can improve food security by reducing poverty and chronic malnutrition, producing food more sustainably, relying more on locally grown food, and cutting food waste.

Use Government Policies to Improve Food Production and Security

Agriculture is a financially risky business. Whether farmers have a good or bad year depends on factors over which they have little control, including weather, crop prices, pests and diseases, interest rates on loans, and global food markets.

Governments use two main approaches to influence food production. First, they can *control food prices* by putting a legally mandated upper limit on them in order to keep them artificially low. This makes consumers happy but makes it harder for farmers to make a living.

Second, they can *provide subsidies* by giving farmers price supports, tax breaks, and other financial support to help them stay in business and to encourage them to increase food production. However, if government subsidies are too generous and the weather is good, farmers and livestock producers may produce more food than can be sold.

Some analysts call for ending such subsidies. They point to New Zealand, which ended farm subsidies in 1984. After the shock wore off, innovation took over and production of some foods such as milk quadrupled. Brazil has also ended most of its farm subsidies. Some analysts call for replacing traditional subsidies for farmers with subsidies that promote more environmentally sustainable farming practices.

Similarly, government subsidies to fishing fleets can promote overfishing and the reduction of aquatic biodiversity. Many analysts call for replacing those harmful subsidies with subsidies that would promote more sustainable fishing and aquaculture.

Other Government and Private Programs Are Increasing Food Security

Studies by the United Nations Children's Fund (UNICEF) indicate that one-half to two-thirds of nutrition-related childhood deaths could be prevented at an average annual cost of $5–$10 per child. This would involve simple measures such as immunizing more children against childhood diseases, preventing dehydration due to diarrhea by giving infants a mixture of sugar and salt in their water, and preventing blindness by giving children an inexpensive vitamin A capsule twice a year.

There are also many private, mostly nonprofit, organizations that are working to help individuals, communities, and nations to improve their food security and produce food more sustainably. For example, Growing Power's Will Allen (**Core Case Study**) argues that instead of trying to transfer complex technologies such as genetic engineering to less-developed countries, we should be helping them to develop simple, sustainable, local food production and distribution systems that will give them more control over their food security.

Sustainable agriculturalists and National Geographic Emerging Explorers Cid Simones and Paola Segura work with small farmers to show them how to grow food more sustainably on small plots in the tropical forests of Brazil. They train one family at a time. In return, each family must teach five other families and thus help to spread more sustainable farming methods. Another person who is working toward this goal in Africa is National Geographic Explorer Jennifer Burney (Individuals Matter 10.1).

REBECCA HALE/National Geographic Creative

Jennifer Burney: Environmental Scientist and National Geographic Emerging Explorer

Environmental scientist and National Geographic Emerging Explorer Jennifer Burney notes that subsistence farmers represent the majority of the world's poorest people and need to boost their productivity for better standards of living and health. She is trying to help such farmers in Africa to grow, distribute, and cook their food using resources like water, fertilizer, and energy as efficiently as possible.

For example, to deal with the problem of seasonal water shortages in many parts of sub-Saharan Africa, Burney has helped organizations connect two technologies—solar irrigation systems and drip irrigation (background photo)—that could serve as a solution. Drip irrigation systems sip water and drip it directly onto plant roots instead of pumping and dumping it. Solar-powered pumps work without the need for batteries or fuel. On sunny days, when crops need water more, the solar panels speed the pumping; on cloudy days when there is less evaporation, the pumping slows down. Thus, only the amount of water that is needed is pumped on most days. This has allowed farmers to grow fruits and vegetables on a larger scale and to improve their incomes and food security.

Background photo: Nhtnog Do

We Can Grow and Buy More Food Locally and Cut Food Waste

One way to increase food security is to grow more of our food locally or regionally, ideally with certified organic farming practices. A growing number of consumers are becoming "locavores" and buying more of their food from local and regional producers in farmers' markets.

In addition, many people are participating in *community-supported agriculture (CSA)* programs in which they buy shares of a local farmer's crop and receive a box of fruits or vegetables each week during the summer and fall. Growing Power (**Core Case Study**) runs such a program that serves people now living in a food desert. In some communities, including Raleigh, North Carolina, mobile *grocers-on-wheels* businesses bring fresh and healthy food to people who live in food deserts. By buying locally, people support local economies and farm families. They also help to reduce fossil fuel energy costs for food producers, as well as the greenhouse gas emissions resulting from refrigeration and transportation of food products over long distances.

An increase in the demand for locally grown food could result in more small, diversified farms that produce organic, minimally processed food from plants and animals. Such eco-farming could be one of this century's challenging new careers for many young people. **GREEN CAREER: Small-scale sustainable agriculture**

According to the USDA, around 15% of the world's food is grown in urban areas, and this percentage could easily be doubled. People are planting gardens and raising chickens in many urban and suburban backyards, growing dwarf fruit trees in large containers of soil, and raising vegetables on rooftops, balconies, and patios. People are also building raised gardening beds in urban parking lots—a growing practice known as *asphalt gardening*. Food columnist Mark Bittman estimated that converting 10% of Americans lawns into food-producing gardens would supply one-third of the country's fresh produce.

Many urban schools, colleges, and universities are benefitting greatly from having gardens on school grounds. Not only do the students have a ready source of fresh produce, but they also learn about where their food comes from and how to grow their own food more sustainably. FoodCorps, a part of the U.S. government AmeriCorps program, is trying to help reduce childhood obesity by promoting nutrition education, farm-to-school food delivery programs, and school gardens.

In the future, much of our food might be grown in cities within specially designed high-rise buildings. Growing Power (**Core Case Study**) has plans to build such a *vertical farm* at its Milwaukee site. This five-story, largely glass-enclosed building with crops growing on every floor would put into practice the three **scientific principles of sustainability**. It would have rooftop solar

GOOD NEWS

SUSTAINABILITY

panels for generating electricity, and the building would capture and recycle rainwater for irrigating its wide diversity of crops.

Finally, people can sharply cut food waste as an important component of improving food security (**Concept 10.5**). A 2011 UN study found that about 33% of all food produced globally is lost during production or thrown away. According to the EPA and a 2012 study by the Natural Resources Defense Council (NRDC), Americans throw away 30% to 40% of the country's food supply each year.

10.6 HOW CAN WE PRODUCE FOOD MORE SUSTAINABLY?

CONCEPT 10.6 We can produce food more sustainably by using resources more efficiently, sharply decreasing the harmful environmental effects of industrialized food production, and eliminating government subsidies that promote such harmful impacts.

Many Farmers Are Reducing Soil Erosion

Land used for food production must have fertile topsoil (Figure 10.A), which takes hundreds of years to form. Thus, sharply reducing topsoil erosion is the single most important component of more sustainable agriculture and one of the most important ways to increase our beneficial environmental impact.

Soil conservation involves using a variety of methods to reduce topsoil erosion and restore soil fertility, mostly by keeping the land covered with vegetation. For example, *terracing* involves converting steeply sloped land into a series of broad, nearly level terraces that run across the land's contours (Figure 10.22a). Each terrace retains water for crops and reduces topsoil erosion by controlling runoff.

On less steeply sloped land, *contour planting* (Figure 10.22b) can be used to reduce topsoil erosion. It involves plowing and planting crops in rows across the slope of the land rather than up and down. Each row acts as a small dam to help hold topsoil by slowing runoff. Similarly, *strip-cropping* (Figure 10.22b) helps to reduce erosion and to restore soil fertility with alternating strips of a row crop (such as corn or cotton) and another crop that completely covers the soil, called a *cover crop* (such as alfalfa, clover, oats, or rye). The cover crop traps topsoil that erodes from the row crop and catches and reduces water runoff.

Alley cropping, or *agroforestry* (Figure 10.22c), is another way to slow the erosion of topsoil and to maintain soil fertility. One or more crops, usually legumes or other crops that add nitrogen to the soil, are planted together in alleys between orchard trees or fruit-bearing shrubs, which provide shade. This reduces water loss by evaporation and helps retain and slowly release soil moisture.

Farmers can also establish *windbreaks*, or *shelterbelts*, of trees around crop fields to reduce wind erosion (Figure 10.22d). The trees retain soil moisture, supply wood for fuel, and provide habitats for birds and insects that help with pest control and pollination.

Another way to greatly reduce topsoil erosion is to eliminate or minimize the plowing and tilling of topsoil and to leave crop residues on the ground. Such *conservation-tillage farming* uses special tillers and planting machines that inject seeds and fertilizer directly through crop residues into minimally disturbed topsoil. Weeds are controlled with herbicides. This type of farming increases crop yields and greatly reduces soil erosion and water pollution from sediment and fertilizer runoff.

In 2011, farmers used conservation tillage on about 63% of U.S. cropland (up from 17% in 1982). However, one drawback is that the greater use of herbicides is promoting the growth of herbicide-resistant weeds that force farmers to use larger doses of weed killers or, in some cases, to return to plowing. The USDA estimates that by using conservation tillage on 80% of U.S. cropland, farmers could reduce topsoil erosion by at least 50%. Conservation tillage is only used on about 10% of the world's cropland, although it is widely used in some countries, including the United States (see the following Case Study), Brazil, Argentina, Canada, and Australia.

Still another way to conserve the earth's topsoil is to retire the estimated one-tenth of the world's highly erodible cropland. The goal would be to identify *erosion hotspots*, withdraw these areas from cultivation, and plant them with grasses or trees, at least until their topsoil has been renewed.

CASE STUDY

Soil Erosion in the United States

In the United States, at least a third of the country's original topsoil is gone and much of the rest is degraded. In the state of Iowa, which has the world's highest concentration of prime farmland, half of the topsoil is gone after a century of industrialized farming. According to the Natural Resources Conservation Service, 90% of American farmland is, on average, losing topsoil 17 times faster than new topsoil is being formed.

In the early 1930s, the Great Plains in the Midwestern United States experienced an extreme drought known as the Dust Bowl, with a sharp drop in crop productivity over a vast area. This occurred because too many fields were overplowed and left open to hot dry winds and severe soil erosion. In 1935, the United States passed the *Soil Erosion Act*, which established the Soil Conservation Service (SCS) as part of the USDA. It gave farmers and ranchers technical assistance for setting up soil conservation programs that did help to reduce erosion. (The SCS is now called the Natural Resources Conservation Service, or NRCS.)

FIGURE 10.22 Soil conservation methods include **(a)** terracing; **(b)** contour planting and strip cropping; **(c)** alley cropping; and **(d)** windbreaks between crop fields.

With the help of the NRCS, U.S. farmers are sharply reducing some of their topsoil losses through a combination of conservation-tillage farming and government-sponsored soil conservation programs (**Concept 10.6**). Under the 1985 Food Security Act (Farm Act), more than 400,000 farmers participating in the Conservation Reserve Program received subsidy payments for taking highly erodible land—totaling an area larger than the U.S. state of New York—out of production and replanting it with grass or trees for 10–15 years. Since 1985, these efforts have cut total topsoil losses on U.S. cropland by 40%.

There is still room for improvement, however. Effective topsoil conservation is practiced today on only half of all U.S. agricultural land. But the United States is currently the only major food-producing nation to be significantly reducing its topsoil losses.

CONSIDER THIS. . .

CONNECTIONS Corn, Ethanol, and Soil Conservation

In recent years, some U.S. farmers took erodible land out of the conservation reserve in order to receive more generous government subsidies for planting corn (which removes nitrogen from the soil and reduces the ability of soil to store carbon by removing carbon dioxide from the atmosphere) to make ethanol for use as a motor vehicle fuel. This led to mounting political pressure to abandon or sharply cut back on the nation's highly successful topsoil conservation reserve program.

We Can Restore Soil Fertility

The best way to maintain soil fertility is through topsoil conservation, especially through methods that keep topsoil covered with vegetation (Figure 10.22). The next best option is to restore some of the lost plant nutrients that

FIGURE 10.23 Chickens add fertilizer to the soil on this Pennsylvania farm. Mobile chicken coops are moved daily to distribute the manure more evenly and reduce runoff of fertilizer.

have been washed, blown, or leached out of topsoil, or that have been removed by repeated crop harvesting. To do this, farmers can use **organic fertilizer** derived from plant and animal materials or **synthetic inorganic fertilizer** made of inorganic compounds that contain *nitrogen*, *phosphorus*, and *potassium* along with trace amounts of other plant nutrients.

There are several types of *organic fertilizers*. One is **animal manure**: the dung and urine of cattle, horses, poultry (Figure 10.23), and other farm animals. It improves topsoil structure, adds organic nitrogen, and stimulates the growth of beneficial soil bacteria and fungi. Another type, called **green manure**, consists of freshly cut or growing green vegetation that is plowed into the topsoil to increase the organic matter and humus available to the next crop. A third type is **compost**, produced when microorganisms break down organic matter such as leaves, crop residues, food wastes, paper, and wood in the presence of oxygen.

The Growing Power farm (**Core Case Study**) depends greatly on its large piles of compost. Will Allen invites local grocers and restaurant owners to send their food wastes to add to the pile. Also, the process of composting generates a considerable amount of heat, which is used to help warm the farm's greenhouses during cold months.

One way to degrade soils is to plant crops such as corn and cotton on the same land several years in a row, a prac-

tice that can deplete nutrients—especially nitrogen—in the topsoil. Crop rotation is one way to reduce such losses. A farmer plants an area with a nutrient-depleting crop one year, and the next year, plants the same area with legumes, whose root nodules add nitrogen to the soil. This method helps to restore topsoil nutrients while reducing erosion by keeping the topsoil covered with vegetation.

Many farmers, especially those in more-developed countries, rely on synthetic inorganic fertilizers. The use of these products has grown more than ninefold since 1950, and it now accounts for about 25% of the world's crop yield. While these fertilizers can replace depleted inorganic nutrients, they do not replace organic matter. Completely restoring topsoil nutrients requires both inorganic and organic fertilizers.

We Can Reduce Soil Salinization and Desertification

We know how to prevent and deal with soil salinization, as summarized in Figure 10.24. The problem is that most of these solutions are costly.

Reducing desertification is not easy. We cannot control the timing and location of prolonged droughts caused by changes in weather and climate patterns. But we can reduce population growth, overgrazing, deforestation, and

Soil Salinization

Prevention	Cleanup
Reduce irrigation	Flush soil (expensive and inefficient)
Use more efficient irrigation methods	Stop growing crops for 2–5 years
Switch to salt-tolerant crops	Install underground drainage systems

© 2016 Cengage Learning

FIGURE 10.24 Ways to prevent and ways to clean up soil salinization (**Concept 10.6**). **Questions:** Which two of these solutions do you think are the best ones? Why?

Photo: USDA Natural Resources Conservation Service

destructive forms of planting and irrigation in dryland areas, which have left much land vulnerable to topsoil erosion and thus desertification. We can also work to decrease the human contribution to projected climate change, which could increase the severity of droughts in larger areas of the world during this century.

It is possible to restore land suffering from desertification by planting trees and other plants that anchor topsoil and hold water (Figure 10.12). We can also grow trees and crops together (alley cropping, Figure 10.22c), and establish windbreaks around farm fields (Figure 10.22d).

Some Producers Practice More Sustainable Aquaculture

Figure 10.25 lists some ways to make aquaculture more sustainable and to reduce its harmful environmental effects. The Aquaculture Stewardship Council (ASC) has developed aquaculture sustainability standards, but it has certified only about 4.6% of the world's aquaculture production operations.

One approach is *open-ocean aquaculture,* which involves raising large carnivorous fish in underwater pens—some as large as a high school gymnasium. Some are located as far as 300 kilometers (190 miles) offshore (see Figure 9.24, p. 207) where rapid currents can sweep away fish wastes and dilute them.

Other fish farmers are reducing coastal damage from aquaculture by raising shrimp and fish species in inland facilities using zero-discharge freshwater ponds and tanks. In such *recirculating aquaculture systems,* the water used to raise the fish is continually recycled.

More Sustainable Aquaculture

- Protect mangrove forests and estuaries
- Improve management of wastes
- Reduce escape of aquaculture species into the wild
- Set up self-sustaining polyaquaculture systems that combine aquatic plants, fish, and shellfish
- Certify sustainable forms of aquaculture

© 2016 Cengage Learning

FIGURE 10.25 We can make aquaculture more sustainable and reduce its harmful effects. **Questions:** Which two of these solutions do you think are the best ones? Why?

In the long run, making aquaculture more sustainable will require some fundamental changes for producers and consumers. One change is for more consumers to choose fish species that eat algae and other vegetation rather than other fish. Raising carnivorous fishes such as salmon, trout, tuna, grouper, and cod contributes to overfishing and population crashes within species used to feed these carnivores, and will eventually be unsustainable. Aquaculture producers can avoid this problem by raising plant-eating fishes such as carp, tilapia, and catfish, as long as they do not try to increase yields by feeding fishmeal to such species, as many of them are doing.

One advocate of sustainable seafood consumption is Barton Seaver, a conservationist and National Geographic Fellow. As a chef at several of Washington, D.C.'s, top restaurants, he brought the concept of sustainable seafood consumption to the nation's capital. The Seafood Choices Alliance has called him a leader in seafood sustainability.

Fish farmers can also emphasize *polyaquaculture,* which has been part of aquaculture for centuries, especially in Southeast Asia. Polyaquaculture operations raise fish or shrimp along with algae, seaweeds, and shellfish in coastal lagoons, ponds, and tanks. The wastes of the fish or shrimp feed the other species. Polyaquaculture applies the chemical cycling and biodiversity **principles of sustainability**. **GREEN CAREER: Sustainable aquaculture**

We Can Produce Meat and Dairy Products More Efficiently

Meat production has a huge environmental impact and meat consumption is the largest factor in the growing ecological footprints of individuals in affluent nations.

A more sustainable form of meat production and consumption would involve shifting from less grain-efficient forms of animal protein, such as beef, pork, and carnivorous fish produced by aquaculture, to more grain-efficient forms, such as poultry and plant-eating farmed fish

(Figure 10.26). We could also shift from buying chicken that has been raised in CAFOs (Figure 10.6), which produce large amounts of manure that is not returned to the soil, to supporting free-range chicken operations that use chicken manure to fertilize the soil (Figure 10.23).

A growing number of people have one or two meatless days per week. Others go further and eliminate most or all meat from their diets, replacing it with a balanced vegetarian diet of fruits, vegetables, and protein-rich foods such as peas, beans, and lentils. According to agricultural science writer Michael Pollan, if all Americans picked one day per week to have no meat, the reduction in greenhouse gas emissions would be equivalent to taking 30 to 40 million cars off the road for a year.

We Can Make a Shift to More Sustainable Food Production

Modern industrialized food production has yielded huge amounts of food at affordable prices, but to a growing number of analysts, it is unsustainable, because it violates the three **scientific principles of sustainability**. It relies heavily on the use of fossil fuels and thus adds greenhouse gases and other air pollutants to the atmosphere and contributes to climate change. It also reduces biodiversity and agrobiodiversity and interferes with the cycling of plant nutrients. These harmful effects are hidden from consumers because most of the harmful environmental and health costs of food production (Figure 10.8) are not included in the market prices of food—a violation of the full-cost pricing **principle of sustainability**.

A more sustainable food production system would have several major components (Figure 10.27) (**Concept 10.6**). One such component is USDA 100% Certified Organic Agriculture (Figure 10.3). Many experts support a shift to organic farming because it sharply reduces the harmful environmental and health effects of industrialized farming, improves the condition of topsoil, and reduces pollution of air and water.

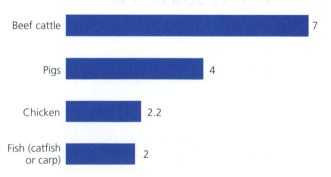

FIGURE 10.26 Kilograms of grain required for each kilogram of body weight added for each type of animal.

Compiled by the authors using data from U.S. Department of Agriculture.

Solutions

More Sustainable Food Production

More		Less
High-yield polyculture		Soil erosion
Organic fertilizers		Soil salinization
Biological pest control		Water pollution
Integrated pest management		Aquifer depletion
Efficient irrigation		Overgrazing
Perennial crops		Overfishing
Crop rotation		Loss of biodiversity and agrobiodiversity
Water-efficient crops		Fossil fuel use
Soil conservation		Greenhouse gas emissions
Subsidies for sustainable farming		Subsidies for unsustainable farming

© 2016 Cengage Learning

FIGURE 10.27 More sustainable, low-input food production has a number of major components. (**Concept 10.6**). **Questions:** For each list in this diagram (left and right), which two items do you think are the most important? Why?

Top: Marko5/Shutterstock.com. Center: Anhong/Dreamstime. Bottom: pacopi/Shutterstock.com.

Another important component of more sustainable agriculture would be to rely less on conventional monoculture and more on organic polyculture. Of particular interest to some scientists is the idea of using polyculture to grow *perennial crops*—crops that grow back year after year on their own (Science Focus 10.2).

Another key to developing more sustainable agriculture is to shift from using fossil fuels to relying more on renewable energy for food production—an important application of the solar energy **principle of sustainability** that has been well demonstrated by the Growing Power farm (**Core Case Study**). To produce the electricity and fuels needed for food production, farmers can make greater use of renewable solar energy (see chapter-opening photo), wind, flowing water, and biofuels produced from farm wastes in tanks called *biogas digesters*.

Analysts suggest five major strategies to help farmers and consumers to make the transition to more sustainable agriculture over the next 50 years (**Concept 10.6**). *First,* greatly increase research on more sustainable organic farming and perennial polyculture, and on improving human nutrition. *Second,* establish education and training

SCIENCE**FOCUS**10.2

PERENNIAL POLYCULTURE AND THE LAND INSTITUTE

Some scientists call for greater reliance on polycultures of perennial crops as a component of more sustainable agriculture. Such crops can live for many years without having to be replanted and are better adapted to regional soil and climate conditions than are most annual crops. One scientist who is exploring perennial agriculture is soil expert Jerry Glover, a National Geographic Emerging Explorer. He studies perennial farming systems around the world, working for the U.S. Agency for International Development.

More than three decades ago, plant geneticist Wes Jackson co-founded the Land Institute in the U.S. state of Kansas. One of the institute's goals has been to grow a diverse mixture (polyculture) of edible perennial plants to

supplement traditional annual monoculture crops and to help reduce the latter's harmful environmental effects.

Perennial crops, which live for several years, help farmers copy nature by better using and conserving natural resources—sunlight, soil, and water. Because there is no need to till the soil and replant perennials each year, this approach produces much less topsoil erosion and water pollution. It also reduces the need for irrigation because the deep roots of such perennials retain more water than do the shorter roots of annuals (Figure 10.B). Often, there is a reduced need for chemical fertilizers and pesticides, and thus little or no pollution from these sources. Perennial crops also remove and store more carbon from the atmosphere, and growing them requires less energy

than does growing annual crops in conventional monocultures.

Critical Thinking

Why do you think large seed companies generally oppose this form of more sustainable agriculture?

FIGURE 10.B The roots of an annual wheat crop plant (left) are much shorter than those of big bluestem (right), a tall-grass prairie perennial plant.

The Land Institute

---

programs in more sustainable agriculture for students, farmers, and government agricultural officials. *Third,* set up an international fund to give farmers in poor countries access to various types of more sustainable agriculture. *Fourth,* replace government subsidies for environmentally harmful forms of industrialized food production with subsidies that encourage more sustainable food production. And *fifth,* mount a massive program to educate consumers about where their food really comes from, how it is produced, and what are the harmful environmental and health effects of industrialized food production.

Figure 10.28 lists ways in which you can promote more sustainable food production. The demand for such production would likely grow if large numbers of people were to adopt a nutritious diet, such as the one suggested by the Harvard School of Public Health in 2014 (Figure 10.29).

You could also grow some of your own food—channeling your inner farmer and getting your hands dirty by raising some organic vegetables in your back yard, in a window box, or in a shared neighborhood garden. You would learn a little about where some of your food comes from, while making a beneficial environmental impact.

What Can You Do?

More Sustainable Food Production

- Eat less meat, no meat, or organically certified meat
- Choose sustainably produced herbivorous fish
- Use organic farming to grow some of your food
- Buy certified organic food
- Eat locally grown food
- Compost food wastes
- Cut food waste

© 2016 Cengage Learning

FIGURE 10.28 Individuals matter: Ways to promote more sustainable food production (**Concept 10.6**). *Questions:* Which three of these actions do you think are the most important ones to take? Why?

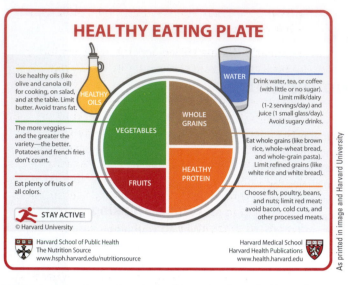

HEALTHY EATING PLATE

Use healthy oils (like olive and canola oil) for cooking, on salad, and at the table. Limit butter. Avoid trans fat.

The more veggies— and the greater the variety—the better. Potatoes and french fries don't count.

Eat plenty of fruits of all colors.

HEALTHY OILS

VEGETABLES

WHOLE GRAINS

FRUITS

HEALTHY PROTEIN

WATER

Drink water, tea, or coffee (with little or no sugar). Limit milk/dairy (1-2 servings/day) and juice (1 small glass/day). Avoid sugary drinks.

Eat whole grains (like brown rice, whole-wheat bread, and whole-grain pasta). Limit refined grains (like white rice and white bread).

Choose fish, poultry, beans, and nuts; limit red meat; avoid bacon, cold cuts, and other processed meats.

STAY ACTIVE!
© Harvard University

Harvard School of Public Health
The Nutrition Source
www.hsph.harvard.edu/nutritionsource

Harvard Medical School
Harvard Health Publications
www.health.harvard.edu

As printed in image and Harvard University

FIGURE 10.29 Suggestions for healthy eating from the Harvard School of Public Health.

TYING IT ALL TOGETHER
Growing Power and Sustainability

This chapter began with a look at how the ecologically based urban Growing Power farm (**Core Case Study**), is providing a diversity of good food to people living in a food desert. Its founder Will Allen, in demonstrating how organic food can be grown more sustainably at affordable prices, is showing how to make the transition to more sustainable food production while applying the three **scientific principles of sustainability**. Modern industrialized agriculture, aquaculture, and other forms of industrialized food production violate all of these principles.

Making the transition to more sustainable food production means relying more on solar (see chapter-opening photo) and other forms of renewable energy and less on fossil fuels. It also means sustaining chemical cycling by conserving topsoil and returning crop residues and animal wastes to the soil. It involves helping to sustain natural, agricultural, and aquatic biodiversity by relying on a greater variety of crop and animal strains and seafood, produced by certified organic methods and sold locally in grocery stores and farmers' markets (see photo at right). Controlling pest populations through broader use of conventional and perennial polyculture and integrated pest management will also help to sustain biodiversity.

Governments could help these efforts by replacing environmentally harmful agricultural and fishing subsidies and tax breaks with more environmentally beneficial ones. Finally, the transition to more sustainable food production would be accelerated for the benefit of the environment as well as current and future generations if we could find ways to include the harmful environmental and health costs of food production in the market prices of food, in keeping with all three **social science principles of sustainability**.

Baloncici/Shutterstock.com

Chapter Review

Core Case Study

1. Summarize the benefits that the Growing Power farm has brought to its community. How does the farm showcase the three **scientific principles of sustainability**?

Section 10.1

2. What are the two key concepts for this section? Define **food security** and **food insecurity**. What is the root cause of food insecurity? Distinguish between **chronic undernutrition (hunger)** and **chronic malnutrition** and describe their harmful effects. Describe the effects of diet deficiencies in vitamin A, iron, and iodine. What is **overnutrition** and what are its harmful effects?

Section 10.2

3. What is the key concept for this section? What three systems supply most of the world's food? Define and distinguish among **industrialized agriculture (high-input agriculture)**, **plantation agriculture**, **traditional subsistence agriculture**, and **traditional intensive agriculture**. Define **polyculture** and summarize its benefits. Define **organic agriculture** and compare its main components with those of conventional industrialized agriculture. What is a **green revolution**? Summarize the story of industrialized food production in the United States.

4. Distinguish between crossbreeding through artificial selection and genetic engineering. Describe the second gene revolution based on genetic engineering. Summarize the growth of industrialized meat production. What are feedlots and CAFOs? What is a **fishery**? What is **aquaculture (fish farming)**? Explain why industrialized food production requires large inputs of energy. Why does it result in a net energy loss?

Section 10.3

5. What is the key concept for this section? List two major benefits of high-yield modern agriculture. Define **soil** and describe its formation and the major layers in mature soils. What is **topsoil** and why is it one of our most important resources? What is **soil erosion** and what are its two major harmful environmental effects? What is **desertification** and what are its harmful environmental effects? Define **soil salinization** and **waterlogging** and explain why they are harmful. What is soil pollution and what are two of its causes?

6. Summarize industrialized agriculture's contribution to projected climate change. Explain how industrialized food production systems have caused losses in biodiversity. What is **agrobiodiversity** and how is it being affected by industrialized food production? List the advantages and disadvantages of using genetic engineering in food production. What factors can limit green revolutions? Compare the benefits and harmful effects of industrialized meat production. Explain the connection between feeding livestock and the formation of ocean dead zones. Compare the benefits and harmful effects of aquaculture.

Section 10.4

7. What is the key concept for this section? What is a **pest**? Define and give two examples of a **pesticide**. Summarize the advantages and disadvantages of using synthetic pesticides. Describe the use of laws and treaties to help protect U.S. citizens from the harmful effects of pesticides. List seven alternatives to conventional pesticides. Define **integrated pest management (IPM)** and list its advantages.

Section 10.5

8. What is the key concept for this section? What are the two main approaches used by governments to influence food production? How have governments used subsidies to influence food production and what have been some of their effects? Describe the system used by Jennifer Burney to help people grow crops in parts of sub-Saharan Africa. What are two other ways in which organizations are improving food security? Explain three of the benefits of buying locally grown food. How can urban farming help to increase food security?

Section 10.6

9. What is the key concept for this section? What is **soil conservation**? Describe six ways to reduce topsoil erosion. Distinguish among **organic fertilizer**, **synthetic inorganic fertilizer**, **animal manure**, **green manure**, and **compost**. How does crop rotation help restore topsoil fertility? What are some ways to prevent and some ways to clean up soil salinization? How can we reduce desertification? Describe three ways to make aquaculture more sustainable. What are some ways to make meat production and consumption more sustainable? Summarize three important components of a more sustainable food production system. List the

advantages of relying more on organic polyculture and perennial crops. What are five strategies that could help farmers and consumers to shift to more sustainable food production? What are three important ways in which individual consumers can help to promote more sustainable food production?

10. What are the three big ideas of this chapter? Explain how making the transition to more sustainable food production such as that promoted by the Growing Power farm (**Core Case Study**) will involve applying the six **principles of sustainability**.

Note: Key terms are in bold type.

Critical Thinking

1. Suppose you got a job with Growing Power, Inc. (**Core Case Study**) and were given the assignment to turn an abandoned suburban shopping center and its large parking lot into an organic farm. Write up a plan for how you would accomplish this.

2. Do you think that the advantages of organic agriculture outweigh its disadvantages? Explain. Do you eat or grow organic foods? If so, explain your reasoning for making this choice. If not, explain your reasoning for some of the food choices you do make.

3. Food producers can now produce more than enough food to feed everyone on the planet a healthy diet. Given this fact, why do you think that nearly a billion people are chronically undernourished or malnourished? Assume you are in charge of solving this problem, and write a plan for how you will accomplish it.

4. Explain why you support or oppose greatly increasing the use of **(a)** genetically modified food production and **(b)** organic perennial polyculture.

5. What might happen to industrialized food production if oil prices rise sharply? How might this affect your life? How will it affect the lives of any children or grandchildren you might eventually have? List two ways in which you would deal with these changes.

6. You are the head of a major agricultural agency in the area where you live. Weigh the advantages and disadvantages of using synthetic pesticides and explain why you would support or oppose the increased use of such pesticides as a way to help farmers raise their yields. What are the alternatives?

7. If the mosquito population in the area where you live were proven to be carrying malaria or some dangerous virus, would you want to spray DDT in your yard, inside your home, or all through the local area to reduce this risk? Explain. What are the alternatives?

8. According to physicist Albert Einstein, "Nothing will benefit human health and increase the chances of survival of life on Earth as much as the evolution to a vegetarian diet." Explain your interpretation of this statement. Are you willing to eat less meat or no meat? Explain.

Doing Environmental Science

For 1 week, weigh the food that is purchased in your home and the food that is thrown out. Also, keep track of the types of food you eat, using categories like fruits,

vegetables, meats, dairy, and even more specific categories if you wish. Record and compare these numbers and other data from day to day. Develop a plan for cutting your household food waste in half. Consider making a similar study for your school cafeteria and reporting the results and your recommendations to school officials.

Global Environment Watch Exercise

In the *Soil Erosion* portal, look for information on causes of soil erosion and how it affects soil fertility. Write a report on your findings. If you were to overhear a group of

farmers complaining about how much money they must spend on fertilizers, what suggestions would you give them for saving money? Include your answer to this question, along with your reasoning, in your report.

Ecological Footprint Analysis

The following table gives the world's fish harvest and population data.

1. Use the world fish harvest and population data in the table to calculate the per capita fish consumption for 1990–2012 in kilograms per person. (*Hints:* 1 million metric tons equals 1 billion kilograms; the human population data are expressed in billions; and per capita consumption can be calculated directly by dividing the total amount consumed by a population figure for any year.)

2. Did per capita fish consumption generally increase or decrease between 1990 and 2012?

3. In what years did per capita fish consumption decrease?

World Fish Harvest

Years	Fish Catch (million metric tons)	Aquaculture (million metric tons)	Total (million metric tons)	World Population (in billions)	Per Capita Fish Consumption (kilograms/person)
1990	84.8	13.1	97.9	5.27	
1991	83.7	13.7	97.4	5.36	
1992	85.2	15.4	100.6	5.44	
1993	86.6	17.8	104.4	5.52	
1994	92.1	20.8	112.9	5.60	
1995	92.4	24.4	116.8	5.68	
1996	93.8	26.6	120.4	5.76	
1997	94.3	28.6	122.9	5.84	
1998	87.6	30.5	118.1	5.92	
1999	93.7	33.4	127.1	6.00	
2000	95.5	35.5	131.0	6.07	
2001	92.8	37.8	130.6	6.15	
2002	93.0	40.0	133.0	6.22	
2003	90.2	42.3	132.5	6.31	
2004	94.6	45.9	140.5	6.39	
2005	94.2	48.5	142.7	6.46	
2006	92.0	51.7	143.7	6.54	
2007	90.1	52.1	142.2	6.61	
2008	89.7	52.5	142.3	6.69	
2009	90.0	55.7	145.7	6.82	
2010	89.0	59.0	148.0	6.90	
2011	93.5	62.7	156.2	7.00	
2012	90.2	66.5	156.7	7.05	

Compiled by the authors using data from UN Food and Agriculture Organization and Earth Policy Institute.

CENGAGEbrain.com To access course materials, including Aplia homework, please visit www.cengagebrain.com.

WWW.CENGAGEBRAIN.COM **247**

11

WATER RESOURCES AND WATER POLLUTION

Our liquid planet glows like a soft blue sapphire in the hard-edged darkness of space. There is nothing else like it in the solar system. It is because of water.

JOHN TODD

The Colorado River Story

© 2016 Cengage Learning

FIGURE 11.1 The *Colorado River basin:* The area drained by this river system is more than one-twelfth of the land area of the lower 48 states. This map shows 6 of the river's 14 dams.

The Colorado River,
the major river of the arid southwestern United States, flows 2,300 kilometers (1,400 miles) through seven states to the Gulf of California (Figure 11.1). Most of its water comes from snowmelt in the Rocky Mountains. During the past 100 years, this once free-flowing river has been tamed by a gigantic plumbing system consisting of 14 major dams and reservoirs (Figure 11.2), and canals that help control flooding and supply water and electricity to farmers, ranchers, industries, and cities.

This system of dams and reservoirs provides water and electricity from hydroelectric plants at the major dams for roughly 40 million people in seven states—about one of every eight people in the United States. The river's water is used to produce about 15% of the nation's crops and livestock. The system also supplies water to some of the nation's driest and hottest cities. Take away this tamed river, and Las Vegas, Nevada, and Phoenix, Arizona, would be a mostly uninhabitable desert areas; San Diego and Los Angeles, California, could not support their present populations; and in California's Imperial Valley,

cactus and mesquite plants would eventually replace vast fields of vegetables.

So much water is withdrawn from this river to grow crops and support cities in this dry, desert-like climate that very little of it reaches the sea. To make matters worse, since 1999, the system has experienced severe **drought**, a prolonged period in which precipitation is at least 70% lower and evaporation is higher than normal. This overuse of the Colorado River illustrates the challenges faced by governments and people living in arid and semiarid regions with shared river systems, as population growth and economic growth place increasing demands on limited or decreasing supplies of surface water.

To many analysts, emerging shortages of water for drinking and irrigation in several parts of the world represent one of the major environmental challenges that we will face during this century.

Pete Mcbride/National Geographic Stock

FIGURE 11.2 The Glen Canyon Dam was built to create the Lake Powell reservoir on the Colorado River.

11.1 WILL WE HAVE ENOUGH USABLE WATER?

CONCEPT 11.1A We are using available freshwater unsustainably by extracting it faster than nature can replace it, and by wasting, polluting, and underpricing this irreplaceable natural resource.

CONCEPT 11.1B Freshwater supplies are not evenly distributed, and one of every nine people on the planet does not have adequate access to clean water.

Freshwater Is an Irreplaceable Resource That We Are Managing Poorly

We live on a planet that is unique in our solar system because of a precious layer of water—most of it saltwater—covering about 71% of its surface (see Figure 14, p. S30, Supplement 4). Look in the mirror. What you see is about 60% water, most of it inside your cells.

Water is an amazing chemical with unique properties that help to keep us and other species alive (see Science Focus 3.2, p. 51). You could survive for several weeks without food, but for only a few days without **freshwater**, or water that contains very low levels of dissolved salts. We have no substitute for this vital form of natural capital (**Concept 11.1A**).

It takes huge amounts of water to supply us with food and with most of the other things that we use to meet our daily needs and wants. Water also plays a key role in determining and moderating the earth's climates and in removing and diluting some of the pollutants and wastes that we produce.

Access to freshwater is a *global health issue*. The World Health Organization (WHO) estimates that each day an average of 4,400 people die from waterborne infectious diseases because they do not have access to safe drinking water.

Access to freshwater is also an *economic issue* because water is vital for producing food and energy and for reducing poverty. Because almost half of the world's people do not have water piped to their homes, people in many less-developed countries, usually women and children, have to find and carry daily supplies of freshwater from distant sources or wells (see chapter-opening photo). Water is also a *national and global security issue* because of increasing tensions within and between some nations over access to limited freshwater resources that they share.

Finally, water is an *environmental issue* because excessive withdrawal of freshwater from rivers and aquifers has resulted in falling water tables, dwindling river flows (**Core Case Study**), shrinking lakes, and disappearing wetlands. This, in combination with water pollution in many areas of the world, has degraded water quality, reduced fish populations, hastened the extinction of some aquatic species, and degraded aquatic ecosystem services (see Figures 7.18, p. 147, 7.24, p. 151, and 7.25, p. 152).

Most of the Earth's Freshwater Is Not Available to Us

Only a tiny fraction—*about 0.024%* of the planet's enormous water supply—is readily available to us as liquid freshwater, which is stored in accessible underground deposits and in lakes, rivers, and streams. The rest is in the salty oceans (about 96.5% of the earth's volume of liquid water), in frozen polar ice caps and glaciers (1.7%), and in deep underground aquifers (1.7%; see Figure 14, p. S30, Supplement 4).

Fortunately, the world's freshwater supply is continually recycled, purified, and distributed in the earth's *hydrologic cycle* (see Figure 3.13, p. 52). This irreplaceable water recycling and purification system works well, unless we alter it, overload it with pollutants, or withdraw freshwater from underground and surface water supplies faster than natural processes replenish it.

We can also alter the hydrologic cycle through our influence on projected climate change. Research indicates that atmospheric warming is altering the water cycle by evaporating more water into the atmosphere. As a result, on average, wet places will get wetter with more frequent and heavier flooding, dry places will get drier with more intense drought, and many storms will be more violent.

On a global basis, we have plenty of freshwater, but it is not distributed evenly (Figure 11.3) (**Concept 11.1B**). For example, Canada, with only 0.5% of the world's population, has 20% of its liquid freshwater, while China, with 19% of the world's people, has only 6.5% of the supply.

Groundwater and Surface Water Are Critical Resources

Some precipitation infiltrates the ground and percolates downward through spaces in soil, gravel, and rock until an impenetrable layer of rock or clay stops it. The freshwater in these spaces underground is called **groundwater**—a key component of the earth's natural capital.

The spaces in soil and rock close to the earth's surface hold little moisture. However, below a certain depth, in the **zone of saturation**, these spaces are completely filled with freshwater. The top of this groundwater zone is the **water table**. It falls in dry weather, or when we remove groundwater from this zone faster than nature can replenish it, and it rises in wet weather.

Deeper down are geological layers called **aquifers**, underground caverns and porous layers of sand, gravel, or rock through which groundwater flows. Some caverns have rivers of groundwater flowing through them. However, the porous layers of sand, gravel, or rock in most aquifers are like large, elongated sponges through which groundwater seeps—typically moving only a meter or so (about 3 feet) per year and rarely more than 0.3 meter (1 foot) per day. Watertight layers of rock or clay below such aquifers keep the freshwater from escaping deeper into the earth.

FIGURE 11.3 Some countries have ample freshwater. The Iguazu River, marking part of the border between Brazil and Argentina, flows over the world's largest series of falls (left). Other countries such as Saudi Arabia (right) have very little freshwater.

Left: jose marques lopes/Shutterstock.com; Right: David Steele/Shutterstock.com

We use pumps to bring this groundwater to the surface for irrigating crops and supplying households and industries. Most aquifers are replenished naturally by precipitation that percolates downward through exposed soil and rock, a process called *natural recharge.* Others are recharged from the side by *lateral recharge* from nearby lakes, rivers, and streams.

Most aquifers recharge extremely slowly, and because so much of the earth's urban area landscapes have been built on or paved over, freshwater can no longer penetrate the ground to recharge aquifers below such areas. In addition, in dry areas of the world, there is little precipitation available to recharge aquifers. Deep aquifers that get very little, if any, recharge are called *nonrenewable aquifers.* Withdrawing freshwater from these aquifers amounts to *mining* a nonrenewable resource—an example of the tragedy of the commons (see Chapter 1, pp. 11–12).

Another crucial resource is **surface water**, the freshwater from rain and melted snow that flows or is stored in lakes, reservoirs, wetlands, streams, rivers, estuaries, and the oceans. Precipitation that does not infiltrate the ground or return to the atmosphere by evaporation is called **surface runoff**. The land from which surface runoff drains into a particular stream, lake, wetland, or other body of water is called its **watershed**, or **drainage basin**. For example, the drainage basin for the Colorado River is shown in yellow and green on the map in Figure 11.1 (**Core Case Study**).

We Are Using Increasing Amounts of the World's Reliable Runoff

According to *hydrologists* (scientists who study water and its movements above, on, and below the earth's surface), two-thirds of the annual surface runoff of freshwater into rivers and streams is lost in seasonal floods and is not available for human use. The remaining one-third is **reliable surface runoff**, which we can generally count on as a source of freshwater from year to year.

During the last century, the human population tripled, global water withdrawals increased sevenfold, and per capita withdrawals quadrupled. As a result, we now withdraw about 34% of the world's reliable runoff. This is a global average. In the arid American Southwest, up to 70% of the reliable runoff is withdrawn for human purposes, mostly for irrigation (**Core Case Study**). Some water experts project that because of a combination of population growth, rising rates of water use per person, longer dry periods in some areas, and failure to reduce unnecessary water losses, we are likely to be withdrawing up to 90% of the reliable freshwater runoff by 2025.

Worldwide, we use 70% of the freshwater we withdraw each year from rivers, lakes, and aquifers to irrigate cropland and raise livestock. (In arid regions, on average, 90% of all water withdrawn is used for food production.) Industry uses roughly another 20% of the water withdrawn globally each year, and cities and residences use the remaining 10%. Our **water footprint** is a rough measure of the volume of freshwater that we use directly and indirectly to stay alive and to support our lifestyles. Agriculture accounts for 92% of humanity's water footprint.

According to the American Water Works Association, each day, the average American (see Case Study that fol-

These items...	take this many bathtubs full of water to produce
Cup of coffee	1
2-liter soda	3
Loaf of bread	4
1 dozen eggs	14
1 pound of cheese	15
Hamburger	16
T-shirt	17
Blue jeans	72
Small car	2,600
Medium-sized house	16,000

FIGURE 11.4 Producing and delivering a single one of each of the products listed here requires the equivalent of at least one and usually many bathtubs full of freshwater, called *virtual water*. *Note:* 1 bathtub = 151 liters (40 gallons).

Compiled by the authors using data from UN Food and Agriculture Organization, UNESCO-IHE Institute for Water Education, World Water Council, and Water Footprint Network.

lows) directly uses about 260 liters (69 gallons) of freshwater—enough water to fill about 1.7 typical bathtubs of water. (A bathtub can contain about 151 liters or 40 gallons of water.) This water is used mostly for flushing toilets (27%), washing clothes (22%), taking showers (17%), and running faucets (16%), or is lost through leaking pipes, faucets, and other fixtures (14%).

We use many times more freshwater indirectly. This water is called **virtual water**, the freshwater that is not directly consumed but is used to produce food and other products. It makes up a large part of our water footprints, especially in more-developed countries. Producing and delivering a typical hamburger, for example, takes about 2,400 liters (630 gallons or about 16 bathtubs) of freshwater—most of which is used to grow grain that is fed to cattle.

Figure 11.4 shows one way to measure the amounts of virtual water used for producing and delivering products. These values can vary depending on how much of the supply chain is included, but they give us a rough estimate of the size of our water footprints.

For some water-short countries, it makes sense to save real freshwater by importing virtual water through food imports, instead of producing food domestically. Such countries include Egypt and other Middle Eastern nations in dry climates with little freshwater. Large exporters of virtual water—mostly in the form of wheat, corn, soybeans, alfalfa, and other foods—are the European Union, the United States, Brazil, and Australia.

Freshwater Resources in the United States

The United States has more than enough renewable freshwater to meet its needs. But it is unevenly distributed and much of it is contaminated by agricultural and industrial practices. The eastern states usually have ample precipitation, whereas many western and southwestern states have little (Figure 11.5).

According to the U.S. Geological Survey (USGS), the major uses of groundwater and surface freshwater in the United States are the cooling of electric power plants (41% of total water use), irrigation (37%), public water supplies (13%), industry (5%), and livestock production (4%). In the eastern United States, most water is used for manufacturing and for cooling power plants. In the arid and semi-arid regions of the western half of the United States (**Core Case Study**), irrigation counts for as much as 85% of freshwater use.

Water tables in many water-short areas, especially in the dry western states, are dropping quickly as farmers and rapidly growing urban areas deplete many aquifers faster than they can be recharged. The USGS and the U.S. Department of the Interior projected that areas of at least 36 states are likely to face freshwater shortages by 2025 because of a combination of drought, rising temperatures, population growth, urban sprawl, and increased per capita water use (Figure 11.6). In addition, Columbia University climate researchers led by Richard Seager used well-tested climate models to project that the southwestern United States and parts of northern Mexico are very likely to have long periods of extreme drought throughout most of the rest of this century.

The Colorado River system (Figure 11.1) will be directly affected by such drought. There are three major

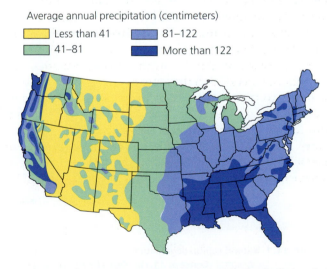

Average annual precipitation (centimeters)

- Less than 41
- 41–81
- 81–122
- More than 122

FIGURE 11.5 Long-term average annual precipitation in the continental United States.

Compiled by the authors using data from U.S. Water Resources Council and U.S. Geological Survey.

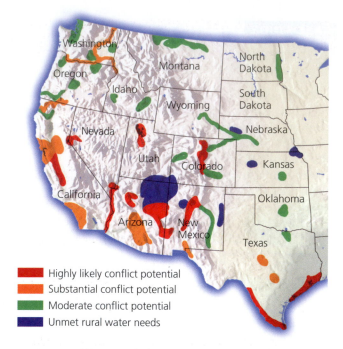

Highly likely conflict potential
Substantial conflict potential
Moderate conflict potential
Unmet rural water needs

FIGURE 11.6 Water scarcity hotspots in 17 western states that, by 2025, could face intense conflicts over scarce water needed for urban growth, irrigation, recreation, and wildlife. **Question:** Which, if any, of these areas are found in the Colorado River basin (**Core Case Study**)?

Compiled by the authors using data from U.S. Department of the Interior and U.S. Geological Survey.

problems associated with the use of freshwater from this river (**Core Case Study**). *First,* the Colorado River basin includes some of the driest lands in the United States and Mexico. *Second,* legal pacts signed in 1922 and 1944 between Mexico and the U.S. states that share the river's water allocated more freshwater for human use than the river can supply—even in rare years when there is no drought. These pacts allocated no water for protecting aquatic and terrestrial wildlife. *Third,* since 1960, the river has rarely flowed all the way to the Gulf of California because of its reduced water flow (due to many dams), increased freshwater withdrawals, and prolonged drought.

Freshwater Shortages Will Grow

The main factors that cause water scarcity in any particular area are a dry climate, drought, too many people using a freshwater supply more quickly than it can be replenished, and inefficient use of freshwater. Figure 11.7 shows the current degree of *freshwater scarcity stress*—a measure based on a comparison of the amount of freshwater available with the amount used by humans—faced by the world's major river systems.

More than 30 countries—most of them in the Middle East and Africa—now face stress from freshwater scarcity.

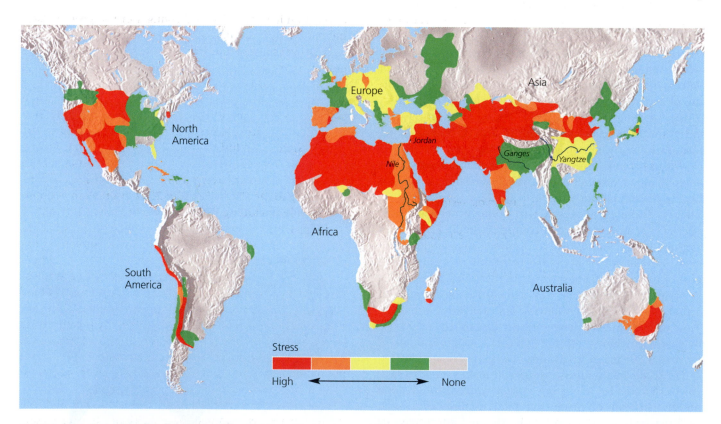

Stress

High ⟷ None

FIGURE 11.7 Natural capital degradation: The world's major river basins differ in their degree of *freshwater scarcity stress* (**Concept 11.1B**). **Questions:** Do you live in a freshwater-stressed area? If so, what signs of stress have you noticed? In what ways, if any, has it affected your life?

Compiled by the authors using data from World Commission on Water Use for the 21st Century, UN Food and Agriculture Organization, and World Water Council.

By 2050, some 60 countries, many of them in Asia, with three-fourths of the world's population, are likely to be suffering from such freshwater scarcity stress.

Currently, about 30% of the earth's land area—a total area roughly 5 times the size of the United States—experiences severe drought. By 2059, as much as 45% of the earth's land surface—about 7 times the area of the United States—could experience extreme drought from a combination of natural drought cycles and projected climate change, according to a study by climate researcher David Rind and his colleagues.

In 263 of the world's water basins, two or more countries share the available freshwater supplies. However, countries in only 158 of those basins have water-sharing agreements. This explains why conflicts among nations over shared freshwater resources are likely to happen more as populations grow, as demand for water increases, and as supplies shrink in many parts of the world.

In 2012, the United Nations (UN) and the WHO reported that about 780 million people—nearly 2.5 times the U.S. population—did not have regular access to enough clean water for drinking, cooking, and washing, mostly due to poverty (**Concept 11.1B**). The report also noted that more than 2 billion people had gained access to clean water between 1990 and 2012. However, many analysts view the likelihood of expanding water shortages in many parts of the world as one of the most serious environmental, health, and economic problems.

11.2 HOW CAN WE INCREASE FRESHWATER SUPPLIES?

CONCEPT 11.2A Groundwater used to supply cities and grow food is being pumped from aquifers in some areas faster than it is renewed by precipitation.

CONCEPT 11.2B Large dam-and-reservoir systems and water transfer projects have greatly expanded water supplies in some areas, but have also disrupted ecosystems and displaced people.

CONCEPT 11.2C We can convert salty ocean water to freshwater, but the cost is high, and the resulting salty brine must be disposed of without harming aquatic or terrestrial ecosystems.

Groundwater Is Being Withdrawn Faster Than It Is Replenished in Some Areas

Aquifers provide drinking water for nearly half of the world's people. In the United States, aquifers supply almost all of the drinking water in rural areas, 20% of that in urban areas, and 37% of the country's irrigation water. Most aquifers are renewable resources unless the groundwater they contain becomes contaminated or is removed

FIGURE 11.8 Withdrawing groundwater from aquifers has advantages and disadvantages. **Questions:** Which two advantages and which two disadvantages do you think are the most important? Why?

Top: Ulrich Mueller/Shutterstock.com

faster than it is replenished. Relying more on groundwater has advantages and disadvantages (Figure 11.8).

Test wells and other data indicate that water tables are falling in many areas of the world because the rate at which most of the world's aquifers are being pumped (mostly to irrigate crops) is greater than the rate of natural recharge from rainfall and snowmelt (**Concept 11.2A**). The world's three largest grain producers—China, the United States, and India—as well as Mexico, Saudi Arabia, Iran, Iraq, Yemen, Israel, Libya, Egypt, Pakistan, South Korea, and Spain are overpumping many of their aquifers. According to the World Bank, in 2012, more than 400 million people (including 190 million in India and 130 million in China) were consuming grain produced through this unsustainable use of groundwater. This number is expected to grow.

Such overpumping can lead to shortages of groundwater needed for irrigation, followed by sharp drops in food production, rising food prices, hunger, and social unrest. For example, much of the Middle East is facing a growing water and food crisis and increasing tensions among its nations, brought on mostly by falling water tables, rapid population growth, and disagreements over access to shared water supplies from the region's rivers.

Saudi Arabia is as water-poor as it is oil-rich. Much of its freshwater is pumped from ancient, nonrenewable deep aquifers to irrigate crops such as wheat grown on desert land (Figure 11.9) and to fill large numbers of fountains and swimming pools, which lose a great deal of water through evaporation into the hot, dry desert air. In 2008,

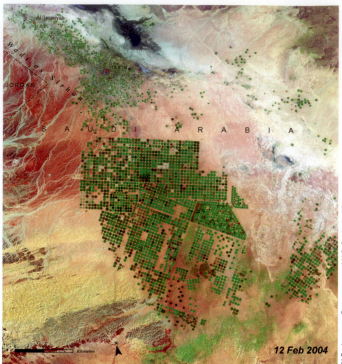

2 Feb 1986

12 Feb 2004

UN Environment Programme and U.S. Geological Survey

UN Environment Programme and U.S. Geological Survey

FIGURE 11.9 Natural capital degradation: Satellite photos of farmland irrigated by groundwater pumped from a deep aquifer in a vast desert region of Saudi Arabia between 1986 (left) and 2004 (right). Irrigated areas appear as green dots (each representing a circular spray system) and brown dots show areas where wells have gone dry and the land has returned to desert.

Saudi Arabia announced that irrigated wheat production had largely depleted its major deep aquifer and said that it would stop producing wheat by 2016 and import grain (virtual water) to help feed its 30 million people.

CASE STUDY

Aquifer Depletion in the United States

In the United States, groundwater is being withdrawn from aquifers, on average, 4 times faster than it is replenished, according to the USGS (**Concept 11.2A**). Figure 11.10 shows the areas of greatest aquifer depletion. One of the most serious overdrafts of groundwater is in the lower half of the Ogallala Aquifer, the world's largest known aquifer, which lies under eight Midwestern states from southern South Dakota to Texas (blowup section of Figure 11.10).

The Ogallala Aquifer supplies about one-third of all the groundwater used in the United States and has helped to turn the Great Plains into one of world's most productive irrigated agricultural regions. The problem is that the Ogallala is essentially a one-time deposit of liquid natural capital with a very slow rate of recharge.

In parts of the southern half of the Ogallala, groundwater is being pumped out 10–40 times faster than the slow natural recharge rate. This has lowered water tables and raised pumping costs, especially in northern Texas.

Such overpumping of aquifers, along with reduced access to Colorado River water (**Core Case Study**) and population growth, have led to the shrinkage of irrigated croplands in Texas, Arizona, Colorado, and California, as well as increasing competition for water among farmers, ranchers, and growing urban areas.

Government *subsidies*—payments or tax breaks designed to increase crop production—have encouraged farmers to grow water-thirsty crops in dry areas, which has accelerated depletion of the Ogallala Aquifer. In particular, corn—a very thirsty crop—has been planted widely on fields watered by the Ogallala. Serious aquifer depletion is also taking place in California's semiarid Central Valley, which supplies half of the country's fruits and vegetables (the long red area in the California portion of Figure 11.10).

CONSIDER THIS. . .

CONNECTIONS Aquifer Depletion in California and Meat Consumption in China

California's Central Valley produces huge amounts of alfalfa, which is used as a supplemental feed for cattle and dairy cows. Alfalfa requires more water than any other crop in California, but very little of this alfalfa is used to feed cattle or dairy cows in California, because alfalfa growers make more money by shipping most of it to China. As a result, every year, alfalfa growers export billions of gallons of virtual water from a drought-ridden area of California to China to support its growing consumption of meat and milk.

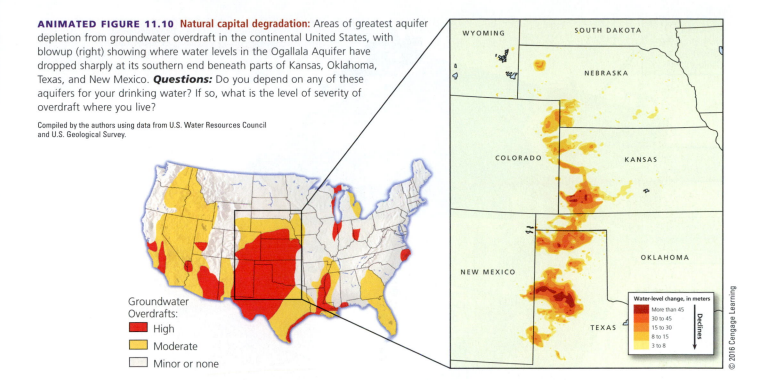

ANIMATED FIGURE 11.10 **Natural capital degradation:** Areas of greatest aquifer depletion from groundwater overdraft in the continental United States, with blowup (right) showing where water levels in the Ogallala Aquifer have dropped sharply at its southern end beneath parts of Kansas, Oklahoma, Texas, and New Mexico. **Questions:** Do you depend on any of these aquifers for your drinking water? If so, what is the level of severity of overdraft where you live?

Compiled by the authors using data from U.S. Water Resources Council and U.S. Geological Survey.

Groundwater Overdrafts:
- High
- Moderate
- Minor or none

Overpumping of Aquifers Can Have Harmful Effects

The overpumping of aquifers can contribute to limits on food production, rising food prices, and widening gaps between the rich and poor in some areas. As water tables drop, the energy and financial costs of pumping the water from lower depths rise sharply as farmers must drill deeper wells, buy larger pumps, and use more electricity to run those pumps. Poor farmers cannot afford to do this and often end up losing their land and working for richer farmers or migrating to cities that are already crowded with poor people struggling to survive.

Withdrawing large amounts of groundwater sometimes causes the sand and rock that is held in place by water pressure in aquifers to collapse. This can cause the land above the aquifer to *subside* or sink, a phenomenon known as *land subsidence*. Extreme, sudden subsidence, sometimes referred to as a *sinkhole*, can swallow cars and houses. Once an aquifer becomes compressed by subsidence, recharge is impossible. In addition, land subsidence can damage roadways, water and sewer lines, and building foundations.

Groundwater overdrafts in coastal areas, where many of the world's largest cities and industrial areas are found, can pull saltwater into freshwater aquifers. The resulting contaminated groundwater is undrinkable and unusable for irrigation. This problem is especially serious in coastal areas of the U.S. states of California, Texas, Florida, Georgia, South Carolina, and New Jersey, as well as in coastal areas of Turkey, Thailand, and the Philippines.

Solutions

Groundwater Depletion

Prevention

Use water more efficiently

Subsidize water conservation

Limit number of wells

Stop growing water-intensive crops in dry areas

Control

Raise price of water to discourage waste

Tax water pumped from wells near surface water

Build rain gardens in urban areas

Use permeable paving material on streets, sidewalks, and driveways

FIGURE 11.11 Ways to prevent or slow groundwater depletion by using freshwater more sustainably. **Questions:** Which two of these solutions do you think are the best ones? Why?

Top: Anhong/Dreamstime. Bottom: Banol2007/Dreamstime.

Figure 11.11 lists ways to prevent or slow the problem of aquifer depletion by using this largely renewable resource more sustainably. The challenge is to educate people about the dangers of depleting vital underground supplies of water that they cannot see. **GREEN CAREER: Hydrogeologist**

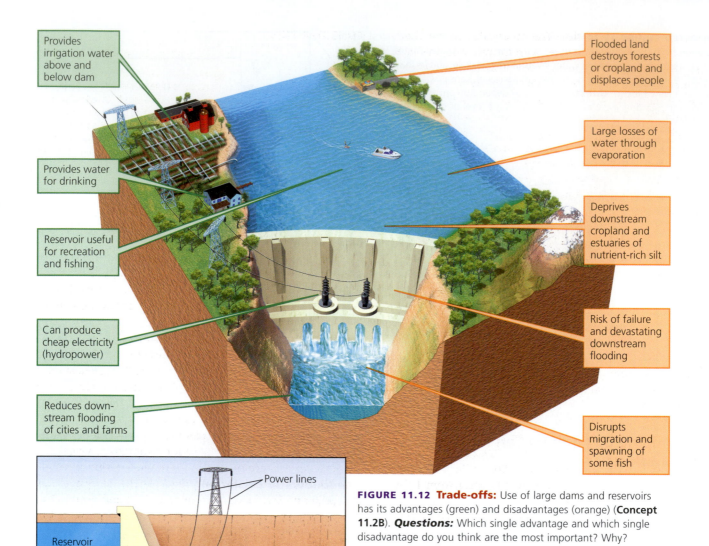

Provides irrigation water above and below dam

Provides water for drinking

Reservoir useful for recreation and fishing

Can produce cheap electricity (hydropower)

Reduces down-stream flooding of cities and farms

Flooded land destroys forests or cropland and displaces people

Large losses of water through evaporation

Deprives downstream cropland and estuaries of nutrient-rich silt

Risk of failure and devastating downstream flooding

Disrupts migration and spawning of some fish

Power lines

Reservoir

Dam

Intake

Powerhouse

Turbine

© 2016 Cengage Learning

FIGURE 11.12 Trade-offs: Use of large dams and reservoirs has its advantages (green) and disadvantages (orange) (**Concept 11.2B**). **Questions:** Which single advantage and which single disadvantage do you think are the most important? Why?

Use of Large Dams Provides Benefits and Creates Problems

A **dam** is a structure built across a river to control its flow. Usually, dammed water creates an artificial lake, or **reservoir**, behind the dam (Figure 11.2). The main goals of a dam-and-reservoir system are to capture and store the surface runoff from a river's watershed, and release it as needed to control floods, to generate electricity (hydropower), and to supply freshwater for irrigation and for

towns and cities. Reservoirs also provide recreational activities such as swimming, fishing, and boating. Large dams and reservoirs provide benefits but have drawbacks (Figure 11.12).

The world's 45,000 large dams capture and store about 14% of the world's surface runoff, provide water for almost half of all irrigated cropland, and supply more than half the electricity used in 65 countries. By using dams, we have increased the annual reliable runoff available for our uses by nearly 33%. On the down side, this engineering approach to river management has displaced 40–80 million people from their homes, flooded a large area of mostly productive land to establish reservoirs, and impaired some of the important ecosystem services that rivers provide (see Figure 7.25, left, p. 152) (**Concept 11.2B**)

A 2007 study by the World Wildlife Fund (WWF) estimated that about one out of five of the world's freshwater fish and plant species are either extinct or endangered, primarily because dams and water withdrawals have sharply decreased the flows of many rivers such as the Colorado (**Core Case Study**). This helps explain why estimated extinction rates for freshwater life are 4–6 times higher than for marine or terrestrial species. The study also found that, because of dams, excessive water withdrawals, and, in some areas, prolonged severe drought, only 21 of the planet's 177 longest rivers consistently run all the way to the sea before running dry (see Case Study that follows).

Within 50 years, the reservoirs behind dams typically fill up with sediments such as mud and silt, which makes them useless for storing water or producing electricity. For example, in the Colorado River system (Core Case Study), the equivalent of roughly 20,000 dump-truck loads of silt are deposited on the bottoms of the Lake Powell and Lake Mead reservoirs every day. Sometime during this century, these two reservoirs will probably be too full of silt to function as designed. About 85% of all U.S. dam-and-reservoir systems will be 50 years old or more by 2025, and some of those aging dams are being removed because their reservoirs have filled with silt.

If climate change occurs as projected during this century, it will intensify shortages of water in many parts of the world. For example, mountain snows that feed the Colorado River system (Core Case Study) will melt faster and earlier, making less freshwater available to the river system when it is needed for irrigation during hot and dry summer months.

Also, as the river's flow has slowed, the water level in Lake Mead has dropped below the intake pipes that are a key part of hydroelectric production. This has forced the city of Las Vegas to spend more than $800 million building lower intake pipes to keep the water flowing from Lake Mead through the Hoover Dam.

If some of the Colorado's largest reservoirs drop dramatically or become filled with silt during this century, the region could experience economic and ecological disruptions with political and legal battles over who will get how much of the region's greatly diminished freshwater supply. Agricultural production would drop sharply and the region's major desert cities such as Las Vegas, Nevada (which depends on the Colorado for more than 90% of its freshwater needs), and Phoenix, Arizona, would be challenged to survive. A 2012 report from the U.S. Bureau of Reclamation concluded that over the next 50 years, the Colorado River will not be able to meet the projected water demands of Arizona, New Mexico, and California.

Nearly 3 billion people in South America, China, India, and other parts of Asia depend on river flows fed by mountain glaciers, which act like aquatic savings accounts. They store precipitation as ice and snow in wet periods and release it slowly during dry seasons for use on farms and in cities. In 2010, according to the World Glacier Monitoring Service, many of these mountain glaciers had been shrinking for 19 consecutive years, mostly due to a warming atmosphere. For a while, their melting will increase water supplies. However, if these glaciers eventually disappear, most of those 3 billion people will likely be short of water for all purposes, including food production.

Severe drought and land use activities such as all-terrain vehicle use in desert areas can expose soil to erosion by wind, which adds dust to the atmosphere. Strong winds can carry this dust to mountain ranges where it lands on mountain snowpacks. The dust-darkened snow melts faster because it absorbs more solar energy than white snow absorbs. These snowpacks then melt earlier and faster, and water supplies for farmers and urban areas are increasingly strained during the warmer summer months.

CASE STUDY

How Dams Can Kill an Estuary

Since 1905, the amount of water flowing to the mouth of the Colorado River (Core Case Study) has dropped dramatically. In most years since 1960, the river has dwindled to a small, sluggish stream by the time it reaches the Gulf of California.

This is the subject of an online short film by National Geographic Explorer Alexandra Cousteau (see Individuals Matter 7.1, p. 156), called *Death of a River: The Colorado River Delta*. In that film, Cousteau explains that the river once emptied into a vast *delta*, the wetland area at the mouth of a river containing the river's estuary. It hosted forests, lagoons, and marshes rich in plant and animal life and supported a thriving coastal fishery for hundreds of years.

Since the damming of the Colorado—within one human lifetime—this biologically diverse delta ecosystem has collapsed and is now covered by mud flats and desert. The dams upstream have cut off the water supply that kept the system alive. The delta and its wildlife are now mostly gone and its coastal fishery that fed many generations of area residents is disappearing.

Historically, about 80% of the water withdrawn from the Colorado has been used to irrigate crops and raise cattle with the help of government subsidies. Such subsidies have led to inefficient use of irrigation water for growing thirsty crops such as rice, cotton, and alfalfa. Also, much of the water in the Colorado's reservoirs is lost through evaporation and seepage of water into porous rock beds under the reservoirs.

According to Cousteau, if just 1% of the river's flow were restored to the delta area, much of it could be partially restored. In 2014, the floodgates of the Morelos Dam near Yuma, Arizona were opened for 2 months to send Colorado River water through the delta to the Gulf of California for the first time in years. Researchers will evaluate the effects of this experiment on the vegetation and soils in the delta.

To deal with the root of this problem, water experts call for the seven states using the Colorado River to enact and enforce strict water conservation measures, including cleaning up and recycling sewage water, as is being done in some urban areas in Southern California. They also call for phasing out state and federal government subsidies for agricul-

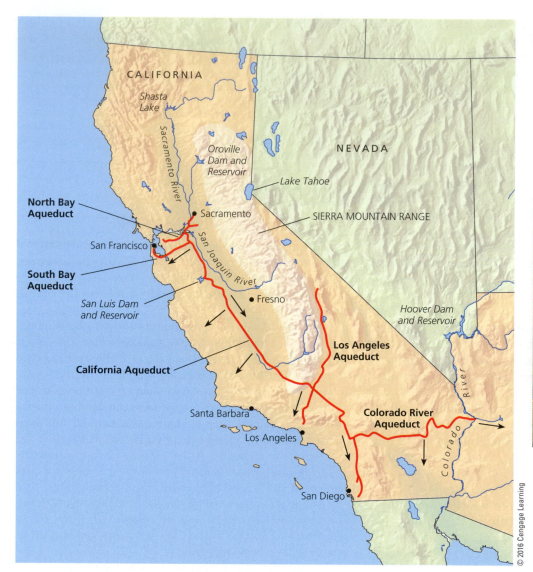

FIGURE 11.13 The California State Water Project transfers huge volumes of freshwater from one watershed to another. The red arrows on the map show the general direction of water flow. The photo shows one of the aqueducts carrying water within the system. **Question:** What effects might this system have on the areas from which the water is taken?

ture in this region, shifting water-thirsty crops to less arid areas, and banning or severely restricting the use of surface water and groundwater to keep golf courses and lawns green in the desert areas of the Colorado River basin. They suggest that the best way to implement such solutions is to sharply raise the historically low price of the river's freshwater over the next decade—another application of the full-cost pricing **principle of sustainability**.

Water Transfers Can Be Inefficient and Environmentally Harmful

In the world's 364 water transfer projects, canals and pipelines are used to transfer large volumes of water from water-rich areas to water-poor areas. For example, when you have lettuce in a salad in the United States, chances are good that it was grown in the arid Central Valley of California, partly with the use of irrigation water from the mountains of northeastern California. The *California State*

Water Project (see map in Figure 11.13) is one of the world's largest freshwater transfer projects. It uses a maze of giant dams, pumps, and lined canals, or *aqueducts* (photo in Figure 11.13), to transport freshwater to the vast farms and heavily populated cities of central and southern California.

This massive water transfer has yielded many benefits. California's heavily irrigated Central Valley supplies half of the United States' fruits and vegetables, and the cities of San Diego and Los Angeles have grown and flourished because of this water transfer. However, the project has also reduced the flow of the Sacramento River, threatening fisheries and reducing the flushing action that helps to cleanse San Francisco Bay of pollutants. The bay has suffered from pollution, and the flow of freshwater to its coastal marshes and other ecosystems has dropped, putting stress on wildlife species that depend on these ecosystems (**Concept 11.2B**).

Water transfers like the California Project are typically subsidized by governments, which has led to inefficient uses such as the irrigation of lettuce, alfalfa, almonds, and other thirsty crops in desert-like areas. In central California, agriculture consumes three-fourths of the water that is transferred, and much of it is lost through inefficient irrigation systems. Studies have shown that making irrigation just 10% more efficient would provide all the water necessary for domestic and industrial uses in southern California. Yet, the inefficient use of water continues because taxpayer subsidies have reduced the price of water to the point where farmers and urban residents have little incentive to invest in more efficient irrigation and water-saving toilets, showerheads, and other water-saving devices. Many hydrologists and economists call for replacing these subsidies with subsidies that would encourage more efficient use of water.

According to several studies, projected climate change will make matters worse by sharply reducing surface water availability in California. In northern California, many people depend on *snowpacks*, bodies of densely packed, slowly melting snow in the High Sierra Mountains, for more than 60% of their freshwater during the hot, dry summer months. Projected atmospheric warming could shrink the snowpacks by as much as 40% by 2050 and by as much as 90% by the end of this century. This will sharply reduce the amount of freshwater available for northern residents and ecosystems, as well as for transfer to arid and semiarid central and southern California.

There are many other examples around the world of water transfers that have resulted in environmental degradation (see the following Case Study).

The Aral Sea Disaster: A Glaring Example of Unintended Consequences

The shrinking of the Aral Sea (Figure 11.14) is the result of a large-scale freshwater transfer project in an area of the former Soviet Union with the driest climate in central Asia. Since 1960, enormous amounts of irrigation water have been diverted from the two rivers that supply water to the Aral Sea. The goal was to create one of the world's largest irrigated areas, mostly for raising cotton and rice. The irrigation canal, the world's longest, stretches more than 1,300 kilometers (800 miles)—roughly the distance between the two U.S. cities of Boston, Massachusetts, and Chicago, Illinois.

This project, coupled with drought and high evaporation rates due to the area's hot and dry climate, has caused a regional ecological and economic disaster. Since 1961, the sea's salinity has risen sevenfold and the average level of its water has dropped by an amount roughly equal to the height of a six-story building. The Southern Aral Sea has lost 90% of its volume of water and most of its lake

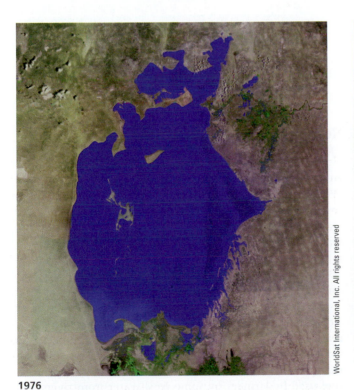

1976

2012

FIGURE 11.14 Natural capital degradation: The *Aral Sea*, straddling the borders of Kazakhstan and Uzbekistan, was one of the world's largest saline lakes. These satellite photos show the sea in 1976 (top) and in 2012 (bottom). ***Question:*** What do you think should be done to help prevent further shrinkage of the Aral Sea?

bottom is now a white salt desert (Figure 11.14, right photo). Water withdrawal for agriculture has reduced the two rivers feeding the sea to mere trickles.

About 85% of the area's wetlands have been eliminated and about half the local bird and mammal species have disappeared. The sea's greatly increased salt concentration—3 times saltier than ocean water—has caused the presumed local extinction of 26 of the area's 32 native fish species. This has devastated the area's fishing industry, which once provided work for more than 60,000 people. Fishing villages and boats once located on the sea's coastline now sit abandoned in a salty desert.

Winds pick up the sand and salty dust and blow it onto fields as far as 500 kilometers (310 miles) away. As the salt spreads, it pollutes water and kills wildlife, crops, and other vegetation. Aral Sea dust settling on glaciers in the Himalayas is causing them to melt at a faster-than-normal rate.

The shrinkage of the Aral Sea has also altered the area's climate. The shrunken sea no longer acts as a thermal buffer to moderate the heat of summer and the extreme cold of winter. Now there is less rain, summers are hotter and drier, winters are colder, and the growing season is shorter. The combination of such climate change and severe salinization has reduced crop yields by 20–50% on almost one-third of the area's cropland—the opposite of the project's intended consequences.

Since 1999, the UN, the World Bank, and the five countries surrounding the lake have worked to improve irrigation efficiency. They have also partially replaced thirsty crops with other crops that require less irrigation water. Because of a dike built to block the flow of water from the Northern Aral Sea into the southern sea, the level of the northern sea has risen by 2 meters (7 feet), its salinity has dropped, dissolved oxygen levels are up, and it supports a healthy fishery.

However, the formerly much larger southern sea is still shrinking. By 2012, its eastern lobe was essentially gone (Figure 11.14, right photo), and the European Space Agency projects that the rest of the Southern Aral Sea could dry up completely by 2020.

Removing Salt from Seawater Is Costly and Has Harmful Effects

Desalination is the process of removing dissolved salts from ocean water or from brackish (slightly salty) water in aquifers or lakes. It is another way to increase supplies of freshwater (**Concept 11.2C**).

The two most widely used methods for desalinating water are distillation and reverse osmosis. *Distillation* involves heating saltwater until it evaporates (leaving behind salts in solid form) and condenses as freshwater. *Reverse osmosis* (or *microfiltration*) uses high pressure to force saltwater through a membrane filter with pores small enough to remove the salt and other impurities. Desalination supplies less than 0.3% of the world's demand and only about 0.4% of the U.S. demand for freshwater.

There are three major problems with the widespread use of desalination. *First* is the high cost, because it takes a lot of increasingly expensive energy to remove salt from seawater. A *second* problem is that pumping large volumes of seawater through pipes requires the use of chemicals to sterilize the water and to keep down algae growth, and this kills many marine organisms and also requires large inputs of energy and money. *Third*, desalination produces huge quantities of salty wastewater that must go somewhere. Dumping it into nearby coastal ocean waters increases the salinity of those waters, which can threaten food resources and aquatic life, especially if it is dumped near coral reefs, marshes, or mangrove forests. Disposing of it on land could contaminate groundwater and surface water (**Concept 11.2C**).

Currently, desalination is practical only for water-short countries and cities that can afford its high cost. However, scientists and engineers are working to develop better and more affordable desalination technologies.

11.3 HOW CAN WE USE FRESHWATER MORE SUSTAINABLY?

CONCEPT 11.3 We can use freshwater more sustainably by cutting water waste, raising water prices, slowing population growth, and protecting aquifers, forests, and other ecosystems that store freshwater.

Reducing Freshwater Losses Can Provide Many Benefits

According to water resource expert Mohamed El-Ashry of the World Resources Institute, about 66% of the freshwater used in the world and about 50% of the freshwater used in the United States is lost through evaporation, leaks, and inefficient use. El-Ashry estimates that it is economically and technically feasible to reduce such losses to 15%, thereby meeting most of the world's freshwater needs for the foreseeable future.

GOOD NEWS

So why do we have such large losses of freshwater? According to water resource experts, there are two major reasons. First, the *cost of freshwater to most users is low.* Such underpricing is mostly the result of government subsidies that provide irrigation water or the electricity and diesel fuel used by farmers to pump freshwater from rivers and aquifers, at below-market prices. People in the United States and China—two of the world's largest water users—respectively pay only 2.9% and 0.5% of their disposable income for water, according to a 2012 study by the Deutsche Bank. Because of this failure to apply the full-cost pricing **principle of sustainability**, users have little or no financial incentive to invest in water-saving technologies.

CONSIDER THIS. . .

THINKING ABOUT Government Freshwater Subsidies

Many argue that government freshwater subsidies promote the farming of unproductive land, stimulate local economies, and help to keep food and electricity prices low. Do you think this is reason enough for governments to continue providing subsidies to farmers and cities to help keep the price of freshwater low? Explain.

Higher prices for freshwater encourage water conservation but make it difficult for low-income farmers and city dwellers to buy enough water to meet their needs. When South Africa raised water prices, it dealt with this problem by establishing *lifeline rates*, which give each household a set amount of free or low-priced water to meet basic needs. When users exceed this amount, they pay increasingly higher prices as their water use increases. This is a *user-pays* approach.

The second major cause of unnecessary losses of freshwater is a *lack of government subsidies for improving the efficiency of water use*. Withdrawing some of the subsidies that encourage inefficient water use and replacing them with subsidies for more efficient water use would sharply reduce water losses. Understandably, farmers and industries that receive subsidies that keep water prices low have vigorously opposed efforts to eliminate or reduce them.

We Can Improve Efficiency in Irrigation

Only about 60% of the world's irrigation water reaches crops, which means that most irrigation systems are highly inefficient. The most inefficient irrigation system, commonly used in less-developed countries, is *flood irrigation,* in which water is pumped from a groundwater or surface water source through unlined ditches where it flows by gravity to the crops being watered (Figure 11.15, left). This method delivers far more water than is needed for crop growth and typically, about 45% of this water is lost through evaporation, seepage, and runoff. With existing irrigation technologies (see Figure 11.15, middle and right), this loss could be reduced to 5–10%.

FIGURE 11.15 Traditional irrigation methods rely on gravity and flowing water (left). Newer systems such as center-pivot, low-pressure sprinkler irrigation (center), and drip irrigation (right) are far more efficient.

© 2016 Cengage Learning

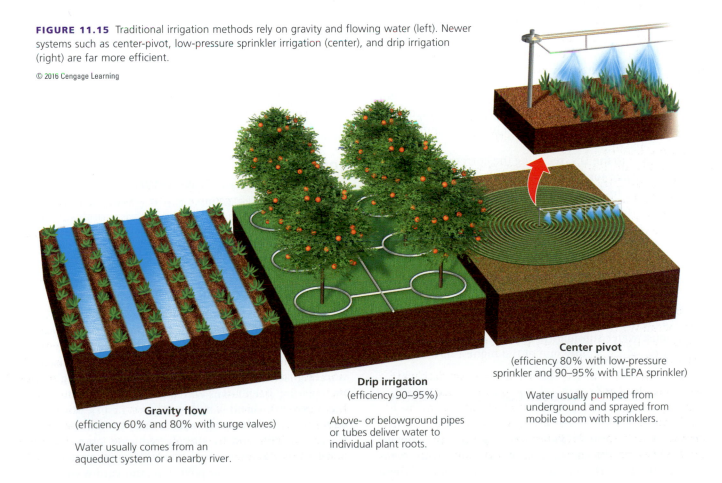

Gravity flow
(efficiency 60% and 80% with surge valves)

Water usually comes from an aqueduct system or a nearby river.

Drip irrigation
(efficiency 90–95%)

Above- or belowground pipes or tubes deliver water to individual plant roots.

Center pivot
(efficiency 80% with low-pressure sprinkler and 90–95% with LEPA sprinkler)

Water usually pumped from underground and sprayed from mobile boom with sprinklers.

Solutions

Reducing Irrigation Water Losses

- Avoid growing thirsty crops in dry areas

- Import water-intensive crops and meat

- Encourage organic farming and polyculture to retain soil moisture

- Monitor soil moisture to add water only when necessary

- Expand use of drip irrigation and other efficient methods

- Irrigate at night to reduce evaporation

- Line canals that bring water to irrigation ditches

- Irrigate with treated wastewater

© 2016 Cengage Learning

FIGURE 11.16 Ways to reduce freshwater losses in irrigation. *Questions:* Which two of these solutions do you think are the best ones? Why?

Since the early 1990s, the global area of cropland on which drip irrigation is used has increased more than six-fold, with most of this growth happening in the United States, China, and India. Even so, drip irrigation is used on less than 4% of the irrigated crop fields in the world and in the United States, largely because most drip irrigation systems are costly. This percentage rises to 13% in the U.S. state of California, 66% in Israel, and 90% in Cyprus. If freshwater were priced closer to the value of the ecosystem services it provides and if government subsidies for inefficient use of water were reduced or eliminated, drip irrigation could more easily be used to irrigate most of the world's crops.

According to the UN, reducing the current global withdrawal of water for irrigation by just 10% would save enough water to grow crops and meet the estimated additional water demands of the earth's cities and industries through 2025. Figure 11.16 summarizes several ways to reduce water losses in crop irrigation. Since 1950, Israel has used many of these techniques to slash irrigation water losses by 84% while irrigating 44% more land. Israel now treats and reuses 30% of its municipal sewage water for crop production and plans to increase this to 80% by 2025. The government also gradually eliminated most water subsidies to raise Israel's price of irrigation water, which is now one of the highest in the world.

We Can Cut Freshwater Losses in Industries and Homes

Producers of chemicals, paper, oil, coal, primary metals, and processed foods consume almost 90% of the freshwater used by industries in the United States. Some of these industries recapture, purify, and recycle water to reduce their water use and water treatment costs. For example, more than 95% of the water used to make steel can be recycled. Even so, most industrial processes could be redesigned to use much less water. **GREEN CAREER: Water conservation specialist**

Flushing toilets with freshwater (most of it clean enough to drink) is the single largest use of domestic freshwater in the United States and accounts for about one-fourth of home water use. Since 1992, U.S. government standards have required that new toilets use no more than 6.1 liters (1.6 gallons) of water per flush. Even at this rate, just two flushes of such a toilet require more than the daily amount of water available for all uses to many of the world's poor people living in arid regions (see chapter-opening photo).

Other water-saving appliances are widely available. Low-flow showerheads can save large amounts of water by cutting the flow of a shower in half. Front-loading clothes washers use 30% less water than top-loading machines use. According to the American Water Works Association, if the typical American household were to stop all water leaks and use these devices, along with low-flow toilets and faucets, it could cut its daily water use by nearly a third.

According to UN studies, 30–60% of the water supplied in nearly all of the world's major cities in less-developed countries is lost, primarily through leakage from water mains, pipes, pumps, and valves. Water experts say that fixing these leaks should be a high priority for water-short countries, because it would increase water supplies and cost less than building dams or importing water. Even in advanced industrialized countries such as the United States, these losses to leakage average 10–30%. However, leakage losses have been reduced to about 3% in Copenhagen, Denmark, and to 5% in Fukuoka, Japan.

CONSIDER THIS. . .

CONNECTIONS Water Leaks and Water Bills

Any water leak wastes freshwater and raises water bills. You can detect a silent toilet water leak by adding a few drops of food coloring to the toilet tank and waiting 5 minutes. If the color shows up in the bowl, you have a leak. Also, a faucet leaking one drop per second can drip about 8,200 liters (3,000 gallons) of water down the drain in a year—enough to fill about 75 bathtubs. This also represents money going down the drain.

Many homeowners and businesses in water-short areas are using drip irrigation on their properties to cut water losses. Some are also using smart sprinkler systems with moisture sensors that have helped to cut water used for watering lawns by up to 40%. Others are copying nature by replacing green lawns with a mix of native plants that need little or no watering (Figure 11.17). Such water-thrifty landscaping saves money by reducing water use by 30–85% and by sharply reducing labor, fertilizer, and fuel requirements. It also can help landowners to reduce polluted runoff, air pollution, and yard wastes.

FIGURE 11.17 This yard in Encinitas, a city in a dry area of southern California (USA), uses a mix of plants that are native to the arid environment and require little watering.

iStockphoto.com/Escaflowne

This example of reconciliation ecology (see Chapter 9, p. 204) also helps to provide habitats and food for threatened honeybee, butterfly, and songbird species. It is an application of the biodiversity **principle of sustainability**, as well as a good way to make a beneficial environmental impact.

However, in some more-developed countries, people who live in arid areas maintain green lawns by watering them heavily. Some communities and housing developments in water-short areas have even passed ordinances that require green lawns and prohibit the planting of native vegetation in place of lawns.

Water used in homes can be reused and recycled. About 50–75% of a typical household's *gray water*—used water from bathtubs, showers, sinks, dishwashers, and clothes washers—could be stored in a holding tank and then reused to irrigate lawns and nonedible plants, to flush toilets, and to wash cars. Such efforts mimic the way nature recycles water, and thus they follow the chemical cycling **principle of sustainability**.

The relatively low cost of water in most communities is one of the major causes of excessive water use in homes and industries. Many water utility and irrigation authorities charge an annual flat fee for water use, and some

Solutions

Reducing Water Losses

- Redesign manufacturing processes to use less water

- Recycle water in industry

- Fix water leaks

- Landscape yards with plants that require little water

- Use drip irrigation on gardens and lawns

- Use water-saving showerheads, faucets, appliances, and toilets (or waterless composting toilets)

- Collect and reuse gray water in and around houses, apartments, and office buildings

- Raise water prices and use meters, especially in dry urban areas

© 2016 Cengage Learning

FIGURE 11.18 Ways to reduce freshwater losses in industries, homes, and businesses (**Concept 11.3**). *Questions:* Which three of these solutions do you think are the best ones? Why?

charge less to the largest users of water. About one-fifth of all U.S. public water systems do not use water meters and charge a single low annual rate for almost unlimited use of high-quality freshwater.

When the U.S. city of Boulder, Colorado, introduced water meters, water use per person dropped by 40%. In some cities in Brazil, people buy *smart cards,* each of which contains a certain number of water credits that entitle their owners to measured amounts of freshwater. Brazilian officials say this approach saves water and typically reduces household water bills by 40%. Figure 11.18 lists various ways to use water more efficiently in industries, homes, and businesses (**Concept 11.3**).

We can also collect and store rainwater for future use. For example, Los Angeles, California, plans to build large ponds to capture and store the water that falls on its streets and roofs during winter storms. The plan is to pass the captured rainwater through newly constructed wetlands for transfer into drinking water aquifers. This would supply more drinking water and reduce flooding from storms.

We Can Use Less Water to Remove Wastes

Currently, we use large amounts of freshwater clean enough to drink to flush away industrial, animal, and household wastes. According to the UN Food and Agriculture Organization (FAO), if current growth trends in population and water use continue, within 40 years, we will need the world's entire reliable flow of river water just to dilute and transport the wastes we produce each year.

We could save much of this freshwater by recycling and reusing gray water from homes and businesses and wastewater from sewage treatment plants for purposes such as cleaning equipment and watering lawns and some crops. Israel reuses 70% of its wastewater to irrigate non-food crops. In Singapore, all sewage water is treated at reclamation plants for reuse by industry. U.S. cities such as Las Vegas, Nevada, and Los Angeles, California, are also beginning to clean up and reuse some of their wastewater.

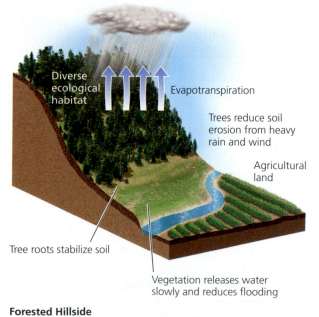

Diverse ecological habitat

Evapotranspiration

Trees reduce soil erosion from heavy rain and wind

Agricultural land

Tree roots stabilize soil

Vegetation releases water slowly and reduces flooding

Forested Hillside

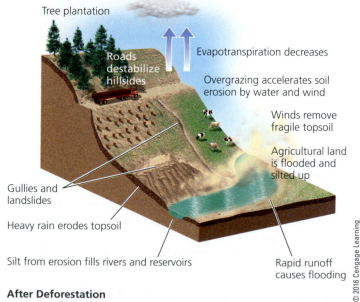

Tree plantation

Roads destabilize hillsides

Evapotranspiration decreases

Overgrazing accelerates soil erosion by water and wind

Winds remove fragile topsoil

Agricultural land is flooded and silted up

Gullies and landslides

Heavy rain erodes topsoil

Silt from erosion fills rivers and reservoirs

Rapid runoff causes flooding

After Deforestation

© 2016 Cengage Learning

ANIMATED FIGURE 11.19 Natural capital degradation: A hillside before and after deforestation. *Question:* How might a drought in this area make these conditions even worse?

However, only about 7% of the water in the United States is recycled, cleaned up, and reused. Sharply raising this percentage would be a way to apply the chemical cycling **principle of sustainability**.

We Can Reduce the Pollution of Freshwater Supplies Due to Flooding

Some areas have too little freshwater, but others sometimes have too much because of natural flooding by streams, caused mostly by heavy rain or rapidly melting snow. We could use water more sustainably by reducing the human contribution to flooding and the resulting pollution of water sources.

Human activities have contributed to flooding and water pollution in several ways. First, in efforts to reduce the threat of flooding on *floodplains,* or land areas adjacent to streams and rivers, some rivers have been narrowed and straightened (or *channelized*), surrounded by protective dikes and *levees* (long mounds of earth along their banks), and dammed to create reservoirs that store and release water as needed (Figure 11.2). However, in the long run, such measures can lead to greatly increased flooding and surface water pollution when heavy snowmelt or prolonged rains overwhelm them.

Second, in developing land, we have *removed a great deal of water-absorbing vegetation,* especially on hillsides (Figure 11.19). Once the trees on a hillside have been cut for timber, fuelwood, livestock grazing, or farming, water from precipitation rushes down the denuded slopes, erodes

precious topsoil, and can increase flooding and pollution in local streams.

The third human activity that increases the severity of flooding is the *draining of wetlands* that naturally absorb floodwaters. These areas are then often covered with pavement and buildings that greatly increase runoff, which contributes to flooding and pollution of surface waters.

Many scientists argue that we could reduce our contribution to such flooding and water pollution by relying less on engineered devices such as dams and levees and more on nature's systems such as wetlands and forests in watersheds. To that end, they call for preserving existing wetlands and restoring degraded wetlands that lie in floodplains to take advantage of the natural flood control they provide. These and other ways to reduce our contribution to flooding are listed in Figure 11.20.

We Can Use Water More Sustainably

Each of us can reduce our water footprints by using less water and using it much more efficiently (Figure 11.21).

Finding more sustainable ways to use freshwater is the subject of some major research efforts. One group that is working on this problem is the Global Water Policy Project, founded by the renowned water supply expert and National Geographic Explorer Sandra Postel (Individuals Matter 11.1).

Solutions

Reducing Flood Damage

Prevention		Control
Preserve forests in watersheds		Strengthen and deepen streams (channelization)
Preserve and restore wetlands on floodplains		
Tax development on floodplains		Build levees or floodwalls along streams
Increase use of floodplains for sustainable agriculture and forestry		Build dams

© 2016 Cengage Learning

FIGURE 11.20 Methods for reducing the harmful effects of flooding. **Questions:** Which two of these solutions do you think are the best ones? Why?

What Can You Do?

Water Use and Waste

- Use water-saving toilets, showerheads, and faucets
- Take short showers instead of baths
- Turn off sink faucets while brushing teeth, shaving, or washing
- Wash only full loads of clothes or use the lowest possible water-level setting for smaller loads
- Repair water leaks
- Wash your car from a bucket of soapy water, use gray water, and use the hose for rinsing only
- If you use a commercial car wash, try to find one that recycles its water
- Replace your lawn with native plants that need little if any watering
- Water lawns and gardens only in the early morning or evening and use gray water
- Use drip irrigation and mulch for gardens and flowerbeds

© 2016 Cengage Learning

FIGURE 11.21 Individuals matter: You can reduce your use and waste of freshwater. **Questions:** Which of these steps have you taken? Which would you like to take?

Sandra Postel: National Geographic Fellow and Freshwater Conservationist

Sandra Postel is regarded as one of the world's most respected authorities on water issues. In 1994 she founded the Global Water Policy Project, which, through research and education, promotes a deeper knowledge of, and more sustainable use of, the earth's finite freshwater supply. Postel has authored or coauthored several influential books and written scores of articles for popular and scholarly publications about using water more sustainably.

In her quest to educate people about water supply issues, Postel has appeared on several television news shows, taken part in numerous environmental documentary films (including the BBC's *Planet Earth*), and addressed the European Parliament. In 2010, she was appointed Freshwater Fellow of the National Geographic Society, where she serves as the lead water expert for the society's freshwater efforts and co-director of *Change the Course*, a national freshwater conservation and restoration campaign being piloted in the Colorado River Basin. In 2002, Postel was named one of the "Scientific American 50" for her contributions to science and technology related to more sustainable use of the earth's freshwater supply.

11.4 HOW CAN WE DEAL WITH WATER POLLUTION?

CONCEPT 11.4 Reducing water pollution requires that we prevent it, work with nature to treat sewage, cut resource use and waste, reduce poverty, and slow population growth.

Water Pollution Comes from Point and Nonpoint Sources

Water pollution is any change in water quality that can harm living organisms or make the water unfit for human uses such as drinking, irrigation, and recreation. It can come from single (point) sources or from larger and dispersed (nonpoint) sources. **Point sources** discharge pollutants into bodies of surface water at specific locations through drain pipes (Figure 11.22), ditches, or sewer lines. Examples include factories, sewage treatment plants (which remove some, but not all, pollutants), underground mines, oil wells, and oil tankers.

Because point sources are located in specific places, they are fairly easy to identify, monitor, and regulate. Most of the world's more-developed countries have laws that help control point-source discharges of harmful chemicals

FIGURE 11.22 Point source of water pollution from an industrial plant.

into aquatic systems. In most of the less-developed countries, there is little control of such discharges.

Nonpoint sources are broad and diffuse areas where rainfall or snowmelt washes pollutants off the land into bodies of surface water. Examples include runoff of eroded soil and chemicals such as fertilizers and pesticides from cropland, feedlots, logged forests, urban streets, parking lots, lawns, and golf courses. We have made little progress in controlling water pollution from nonpoint sources because of the difficulty and expense of identifying and controlling discharges from so many diffuse sources. According to the U.S. Environmental Protection Agency (EPA), nonpoint source pollution is the main reason why 40% of all U.S. rivers, lakes, and estuaries are still not clean enough for uses such as fishing and swimming, despite the enactment of major water pollution control laws 40 years ago.

Agricultural activities are by far the leading cause of water pollution. Sediment eroded from agricultural lands (see Figure 10.9, p. 224) is the most common pollutant. Other major agricultural pollutants include fertilizers and pesticides, bacteria from livestock and food-processing wastes, and excess salts from soils of irrigated cropland. *Industrial facilities,* which emit a variety of harmful chemicals, are a second major source of water pollution. *Mining* is the third biggest source of water pollution. Surface mining disturbs the land, creating major erosion of sediments and runoff of toxic chemicals.

Another form of water pollution is caused by the widespread use of human-made materials such as plastics used to make millions of products. Much of the plastic that is improperly discarded eventually winds up in waterways (see Figure 1.9, p. 11) and in the oceans. Such discarded plastic products can harm various forms of wildlife (Figure 11.23).

TABLE 11.1 Major Water Pollutants and Their Sources

Type/Effects	Examples	Major Sources
Infectious agents (pathogens) *Cause diseases*	Bacteria, viruses, protozoa, and parasites	Human and animal wastes
Oxygen-demanding wastes *Deplete dissolved oxygen needed by aquatic species*	Biodegradable animal wastes and plant debris	Sewage, animal feedlots, food-processing facilities, paper mills
Plant nutrients *Cause excessive growth of algae and other species*	Nitrates (NO_3^-) and phosphates (PO_4^{3-})	Sewage, animal wastes, inorganic fertilizers
Organic chemicals *Add toxins to aquatic systems*	Oil, gasoline, plastics, pesticides, cleaning solvents	Industry, farms, households
Inorganic chemicals *Add toxins to aquatic systems*	Acids, bases, salts, and metal compounds	Industry, households, surface runoff
Sediments *Disrupt photosynthesis, food webs, and other processes*	Soil, silt	Land erosion
Heavy metals *Cause cancer, disrupt immune and endocrine systems*	Lead, mercury, arsenic	Unlined landfills, household chemicals, mining refuse, and industrial discharges
Thermal *Make some species vulnerable to disease*	Heat	Electric power and industrial plants

© 2016 Cengage Learning

CONSIDER THIS. . .

CONNECTIONS Atmospheric Warming and Water Pollution

Projected climate change from atmospheric warming will likely contribute to water pollution in some areas of the globe. In a warmer world, some regions will get more precipitation and other areas will get less. More intense downpours will flush more harmful chemicals, plant nutrients, and disease-causing microorganisms into some waterways. In other areas, prolonged drought will reduce river flows that dilute wastes.

One of the major water pollution problems that we face is exposure to infectious bacteria, viruses, and parasites that can be transferred into water from the wastes of humans and other animals. The WHO estimates that each year, more than 1.6 million people die from largely preventable waterborne infectious diseases that they get by drinking contaminated water or by not having enough clean water for adequate hygiene. Every 18 seconds, a young child somewhere in the world dies from diarrhea caused by drinking contaminated water, according to the WHO. Table 11.1 lists the major types of water pollutants along with examples of each and their harmful effects and sources.

Doris Alcorn/U.S. National Maritime Fisheries

FIGURE 11.23 This Hawaiian monk seal was slowly starving to death before a discarded piece of plastic was removed from its snout.

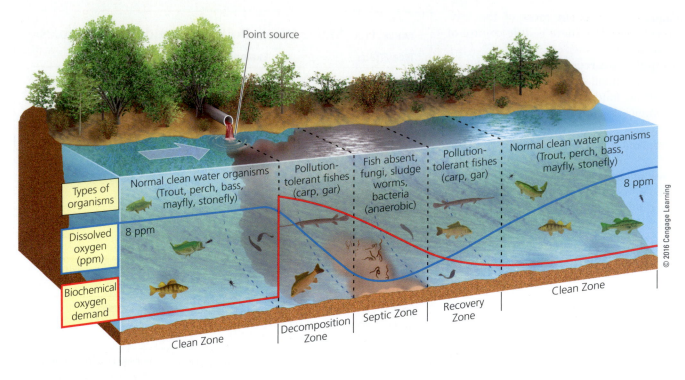

ANIMATED FIGURE 11.24 Natural capital: A stream can dilute and decay degradable, oxygen-demanding wastes, and it can also dilute heated water. This figure shows the oxygen sag curve (blue) and the curve of oxygen demand (red). Streams recover from oxygen-demanding wastes and from the injection of heated water if they are given enough time and are not overloaded. **_Question:_** What would be the effect of putting another discharge pipe emitting biodegradable waste to the right of the one in this picture?

A 2009 study by the National Academy of Sciences ranked the following pollutants as the most serious threats to U.S. stream and lake water quality. In decreasing order, they are: mercury, pathogens from leaking sewer pipes, sediment from land disturbance and stream erosion, metals other than mercury, and nutrients that cause oxygen depletion.

Streams Can Cleanse Themselves, If We Do Not Overload Them

Flowing rivers and streams can recover rapidly from moderate levels of degradable, oxygen-demanding wastes through a combination of dilution and bacterial biodegradation of such wastes. But this natural recovery process does not work when a stream becomes overloaded with such pollutants or when drought, damming, or water diversion reduces its flow (**Concept 11.4**). Also, while this process can remove biodegradable wastes, it does not eliminate slowly degradable and nondegradable pollutants.

In a flowing stream, the breakdown of biodegradable wastes by bacteria depletes dissolved oxygen and creates an *oxygen sag curve* (Figure 11.24). This reduces or eliminates populations of organisms with high oxygen requirements until the stream is cleansed of oxygen-demanding wastes.

Laws enacted in the 1970s to control water pollution have led to a greatly increased number of facilities that treat **wastewater**—water that contains sewage and other wastes from homes and industries—in the United States and in most other more-developed countries. Such laws also require industries to reduce or eliminate their point-source discharges of harmful chemicals into surface waters. This is an impressive accomplishment given the country's increased economic activity, resource consumption, and population growth since the passage of these laws.

In most less-developed countries, stream pollution from discharges of untreated sewage, industrial wastes, and discarded trash is a serious and growing problem. According to the Global Water Policy Project, most cities in less-developed countries discharge 80–90% of their untreated sewage directly into rivers, streams, and lakes whose waters are often used also for drinking, bathing, and washing clothes.

According to the World Commission on Water for the 21st Century, half of the world's 500 major rivers are heavily polluted, with most of these polluted waterways running through less-developed countries. A majority of these countries cannot afford to build waste treatment plants and do not have, or do not enforce, laws for controlling water pollution.

FIGURE 11.25 Severe cultural eutrophication of Chao Lake in China.

Yang Xiaoyuan/Xinhua Press/Corbis Wire/Corbis

Industrial wastes and sewage pollute more than two-thirds of India's water resources as well as 54 of the 78 rivers and streams monitored in China. According to the Ministry of Environmental Protection, some 380 million Chinese people drink unsafe water and nearly half of China's rivers carry water that is too toxic to touch, much less drink.

More than half of the lakes in China suffer from cultural eutrophication.

Lakes and Reservoirs Are Especially Vulnerable to Water Pollution

Lakes and reservoirs are generally less effective at diluting pollutants than streams are, for two reasons. *First,* lakes and reservoirs often contain stratified layers (see Figure 7.26, p. 152) that undergo little vertical mixing. *Second,* they have low flow rates or no flow at all. The flushing and changing of water in lakes and large artificial reservoirs can take from 1 to 100 years, compared with several days to several weeks for streams.

As a result, lakes and reservoirs are more vulnerable than streams are to contamination by runoff or discharges of plant nutrients, oil, pesticides, and nondegradable toxic substances, such as lead, mercury, and arsenic. Many toxic chemicals and acids also enter lakes and reservoirs from the atmosphere.

Eutrophication is the name given to the natural nutrient enrichment of a shallow lake, a coastal area at the mouth of a river, or a slow-moving stream. It is caused mostly by runoff of plant nutrients such as nitrates and phosphates from land bordering such bodies of water. An *oligotrophic lake* is low in nutrients and its water is clear (see Figure 7.27, p. 153). Over time, some lakes become more eutrophic (see Figure 7.28, p. 154) as nutrients are added from natural and human sources in the surrounding watersheds.

Near urban or agricultural areas, human activities can greatly accelerate the input of plant nutrients to a lake—a process called **cultural eutrophication**. Such inputs involve mostly nitrate- and phosphate-containing effluents from various sources, including farmland, feedlots, urban streets and parking lots, chemically fertilized lawns, mining sites, and municipal sewage treatment plants. Some nitrogen also reaches lakes by deposition from the atmosphere.

During hot weather or drought, this nutrient overload can produce dense growths, or "blooms," of organisms such as algae and cyanobacteria (see Figure 7.28, p. 154). When the algae die, they are decomposed by swelling populations of oxygen-consuming bacteria, which deplete dissolved oxygen in the surface layer of water near the shore as well as in the bottom layer of a lake or coastal area. This can kill fish, shellfish, and other aerobic aquatic animals that cannot move to safer waters. If excess nutrients continue to flow into a lake, bacteria that don't require oxygen take over and produce gaseous products such as smelly, highly toxic hydrogen sulfide and flammable methane.

According to the EPA, about one-third of the 100,000 medium to large lakes and 85% of the large lakes near major U. S. population centers have some degree of cultural eutrophication. The International Water Association estimates that more than half of the lakes in China suffer from cultural eutrophication (Figure 11.25).

There are several ways to *prevent* or *reduce* cultural eutrophication. We can use advanced (but expensive) waste treatment processes to remove nitrates and phosphates from wastewater before it enters a body of water. Then we can mimic the earth's natural cycling of nutrients by recycling them to the soil instead of dumping them into waterways—a very important way to increase our beneficial environmental impact. We can also use a preventive approach by banning or limiting the use of phosphates in household detergents and other cleaning agents, and by employing soil conservation (see p. 238) and land-use control to reduce nutrient runoff.

We can *clean up* lakes suffering from cultural eutrophication by mechanically removing excess weeds, controlling undesirable plant growth with herbicides and algaecides, and pumping air into lakes and reservoirs to prevent oxygen depletion—all of which are expensive and energy-intensive methods. Most lakes and other surface waters can recover from cultural eutrophication, if excessive inputs of plant nutrients are stopped.

GOOD NEWS

Groundwater Cannot Cleanse Itself Very Well

According to many scientists, groundwater pollution is a serious threat to human health. Common pollutants such as fertilizers, pesticides, gasoline, and organic solvents can seep into groundwater from numerous sources (Figure 11.26).

FIGURE 11.26 Natural capital degradation: Principal sources of groundwater contamination in the United States. Another source in coastal areas is saltwater intrusion from excessive groundwater withdrawal. (Figure is not drawn to scale.) *Question:* What are three sources shown in this figure that might be affecting groundwater in your area?

© 2016 Cengage Learning

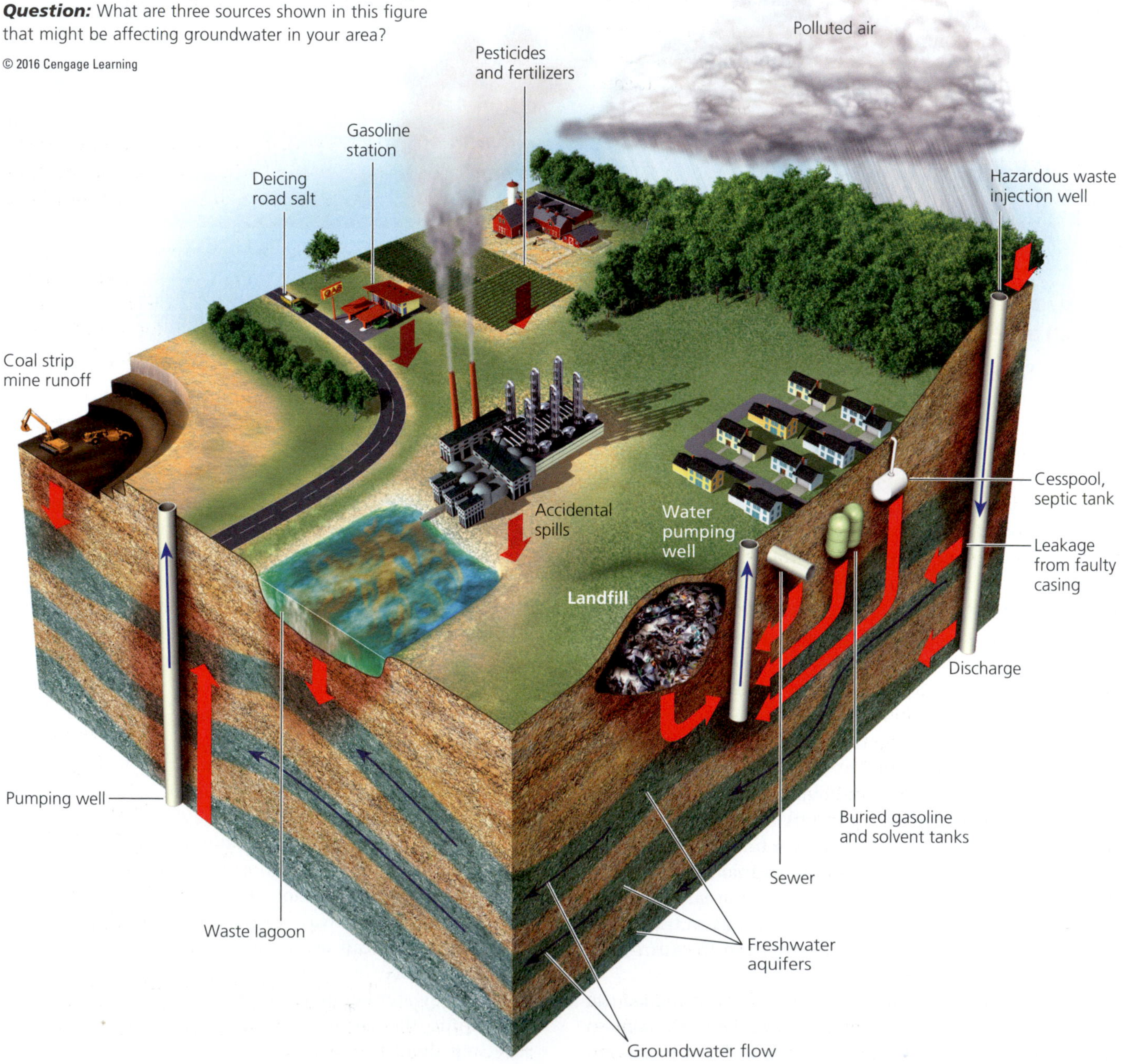

Polluted air

Pesticides and fertilizers

Gasoline station

Deicing road salt

Hazardous waste injection well

Coal strip mine runoff

Accidental spills

Water pumping well

Cesspool, septic tank

Landfill

Leakage from faulty casing

Discharge

Pumping well

Waste lagoon

Buried gasoline and solvent tanks

Sewer

Freshwater aquifers

Groundwater flow

People who dump or spill gasoline, oil, and paint thinners and other organic solvents onto the ground also risk contaminating groundwater.

There is increasing concern about a new and growing potential threat to groundwater—the drilling of thousands of new oil and natural gas wells in parts of the United States involving a process called *hydraulic fracturing,* or *fracking,* which is discussed in detail in Chapter 13. Groundwater contamination could result from leaky gas well pipes and pipefittings and from contaminated wastewater brought to the surface during fracking operations.

When groundwater becomes contaminated, it cannot cleanse itself of degradable wastes as quickly as flowing surface water can. Groundwater flows so slowly that contaminants are not diluted and dispersed effectively. In addition, groundwater usually has much lower concentrations of dissolved oxygen (which helps decompose many contaminants) and smaller populations of decomposing bacteria. And the usually cold temperatures of groundwater slow down chemical reactions that decompose wastes.

Thus, it can take decades to thousands of years for contaminated groundwater to cleanse itself of slowly degradable wastes (such as DDT). On a human time scale, nondegradable wastes (such as toxic lead and arsenic) remain in the water permanently.

Groundwater Pollution Is a Serious Hidden Threat in Some Areas

On a global scale, we do not know much about groundwater pollution because few countries go to the great expense of locating and tracking pollutants and testing aquifers. But the results of scientific studies in scattered parts of the world are alarming.

For example, in 2010, about 90% of China's shallow groundwater was polluted with chemicals such as toxic heavy metals, organic solvents, nitrates, petrochemicals, and pesticides, according to the Chinese Ministry of Land and Resources. About 37% of this groundwater is so polluted that it cannot even be treated for use as drinking water. China has more than 200 "cancer villages" located near mines and factories where cancer rates are far above the national average. Some researchers say there may be more than 400 such villages.

In the United States, an EPA survey of 26,000 industrial waste ponds and lagoons found that one-third of them had no liners to prevent toxic liquid wastes from seeping into aquifers. One-third of these sites are within 1.6 kilometers (1 mile) of a drinking water well. In addition, almost two-thirds of the country's liquid hazardous wastes are injected into the ground in disposal wells (Figure 11.26). Leaking injection pipes and seals in such wells can contaminate aquifers used as sources of drinking water.

The EPA has overseen the cleanup of about 357,000 of the more than 479,000 underground tanks in the United States that were leaking gasoline, diesel fuel, home heating oil, or toxic solvents into groundwater. During this century, scientists expect many of the millions of such tanks, which have been installed around the world, to become corroded and leaky, possibly contaminating groundwater and becoming a major global health problem. Determining the extent of a leak from a single underground tank can cost $25,000–$250,000, and cleanup costs range from $10,000 to more than $250,000.

If the chemical reaches an aquifer, effective cleanup is often not possible or is too costly. Although there are ways to clean up contaminated groundwater (Figure 11.27, right), such methods are very expensive. Cleaning up a single contaminated aquifer can cost anywhere from $10 million to $10 billion. Thus, preventing groundwater contamination (Figure 11.27, left) is the only effective way to deal with this serious water pollution problem.

There Are Many Ways to Purify Drinking Water

Most of the more-developed countries have laws establishing drinking water standards. But most of the less-developed countries do not have such laws or, if they have them, they do not enforce them.

In more-developed countries, wherever people depend on surface water sources, drinking water is typically stored in a reservoir for several days. This improves clarity and taste by increasing dissolved oxygen content and allowing suspended matter to settle. The water is then pumped to a pu-

Solutions

Groundwater Pollution

Prevention	Cleanup
Find substitutes for toxic chemicals	Pump to surface, clean, and return to aquifer (very expensive)
Keep toxic chemicals out of the environment	Inject microorganisms to clean up contamination (less expensive but still costly)
Require leak detectors on underground tanks	
Ban hazardous waste disposal in landfills and injection wells	Pump nanoparticles of inorganic compounds to remove pollutants (still being developed)
Store harmful liquids in aboveground tanks with leak detection and collection systems	

© 2016 Cengage Learning

FIGURE 11.27 There are ways to prevent and ways to clean up contaminated groundwater, but prevention is the only effective approach. **Question:** Which two of these preventive solutions do you think are the most important? Why?

rification plant and treated to meet government drinking water standards. In areas with very pure groundwater or surface water sources, little treatment is necessary. Several major U.S. cities, including New York City, Boston, Seattle, and Portland, Oregon, have avoided building expensive water treatment facilities and increased their beneficial environmental impact by investing in protection of the forests and wetlands in the watersheds that provide their water supplies.

We have the technology to convert sewer water into pure drinking water. One such process involves microfiltration to remove bacteria and suspended solids; reverse osmosis to remove minerals, viruses, and various organic compounds; and hydrogen peroxide and ultraviolet light to remove additional organic compounds. In a world where we will face increasing shortages of drinking water, wastewater purification is likely to become a major growth business. **GREEN CAREER: Wastewater purification**

In most places, the reclamation of wastewater still faces opposition from citizens and from some health officials who are unaware of the advances in this technology. However, the California cities of Los Angeles and San Diego are purifying wastewater to the point where it is fit to drink

and putting this water into underlying aquifers, thereby reducing their dependence on water imported from Northern California (Figure 11.13) and from the Colorado River (**Core Case Study**).

We can also use simpler measures to purify drinking water. In tropical countries that lack centralized water treatment systems, the WHO urges people to purify drinking water by exposing a clear plastic bottle filled with contaminated water to intense sunlight. The sun's heat and ultraviolet (UV) rays can kill infectious microbes in as little as 3 hours. Painting one side of the bottle black can improve heat absorption in this simple solar disinfection method, which applies the solar energy **principle of sustainability**. Where this simple measure has been used, the incidence of dangerous childhood diarrhea has decreased by 30–40%. Researchers have found that they can speed up this disinfection process by adding lime juice to the bottles of water.

Danish inventor Torben Vestergaard Frandsen has developed the *LifeStraw*™, an inexpensive, portable water filter that eliminates many viruses and parasites from water that is drawn through it (Figure 11.28). This filter has been particularly useful in Africa, where aid agencies are

FIGURE 11.28 The *LifeStraw*™ is a personal water purification device that gives many poor people access to safe drinking water. Here, four young men in Uganda demonstrate its use. **Question:** Do you think the development of such devices should make prevention of water pollution less of a priority? Explain.

Vestergaard Frandsen

distributing it. Another option being used by more and more people around the world is bottled water, which has created or worsened some environmental problems (see the following Case Study).

Is Bottled Water a Good Option?

Bottled water can be a useful (but expensive) option in countries and areas where people do not have access to safe and clean drinking water. However, despite some problems, experts say the United States has some of the world's cleanest drinking water. Municipal water systems in the United States are required to test their water regularly for a number of pollutants and to make the results available to citizens.

Yet about half of all Americans worry about getting sick from tap water contaminants, and many drink high-priced bottled water or install expensive water purification systems. Studies by the Natural Resources Defense Council (NRDC) reveal that in the United States, a bottle of water costs between 240 and 10,000 times as much as the same volume of tap water. Water expert Peter Gleick has estimated that more than 40% of the expensive bottled water that Americans drink is really bottled tap water. And a 4-year study by the NRDC found bacteria and synthetic organic chemicals in one-third of the bottles tested. EPA contamination standards that apply to public water supplies do not apply to bottled water.

Use of bottled water also causes environmental problems. In the United States, according to the Container Recycling Institute, more than 67 million plastic water bottles are discarded every day—enough bottles in a year to wrap around the planet at its equator about 280 times, if lined up end-to-end. Most water bottles are made of recyclable plastic, but in the United States, only about 29% of these bottles get recycled. The billions of discarded bottles end up in landfills, where they can remain for hundreds of years, or are burned in incinerators, which release some of their harmful chemicals into the atmosphere. Even worse, millions get scattered on the land and end up in rivers, lakes, and oceans. By contrast, in Germany most bottled water is sold in returnable and reusable glass bottles.

It takes huge amounts of energy to manufacture bottled water and to transport it across countries and around the world, as well as to refrigerate much of it in stores. Toxic gases and liquids are released during the manufacture of plastic water bottles, and greenhouse gases and other air pollutants are emitted by the fossil fuels burned to make them and to deliver bottled water to suppliers. According to the Pacific Institute, the oil used to pump, process, bottle, transport, and refrigerate the bottled water used annually in the United States would be enough to fuel 3 million cars for a year. In addition, withdrawing groundwater for bottling is helping to deplete some aquifers.

Because of these harmful environmental impacts and the high cost of bottled water, there is a growing *back-to-the-tap* movement. From San Francisco to New York to Paris, city governments, restaurants, schools, religious groups, and many consumers are refusing to buy bottled water. People are also refilling portable bottles with tap water and using simple filters to improve the taste and color of water where necessary. Some health officials suggest that before drinking expensive bottled water or buying costly home water purifiers, consumers have their water tested by local health departments or private labs (but not by companies trying to sell water purification equipment).

Ocean Pollution Is a Growing and Poorly Understood Problem

Why should we care about the oceans? Short answer: Because they keep us alive. Oceans help to provide and recycle the planet's freshwater through the water cycle (see Figure 3.13, p. 52). They also strongly affect weather and climate, help to regulate the earth's temperature, and absorb some of the massive amounts of carbon dioxide that we emit into the atmosphere. As oceanographer and explorer Sylvia A. Earle (see Individuals Matter 9.1, p. 209) reminds us: "Even if you never have the chance to see or touch the ocean, the ocean touches you with every breath you take, every drop of water you drink, every bite you consume. Everyone, everywhere is inextricably connected to and utterly dependent upon the existence of the sea." Despite its importance, we treat the ocean as the world's largest dump for the massive and growing amount of wastes and pollutants that we produce.

Coastal areas—especially wetlands, estuaries, coral reefs, and mangrove swamps—bear the brunt of our inputs of pollutants and wastes into the ocean (Figure 11.29). Roughly 40% of the world's people (53% in the United States) live on or near coastlines, which helps explain why 80% of marine pollution originates on land; and coastal populations are projected to double by 2050.

According to a study by the U.N. Environment Programme (UNEP), 80–90% of the municipal sewage from coastal areas of less-developed countries is dumped into oceans without treatment. This often overwhelms the ability of the coastal waters to degrade the wastes. By adding excessive amounts of nitrates and phosphates to the ocean instead of recycling these vital plant nutrients to the soil, this activity alters the nitrogen and phosphorus cycles and upsets marine ecosystems. For example, many areas of China's coastline are so choked with algae growing on the nutrients provided by sewage that some scientists have concluded that large areas of China's coastal waters can no longer sustain marine ecosystems. The dumping of wastes in coastal waters is a serious violation of the chemical cycling **principle of sustainability**.

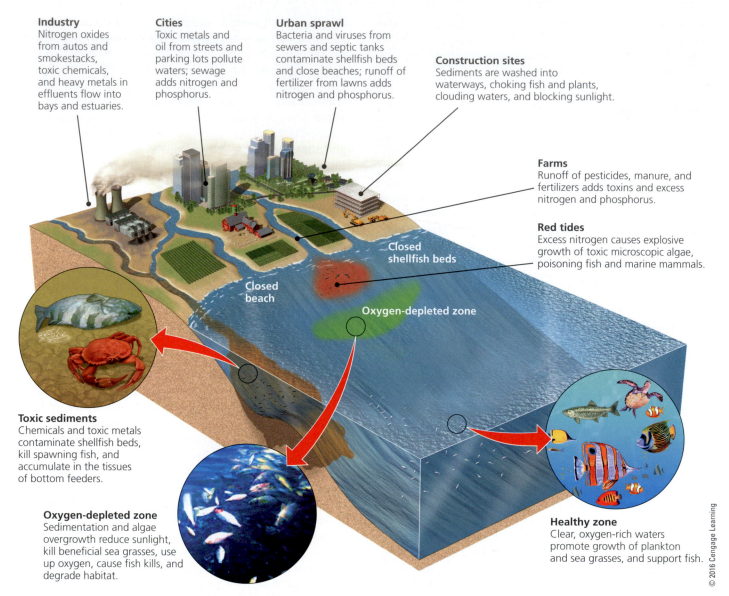

Industry
Nitrogen oxides from autos and smokestacks, toxic chemicals, and heavy metals in effluents flow into bays and estuaries.

Cities
Toxic metals and oil from streets and parking lots pollute waters; sewage adds nitrogen and phosphorus.

Urban sprawl
Bacteria and viruses from sewers and septic tanks contaminate shellfish beds and close beaches; runoff of fertilizer from lawns adds nitrogen and phosphorus.

Construction sites
Sediments are washed into waterways, choking fish and plants, clouding waters, and blocking sunlight.

Farms
Runoff of pesticides, manure, and fertilizers adds toxins and excess nitrogen and phosphorus.

Red tides
Excess nitrogen causes explosive growth of toxic microscopic algae, poisoning fish and marine mammals.

Closed shellfish beds

Closed beach

Oxygen-depleted zone

Toxic sediments
Chemicals and toxic metals contaminate shellfish beds, kill spawning fish, and accumulate in the tissues of bottom feeders.

Oxygen-depleted zone
Sedimentation and algae overgrowth reduce sunlight, kill beneficial sea grasses, use up oxygen, cause fish kills, and degrade habitat.

Healthy zone
Clear, oxygen-rich waters promote growth of plankton and sea grasses, and support fish.

© 2016 Cengage Learning

FIGURE 11.29 Natural capital degradation: Residential areas, factories, and farms all contribute to the pollution of coastal waters. **Questions:** What do you think are the three worst pollution problems shown here? For each one, how does it affect two or more of the ecosystem and economic services listed in Figure 7.18 (p. 147)?

In deeper waters, the oceans can dilute, disperse, and degrade large amounts of raw sewage and other types of degradable pollutants. Some scientists suggest that it is safer to dump sewage sludge, toxic mining wastes, and most other harmful wastes into the deep ocean than to bury them on land or burn them in incinerators. Other scientists disagree, pointing out that dumping harmful wastes into the ocean would delay urgently needed pollution prevention measures and promote further degradation of this vital part of the earth's life-support system.

Recent studies of some U.S. coastal waters have found vast colonies of viruses thriving in raw sewage and in effluents from sewage treatment plants (which do not remove viruses) and leaking septic tanks. According to one study, one-fourth of the people using coastal beaches in the United States develop ear infections, sore throats, eye irritations, respiratory diseases, or gastrointestinal diseases from swimming in seawater containing infectious viruses and bacteria.

Scientists also point to the underreported problem of pollution from cruise ships. A cruise liner can carry as many as 6,300 passengers and 2,300 crewmembers, and it can generate as much waste (toxic chemicals, garbage, sewage, and waste oil) as a small city. Many cruise ships dump these wastes at sea. In U.S. waters, such dumping is illegal, but some ships continue dumping secretively, usually at night. Some environmentally aware vacationers are refusing to go on cruise ships that do not have sophisticated systems for dealing with the wastes they produce.

Runoff of sewage and agricultural wastes into coastal waters introduces large quantities of nitrate (NO_3^-) and

U.S. Coast Guard

Joel Sartore/National Geographic Creative

FIGURE 11.30 The Deepwater Horizon drilling platform exploded, burned, and sank in the Gulf of Mexico on April 20, 2010. The accident killed 11 of the rig's crewmembers and, during the following 3 months, released about 206 million barrels of crude oil into gulf waters before the leaking well was capped. The photo below shows a brown pelican that was severely oiled by the accident.

phosphate (PO_4^{3-}) plant nutrients, which can cause explosive growths of harmful algae. These *harmful algal blooms*—also called red, brown, or green toxic tides—can release waterborne and airborne toxins that poison seafood, damage fisheries, kill some fish-eating birds, and reduce tourism. Each year, harmful algal blooms lead to the poisoning of about 60,000 Americans who eat shellfish contaminated by the algae.

Harmful algal blooms occur annually in several hundred *oxygen-depleted zones* around the world, mostly in temperate coastal waters and in large bodies of water with restricted outflows, such as the Baltic and Black seas. The largest such zone in U.S. coastal waters forms each year in the northern Gulf of Mexico (Science Focus 11.1). A study by Luan Weixin, of China's Dalain Maritime University, found that nitrates and phosphates have seriously contaminated about half of China's shallow coastal waters.

Ocean Pollution from Oil

Crude petroleum (oil as it comes out of the ground) and *refined petroleum* (fuel oil, diesel, gasoline, and other processed petroleum products) reach the ocean from a number of sources and become highly disruptive pollutants. The most visible sources are tanker accidents, such as the huge *Exxon Valdez* oil spill in the U.S. state of Alaska in 1989, and blowouts at offshore oil drilling rigs, such as that of the BP Deepwater Horizon rig in the Gulf of Mexico in 2010 (Figure 11.30)—the worst-ever oil spill in U.S. waters, with damages and fines exceeding $15 billion.

THE GULF OF MEXICO'S ANNUAL DEAD ZONE

The Mississippi River basin (Figure 11.A, map) lies within 31 states and contains almost two-thirds of the continental U.S. land area. With more than half of all U.S. croplands, it is one of the world's most productive agricultural regions. However, water draining into the Mississippi River and its tributaries from farms, cities, factories, and sewage treatment plants in this huge basin contains sediments and other pollutants that end up in the Gulf of Mexico (Figure 11.A, photo)—a major supplier of the country's fish and shellfish.

Each spring and summer, this huge input of plant nutrients, mostly nitrates from crop fertilizers, enters the northern Gulf of Mexico and overfertilizes the coastal waters of the U.S. states of Mississippi, Louisiana, and Texas. One result is an explosion of populations of phytoplankton (mostly algae) that eventually die and fall to the sea floor. Hordes of oxygen-consuming bacteria that decompose the phytoplankton remains deplete the dissolved oxygen in the Gulf's bottom layer of water.

The huge volume of water affected by this seasonal event is called a dead zone because it contains little animal marine life. Its low oxygen levels (Figure 11.B) drive away faster-swimming marine organisms and suffocate bottom-dwelling fish, crabs, oysters, and shrimp that cannot move to less polluted areas. Large amounts of sediment, mostly from soil eroded from the

FIGURE 11.A Water containing sediment, dissolved nitrate fertilizers, and other pollutants drains from the Mississippi River basin (map) into the Mississippi River and from there into the northern Gulf of Mexico (photo).

Missouri River
Mississippi River Basin
Ohio River
Mississippi River

Mississippi River

GULF OF MEXICO

LIAM GUNLEY/NASA

However, studies show that the largest source of ocean oil pollution is urban and industrial runoff from land, much of it from leaks in pipelines, refineries, and other oil-handling and storage facilities. An estimated one-third to one-half of all ocean oil pollution comes from oil and oil products that are intentionally dumped or accidentally spilled or leaked onto the land or into sewers by homeowners and industries.

Volatile organic hydrocarbons in oil and other petroleum products kill many aquatic organisms immediately upon contact, especially if these animals are in their vulnerable larval forms. Other chemicals in oil form tarlike globs that float on the surface and coat the feathers of seabirds and the fur of marine mammals (Figure 11.30), which destroys their natural heat insulation and buoyancy, causing many of them to drown or die from loss of body heat.

Heavy oil components that sink to the ocean floor or wash into estuaries and coastal wetlands can smother bottom-dwelling organisms such as crabs, oysters, mussels, and clams, or make them unfit for human consumption. Some oil spills have killed coral reefs.

Research shows that populations of many forms of marine life can recover from exposure to large amounts of *crude oil* in warm waters with fairly rapid currents within about 3 years. But in cold and calm waters, recovery can take decades. In addition, recovery from exposure to *refined oil*, especially in estuaries and salt marshes, can take 10–20 years or longer. Oil slicks that wash onto beaches can have a serious economic impact on coastal residents, who lose income normally gained from fishing and tourist activities.

Some oil spills that are not too large can be partially cleaned up by mechanical means, including floating

Mississippi River basin, can also kill bottom-dwelling forms of animal aquatic life. The dead zone appears each spring and grows until fall when storms churn the water and redistribute dissolved oxygen to the Gulf bottom.

The size of the Gulf of Mexico's annual dead zone depends primarily on the amount of water flowing into the Mississippi River each year. In years with ample rainfall and snowmelt, such as 2003, it has covered an area as large as the state of Massachusetts—27,300 square kilometers (10,600 square miles). In 2013, a severe drought year in parts of the country, it covered a smaller area of about 15,125 square kilometers (5,840 square miles)—about the size of the state of Connecticut.

The annual Gulf of Mexico dead zone (one of about 400 dead zones found throughout the world) represents a disruption of the nitrogen cycle (see Figure 3.15, p. 54) caused primarily by human activities. It happens because huge quantities of nitrogen from nitrate fertilizers are added to ecosystems such as the Mississippi River and

FIGURE 11.B The colorized area in this satellite image of the northern Gulf of Mexico represents the seasonal dead zone of 2012 with the red area having the lowest oxygen level.

Gulf of Mexico much faster than the nitrogen cycle can remove them. As a result, by producing crops for food and ethanol to fuel cars in the vast Mississippi basin, we end up disrupting coastal aquatic life and seafood production in the Gulf of Mexico.

Because of the size and agricultural importance of the Mississippi River basin, there is no easy solution to the problem of severe cultural eutrophication of this coastal zone. The best hope lies in preventive measures, including applying less fertilizer on farms upstream, injecting fertilizer below the soil surface, and using controlled-release fertilizers that have water-

insoluble coatings. Other such measures include planting strips of forests and grasslands along waterways to soak up excess nitrogen, restoring Gulf Coast wetlands that once filtered some of the plant nutrients, and creating wetlands between crop fields and streams emptying into the Mississippi River. Another measure would be to reduce or eliminate government subsidies for growing corn to make ethanol.

Critical Thinking

Which three of the preventive measures described above do you believe would be the most effective? Explain.

booms, skimmer boats, and absorbent devices such as giant pillows filled with feathers or hair. But scientists estimate that current cleanup methods can recover typically no more than 15% of the oil from a major spill.

Thus, *preventing* oil pollution is the most effective and, in the long run, the least costly approach. One of the best ways to prevent tanker spills is to use oil tankers with double hulls. Stricter safety standards and inspections could help to reduce oil well blowouts at sea. Most important, businesses, institutions, and citizens living in coastal areas must take care to prevent leaks and spillage of even the smallest amounts of oil and oil products such as paint thinners and gasoline.

Most ocean pollution occurs in coastal waters and comes from human activities on land. Figure 11.31 lists ways to prevent pollution of coastal waters and ways to reduce it.

CONSIDER THIS. . .
THINKING ABOUT Ocean Oil Pollution

What are three ways in which you might be contributing to ocean oil pollution? How could you reduce your contribution to this environmental problem?

Reducing Surface Water Pollution from Nonpoint Sources

There are a number of ways to reduce nonpoint sources of surface water pollution, most of which come from agricultural practices. Farmers can reduce soil erosion by keeping cropland covered with vegetation and using conservation tillage and other soil conservation methods (see Chapter 10, pp. 238–239). They can also reduce the amount of fertilizer that runs off into surface waters by using slow-release

Solutions

Coastal Water Pollution

Prevention		Cleanup
Separate sewage and storm water lines	Ban dumping of wastes and sewage by ships in coastal waters	Improve oil-spill cleanup capabilities
Require secondary treatment of coastal sewage	Strictly regulate coastal development, oil drilling, and oil shipping	Use nanoparticles on sewage and oil spills to dissolve the oil or sewage (still under development)
Use wetlands and other natural methods to treat sewage	Require double hulls for oil tankers	

FIGURE 11.31 Methods for preventing excessive pollution of coastal waters and methods for cleaning it up (**Concept 11.4**). **Questions:** Which two of these solutions do you think are the most important? Why?

Top: Rob Byron/Shutterstock.com. Bottom: Igor Karasi/Shutterstock.

fertilizer, using no fertilizer on steeply sloped land, and planting buffer zones of vegetation between cultivated fields and nearby surface waters.

Organic farming (see Chapter 10, p. 219) and other forms of more sustainable food production (see Figure 10.27, p. 242) can also help to prevent water pollution caused by nutrient overload because they use little if any synthetic inorganic fertilizers and pesticides. Farmers can reduce pesticide runoff by applying pesticides only when needed and by relying more on integrated pest management (see Chapter 10, p. 235). In addition, they can control runoff and infiltration of manure from animal feedlots by planting buffer zones and by locating feedlots, pastures, and animal waste storage sites away from steeply sloped land, surface water, and flood zones. All of these are ways to have a beneficial environmental impact.

Laws Can Be Used to Reduce Water Pollution from Point Sources

The Federal Water Pollution Control Act of 1972 (renamed the Clean Water Act when it was amended in 1977) and the 1987 Water Quality Act form the basis of U.S. efforts to control pollution of the country's surface waters. The Clean Water Act sets standards for allowed levels of 100 key water pollutants and requires polluters to get permits that limit how much of these various pollutants they can discharge into aquatic systems.

The EPA has also been experimenting with a *discharge trading policy,* which uses market forces to reduce water pollution in the United States. Under this program, a per-mit holder can pollute at higher levels than allowed in its permit if it buys credits from permit holders who are polluting below their allowed levels.

Environmental scientists warn that the effectiveness of such a system depends on how low the cap on total pollution levels in any given area is set and on how regularly the cap is lowered. They also warn that discharge trading could allow water pollutants to build up to dangerous levels in areas where credits are bought. Neither adequate scrutiny of the cap levels nor gradual lowering of caps is a part of the current EPA discharge trading system.

Some environmental and health scientists call for strengthening the U.S. Clean Water Act. Suggested improvements include shifting the focus of the law from end-of-pipe removal of specific pollutants to water pollution prevention; greatly increased monitoring for violations of the law and much larger mandatory fines for violators; and regulating irrigation water quality (for which there is no federal regulation). Another suggestion is to expand the rights of citizens to bring lawsuits to ensure that water pollution laws are enforced. Still another suggestion is to rewrite the Clean Water Act to clarify that it covers all waterways. This was the original intent of the law, but it has since been muddled by court decisions, and some polluters have taken advantage of the resulting confusion to keep polluting in many areas.

Many people oppose these proposals, contending that the Clean Water Act's regulations are already too restrictive and costly. Some state and local officials argue that in many communities, it is unnecessary and too expensive to test all the water for pollutants as required by federal law.

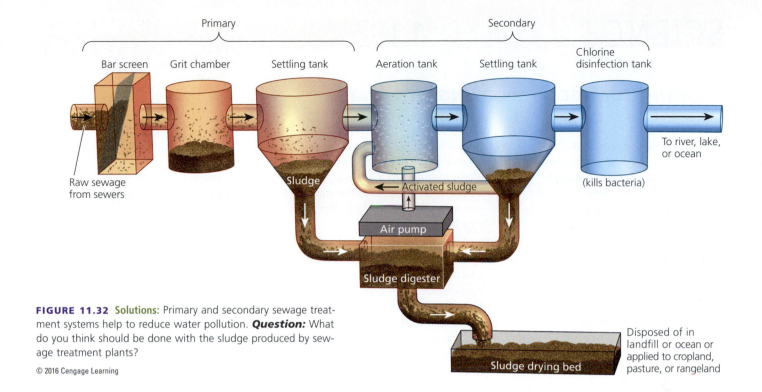

Bar screen   Grit chamber   Settling tank   Aeration tank   Settling tank   Chlorine disinfection tank

Raw sewage from sewers

Sludge

Activated sludge

To river, lake, or ocean

(kills bacteria)

Air pump

Sludge digester

Sludge drying bed

Disposed of in landfill or ocean or applied to cropland, pasture, or rangeland

FIGURE 11.32 **Solutions:** Primary and secondary sewage treatment systems help to reduce water pollution. **Question:** What do you think should be done with the sludge produced by sewage treatment plants?

© 2016 Cengage Learning

Some members of Congress, under pressure from regulated industries, go further and want to seriously weaken or repeal the Clean Water Act and other government environmental regulations, arguing that such regulations hinder economic growth. In 2012, William K. Reilly, who headed the EPA from 1989 to 1993 and served as the co-chairman of a presidential commission on offshore drilling, said: "If we buy into the misguided notion that reducing protection of our waters will somehow ignite the economy, we will shortchange our health, environment, and economy."

Sewage Treatment Reduces Water Pollution

In rural and suburban areas with suitable soils, sewage from each house usually is discharged into a **septic tank** with a large drainage field. In such a system, household sewage and wastewater is pumped into a settling tank, where grease and oil rise to the top, and solids fall to the bottom and are decomposed by bacteria. The partially treated wastewater that results is discharged into a large drainage (absorption) field through small holes in perforated pipes embedded in porous gravel or crushed stone just below the soil's surface. As these wastes drain from the pipes and percolate downward, the soil filters out some potential pollutants and soil bacteria decompose biodegradable materials. About one-fourth of all homes in the United States are served by septic

tanks. They work well as long as they are not overloaded and their solid wastes are regularly pumped out.

In urban areas in the United States and other more-developed countries, most waterborne wastes from homes, businesses, and storm runoff flow through a network of sewer pipes to *wastewater* or *sewage treatment plants*. Raw sewage reaching a treatment plant typically undergoes one or two levels of wastewater treatment. The first is **primary sewage treatment**—a *physical* process that uses screens and a grit tank to remove large floating objects and to allow solids such as sand and rock to settle out. Then the waste stream flows into a primary settling tank where suspended solids settle out as sludge (Figure 11.32, left).

The second level is **secondary sewage treatment**—a *biological* process in which aerobic bacteria remove as much as 90% of dissolved and biodegradable, oxygen-demanding organic wastes (Figure 11.32, right). A combination of primary and secondary treatment removes 95–97% of the suspended solids and oxygen-demanding organic wastes, 70% of most toxic metal compounds and nonpersistent synthetic organic chemicals, 70% of the phosphorus, and 50% of the nitrogen. However, this process removes only a tiny fraction of persistent and potentially toxic organic substances found in some pesticides and in discarded medicines that people put into sewage systems, and it does not kill pathogens.

Before discharge, water from sewage treatment plants usually undergoes *bleaching*, to remove water coloration, and *disinfection* to kill disease-carrying bacteria and some

TREATING SEWAGE BY WORKING WITH NATURE

Some communities and individuals are seeking better ways to purify sewage by working with nature (**Concept 11.4**). Biologist John Todd has developed an ecological approach to treating sewage, which he calls *living machines* (Figure 11.C).

This purification process begins when sewage flows into a passive solar greenhouse or outdoor site containing rows of large open tanks populated by an increasingly complex series of organisms. In the first set of tanks, algae and microorganisms decompose organic wastes, with sunlight speeding up the process. Water hyacinths, cattails, bulrushes, and other aquatic plants growing in the tanks take up the resulting nutrients.

After flowing though several of these natural purification tanks, the water passes through an artificial marsh made of sand, gravel, and bulrushes, which filters out algae and remaining organic waste. Some of the plants also absorb, or *sequester,* toxic metals such as lead and mercury and secrete natural antibiotic compounds that kill pathogens.

Next, the water flows into aquarium tanks, where snails and zooplankton consume microorganisms and are in turn consumed by crayfish, tilapia, and other fish that can be eaten or sold as bait. After 10 days, the clear water flows into a second artificial marsh for final filtering and cleansing. The water can be made pure enough to drink by treating it with ultraviolet light or by passing the water through an ozone generator, usually immersed out of sight in an attractive pond or wetland habitat. Operating costs are about the same as those of a conventional sewage treatment plant. These systems are widely used on a small scale. However, they have been difficult to maintain on a scale large enough to handle the typical variety of chemicals in the sewage wastes from more-developed urban areas.

More than 800 cities and towns around the world (150 in the United States) use natural or artificially created wetlands to treat sewage as a lower-cost alternative to expensive waste treatment plants. For example, in Arcata, California—a coastal town of 18,000 people—scientists and workers created some 65 hectares (160 acres) of wetlands between the town and the adjacent Humboldt Bay. The marshes and ponds, developed on land that was once a dump, act as a natural waste treatment plant. The project cost was less than half the estimated price of a conventional treatment plant.

This system returns purified water to Humboldt Bay, and the sludge that is removed is processed for use as fertilizer. The marshes and ponds also serve as an Audubon Society bird sanctuary, which provides habitats for thousands of seabirds, otters, and other marine animals. The town has even celebrated its natural sewage treatment system with an annual "Flush with Pride" festival.

This approach and the living machine system developed by John Todd apply all three **scientific principles of sustainability**: using solar energy, employing natural processes to remove and recycle nutrients and other chemicals, and relying on a diversity of organisms and natural processes.

Critical Thinking

Can you think of any disadvantages of using such a nature-based system instead of a conventional sewage treatment plant? Do you think any such disadvantages outweigh the advantages? Why or why not?

FIGURE 11.C Solutions: The Solar Sewage Treatment Plant in the U.S. city of Providence, Rhode Island, is an ecological wastewater purification system, also called a *living machine.* Biologist John Todd is demonstrating this ecological process he invented for purifying wastewater by using the sun and a series of tanks containing living organisms.

Ocean Arks International

(but not all) viruses. The usual method for accomplishing this is *chlorination*. But chlorine can react with organic materials in water to form small amounts of chlorinated hydrocarbons. Some of these chemicals cause cancers in test animals, can increase the risk of miscarriages, and can damage the human nervous, immune, and endocrine systems. Use of other disinfectants, such as ozone and ultraviolet light, is increasing, but they cost more and their effects do not last as long as those of chlorination.

We Can Improve Conventional Sewage Treatment and Emphasize Nutrient Recovery

Some scientists call for redesigning the conventional sewage treatment system shown in Figure 11.32. The idea is to prevent toxic and hazardous chemicals from reaching sewage treatment plants and thus from getting into sludge and water discharged from such plants.

They suggest several ways to implement this pollution prevention approach. One is to require industries and businesses to remove toxic and hazardous wastes from water sent to municipal sewage treatment plants, which would help to implement the full-cost pricing **principle of sustainability** by increasing the cost of creating waste and pollution. We could also encourage industries to reduce or eliminate their use and waste of toxic chemicals, which would reduce their expense in complying with water pollution control laws.

Another suggestion is to require or encourage more households, apartment buildings, and offices to eliminate sewage outputs by switching to waterless, odorless *composting toilet systems*, to be installed, maintained, and managed by professionals. These systems, pioneered several decades ago in Sweden, convert nutrient-rich human fecal matter into a soil-like humus that can be used as a fertilizer supplement. This process returns plant nutrients in human waste to the soil, mimicking the chemical cycling **principle of sustainability**.

On a larger scale, such systems would be cheaper to install and maintain than current sewage systems are, because they do not require vast systems of underground pipes connected to centralized sewage treatment plants. They also save large amounts of water, reduce water bills, and decrease the amount of energy used to pump and purify water. This more environmentally sustainable replacement for the conventional toilet is now being used in parts of more than a dozen countries, including China, India, Mexico, Syria, and South Africa.

A Swedish entrepreneur has developed a biodegradable single-use plastic bag that can be used as a toilet in urban slums and in other areas where many of the world's people do not have access to toilets. After it is used, the bag is knotted and buried. A thin layer of urea in this bag kills the disease-producing pathogens in the feces and helps break down the waste into plant nutrients that are then recycled. This is a simple and inexpensive, low-tech application of the chemical cycling **principle of sustainability**.

Some communities are also using unconventional, but highly effective, *ecological sewage treatment systems*, which work with nature (Science Focus 11.2).

There Are Sustainable Ways to Reduce and Prevent Water Pollution

It is encouraging that since 1970, most of the world's more-developed countries have enacted laws and regulations that have significantly reduced point-source water pollution. These improvements were largely the result of *bottom-up* political pressure on elected officials from individuals and groups. On the other hand, little has been done to reduce water pollution in most of the less-developed countries.

Many environmental and health scientists argue that the next step is to increase efforts to reduce and prevent water pollution in both more- and less-developed countries as an important way to increase our beneficial environmental impact. They would begin by asking the question: *How can we avoid producing water pollutants in the first place?* (**Concept 11.4**). Figure 11.33 lists ways to achieve this goal over the next several decades.

This shift to pollution prevention will not take place unless citizens put political pressure on elected officials and also take actions to reduce their own daily contributions to water pollution. Figure 11.34 lists some actions you can take to help reduce water pollution.

Solutions

Water Pollution

- Prevent groundwater contamination
- Reduce nonpoint runoff
- Work with nature to treat sewage and reuse treated wastewater
- Find substitutes for toxic pollutants
- Practice the four Rs of resource use (refuse, reduce, reuse, recycle)
- Reduce air pollution
- Reduce poverty
- Slow population growth

© 2016 Cengage Learning

FIGURE 11.33 Ways to prevent or reduce water pollution (**Concept 11.4**). *Questions:* Which two of these solutions do you think are the most important? Why?

What Can You Do?

Reducing Water Pollution

- Fertilize garden and yard plants with manure or compost instead of commercial inorganic fertilizer
- Minimize use of pesticides, especially near bodies of water
- Prevent yard wastes from entering storm drains
- Do not use water fresheners in toilets
- Do not flush unwanted medicines down the toilet
- Do not pour pesticides, paints, solvents, oil, antifreeze, or other harmful chemicals down the drain or onto the ground

© 2016 Cengage Learning

FIGURE 11.34 Individuals matter: You can help reduce water pollution. **Questions:** Which three of these steps do you think are the most important ones to take? Why?

TYING IT ALL TOGETHER
The Colorado River, Sustainability, and Reducing Water Pollution

The Core Case Study that opens this chapter discusses the problems and tensions that can occur when a large number of U.S. states in a water-short region share a limited river water resource. Such problems are representative of those faced in many other dry regions of the world, especially areas where the population is growing rapidly and water resources are dwindling for various reasons.

Large dams, river diversions, levees, water transfers, and other big engineering schemes have helped to provide much of the world with electricity, food from irrigated crops, drinking water, and flood control. However, they have also degraded the aquatic natural capital necessary for long-term economic and ecological sustainability by seriously disrupting rivers, streams, wetlands, aquifers, and other aquatic systems. We are also degrading aquatic natural capital by polluting much of the world's rivers, lakes, oceans, and aquifers.

The three **scientific principles of sustainability** can guide us in using water resources more sustainably and in reducing and preventing water pollution. Scientists hope to use solar energy to desalinate water and expand freshwater supplies and to purify much of the water we use. Purifying and recycling more water will help us to reduce water waste. We can use natural nutrient cycling to treat our wastes in wetland-based sewage treatment systems and to remove nitrate and phosphate plant nutrients from sewage and recycle them to the soil. By sustaining water supplies and avoiding the pollution of aquatic systems and their bordering terrestrial systems, we can preserve biodiversity and its ecosystem services—a key factor in maintaining water supplies and water quality.

Atm2003/Shutterstock.com

Chapter Review

Core Case Study

1. Summarize the importance of the Colorado River basin in the United States and how human activities are stressing this system. Define **drought** and explain how it has affected the Colorado River system.

Section 11.1

2. What are the two key concepts for this section? Define **freshwater**. Explain why access to water is a health issue, an economic issue, a national and global security issue, and an environmental issue. What percentage of the earth's freshwater is available to us? Explain how water is recycled by the hydrologic cycle and how human activities can interfere with this cycle. Define **groundwater**, **zone of saturation**, **water table**, **aquifer**, **surface water**, **surface runoff**, **watershed (drainage basin)**, and **reliable surface runoff**.

3. What percentage of the world's reliable runoff are we using and what percentage are we likely to be using by 2025? How is most of the world's water used? Define **water footprint** and **virtual water** and give two examples of each. Describe the availability and use of freshwater resources in the United States and the water shortages that could occur during this century. What are four major problems resulting from the way people are using water from the Colorado River basin? What percentage of the earth's land suffers from severe drought today and how might this change by 2059? How many people in the world lack regular access to clean water today?

Section 11.2

4. What are the three key concepts for this section? What are the advantages and disadvantages of withdrawing groundwater? Summarize the problem of groundwater depletion in the world and in the United States, especially in the Ogallala Aquifer. List two problems that result from the overpumping of aquifers. List some ways to prevent or slow groundwater depletion.

5. What is a **dam**? What is a **reservoir**? What are the advantages and disadvantages of using large dams and reservoirs? What has happened to water flows in the Colorado River (**Core Case Study**) since 1960? List three possible solutions to the supply problems in the Colorado River basin. List the pros and cons of the California Water Project. Describe the environmental and health disaster caused by the Aral Sea water transfer project. Define **desalination** and distinguish between distillation and reverse osmosis as methods for desalinating water. What are three limitations of desalination?

Section 11.3

6. What is the key concept for this section? What percentage of available freshwater is lost through inefficient use and other causes in the world and in the United States? What are two major reasons for those losses? Describe three major irrigation methods and list ways to reduce water losses in irrigation. List four ways to reduce water waste in industries and homes and three ways to use less water to remove wastes. How does flooding affect water supplies and what are three ways in which we can reduce our contribution to flooding? List four ways in which each of us can reduce our water footprint.

Section 11.4

7. What is the key concept for this section? What is **water pollution**? Distinguish between **point sources** and **nonpoint sources** of water pollution and give an example of each. List nine major types of water pollutants and give an example of each. Summarize the relationship between atmospheric warming and water pollution.

8. Explain how streams can cleanse themselves and how these cleansing processes can be overwhelmed. What is **wastewater**? Describe the state of stream pollution in more- and less-developed countries. Give two reasons why lakes cannot cleanse themselves as readily as streams can. Distinguish between **eutrophication** and **cultural eutrophication**. List ways to prevent or reduce cultural eutrophication. Explain why groundwater cannot cleanse itself very well. What are the major sources of groundwater contamination in the United States? List ways to prevent or clean up groundwater contamination. List some ways to purify drinking water. Describe the environmental problems caused by the widespread use of bottled water.

9. Why should we care about the oceans? How are coastal waters and deeper ocean waters polluted? What causes algal blooms and what are their negative effects? Describe oxygen depletion in the northern Gulf of Mexico. What are the effects of oil pollution of the oceans and what can be done to reduce such pollution? How can we prevent or reduce pollution of coastal waters?

10. List ways to reduce water pollution from **(a)** non-point sources and **(b)** point sources. Describe the U.S. experience with reducing point-source water pollution and list ways to improve such efforts. What is a **septic tank** and how does it work? Explain how **primary sewage treatment** and **secondary sewage treatment** are used to help purify water. How could we improve conventional sewage treatment? What is a composting toilet system? Describe John Todd's use of living machines to treat sewage. Explain how wetlands can be used to treat sewage. List six ways to prevent and reduce water pollution. List five things you can do to reduce water pollution. What are this chapter's *three big ideas*? Explain how the three **scientific principles of sustainability** can guide us in using water resources more sustainably and in reducing and preventing water pollution.

Note: Key terms are in bold type.

Critical Thinking

1. What do you think are the three most important priorities for dealing with the water resource problems of the Colorado River basin, as discussed in this chapter's **Core Case Study**? Explain your choices.

2. What role does population growth play in water supply problems? Relate this to water supply problems of the Colorado River basin (**Core Case Study**).

3. Explain why you are for or against **(a)** raising the price of water while providing lower lifeline rates for poor consumers, **(b)** withdrawing government subsidies that provide farmers with water at low cost, and **(c)** providing government subsidies to farmers for improving irrigation efficiency.

4. Calculate how many liters (and gallons) of water are lost in 1 month by a toilet or faucet that leaks 2 drops of water per second. (One liter of water equals about 3,500 drops and 1 liter equals 0.265 gallon.) How many bathtubs (each containing about 151 liters or 40 gallons) could be filled with this lost water?

5. List the three most important ways in which you could use water more efficiently. Which, if any, of these measures do you already take?

6. How might you be contributing directly or indirectly to the annual dead zone that forms in the Gulf of Mexico? What are three things you could do to reduce your contribution?

7. How might you be contributing directly or indirectly to groundwater pollution? What are three things you could do to reduce your contribution?

8. When you flush your toilet, where does the wastewater go? Trace the actual flow of this water in your community from your toilet through sewers to a wastewater treatment plant (or to a septic system) and from there to the environment. Try to visit a local sewage treatment plant to see what it does with wastewater. Compare the processes it uses with those shown in Figure 11.32. What happens to the sludge produced by this plant? What improvements, if any, would you suggest for this plant?

Doing Environmental Science

Do some research on the water resources in your community and write a report answering the following questions:

a. What are the principle sources of your community's drinking water?

b. How is your drinking water treated?

c. What are your community's principal nonpoint sources of contamination of surface water and groundwater?

d. What problems related to drinking water, if any, have arisen in your community? What actions, if any, has your local government taken to solve such problems?

e. Is groundwater contamination a problem? If so, where, and what has been done about the problem?

Global Environment Watch Exercise

Search for *Ogallala Aquifer* and research the decline of this aquifer. Plot the decline on a graph and list the three areas over the aquifer where the decline is the worst. Look for projections on how much more the aquifer could decline in the future and take notes on this. Find information on the causes of this decline and determine which are the three largest causes. Learn what is being done to address each of these causes and write a report explaining the causes, projections, and possible ways to slow the decline of the Ogallala Aquifer.

Ecological Footprint Analysis

In 2005 (the latest year for which these data were available), the population of the U.S. state of Florida consumed 24.5 billion liters (6.5 billion gallons) of freshwater daily. It is projected that in 2025, the daily consumption will increase to 32.1 billion liters (8.5 billion gallons) per day. Between 2005 and 2025, the population of Florida was projected to increase from 17.5 million to 25.9 million.

1. Based on total freshwater use:
   a. Calculate the per capita consumption of water per day in Florida in 2005 and the projected per capita consumption per day for 2025.
   b. Calculate the per capita consumption of water for the year in Florida in 2005 and the projected per capita consumption per year for 2025.

2. In 2005, how did the Florida *average water footprint* (consumption per person per year, in this case, based only on water used within the state) compare with (a) the average U.S. water footprint of approximately 249,000 liters (66,000 gallons) per person per year, and (b) with the global average water footprint of 123,770 liters (32,800 gallons) per person per year?

CENGAGE**brain**.com To access course materials, including Aplia homework, please visit www.cengagebrain.com.

WWW.CENGAGEBRAIN.COM **287**

12

GEOLOGY AND NONRENEWABLE MINERAL RESOURCES

Civilization exists by geological consent, subject to change without notice.

WILL DURANT

12.3 What are the environmental effects of using nonrenewable mineral resources?

12.4 How can we use mineral resources more sustainably?

12.5 What are the earth's major geological hazards?

Open-pit copper mine in Utah. It is almost 5 kilometers (3 miles) wide and 1,200 meters (4,000 feet) deep, and is getting deeper.

Lee Prince/Shutterstock.com

The Crucial Importance of Rare Earth Metals

Mineral resources

extracted from the earth's crust are processed into an amazing variety of products that can make our lives easier and provide us with economic

FIGURE 12.1 The highly reactive rare earth metal lanthanum (La) and its compounds are used in a variety of products, including batteries in electric and hybrid-electric cars (Figure 12.2) and lenses in cameras and video cameras.

SPL/Science Source

benefits and jobs. You are probably familiar with mineral resources such as gold, copper, aluminum, sand, and gravel. But people are generally much less aware of the *rare earth metals and oxides,* some of which are crucial to the major technologies that support today's lifestyles and economies.

The 17 rare earth metals, also known as *rare earths,* include scandium, yttrium, and 15 lanthanide chemical elements, including lanthanum (see Figure 12.1 and the Periodic Table in Figure 1, p. S5, Supplement 3). Because of their superior magnetic strength and other unique properties, these elements and their compounds are very important for a number of widely used technologies.

Rare earths are used to make liquid crystal display flat screens for computers and television sets, energy-efficient compact fluorescent and LED light bulbs, solar cells, fiber-optic cables, cell phones, and digital cameras. They are

also important in the manufacture of batteries and motors for electric and hybrid-electric cars (Figure 12.2), catalytic converters in car exhaust systems, jet engines, the powerful magnets in wind turbine generators, and solar cells. Rare earths also go into missile guidance systems, smart bombs, aircraft electronics, and satellites.

Without affordable supplies of these metals, industrialized nations could not develop the current versions of cleaner energy technology and other high-tech products that will be major sources of economic growth during this century. Nations also will need these metals to maintain their military strength.

In this chapter we look at the nature and supplies of mineral resources, the rock cycle that produces them, the harmful environmental impacts of our using them, and how we can use these resources more sustainably. We also look at the earth's natural geologic processes and major geologic hazards.

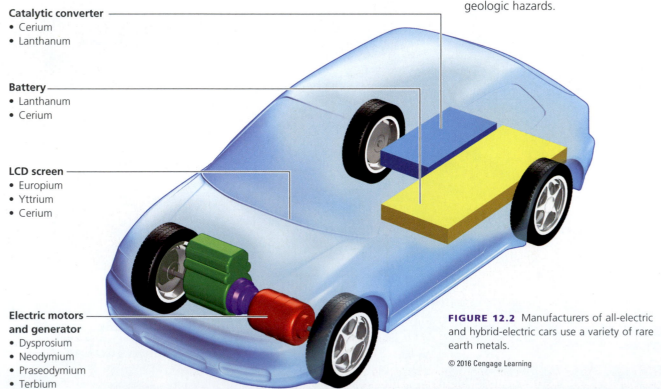

Catalytic converter
• Cerium
• Lanthanum

Battery
• Lanthanum
• Cerium

LCD screen
• Europium
• Yttrium
• Cerium

Electric motors and generator
• Dysprosium
• Neodymium
• Praseodymium
• Terbium

FIGURE 12.2 Manufacturers of all-electric and hybrid-electric cars use a variety of rare earth metals.

© 2016 Cengage Learning

12.1 WHAT ARE THE EARTH'S MAJOR GEOLOGICAL PROCESSES AND WHAT ARE MINERAL RESOURCES?

CONCEPT 12.1A Dynamic processes within the earth and on its surface produce the mineral resources we depend on.

CONCEPT 12.1B Mineral resources are nonrenewable because they are produced and renewed over millions of years largely by the earth's rock cycle.

The Earth Is a Dynamic Planet

Geology is the scientific study of dynamic processes taking place on the earth's surface and in its interior. Scientific evidence indicates that the earth formed about 4.6 billion years ago. As the primitive earth cooled over millions of years, its interior separated into three major concentric zones: the *core,* the *mantle,* and the *crust* (Figure 12.3).

The **core** is the earth's innermost zone. It is extremely hot and has a solid inner part, surrounded by a thick layer of *molten rock,* or hot liquid rock, and semisolid material. Surrounding the core is a thick zone called the **mantle**—

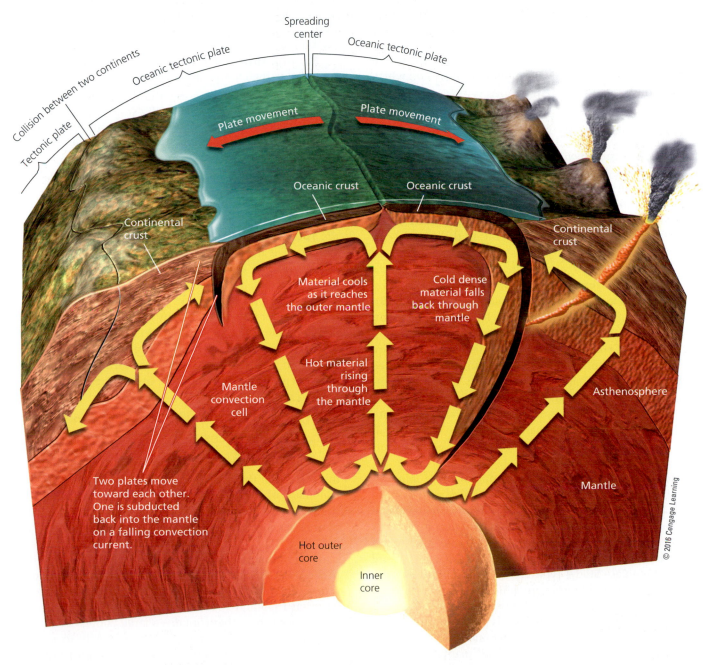

ANIMATED FIGURE 12.3 The earth has a core, mantle, and crust. Within the core and mantle, dynamic forces have major effects on what happens in the crust and on the surface.

© 2016 Cengage Learning

a zone made mostly of solid rock that can be soft and pliable at very high temperatures. The outermost part of the mantle is solid rock, and under that part is a zone called the **asthenosphere**—a volume of hot, partly melted rock that flows. Tremendous heat within the core and mantle generate *convection cells* or *currents* that slowly move large volumes of rock and heat in loops within the mantle like gigantic conveyer belts (Figure 12.3). Some of the molten rock in the asthenosphere flows upward into the crust, where it is referred to as *magma*.

The outermost and thinnest zone of solid material is the earth's **crust**. It consists of the *continental crust,* which underlies the continents (including the continental shelves extending into the oceans), and the *oceanic crust,* which underlies the ocean basins and makes up 71% of the earth's crust. The combination of the crust and the rigid, outermost part of the mantle is called the **lithosphere**. It is in this zone that we find the mineral resources on which we depend (**Concept 12.1A**).

What Are Minerals and Rocks?

The earth's crust beneath our feet consists mostly of minerals and rocks. A **mineral** is a naturally occurring chemical element or inorganic compound that exists as a solid with a regularly repeating internal arrangement of its atoms or ions (a *crystalline solid*). A **mineral resource** is a concentration of one or more minerals in the earth's crust that we can extract and process into raw materials and useful products at an affordable cost. Because minerals take millions of year to form, they are *nonrenewable resources,* and their supplies can be depleted (**Concept 12.1B**).

A few minerals consist of a single chemical element. They include gold (see Figure 2.3, right, p. 29) and rare earth metals (**Core Case Study**) such as lanthanum (Figure 12.1). However, most of the more than 2,000 identified mineral resources that we use occur as inorganic compounds formed by various combinations of elements. Examples include salt (sodium chloride, or NaCl; see Figure 2, p. S6, in Supplement 3) quartz (silicon dioxide, or SiO_2), and rare earth oxides (formed when rare earth metals combine with oxygen).

Rock is a solid combination of one or more minerals found in the earth's crust. Some kinds of rock such as limestone (calcium carbonate, or $CaCO_3$) and quartzite (silicon dioxide, or SiO_2) contain only one mineral. But most rocks consist of two or more minerals. For example, granite is a mixture of mica, feldspar, and quartz crystals. Deposits of rare earths (**Core Case Study**) typically contain a mixture of the metals and their oxides that are difficult and costly to separate from one another and to purify to an acceptable level.

Based on the way it forms, rock is placed in three broad classes: sedimentary, igneous, or metamorphic. **Sedimentary rock** is made of *sediments*—dead plant and animal remains and tiny particles of weathered and eroded rocks. These sediments are transported by water, wind, or gravity to downstream, downwind, downhill, or underwater sites. There they are deposited in layers that accumulate over time. Eventually, the increasing weight and pressure on the underlying layers convert the sedimentary layers to rock. Examples include *sandstone* and *shale* (formed from pressure created by deposited layers made primarily of sand), *dolomite* and *limestone* (formed from the compacted shells, skeletons, and other remains of dead aquatic organisms), and *lignite* and *bituminous coal* (derived from compacted plant remains).

Igneous rock forms below or on the earth's surface under intense heat and pressure when magma wells up from the earth's mantle and then cools and hardens. Examples include *granite* (formed underground) and *lava rock* (formed aboveground). Igneous rock forms the bulk of the earth's crust but is usually covered by sedimentary rock.

Metamorphic rock forms when an existing rock is subjected to high temperatures (which may cause it to melt partially), high pressures, chemically active fluids, or a combination of these agents. Examples include *slate* (formed when shale and mudstone are heated) and *marble* (produced when limestone is exposed to heat and pressure).

The Earth's Rocks Are Recycled Very Slowly

The interaction of physical and chemical processes that change the earth's rocks from one type to another is called the **rock cycle** (Figure 12.4 and **Concept 12.1B**). Rocks are recycled over millions of years by three processes—*erosion, melting,* and *metamorphism*—which produce *sedimentary, igneous,* and *metamorphic* rocks, respectively.

In these processes, rocks are broken down, melted, fused together into new forms by heat and pressure, cooled, and sometimes recrystallized within the earth's interior and crust. The rock cycle is the slowest of the earth's cyclic processes and plays the major role in the formation of concentrated deposits of nonrenewable mineral resources.

We Depend on a Variety of Nonrenewable Mineral Resources

We know how to find and extract more than 100 different minerals from the earth's crust. They include *metallic minerals* made of single elements, such as aluminum, gold, and the rare earths (**Core Case Study**), and those made of compounds such as rare earth oxides. Other mineral resources, including sand and limestone, are *nonmetallic minerals* made of nonmetallic elements and compounds. According to the U.S. Geological Survey (USGS), the quantity of nonrenewable minerals extracted globally increased threefold between 1995 and 2010.

An **ore** is rock that contains a large enough concentration of a particular mineral—often a metal—to make it profitable for mining and processing. A **high-grade ore** contains a high concentration of the desired mineral, whereas a **low-grade ore** contains a lower concentration.

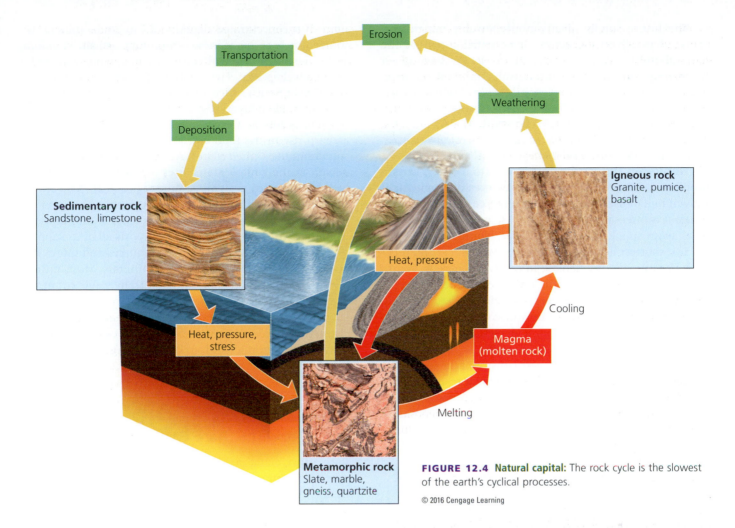

Sedimentary rock
Sandstone, limestone

Igneous rock
Granite, pumice, basalt

Erosion

Transportation

Weathering

Deposition

Heat, pressure

Heat, pressure, stress

Cooling

Magma (molten rock)

Melting

Metamorphic rock
Slate, marble, gneiss, quartzite

FIGURE 12.4 Natural capital: The rock cycle is the slowest of the earth's cyclical processes.

© 2016 Cengage Learning

Nonrenewable mineral resources are used for many purposes. Today, about 60 of the chemical elements in the periodic table (see Figure 1, p. S5, Supplement 3) are used for making computer chips. *Aluminum* (Al) is used as a structural material in beverage cans, motor vehicles, aircraft, and buildings. *Steel,* an essential material used in buildings, machinery, and motor vehicles, is a mixture (or *alloy*) of iron (Fe) and other elements, including *manganese* (Mn), *cobalt* (Co), and *chromium* (Cr). *Copper* (Cu), a good conductor of electricity, is used to make electrical and communications wiring and plumbing pipes. *Gold* (Au) is a component of electrical equipment, tooth fillings, jewelry, coins, and some medical implants. And many of today's important new technologies make use of various rare earth elements (**Core Case Study**).

There are several widely used nonmetallic mineral resources. *Sand,* which is mostly silicon dioxide (SiO_2), is used to make glass, bricks, and concrete for the construction of roads and buildings. *Gravel* is used for roadbeds and to make concrete. Another common nonmetallic mineral is *limestone* (mostly calcium carbonate, or $CaCO_3$), which is crushed to make concrete and cement. Still another is *phosphate salts,* used to make inorganic fertilizers and certain detergents.

12.2 HOW LONG MIGHT SUPPLIES OF NONRENEWABLE MINERAL RESOURCES LAST?

CONCEPT 12.2A Nonrenewable mineral resources exist in finite amounts and can become economically depleted when it costs more than it is worth to find, extract, and process the remaining deposits.

CONCEPT 12.2B There are several ways to extend supplies of mineral resources, but each of them is limited by economic and environmental factors.

Supplies of Nonrenewable Mineral Resources Can Be Economically Depleted

Most published estimates of the supply of a given nonrenewable mineral resource refer to its **reserves**: identified deposits from which we can extract the mineral profitably at current prices. Reserves can be expanded when we find new, profitable deposits or when higher prices or improved mining technologies make it profitable to extract deposits that previously were too expensive to remove.

The future supply of any nonrenewable mineral resource depends on the actual or potential supply of the mineral and the rate at which we use it. We have never completely run out of a nonrenewable mineral resource, but a mineral becomes *economically depleted* when it costs more than it is worth to find, extract, transport, and process the remaining deposits (**Concept 12.2A**). At that point, there are five choices: *recycle or reuse existing supplies, waste less, use less, find a substitute,* or *do without.*

Depletion time is the time it takes to use up a certain proportion—usually 80%—of the reserves of a mineral at a given rate of use. When experts disagree about depletion times, it is often because they are using different assumptions about supplies and rates of use (Figure 12.5).

The shortest depletion-time estimate assumes no recycling or reuse and no increase in reserves (curve A, Figure 12.5). A longer depletion-time estimate assumes that recycling will stretch existing reserves and that better mining technology, higher prices, or new discoveries will increase reserves (curve B). The longest depletion-time estimate (curve C) makes the same assumptions as A and B, but also includes reuse and reduced consumption to further expand reserves. Finding a substitute for a resource leads to a new set of depletion curves for the new resource.

The earth's crust contains fairly abundant deposits of nonrenewable mineral resources such as iron and alumi-

num. But concentrated deposits of important mineral resources such as manganese, chromium, cobalt, platinum, and some of the rare earths (see the following Case Study) are relatively scarce. Five nations—the United States, Canada, Russia, South Africa, and Australia—supply most of the nonrenewable mineral resources used by modern societies.

Since 1900, and especially since 1950, there has been a sharp rise in the total and per capita use of mineral resources in the United States. According to the USGS, each person in the United States uses an average of 22 metric tons (24 tons) of mineral resources per year. As a result, the United States has economically depleted some of its once-rich deposits of metals such as lead, aluminum, and iron. Currently, the United States imports all of its supplies of 20 key nonrenewable mineral resources and more than 90% of its supplies of 4 other key minerals. Most of these imports come from reliable and politically stable countries. But there are serious concerns about access to adequate supplies of four *strategic metal resources*—manganese, cobalt, chromium, and platinum—which are essential for the country's economic and military strength. The United States has little or no reserves of these metals.

CASE STUDY

Global and U.S. Rare Earth Supplies

Rare earth elements (**Core Case Study**) are not actually rare, for the most part, but they are hard to find in concentrations high enough to extract and process at an affordable price. According to the USGS, in 2010, China had roughly 50% of the world's known rare earth reserves, Russia had 15%, and the United States 13%.

In 2013, China produced about 80% of the world's rare earth metals and oxides. This is partly because China does not strictly regulate the environmentally disruptive mining and processing of rare earths. This means that Chinese companies have lower production costs than do companies in countries that regulate rare earth mining and processing more strictly.

The United States and Japan are heavily dependent on rare earths and their oxides. Japan has no rare earth reserves. In the United States, the only rare earth mine, located in California, used to be the world's largest supplier of rare earth metals. However, it closed down because of the expense of meeting pollution regulations, and because China had driven the prices of rare earth metals down to a point where the mine was too costly to operate. It has reopened, but it contains mostly lighter rare earths (with lower atomic numbers), which are more abundant and thus are not as valuable.

China dominates the world in converting rare earth minerals to individual metals and oxides—a complex, lengthy, and environmentally harmful chemical process. Since 2010, China has been reducing its exports of rare earth metals and their oxides to other countries and has sharply raised its prices on such exports.

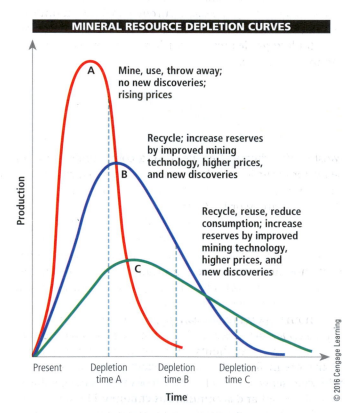

MINERAL RESOURCE DEPLETION CURVES

A Mine, use, throw away; no new discoveries; rising prices

Recycle; increase reserves by improved mining technology, higher prices, and new discoveries

B

Recycle, reuse, reduce consumption; increase reserves by improved mining technology, higher prices, and new discoveries

C

Production

Present | Depletion time A | Depletion time B | Depletion time C

Time

© 2016 Cengage Learning

FIGURE 12.5 *Natural capital depletion:* Each of these *depletion curves* for a mineral resource is based on a different set of assumptions. Dashed vertical lines represent the times at which 80% depletion occurs.

Market Prices Affect Supplies of Nonrenewable Mineral Resources

Geological processes determine the quantity and location of a nonrenewable mineral resource in the earth's crust, but economics determines what part of the known supply is extracted and used. An increase in the price of a scarce mineral resource can often lead to increased supplies and can encourage more efficient use, but there are limits to this effect (**Concept 12.2B**).

According to standard economic theory, in a competitive market system when a resource becomes scarce, its price rises. This can encourage exploration for new deposits, stimulate development of better mining technology, and make it profitable to mine lower-grade ores. It can also encourage a search for substitutes and promote resource conservation.

CONSIDER THIS. . .

CONNECTIONS Metal Prices and Thievery

Resource scarcity can also promote thievery. For example, because of increasing demand, copper prices have risen sharply in recent years. As a result, in several U.S. communities, people have been stealing copper to sell it—stripping abandoned houses of copper pipe and wiring and stealing outdoor central air conditioning units for their copper coils. They have also stolen electrical wiring from beneath city streets and copper piping from farm irrigation systems. In one Oklahoma town, someone cut down several utility poles and stole the copper electrical wiring, causing a blackout.

According to some economists, this price effect may no longer apply very well in most of the more-developed countries. Governments in such countries often use subsidies, tax breaks, and import tariffs to control the supply, demand, and prices of key mineral resources to such an extent that a truly competitive free market does not exist. In the United States, for instance, mining companies get various types of government subsidies, including *depletion allowances*—permission to deduct from their taxable incomes the costs of developing and extracting mineral resources. These allowances amount to 5–22% of their gross income gained from selling the mineral resources.

Mining company representatives insist that they need taxpayer subsidies and tax breaks to keep the prices of minerals low for consumers. They also claim that, without these subsidies and tax breaks, their companies might move their operations to other countries where they would not have to pay taxes or comply with strict mining and pollution control regulations.

Can We Expand Reserves by Mining Lower-Grade Ores?

Some analysts contend that we can increase supplies of some minerals by extracting them from lower-grade ores. They point to the development of new earth-moving equipment, improved techniques for removing impurities from ores, and other technological advances in mineral extraction and processing that can make lower-grade ores accessible, sometimes at lower costs. For example, in 1900, the copper ore mined in the United States was typically about 5% copper by weight. Today, it is typically about 0.5%, yet copper costs less (when prices are adjusted for inflation).

However, several factors can limit the mining of lower-grade ores (**Concept 12.2B**). For one, it requires mining and processing larger volumes of ore, which takes much more energy and costs more. Another factor is the dwindling supplies of freshwater needed for the mining and processing of some minerals, especially in arid and semi-arid areas. A third limiting factor is the growing environmental impacts of land disruption, along with waste material and pollution produced during mining and processing. We discuss this further in the next section of this chapter.

One way to improve mining technology and reduce its environmental impact is to use a biological approach, sometimes called *biomining*. Miners use natural or genetically engineered bacteria to remove desired metals from ores through wells bored into the deposits. This leaves the surrounding environment undisturbed and reduces the air and water pollution associated with removing the metal from metal ores. On the down side, biomining is slow. It can take decades to remove the same amount of material that conventional methods can remove within months or years. So far, biomining methods are economically feasible only for low-grade ores for which conventional techniques are too expensive.

Can We Get More Minerals from the Oceans?

Most of the chemical elements and compounds found in seawater occur in such low concentrations that recovering these mineral resources takes more energy and money than they are worth. Currently, only magnesium, bromine, and sodium chloride are abundant enough to be extracted profitably from seawater. On the other hand, in sediments along the shallow continental shelf and adjacent shorelines, there are significant deposits of minerals such as sand, gravel, phosphates, copper, iron, tungsten, silver, titanium, platinum, and diamonds.

Another potential ocean source of some minerals is *hydrothermal ore deposits* that form when superheated, mineral-rich water shoots out of vents in volcanic regions of the ocean floor. As the hot water comes into contact with cold seawater, black particles of various metal sulfides precipitate out and accumulate as chimney-like structures, called *black smokers,* near the hot water vents (Figure 12.6). These deposits are especially rich in minerals such as copper, lead, zinc, silver, gold, and some of the rare earth metals. A variety of exotic forms of life—including giant clams,

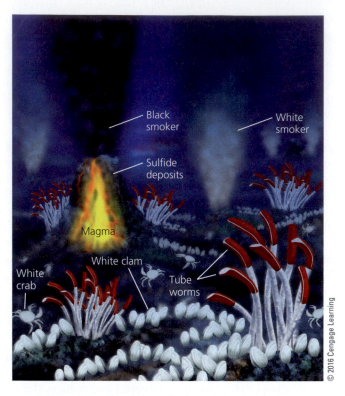

FIGURE 12.6 **Natural capital:** *Hydrothermal deposits*, or *black smokers*, are rich in various minerals.

six-foot tubeworms, and eyeless shrimp—live in the dark depths around these black smokers, supported by bacteria that produce food through chemosynthesis using sulfur compounds discharged by the vents.

Because of the rapidly rising prices of many of these metals, there is growing interest in deep-sea mining. In 2011, China began using remote-controlled underwater equipment and a manned deep-sea craft to evaluate the mining of mineral deposits around black smokers in the Indian Ocean near Madagascar as well as in other deep-sea areas. Also in that year, Japanese explorers found large deposits of rare-earth minerals (**Core Case Study**) on the floor of the Pacific Ocean. The UN International Seabed

Authority, established to manage seafloor mining in international waters, began issuing mining permits in 2011.

Some analysts say that seafloor mining is less environmentally harmful than mining on land. However, marine biologists are concerned that the sediment stirred up by such mining could harm or kill organisms that feed by filtering seawater. Proponents of the mining say that the number of potential mining sites is quite small and that many of these organisms can live elsewhere.

Another possible source of metals from the ocean is the potato-size *manganese nodules* that cover large areas of the Pacific Ocean floor and smaller areas of the Atlantic and Indian Ocean floors. They also contain some low concentrations of various rare earth minerals (**Core Case Study**). These modules could be sucked up by giant vacuum pipes or scooped up by underwater mining machines.

So far, mining on the ocean floor has been hindered by the high costs involved, the potential threat to marine ecosystems, and arguments over rights to the minerals in deep ocean areas that belong to no one country.

12.3 WHAT ARE THE ENVIRONMENTAL EFFECTS OF USING NONRENEWABLE MINERAL RESOURCES?

CONCEPT 12.3 Extracting minerals from the earth's crust and converting them to useful products can disturb the land, erode soils, produce large amounts of solid waste, and pollute the air, water, and soil.

Removal and Use of Mineral Deposits Have Harmful Environmental Effects

Every metal product has a *life cycle* that includes mining the mineral, processing it, manufacturing the product, and disposal or recycling of the product (Figure 12.7). This

Mining → Metal ore → Separation of ore from waste material → Smelting → Melting metal → Conversion to product → Discarding of product

Recycling

FIGURE 12.7 Each metal product that we use has a *life cycle*.

Left: kaband/Shutterstock.com. Second to left: Andrey N Bannov/Shutterstock.com. Center left: Vladimir Melnik/Shutterstock.com. Center: mares/Shutterstock.com. Center right: Zhu Difeng/Shutterstock.com. Second to right: Michael Shake/Shutterstock.com. Right: Pakhnyushcha/Shutterstock.com.

FIGURE 12.8 Natural capital degradation: This spoils pile in Zielitz, Germany, is made up of waste material from the mining of potassium salts used to make fertilizers.

FIGURE 12.9 Natural capital degradation: Area strip mining for coal in the U.S. state of Wyoming.

process makes use of large amounts of energy and water, and results in pollution and waste (**Concept 12.3**).

The environmental impacts of mining a metal ore are determined partly by the ore's percentage of metal content, or *grade*. The more accessible higher-grade ores are usually exploited first. Mining lower-grade ores takes more money, energy, water, and other resources, and leads to more land disruption, mining waste, and pollution.

Several different mining techniques are used to remove mineral deposits. Shallow mineral deposits are removed by **surface mining**, in which vegetation, soil, and rock overlying a mineral deposit are cleared away. The soil and rock, called **overburden**, are usually deposited in piles of waste material called **spoils** (Figure 12.8). Surface mining is used to extract about 90% of the nonfuel mineral resources and 60% of the coal used in the United States.

Different types of surface mining can be used, depending on two factors: the resource being sought and the local topography. In **open-pit mining**, machines are used to dig very large holes and remove metal ores containing copper (see chapter-opening photo), gold (see Case Study that follows), or other metals, or sand, gravel, or stone.

Strip mining is any form of mining involving the extraction of mineral deposits that lie in large horizontal beds close to the earth's surface. In **area strip mining**, used where the terrain is fairly flat, a gigantic earthmover strips away the overburden, and a power shovel—which can be as tall as a 20-story building—removes the mineral deposit or an energy resource such as coal (Figure 12.9). The resulting trench is filled with overburden, and a new cut is made parallel to the previous one. This process is repeated over the entire site.

Contour strip mining (Figure 12.10) is used mostly to mine coal and various mineral resources on hilly or mountainous terrain. Huge power shovels and bulldozers cut a series of terraces into the side of a hill. Then, earthmovers remove the overburden, an excavator or power

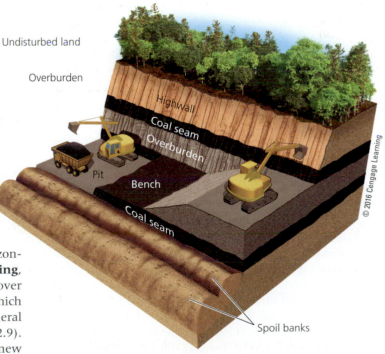

Undisturbed land

Overburden

Highwall

Coal seam

Overburden

Pit

Bench

Coal seam

Spoil banks

FIGURE 12.10 Natural capital degradation: Contour strip mining is used in hilly or mountainous terrain.

Several hundred mountaintops have been blown apart to mine coal in the United States.

Jim West/Age Fotostock

FIGURE 12.11 **Natural capital degradation:** Mountaintop removal coal mining near Whitesville, West Virginia.

shovel extracts the coal, and the overburden from each new terrace is dumped onto the one below. Unless the land is restored, what is left are a series of spoils banks and a highly erodible bank of soil and rock called a *highwall*.

Another surface mining method is **mountaintop removal**, in which the top of a mountain is removed to expose seams of coal, which are then extracted (Figure 12.11). This method is commonly used in the Appalachian Mountains of the United States. With this method, after a mountaintop is blown apart, enormous machines plow waste rock and dirt into valleys below the mountaintops. This destroys forests, buries mountain streams, and increases the risk of flooding. Wastewater and toxic sludge, produced when the coal is processed, are often stored behind dams in these valleys, which can overflow or collapse and release toxic substances such as arsenic and mercury.

In the United States, more than 500 mountaintops in West Virginia and other Appalachian states have been removed to extract coal. According to the U.S. Environmental Protection Agency (EPA), the resulting spoils have buried more than 1,100 kilometers (700 miles) of streams—a total roughly equal in length to the distance between the two U.S. cities of New York and Chicago.

The U.S. Department of the Interior estimates that at least 500,000 surface-mined sites dot the U.S. landscape, mostly in the West. Such sites can be cleaned up and restored, but it is costly and is rarely done even when required by law.

Deep deposits of minerals are removed by **subsurface mining**, in which underground mineral resources are removed through tunnels and shafts. This method is used to remove metal ores and coal that are too deep to be extracted by surface mining. Miners dig a deep, vertical shaft, blast open subsurface tunnels and chambers to reach the deposit, and use machinery to remove the resource and transport it to the surface.

Subsurface mining disturbs less than one-tenth as much land as surface mining disturbs, and it usually produces less waste material. However, it creates hazards such as cave-ins, explosions, and fires for miners. Miners often get lung diseases caused by prolonged inhalation of mineral or coal dust in subsurface mines. Another problem is *subsidence*—the collapse of land above some underground mines. It can damage houses, crack sewer lines, break gas mains, and disrupt groundwater systems.

FIGURE 12.12 Gold mine in New Zealand.

Surface and subsurface mining operations also produce large amounts of solid waste—three-fourths of all U.S. solid waste—and cause major water and air pollution. For example, *acid mine drainage* occurs when rainwater that seeps through an underground mine or a spoils pile from a surface mine carries sulfuric acid (H_2SO_4, produced when aerobic bacteria act on minerals in the spoils) to nearby streams and groundwater.

According to the EPA, mining has polluted mountain streams in about 40% of the western watersheds in the United States, and it accounts for 50% of all the country's emissions of toxic chemicals into the atmosphere. In fact, the mining industry produces more of such toxic emissions than any other U.S. industry.

Where environmental regulations and enforcement have not been strict, the mining and processing of rare earths (**Core Case Study**) has had a very harmful environmental impact. For example, in China, the extraction and processing of rare earth metals and oxides has stripped much land of its vegetation and topsoil while polluting the air, acidifying streams, and leaving toxic and radioactive waste piles.

The Real Cost of Gold

Many newlyweds would be surprised to know that mining enough gold to make their wedding rings produced roughly enough mining waste to equal the total weight of more than three mid-size cars. This waste is usually left piled near the mine site and can pollute the air and nearby surface water.

In 2012, the world's top five gold-producing countries were, in order, China, Australia, the United States, South Africa, and Russia. Many mining companies dig up massive amounts of rock (Figure 12.12) containing only small concentrations of gold. At about 90% of the world's gold mines, the mineral is extracted with the use of a solution of highly toxic cyanide salts sprayed onto piles of crushed rock. The solution reacts with the gold and then drains off the rocks, pulling some gold with it, into settling ponds. After the solution is recirculated a number of times, the gold is removed from the ponds.

Until sunlight breaks down the cyanide, it is extremely toxic to birds and mammals drawn to these ponds in search of water. The ponds can also leak or overflow, which poses

SCIENCE**FOCUS**12.1

THE NANOTECHNOLOGY REVOLUTION

Nanotechnology, or *tiny tech,* uses science and engineering to manipulate and create materials out of atoms and molecules at the ultra-small scale of less than 100 nanometers. The diameter of the period at the end of this sentence is about a half million nanometers.

At the nanometer level, conventional materials have unconventional and unexpected properties. For example, scientists have learned to link carbon atoms together to form carbon nanotubes that are 60 times stronger than high-grade steel. An incredibly thin and nearly invisible thread of this material is strong enough to suspend a pickup truck. Using carbon nanotubes to build cars would make them stronger and safer and would improve gas mileage by making them up to 80% lighter.

Currently, nanomaterials are used in more than 1,300 consumer products and the number is growing rapidly.

Such products include certain batteries, stain-resistant and wrinkle-free clothes, odor-eating socks, self-cleaning glass surfaces and exterior surfaces on buildings, self-cleaning sinks and toilets, sunscreens, waterproof coatings for cell phones and other electrical devices, some cosmetics, some processed foods, and food containers that release nanosilver ions to kill bacteria, molds, and fungi. A new one-atom thick nanotextile material that can conduct electricity could be incorporated into clothing, and this could allow you to charge your cellphone or start your laptop using solar energy by plugging it into your jeans or your tee shirt. In 2014, Swedish scientists developed a paper filter made of cellulose nanofibers that can remove viruses from drinking water.

Nanotechnologists envision technological innovations such as a supercomputer smaller than a grain of rice, thin and flexible solar cell films that could

be attached to or painted onto almost any surface, motors smaller than a human cell, biocomposite materials that would make our bones and tendons super strong, nanovessels filled with medicines that could be delivered to cells anywhere in the body, and nanomolecules specifically designed to seek out and kill cancer cells. Computer-controlled tabletop nanofactories could produce items from computerized blueprints within minutes. Nanotechnology would allow us make materials from the bottom up, using atoms of abundant elements (primarily hydrogen, oxygen, nitrogen, carbon, silicon, and aluminum) as substitutes for scarcer elements, such as copper, cobalt, nickel, and tin, to make a wide variety of products.

Nanotechnology has many potential environmental benefits. Designing and building existing and new products from the molecular level up would greatly reduce the need to mine many

threats to underground drinking water supplies and to fish and other forms of life in nearby lakes and streams. Special liners in the settling ponds can help prevent leaks, but some have failed. According to the EPA, all such liners are likely to leak, eventually.

After extracting the gold from a mine, some mining companies have declared bankruptcy. This has allowed them to walk away from cleaning up their mining operations, leaving behind holding ponds full of cyanide-laden water.

Since 1980, millions of poverty-stricken miners in less-developed countries have streamed into tropical forests in search of gold, and this influx has increased greatly since 2001 when the price of gold began rising sharply. As a result, illegal deforestation to make way for mines has also increased rapidly in parts of the Amazon Basin. These miners use toxic mercury illegally to separate gold from its ore, and they heat the mixture of gold and mercury to vaporize the mercury and leave the gold, causing dangerous air and water pollution. Many of these miners and

villagers living near the mines eventually inhale toxic mercury vapor, drink mercury-laden water, or eat fish contaminated by mercury.

There are nontoxic methods for extracting gold, but the cyanide process is strongly entrenched in the industry. For gold and other mineral resource use to become less environmentally harmful, consumers will have to start demanding more sustainably produced mineral products and substitutes, and producers will have to find ways to produce them—a subject of the next section of this chapter.

Removing Metals from Ores Has Harmful Environmental Effects

Ore extracted by mining typically has two components: the ore mineral, containing the desired metal, and waste material. Removing the waste material from ores produces **tailings**—rock wastes that are left in piles or put into

materials. It would also require less material and energy and reduce waste production. We may be able to use nanoparticles to remove industrial pollutants in contaminated air, soil, and groundwater. Nanofilters might someday be used to desalinate and purify seawater at an affordable cost, thereby helping to increase drinking water supplies. **GREEN CAREER: Environmental nanotechnology**

So what's the catch? The main problem is serious concerns about the possible harmful health effects of nanotechnology on humans. The tiny size and large combined surface area of the huge numbers of these nanoparticles involved in any particular application make them more chemically reactive and potentially more toxic to humans and other animals than are conventional materials composed of much larger particles. Laboratory studies involving mice and other test animals reveal that nanoparticles can

- be inhaled deeply into the lungs and absorbed into the bloodstream and

can penetrate cell membranes, including those in the brain;

- move across the placenta from a mother to her fetus and can move from the nasal passage to the brain;
- result in lung damage similar to that caused by mesothelioma, a deadly cancer resulting from the inhalation of asbestos particles; and
- cause changes in the rate and rhythm of a rodent's heartbeat, similar to such changes caused by heart disease.

We know far too little about these and other risks at a time when the use of untested, unregulated, and unlabeled nanoparticles is growing rapidly. An expert panel from the U.S. National Academy of Sciences has said that the U.S. government is not doing enough to evaluate the potential health and environmental risks of using engineered nanomaterials. For example, the U.S. Food and Drug Administration does not even maintain a list of the food products and cosmetics that contain nanomaterials.

By contrast, the European Union (EU) takes a precautionary approach to the use of nanomaterials and other untested chemicals. While U.S. regulators generally assume that nanoparticles and chemicals are innocent until shown to be harmful, EU laws require that manufacturers demonstrate the safety of their products before they can enter the marketplace.

Many analysts say we need to take three steps before unleashing nanotechnology more broadly. *First,* greatly increase research on the potential harmful health effects of nanoparticles. *Second,* develop guidelines and regulations for controlling its growing applications until we know more about the potentially harmful effects of this new technology. *Third,* require labeling of all products containing nanoparticles.

Critical Thinking

Do you think the potential benefits of nanotechnology products outweigh their potentially harmful effects? Explain.

---

ponds where they settle out. Particles of toxic metals in tailings piles can be blown by the wind or washed out by rainfall and can contaminate surface water and groundwater. Tailings ponds can leak and contaminate surface water and groundwater.

After the waste material is removed, heat or chemical solvents are used to extract the metals from mineral ores. Heating ores to release metals is called **smelting** (Figure 12.7). Without effective pollution control equipment, a smelter emits large quantities of air pollutants, including sulfur dioxide and suspended toxic particles, which damage vegetation and acidify soils in the surrounding area. Smelters also cause water pollution and produce liquid and solid hazardous wastes that require safe disposal. A 2012 study by Blacksmith Institute and Green Cross Switzerland found that lead smelting is the world's second most toxic industry after the recycling of lead-acid batteries. Using chemicals to extract metals from their ores can also create numerous problems, as we saw in the case of using cyanide to remove gold.

12.4 HOW CAN WE USE MINERAL RESOURCES MORE SUSTAINABLY?

CONCEPT 12.4 We can try to find substitutes for scarce resources, reduce resource waste, and recycle and reuse minerals.

We Can Find Substitutes for Some Scarce Mineral Resources

Some analysts believe that even if supplies of key minerals become too expensive or too scarce due to unsustainable use, human ingenuity will find substitutes (**Concept 12.4**). They point to the current *materials revolution* in which silicon and other materials are replacing some metals for common uses. They also point out the possibilities of finding substitutes for scarce minerals through nanotechnology (Science Focus 12.1), as well as through other emerging technologies.

GRAPHENE: A REVOLUTIONARY MATERIAL

Graphene is made from graphite—a form of carbon that occurs as a mineral in some rocks. Ultrathin graphene consists of a single layer of carbon atoms packed into a two-dimensional hexagonal lattice (somewhat like chicken wire) that can be applied as a transparent film to surfaces (Figure 12.A).

Graphene is one of the world's thinnest and strongest materials and is light, flexible, and stretchable. A single layer of graphene is 150,000 times thinner than a human hair and 200 times stronger than structural steel. A sheet of this amazing material stretched over a coffee mug could support the weight of a car. It is also a good conductor of electricity and conducts heat better than any known material.

The use of graphene could revolutionize the electric car industry by leading to the production of batteries that can be recharged 10 times faster and hold 10 times more power than current car batteries. Graphene composites can also be used to make stronger and lighter plastics, lightweight aircraft and motor vehicles, flexible computer

tablets, and TV screens as thin as a magazine. Within 5 years, it might also be used to make flexible, more efficient, less costly solar cells that can be attached to almost anything. Graphene could also replace the silicon transistors that are used as chips in computers and other electronic devices, and this could make almost any electronic device run much faster and use less power. Indeed, some scientists contend that the use of graphene will change the world more than any technological development since the invention of the silicon chip.

Graphene is made from very high purity and expensive graphite. According to the USGS, China controls about 73% of the world's high-purity graphite production. As with rare earth metals (**Core Case Study**), the United States mines very little natural graphite and imports most of its graphite from China, which may restrict U.S. product exports as the use of graphene grows.

Geologists are looking for deposits of graphite in the United States. However, in 2011, a team of Rice University

FIGURE 12.A Sheet of graphene, which consists of multiple layers of carbon atoms linked together in a hexagonal lattice, is a revolutionary new material.

chemists, led by James M. Tour, found ways to make large sheets of high-quality graphene from inexpensive materials found in garbage and from dog feces. If such a process becomes economically feasible, concern over supplies of graphite could vanish, along with any harmful environmental effects of the mining and processing of graphite.

Critical Thinking

Would you invest money in a company that mines and purifies graphite and converts it to graphene? Explain.

For example, fiber-optic glass cables that transmit pulses of light are replacing copper and aluminum wires in telephone cables, and nanowires may eventually replace fiber-optic glass cables. High-strength plastics and materials, strengthened by lightweight carbon, hemp, and glass fibers, are beginning to transform the automobile and aerospace industries. These new materials do not need painting (which reduces pollution and costs), can be molded into any shape, and can increase fuel efficiency by greatly reducing the weights of motor vehicles and airplanes. Such new materials are even being used to build bridges. One new material, called *graphene,* promises to be another new breakthrough material (Science Focus 12.2).

But resource substitution is not a cure-all. For example, platinum is currently unrivaled as a catalyst and is used in industrial processes to speed up chemical reac-

tions, and chromium is an essential ingredient of stainless steel. We can try to find substitutes for such scarce resources, but this may not always be possible.

We Can Use Mineral Resources More Sustainably

Figure 12.13 lists several ways to use mineral resources more sustainably (**Concept 12.4**). One strategy is to focus on recycling and reuse of nonrenewable mineral resources, especially valuable or scarce metals such as gold, iron, copper, aluminum, and platinum. Recycling, an application of the chemical cycling **principle of sustainability**, has a much lower environmental impact than that of mining and processing metals from ores. For example, recycling aluminum beverage cans and scrap

Yu-Guo Guo: Designer of Nanotechnology Batteries and National Geographic Emerging Explorer

Yu-Guo Guo is a professor of chemistry and a nanotechnology researcher at the Chinese Academy of Sciences in Beijing. He has invented nanomaterials that can be used to make lithium-ion battery packs smaller, more powerful, and less costly, which makes them more useful for powering electric cars and electric bicycles. This is an important scientific advance, because the battery pack is the most important and expensive part of any electric vehicle.

Guo's innovative use of nanomaterials has greatly increased the power of lithium-ion batteries by enabling electric current to flow more efficiently through what he calls "3-D conducting nanonetworks." With this promising technology, lithium-ion battery packs in electric vehicles can be fully charged in a few minutes. They also have twice the energy storage capacity of today's batteries, and thus will extend the range of electric vehicles by enabling them to run longer.

Guo predicts that within 5 years, the electric-vehicle market will be well established and, in some cities, will account for 10% of all cars. He is also interested in developing nanomaterials for use in solar cells and fuel cells that could be used to generate electricity and to power vehicles.

Solutions

Sustainable Use of Nonrenewable Minerals

- Reuse or recycle metal products whenever possible

- Redesign manufacturing processes to use less mineral resources

- Reduce mining subsidies

- Increase subsidies for reuse, recycling, and finding substitutes

FIGURE 12.13 We can use nonrenewable mineral resources more sustainably (**Concept 12.4**). **Questions:** Which two of these solutions do you think are the most important? Why?

aluminum produces 95% less air pollution and 97% less water pollution, and uses 95% less energy, than mining and processing aluminum ore. Cleaning up and reusing items instead of recycling them has an even lower environmental impact.

Researchers are working hard to try to ensure adequate supplies of rare earths for the short term and to find alternatives to these materials for the long term (**Core Case Study**). One way to increase supplies is to extract and recycle rare earth metals from the massive amounts of electronic wastes that are being produced. So far, however, less than 1% of rare earth metals are recovered and recycled. The USGS and Department of Energy are also evaluating mining wastes as a potential source of rare earths.

Companies that build batteries for electric cars are beginning to switch from making nickel-metal-hydride batteries, which require the rare earth metal lanthanum, to manufacturing lighter-weight lithium-ion batteries, which researchers are now trying to improve (Individuals Matter 12.1).

Lithium (Li), the world's lightest metal, is a vital component of lithium-ion batteries, which are used in cell phones, iPads, laptop computers, and a growing number of other products. The problem is that some countries, including the United States, do not have large supplies of lithium. The South American countries of Bolivia, Chile, and Argentina have about 80% of the global reserves of lithium. Bolivia alone has about 50% of these reserves, while the United States holds only about 3%.

As a result of this, Japan, China, South Korea, and the United Arab Emirates are buying up access to global lithium reserves to ensure their ability to sell lithium or batteries to the rest of the world. Within a few decades, the United States may be heavily dependent on expensive imports of lithium and lithium batteries. However, in 2014, the company Simbol, Inc. began building a plant in California that is designed to extract lithium from brine waste produced by geothermal power plants. If it is successful, this process could lessen some U.S. dependence on imported lithium.

Scientists are also searching for substitutes for rare earth metals that could be used to make increasingly important powerful magnets and related devices. In Japan and the United States, researchers are developing a variety of such devices that require no rare earth minerals, are light and compact, and can deliver more power with greater efficiency at a reduced cost. These include electric motors for hybrid-electric vehicles and electromagnets for use in wind turbines.

12.5 WHAT ARE THE EARTH'S MAJOR GEOLOGICAL HAZARDS?

CONCEPT 12.5 Dynamic processes move matter within the earth and on its surface and can cause volcanic eruptions, earthquakes, tsunamis, erosion, and landslides.

The Earth Beneath Your Feet Is Moving

We tend to think of the earth's crust as solid and unmoving. However, according to geologists, the flows of energy and heated material within the earth's convection cells

(Figure 12.3) are so powerful that they have caused the lithosphere to break up into a dozen or so huge rigid plates, called **tectonic plates**, which move extremely slowly atop the asthenosphere (Figure 12.14).

These gigantic plates are somewhat like the world's largest and slowest-moving surfboards on which we ride without noticing their movement. Their typical speed is about the rate at which your fingernails grow. Throughout the earth's history, continents have split apart and joined as tectonic plates shifted around atop the earth's asthenosphere (Figure 4.D, p. 76).

Much of the geological activity at the earth's surface takes place at the boundaries between tectonic plates as they separate, collide, or grind along against each other (Figure 12.15). The tremendous forces produced at these plate boundaries can cause mountains or deep rifts to form, earthquakes to shake parts of the crust, and volcanoes to erupt.

Volcanoes Release Molten Rock from the Earth's Interior

An active **volcano** occurs where magma rising in a plume through the lithosphere reaches the earth's surface through a central vent or a long crack, called a *fissure* (Figure 12.16). Magma that reaches the earth's surface is called *lava* and often builds into a cone.

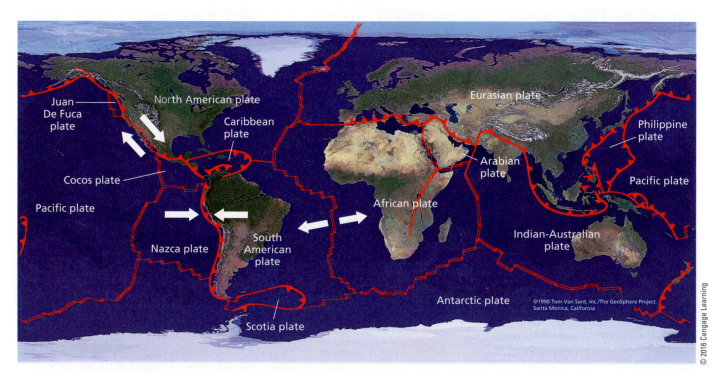

ANIMATED FIGURE 12.14 The earth's crust has been fractured into several major tectonic plates.
Question: Which plate are you riding on?

© 2016 Cengage Learning

North American plate

San Francisco

San Andreas Fault

Pacific plate

Los Angeles

Kevin Schafer/Age Fotostock

FIGURE 12.15 The North American Plate and the Pacific Plate (Figure 12.14) slide very slowly against each other in opposite directions along the San Andreas fault (shown here) which runs almost the full length of California and is responsible for earthquakes of various magnitudes.

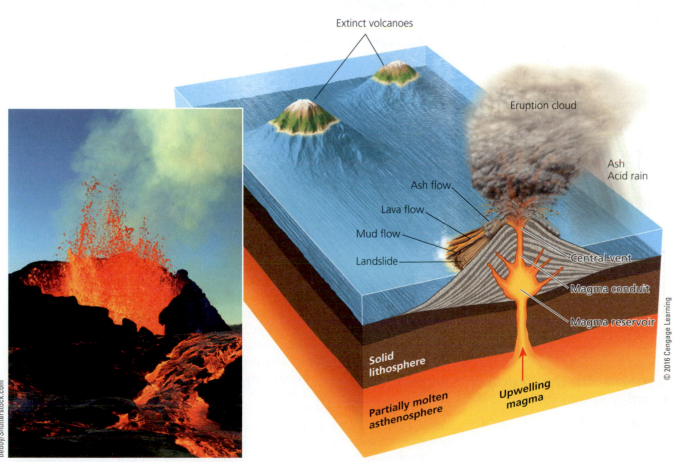

Extinct volcanoes

Eruption cloud

Ash
Acid rain

Ash flow

Lava flow

Mud flow

Landslide

Central vent

Magma conduit

Magma reservoir

Solid lithosphere

Partially molten asthenosphere

Upwelling magma

beboy/Shutterstock.com

© 2016 Cengage Learning

FIGURE 12.16 Sometimes, the internal pressure in a volcano is high enough to cause lava, ash, and gases to be ejected into the atmosphere (photo inset) or to flow over land, causing considerable damage.

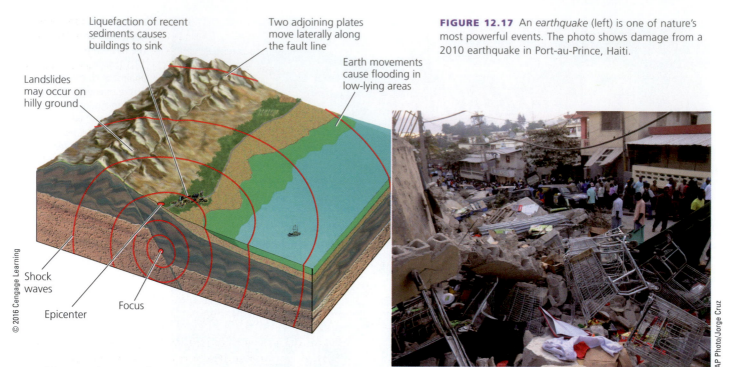

Liquefaction of recent sediments causes buildings to sink

Two adjoining plates move laterally along the fault line

Landslides may occur on hilly ground

Earth movements cause flooding in low-lying areas

Shock waves

Epicenter

Focus

© 2016 Cengage Learning

FIGURE 12.17 An *earthquake* (left) is one of nature's most powerful events. The photo shows damage from a 2010 earthquake in Port-au-Prince, Haiti.

AP Photo/Jorge Cruz

Many volcanoes form along the boundaries of the earth's tectonic plates when one plate slides under or moves away from another plate. Volcanoes can *erupt*, releasing large chunks of lava rock, glowing hot ash, liquid lava, and gases (including water vapor, carbon dioxide, and sulfur dioxide) into the environment (**Concept 12.5**). Eruptions can be extremely destructive, causing loss of life and obliterating ecosystems and human communities.

While volcanic eruptions can be destructive, they do provide some benefits. They can result in the formation of majestic mountains and lakes and the weathering of lava contributes to fertile soils. Hundreds of volcanoes have erupted on the ocean floor, building cones that have reached the ocean's surface, eventually to form islands that have become suitable for human settlement.

We can reduce the loss of human life and some of the property damage caused by volcanic eruptions by using historical records and geological measurements to identify high-risk areas, so that people can avoid living in those areas. We also use monitoring devices that warn us when volcanoes are likely to erupt, and in some areas that are prone to volcanic activity, evacuation plans have been developed.

Earthquakes Are Geological Rock-and-Roll Events

Forces inside the earth's mantle put tremendous stress on rock within the crust. Such stresses can be great enough to cause sudden breakage and shifting of the rock, producing a *fault*, or fracture in the earth's crust (Figure 12.15). When a fault forms, or when there is abrupt movement on an existing fault, energy that has accumulated over time is released in the form of vibrations, called *seismic waves*, that move in all directions through the surrounding rock—an event called an **earthquake** (Figure 12.17 and **Concept 12.5**). Most earthquakes occur at the boundaries of tectonic plates.

Seismic waves move upward and outward from the earthquake's *focus* like ripples in a pool of water. Scientists measure the severity of an earthquake by the *magnitude* of its seismic waves. The magnitude is a measure of ground motion (shaking) caused by the earthquake, as indicated by the *amplitude*, or size of the seismic waves when they reach a recording instrument, called a *seismograph*.

Scientists use the *Richter scale*, on which each unit has an amplitude 10 times greater than the next smaller unit. Seismologists rate earthquakes as *insignificant* (less than 4.0 on the Richter scale), *minor* (4.0–4.9), *damaging* (5.0–5.9), *destructive* (6.0–6.9), *major* (7.0–7.9), and *great* (over 8.0). The largest recorded earthquake occurred in Chile on May 22, 1960, and measured 9.5 on the Richter scale. Each year, scientists record the magnitudes of more than 1 million earthquakes, most of which are too small to feel.

The primary effects of earthquakes include shaking and sometimes a permanent vertical or horizontal displacement of a part of the crust. These effects can have serious consequences for people and for buildings, bridges, freeway overpasses, dams, and pipelines. A major earthquake is a very large rock-and-roll geological event.

One way to reduce the loss of life and property damage from earthquakes is to examine historical records and make geological measurements to locate active fault zones.

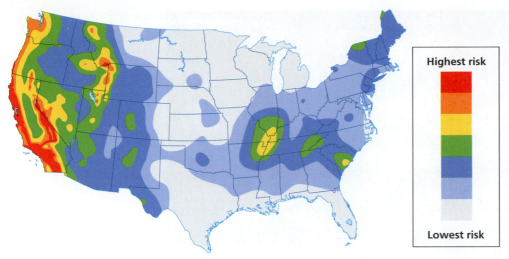

FIGURE 12.18 Comparison of degrees of earthquake risk across the continental United States.

Compiled by the authors using data from the U.S. Geological Survey.

Highest risk

Lowest risk

We can then map high-risk areas (Figure 12.18) and establish building codes that regulate the placement and design of buildings in such areas. Then people can evaluate the risk and factor it into their decisions about where to live. Also, engineers know how to make homes, large buildings, bridges, and freeways more earthquake resistant, although this is costly.

Earthquakes on the Ocean Floor Can Cause Huge Waves Called Tsunamis

A **tsunami** is a series of large waves generated when part of the ocean floor suddenly rises or drops (Figure 12.19). Most large tsunamis are caused when certain types of faults in the ocean floor move up or down as a result of a

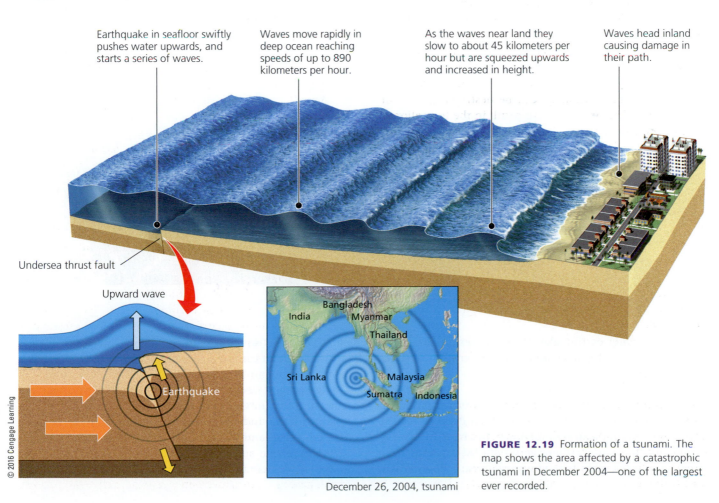

Earthquake in seafloor swiftly pushes water upwards, and starts a series of waves.

Waves move rapidly in deep ocean reaching speeds of up to 890 kilometers per hour.

As the waves near land they slow to about 45 kilometers per hour but are squeezed upwards and increased in height.

Waves head inland causing damage in their path.

Undersea thrust fault

Upward wave

Earthquake

© 2016 Cengage Learning

December 26, 2004, tsunami

FIGURE 12.19 Formation of a tsunami. The map shows the area affected by a catastrophic tsunami in December 2004—one of the largest ever recorded.

FIGURE 12.20 The Banda Aceh Shore near Gleebruk, Indonesia on June 23, 2004 (left), and on December 28, 2004 (right), after it was stuck by a tsunami.

large underwater earthquake. Other causes are landslides generated by earthquakes and volcanic eruptions (**Concept 12.5**).

Tsunamis are often called *tidal waves,* although they have nothing to do with tides. They can travel far across the ocean at the speed of a jet plane. In deep water the waves are very far apart—sometimes hundreds of kilometers—and their crests are not very high. As a tsunami approaches a coast with its shallower waters, it slows down,

its wave crests squeeze closer together, and their heights grow rapidly. It can hit a coast as a series of towering walls of water that can level buildings.

The largest recorded loss of life from a tsunami occurred in December 2004 when a great underwater earthquake in the Indian Ocean with a magnitude of 9.15 caused a tsunami with huge waves that killed around 279,900 people and devastated many coastal areas of Indonesia (see map in Figure 12.19 and Figure 12.20), Thailand, Sri Lanka, South India, and even eastern Africa. It also displaced about 1.8 million people (1.3 million of them in India and Indonesia), and destroyed or damaged about 470,000 buildings and houses. There were no recording devices in place to provide an early warning of this tsunami.

In 2011, a large tsunami caused by a powerful earthquake off the coast of Japan generated 3-story high waves that killed almost 19,000 people, displaced more than 300,000 people, and destroyed or damaged 125,000 buildings. It also heavily damaged three nuclear reactors, which then released dangerous radioactivity into the surrounding environment.

In some areas, scientists have built networks of ocean buoys and pressure recorders on the ocean floor to collect data that can be relayed to tsunami emergency warning centers. However, these networks are far from complete.

- Dynamic forces that move matter within the earth and on its surface recycle the earth's rocks, form deposits of mineral resources, and cause volcanic eruptions, earthquakes, and tsunamis.

- The available supply of a mineral resource depends on how much of it is in the earth's crust, how fast we use it, the mining technology used to obtain it, its market prices, and the harmful environmental effects of removing and using it.

- We can use mineral resources more sustainably by trying to find substitutes for scarce resources, reducing resource waste, and reusing and recycling nonrenewable minerals.

TYING IT ALL TOGETHER

Rare Earth Metals and Sustainability

M. Dykstra/Shutterstock.com

In the Core Case Study that opened this chapter, we learned about the importance of various rare earth metals that are used in a variety of modern technologies, including electric motor vehicles and energy-efficient compact fluorescent light bulbs (photo to left).

In this chapter, we looked at technological developments that could help us to expand supplies of mineral resources and to use them more sustainably. For example, if we develop it safely, we could use nanotechnology to make new materials that could replace scarce mineral resources and greatly reduce the environmental impacts of mining and processing such resources. We looked at how biomining makes use of microbes to extract mineral resources without disturbing the land or polluting air and water as much as conventional mining operations do. Another emerging technology uses graphene to replace conventional transistors and to produce more efficient and affordable solar cells to generate electricity—an application of the solar energy **principle of sustainability**.

We can also use mineral resources more sustainably by reusing and recycling them, and by reducing unnecessary resource use and waste—applying the chemical cycling **principle of sustainability**. In addition, industries can mimic nature by using a diversity of ways to reduce the harmful environmental impacts of mining and processing mineral resources, thus applying the biodiversity **principle of sustainability**.

Chapter Review

Core Case Study

1. Explain the importance of the rare earth metals.

Section 12.1

2. What are the two key concepts for this section? Define **geology**, **core**, **mantle**, **asthenosphere**, **crust**, and **lithosphere**. Define **mineral**, **mineral resource**, and **rock**. Define and distinguish among **sedimentary rock**, **igneous rock**, and **metamorphic rock** and give an example of each. Define the **rock cycle** and explain its importance. Define **ore** and distinguish between a **high-grade ore** and a **low-grade ore**. List five important nonrenewable mineral resources and their uses.

Section 12.2

3. What are the two key concepts for this section? What are the **reserves** of a mineral resource and how can they be expanded? What two factors determine the future supply of a nonrenewable mineral resource? Explain how the supply of a nonrenewable mineral resource can be economically depleted and list the five choices we have when this occurs. What is **depletion time** and what factors affect it?

4. What five nations supply most of the world's nonrenewable mineral resources? How dependent is the United States on other countries for important nonrenewable mineral resources? Explain the concern over U.S. access to rare earth mineral resources. Describe the conventional view of the relationship between the supply of a mineral resource and its market price. Explain why some economists believe this relationship no longer applies.

5. Summarize the opportunities and limitations of expanding mineral supplies by mining lower-grade ores. What are the advantages and disadvantages of biomining? Describe the opportunities and possible problems that could result from deep-sea mining.

Section 12.3

6. What is the key concept for this section? Summarize the life cycle of a metal product.

7. What is **surface mining**? Define **overburden** and **spoils**. Define **open-pit mining** and **strip mining**, and distinguish among **area strip mining**, **contour strip mining**, and **mountaintop removal mining**. Describe three harmful environmental effects of surface mining. What is **subsurface mining**? Summarize the harmful effects of gold mining. Define **tailings** and explain why they can be hazardous. What is **smelting** and what are its major harmful environmental effects?

Section 12.4

8. What is the key concept for this section? Give two examples of promising substitutes for key mineral resources. What is **nanotechnology** and what are some of its potential environmental and other benefits? What are some problems that could arise from the widespread use of nanotechnology? Describe the potential of using graphene as a new resource. Explain the benefits of recycling and reusing valuable metals. List five ways to use nonrenewable mineral resources more sustainably. What are two examples of research into substitutes for rare earth metals? Explain why uneven distribution of lithium among various countries is a concern.

Section 12.5

9. What is the key concept for this section? What are **tectonic plates**, and what typically happens when they collide, move apart, or grind against one another? Define **volcano** and describe the nature and major effects of a volcanic eruption. Define **earthquake** and describe its nature and major effects. What is a **tsunami** and what are its major effects?

10. What are the three big ideas of this chapter? Explain how we can apply the three **scientific principles of sustainability** to obtain and use rare earth metals and other nonrenewable mineral resources in more sustainable ways.

Note: Key terms are in bold type.

Critical Thinking

1. Give three reasons why rare earth metals (**Core Case Study**) are important to your lifestyle.

2. Would you favor giving the owners of the California rare earth metals mine significant government tax breaks and subsidies to put the mine back into full production? Explain. Would you favor reducing the environmental regulations for the mining and processing of these metals? Explain.

3. You are an igneous rock. Describe what you experience as you move through the rock cycle. Repeat this exercise, assuming you are a sedimentary rock and again assuming you are a metamorphic rock.

4. What are three ways in which you benefit from the rock cycle?

5. Use the second law of thermodynamics (see Chapter 2, p. 35) to analyze the scientific and economic feasibility of each of the following processes:
   a. Extracting certain minerals from seawater
   b. Mining increasingly lower-grade deposits of minerals
   c. Continuing to mine, use, and recycle minerals at increasing rates

6. Suppose you were told that mining deep-ocean mineral resources would mean severely degrading ocean bottom habitats and life forms such as giant tubeworms and giant clams. Do you think that such information should be used to prevent ocean bottom mining? Explain.

7. List three ways in which a nanotechnology revolution could benefit you and three ways in which it could harm you. Do you think the benefits outweigh the harms? Explain.

8. What are three ways to reduce the harmful environmental impacts of the mining and processing of nonrenewable mineral resources? What are three aspects of your lifestyle that contribute to these harmful impacts?

Doing Environmental Science

Do research to determine which mineral resources go into the manufacture of each of the following items and how much of each of these resources are required to make each item: **(a)** a cell phone, **(b)** a wide-screen TV, and **(c)** a large pickup truck. Pick three of the lesser-known mineral materials that you have learned about in this exercise and do more research to find out where in the world most of that material comes from. For each of the three minerals you chose, try to find out what kinds of environmental effects have resulted from the mining of the mineral in at least one of the places where it is mined. You might also find out about steps that have been taken to deal with those effects. Write a report summarizing all of your findings.

Global Environment Watch Exercise

Use the Global Environment Watch site to find and read an article that deals with U.S. rare earth metal supplies (**Core Case Study**). Summarize the conclusions expressed in the article. Is there scientific information cited in the article to support the author's conclusions? Give specific examples. Do you think there are any types of supporting scientific data not mentioned in the article that would strengthen the author's conclusions? For example, would you add statistical data to support a point, or data in a graph indicating possible cause-and-effect relationships? Be specific and give reasons for your suggestions.

Data Analysis

Rare earth metals are widely used in a variety of important products (**Core Case Study**). According to the U.S. Geological Survey, China has about 50% of the world's reserves of rare earth metals. Use this information to answer the following questions.

1. In 2010, China had 55 million metric tons of rare earth metals in its reserves and produced 130,000 metric tons of these metals. At this rate of production how long will China's rare earth reserves last?

2. In 2010, the global demand for rare earth metals was about 133,600 metric tons. At this annual rate of use, if China were to produce all of the world's rare earth metals, how long would their reserves last?

3. The annual global demand for rare earth metals is projected to rise to at least 185,000 metric tons by 2015. At this rate, if China were to produce all of the world's rare earth metals, how long would its reserves last?

CENGAGEbrain.com To access course materials, including Aplia homework, please visit www.cengagebrain.com.

WWW.CENGAGEBRAIN.COM **311**

13
ENERGY RESOURCES

KEY QUESTIONS

13.1 What is net energy and why is it important?

13.2 What are the advantages and disadvantages of using fossil fuels?

Tractor-trailer hauling two wind turbine blades to a wind farm in China.

Greg Girard/National Geographic Creative

The Astounding Potential for Wind Power in the United States

Between the earth's equator and its polar regions, the sun's rays strike the earth at different angles. This results in more solar energy at the equator and less at the poles. Together with the earth's rotation, these differences create flows of air called *wind* (see Figure 7.3, p. 132). Because wind is an indirect form of solar energy, relying more on wind is one way to apply the solar energy **principle of sustainability**.

We can capture this energy with groups of wind turbines called *wind farms* that convert wind energy to electrical energy (Figure 13.1, left) that can be fed into electrical grids. We can locate wind farms on land or at sea (Figure 13.1, right).

Because today's wind turbine towers can be as tall as a 30-story building and typically have blades as long as 60 meters (200 feet) (see chapter-opening photo), they can tap into the stronger and more reliable winds found at higher altitudes on land and at sea. This allows them to produce more

electricity at a lower cost. The U.S. Department of Energy (DOE) estimates that wind farms at favorable sites in just three U.S. states—North Dakota, Kansas, and Texas—could more than meet the electricity needs of the lower 48 states. In addition, the National Renewable Energy Laboratory (NREL) estimates that offshore winds—off the Atlantic and Pacific coasts, and off the shores of the Great Lakes—could generate four times the electricity currently used in the lower 48 states.

Texas, the nation's leading oil-producing state, is also the nation's leading producer of electricity from wind power, followed by California. These two states produce more electricity from wind than all but five of the world's countries. To some analysts, wind power has more advantages and fewer serious disadvantages than all other energy resources except for energy efficiency, another topic covered in this chapter.

Wind power proponents call for a crash program to develop land-based and offshore wind farms in the United States and to build an updated and expanded smart electrical grid to transmit the electricity they produce to consumers. Over time, this would help reduce the country's dependence on coal, which, when burned to produce electricity, emits air pollutants that kill at least 24,000 Americans a year and adds huge quantities of climate-changing CO_2 to the atmosphere. Greatly expanding the U.S. wind turbine industry would also create many thousands of new jobs and boost the American economy.

In this chapter, we explore and compare the benefits and drawbacks of the key energy resources on which we now depend, including nonrenewable resources such as oil, coal, and nuclear power, and renewable resources such as wind, solar energy, flowing water, and the earth's internal heat.

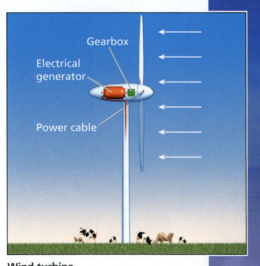

Wind turbine

FIGURE 13.1 Wind turbines convert the kinetic energy in wind to electricity, another form of kinetic energy (moving electrons). Wind power is an indirect form of solar energy.

WHAT IS NET ENERGY AND WHY IS IT IMPORTANT?

CONCEPT 13.1 Energy resources vary greatly in their *net energy yields*—the amount of energy available from a resource minus the amount of energy needed to make it available.

Net Energy Is the Only Energy That Really Counts

It takes energy to produce energy. For example, before oil becomes useful to us, it must be found, pumped up from beneath the ground or ocean floor, transferred to a refinery, converted to gasoline and other fuels and chemicals, and delivered to consumers. Each of these steps uses high-quality energy, obtained mostly by burning fossil fuels, especially gasoline and diesel fuel.

The usable amount of high-quality energy available from an energy resource is its **net energy yield**. It is the total amount of high-quality energy available from a given quantity of an energy resource minus the high-quality energy needed to make the energy available (**Concept 13.1**). Suppose that it takes about 9 units of high-quality energy to produce 10 units of high-quality energy from an energy resource. Then the net energy yield is 1 unit of energy.

Net energy is like the net profit earned by a business after it deducts its expenses. If a business has $1 million in sales and $900,000 in expenses, its net profit is $100,000.

Figure 13.2 shows generalized net energy yields for energy resources and systems that generate electricity, heat homes and other buildings, produce high-temperature heat for industrial processes, and provide transportation. It is based on several sources of scientific data and classifies estimated net energy yields as high, medium, low, or negative (negative being a net energy loss). Many analysts view net energy yield as the single most important measure that we can use to evaluate the long-term economic usefulness of different energy resources.

CONSIDER THIS...

CONNECTIONS Energy Resources and the Laws of Thermodynamics

Net energy yields are based on the first and second laws of thermodynamics (see pp. 34–36)—two scientific laws that we cannot violate no matter how clever we are. Technology can help us exploit an energy resource more cheaply, cleanly, and efficiently, but because of the first law we cannot create energy. And because of the second law, whenever we use any type of high-quality energy, it is always degraded into lower-quality energy, mostly heat that ends up in the environment. This means that we cannot recycle high-quality energy.

Electricity	Net Energy Yield
Energy efficiency	High
Hydropower	High
Wind	High
Coal	High
Natural gas	Medium
Geothermal energy	Medium
Solar cells	Low to medium
Nuclear fuel cycle	Low
Hydrogen	Negative (Energy loss)

Space Heating	Net Energy Yield
Energy efficiency	High
Passive solar	Medium
Natural gas	Medium
Geothermal energy	Medium
Oil	Medium
Active solar	Low to medium
Heavy shale oil	Low
Heavy oil from tar sands	Low
Electricity	Low
Hydrogen	Negative (Energy loss)

High-Temperature Industrial Heat	Net Energy Yield
Energy efficiency (cogeneration)	High
Coal	High
Natural gas	Medium
Oil	Medium
Heavy shale oil	Low
Heavy oil from tar sands	Low
Direct solar (concentrated)	Low
Hydrogen	Negative (Energy loss)

Transportation	Net Energy Yield
Energy efficiency	High
Gasoline	High
Natural gas	Medium
Ethanol (from sugarcane)	Medium
Diesel	Medium
Gasoline from heavy shale oil	Low
Gasoline from heavy tar sand oil	Low
Ethanol (from corn)	Low
Biodiesel (from soy)	Low
Hydrogen	Negative (Energy loss)

Top left: Yegor Korzh/Shutterstock.com. Bottom left: Donald Aitken/National Renewable Energy Laboratory. Top right: Serdar Tibet/Shutterstock.com. Bottom right: Michel Stevelmans/Shutterstock.com.

FIGURE 13.2 Generalized net energy yields for various energy resources and systems (**Concept 13.1**).
Question: Based only on these data, which two resources in each category should we be using?

Compiled by the authors using data from the U.S. Department of Energy; U.S. Department of Agriculture; Colorado Energy Research Institute, *Net Energy Analysis*, 1976; Howard T. Odum and Elisabeth C. Odum, *Energy Basis for Man and Nature*, 3rd ed., New York: McGraw-Hill, 1981; and Charles A. S. Hall and Kent A. Klitgaard, *Energy and the Wealth of Nations*, New York: Springer, 2012.

Some Energy Resources Need Help to Compete in the Marketplace

The following general rule can help us to evaluate the long-term economic usefulness of an energy resource based on its net energy yield: *It is very difficult for an energy resource with a low or negative net energy yield to compete in the marketplace with other energy alternatives that have medium to high net energy yields unless it receives subsidies or tax breaks from the government (taxpayers) or other outside sources.*

For example, electricity produced by nuclear power has a low net energy yield because large amounts of high-quality energy are needed for each step in the *nuclear power fuel cycle:* to extract and process uranium ore, convert it into nuclear fuel, build and operate nuclear power plants, safely store the resulting highly radioactive wastes for thousands of years, dismantle each highly radioactive plant after its useful life (typically 40–60 years), and safely store its high-level radioactive materials for thousands of years. The low net energy yield for the whole fuel cycle is one reason why governments (taxpayers) throughout the world must heavily subsidize nuclear-generated electricity to make it available to consumers at an affordable price.

13.2 WHAT ARE THE ADVANTAGES AND DISADVANTAGES OF USING FOSSIL FUELS?

CONCEPT 13.2 Oil, natural gas, and coal are currently abundant and relatively inexpensive, but using them causes air and water pollution, degrades large areas of land, and releases climate-changing greenhouse gases to the atmosphere.

Fossil Fuels Supply Most of Our Commercial Energy

The energy that heats the earth and makes it livable for us comes from the sun at no cost to us—in keeping with the solar energy **principle of sustainability**. Without this essentially inexhaustible input of solar energy, the earth's average temperature would be −240 °C (−400 °F) and life as we know it would not exist.

This direct input of solar energy produces several other forms of renewable energy resources that can be thought of as indirect solar energy: *wind* (moving air masses heated by the sun, see **Core Case Study**), *flowing water* (made possible by solar energy, which evaporates water that returns to the earth as precipitation and flows in rivers), and *biomass* (solar energy converted to chemical energy and stored in the tissues of trees and other plants that can be burned as a source of energy).

In 2012, the world's three largest users of *commercial energy*—energy sold in the marketplace—were China (22% of the global total), the United States (18%), and Russia (6%). Most of this energy comes from extracting and burning *nonrenewable energy resources* obtained from the earth's crust. About 91% of the world's commercial energy comes from nonrenewable resources—87% from carbon-containing fossil fuels (oil, natural gas, and coal) and 4% from nuclear power (Figure 13.3, left). The remaining 9% of the commercial energy we use comes from *renewable energy resources* such as biomass, hydropower, geothermal, wind, and solar energy. (In Supplement 5, see Figure 1, p. S39, showing total and per capita U.S. energy use from 1950 to 2012; and Figure 2, p. S39, showing U.S. energy consumption by fuel type between 1980 and 2012, with projections to 2040.)

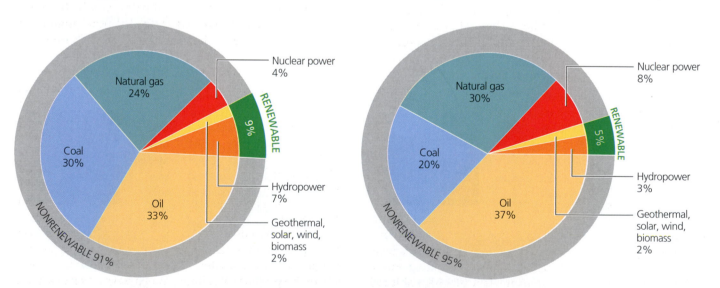

FIGURE 13.3 Energy use by source throughout the world (left) and in the United States (right) in 2012.

Compiled by the authors using data from British Petroleum, U.S. Energy Information Administration, and International Energy Agency.

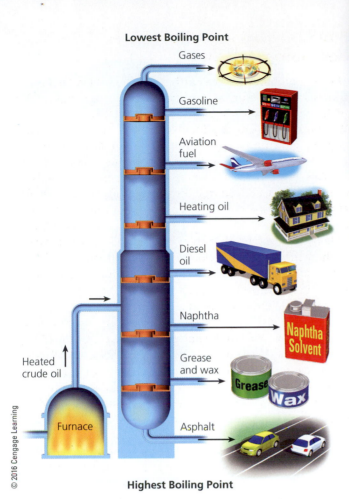

Lowest Boiling Point

Gases

Gasoline

Aviation fuel

Heating oil

Diesel oil

Naphtha

Grease and wax

Asphalt

Heated crude oil

Furnace

© 2016 Cengage Learning

Highest Boiling Point

Natalia Bratslavsky/Shutterstock.com

FIGURE 13.4 When crude oil is refined, many of its components are removed at various levels of a distillation column, depending on their boiling points. The most volatile components with the lowest boiling points are removed at the top of the column, which can be as tall as a nine-story building. The photo shows an oil refinery in Texas.

We Depend Heavily on Oil

Crude oil, or **petroleum**, is a black, gooey liquid containing a mixture of combustible hydrocarbons (containing hydrogen and carbon atoms) along with small amounts of sulfur, oxygen, and nitrogen impurities. It is also known as *conventional* or *light crude oil*. It was formed from the decayed remains of ancient organisms that were crushed beneath layers of rock for millions of years.

Deposits of conventional crude oil and natural gas often are trapped together beneath layers of impermeable rock within the earth's crust on land or under the seafloor. The crude oil is dispersed in pores and cracks in underground rock formations, somewhat like water saturating a sponge.

Geologists identify potential oil deposits by using large machines to pound the earth, sending shock waves deep underground, and they measure how long it takes for the waves to be reflected back. This information is fed into computers and converted into *3-D seismic maps* of the underground that show the locations and sizes of various rock formations. In 2014, scientists were evaluating a new technology in which a fast-moving plane at a high altitude seeks to detect changes in the gravitational field deep underground to pinpoint perspective oil and natural gas deposits.

When oil is found, oil companies drill holes and remove rock cores from potential oil deposit areas to learn whether there is enough oil to be extracted profitably. If there is, one or more wells are drilled and the oil, drawn by gravity out of the rock pores, flows to the bottom of the well and is pumped from there to the surface.

After years of pumping, usually a decade or so, the pressure in a well drops and its rate of conventional crude oil production starts to decline. This point in time is referred to as **peak production** for the well. The same thing can happen to a large oil field when the overall rate of production from its numerous wells begins to decline.

Global peak production occurs when the rate of global production of crude oil begins to decline faster than new oil fields are found and put into production. There is disagreement among experts about whether we have reached this point and when we might reach it if we have not. When this happens we can expect net energy yields from crude oil to decline and oil prices to rise as the largest, cheapest, and most accessible oil fields with high net energy yields become depleted. At global peak production, we will start to use the latter half of the world's estimated crude oil reserves.

Crude oil from a well cannot be used as it is. It is transported to a refinery by pipeline, truck, rail, or ship (oil tanker) where it is heated to separate it into various fuels and other components with different boiling points (Figure 13.4) in a complex process called **refining**. This process, like all other steps in the cycle of oil production and use, requires an input of high-quality energy and decreases the net energy yield of oil.

Oil refining also produces large quantities of petroleum coke or *petcoke,* a black powder that often accumulates in huge piles. Winds can blow harmful fine particles from these piles into the atmosphere, unless they are covered or the powder is stored safely to prevent such pollution. About 2% of the products of refining, called **petrochemicals**, are used as raw materials to make industrial organic chemicals, cleaning fluids, pesticides, plastics, synthetic fibers, paints, medicines, cosmetics, and many other products.

REMOVING TIGHTLY HELD OIL AND NATURAL GAS BY DRILLING SIDEWAYS AND FRACKING

Geologists have known for decades about vast deposits of oil and natural gas that are widely dispersed and trapped between compressed layers of shale rock formations found deep underground in many areas of the United States, including North Dakota, Texas, and Pennsylvania.

Until recently, it cost too much to extract such oil (called *tight oil*) and natural gas from shale rock. This situation has changed because of high oil prices along with the growing use of two extraction technologies (Figure 13.A). One is **horizontal drilling**, a method that involves drilling first vertically to a certain point, then turning the flexible shaft of the drill and drilling horizontally to gain access to the tightly held oil and natural gas deposits.

The second technology, called **hydraulic fracturing** or **fracking**, is then used to free this trapped oil and natural gas. In a fracking operation, high-pressure pumps blast a huge volume of a slimy mixture of water, sand, and various chemicals into a well to fracture the porous rock and create cracks. The sand becomes wedged in the cracks and keeps them open to allow the oil or natural gas and about half the water-chemical mixture to flow out and be pumped to the surface. This hazardous slurry contains a mix of naturally occurring salts, toxic heavy metals, and radioactive materials

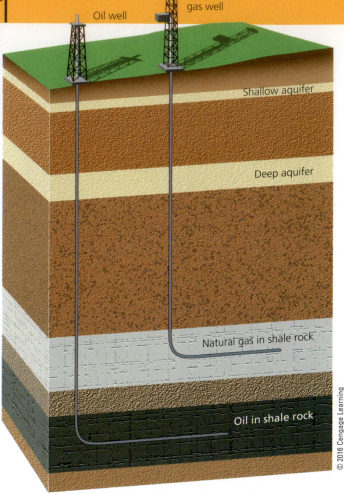

Oil well

Natural gas well

Shallow aquifer

Deep aquifer

Natural gas in shale rock

Oil in shale rock

© 2016 Cengage Learning

FIGURE 13.A Horizontal drilling and hydraulic fracturing, or fracking, are being used to release large amounts of oil and natural gas that are tightly held in shale rock formations.

leached from the rock, along with some potentially harmful drilling chemicals that natural gas companies are not required to identify or publicize. After fracking, the hazardous wastewater is injected into deep underground hazardous waste wells, stored in holding ponds, or cleaned up and reused in the fracking process.

Each day, energy companies drill and frack about 100 new oil or natural gas wells in parts of the United States. They drill a well, frack it several times, and then drill a new well and repeat the process. The growing use of these extraction technologies has brought about what some experts are calling a new era of oil and natural gas production in the United States. But such production also results in a lower net energy yield, compared to conventional oil and natural gas production, which means that market prices of oil and natural gas must remain high enough to make it profitable to produce oil and natural gas from these deposits.

Critical Thinking

Why do you think horizontal drilling allows better access to tightly held oil and natural gas deposits than does drilling vertically into such deposits?

Is the World Running Out of Crude Oil?

We use an astonishing amount of oil—currently, the lifeblood of most economies. Laid end to end, the roughly 32 billion barrels of crude oil used worldwide in 2012 would stretch to about 28 million kilometers (18 million miles)—far enough to reach to the moon and back about 37 times. (One barrel of oil contains 159 liters or 42 gallons of oil.)

How much crude oil is there? No one knows, although geologists have estimated the amounts existing in identified oil deposits. Available deposits are called **proven oil reserves**—deposits from which the oil can be extracted profitably at current prices using current technology. Proven oil reserves are not fixed. For example, a combination of oil extraction technologies (Science Focus 13.1) and high oil prices have made it profitable to extract light crude oil that is tightly held in layers of dense shale rock and was once inaccessible.

The world is not about to run out of crude oil in the near future. We can produce more conventional light crude oil from far offshore in deep ocean seabed deposits and from remote areas with severe climates such as the Arctic Circle. We can also rely more on unconventional heavy oil—a thick type of crude oil that does not flow as easily as light oil—from depleted oil wells and other sources. But the use of these sources of oil results in lower net energy yields, higher environmental impacts, higher production costs, and probably higher oil prices.

The 12 countries that make up the Organization of Petroleum Exporting Countries (OPEC) have about 72% of the world's proven crude oil reserves and thus are likely to control most of the world's conventional oil supplies for many years to come. In 2014, OPEC's members were Algeria, Angola, Ecuador, Iran, Iraq, Kuwait, Libya, Nigeria, Qatar, Saudi Arabia, United Arab Emirates, and Venezuela. According to BP and the U.S. Energy Information Administration (EIA), in 2012:

- the three countries with the world's largest shares of the world's proven crude oil reserves were Venezuela (18%), Saudi Arabia (16%), and Canada (10%, including heavy oil from tar sands, covered later in chapter);

- the three largest producers of crude oil were Saudi Arabia (13.3% of global production), Russia (12.8%), and the United States (9.6%, see the following Case Study); and

- the world's three largest consumers of crude oil—the United States (20% of global total), China (12%), and Japan (5%)—had, in order, only about 2.1%, 1%, and 0.003% of the world's proven crude oil reserves.

If global consumption of crude oil continues to grow at about 2% per year, we will need to expand global proven crude oil reserves with moderate to high net energy yields by an amount equal to Saudi Arabia's current reserves—the world's second largest—every 7 years, according to estimates by the DOE and the U.S. Geological Survey (USGS). Most oil geologists say this is highly unlikely.

CASE STUDY

Oil Production and Consumption in the United States

The United States gets about 87% of its commercial energy from fossil fuels, with 33% coming from crude oil (Figure 13.3, right). Currently, oil production in the United States, especially from shale rock deposits, is growing enough to make the United States the world's top producer of crude oil by 2017, according to a 2013 report by the International Energy Agency. (See Figure 15, p. S32, in Supplement 4 for a map of most of North America's proven and potential fossil fuel reserves; and Figure 2, p. S39, in

Supplement 5 for a graph of U.S. oil consumption between 1980 and 2012, with projections to 2040.)

Since 1982, oil consumption in the United States has greatly exceeded domestic production and thus the country has had to import some of the oil it uses. This gap has been narrowed by rising domestic production of tight oil. The United States imported about 40% of its crude oil in 2012, compared to 60% in 2005. According to the DOE, the United States spent an average of about $824,000 per minute on imported oil in 2012. The four largest suppliers of imported oil for the United States in 2012 were Canada (34%), Saudi Arabia (18%), Venezuela (12%), and Russia (10%).

Can the United States significantly reduce its dependence on oil imports by producing its own oil faster than its current oil supply is being depleted? Some say "yes" and project that domestic oil production will increase dramatically over the next few decades—especially from oil found in shale rock.

However, production of tightly held shale rock oil has a lower net energy yield, higher costs, and a higher environmental impact than does production from more accessible conventional oil deposits such as those in Saudi Arabia. In addition, experience indicates that production of oil from shale rock beds drops off about twice as fast as it does in most conventional oil fields.

Thus, according to a 2013 report by the International Energy Agency, oil produced from shale rock in the United States is likely to peak around 2020 and then decline for 2 to 3 decades as the richest deposits (called sweet spots) are depleted. This could impact the nation, but is not a major concern for oil producers who can develop a well for $7 million and make $35 million (or more) over the life of the well. The long-term problem for the United States is that it uses about 20% of the oil produced globally, produces about 10% of the world's oil, and has only 2% of the world's proven crude oil reserves.

Using Crude Oil Has Advantages and Disadvantages

Figure 13.5 lists major advantages and disadvantages of using crude oil as an energy resource. A critical and growing problem is that burning oil or any carbon-containing fossil fuel releases the greenhouse gas CO_2 into the atmosphere. According to most of the world's top climate scientists, this has played an important role in warming the atmosphere, which will contribute to projected climate change during this century, as discussed in Chapter 15.

Heavy Oil Is a Potential Resource with a High Environmental Impact

A potential supply of heavy oil is **shale oil**—oil that is *integrated within* bodies of shale rock (as opposed to *trapped between* layers of rock; see Science Focus 13.1). Producing

Trade-Offs

Conventional Oil

Advantages	Disadvantages
Ample supply for several decades	Water pollution from oil spills and leaks
Net energy yield is medium but decreasing	Environmental costs not included in market price
Low land disruption	Releases CO_2 and other air pollutants when burned
Efficient distribution system	Vulnerable to international supply interruptions

FIGURE 13.5 Using conventional light oil as an energy resource has advantages and disadvantages. **Questions:** Which single advantage and which single disadvantage do you think are the most important? Why? Do the advantages outweigh the disadvantages? Explain.

Photo: Richard Goldberg/Shutterstock.com

FIGURE 13.6 Heavy shale oil (right) can be extracted from oil shale rock (left).

it involves mining, crushing, and heating oil shale rock (Figure 13.6, left) to extract a mixture of hydrocarbons called *kerogen* that can be distilled to produce shale oil (Figure 13.6, right). Before the thick shale oil is pumped through a pressurized pipeline to a refinery, it must be heated to increase its flow rate and processed to remove sulfur, nitrogen, and other impurities.

CONSIDER THIS. . .

THINKING ABOUT Shale Oil Production

How do you think the process described here affects the net energy yield of shale oil? Explain.

About 72% of the world's estimated oil shale rock reserves are buried deep in rock formations located primarily under government-owned land in the U.S. states of Colorado, Wyoming, and Utah. However, the net energy yield for this energy resource is low (Figure 13.2) and the process has a large and harmful environmental impact. In 2011, the U.S. Bureau of Land Management stated that unless oil prices rise sharply, "there are no economically viable ways yet known to extract and process oil shale for commercial purposes."

A growing source of heavy oil is **tar sands**, or **oil sands**, which are a mixture of clay, sand, water, and a combustible organic material called *bitumen*—a thick, sticky, tar-like heavy oil with a high sulfur content. Northeastern Alberta in Canada has three-fourths of the world's tar sands resource in sandy soil under a vast area of remote boreal forest. If we include its conventional light oil and its heavy oil from tar sands, Canada has the world's third largest proven oil reserves.

The two big drawbacks to producing oil from tar sands are its low net energy yield (Figure 13.2) and its major harmful impacts on the land (Figure 13.7), air, water, wildlife, and climate. According to a 2014 Canadian government report, the largest producer of Canada's greenhouse gases is its energy sector, with over 70% of the emissions coming from its tar sands industry. And the volume of tar sands mining is expected to nearly double by 2021. Figure 13.8 lists the major advantages and disadvantages of producing heavy oil from oil shale rock and tar sands.

Since 2005, there has been a controversy over building the Keystone XL pipeline, which would allow Canada to send its heavy crude oil (bitumen) through the United States to Gulf Coast refineries. Keystone proponents argue that it would create thousands of jobs while adding to oil supplies, but opponents say it would result in less than 100 permanent jobs, once the pipeline is built. Opponents also contend that the pipeline will promote the development of the world's most environmentally harmful form of oil production, and that it would be built over a part of the Ogallala Aquifer (see p. 256), on which millions of Americans depend for irrigation and drinking water. Furthermore, they say, it would not boost U.S. oil reserves because most of the resulting oil would be sold overseas.

Proponents argue that if this pipeline is not built, Canada could build a pipeline to the refineries on its west coast and then, after refining it, ship most of its heavy oil to China. It might also ship much of its heavy oil to other refineries by truck and rail, which pipeline proponents say, would add to the risks of oil spills and higher greenhouse gas emissions. In response, opponents say that just one spill from the pipeline could permanently damage some part of the irreplaceable Ogallala Aquifer.

In 2014, this controversy was mounting, with heavy pressure being applied by both sides to decision makers in the U.S. government.

Producing heavy oil from tar sands has a high environmental impact.

Christopher Kolaczan/Shutterstock.com

FIGURE 13.7 Tar sands mining operation in Alberta, Canada.

Trade-Offs

Heavy Oils from Oil Shale and Tar Sand

Advantages	Disadvantages
Large potential supplies	Low net energy yield
Easily transported within and between countries	Releases CO_2 and other air pollutants when produced and burned
Efficient distribution system in place	Severe land disruption and high water use

© 2016 Cengage Learning

FIGURE 13.8 Using heavy oil from tar sands and from oil shale rock as an energy resource has advantages and disadvantages (**Concept 13.2**). **Questions:** Which single advantage and which single disadvantage do you think are the most important? Why? Do the advantages outweigh the disadvantages? Explain.

Photo: Christopher Kolaczan/Shutterstock.com

Natural Gas Is a Versatile and Widely Used Fuel

Natural gas is a mixture of gases of which 50–90% is methane (CH_4). It also contains smaller amounts of heavier gaseous hydrocarbons such as propane (C_3H_8) and butane (C_4H_{10}), and small amounts of highly toxic hydrogen sulfide (H_2S). This versatile fuel has a medium net energy yield (Figure 13.2) and is widely used for cooking, heating, and industrial purposes. It can also be used as a fuel for cars and trucks and for natural gas turbines used to produce electricity in power plants.

This versatility helps to explain why natural gas provided about 28% of the energy (Figure 13.3, right) and 27% of the electricity consumed in the United States in 2013 (up from 17% in 2003). It burns cleaner than oil and much cleaner than coal, and when burned completely, it emits about 30% less CO_2 than oil and about 50% less than coal.

Conventional natural gas is often found in deposits lying above deposits of conventional oil. It also exists in tightly held deposits between layers of shale rock and can be extracted through horizontal drilling and fracking (Science Focus 13.1). (See Figure 17, p. S34, in Supplement 4 for a map of U.S. natural gas shale rock deposits.)

When a natural gas deposit is tapped, propane and butane gases can be liquefied under high pressure and removed as **liquefied petroleum gas (LPG)**. LPG is stored in pressurized tanks for use mostly in rural areas not served by natural gas pipelines. The rest of the gas (mostly methane) is purified and pumped into pressurized pipelines for distribution across land areas.

Natural gas can also be transported across oceans, by converting it to **liquefied natural gas (LNG)** at a high pressure and at a very low temperature. This highly flammable liquid is transported in refrigerated tanker ships. At its destination port, it is heated and converted back to the gaseous state and then distributed by pipeline. The liquefying of natural gas greatly reduces its net energy yield and increases its cost.

In 2012, the three countries with world's largest proven natural gas reserves were Iran (with 18.0% of the global total), Russia (17.6%), and Qatar (13.4%), according to British Petroleum (BP) and the International Energy Agency. The United States and China had 4.5% and 1.7%, respectively, of the world's proven natural gas reserves. Japan has no significant natural gas reserves and depends heavily on imports of expensive LNG.

In 2012, the world's three largest producers of natural gas were the United States (20% of global production), Russia (18%), and Iran (5%), and the three largest consumers were the United States (using 22% of the global total), Russia (11%), and Iran (5%). (See Figure 2 in Supplement 5 for a graph showing U.S. natural gas consumption between 1980 and 2012, with projections to 2040.) Currently, the United States does not have to rely on natural gas imports because its production has been increasing rapidly, mostly because of the growing use of fracking in shale rock beds (Science Focus 13.1), which accounted for about 30% of U.S. natural gas production in 2012 (up from 2% in 2000).

The demand for natural gas in the United States is projected to more than double between 2010 and 2050. If much of this demand is met by increased production of natural gas from shale rock, the United States could continue to meet its needs for natural gas from domestic resources. However, U.S. natural gas reserves might have been overestimated. In 2011, the USGS cut its nationwide estimate of recoverable natural gas from shale rock by 50% and pointed out that natural gas production from shale rock tends to peak and drop off much faster than does production from conventional natural gas wells.

In addition, extracting and producing natural gas from shale rock reduces the net energy yield and, without effective regulation, can increase the harmful environmental impacts of production (see Case Study that follows). Figure 13.9 lists the advantages and disadvantages of using conventional natural gas as an energy resource.

Natural Gas Production and Fracking in the United States

The U.S. Energy Information Administration projects that, within a decade or two, at least 100,000 more natural gas wells will be drilled and fracked in the United States. If this projection holds, such wells could be producing 50% of the country's natural gas by 2040, up from 30% in 2012. Some analysts warn that without more monitoring and regulation of the entire natural gas production and distribution process, including fracking, the greatly increased production of natural gas (and oil) from shale rock could have several harmful environmental effects:

- Fracking requires enormous volumes of water. A typical fracking operation pumps about 18.9 million liters (5 million gallons) of water and chemicals underground. At least half of this water remains deep underground, permanently removed from the water cycle. In water-short areas this could help to reduce available surface water, deplete aquifers, degrade aquatic habitats, and diminish the availability of water for irrigation and other purposes. About half of the fracked wells in the United States are in water-stressed areas.

- Each fracked well produces huge volumes of hazardous wastewater that are brought to the surface along with the released natural gas. Many scientists warn that the potentially harmful chemicals in fracking slurries, including arsenic and naturally occurring radioactive elements such as radium, could contaminate groundwater and surface waters.

- Groundwater that is used in homes can become contaminated by natural gas leaking from faulty well casings and valves, and when the water is drawn from a tap, this natural gas can catch fire (Figure 13.10). In 2013, Rob Jackson and other Duke University environmental scientists studied 141 drinking water wells in parts of Pennsylvania and New York. They found that wells located near fracking sites were six times more likely to be contaminated with methane than were those located farther away from such sites. The researchers also concluded that most of the dissolved methane found in these wells was likely the result of inadequately sealed natural gas well pipes.

Trade-Offs

Conventional Natural Gas

Advantages	Disadvantages
Ample supplies	Low net energy yield for LNG
Versatile fuel	Production and delivery may emit more CO_2 and CH_4 per unit of energy produced than coal
Medium net energy yield	
Emits less CO_2 and other air pollutants than other fossil fuels when burned	Fracking uses and pollutes large volumes of water
	Potential groundwater pollution from fracking

© 2016 Cengage Learning

FIGURE 13.9 Use of conventional natural gas as an energy resource has advantages and disadvantages. **Questions:** Which single advantage and which single disadvantage do you think are the most important? Why? Do you think that the advantages outweigh the disadvantages? Explain.

Photo: Werner Muenzker/Shutterstock.com

- A 2014 study led by scientists Paul Shepson and Jed Sparks used aircraft to measure levels of the potent greenhouse gas methane in the air above natural gas wells and found that methane levels during the drilling phase were 100 to 1,000 times higher than the EPA had estimated.

- Leaks of methane, along with exhaust fumes from heavy truck traffic and electrical generators on fracking sites, have raised local levels of air pollution. For example, numerous fracking wells near the small Wyoming town of Pinedale have led to smog levels equivalent to those typical of Los Angeles, California.

- According to a 2012 study by the National Academy of Sciences and another study by the USGS, in recent years, one of the causes of hundreds of small earthquakes in 13 states has been the shifting of bedrock resulting from the high-pressure injection of fracking wastewater into deep underground hazardous waste wells. In some areas, the fracking process itself is suspected of causing small earthquakes. Such earthquakes could release hazardous wastewater into aquifers and cause breaks in the steel linings and cement seals of oil and gas well pipes. Between 2012 and 2014, the U.S. state of Ohio shut down some fracking operations and deep disposal wells after several small earthquakes occurred near these sites.

Producers point out that increased natural gas production from fracking has lowered U.S. natural gas prices and benefitted those who use natural gas (55% of all U.S. consumers). In addition, the natural gas fracking boom in Pennsylvania created about 18,000 jobs, and producers paid millions of dollars to land owners for signing leases to the gas under their land, along with royalties on the gas removed. And with more electricity generated by use of natural gas, the resulting drop in the use of coal to produce electricity has reduced emissions of CO_2 and other air pollutants.

Producers also argue that no groundwater contamination directly due to fracking has ever been recorded, mostly because the fracking takes place far below drinking-water aquifers (Figure 13.A). However, critics contend that natural gas producers have squelched numerous reports of drinking water contamination near fracking sites by offering financial settlements to people who make such claims. To receive payments, those people must promise not to reveal any information about the alleged contamination. In 2014, documents obtained from the state governments of Pennsylvania, Texas, Ohio, and West Virginia revealed hundreds of complaints since 2003 about contamination and diminished water flow from private water wells due to oil and natural gas production. Most of the confirmed examples of pollution involved methane gas in drinking water (Figure 13.10).

According to the U.S. Environmental Protection Agency (EPA), the natural gas and petroleum industry is the largest source of methane emissions into the atmosphere from hu-

FIGURE 13.10 Natural gas fizzing from this faucet in a Pennsylvania home can be lit like a natural gas stove burner. This began to happen in the area after an energy company drilled a fracking well, but the company denies responsibility. The homeowners have to keep their windows partly open year-round to keep the lethal and explosive gas from building up in the house.

man activities, followed by releases from livestock (mostly their belching), landfills, and coal mining. About two-thirds of global emissions of methane—a potent greenhouse gas—are caused by human activities, and emissions have been increasing since 2007, according to National Oceanic and Atmospheric Administration (NOAA).

Methane can also reach the air, and possibly aquifers, from the large number of wells abandoned by small producers that declared bankruptcy without sealing the wells properly and reclaiming the disturbed land. Such problems indicate that inspection and regulation of the entire natural gas production process by the state and federal governments is inadequate.

Currently, people who rely on aquifers and streams for their drinking water in areas undergoing rapid increases in shale gas production have little protection from pollution of their air and water supplies that might result from natural gas production. This is because, under political pressure from natural gas suppliers, Congress in 2005 excluded natural gas fracking from EPA regulations under the federal Safe Drinking Water Act. Other loopholes have also exempted natural gas production from parts of several other federal environmental laws, including the Clean Water Act, the Clean Air Act, and the National Environmental Policy Act.

Without stricter regulation and monitoring, drilling another 100,000 natural gas wells during the next 10–20 years will likely increase the risk of air and water pollution from the entire natural gas production process and its distribution by pipelines. This could cause a public backlash against this technology. To avoid this, some analysts call for **(1)** getting better data on the impacts of fracking and on leaks from well-casings, valves, storage tanks, compressors, and pipelines, **(2)** setting higher standards for quality control in the building of natural gas wells, **(3)** fixing leaks in the natural gas production and distribution system, and **(4)** pressuring Congress to revoke all exemptions from U.S. environmental laws for the natural gas industry.

Coal Is a Plentiful but Dirty Fuel

Coal is a solid fossil fuel formed from the remains of land plants that were buried and exposed to intense heat and pressure for 300–400 million years (Figure 13.11).

In 2013, coal-burning power plants (Figure 13.12) generated about 40% of the world's electricity, 39% of the electricity used in the United States (down from 51% in 2003), and 65% of that used in China. Coal is also burned in industrial plants to make steel, cement, and other products. In 2012, the three largest producers of coal were China (48% of global total), the United States (13%), and Australia (6%), and the three largest consumers were China (50%), the United States (12%), and India (8%). In 2013, China consumed roughly as much coal as the rest of the world combined. (See Figure 3, p. S40, in Supplement 5 for a graph showing global coal consumption between 1950 and 2012 and Figure 4, p. S40, for graphs showing coal consumption in the United States and China between 1965 and 2012.)

In 2012, the three countries with the world's largest proven coal reserves were the United States (28% of global total), Russia (18%), and China (13%). The USGS estimates that identified U.S. coal reserves (see Figures 15 and 16, pp. S32 and S33, in Supplement 4) could last about 250 years at the current consumption rate.

The problem is that coal is by far the dirtiest of all fossil fuels, starting with the mining of coal, which severely degrades land (see Figures 12.9 through 12.11, pp. 297–298) and pollutes water and air. Coal is mostly carbon but contains small amounts of sulfur, which are converted to the air pollutant sulfur dioxide (SO_2) when the coal burns.

Currently, China is the world's leading emitter of SO_2. These emissions contribute to the problem of acid precipitation and to serious human health problems.

The burning of coal also releases large amounts of black carbon particulates, or soot (Figure 13.13), and much smaller, fine particles of air pollutants such as toxic mercury. The fine particles can get past our bodies' natural defenses and into our lungs, causing various severe illnesses. According to a 2010 study by the Clean Air Task Force, fine-particle pollution in the United States, mostly from the older coal-burning power plants without the latest air pollution control technology, kills at least 13,000 people a year—an average of 3 people every 2 hours. In China, outdoor air pollution from the burning of coal contributes to 1.2 million premature deaths per year (137 per hour), according to a 2010 study by University of Washington scientists and the World Health Organization.

Coal-burning power and industrial plants are among the largest emitters of the greenhouse gas CO_2 (Figure 13.14), which is contributing to atmospheric warming and projected climate change (covered in Chapter 15). Coal combustion produces about 39% of global CO_2 emissions. China leads the world in such emissions, followed by the United States. Another problem with coal combustion is that it emits trace amounts of radioactive materials as well as toxic mercury into the atmosphere.

In addition, many coal-burning plants employ scrubbers to remove some of these pollutants before they leave the smokestacks, and this produces *coal ash* (Figure 13.12)—a toxic mix that can contain dangerous, indestructible chemical elements such as arsenic, lead, mercury, cadmium, and radioactive radium. It must be stored safely, essentially forever. In the United States, some of this ash is buried in landfills or in active or abandoned mines and some is sold for use in building products like concrete, cement, and wallboard. Some is also made into a wet slurry and stored in holding ponds. Some of these

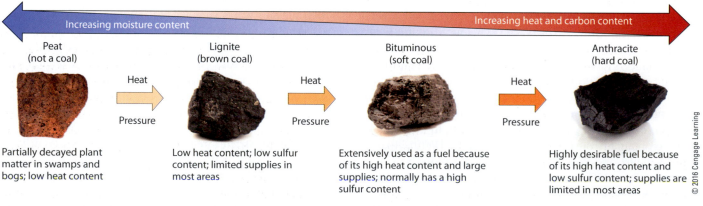

Increasing moisture content ← → **Increasing heat and carbon content**

Peat (not a coal) → Heat / Pressure → **Lignite** (brown coal) → Heat / Pressure → **Bituminous** (soft coal) → Heat / Pressure → **Anthracite** (hard coal)

Partially decayed plant matter in swamps and bogs; low heat content

Low heat content; low sulfur content; limited supplies in most areas

Extensively used as a fuel because of its high heat content and large supplies; normally has a high sulfur content

Highly desirable fuel because of its high heat content and low sulfur content; supplies are limited in most areas

© 2016 Cengage Learning

FIGURE 13.11 Over millions of years, several different types of coal have formed. Peat is a soil material made of moist, partially decomposed organic matter, similar to coal. It is not classified as a coal, although it is used as a fuel. These different major types of coal vary in the amounts of heat, carbon dioxide, and sulfur dioxide released per unit of mass when they are burned.

Left: Siim Sepp/Shutterstock. Left center: JIANG HONGYAN/Shutterstock. Right center: farbled/Shutterstock. Right: anat chant/Shutterstock.

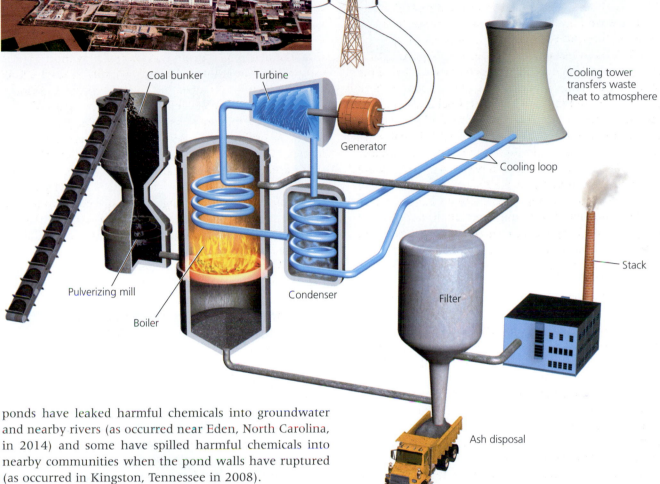

FIGURE 13.12 This power plant burns pulverized coal to boil water and produce steam that spins a turbine to produce electricity. The largest of the more than 550 coal-burning power plants in the United States, located in Indiana, burns three 100-car train-loads of coal per day. **Question:** Does the electricity that you use come from a coal-burning power plant?

Waste heat

Cooling tower transfers waste heat to atmosphere

Coal bunker

Turbine

Generator

Cooling loop

Stack

Pulverizing mill

Condenser

Filter

Boiler

Ash disposal

ponds have leaked harmful chemicals into groundwater and nearby rivers (as occurred near Eden, North Carolina, in 2014) and some have spilled harmful chemicals into nearby communities when the pond walls have ruptured (as occurred in Kingston, Tennessee in 2008).

In 2013, there were about 600 coal-ash ponds and 300 landfills storing coal ash in the United States. Some 45 coal-ash ponds were classified as having a "high hazard potential," according to the EPA. There is growing pressure to transfer wet coal ash slurry from holding ponds, which are viewed by some as disasters waiting to happen, to double-lined landfills. However, this would be costly and it would likely raise electricity prices.

The primary reason that coal is a relatively cheap way to produce electricity is that most of its harmful environmental and health costs are not included in the market price of coal-generated electricity. There are ways to include such costs, and they include requiring stricter pollution controls, regulating coal ash as a hazardous waste, storing it in dry, double-lined landfills, and putting a tax on CO_2 emissions from the burning of coal and other fossil fuels.

For decades, the U.S. coal and electric utility industries have fought such regulations and taxes, arguing that the potentially harmful chemicals found in coal ash are not at

high enough levels to harm humans. In addition, regulating coal ash as a hazardous waste and taxing CO_2 emissions would increase the cost of using coal as an energy resource and make it less competitive with other, cleaner ways to produce electricity such as with natural gas, wind, and solar cells.

Since 2008 the coal and electric utility industries have mounted a highly effective and well-financed publicity campaign built around the notion of *clean coal*, focused on the fact that coal can be burned more cleanly. However, critics point out that the concept of clean coal is misleading. Even with stricter air pollution controls, the burning of coal will always involve some emissions of health-damaging air pollutants and climate-changing CO_2. It will always create indestructible and hazardous coal ash, that must be stored safely, essentially forever, especially if scrubbers are used more widely to reduce air pollution from burning coal. And mining and transporting coal will

FIGURE 13.13 This coal-burning industrial plant produces large amounts of air pollution because it has inadequate air pollution controls.

Coal-fired electricity — 286%
Synthetic oil and gas produced from coal — 150%
Coal — 100%
Tar sand — 92%
Oil — 86%
Natural gas — 58%
Nuclear power fuel cycle — 17%
Geothermal — 10%

Compiled by the authors using data from U.S. Department of Energy.

FIGURE 13.14 Carbon dioxide emissions, expressed as percentages of emissions released by burning coal directly, vary with different energy resources. **Question:** Which of these produces more CO_2 emissions per kilogram: burning coal to heat a house, or heating with electricity generated by coal?

Trade-Offs

Coal

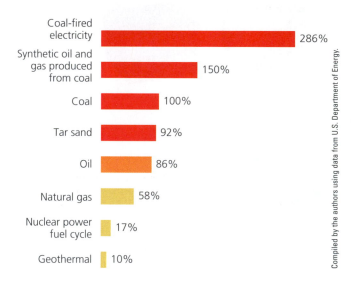

Advantages	Disadvantages
Ample supplies in many countries	Severe land disturbance and water pollution
Medium to high net energy yield	Fine particle and toxic mercury emissions threaten human health
Low cost when environmental costs are not included	Emits large amounts of CO_2 and other air pollutants when produced and burned

© 2016 Cengage Learning

FIGURE 13.15 Using coal as an energy resource has advantages and disadvantages. **Questions:** Which single advantage and which single disadvantage do you think are the most important? Why? Do you think that the advantages outweigh the disadvantages? Explain.

Photo: El Greco/Shutterstock.com

always involve disrupting land—in many cases, vast areas of land—and polluting water and air.

Because of increasing competition from cleaner-burning natural gas, few new coal-burning power plants are being built in the United States and some older ones are being closed. According to the U.S. Energy Information Agency, natural gas should overtake coal as the nation's largest source of electricity by the 2030s, and sooner if the government regulates greenhouse gas emissions from fossil fuel combustion. In response to these changes, U.S. coal producers are exporting increasing amounts of coal to the United Kingdom, the Netherlands, Brazil, South Korea, China, and other countries. Figure 13.15 lists the advantages and disadvantages of using coal as an energy resource.

13.3 WHAT ARE THE ADVANTAGES AND DISADVANTAGES OF USING NUCLEAR POWER?

CONCEPT 13.3 Nuclear power has a low environmental impact and a very low accident risk, but its use has been limited by a low net energy yield, high costs, fear of accidents, long-lived radioactive wastes, and its role in the spread of nuclear weapons technology.

How Does a Nuclear Fission Reactor Work?

To evaluate the advantages and disadvantages of nuclear power, we must know how a nuclear power plant and its accompanying nuclear fuel cycle work. A nuclear power plant is a highly complex and costly system designed to perform a relatively simple task: to boil water to produce steam that spins a turbine and generates electricity.

What makes nuclear power complex and costly is the use of a controlled nuclear fission reaction to provide the heat. (See p. 33 and Figure 14 (middle), p. S13, in Supplement 3.) The fission reaction takes place in a *reactor*. The most common reactors, called *light-water reactors* (LWRs, see Figure 13.16), produce 85% of the world's nuclear-generated electricity (100% in the United States).

Ron Chapple/Dreamstime.com

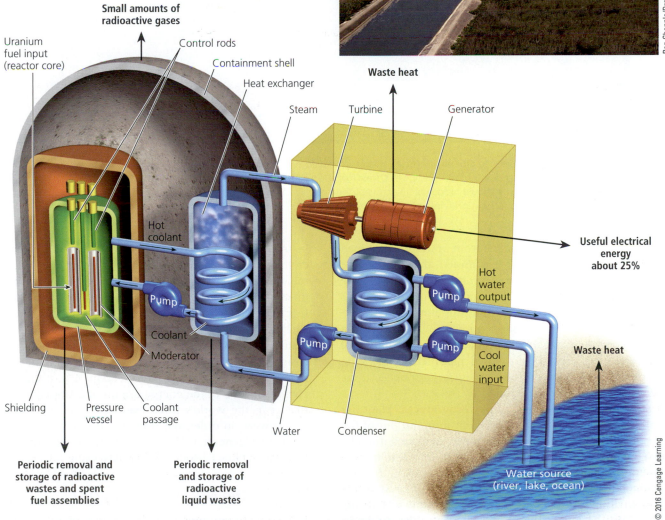

© 2016 Cengage Learning

FIGURE 13.16 This water-cooled nuclear power plant, with a pressurized water reactor, produces intense heat that is used to convert water to steam, which spins a turbine that generates electricity. **Question:** How does this plant differ from the coal-burning plant in Figure 13.12?

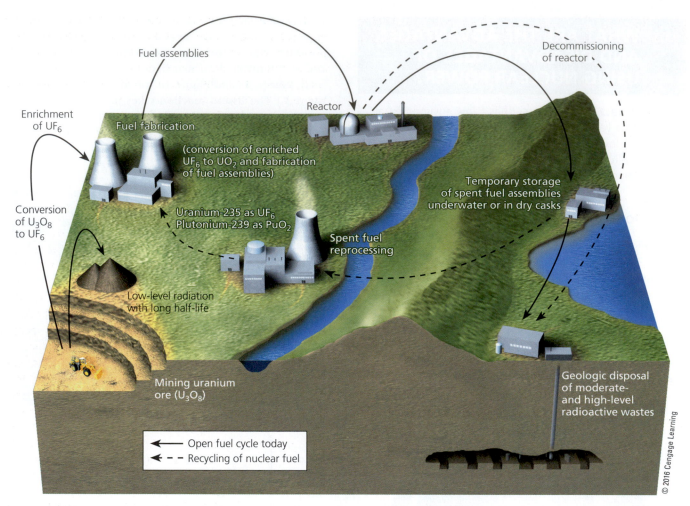

Fuel assemblies

Decommissioning of reactor

Reactor

Enrichment of UF₆

Fuel fabrication

(conversion of enriched UF₆ to UO₂ and fabrication of fuel assemblies)

Conversion of U₃O₈ to UF₆

Uranium-235 as UF₆
Plutonium-239 as PuO₂

Temporary storage of spent fuel assemblies underwater or in dry casks

Spent fuel reprocessing

Low-level radiation with long half-life

Mining uranium ore (U₃O₈)

Geologic disposal of moderate- and high-level radioactive wastes

⟶ Open fuel cycle today
⟵ - - Recycling of nuclear fuel

© 2016 Cengage Learning

FIGURE 13.17 Using nuclear power to produce electricity involves a sequence of steps and technologies that together are called the *nuclear power fuel cycle*. **Question:** Do you think the market price of nuclear-generated electricity should include all the costs of the nuclear power fuel cycle, in keeping with the full-cost pricing **principle of sustainability**? Explain.

The fuel for a reactor is made from uranium ore mined from the earth's crust. After it is mined, the ore must be enriched to increase the concentration of its fissionable uranium-235 by 1% to 5%. The enriched uranium-235 is processed into small pellets of uranium dioxide. Each pellet, about the size of an eraser on a pencil, contains the energy equivalent of about a ton of coal. Large numbers of the pellets are packed into closed pipes, called *fuel rods,* which are then grouped together in *fuel assemblies,* to be placed in the core of a reactor.

To regulate the amount of power produced, plant operators use *control rods,* moving them into and out of the reactor core to absorb more or fewer neutrons, thereby slowing or speeding the fission reaction. A *coolant,* usually water, circulates through the reactor's core to remove heat and keep the fuel rods and other reactor components from melting and releasing massive amounts of radioactivity into the environment. An LWR includes an emergency core cooling system as a backup to help prevent such meltdowns.

A *containment shell* made of thick, steel-reinforced concrete surrounds the reactor core. It is designed to help keep radioactive materials from escaping into the environment, in case there is an internal explosion or a core meltdown. It is also intended to protect the core from external threats such as tornadoes and plane crashes. These essential safety features help to explain why a new nuclear power plant costs as much as $10 billion and why that cost is on the rise. In 2013, nuclear power produced about 4% of the world's energy and 8% of the energy used in the United States. In that year, the world's three leading producers of nuclear power were, in order, the United States, France, and Russia. France generates 75% of its electricity and the United States generates 19% of its electricity using nuclear power.

What Is the Nuclear Fuel Cycle?

Building and running a nuclear power plant is only one part of the **nuclear fuel cycle** (Figure 13.17), which also includes the mining of uranium, processing and enriching

Conventional Nuclear Fuel Cycle

Advantages	Disadvantages
Low environmental impact (without accidents)	Low net energy yield
Emits 1/6 as much CO_2 as coal	High overall cost
Low risk of accidents in modern plants	Produces long-lived, harmful radioactive wastes
	Promotes spread of nuclear weapons

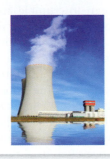

© 2016 Cengage Learning; Photo: Kletr/Shutterstock.com

FIGURE 13.18 Using the nuclear power fuel cycle (Figure 13.17) to produce electricity has advantages and disadvantages. **Questions:** Which single advantage and which single disadvantage do you think are the most important? Why? Do you think that the advantages outweigh the disadvantages? Explain.

the uranium to make fuel, using it in a reactor, safely storing the resulting highly radioactive wastes for thousands of years until their radioactivity falls to safe levels, and retiring the worn-out plant by taking it apart and storing its high- and moderate-level radioactive parts safely for thousands of years.

As long as a reactor is operating safely, the power plant itself has a fairly low environmental impact and a very low risk of an accident. However, considering the entire nuclear fuel cycle, the potential environmental impact increases.

In evaluating the safety, economic feasibility, net energy yield, and overall environmental impact of nuclear power, energy experts and economists caution us to look at the entire nuclear fuel cycle, not just the power plant itself. Figure 13.18 lists the major advantages and disadvantages of producing electricity by using the nuclear power fuel cycle (**Concept 13.3**).

Let's look more closely at some of the challenges involved in using nuclear power.

Dealing with Radioactive Nuclear Wastes Is a Difficult Scientific and Political Problem

The high-grade uranium fuel in a typical nuclear reactor lasts for 3–4 years, after which it becomes *spent*, or useless, and must be replaced. The spent-fuel rods are so thermally hot and highly radioactive that they cannot be simply thrown away. Researchers have found that 10 years after being removed from a reactor, a single spent-fuel rod assembly can still emit enough radiation to kill a person standing 1 meter (39 inches) away in less than 3 minutes.

After spent-fuel rod assemblies are removed from reactors, they are stored in *water-filled pools* (Figure 13.19, left). After several years of cooling and decay of some of their radioactivity, they can be transferred to *dry casks* made of heat-resistant metal alloys and concrete and filled with inert helium gas (Figure 13.19, right). These casks are licensed for 20 years and could last for 100 or more years—only a tiny fraction of the thousands of years that the radioactive waste must be safely stored. In 2005, a study by the U.S. National Academy of Sciences warned that the

U.S. Department of Energy/Nuclear Regulatory Commission

FIGURE 13.19 After 3 or 4 years in a reactor, spent-fuel rods are removed and stored in a deep pool of water contained in a steel-lined concrete basin (left) for cooling. After about 5 years of cooling, the fuel rods can be stored upright on concrete pads (right) in sealed dry-storage casks made of heat-resistant metal alloys and thick concrete. **Questions:** Would you be willing to live within a block or two of these casks or have them transported through the area where you live in the event that they were transferred to a long-term storage site? Explain. What are the alternatives?

intensely radioactive waste storage pools and dry casks at 68 of the country's 100 commercial nuclear reactors were especially vulnerable to sabotage or terrorist attack because they were stored outside of the heavily protected reactor containment buildings.

Spent nuclear fuel rods can be processed to remove radioactive plutonium, which can then be used as nuclear fuel or for making nuclear weapons, thus closing the nuclear fuel cycle (Figure 13.17). This reprocessing reduces the storage time for the remaining wastes from up to 240,000 years (longer than the current version of the human species has been around) to about 10,000 years.

However, reprocessing is very costly and produces bomb-grade plutonium that can be used by nations or terrorists to make nuclear weapons, as India did in 1974. This is mainly why the U.S. government, after spending billions of dollars, abandoned this fuel recycling approach in 1977. Currently, France, Russia, Japan, India, the United Kingdom, and China reprocess some of their nuclear fuel.

Most scientists and engineers agree in principle that deep burial in an underground repository is the safest and cheapest way to store high-level radioactive wastes for thousands of years. However, some scientists contend that it is not possible to demonstrate that this or any storage method will work for thousands of years.

Between 1987 and 2009, the DOE spent $12 billion on research and testing for a long-term underground nuclear waste storage site in the Yucca Mountain desert region of Nevada. In 2010, this project was abandoned for scientific and political reasons. A government panel is looking for alternative solutions and sites. Some geologists have suggested that nuclear waste might be stored in impermeable shale rock formations, like those used to produce oil and natural gas. Research will be needed to evaluate this option. Meanwhile these deadly wastes are building up, with about 78% stored in pools and 22% stored in dry casks (Figure 13.19).

Another radioactive waste problem arises when a nuclear power plant reaches the end of its useful life after about 40 to 60 years and must be *decommissioned,* or closed and torn down. Around the world, 285 of the 430 commercial nuclear reactors operating in 2014 will have to be decommissioned by 2025.

Eventually all nuclear plants will have to be dismantled and their high-level radioactive materials will have to be stored safely. Scientists have proposed three ways to do this. One strategy is to remove and store the highly radioactive parts in a permanent, secure repository. A second approach is to install a physical barrier around the plant and set up full-time security for 30–100 years, until the plant can be dismantled after its radioactivity has reached safer levels. These levels would still be high enough to require long-term safe storage of leftover materials.

A third option is to enclose the entire plant in a concrete and steel-reinforced tomb, called a *containment structure.* This is what was done with a reactor at Chernobyl, Ukraine, that exploded and nearly melted down in 1986, due to a combination of poor reactor design and human operator error. The explosion and the radiation released over large areas have killed hundreds and perhaps thousands of people and contaminated a vast area of land with long-lasting radioactive fallout. A few years after the containment structure was built, it began to crumble and leak radioactive wastes, due to the corrosive nature of the radiation inside the damaged reactor. The structure is being rebuilt at great cost and is unlikely to last even several hundred years.

Regardless of the method chosen, the high costs of retiring nuclear plants adds to the enormous costs of the nuclear power fuel cycle and reduces its already low net energy yield. Even if all the nuclear power plants in the world were shut down tomorrow, we would still have to find a way to protect ourselves from their high-level radioactive components for thousands of years.

Can Nuclear Power Help to Slow Projected Climate Change?

Nuclear power advocates contend that by increasing our use of nuclear power, we could greatly reduce CO_2 emissions that are contributing to climate change. Scientists point out that this is only partially correct. While nuclear plants are operating, they do not emit CO_2. However, during the 10 years that it typically takes to build a plant, especially in the manufacturing of huge quantities of construction cement, large amounts of CO_2 are emitted. Every other step in the nuclear power fuel cycle (Figure 13.17) also involves CO_2 emissions. Such emissions are much lower than those from coal-burning power plants (Figure 13.14) but they still contribute to atmospheric warming and projected climate change.

Experts Disagree about the Future of Nuclear Power

In the 1950s, researchers predicted that by the year 2000, at least 1,800 nuclear power plants would supply 21% of the world's commercial energy (25% of that in the United States) and most of the world's electricity. After almost 60 years of development, a huge financial investment, and enormous government subsidies, some 430 commercial nuclear reactors in 31 countries produced only 4% of the world's commercial energy and 15% of its electricity in 2012. In the United States, 100 licensed commercial nuclear power reactors generated about 8% of the country's overall energy and 19% of its electricity in 2012.

Globally in 2013, 15 new nuclear reactors were under construction (8 of them in China, which had 17 reactors in 2012). Another 156 reactors are planned (most of them in China), but even if they are completed after a decade or two, they will not begin to replace the 285 aging reactors that must be decommissioned by 2025. This helps to explain why the production of electricity from nuclear power has essentially leveled off since 1985 (see Figure 5, p. S40, in Supplement 5) and is now the world's slowest-growing form of commercial energy. In the United States, the amount of electricity produced from nuclear power leveled off between 2000 and 2010 and has dropped since then.

The future of nuclear power is a subject of debate. Critics argue that the most serious problem with the nuclear power fuel cycle is that it is uneconomical. They contend that the nuclear power industry could not exist without high levels of financial support from governments and taxpayers, because of the extraordinarily high cost of ensuring safety and the low net energy yield of the nuclear fuel cycle (Figure 13.2).

For example, the U.S. government has provided large research and development subsidies, tax breaks, and loan guarantees to the industry (with taxpayers accepting the risk of any debt defaults) for more than 50 years. It also assumes most of the financial burden of finding ways to store radioactive wastes. In addition, the government provides accident insurance guarantees (under the Price-Anderson Act passed by Congress in 1957), because insurance companies have refused to fully insure any nuclear reactor against the consequences of a catastrophic accident.

According to the nonpartisan Congressional Research Service, since 1948, the U.S. government has spent more than $95 billion (in 2011 dollars) on nuclear energy research and development (R & D)—more than four times the amount spent on R & D for solar, wind, geothermal, biomass, biofuels, and hydropower combined. Many analysts question the need for continuing such taxpayer support for nuclear power.

Because of the multiple built-in safety features, the risk of exposure to radioactivity from nuclear power plants in the United States and in most other more-developed countries is very low. However, several explosions and partial or complete meltdowns have occurred (see the following Case Study). These accidents have dampened public and investor confidence in nuclear power.

Another serious safety concern related to commercial nuclear power is the spread of nuclear weapons technology around the world. In the international marketplace, the United States and 14 other countries have been selling commercial and experimental nuclear reactors and uranium fuel-enrichment and purification technology for decades. Much of this information and equipment can be used to produce bomb-grade uranium and plutonium for use in nuclear weapons. Energy expert John Holdren pointed out that the 60 countries that have nuclear weapons or the knowledge to develop them have gained most of such information by using civilian nuclear power technology. Some critics see this as the single most important reason for not building more nuclear power plants.

Proponents of nuclear power argue that governments should continue funding research, development, and pilot-plant testing of potentially safer and less costly new types of reactors. The nuclear industry claims that hundreds of

Solutions

- Reactors must be built so that a runaway chain reaction is impossible.

- The reactor fuel and methods of fuel enrichment and fuel reprocessing must be such that they cannot be used to make nuclear weapons.

- Spent fuel and dismantled structures must be easy to dispose of without burdening future generations with harmful radioactive waste.

- Taking its entire fuel cycle into account, nuclear power must generate a net energy yield high enough so that it does not need government subsidies, tax breaks, or loan guarantees to compete in the open marketplace.

- Its entire fuel cycle must generate fewer greenhouse gas emissions than other energy alternatives.

© 2016 Cengage Learning

FIGURE 13.20 Some energy analysts say that any new generation of nuclear power plants should meet all of these five criteria. **Question:** Do you agree or disagree with these analysts? Explain.

new *advanced light-water reactors (ALWRs)* could be built in just a few years. ALWRs have built-in safety features designed to make meltdowns and releases of radioactive emissions almost impossible. The industry is also evaluating the development of smaller modular light-water reactors—about the size of a school bus—that could be built in a factory, delivered to a site, and installed underground.

Some scientists call for replacing today's uranium-based reactors with new ones to be fueled by thorium. They argue that such reactors would be much less costly and safer because they cannot melt down. Also, the nuclear waste they produce cannot be used to make nuclear weapons. China plans to explore this option.

Some analysts believe that, in order to be environmentally and economically acceptable, any new-generation nuclear technology should meet the five criteria listed in Figure 13.20. In the United States, even with considerable government subsidies and loan guarantees, most utility companies and money lenders are unwilling to take on the financial risk of building new nuclear plants of any design as long as electricity can be produced more cheaply with the use of natural gas and wind power (and solar cells if solar prices keep falling).

CASE STUDY

The Fukushima Daiichi Nuclear Power Plant Accident in Japan

A major accident occurred on March 11, 2011, at the Fukushima Daiichi Nuclear Power Plant on the northeast coast of Japan. A strong offshore earthquake that caused a severe tsunami (see Figure 12.19, p. 307) devastated coastal communities and triggered the worst nuclear accident since the Chernobyl explosion in 1986. An immense wave of seawater washed over the nuclear plant's protective seawalls and knocked out the circuits and backup diesel generators of the emergency core cooling systems for three of the reactors. Then, explosions (presumably from the buildup of hydrogen gas) blew the roofs off three of the reactor buildings and released radioactivity into the atmosphere and nearby coastal waters.

Evidence indicates that the cores of these three reactors suffered full meltdowns and contaminated a large area with low to moderate levels of radioactivity. In 2013, radioactivity from contaminated groundwater and from one of the plant's 1,000 wastewater storage tanks was leaking into the coastal waters near the plant. Officials say the costly cleanup and decommissioning of the damaged reactors will take several decades.

The tsunami killed about 8,200 people and coupled with the nuclear plant disaster created 270,000 refugees, with 3,000 refugees dying from medical problems and suicides. This event greatly damaged the confidence of Japanese citizens in the safety of nuclear power and led the government to shut down all of the country's nuclear reactors and consider abandoning its use of nuclear power. By 2013 Japan was relying more on imports of expensive liquefied natural gas (LNG) and much cheaper coal to produce electricity. In 2013, Japan was the world's second largest importer of coal (much of it from Australia) after China. Japan's serious nuclear power accident also prompted Germany, Switzerland, and Belgium to announce plans for phasing out nuclear power.

Is Nuclear Fusion the Answer?

Other proponents of nuclear power hope to develop **nuclear fusion**—a nuclear change at the atomic level in which the nuclei of two isotopes of a light element such as hydrogen are forced together at extremely high temperatures until they fuse to form a heavier nucleus, releasing energy in the process (see Figure 14, bottom, p. S13, in Supplement 3). Some scientists hope that controlled nuclear fusion will provide an almost limitless source of energy.

With nuclear fusion, there would be no risk of a meltdown or of a release of large amounts of radioactive materials, and little risk of the additional spread of nuclear weapons. Fusion power might also be used to destroy toxic wastes and to supply electricity for desalinating water and for decomposing water to produce hydrogen fuel as a very clean-burning energy source.

However, in the United States, after more than 50 years of research and a $25 billion investment (mostly by the government), controlled nuclear fusion is still in the laboratory stage. None of the approaches tested so far has produced more energy than they used. In 2006, the United States, China, Russia, Japan, South Korea, India, and the European Union agreed to spend at least $12.8

billion in a joint effort to build a large-scale experimental nuclear fusion reactor by 2026 to determine if it can produce a net energy yield at an affordable cost. By 2012, the estimated cost of this project had doubled and it was far behind schedule. Unless there is an unexpected scientific breakthrough, some skeptics say, "Nuclear fusion is the power of the future and always will be."

13.4 WHY IS ENERGY EFFICIENCY AN IMPORTANT ENERGY RESOURCE?

CONCEPT 13.4 Improvements in energy efficiency could save at least a third of the energy used in the world and up to 43% of the energy used in the United States.

We Use Energy Inefficiently

Energy efficiency is a measure of how much useful work we can get from each unit of energy we use. Improving energy efficiency means using less energy to provide the same amount of work.

You may be surprised to learn that roughly 84% of all commercial energy used in the United States is wasted (Figure 13.21). About 41% of this energy unavoidably ends up as low-quality waste heat in the environment

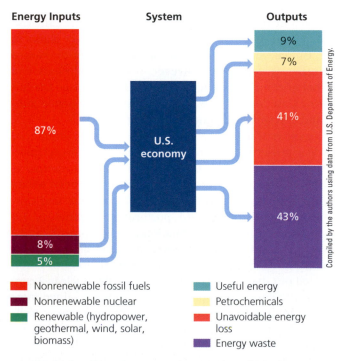

FIGURE 13.21 Flow of commercial energy through the U.S. economy. Only 16% of the country's high-quality energy ends up performing useful tasks. **Question:** What are two examples of unnecessary energy waste?

because of the degradation of energy quality imposed by the second law of thermodynamics. The other 43% is wasted unnecessarily, mostly due to the inefficiency of industrial motors, most motor vehicles, power plants, and numerous other energy-consuming devices.

Another reason for our inefficient use of energy is that many people live and work in poorly insulated, badly designed buildings that require excessive heating during cold weather and excessive cooling during hot weather. Also, many Americans live in ever-expanding suburban areas around large cities where they must depend on cars for getting around. Roughly three of every four Americans commute to work, mostly in energy-inefficient vehicles, and only 5% rely on more energy-efficient mass transit.

We waste large amounts of high-quality energy and money by relying on various energy-inefficient devices. One example is the huge *data centers* that process information flowing on the Internet and provide space for cloud-based data storage services. Typically they use only about 10% of the electrical energy they pull off of the grid. The other 90% ends up as low-quality heat that flows into the environment. Most of these centers run 24 hours a day at their maximum capacities regardless of the demand, and they require large amounts of energy for cooling. Another example is the *internal combustion engine*, which propels most motor vehicles. Only about 20% of the money people spend on gasoline provides them with transportation. The other 80% of the energy released by burning gasoline ends up as waste heat in the atmosphere.

By improving energy efficiency, we can gain numerous economic and environmental benefits (Figure 13.22). To most energy analysts, *it is the quickest, cleanest, and usually the cheapest way to provide more energy, reduce pollution and environmental degradation, and slow projected climate change and ocean acidification.*

We Can Improve Energy Efficiency in Industries and Utilities

Industry accounts for about 30% of the world's energy consumption and 33% of U.S. energy consumption, mostly for producing metals, chemicals, petrochemicals, cement, and paper. There are several ways in which industries can improve their energy efficiency and save money in the process (**Concept 13.4**).

One way to save energy is to use **cogeneration**, which involves using a *combined heat and power (CHP)* system. For example, the steam used for generating electricity in a CHP system can be captured and used again to heat the plant or other nearby buildings, rather than released into the environment as waste heat. The energy efficiency of these systems is 75–90%. Denmark leads the world by getting 52% of its electricity from CHP systems, compared to 8% in the United States.

Industries can also use more energy-efficient *electric motors*. Most of the widely used inefficient motors run only at full speed with their output throttled to match the task—somewhat like using one foot to push the gas pedal to the floorboard of your car and putting your other foot on the brake pedal to control its speed. They can be replaced with more energy-efficient variable-speed motors, which run at the minimum rate needed for each job.

Recycling materials such as steel and other metals is a third way for industry to save energy and money. For example, producing steel from recycled scrap iron uses 75% less high-quality energy than does producing steel from virgin iron ore, and it emits 40% less CO_2. A fourth way is to use more energy-efficient *compact fluorescent* and *LED lighting*.

Yet another way to save energy would be to redesign the software and cooling systems in electronic data processing systems. Servers could also be powered down when they are not in use and throttled back automatically when data traffic is light.

A growing number of major corporations are saving money by improving energy efficiency. For example, the CEO of Dow Chemical Company, which operates 165 manufacturing plants in 37 countries, estimated that between 1996 and 2009, energy efficiency improvements cost Dow about $1.1 billion, but resulted in savings of about $9.4 billion.

Solutions

Improving Energy Efficiency

- Prolongs fossil fuel supplies

- Reduces oil imports and improves energy security

- Very high net energy yield

- Low cost

- Reduces pollution and environmental degradation

- Buys time to phase in renewable energy

- Creates local jobs

© 2016 Cengage Learning

FIGURE 13.22 Improving energy efficiency has several benefits. **Questions:** Which two of these benefits do you think are the most important? Why?

Top: Dmitriy Raykin/Shutterstock.com. Center: V. J. Matthew/Shutterstock.com. Bottom: andrea lehmkuhl/Shutterstock.com.

Building a Smarter and More Energy-Efficient Electrical Grid

We can also improve the efficiency of transmitting electricity to industries and communities. Grid systems of high-voltage transmission lines carry electricity from power plants, wind farms (**Core Case Study**), and other electricity producers to users. In the United States, many energy experts place top priority on converting and expanding today's outdated electrical grid system into a regional and national *smart grid*. This would be an energy-efficient, digitally controlled, ultra-high-voltage (UHV) system with superefficient transmission lines. According to the DOE, building such a grid would cost the United States up to $800 billion over the next 20 years but would save the U.S. economy up to $2 trillion during that period. Much of this savings would go directly to consumers, because the system would include smart electric meters to help consumers use their energy as efficiently as possible.

Solar and wind power, used to generate electricity, are the fastest growing energy resources in the world and in the United States (see Figures 6 and 7, p. S41, in Supplement 5). However, their growth will be limited unless the new wind farms and solar cell power plants being built in sparsely populated areas can be connected to a smart grid. Such a grid could quickly adjust for a major power loss in one area by automatically bringing in electricity from other parts of the country. It would also make it easier for power companies and consumers to buy electricity produced from wind, solar, and other renewable forms of energy in areas where they are not directly available.

We Can Improve Energy Efficiency and Save Money in Transportation

Between 1973 and 2013, the average fuel economy for new cars and light trucks in the United States increased from 5 kilometers per liter, or *kpl* (11.9 miles per gallon, or *mpg*) to 10.6 kpl (24.9 mpg). The government goal is for such vehicles to get 23.3 kpl (54.5 mpg) by 2025. Energy experts such as Joseph Romm argue that all new cars and light trucks sold in the United States could get more than 43 kpl (100 mpg), using available technology, by 2040, and they call for such government standards. Fuel economy standards for new vehicles in Europe, Japan, China, and Canada are much higher than are those in the United States.

One reason why many consumers buy large, inefficient motor vehicles is that they do not realize that gasoline costs them much more than the price they pay at the pump. The *hidden costs* not included in the market price of gasoline include government subsidies and tax breaks for oil companies, car manufacturers, and road builders; costs of pollution control and cleanup; and higher medical bills and health insurance premiums resulting from illnesses caused by air and water pollution from the production and use of motor vehicles.

Consumers pay these hidden costs, but not at the gas pump. According to a study by the International Center for Technology Assessment, the hidden costs of gasoline for U.S. consumers are about $3.18 per liter ($12 per gallon). Thus, when gas costs $3 a gallon, U.S. consumers are really paying about $15 a gallon.

One way to include more of these hidden costs in the market price, and thus implement the full-cost pricing **principle of sustainability**, is through higher gasoline taxes, which are widely used in Europe but are politically unpopular in the United States. Some analysts call for increasing U.S. gasoline taxes and reducing payroll and income taxes to balance such increases, thereby relieving consumers of any additional financial burden. Another way for governments to encourage higher energy efficiency in transportation is to give consumers significant tax breaks or other economic incentives to encourage them to buy more fuel-efficient vehicles.

Other ways to save energy and money in transportation include building or expanding mass transit systems within cities, constructing high-speed rail lines between cities (as is done in Japan, much of Europe, and China), and carrying more freight by rail instead of in heavy trucks. Another approach is to encourage bicycle use by building bike lanes along highways and city streets.

Energy-Efficient Vehicles Are Available

There is growing interest in developing superefficient, ultralight, and ultrastrong cars using existing technology. One of these vehicles is the energy-efficient, gasoline-electric *hybrid car* (Figure 13.23, left). These cars have a small, traditional gasoline-powered engine and a battery-powered electric motor used to provide the energy needed for acceleration and hill climbing. The most efficient current models of these cars get a combined city/highway mileage of up to 21 kpl (50 mpg) and emit about 65% less CO_2 per kilometer driven than do comparable conventional cars.

Another option is the *plug-in hybrid electric vehicle* (Figure 13.23, right). Such vehicles can travel 48–97 kilometers (30–60 miles) on electricity alone. Then the small gasoline motor kicks in, recharges the battery, and extends the driving range to 600 kilometers (370 miles) or more. The battery can be plugged into a conventional 110-volt outlet and fully charged in 6 to 8 hours, or in a much shorter time using a 220-volt outlet. Still another option is an all-electric vehicle that runs on a battery only. The problem for the average consumer is that the prices on hybrid, plug-in hybrid, and all-electric cars are high because of the high cost of their batteries.

According to a DOE study, replacing most of the current U.S. vehicle fleet with plug-in hybrid vehicles over 3 decades would cut U.S. oil consumption by 70–90%, eliminate the need for costly oil imports, save consumers money, and reduce CO_2 emissions by 27%. If the batteries in these cars were recharged mostly by electricity gener-

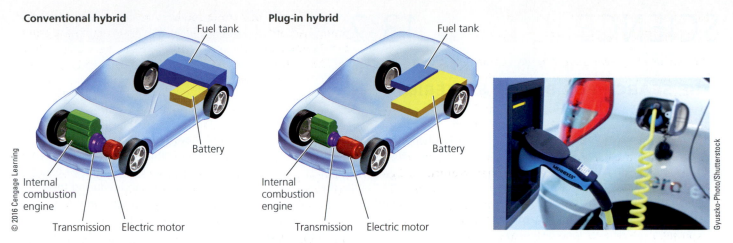

Conventional hybrid | Plug-in hybrid

Fuel tank

Battery

Internal combustion engine

Transmission   Electric motor

Fuel tank

Battery

Internal combustion engine

Transmission   Electric motor

© 2016 Cengage Learning

Gyuszko-Photo/Shutterstock

FIGURE 13.23 **Solutions:** A *conventional gasoline-electric hybrid* vehicle (left) is powered mostly by a small internal combustion engine with an assist from a strong battery. A *plug-in hybrid electric* vehicle (right) typically has a smaller internal combustion engine with a second and more powerful battery that can be plugged into a 110-volt or 220-volt outlet and recharged (see photo). An all-electric vehicle (not shown) runs completely on a rechargeable battery. **Question:** Would you buy one of these vehicles? Explain.

ated by renewable resources such as wind (**Core Case Study**), solar cells, or hydroelectric power, U.S. emissions of CO_2 would drop by 80–90%. This would help to reduce projected climate change and save thousands of lives by reducing air pollution from motor vehicles and from coal-burning power plants.

The key to greatly increasing the use of hybrid, plug-in hybrid, and all-electric motor vehicles is to ramp up research and development of improved, more affordable batteries (Science Focus 13.2). Another important factor will be to build a network of recharging stations in many convenient locations within and between communities.

A future stage in the development of superefficient cars might be an electric vehicle that uses a *fuel cell*—a device that uses hydrogen gas (H_2) as a fuel to produce electricity when it reacts with oxygen gas (O_2) in the atmosphere and emits harmless water vapor. Fuel cells are at least twice as efficient as internal combustion engines, have no moving parts, and require little maintenance. Their H_2 fuel is usually produced with the use of electricity, and if such electricity could be generated by wind or solar power, the widespread use of fuel cells would drastically reduce air pollution and CO_2 emissions that contribute to projected climate change—two of the world's most serious environmental, economic, and health problems. However, in the short run, such cars will probably be very expensive, in part because H_2 has a negative net energy yield, as discussed later in this chapter. **GREEN CAREER: Fuel-cell technology**

We can also improve a vehicle's fuel efficiency by reducing its weight. One way to do this is to make car bodies out of *ultralight* and *ultrastrong* composite materials such as fiberglass, carbon fiber, hemp-fiber, and graphene. The current cost of making such car bodies is high, but mass production would likely bring these costs down.

We Can Design Buildings That Save Energy and Money

We know how to make a transition to living and working in more sustainable buildings over the next few decades. For example, if a building is oriented to face the sun, it can get more of its heat from solar energy, which can cut heating costs by up to 20%. This is a simple application of the solar energy **principle of sustainability** that people have been using for centuries. Heating costs can be cut by as much as 75% when the building is well insulated and airtight.

Green architecture, based on energy-efficient, resource-efficient, and money-saving designs, employs several old and new technologies. Along with natural lighting, direct solar heating, insulated windows, and energy-efficient appliances and lighting, it makes use of solar hot water heaters, electricity from solar cells, windows that darken automatically to deflect heat from the sun, thin sheets of aerogel insulation, and recycling of wastewater. Some homes and urban buildings also have *living roofs,* or *green roofs,* covered with specially designed soil and vegetation (Figure 13.24). Such a roof can reduce the costs of cooling and heating a building by absorbing heat from the summer sun and helping to insulate the structure and retain heat in the winter. **GREEN CAREERS: Sustainable environmental design and architecture**

Superinsulation is very important in energy-efficient design. A house can be so heavily insulated and airtight that heat from direct sunlight, appliances, and human bodies can warm it with little or no need for a backup heating system, even in extremely cold climates. Superinsulated houses in Sweden use 90% less energy for heating and cooling than do typical American homes of the same size. One example of superinsulation is straw-bale con-

THE SEARCH FOR BETTER BATTERIES

The major obstacle standing in the way of wider use of plug-in hybrid-electric and all-electric vehicles is the need for an affordable, small, light-weight, and easily rechargeable car battery that can store enough energy to power a vehicle over long distances.

One promising type of battery is the *lithium-ion battery*, commonly used in laptop computers and cell phones. These batteries are light (because lithium is the lightest solid chemical element) and can pack a lot of energy into a small space. They can be hooked together to power hybrid, plug-in hybrid, and all-electric motor vehicles.

Much of the current research on batteries involves nanotechnology (see Science Focus 12.1, p. 300). For example, researchers at the Massachusetts Institute of Technology (MIT) have developed a new type of lithium-ion battery using such technology. It is less expensive and can be charged 40 times faster than the batteries that are used to power many of today's hybrid vehicles.

MIT researcher Angela Belcher is working on another new type of battery. She has genetically engineered a virus that can coat itself with electricity-conducting materials to form a tiny nanowire. Belcher is trying to find ways to link these wires up to form the components of a battery far more compact and powerful than any yet developed. Such viral batteries would essentially grow themselves without producing the toxic wastes often associated with battery production and recycling.

Scientists have also developed *ultra-capacitors*, which are small batteries that quickly store and release large amounts of electrical energy, thus providing the power needed for quick acceleration. They can be recharged in minutes, can hold a charge much longer than conventional batteries, and don't have to be replaced as frequently as conventional batteries. Researchers are working on increasing their energy storage capacity and reducing their costs per unit of energy produced. A promising candidate is a graphene (see Science Focus 12.2, p. 302) supercapacitor with an energy storage capacity comparable to that of a long-life battery.

If any one or a combination of these or other new battery technologies can be mass-produced at an affordable cost, plug-in hybrid and all-electric vehicles could take over the car and truck market within a few decades, which would greatly reduce air pollution and climate-changing CO_2 emissions.
GREEN CAREER: Battery engineer

Critical Thinking

How might your life change if one or more of the new battery technologies discussed above become a reality?

---

struction, in which a house's walls are built of straw bales that are covered on the inside and outside with adobe (Figure 13.25). Such walls can have insulating values of 2 to 6 times those of conventional walls. (See the online guest essay by Nancy Wicks on this topic.)

Green building certification standards have been adopted in 21 countries, thanks to the efforts of the World Green Building Council. Between 1999 and 2012, the U.S. Green Building Council's Leadership in Energy and Environmental Design (LEED) program has awarded silver, gold, and platinum standard certificates to more than 44,200 U.S. buildings that meet certain efficiency and environmental standards—the largest number by far of any country. The Sustainable Building Coalition has proposed that we go further and evaluate the sustainability of buildings over their lifetimes according to measurements of energy use, water use, CO_2 emissions, outputs of solid and hazardous wastes, thermal comfort, and indoor air quality.

We Can Save Money and Energy in Existing Buildings

Here are some ways to save energy and money in existing buildings:

- *Insulate the building and plug leaks.* About one-third of the heated air in typical U.S. homes and other buildings escapes through holes, cracks, and single-pane windows (Figure 13.26). During hot weather, these windows and cracks let heat in, which increases the need for air conditioning.

- *Use energy-efficient windows.* They can cut heat losses by two-thirds in winter, and cut summer cooling costs.

- *Stop other heating and cooling losses.* Seal leaky heating and cooling ducts in attics and unheated basements.

- *Heat houses and other buildings more efficiently.* In order, the most energy-efficient ways to heat indoor space or to conserve indoor heat are: superinsulation (including plugging leaks); a geothermal heat pump that transfers

FIGURE 13.24 Green roof on Chicago's City Hall.

FIGURE 13.25 Solutions: Energy-efficient, Victorian-style straw-bale house in Crested Butte, Colorado, during construction (left) and after it was completed (right).

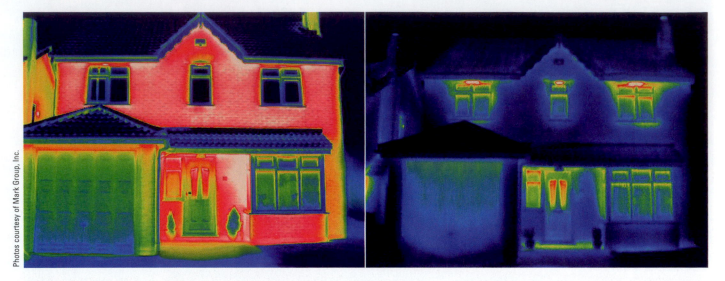

FIGURE 13.26 A *thermogram*, or infrared photo, of a house in Great Britain before it was well insulated (left) and after (right). (Red hues indicate heat loss.) Many homes are so full of leaks that their heat loss in cold weather and heat gain in hot weather are equivalent to what would be lost through a large, window-sized hole in a wall of the house. **Question:** How do you think the place where you live would compare to the house on the left in terms of heat loss?

heat stored from underground into a home; passive solar heating; a high-efficiency, conventional heat pump (in warm climates only); and a high-efficiency natural gas furnace.

- *Heat water more efficiently.* One approach is to use a roof-mounted solar hot water heater. Another option is a *tankless instant water heater* fired by natural gas or LPG (but not an electric heater, which is inefficient) that can provide hot water for as long as it is needed.

- *Use energy-efficient appliances.* A refrigerator with its freezer in a drawer on the bottom uses about half as much energy as one with the freezer on top or on the side, which allows dense cold air to flow out quickly whenever the door is opened. Microwave ovens use 25–50% less electricity than electric stoves do for cooking and 20% less than convection ovens use. Clothes dryers with moisture sensors cut energy use by 15%. Front-loading clothes washers use 55% less energy and 30% less water than top-loading models use and cut operating costs in half.

- *Stop using the standby mode.* According to the DOE, keeping TVs and other electronic devices on standby when they are not being used consumes about 10% of the electricity used by a typical household. Consumers can reduce their energy use and their monthly power bills by plugging their standby electronic devices into smart power strips that cut off power to a device when it detects that the device has been turned off.

- *Use energy-efficient computers.* According to the EPA, if all computers sold in the United States met its Energy Star requirements, consumers would save $1.8 billion a year in energy costs and reduce greenhouse gas emissions by an amount equal to that of taking about 2 million cars off the road.

- *Use energy-efficient lighting.* As incandescent lightbulbs are being phased out, homeowners, colleges and universities, and businesses are switching to more energy-efficient and longer-lasting compact fluorescent light (CFL) and LED bulbs. LED bulbs use 85% less energy, save consumers money, and last up to 25 times as long as incandescent bulbs and 4 times as long as CFLs. When LED bulbs first came out, they cost $100 a bulb. In 2014, prices were down to $3–$6 a bulb and still falling. The DOE estimates that rapid adoption of LED bulbs over the next 20 years in the United States would save consumers money and eliminate the need to build 40 new power plants. Another way to make lighting more efficient is to use automatic motion sensors to turn lights on and off as needed.

Figure 13.27 summarizes several ways in which you can cut your energy use and save money in the place where you live.

Why Are We Still Wasting So Much Energy and Money?

Considering its impressive array of benefits (Figure 13.22), why is there so little emphasis on improving energy efficiency? One reason is that energy resources such as fossil fuels and nuclear power are artificially cheap, primarily because of the government subsidies and tax breaks they

Attic
- Hang reflective foil near roof to reflect heat.
- Use house fan.
- Be sure attic insulation is at least 30 centimeters (12 inches).

Bathroom
- Install water-saving toilets, faucets, and shower heads.
- Repair water leaks promptly.

Kitchen
- Use microwave rather than stove or oven as much as possible.
- Run only full loads in dishwasher and use low- or no-heat drying.
- Clean refrigerator coils regularly.

Outside
Plant deciduous trees to block summer sun and let in winter sunlight.

Other rooms
- Use compact fluorescent lightbulbs or LEDs and avoid using incandescent bulbs.
- Turn off lights, computers, TV, and other electronic devices when they are not in use.
- Use high-efficiency windows; use insulating window covers and close them at night and on sunny, hot days.
- Set thermostat as low as you can in winter and as high as you can in summer.
- Weather-strip and caulk doors, windows, light fixtures, and wall sockets.
- Keep heating and cooling vents free of obstructions.
- Keep fireplace damper closed when not in use.
- Use fans instead of, or along with, air conditioning.

Basement or utility room
- Use front-loading clothes washer. If possible run only full loads with warm or cold water.
- If possible, hang clothes on racks for drying.
- Run only full loads in clothes dryer and use lower heat setting.
- Set water heater at 140°F if dishwasher is used and 120°F or lower if no dishwasher is used.
- Use water heater thermal blanket.
- Insulate exposed hot water pipes.
- Regularly clean or replace furnace filters.

© 2016 Cengage Learning

FIGURE 13.27 **Individuals matter:** You can save energy and money where you live. **Questions:** Which of these things do you already do? Which ones could you do tomorrow?

receive and because their market prices do not include the harmful environmental and health costs of producing and using them. According to conservative estimates by the International Energy Agency and the Global Subsidies Initiative, in 2011, governments around the world spent more than $620 billion on subsidies for the highly profitable fossil fuels industry compared to $88 billion on subsidies for renewable energy. This distortion of the energy marketplace violates the full-cost pricing **principle of sustainability**.

Another reason for continuing energy waste is that there are too few long-lasting government tax breaks, rebates, low-interest long-term loans, and other economic incentives for investing in energy efficiency. A third reason is that some governments have not put a high priority on educating the public about the environmental and money-saving advantages of improving energy efficiency.

In 2011, when Japan was getting 30% of its electricity from nuclear power, an earthquake and tsunami shut down three of its nuclear reactors, and eventually all of them were closed, at least temporarily (Case Study, p. 331). Within three years, Japan was able to replace half of its lost nuclear power by reducing its use of electricity through conservation and by improvements in energy efficiency. This is an important lesson for the United States, China, and other countries.

GOOD NEWS

13.5 WHAT ARE THE ADVANTAGES AND DISADVANTAGES OF USING RENEWABLE ENERGY RESOURCES?

CONCEPT 13.5 By using a mix of renewable energy resources, we can satisfy our energy needs while drastically reducing pollution, greenhouse gas emissions, and biodiversity losses.

We Can Use Renewable Energy for Many Purposes

The lesson from one of nature's three **scientific principles of sustainability** is to *rely mostly on solar energy*. We can get renewable solar energy directly from the sun or indirectly from wind (**Core Case Study**), flowing water, and biomass. Another form of renewable energy is geothermal energy, or heat from the earth's interior.

Studies show that with increased and consistent government backing in the form of research and development funds, subsidies, and tax breaks, renewable energy could provide 20% of the world's electricity by 2025 and 50% by 2050. In 2012, thirteen countries got more than 30% of their electricity from renewable energy, compared to 13% in the United States. In that year, the National Renewable Energy Laboratory projected that, with a crash program, the United States could get 50% of its electricity from renewable energy sources by 2050.

China is by far the world's largest user of coal and leads the world in climate-changing CO_2 emissions. China also has the world's largest capacity for electricity from wind power and solar cells and plans to become the largest user and seller of wind turbines and solar cells—projected to be two of the world's fastest growing businesses over the next few decades. China's goal is to greatly expand its production of electricity from renewable wind, sun, and flowing water (hydropower) to help reduce its use of coal and the resulting outdoor air pollution that kills about 1.2 million of its citizens each year.

According to a 2014 report by the U.S. National Science Foundation, in 2012, China spent $60 billion on clean energy investments compared to $20 billion spent by the United States and $20 billion spent by the European Union. However, between 2005 and 2013, China increased its production of coal by 1 billion metric tons—an amount that took the United States 150 years to produce, according to energy expert Vaclav Smil. And by 2020, China plans to produce another billion metric tons of coal.

Engineer, renewable energy advocate, and National Geographic Emerging Explorer Ibrahim Togola is promoting the use of renewable energy resources like solar, wind, and biofuel to produce electricity in rural areas of his native country of Mali, Africa. His goal is to reduce deforestation caused by reliance on wood and charcoal for 80% of the country's energy supply and contribute to wealth creation in the country by promoting entrepreneurship. He has also helped villagers establish tree-planting programs to help replenish cleared forests.

If renewable energy is so great, why does it provide only 8% of the world's energy (Figure 13.3) and 5% of the energy used in the United States? There are four major reasons. *First,* since 1950, government tax breaks, subsidies, and funding for research and development of renewable energy resources have been much lower than those for fossil fuels (especially oil) and nuclear power, although subsidies and tax breaks for renewables have increased in recent years.

Second, although government subsidies and tax breaks for fossil fuels and nuclear power have essentially been guaranteed for many decades, those for renewable energy in the United States have to be renewed by Congress every few years. This measure is mostly a result of political pressure from the fossil fuel industry trying to head off competition from these rapidly developing technologies. The resulting financial uncertainty makes it risky for entrepreneurs, companies, and homeowners to invest in renewable energy.

Third, the prices we pay for nonrenewable fossil fuels and nuclear power do not include most of the harmful environmental and human health costs of producing and using them. This helps to shield them from free-market competition with renewable sources of energy.

Fourth, history shows that it has typically taken about 50–60 years to make the transition from one dominant fuel to another, such as from wood to coal and from coal to oil and natural gas. Renewable wind and solar energy are the world's fastest growing sources of energy, but it will likely take decades for them to supply 25% or more of the world's energy or electricity.

Fossil fuel and utility companies fear losing business to wind- and solar-powered electricity, and they argue that both of these energy technologies are too expensive. However, the prices for electricity produced by the wind and sun are falling very rapidly because of the development of new technologies and cost savings from mass production. **GOOD NEWS**

We Can Heat Buildings and Water with Solar Energy

A building that has enough access to sunlight can get all or most of its heat through a **passive solar heating system** (Figure 13.28, left). Such a system absorbs and stores heat from the sun directly within a well-insulated structure. Water tanks and walls and floors of concrete, adobe, brick, or stone can store much of the collected solar energy as heat and release it slowly throughout the day and night. In a passively heated building, a small backup heating system such as a vented natural gas or propane heater can be used, but is usually not necessary in a well-designed and well-insulated building. (See the online guest essay by Nancy Wicks on this topic.)

An **active solar heating system** (Figure 13.28, right) captures energy from the sun by pumping a heat-absorbing

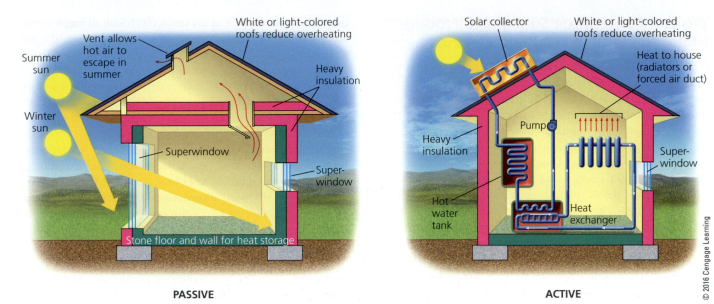

FIGURE 13.28 **Solutions:** Passive (left) and active (right) solar home heating systems.

Trade-Offs

Passive or Active Solar Heating

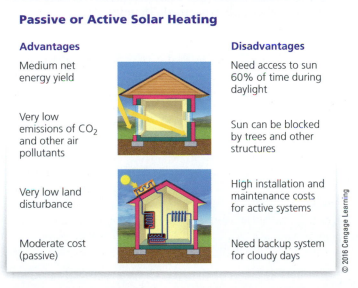

Advantages	Disadvantages
Medium net energy yield	Need access to sun 60% of time during daylight
Very low emissions of CO_2 and other air pollutants	Sun can be blocked by trees and other structures
Very low land disturbance	High installation and maintenance costs for active systems
Moderate cost (passive)	Need backup system for cloudy days

FIGURE 13.29 Heating a house with a passive or active solar energy system has advantages and disadvantages. **Questions:** Which single advantage and which single disadvantage do you think are the most important? Why? Do you think that the advantages outweigh the disadvantages?

fluid (such as water or an antifreeze solution) through special collectors, usually mounted on a roof or on special racks to face the sun. Some of the collected heat can be used directly. The rest can be stored in large insulated containers filled with gravel, water, clay, or a heat-absorbing chemical, and used as needed.

Rooftop active solar collectors are used to heat water in many homes and apartment buildings. With systems that cost the equivalent of as little as $200, about one in ten houses and apartment buildings in China use the sun to

provide hot water. Once the fairly low initial cost is paid, the hot water is heated for free. In Cairo, Egypt, urban planner and Emerging National Geographic Explorer Thomas Culhane has established a nongovernmental organization that works with the residents of the poorest neighborhoods to install rooftop solar water heaters. According to the UN Development Programme, solar water heaters could be used to provide half of the world's hot water.

Passive and active solar systems can be used to heat new homes in areas with adequate sunlight. (See Figures 19 and 20, pp. S36 and S37, in Supplement 4 for maps showing solar energy availability throughout the world and in the United States, respectively.) Figure 13.29 lists the major advantages and disadvantages of using passive or active solar systems for heating buildings.

We Can Cool Buildings Naturally

Direct solar energy works against us when we want to keep a building cool, but we can use indirect solar energy (mainly wind) to help cool buildings. For example, we can open windows to take advantage of breezes and use fans to keep the air moving. When there is no breeze, superinsulation and high-efficiency windows help to keep hot air outside. Here are three other ways to keep cool:

- Block the high summer sun with shade trees, broad overhanging eaves, window awnings, or shades (Figure 13.28, top).

- In warm climates, use a light-colored roof to reflect as much as 90% of the sun's heat (compared to only 10–15% for a dark-colored roof), or use a living or green roof.

- Use geothermal heat pumps to pump cool air from underground into a building during summer.

We Can Concentrate Sunlight to Produce High-Temperature Heat and Electricity

One of the problems with direct solar energy is that it is dispersed. **Solar thermal systems**, also known as *concentrated solar power* (CSP), use different methods to collect and concentrate solar energy in order to boil water and produce steam for generating electricity. These systems are used mostly in desert areas with ample sunlight.

One such system uses troughs of curved collectors that concentrate solar energy and use it to heat synthetic oil in a pipe that runs through the center of each trough (Figure 13.30, left). This concentrated heat—as hot as 400°C (750°F)—is used to boil water and produce steam that powers a turbine that drives a generator to produce electricity. Another system (Figure 13.30, right) uses an array of computer-controlled mirrors to track the sun and focus its energy on a central power tower to provide enough heat to boil water that is used to produce electricity. The heat produced by either of these systems can also be used to melt a certain kind of salt stored in a large insulated container. The heat stored in this *molten salt system* can then be released as needed to produce electricity at night or on cloudy days.

In 2014, the world's largest solar thermal plant using mirrors opened in California's Mojave Desert. This $2.2 billion plant has 350,000 mirrors focused on three 40-story towers. It can produce enough electricity to power 140,000 homes and eliminate annual CO_2 emissions equivalent to taking 88,000 cars off the road.

One problem with solar thermal systems is that their net energy yield is low, which means that they need large government subsidies or tax breaks in order to compete in the marketplace with alternatives that have higher net energy yields. Figure 13.31 summarizes the major advantages and disadvantages of concentrating solar energy to produce high-temperature heat or electricity.

We can use concentrated solar energy on a smaller scale, as well. In some sunny rural areas, people use inexpensive *solar cookers* to focus and concentrate sunlight for boiling and sterilizing water (Figure 13.32, left) and for cooking food (Figure 13.32, right). Solar cookers can replace wood and charcoal fires, thereby helping to reduce indoor air pollution, a major killer of many of the world's poor people. They also help to reduce deforestation by reducing the need for firewood and charcoal made from firewood.

FIGURE 13.30 *Solar thermal power:* This solar power plant (left) in a California desert uses curved (parabolic) solar collectors to concentrate solar energy to provide enough heat to boil water and produce steam for generating electricity. In another type of system (right), an array of mirrors tracks the sun and focuses reflected sunlight on a central receiver to boil the water for producing electricity.

Left: National Renewable Energy Laboratory; Right: Sandia National Laboratories/National Renewable Energy Laboratory

We Can Use Solar Cells to Produce Electricity

We can convert solar energy directly into electrical energy using **photovoltaic (PV) cells**, commonly called **solar cells**. Most solar cells are thin wafers of purified silicon (Si) or polycrystalline silicon with trace amounts of metals that allow them to produce electricity.

Solar cells have no moving parts and they operate safely and quietly with no emissions of pollutants or greenhouse gases. A typical solar cell has a thickness ranging from less than that of a human hair to that of a sheet of paper. When sunlight strikes these transparent cells, they produce electricity (a flow of electrons), and many cells wired together in a panel can produce large amounts of electrical power. Such systems can be connected to existing electrical grid systems or to batteries that store the electrical energy until it is needed.

We can mount arrays of solar cells on rooftops and incorporate them into almost any type of roofing material. Nanotechnology and other emerging technologies will likely allow the manufacturing of solar cells in paper-thin, rigid or flexible sheets (see Figure 12.A, p. 302) that can be printed like newspapers and attached to or embedded in a variety of surfaces such as outdoor walls, windows, drapes, and clothing. Large solar-cell power plants are operating in Portugal, Spain, Germany, South Korea, the southwestern United States, and China (Figure 13.33).

Nearly 1.3 billion people, almost one of every five in the world, do not have access to electricity that most of us take for granted. They live in energy poverty, mostly in less-developed countries in rural villages that are not connected to an electrical grid. A growing number of these individuals and villages now use solar cells to generate electricity (Figure 13.34) and to power highly efficient LED lamps that can replace polluting kerosene lamps.

As these small, off-grid systems reach more rural villages they will help hundreds of millions of people to lift themselves out of poverty. Eventually, new local microgrids of thin-film and other new types of solar cells are expected to drastically lower the cost of providing electricity to these areas. This could greatly reduce the need to build expensive and centralized coal and nuclear power plants and electrical grids to distribute power throughout countries such as India.

FIGURE 13.32 Solutions: Solar cooker (left) in Costa Rica and simple solar oven (right).

chriss73/Shutterstock.com

M. Cornelius/Shutterstock.com

Solar cell power plants are generating electricity in some sunny areas of the world.

Greg Girard/National Geographic Creative

FIGURE 13.33 Solar-cell power plant located near the city of Shizuishan, China. Solar cell power plants need no water for cooling.

Solar cells emit no greenhouse gases, although they are not a carbon-free option, because fossil fuels are used to produce and transport the panels. However, these emissions are small compared to those generated by the use of fossil fuels and the nuclear power fuel cycle. Conventional solar cells also contain toxic materials that must be recovered when the cells wear out after 20–25 years of use, or when they are replaced by new systems.

One problem with solar cells is that they have a low-to-medium net energy yield, depending on the design and the production process. Solar cells typically convert only 20% of the incoming solar energy into electricity. Scientists and engineers are rapidly improving the efficiency of solar cells, which will increase their net energy yield. In 2014, researchers at Germany's Fraunhofer Institute for Solar Energy Systems developed a solar cell with an efficiency of 45%. They plan to scale up this prototype cell for commercial use within 3 years. Also in 2014, Swedish scientist David Barbero and his colleagues found a way to use nanotubes that has the potential to make next-generation solar cells vastly more efficient. Figure 13.35 lists the major advantages and disadvantages of using solar cells.

Between 2008 and 2012, the cost per watt of electricity produced by solar cells in the United States fell by more than 75%, according to the DOE, and prices are expected to keep falling. As a result of government subsidies and tax breaks for solar cell producers and users, along with declining prices, solar cells have become the world's fastest growing way to produce electricity (see Figure 6, p. S41, in Supplement 5). Production is likely to grow much more. Within the next decade or two, next generation thin-film nanotechnology and graphene solar cells (see Science Focus 12.2, p. 302), along with solar cells made with a class of mineral materials known as perovskites, are expected to become inexpensive enough to compete with fossil fuels, especially coal.

Jim Lyons, former chief engineer for General Electric, contends that solar cells will be the world's number-one source of electricity by the end of this century. If that happens, it will represent a huge global application of the solar energy **principle of sustainability** that would sharply reduce air pollution and help to slow climate change. **GREEN CAREER: Solar-cell technology**

Jim Welch/National Renewable Energy Laboratory

FIGURE 13.34 **Solutions:** A solar cell panel provides electricity for lighting this hut in rural West Bengal, India. **Question:** Do you think your government should provide aid to poor countries for obtaining solar-cell systems? Explain.

We Can Produce Electricity from Falling and Flowing Water

Hydropower is any technology that uses the kinetic energy of flowing or falling water to produce electricity. It is an indirect form of solar energy because it depends on the evaporation of water, which is deposited as rain or snow at higher elevations where it can flow to lower elevations in rivers as part of the earth's solar-powered water cycle (see Figure 3.13, p. 52).

The most common approach to harnessing hydropower is to build a high dam across a large river to create a reservoir. Some of the water stored in the reservoir is allowed to flow through large pipes at controlled rates to spin turbines that produce electricity (see Figure 11.12, p. 258).

Hydropower, the leading renewable energy resource, produced about 16% of the world's electricity in 160 countries in 2013. In order, the world's top three producers of hydropower are China, Brazil, and the United States. In 2013, hydropower supplied about 7% of the electricity used in the United States (but about 50% of that used on the West Coast). (See Figure 9, p. S42, in Supplement 5 for a graph showing the global growth of hydropower.)

According to the UN, only about 13% of the world's potential for hydropower has been developed. Much of this untapped potential is in China, India, South America, Central Africa, and parts of the former Soviet Union. China, with the world's largest hydropower output, has plans to triple it by 2020 and is also building or funding more than 200 hydropower dams around the world.

However, some analysts expect that the use of large-scale hydropower plants will fall slowly over the next several decades as many existing reservoirs fill with silt and become useless faster than new systems are built. Also, there is concern over emissions of methane, a potent greenhouse gas, from the decomposition of submerged vegetation in reservoirs, especially in warm climates.

In addition, if atmospheric temperatures continue to rise and contribute to climate change as projected, the electrical output of many of the world's large dams is likely to drop as mountain glaciers, a primary source of their water, will melt. Figure 13.36 lists the major advantages and disadvantages of using large-scale hydropower plants to produce electricity.

The use of *microhydropower generators* may become an increasingly important way to produce electricity. These are floating turbines, each about the size of an overnight suitcase. They can be placed in any stream or river without altering its course to provide electricity at a very low cost with a very low environmental impact. Most of these systems can generate enough electricity to power a moder-

Trade-Offs

Solar Cells

Advantages	Disadvantages
Medium net energy yield	Need access to sun
Little or no direct emissions of CO_2 and other air pollutants	Some designs have low net energy yield
Easy to install, move around, and expand as needed	Need electricity storage system or backup
Competitive cost for newer cells	Costs high for older systems but dropping rapidly
	Solar-cell power plants could disrupt desert ecosystems

FIGURE 13.35 Using solar cells to produce electricity has advantages and disadvantages (**Concept 13.5**). **Questions:** Which single advantage and which single disadvantage do you think are the most important? Why? Do you think that the advantages outweigh the disadvantages? Why?

Top: Martin D. Vonka/Shutterstock.com. Bottom: pedrosala/Shutterstock.com.

Large-Scale Hydropower

Advantages	Disadvantages
High net energy yield	Large land disturbance and displacement of people
Large untapped potential	
Low-cost electricity	High CH_4 emissions from rapid biomass decay in shallow tropical reservoirs
Low emissions of CO_2 and other air pollutants in temperate areas	Disrupts downstream aquatic ecosystems

© 2016 Cengage Learning

FIGURE 13.36 Advantages and disadvantages of using large dams and reservoirs to produce electricity (**Concept 13.5**). *Questions:* Which single advantage and which single disadvantage do you think are the most important? Why? Do you think that the advantages outweigh the disadvantages? Why?

Photo: Andrew Zarivny/Shutterstock.com

ately sized home as far as 1.6 kilometers (1 mile) from where they are located.

We can also produce electricity from flowing water by tapping into the energy from *ocean tides* and *waves*. In some coastal bays and estuaries, water levels can rise or fall by 6 meters (20 feet) or more between daily high and low tides. Dams can be built across the mouths of such bays and estuaries to capture the energy in these flows for hydropower, but sites with large tidal flows are rare. Two large tidal energy dams are currently operating, one at La Rance on the northern coast of France, and the other in Nova Scotia's Bay of Fundy.

For decades, scientists and engineers have also been trying to produce electricity by tapping wave energy along seacoasts where there are almost continuous waves. Scientists estimate that learning how to tap into the world's wave power at an affordable cost could provide more than twice the amount of electricity that the world uses.

However, production of electricity from tidal and wave systems is limited because of a lack of suitable sites, citizen opposition at some sites, high costs, and equipment damage from saltwater corrosion and storms. Proponents hope that improved technology can greatly increase the production of electricity from tides and waves sometime during this century.

In 2014, China was building a pilot plant to evaluate the feasibility of producing electricity by using the difference in temperature between warm surface water and cold deep water in parts of the world's tropical oceans to generate a flow of electrons. The United States experimented with this approach, called ocean thermal-energy conversion (OTEC) in the 1980s, but abandoned it because of its high cost.

Using Wind to Produce Electricity Is an Important Step toward Sustainability

In a 2009 study, Harvard University researcher Xi Lu estimated that wind power (**Core Case Study**) has the potential to produce 40 times the world's current use of electricity. Most of the rapidly growing number of wind farms have been built on land in parts of Europe, China, and the United States. However, the frontier for wind energy is offshore wind farms (Figure 13.1) because winds are generally much stronger and steadier over coastal waters than on land.

Since 1990, wind power has been the world's second fastest-growing source of electricity after solar cells. (See Figure 7, p. S41, in Supplement 5 for a graph showing the rapid exponential growth in electricity production from wind.) In order, the three countries with the largest installed wind power capacity in 2012 were China, the United States, and Germany. In fewer than 10 years, China built the world's largest wind power capacity and plans to triple it by 2020. However, many of China's wind farms are in remote areas and produce no electricity because some of the electrical transmission lines for China's proposed national smart grid have not been built.

In 2013, wind farms in more than 85 countries produced about 3.5% of the world's electricity—enough to provide electricity for more than 500 million people, according to the Earth Policy Institute. By 2050, wind farms could produce about 31% of the world's electricity, according to a 2013 analysis by the market research firm iSUPPLi and the Global Wind Energy Council. In 2013, Denmark produced 34% of its electricity using wind power and plans to increase this to 50% by 2020. In 2013, wind provided 50% of Portugal's electricity, 27% of the electricity used by the U.S. state of Iowa, and 26% of South Dakota's electricity. By 2013, wind turbines in the United States were producing as much electricity as 60 large nuclear reactors with the capacity of land-based wind turbines increasing 12-fold between 2005 and 2012. Texas leads the nation in wind energy production, followed by California, Iowa, and South Dakota.

Even though offshore wind farms are more costly to install, analysts expect to see increasing use of them because they can harness stronger winds and thus reduce the cost of producing electricity. When located far enough offshore they are not visible from the land. Putting them offshore also lessens the need for negotiations among multiple landowners over the locations of turbines and electrical transmission lines. Japan is installing three of the world's largest offshore wind turbines. In 2013 the leaders in building offshore wind farms were, in order, Great Britain, Denmark, Belgium, Germany, China, the Netherlands, Sweden, and Japan. The United States is at the bottom of the list, despite its great offshore wind energy potential.

The United States has an enormous potential for producing electricity from wind power on land and at sea (**Core Case Study** and Figure 21, p. S38, in Supplement 4).

A 2009 study published in the *Proceedings of the U.S. National Academy of Sciences* estimated that the United States has enough wind potential to meet an estimated 16 to 22 times its current electricity needs. According to a study by the U.S. Department of the Interior, with expanded and sustained subsidies and a smart grid, wind farms off the Atlantic and Gulf coasts could generate enough electricity to more than replace all of the country's coal-fired power plants. Many Atlantic and Gulf coast states are making plans to tap into this vast source of energy and create jobs and revenues in the process.

Wind is abundant, widely distributed, and inexhaustible, and wind power is mostly carbon-free and pollution-free. A wind farm can be built within 9 to 12 months and expanded as needed. Although wind farms can cover large areas of land, the turbines occupy only a small total area of land. For each wind turbine located on the land of a farmer or rancher, the landowner typically receives $3,000 to $10,000 a year in royalties, and can still use the land for growing crops or grazing cattle.

A typical wind turbine can generate enough electricity to meet the needs of more than 1,000 homes. An acre of land in northern Iowa planted in corn can produce about $1,000 worth of ethanol car fuel. The same site used for a single wind turbine can produce $300,000 worth of electricity per year, which is why many landowners in favorable wind areas are investing in wind farms.

In addition, wind power has a medium-to-high net energy yield. The DOE and the Worldwatch Institute estimate that, if we were to apply the full-cost pricing **principle of sustainability** by including the harmful environmental and health costs of various energy resources in comparative cost estimates, wind energy would be the least costly way to produce electricity. Between 1990 and 2012, the average cost per kilowatt-hour of electricity from wind power in the United States dropped from 55¢ to 5¢, according to the U.S. Department of Energy.

Like any energy source, wind power has some drawbacks. For example, areas with the greatest wind power potential are often sparsely populated and located far from cities. Thus, to take advantage of its huge potential for using wind energy, the United States will have to invest in replacing and expanding its outdated electrical grids with smart grid systems. Another problem is that winds can die down and thus require a backup source of power, such as natural gas, for generating electricity. However, analysts calculate that a large number of wind farms in different areas connected to a smart grid could take up the slack when winds die down in any one area. This could make wind power a very stable and reliable source of electricity, especially if offshore wind farms were included on the grid.

Scientists are working on ways to store wind energy. Electricity produced by wind can be passed through water to produce hydrogen fuel, which could be thought of as stored wind power. Another option is to use wind-generated electricity to pump pressurized air deep underground into aquifers, caverns, and abandoned natural gas wells. The energy stored in the compressed air could then be released as needed to spin turbines and generate electricity when wind power is not available. Excess electricity from wind farms can also be stored in the millions of hybrid-electric motor vehicles (Figure 13.23) when they are recharging. Together, all these batteries could amount to a gigantic national storage battery.

CONSIDER THIS...

CONNECTIONS Bird and Bat Deaths and Wind Turbines

Wildlife ecologists and bird experts have estimated that collisions with wind turbines kill as many as 573,000 birds and thousands of bats each year in the United States, although some estimates put the figure at 70,000 to 100,000 birds. Compare this to some much larger annual estimates reported in 2013 and 2014 by Defenders of Wildlife, the U.S. Fish and Wildlife Service, and the Smithsonian Conservation Biology Institute: domestic and feral cats kill 1.4 billion to 3.7 billion birds a year; collisions with commercial buildings and homes, at least 1 billion every year; hunters, more than 100 million; cars and trucks, about 340 million; and pesticide poisoning, 67 million. Most of the wind turbines involved in bird and bat deaths were built using outdated designs, and some were built in bird migration corridors. Wind power developers now avoid such corridors, as well as areas with large bat colonies, when building wind farms. Newer turbine designs reduce bird deaths considerably by using slower blade rotation speeds and by not providing places for birds to perch or nest.

Figure 13.37 lists the major advantages and disadvantages of using wind to produce electricity. According to many energy analysts, wind power has more benefits and fewer serious drawbacks than any other energy resource, except for energy efficiency (Figure 13.22). **GREEN CAREER: Wind-energy engineering**

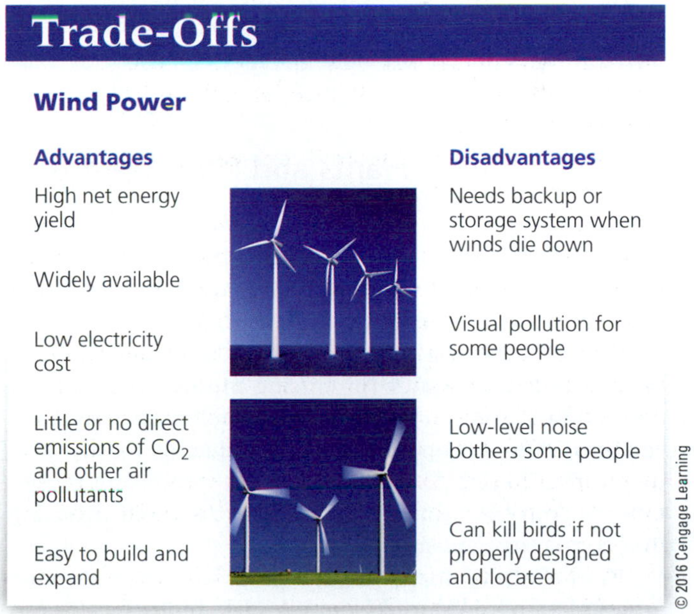

Trade-Offs

Wind Power

Advantages	Disadvantages
High net energy yield	Needs backup or storage system when winds die down
Widely available	
Low electricity cost	Visual pollution for some people
Little or no direct emissions of CO_2 and other air pollutants	Low-level noise bothers some people
Easy to build and expand	Can kill birds if not properly designed and located

© 2016 Cengage Learning

FIGURE 13.37 Using wind to produce electricity has advantages and disadvantages (**Concept 13.5**). *Questions:* Which single advantage and which single disadvantage do you think are the most important? Why? Do you think the advantages outweigh the disadvantages? Why?

Top: TebNad/Shutterstock.com. Bottom: Yegor Korzh/Shutterstock.com.

We Can Produce Energy by Burning Solid Biomass

Biomass consists of plant materials that can be burned as a solid fuel or converted into gaseous or liquid biofuels. Examples are wood, wood wastes, charcoal made from wood, and agricultural wastes (such as sugarcane stalks, rice husks, and corn cobs). Solid biomass is burned mostly for heating and cooking, but also for industrial processes and for generating electricity.

Wood is a renewable fuel only if it is not harvested faster than it is replenished. The problem is that about 2.7 billion people in 77 less-developed countries face a *fuelwood crisis* and are often forced to meet their fuel needs by harvesting wood faster than new vegetation can grow to replace the trees that are removed. One way to deal with this problem is to plant fast-growing trees, shrubs, and perennial grasses in *biomass plantations*. But repeated cycles of growing and harvesting these plantations can deplete the soil of key nutrients. Also, some plantation tree species such as European poplar and American mesquite are invasive species and can spread from plantations to take over nearby natural areas.

Another problem with depending on solid biomass as a fuel is that clearing forests and grasslands to provide the fuel reduces biodiversity and the amount of vegetation that would otherwise capture climate-changing CO_2. At the same time, the burning of wood and other forms of biomass produces CO_2 and other pollutants such as fine particles in smoke. In 2014, the EPA proposed phasing in stricter regulations to curb such pollution from new residential wood-burning stoves in the United States, beginning in 2015. Figure 13.38 lists the general advantages and disadvantages of burning solid biomass as a fuel.

We Can Convert Plants and Plant Wastes to Liquid Biofuels

Liquid biofuels such as *ethanol* (ethyl alcohol produced from plants and plant wastes) and *biodiesel* (produced from vegetable oils) are being used increasingly to fuel motor vehicles. The biggest producers of liquid biofuels are, in order, the United States (producing mostly ethanol from corn), Brazil (producing mostly ethanol from sugarcane residues), the European Union (producing mostly biodiesel from vegetable oils), and China (producing mostly ethanol from non-grain plant sources to avoid diverting grains from its food supply).

Biofuels have three major advantages over gasoline and diesel fuel produced from oil. *First,* biofuel crops can be grown throughout much of the world, and thus they can help countries to reduce their dependence on imported oil. *Second,* if these crops are not used faster than they are replenished by new plant growth, there is no net increase in CO_2 emissions, unless existing grasslands or

---

Trade-Offs

Solid Biomass

Advantages		Disadvantages
Widely available in some areas		Contributes to deforestation
Moderate costs		Clear-cutting can cause soil erosion, water pollution, and loss of wildlife habitat
Medium net energy yield		
No net CO_2 increase if harvested, burned, and replanted sustainably		Can open ecosystems to invasive species
Plantations can help restore degraded lands		Increases CO_2 emissions if harvested and burned unsustainably

FIGURE 13.38 Burning solid biomass as a fuel has advantages and disadvantages (**Concept 13.5**). ***Questions:*** Which single advantage and which single disadvantage do you think are the most important? Why? Do you think the advantages outweigh the disadvantages? Why?

Top: Fir4ik/Shutterstock.com. Bottom: Eppic/Dreamstime.com.

---

forests are cleared to plant biofuel crops. *Third,* biofuels are easy to store and transport through existing fuel networks and can be used in motor vehicles at little additional cost.

Globally, ethanol production rose rapidly, especially in the United States and Brazil, between 1975 and 2012 (see the graph in Figure 8, p. S42, in Supplement 5). Brazil makes ethanol from *bagasse,* a residue produced when sugarcane is crushed. This sugarcane ethanol has a medium net energy yield that is about 8 times higher than that of ethanol produced from corn. About 45% of Brazil's motor vehicles run on ethanol or ethanol–gasoline mixtures produced from sugarcane grown on only 1% of the country's arable land.

In the United States, most ethanol is made from corn. In 2013, about 43% of the corn produced in the United States was used to produce ethanol. However, studies indicate that corn-based ethanol has a low net energy yield because fossil fuels are used heavily for growing the corn and converting it to ethanol. This is one reason why it has received big government subsidies. It also helps to explain why scientists calculate that using this ethanol results in a net increase in greenhouse gas emissions. Also, raising corn requires a great deal of water and thus adds to stresses on diminishing water supplies in many areas.

An alternative to corn-based ethanol is *cellulosic ethanol,* which is produced from the inedible cellulose that makes up most of the biomass of plants—material such as leaves, stalks, and wood chips. Plants that could be used for cellulosic ethanol production are switchgrass, which grows faster than corn, and *Arundo donax*, a giant reed that yields about three times as much ethanol per acre as corn does. Ecologist David Tilman estimates that the net energy yield for cellulosic ethanol is about 5 times higher than that for corn ethanol.

Another approach, being evaluated by scientists at the DOE's Pacific Northwest National Laboratory, involves converting a mixture of algae and water to crude oil in less than an hour by subjecting the mixture to high temperature and pressure in a cooker. Learning how to do this on a large scale at an affordable cost and with an acceptable net energy yield will take more research.

In a UN report on bioenergy, and in another study by R. Zahn and his colleagues, scientists warned that large-scale biofuel crop farming could diminish biodiversity by expanding the clearing of natural forests and grasslands; increase soil degradation, erosion, and nutrient leaching; push small farmers off their land; and raise food prices if food producers can make more money growing corn and other crops to fuel cars than they can raising corn to feed livestock and people.

Recent studies confirm most of the warnings about using corn to produce ethanol. For example, in 2013, the Environmental Working Group reported on a study of the corn-based ethanol program subsidized generously by the U.S. government. The program was judged an ecological disaster that has taken more than 2 million hectares (5 million acres) of land out of the soil conservation reserve, an important topsoil preservation program. It has also led to the filling of wetlands, increased soil erosion, increased contamination of rivers and of the Gulf of Mexico (see Figure 11.A, p. 278) with excess fertilizer, and significantly increased CO_2 emissions.

CONSIDER THIS. . .

CONNECTIONS Biofuels and Climate Change

Nobel Prize–winning chemist Paul Crutzen has warned that intensive farming of biofuel crops could speed up atmospheric warming and projected climate change by producing more greenhouse gases than would be produced by burning fossil fuels instead of biofuels. This would happen if nitrogen fertilizers were used to grow corn and other biofuel crops. Such fertilizers, when applied to the soil, release large amounts of the potent greenhouse gas nitrous oxide. A study by Finn Danielsen and a team of other scientists concluded that keeping tropical rain forests intact is a better way to slow projected climate change than burning and clearing such forests and replacing them with biofuel plantations.

The challenge is to grow crops for food and biofuels by using more sustainable agriculture (see Figure 10.27, p. 242) with less irrigation, land degradation, air and water pollution, greenhouse gas emissions, and degradation of biodiversity—an important way to increase our beneficial environmental impact. Also, any system for producing a biofuel should have a moderate-to-high net energy yield

Trade-Offs

Liquid Biofuels

Advantages		Disadvantages
Reduced CO_2 emissions for some crops		Fuel crops can compete with food crops for land and raise food prices
Medium net energy yield for biodiesel from oil palms		Fuel crops can be invasive species
		Low net energy yield for corn ethanol and for biodiesel from soybeans
Medium net energy yield for ethanol from sugarcane		Higher CO_2 emissions from corn ethanol

© 2016 Cengage Learning

FIGURE 13.39 Use of liquid biofuels has its advantages and disadvantages (**Concept 13.5**). *Questions:* Which single advantage and which single disadvantage do you think are the most important? Why? Do you think the advantages outweigh the disadvantages? Why?

Top: tristan tan/Shutterstock

so that it can compete in the energy marketplace without large government subsidies.

Figure 13.39 compares the advantages and disadvantages of using biodiesel and ethanol liquid biofuels.

We Can Get Energy by Tapping the Earth's Internal Heat

Geothermal energy is heat stored in soil, underground rocks, and fluids in the earth's mantle. We can tap into this stored energy to heat and cool buildings and to produce electricity.

One way to capture geothermal energy is by using a *geothermal heat pump* system (Figure 13.40). It can heat and cool a house by exploiting the temperature difference, almost anywhere in the world, between the earth's surface and underground at a depth of 3–6 meters (10–20 feet), where the earth's temperature typically is 10–20°C (50–60°F) year round. In winter, a closed loop of buried pipes circulates a fluid, which extracts heat from the ground and carries it to a heat pump, which transfers the heat to a home's heat distribution system. In summer, this system works in reverse, removing heat from a home's interior and storing it in the ground. The EPA estimates that such a system can heat or cool a 190-square-meter (2,000-square-foot) house for as little as $1 a day.

According to the EPA, a well-designed geothermal heat pump system is the most energy-efficient, reliable, environmentally clean, and cost-effective way to heat or cool a space. Installation costs can be high but are generally

Geothermal heating

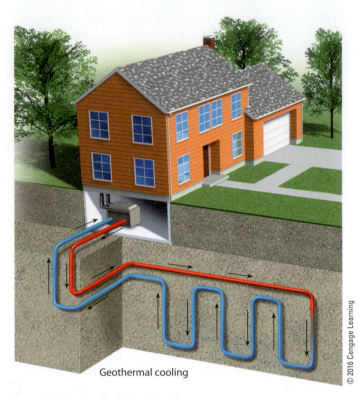

Geothermal cooling

© 2016 Cengage Learning

FIGURE 13.40 Natural capital: A geothermal heat pump system can heat or cool a house almost anywhere.

recouped within 3–5 years, after which these systems save money for their owners.

We can also tap into deeper, more concentrated *hydrothermal reservoirs* of geothermal energy. This is done by drilling wells into the reservoirs to extract their dry steam (with a low water content), wet steam (with a high water content), or hot water, which are then used to heat homes and buildings, provide hot water, grow vegetables in greenhouses, raise fish in aquaculture ponds, and spin turbines to produce electricity. This source of energy is used in 24 countries and provides about 0.3% of the world's electricity and a large percentage of the electricity in some countries such as Iceland (36%) and Nicaragua (12%). In Peru, a National Geographic Young Explorer is carrying out research to develop that country's geothermal resources (Individuals Matter 13.1).

The United States is the world's largest producer of geothermal electricity from hydrothermal reservoirs (see Figure 22, p. S38, in Supplement 4). Most of it is produced in California, Nevada, Utah, and Hawaii. It meets the electricity needs of about 6 million Americans and supplies almost 6% of California's electricity. Geothermal energy is available around the clock.

Another source of geothermal energy is *hot, dry rock* found 5 or more kilometers (3 or more miles) underground almost everywhere. Water can be injected through wells drilled into this rock. Some of this water, absorbing some of the intense heat, becomes steam that is brought to the surface and used to spin turbines to generate electricity. According to the USGS, tapping just 2% of this source of geothermal energy in the United States could produce more than 2,000 times the amount of electricity currently used in the country. The limiting factor is cost, which could be brought down by more research and by improved technology. Some call for the use of geothermal fracking, based on the fracking technology used to produce oil and natural gas (Science Focus 13.1). However, there is concern that tapping this deep source of geothermal energy could increase the frequency of earthquakes in the drilled areas. **GREEN CAREER: Geothermal engineer**

Figure 13.41 lists the major advantages and disadvantages of using geothermal energy. The biggest problem limiting the widespread use of geothermal energy is the lack of hydrothermal sites with concentrations of heat high enough to make it affordable.

Will Hydrogen Save Us?

Some scientists say that the fuel of the future is hydrogen gas (H_2). Most of their research has been focused on using fuel cells that combine H_2 and oxygen gas (O_2) to produce electricity and water vapor ($2 H_2 + O_2 \rightarrow 2 H_2O + energy$), a harmless chemical that is emitted into the atmosphere.

Widespread use of hydrogen as a fuel for running motor vehicles, heating buildings, and producing electricity would eliminate most of the outdoor air pollution that

Sofia Ruzo

Background photo: N.Minton/Shutterstock.com

Andrés Ruzo—Geothermal Energy Sleuth and National Geographic Young Explorer

Andrés Ruzo is a geophysicist with a driving passion to learn about geothermal energy and to show how this renewable and clean energy resource can help us to solve some of the world's energy problems.

This National Geographic Young Explorer and his wife and field assistant, Sofia, spent most of 2012 gathering data to develop the first detailed map of the geothermal energy resources of northern Peru. Their fieldwork involved lowering temperature-measuring equipment down into abandoned water wells and oil wells in the desert area of northern Peru where daytime temperatures can exceed 54°C (130°F). These data will show how heat flows through the upper crust of the earth and how and where this heat can be tapped as a source of energy.

Ruzo believes that geothermal energy is a hidden sleeping giant that can be tapped as an important renewable source of heat and electricity. He says that his goal in life is "to be a force of positive change in the world."

Trade-Offs

Geothermal Energy

Advantages	Disadvantages
Medium net energy yield and high efficiency at accessible sites	High cost except at concentrated and accessible sources
Lower CO_2 emissions than fossil fuels	Scarcity of suitable sites
Low cost at favorable sites	Noise and some CO_2 emissions

© 2016 Cengage Learning

FIGURE 13.41 Using geothermal energy for space heating and for producing electricity or high-temperature heat for industrial processes has advantages and disadvantages (**Concept 13.5**). *Questions:* Which single advantage and which single disadvantage do you think are the most important? Why? Do you think the advantages outweigh the disadvantages? Why?

Photo: N.Minton/Shutterstock.com

comes from motor vehicles and coal-burning power and industrial plants. It would also greatly reduce the threat of projected climate change, because in using it, we emit no CO_2 as long as the H_2 is not produced with the use of fossil fuels or nuclear power. Hydrogen also provides more energy per gram than does any other fuel, making it a lightweight fuel ideal for aviation.

So what's the catch? There are three challenges in turning the vision of hydrogen as a major fuel into reality. *First*, there is hardly any hydrogen gas (H_2) in the earth's atmosphere. We can produce H_2 by heating water or passing electricity through it; by stripping it from the methane (CH_4) found in natural gas and from gasoline molecules; and through a chemical reaction involving coal, oxygen, and steam. As a result, hydrogen has a *negative net energy yield* because it takes more high-quality energy to produce H_2 using these methods than we get by burning it.

Second, fuel cells are the best way to use H_2 to produce electricity, but current versions of fuel cells are expensive. However, progress in the development of nanotechnology (see Science Focus 12.1, p. 300) could lead to less expensive fuel cells.

Third, whether or not a hydrogen-based energy system produces less outdoor air pollution and CO_2 than a fossil fuel system depends on how the H_2 fuel is produced. We could use electricity from coal-burning and nuclear power plants to decompose water into H_2 and O_2. But this approach does not avoid the harmful environmental effects associated with using coal and the nuclear fuel cycle. Also, making H_2 from coal or stripping it from methane or gasoline adds much more CO_2 to the atmosphere per unit of heat generated than does burning the coal or methane directly.

For now, hydrogen's negative net energy yield is a serious limitation and means that this fuel will have to be subsidized in order for it to compete in the open marketplace with fuels that have medium-to-high net energy yields. This could change. In 2011, chemist Daniel Nocera developed an "artificial leaf." This credit-card-sized silicon wafer produces H_2 and O_2 when placed in a glass of tap water and exposed to sunlight, and the hydrogen can be extracted and used to power fuel cells. In 2014, a team of researchers led by chemist Tom Meyer developed an electrochemical cell that mimics photosynthesis by using the sun's energy to split water molecules and to produce H_2 fuel. Scaling up either of these processes to produce large amounts of H_2 at an affordable price with an acceptable net energy yield could represent a tipping point for the future of solar energy and hydrogen that would help implement the solar energy **principle of sustainability** on a global scale.

Figure 13.42 lists the major advantages and disadvantages of using hydrogen as an energy resource. **GREEN CAREER: Hydrogen energy development**

Trade-Offs

Hydrogen

Advantages		Disadvantages
Can be produced from plentiful water at some sites		Negative net energy yield
No CO_2 emissions if produced with use of renewables		CO_2 emissions if produced from carbon-containing compounds
Good substitute for oil		High costs create need for subsidies
High efficiency in fuel cells		Needs H_2 storage and distribution system

Fuel cell

© 2016 Cengage Learning

FIGURE 13.42 Using hydrogen as a fuel for vehicles and for providing heat and electricity has advantages and disadvantages (**Concept 13.5**). *Questions:* Which single advantage and which single disadvantage do you think are the most important? Why? Do you think that the advantages outweigh the disadvantages? Why?

Photo: LovelaceMedia/Shutterstock.com

CONCEPT 13.6 We can make the transition to a more sustainable energy future by greatly improving energy efficiency, using a mix of renewable energy resources, and including the environmental and health costs of energy resources in their market prices.

Choosing Energy Paths

In considering possible energy futures, scientists and energy experts who have evaluated energy alternatives have come to three general conclusions.

First, *during this century, there will likely be a gradual shift from dependence on nonrenewable fossil fuels to a mix of renewable energy from the sun, wind, flowing water (hydropower), and the earth's interior (geothermal energy)*. The use of renewable resources is already enabling many people to depend less on large centralized power systems and more on decentralized systems such as rooftop water heaters, single wind turbines, and solar cell panels. Experts also project a shift from gasoline-powered motor vehicles to hybrid and plug-in electric cars and perhaps to all-electric cars, if there are major improvements in battery technology.

The second general conclusion of experts about the future of energy use is that *a combination of improved energy efficiency and carefully regulated use of natural gas will be the best way to make the transition to using mostly renewable energy resources during this century* (**Concept 13.6**). However, this will have to include much tighter controls on emissions of methane and other greenhouse gases throughout the entire natural gas production and distribution system.

The third general conclusion is that *because fossil fuels are still abundant and artificially cheap, we will continue to use them in large quantities*. This presents two major challenges. One is to find ways to reduce the harmful environmental and health impacts of widespread use of fossil fuels, with special emphasis on reducing outdoor air pollution and greenhouse gas emissions. The second is to find ways to include more of the harmful environmental and health costs of using fossil fuels in their market prices and thus to implement the full-cost pricing **principle of sustainability**. Figure 13.43 summarizes several strategies for making the transition to a more sustainable energy future over the next 50 years (**Concept 13.6**). This shift is gradually taking place.

We have the creativity, the wealth, and most of the technology needed to make the transition to a cleaner and more sustainable energy future within your lifetime, thereby greatly increasing our beneficial environmental im-

GOOD NEWS

Solutions

Making the Transition to a More Sustainable Energy Future

Improve Energy Efficiency

Increase fuel-efficiency standards for vehicles, buildings, and appliances

Provide large tax credits for buying efficient cars, houses, and appliances

Reward utilities for reducing demand for electricity

Greatly increase energy efficiency research and development

More Renewable Energy

Greatly increase use of renewable energy

Provide large subsidies and tax credits for use of renewable energy

Greatly increase renewable energy research and development

Reduce Pollution and Health Risk

Phase out coal subsidies and tax breaks

Levy taxes on coal and oil use

Phase out nuclear power subsidies, tax breaks, and loan guarantees

© 2016 Cengage Learning

FIGURE 13.43 Energy analysts have made a number of suggestions for helping us make the transition to a more sustainable energy future (**Concept 13.6**). *Questions:* Which five of these solutions do you think are the best ones? Why?

Top: andrea lehmkuhl/Shutterstock.com. Bottom: pedrosala/Shutterstock.com.

What Can You Do?

Shifting to More Sustainable Energy Use

- ☒ Walk, bike, or use mass transit or a car pool to get to work or school

- ☒ Drive only vehicles that get at least 17 kpl (40 mpg)

- ☒ Have an energy audit done in the place where you live

- ☒ Superinsulate the place where you live and plug all air leaks

- ☒ Use passive solar heating wherever possible

- ☒ For cooling, open windows and use fans

- ☒ Use a programmable thermostat and energy-efficient heating and cooling systems, lights, and appliances

- ☒ Turn down your water heater's thermostat to 43–49°C (110–120°F) and insulate hot water pipes

- ☒ Turn off lights, TVs, computers, and other electronics when they are not in use

- ☒ Wash laundry in cold water and air dry it on racks

© 2016 Cengage Learning

FIGURE 13.44 Individuals matter: You can make a shift in your own life toward using energy more sustainably. *Questions:* Which three of these measures do you think are the most important ones to take? Why? Which of these steps have you already taken and which do you plan to take?

pact. Making this transition depends primarily on *education, economics,* and *politics*—on how well individuals understand environmental and energy problems and their possible solutions, and on how they vote and then influence their elected officials. People can also vote with their pocketbooks by refusing to buy energy-inefficient and environmentally harmful products and services, and by letting company executives and elected officials know about their choices.

Figure 13.44 lists some ways in which you can take part in the transition to a more environmentally and economically sustainable energy future.

BIG IDEAS

- In evaluating energy resources, we should consider their net energy yields and the environmental and health impacts of using them, as well as their potential supplies.

- By phasing in a mix of renewable energy sources—especially solar, wind, flowing water, sustainable biofuels, and geothermal energy—we could drastically reduce pollution, greenhouse gas emissions, and biodiversity losses.

- Making the transition to a more sustainable energy future will require sharply increasing energy efficiency, using a mix of renewable energy resources, and including the harmful environmental and health costs of energy resources in their market prices.

In the Core Case Study that opens this chapter, we considered the immense potential of wind—an indirect form of solar energy—as an energy resource, particularly in the United States. We have also looked at the advantages and disadvantages of an array of nonrenewable and renewable energy resources.

By relying mostly on nonrenewable fossil fuels, we violate the three **scientific principles of sustainability** (see Supplement 7, p. S50). The technologies we use to obtain energy from these resources disrupt the earth's chemical cycles by diverting huge amounts of water, degrading or destroying terrestrial and aquatic ecosystems, and emitting large quantities of greenhouse gases and other air pollutants. Using these technologies also destroys and degrades biodiversity and ecosystem services.

By relying more on a diversity of direct and indirect forms of renewable solar energy and by greatly improving energy efficiency, we could implement the three **scientific principles of sustainability** during this century and greatly increase our beneficial environmental impact. Making this shift would also follow the three **social science principles of sustainability** (see Supplement 7, p. S51). By including the harmful environmental and health costs of fossil fuels, nuclear energy, and other energy resources in their market prices, we would be applying the *full-cost pricing principle*. This could be accomplished with the aid of compromise and trade-offs in the political arena by applying the *win-win solutions principle*. And many analysts argue that such solutions would greatly benefit life on earth immediately and in the long run—a classic application of the ethical principle of *responsibility to current and future generations*.

Ulrich Mueller/Shutterstock.com

Chapter Review

Core Case Study

1. Describe the potential for using renewable energy from the wind (wind power) to produce most of the electricity used in the United States.

Section 13.1

2. What is the key concept for this section? What is **net energy yield** and why is it important for evaluating energy resources? Use the net energy concept to explain why some energy resources need to be subsidized, and give an example of such a resource.

Section 13.2

3. What is the key concept for this section? What is the earth's main source of energy? What are three forms of indirect solar energy? What is commercial energy and where does most of it come from in the world and in the United States? What is **crude oil (petroleum)** and how is it extracted from the earth? Define **peak production** for an oil well or field. What is **refining**? What are **petrochemicals** and why are they important? What are **proven oil reserves**? Describe the process of removing tightly held oil and natural gas through **horizontal drilling** and **hydraulic fracturing**, or **fracking**. What three countries have the world's largest proven oil reserves and what three countries are the largest users of oil? What percentage of the world's proven oil reserves do the United States and China have? Summarize the story of oil production and consumption in the United States. What are the major advantages and disadvantages of using crude oil as an energy resource? What is **shale oil** and how is this heavy oil produced? What are **tar sands**, or **oil sands**? What is bitumen, and how is it extracted and converted to heavy oil? What are the major advantages and disadvantages of using shale oil and heavy oils produced from tar sands as energy resources?

4. Define **natural gas, liquefied petroleum gas (LPG)**, and **liquefied natural gas (LNG)**. What two countries are the world's largest producers of natural gas and what two countries are the world's largest users of natural gas? Describe the potential for greatly increasing natural gas production in the United States by fracking. List four problems that have resulted from this trend. What are the major advantages and disadvantages of using natural gas as an energy resource? What is **coal**, how is it formed, and how do the various types of coal differ? How does a coal-burning power plant work? What three countries are the largest producers of coal and what two countries are the largest users of coal? Explain why there is no such thing as clean coal. Summarize the major environmental and health problems caused by the use of coal. What are the major advantages and disadvantages of using coal as an energy resource?

Section 13.3

5. What is the key concept for this section? How does a nuclear fission reactor work and what are its major safety features? Describe the **nuclear fuel cycle**. Explain how highly radioactive spent fuel rods are stored and what risks this presents. How has the United States dealt with the nuclear waste problem? What can we do with worn-out nuclear power plants? Summarize the arguments over whether or not the widespread use of nuclear power could help to slow projected climate change during this century. Summarize the arguments of experts who disagree over the future of nuclear power. What is the relationship between nuclear power plants and the spread of nuclear weapons? What are five criteria that any new design for a nuclear power plant should meet, according to some analysts? Describe the Fukushima Daiichi nuclear power plant accident and its effects on the environment. What is **nuclear fusion** and what is its potential as an energy resource?

Section 13.4

6. What is the key concept for this section? What is **energy efficiency**? What percentage of the energy used in the United States is unnecessarily wasted? Briefly describe two widely used energy-inefficient technologies. What are the major benefits of improving energy efficiency? Define and give an example of **cogeneration**. List three other ways to save energy and money in industry. How could an energy-efficient smart electrical grid help us to save energy and money? List three ways to save energy and money in transportation. Explain why the true cost of gasoline is much higher than what consumers pay at the pump. Distinguish among hybrid, plug-in hybrid, all-electric, and fuel-cell motor vehicles.

Summarize the story of the search for better batteries. List four ways to save energy and money **(a)** in new buildings, and **(b)** in existing buildings. List three ways in which you can save energy and money. Give three reasons why we are still wasting so much energy and money.

Section 13.5

7. What is the key concept for this section? List four reasons for why renewable energy is not more widely used than it is. Distinguish between a **passive solar heating system** and an **active solar heating system** and discuss the major advantages and disadvantages of using such systems for heating buildings. What are three ways to cool buildings naturally? Define **solar thermal system** and give an example of a centralized system and a smaller-scale system. List the major advantages and disadvantages of using the centralized systems. What is a **solar cell (photovoltaic** or **PV cell)** and what are the major advantages and disadvantages of using such devices to produce electricity?

8. Define **hydropower** and summarize the potential for expanding it. What are the major advantages and disadvantages of using hydropower to produce electricity? What is the potential for using tides and waves to produce electricity? Summarize the potential for using wind power **(a)** globally, and **(b)** in the United States. What are the major advantages and disadvantages of using wind to produce electricity? What is **biomass** and what are the major advantages and disadvantages of using it to provide heat and electricity? What are the major advantages and disadvantages of using biodiesel fuel and ethanol to power motor vehicles? What is **geothermal energy** and what are three major sources of such energy? What are the major advantages and disadvantages of using geothermal energy as a source of heat and to produce electricity? List the major advantages and disadvantages of using hydrogen as a fuel. Why is it important to compare the various methods of producing hydrogen gas?

Section 13.6

9. What is the key concept for this section? List three of the general conclusions of energy experts with regard to possible energy futures. List five major strategies recommended by such experts for making the transition to a more sustainable energy future.

10. What are this chapter's *three big ideas*? Explain how we can apply each of the six **principles of sustainability** in working to make a transition to a more sustainable energy future.

Note: Key terms are in bold type.

Critical Thinking

1. Suppose that a developer has proposed building a wind farm near where you live (**Core Case Study**). Would you be in favor of the project or opposed to it? Write a letter to your local newspaper or a blog for a website explaining your position and your reasoning. Include the concept of *net energy yield* in your arguments. As part of your research, determine how the electricity you use now is generated and where the power plant is located, and include this information in your arguments.

2. Should governments give a high priority to considering net energy yields when deciding what energy resources to support? What are other factors that should be considered? Which factor or factors should get the most weight in decision making? Explain your thinking.

3. Some analysts argue that in order to continue using oil at the current rate, we must discover and add to global oil reserves the equivalent of two new Saudi Arabian reserves every 7 years. Do you think this is possible? If not, what effects might the failure to find such supplies have on your life and on the lives of any children and grandchildren that you might eventually have?

4. During much of the time since the beginning of the Industrial Revolution, the United States and European nations fueled their economic growth by burning coal, with little effort to control the resulting air pollution. Eventually, as they became more affluent, they established air pollution standards and sought cleaner energy sources. Now China, which has been fueling its rapid economic growth largely by burning coal, says it is being asked to shift to cleaner energy resources before it becomes affluent enough to do so, even though most countries have not made such a shift. Do you think this is a fair expectation? Explain.

5. List five ways in which you unnecessarily waste energy during a typical day, and explain how these actions violate each of the three **scientific principles of sustainability**.

6. What do you think should be the top three energy resources? Explain. What do you think should be the three least-used energy resources? Explain.

7. Explain why you would support or oppose each of the following proposals made by various energy analysts:
   a. Government subsidies for all energy alternatives should be eliminated so that all energy choices can compete on a level playing field in the marketplace.
   b. All government tax breaks and other subsidies for conventional fossil fuels, synthetic natural gas and oil, and nuclear power (fission and fusion) should be eliminated. They should be replaced with subsidies and tax breaks for improving energy efficiency and developing renewable energy resources.
   c. Development of renewable energy resources should be left to private enterprise and should receive little or no help from the federal government, but the nuclear power and fossil fuels industries should continue to receive large federal government subsidies and tax breaks.

8. Congratulations! You are in charge of the world. List the five most important features of your energy policy and explain why each of them is important and how they relate to each other.

Doing Environmental Science

Do a survey of energy use at your school, based on the following questions: How is the electricity generated? How are most of the buildings heated? How is water heated? How are most of the vehicles powered? How is the computer network powered? How could energy efficiency be improved, if at all, in each of these areas? If it does not already do so, how could your school make use of solar, wind, biomass, and other forms of renewable energy? Write up a proposal for using energy more efficiently and sustainably at your school and submit it to school officials.

Global Environment Watch Exercise

Search the term *fracking* and find information on the use of fracking to produce oil and natural gas from shale rock in the United States. Try to answer the following questions: **(1)** In what areas are fracking for either fuel on the rise? **(2)** In one area of your choice, how many fracking wells have been drilled in the past year? **(3)** In that area, how much fuel has been produced in the past year? **(4)** In that area, has there been any opposition to the use of fracking? **(5)** Has the opposition, if any, resulted in any changes to the process or to regulations over the process? **(6)** Is fracking projected to continue in your chosen area, and to what extent? Write a report on your findings.

Ecological Footprint Analysis

Study the table below and then answer these questions by filling in the blank columns in the table.

1. Using Supplement 1 (Measurement Units, p. S1), convert the miles per gallon figures in the table to kilometers per liter (kpl).

2. How many liters (and how many gallons) of gasoline would each type of car use annually if it were driven 19,300 kilometers (12,000 miles) per year?

3. How many kilograms (and how many pounds) of carbon dioxide would be released into the atmosphere annually by each car, based on the fuel consumption calculated in question 2? Assume that the combustion of gasoline releases 2.3 kilograms of CO_2 per liter (19 pounds per gallon).

Combined City/Highway Fuel Efficiency for 2014 Cars (mpg)

Model	Miles per Gallon (mpg)	Kilometers per Liter (kpl)	Annual Liters (Gallons) of Gasoline	Annual CO_2 Emissions
Chevrolet All-Electric Spark	119			
Nissan All-Electric Leaf	114			
Toyota Prius Plug-in Hybrid	58			
Toyota Prius—Hybrid	50			
Chevrolet Cruze S6	30			
Honda Accord S6	25			
Jeep Cherokee 4WD SUV	22			
Ford F150 S6 Pickup	15			
Chevrolet Camaro 8 cyl	14			
Ferrari FF	13			

Compiled by the authors using data from the U.S. Environmental Protection Agency and the U.S. Department of Energy.

CENGAGE**brain**.com To access course materials, including Aplia homework, please visit www.cengagebrain.com.

WWW.CENGAGEBRAIN.COM   **357**

The dose makes the poison.

PARACELSUS, 1540

14

ENVIRONMENTAL HAZARDS
AND HUMAN HEALTH

14.3 What types of chemical hazards do we face?

14.4 How can we evaluate chemical hazards?

14.5 How do we perceive risks and how can we avoid the worst of them?

Many coal-burning factories and power plants release toxic mercury and other air pollutants into the atmosphere.

Dudarev Mikhail/Shutterstock.com

Mercury's Toxic Effects

Mercury (Hg) and its compounds are all toxic. Research indicates that long-term exposure to high levels of mercury can permanently damage the human nervous system, brain, kidneys, and lungs. Fairly low levels of mercury can also harm fetuses and cause birth defects. Pregnant women, nursing mothers and their babies, women of childbearing age, and young children are especially vulnerable to the harmful effects of mercury.

This toxic metal is released into the air from rocks, soil, and volcanoes and by vaporization from the oceans. Such natural sources account for about one-third of the mercury reaching the atmosphere each year. According to the U.S. Environmental Protection Agency (EPA) and the U.N. Environment Programme (UNEP), the remaining two-thirds come from human activities. The largest source of mercury air pollution is the millions of small-scale gold miners in Africa, Latin America, and Asia who use mercury to separate gold from its ore and then heat the mixture of gold and mercury to release the gold (see Case Study,

p. 299). The second largest input of mercury into the atmosphere comes from the smokestacks of coal-burning power and industrial plants (chapter-opening photo), cement kilns, smelters, and solid-waste incinerators.

Because mercury is an element, it cannot be broken down or degraded. Therefore, this indestructible global pollutant can accumulate in soil, in water, and in the tissues of humans and other animals. In the atmosphere, some elemental mercury is converted to more toxic inorganic and organic mercury compounds that can be deposited in lakes and other aquatic environments and on land. Under certain conditions in aquatic systems, bacteria can convert inorganic mercury compounds to highly toxic *methylmercury* (CH_3Hg^+). These compounds are typically deposited from the air or dumped into streams by small-scale gold miners. Like DDT (see Figure 8.9, p. 171), methylmercury can be biologically magnified in food chains and food webs, and high levels of methylmercury are often found in the tissues of large fishes—such as tuna, swordfish, shark, and marlin—that feed at high trophic levels in aquatic systems.

Shrimp and salmon generally have low levels of mercury.

Humans are exposed to mercury in two major ways. *First,* we eat fish and shellfish that are contaminated with methylmercury (Figure 14.1). *Second,* we inhale vaporized elemental mercury or particles of inorganic mercury salts in the air.

The greatest risk from exposure to low levels of methylmercury is that of brain damage in fetuses and young children. Studies estimate that 30,000–60,000 of the children born each year in the United States are likely to have reduced IQs and possible nervous system damage because of such exposure.

This problem raises two important questions: How do scientists determine the potential harm from exposure to mercury and other chemicals? And how serious is the risk of harm from a particular chemical compared to other risks?

In this chapter, we will look at how scientists try to answer these and other questions about our exposure to chemicals. We will also examine health threats from disease-causing bacteria, viruses, and protozoa, and from other hazards that kill millions of people every year.

FIGURE 14.1 This fish (left) is contaminated with mercury, as are the fish in many lakes, including this one in Wisconsin (right).

Left: Peter Casolino/Alamy; Right: Drake Fleege/Alamy

WHAT MAJOR HEALTH HAZARDS DO WE FACE?

CONCEPT 14.1 We face health hazards from biological, chemical, physical, and cultural factors, and from the lifestyle choices we make.

We Face Many Types of Hazards

A **risk** is the probability of suffering harm from a hazard that can cause injury, disease, death, economic loss, or damage. Scientists often state a probability in terms such as, "The lifetime probability of developing lung cancer from smoking one pack of cigarettes per day is 1 in 250." This means that 1 of every 250 people who smoke a pack of cigarettes every day will likely develop lung cancer over a typical lifetime (usually considered to be 70 years). Probability can also be expressed as a percentage, as in a 30% chance of rain.

Risk assessment is the process of using statistical methods to estimate how much harm a particular hazard can cause to human health or to the environment. **Risk management** involves deciding whether and how to reduce a particular risk to a certain level and at what cost. Figure 14.2 summarizes how risks are assessed and managed.

We can suffer harm from five major types of hazards (**Concept 14.1**):

- *Biological hazards* from more than 1,400 **pathogens**, or organisms that can cause disease in other organisms. Examples are bacteria, viruses, parasites, protozoa, and fungi.

- *Chemical hazards* from harmful chemicals in our air, water, soil, food, and human-made products (**Core Case Study**).

FIGURE 14.2 Risk assessment and risk management are used to estimate the seriousness of various risks and to help reduce such risks.
Question: What is an example of how you have applied this process in your daily living?

- *Natural hazards* such as fire, earthquakes, volcanic eruptions, floods, and storms.

- *Cultural hazards* such as unsafe working conditions, criminal assault, and poverty.

- *Lifestyle choices* such as smoking, making poor food choices, and having unsafe sex.

WHAT TYPES OF BIOLOGICAL HAZARDS DO WE FACE?

CONCEPT 14.2 The most serious biological hazards we face are infectious diseases such as flu, AIDS, tuberculosis, diarrheal diseases, and malaria.

Some Diseases Can Spread from One Person to Another

An **infectious disease** is a disease caused by a pathogen such as a bacterium, virus, or parasite invading the body and multiplying in its cells and tissues. **Bacteria** are single-cell organisms that are found everywhere and that can multiply very rapidly on their own. Most bacteria are harmless or beneficial but some cause diseases such as strep throat and tuberculosis (see Case Study that follows). **Viruses** are pathogens that are smaller than bacteria and work by invading a cell and taking over its genetic machinery to copy themselves and then spreading throughout a body. They cause diseases such as flu and acquired immune deficiency syndrome (AIDS). **Parasites,** organisms that live on or inside other organisms and feed on them, can also cause serious infectious diseases, including malaria.

A **transmissible disease** is an infectious disease that can be transmitted from one person to another. Some are bacterial diseases such as tuberculosis and gonorrhea. Others are viral diseases such as the common cold, flu, and AIDS. A **nontransmissible disease** is caused by something other than a living organism and does not spread from one person to another. Such diseases include cardiovascular (heart and blood vessel) diseases, most cancers, asthma, and diabetes.

In 1900, infectious disease was the leading cause of death in the world. Since then, and especially since 1950, the incidences of infectious diseases and the death rates from them have been greatly reduced. This has been achieved mostly by a combination of better health care, the use of antibiotics to treat diseases caused by bacteria, and the development of vaccines to prevent the spread of some viral diseases.

Despite the declining risk of harm from infectious diseases, they remain serious health threats, especially in less-developed countries. Such diseases can be spread through air, water, food, and body fluids such as feces, urine, blood, and droplets sprayed by sneezing and coughing.

GENETIC RESISTANCE TO ANTIBIOTICS IS INCREASING

Antibiotics are chemicals that can kill bacteria. They have played an important role in the 30-year increase in American life expectancy since 1950.

In 2014, the World Health Organization issued a report warning that the age of antibiotics may be coming to an end, because disease-causing bacteria are becoming genetically resistant to the antibiotics that we have long used to fight infectious diseases such as ear infections, bloodstream infections, pneumonia, urinary tract infections, diarrhea, and gonorrhea. We are falling behind in our efforts to prevent infectious bacterial diseases because of the astounding reproductive rate of bacteria, some of which can grow from a population of 1 to well over 16 million in 24 hours. This allows bacteria to

quickly become genetically resistant to an increasing number of antibiotics through natural selection (see Figure 4.9, p. 74).

Other factors can promote such genetic resistance to antibiotics. One is the spread of bacteria around the globe by human travel and international trade. Another is the overuse of pesticides, which leads to growing populations of pesticide-resistant insects and other carriers of bacterial diseases. In addition, some drug-resistant bacteria can quickly transfer their resistance to nonresistant bacteria by exchanging genetic material.

Yet another factor is the overuse of antibiotics for colds, flu, and sore throats, most of which are viral diseases that cannot be treated with antibiotics.

In many countries, antibiotics are available without a prescription, which promotes excessive and unnecessary use. Also, the growing use of antibacterial hand soaps and other antibacterial cleansers could be promoting antibiotic resistance in bacteria.

Bacterial resistance to some antibiotics has also increased because these drugs are widely used to control disease and to promote growth among dairy and beef cattle, poultry, hogs, and other livestock that are raised in large numbers in crowded conditions. In 2011, the U.S. Food and Drug Administration (FDA) estimated that about 80% of all antibiotics used in the United States and 50% of those used worldwide are added to the feed of healthy livestock. The FDA found that

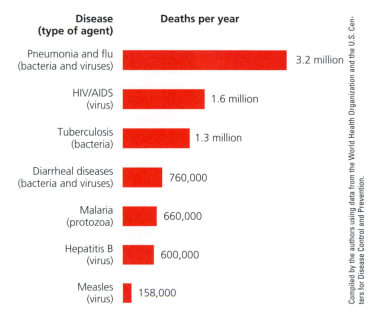

Disease (type of agent) | **Deaths per year**

- Pneumonia and flu (bacteria and viruses) — 3.2 million
- HIV/AIDS (virus) — 1.6 million
- Tuberculosis (bacteria) — 1.3 million
- Diarrheal diseases (bacteria and viruses) — 760,000
- Malaria (protozoa) — 660,000
- Hepatitis B (virus) — 600,000
- Measles (virus) — 158,000

Compiled by the authors using data from the World Health Organization and the U.S. Centers for Disease Control and Prevention.

FIGURE 14.3 The World Health Organization has estimated that the world's seven deadliest infectious diseases kill about 8.3 million people per year—most of them poor people in less-developed countries (**Concept 14.2**). This averages about 22,700 mostly preventable deaths every day. **Question:** How many people, on average, die prematurely from these diseases every hour?

A large-scale outbreak of an infectious disease in an area or a country is called an *epidemic*. A global epidemic such as tuberculosis or AIDS is called a *pandemic*. Figure 14.3 shows the annual death tolls from the world's seven deadliest infectious diseases (**Concept 14.2**).

One reason why infectious disease is still a serious threat is that many disease-carrying bacteria have developed genetic immunity to widely used antibiotics (Science Focus 14.1). Also, many disease-transmitting species of insects such as mosquitoes have become immune to widely used pesticides such as DDT that once helped to control their populations.

CASE STUDY

The Global Threat from Tuberculosis

Since 1990, one of the world's most underreported stories has been the spread of tuberculosis (TB), an extremely contagious bacterial infection of the lungs. Many TB-infected people do not appear to be sick, and most of them do not know they are infected. Left untreated, each person with active TB typically infects a number of other people. Without treatment, about half of the people with active TB

18 of 30 antibiotics used on livestock between 2001 and 2010 posed a high risk to human health through exposure to antibiotic-resistant bacteria in food.

As a result of these factors acting together, every major disease-causing bacterium has now developed strains that resist at least 1 of the roughly 200 antibiotics used to treat bacterial infections, and a growing number of bacteria have developed genetic resistance to more than one antibiotic. We are now seeing the emergence of *superbugs*, bacteria that resist all available antibiotics. In 2013, the Centers for Disease Control and Prevention (CDC) estimated that each year, at least 2 million Americans get infectious diseases from superbugs, and at least 23,000 die. Also, 1 of every 25 U.S. hospital patients picks up such an infection while in the hospital.

For example, a bacterium known as *methicillin-resistant Staphylococcus aureus,* commonly known as MRSA (or "mersa"), has become resistant to most common antibiotics. This type of staph infection first appears on the skin as a red, swollen, and sometimes-painful pimple or boil that will not heal. MRSA can cause a vicious type of pneumonia, flesh-eating wounds, and a quick death if it gets into the bloodstream. MRSA can be found in hospitals, nursing homes, dialysis centers, schools, meeting rooms, gyms, and college dormitories. It can be spread through skin contact, unsanitary use of tattoo needles, and contact with poorly laundered clothing and shared items such as towels, bed linens, athletic equipment, and razors.

Health officials warn that we could be moving into a post-antibiotic era of higher death rates. No major new antibiotics have been developed in recent years, mostly because drug companies lose millions of dollars developing new antibiotics that are used for only a short time to treat various infections, compared to more profitable drugs that users take daily for chronic health problems such as diabetes or high blood pressure.

Researchers are evaluating the use of trace amounts of silver to boost an antibiotic's power. They are also using computer analysis to look for gene patterns in microbes that could be used to develop new antibiotics. Researchers are also looking for ways reduce the ability of germs to cause infections by disrupting the chemical signals they use to release a toxin or to attack.

Critical Thinking

What are three steps that you think we could take to slow the rate at which disease-causing bacteria are developing resistance to antibiotics?

die from bacterial destruction of their lung tissue (Figure 14.4). According to the World Health Organization (WHO), TB strikes about 8.6 million people per year and kills 1.3 million. More than 95% of these deaths occur in less-developed countries, led (in order) by India, Bangladesh, Indonesia, and Pakistan.

Several factors account for the spread of TB since 1990. One is a lack of TB screening and control programs, especially in less-developed countries, where 95% of the new cases occur. However, researchers are developing new and easier ways to detect TB in its victims and to monitor victims for loss of liver function as a possible side effect of taking anti-TB drugs (Individuals Matter 14.1).

A second problem is that most strains of the TB bacterium have developed genetic resistance to the majority of the effective antibiotics. Also, population growth, urbanization, and air travel have greatly increased person-to-person contacts, and TB is spreading faster in areas where large numbers of poor people crowd together, especially in the rapidly growing slums of less-developed countries. A person with active TB might infect several people during a single bus or plane ride.

Slowing the spread of the disease requires early identification and treatment of people with active TB, especially those with a chronic cough, which is the primary way in which the disease is spread from person to person. Treatment with a combination of four inexpensive drugs can cure 90% of individuals with active TB. To be effective, the drugs must be taken every day for 6–9 months. Because the symptoms disappear after a few weeks, many patients, thinking they are cured, stop taking the drugs, allowing the disease to recur, possibly in drug-resistant forms, and to spread to other people.

In recent years, a deadly and apparently incurable form of tuberculosis, known as *multidrug-resistant TB,* has been spreading, especially in India, China, and Russia, and killing about 150,000 people a year. Because this disease cannot be treated effectively with antibiotics, victims must be permanently isolated from the rest of society, and they pose a threat to health workers.

Viral Diseases and Parasites Kill Large Numbers of People

Viruses are not affected by antibiotics and can be very deadly. The biggest viral killer is the *influenza* or *flu virus* (**Concept 14.2**), which is transmitted by the body fluids or airborne emissions of an infected person and often leads to

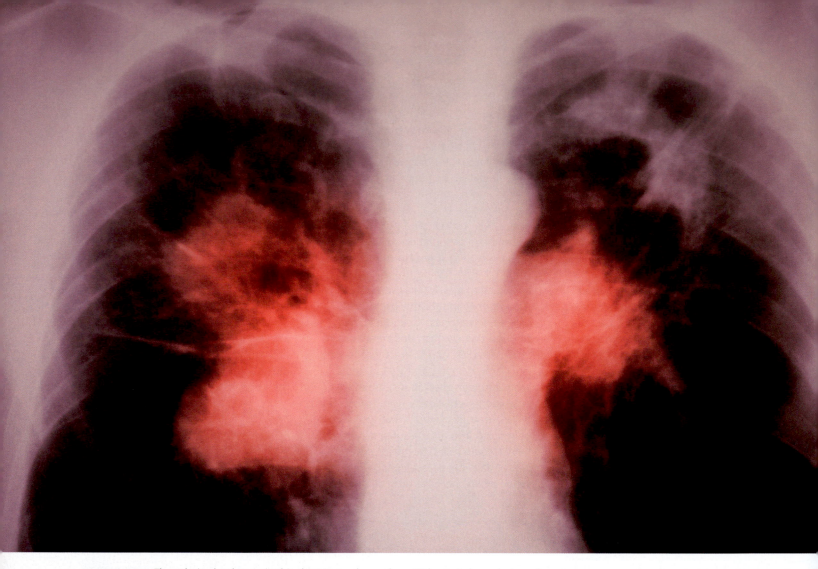

FIGURE 14.4 The colorized red areas in this chest X-ray show where TB bacteria have destroyed lung tissue.

Simon Fraser/Science Source

fatal pneumonia. Flu viruses are so easily transmitted that an especially potent flu virus could spread around the world in a pandemic that could kill millions of people in only a few months.

The second biggest viral killer is the *human immunodeficiency virus, or HIV* (see the following Case Study). On a global scale, according to the WHO, HIV infects about 2.5 million people every year, and the complications resulting from AIDS kill about 1.6 million people annually. HIV is transmitted by unsafe sex, the sharing of needles by drug users, infected mothers who pass the virus to their offspring before or during birth, and exposure to infected blood.

The third largest viral killer is the *hepatitis B virus (HBV),* which damages the liver. According to the WHO, it kills about 600,000 people each year. It is spread in the same ways that HIV is spread.

Scientists estimate that throughout history, more than half of all infectious diseases were originally transmitted to humans from wild or domesticated animals. The West Nile virus (which is transmitted by a mosquito bite and has killed more than 1,500 Americans since 2012), HIV, and a flu strain from birds, called *avian flu,* all fall in this category. The development of such diseases has spurred the growth of the relatively new field of *ecological medicine*—devoted to tracking down disease connections between animals and humans. Scientists in this field have identified several human practices that encourage the spread of diseases among animals and people:

- The clearing or fragmenting of forests to make way for settlements, farms, and expanding cities.

- The hunting of wild game for food. In parts of Africa and Asia, local people who kill monkeys and other animals for bushmeat (see Figure 8.12, p. 175) regularly come in contact with primate blood and can be exposed to a simian (ape or monkey) strain of HIV, which causes AIDS.

- The illegal international trade in wild species.

Hayat Sindi: Science Entrepreneur and National Geographic Emerging Explorer

Growing up in a home of humble means in Saudi Arabia, Hayat Sindi was determined to get an education, become a scientist, and do something for humanity. She was the first Saudi woman to be accepted at Cambridge University. She earned a PhD and taught in the Cambridge's international medical program, and in 2012, she was named a National Geographic Emerging Explorer.

As a visiting scholar, Sindi worked with a team of scientists at Harvard University to co-found a nonprofit marketing company called *Diagnostics for All* to bring low-cost health monitoring to remote, poor areas of the world. The Harvard team had developed an inexpensive diagnostic tool that can be used to detect certain illnesses and medical problems, such as declining liver function. It is a piece of paper the size of a postage stamp, with tiny channels and wells etched into it. Technicians load the paper with certain chemicals and then put a drop of a patient's blood, urine, or saliva onto this paper where the chemicals react with the fluid to change its color. The colors can be read with the aid of a color scale that indicates different sorts of infections and conditions such as declining liver function. The test takes less than a minute, can be conducted by a technician with minimal training, and requires no electricity, clean water, or special equipment. After the paper-testing tool is used, it can be burned on the spot to prevent the spread of any infectious agents.

Sindi has also launched a nonprofit foundation known as the *Institute for Imagination and Ingenuity* to help young people in poor areas of the world to do what she did. It will give funding and guidance to those who wish to attend a university and then return to their home countries to use their talents in helping others.

HAYAT SINDI

Background photo: Antonprado/iStockphoto.com

- Industrialized meat production. For example, a deadly form of *E. coli* bacteria sometimes spreads from livestock to humans when people eat meat contaminated by animal manure. Salmonella bacteria found on animal hides and in poorly processed, contaminated meat also can cause food-borne disease.

Epidemiologist and National Geographic Emerging Explorer Nathan Wolfe is working with partner organizations to create an early warning system for forecasting and containing emerging diseases before they can spread and kill millions. They have begun to create a global network of field sites to identify where people are being exposed to diseases that can be transferred from animals to humans. **GREEN CAREER: Ecological medicine**

You can greatly reduce your chances of getting infectious diseases by practicing good old-fashioned hygiene. Wash your hands thoroughly (for at least 20 seconds) and frequently with plain soap, avoid sharing personal items such as razors or towels, keep all cuts and scrapes covered with bandages until healed, stay away from people who have flu or other viral diseases, and try not to touch your eyes, nose, or mouth.

Yet another growing health hazard is infectious diseases caused by parasites, especially malaria (see the second Case Study that follows).

CASE STUDY

The Global HIV/AIDS Epidemic

The global spread of *acquired immune deficiency syndrome (AIDS)*, caused by infection with HIV, is a major global health threat. This virus cripples the immune system and leaves the body vulnerable to infections such as tuberculosis (TB) and rare forms of cancer such as *Kaposi's sarcoma*. A person infected with HIV can live a normal life, but if the infection develops into AIDS, death is likely.

Since HIV was identified in 1981, this viral infection has spread around the globe. According to the WHO, in 2012, a total of about 35.3 million people worldwide (about 1.1 million in the United States) were living with HIV. In 2012, there were about 2.3 million new cases of AIDS (about 32,000 in the United States)—half of them in people aged 15 to 24. This is down from an estimated 3.2 million new cases in 1997.

The treatment for an HIV infection includes a combination of antiviral drugs that can slow the progress of the virus. However, such drugs cost too much to be used widely in the less-developed countries where AIDS infections are widespread. Also, this treatment can result in a loss of liver function in a victim (Individuals Matter 14.1). It is estimated that about one of every five people infected with HIV is not aware of the infection and can spread the virus for years.

Between 1981 and 2012, more than 36 million people (636,000 in the United States) died of AIDS-related diseases, according to the CDC. Each year, AIDS kills about 1.6 million people (about 15,500 in the United States)—down from a peak of 2.3 million in 2005. AIDS has reduced the life expectancy of the 750 million people living in sub-Saharan Africa from 62 to 47 years, on average, and to 40 years in the seven countries most severely affected by AIDS.

Worldwide, AIDS is the leading cause of death for people of ages 15–49. This affects the population age structures in several African countries, including Botswana (Figure 14.5), where, in 2012, 23% of all people between ages 15 and 49 were infected with HIV, mostly from heterosexual contact. The premature deaths of many teachers, health-care workers, farmers, and other young, productive adults has also contributed to declining education, health care, food production, economic development, and political stability in these countries, along with the disintegration of many families and a large number of orphaned children.

CASE STUDY

Malaria—The Spread of a Deadly Parasite

Almost half of the world's people—most of them living in poor African countries—are at risk of getting malaria (Figure 14.6). So is anyone traveling to malaria-prone areas, because there is no vaccine that can prevent this disease.

Malaria is caused by a parasite that is spread by certain mosquito species. The parasite infects and destroys its victim's red blood cells, causing intense fever, chills, drenching sweats, severe abdominal pain, vomiting, headaches, and increased susceptibility to other diseases. According to the WHO, malaria kills about 660,000 people a year. However, some experts contend this total could be as high as 1.2 million, because public health records are incomplete in many areas. More than 90% of all malaria victims live in sub-Saharan Africa, the area south of the Sahara Desert,

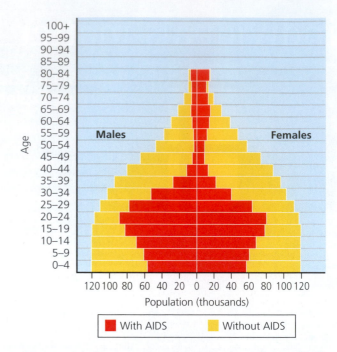

FIGURE 14.5 In Botswana, more than 25% of people ages 15–49 were infected with HIV in 2011. This figure shows two projected age structures for Botswana's population in 2020—one including the possible effects of the AIDS epidemic (red bars), and the other not including those effects (yellow bars). **Question:** How might this affect Botswana's economic development?

Compiled by the authors using data from the U.S. Census Bureau, UN Population Division, and World Health Organization.

and most of them are children younger than age 5. Many of the children who survive suffer brain damage or impaired learning ability.

Four species of protozoan parasites in the genus *Plasmodium* cause malaria. The cycle of infection begins when an uninfected female of any of about 60 *Anopheles* mosquito species bites a person (usually at night) who is infected with the *Plasmodium* parasite, ingests blood that contains the parasite, and later bites an uninfected person. The parasites then move out of the mosquito and into the human's bloodstream and liver, where they multiply. Malaria can also be transmitted by contaminated blood transfusions and by drug users sharing needles.

Over the course of human history, malarial protozoa probably have killed more people than all the wars ever fought. During the 1950s and 1960s, the spread of malaria was slowed when the swamplands and marshes where mosquitoes breed were drained or sprayed with insecticides, and drugs were used to kill the parasites in victims' bloodstreams. Since 1970, however, malaria has come roaring back. Most species of the *Anopheles* mosquito have become genetically resistant to most insecticides and the *Plasmodium* parasites have become genetically resistant to common antimalarial drugs. During this century, as the average atmospheric temperature rises, populations of malaria-carrying mos-

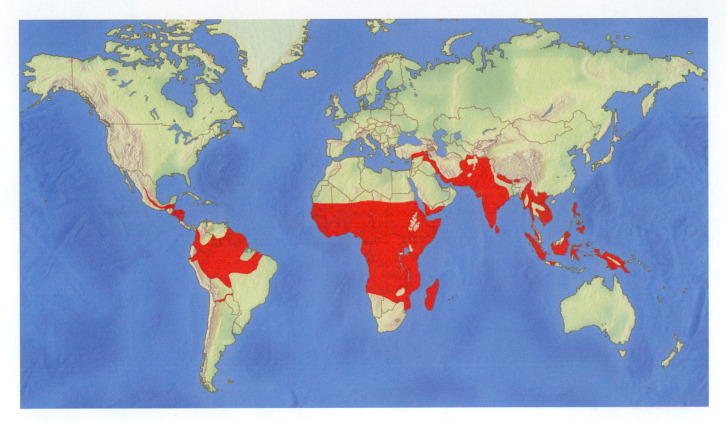

FIGURE 14.6 About 47% of the world's population live in areas in which malaria is prevalent.

Compiled by the authors using data from the World Health Organization and U.S. Centers for Disease Control and Prevention.

quitoes will likely spread from tropical areas to warmer temperate areas of the earth.

Researchers are working to develop new antimalarial drugs and vaccines, as well as biological controls for *Anopheles* mosquitoes. Another approach is to provide poor people in malarial regions with free or inexpensive insecticide-treated bed nets (Figure 14.7) and window screens. Also, zinc and vitamin A supplements could be given to children to boost their resistance to malaria. In addition, we can greatly reduce the incidence of malaria at a low cost by spraying the insides of homes with low concentrations of the pesticide DDT twice a year. While DDT is being phased out in most countries, the WHO supports limited use of DDT for malaria control. In 2014, researchers developed a rapid laser pulse test for malaria that eliminates the need to draw blood. It takes only 20 seconds at half the cost of the conventional test, and because no blood is withdrawn, it cuts the risk of infection.

We Can Reduce the Incidence of Infectious Diseases

According to the WHO, the percentage of all deaths worldwide resulting from infectious diseases dropped from 35% to 15% between 1970 and 2010, primarily because a growing number of children were immunized against the major infectious diseases. Between 1990 and 2010, the estimated annual number of children younger than age 5 who died from infectious diseases dropped from nearly 12 million to 4.4 million. **GOOD NEWS**

Figure 14.8 lists measures promoted by health scientists and public health officials to help prevent or reduce the incidence of infectious diseases—especially in less-developed countries. An important breakthrough has been the development of simple *oral rehydration therapy* to help prevent death from dehydration for victims of severe diarrhea, which causes about one-fourth of all deaths of children younger than age 5. This therapy involves administering a simple solution of boiled water, salt, and sugar or rice at a cost of only a few cents per person. **GREEN CAREER: Infectious disease prevention**

CONSIDER THIS. . .

CONNECTIONS Drinking Water, Latrines, and Infectious Diseases

More than a third of the world's people—2.6 billion—do not have sanitary bathroom facilities, and more than 1 billion get their water for drinking, washing, and cooking from sources polluted by animal or human feces. A key to reducing sickness and premature death due to infectious disease is to focus on providing these people with simple latrines and access to safe drinking water.

Mosquito nets can help to prevent the spread of malaria.

FIGURE 14.7 This baby in Senegal, Africa, is sleeping under an insecticide-treated mosquito net to reduce the risk of being bitten by malaria-carrying mosquitoes.

Olivier Asselin/Alamy

Solutions

Infectious Diseases

- Increase research on tropical diseases and vaccines

- Reduce poverty and malnutrition

- Improve drinking water quality

- Reduce unnecessary use of antibiotics

- Sharply reduce use of antibiotics on livestock

- Immunize children against major viral diseases

- Provide oral rehydration for diarrhea victims

- Conduct global campaign to reduce HIV/AIDS

FIGURE 14.8 Some ways to prevent or reduce the incidence of infectious diseases, especially in less-developed countries. ***Questions:*** Which three of these approaches do you think are the most important? Why?

Top: Omer N Raja/Shutterstock.com. Bottom: Rob Byron/Shutterstock.com.

© 2016 Cengage Learning

WHAT TYPES OF CHEMICAL HAZARDS DO WE FACE?

CONCEPT 14.3 There is growing concern about chemicals in the environment that can cause cancers and birth defects and disrupt the human immune, nervous, and endocrine systems.

Some Chemicals Can Cause Cancers, Mutations, and Birth Defects

A **toxic chemical** is an element or compound that can cause temporary or permanent harm or death to humans. The U.S. Environmental Protection Agency (EPA) has listed arsenic, lead, mercury (**Core Case Study**), vinyl chloride (used to make PVC plastics), and polychlorinated biphenyls (PCBs) as the top five toxic substances in terms of human and environmental health.

There are three major types of potentially toxic agents. **Carcinogens** are chemicals, some types of radiation, and certain viruses that can cause or promote *cancer*—a disease in which malignant cells multiply uncontrollably and create tumors that can damage the body and often lead to premature death. Examples of carcinogens are arsenic, benzene, formaldehyde, gamma radiation, PCBs, radon, ultraviolet (UV) radiation, vinyl chloride, and certain chemicals in tobacco smoke.

Typically, 10–40 years may elapse between the initial exposure to a carcinogen and the appearance of detectable cancer symptoms. Partly because of this time lag, many healthy teenagers and young adults have trouble believing that their smoking, drinking, eating, and other habits today could lead to some form of cancer before they reach age 50.

The second major type of toxic agent, **mutagens**, includes chemicals or forms of radiation that cause or increase the frequency of *mutations*, or changes, in the DNA molecules found in cells. Most mutations cause no harm, but some can lead to cancers and other disorders. For example, nitrous acid (HNO_2), formed by the digestion of nitrite (NO_2^-) preservatives in foods, can cause mutations linked to increases in stomach cancer in people who consume large amounts of processed foods and wine containing such preservatives. Harmful mutations occurring in reproductive cells can be passed on to offspring and to future generations.

Teratogens, a third type of toxic agent, are chemicals that harm or cause birth defects in a fetus or embryo. Ethyl alcohol is a teratogen. Women who drink during pregnancy increase their risk of having babies with low birth weight and a number of physical, developmental, behavioral, and mental problems. Other teratogens are angel dust, benzene, formaldehyde, lead, mercury (**Core Case Study**), PCBs, phthalates, thalidomide, and vinyl chloride. (We discuss some of these hazardous chemicals further on in this chapter.)

Some Chemicals Can Affect Important Body Systems

Since the 1970s, research on wildlife and laboratory animals along with some studies of humans suggest that long-term exposure to some chemicals in the environment can disrupt important body systems, including immune and nervous systems (**Concept 14.3**).

The *immune system* consists of specialized cells and tissues that protect the body against disease and harmful substances by forming *antibodies,* or specialized proteins that render invading agents harmless. Some chemicals such as arsenic, methylmercury (**Core Case Study**), and dioxins can weaken the human immune system and leave the body vulnerable to attacks by allergens and infectious bacteria, viruses, and protozoa.

Some natural and synthetic chemicals in the environment, called *neurotoxins,* can harm the human *nervous system* (brain, spinal cord, and peripheral nerves). Effects can include behavioral changes, learning disabilities, attention-deficit disorder, paralysis, and death. Examples of neurotoxins are PCBs, arsenic, lead, and certain pesticides.

Methylmercury (**Core Case Study**) is an especially dangerous neurotoxin because it is persists in the environment and, like DDT, can be biologically magnified in food chains and food webs (Figure 8.9, p. 171). According to the Natural Resources Defense Council, predatory fish such as tuna, marlin, orange roughy, swordfish, mackerel, Chilean sea bass, grouper, and sharks can have mercury concentrations in their bodies that are 10,000 times higher than the levels in the water around them. A 2009 EPA study found that almost half of the fish tested in 500 lakes and reservoirs across the United States had levels of mercury that exceeded safe levels (Figure 14.1). Similarly, a 2009 study by the U.S. Geological Survey of nearly 300 streams across the United States found mercury-contaminated fish in all of the streams surveyed, with one-fourth of the fish exceeding the safe levels determined by the EPA.

The symptoms of mercury poisoning in adults include poor balance and coordination, muscle weakness, tremors, memory problems, insomnia, hearing loss, loss of hair, and loss of peripheral vision. The EPA estimates that about 1 of every 12 women of childbearing age in the United States has enough mercury in her blood to harm a developing fetus. Figure 14.9 lists ways to prevent or reduce human inputs of mercury (**Core Case Study**) into the environment.

Some Chemicals Affect the Human Endocrine System

The *endocrine system* is a complex network of glands that release tiny amounts of *hormones* into the bloodstreams of humans and other vertebrate animals. Very low levels of these chemical messengers (often measured in parts per

THE CONTROVERSY OVER BPA

The estrogen mimic *bisphenol A (BPA)* serves as a hardening agent in certain plastics (especially shatter-proof polycarbonate) that are used in a variety of products, including some baby bottles, sipping cups, and pacifiers, as well as some reusable water bottles, sports drink and juice bottles, microwave dishes, and food storage containers. BPA is also used to make some dental sealants, as well as the plastic resins that line nearly all food and soft drink cans and cans holding baby formulas and foods. This type of liner allows containers to withstand extreme temperatures, keeps canned food from interacting with the metal in the cans, prevents rust in the cans, and helps to preserve the canned food. In addition, individuals can be exposed to BPA when their hands touch the thermal paper used to produce some cash register receipts.

A study by the CDC indicated that 93% of Americans older than age 6 had trace levels of BPA in their urine. While these levels were well below the acceptable level set by the EPA, the EPA level was established in the late 1980s, long before we knew much about the potential effects of BPA on human health. The CDC study also found that children and adolescents had generally higher urinary BPA levels than adults had.

Research indicates that the BPA in plastics can leach into water or food when the plastic is heated to high temperatures, microwaved, or exposed to acidic liquids. A 2009 Harvard University Medical School study found that there was a 66% increase in BPA levels in the urine of participants who drank from polycarbonate bottles regularly for just 1 week.

By 2013, more than 90 published studies by independent laboratories had found a number of significant adverse effects on test animals from exposure to very low levels of BPA. These effects include brain damage, early puberty, decreased sperm quality, certain

Solutions

Mercury Pollution

Prevention		Control
Phase out waste incineration		Sharply reduce mercury emissions from coal-burning plants and incinerators
Remove mercury from coal before it is burned		Label all products containing mercury
Switch from coal to natural gas and renewable energy resources		Collect and recycle batteries and other products containing mercury

© 2016 Cengage Learning

FIGURE 14.9 Some ways to prevent or control inputs of mercury (**Core Case Study**) into the environment from human sources—mostly coal-burning power plants and incinerators. ***Questions:*** Which four of these solutions do you think are the most important? Why?

Top: Mark Smith/Shutterstock.com. Bottom: tuulijumala/Shutterstock.com.

billion or parts per trillion) regulate bodily systems that control sexual reproduction, growth, development, learning ability, and behavior. Each type of hormone has a unique molecular shape that allows it to attach to certain parts of cells called *receptors,* and to transmit its chemical message.

Molecules of certain pesticides and other synthetic chemicals such as *bisphenol A,* or *BPA* (Science Focus 14.2), have shapes similar to those of natural hormones. This allows them to attach to the molecules of natural hormones and to disrupt the endocrine systems in humans and in some other animals (**Concept 14.3**). These molecules are called *hormonally active agents (HAAs)* or *endocrine disruptors.*

Examples of HAAs include Atrazine™ and several other widely used herbicides, organophosphate pesticides, dioxins, lead, mercury (**Core Case Study**), phthalates, various fire retardants, and BPA. Some HAAs, including BPA and bisphenol S (BPS), act as hormone imposters, or *hormone mimics.* They are chemically similar to estrogens (female sex hormones) and can disrupt the endocrine system by attaching to estrogen receptor molecules. Other HAAs, called *hormone blockers,* disrupt the endocrine system by preventing natural hormones such as androgens (male sex hormones) from attaching to their receptors.

cancers, heart disease, obesity, liver damage, impaired immune function, type 2 diabetes, hyperactivity, impaired learning, impotence in males, and obesity in unborn test animals exposed to BPA. On the other hand, 12 studies funded by the chemical industry found no evidence or only weak evidence of adverse effects from low-level exposure to BPA in test animals.

In the face of these conflicting and possibly confusing research findings, consumers now have more choices, since most makers of baby bottles, sipping cups, and sports water bottles offer BPA-free alternatives. Also, many consumers are avoiding plastic containers with a #7 recycling code (which indicates that BPA can be present). People are also using powdered infant formula instead of liquid formula from metal cans, choosing glass bottles and food containers instead of those made of plastic or lined with plastic resins, and using glass, ceramic, or stainless steel coffee mugs rather than plastic cups.

Many manufacturers have replaced BPA with bisphenol-S (BPS). However, studies indicate that BPS can have effects similar to those of BPA, and BPS is now showing up in human urine at levels similar those of BPA. There are substitutes for the plastic resins containing BPA or BPS that line 90% of the food cans used in the United States. However, these replacements are more expensive, and the potential health effects of some chemicals they contain also need to be evaluated. Some companies, including Campbell Soup and H.J. Heinz, are phasing out the use of BPA in can liners.

Canada, the European Union, and six states have banned the sale of plastic baby bottles that contain BPA. In 2008, the U.S. Food and Drug Administration (FDA) concluded that BPA in food and drink containers was not a health hazard. However, in 2012, the FDA, while not reversing its general position, banned the use of BPA in baby bottles and sipping cups.

Critical Thinking

Should plastics that contain BPA or BPS be banned from use in all children's products? Explain. Should such plastics be banned from use in the liners of canned food containers? Explain. What are the alternatives?

In 2013, the FDA indicated that the chemicals triclosan and triclocarban, widely used in antibacterial soaps and some deodorants, could be hormone disrupters and could be contributing to bacterial resistance to antibiotics (see Figure 4.9, p. 74). The FDA also said that there is no evidence that using these chemicals is any more effective in preventing bacterial infections than is thoroughly washing your hands with plain soap and water. Since 2000, several European countries have restricted the use of triclosan in consumer products.

There is growing concern about another group of HAAs—pollutants that can act as *thyroid disrupters* and cause growth, weight, brain, and behavioral disorders. Scientists are also increasingly concerned about certain HAAs called *phthalates* (pronounced THALL-eights). These chemicals are used to make plastics more flexible and to make cosmetics easier to apply to the skin. They are found in a variety of products, including many detergents, perfumes, cosmetics, baby powders, body lotions for adults and babies, sunscreens, hair sprays, deodorants, soaps, nail polishes, shampoos for adults and babies, and the coatings on many timed-release drugs. They are also found in polyvinyl chloride (PVC) plastic products such as soft vinyl toys, teething rings, blood storage bags, intravenous drip bags, medical tubing, shower curtains, and some plastic food and drink containers that can leach phthalates when heated.

Exposure of laboratory animals to high doses of various phthalates has caused birth defects, liver cancer, kidney and liver damage, premature breast development, immune system suppression, and abnormal sexual development in these animals. Studies at the Mount Sinai School of Medicine and the National Institute of Child Health and Human Development have linked exposure of human babies to phthalates with early puberty in girls and infertility problems in men. A 2014 study led by biologist Heather Patisaul and her colleagues suggested that phthalate exposure might lead to sperm damage in men by interfering with the creation of the male hormone testosterone. The European Union and at least 14 other countries have banned several phthalates. But scientists, government regulators, and manufacturers in the United States are divided on its risks to human health and reproductive systems.

Some scientists hypothesize that rising levels of hormone disrupters in our bodies may be related to more frequent health problems, including sharp drops in male sperm counts and sperm motility; rising rates of testicular cancer and genital birth defects in men; and rising rates of breast cancer in women. Some scientists argue that there

What Can You Do?

Exposure to Hormone Disrupters

- Eat certified organic produce and meats

- Avoid processed, prepackaged, and canned foods

- Use glass and ceramic cookware

- Store food and drinks in glass containers

- Use only natural cleaning and personal care products

- Use natural fabric shower curtains, not vinyl

- Avoid artificial air fresheners, fabric softeners, and dryer sheets

- Use only glass baby bottles and BPA-free, phthalate-free sipping cups, pacifiers, and toys

FIGURE 14.10 Individuals matter: You can reduce your exposure, and that of any children you care for, to hormone disrupters. **Questions:** Which three of these steps do you think are the most important ones to take? Why?

are not enough scientific study results or statistical evidence to strongly link these medical problems with HAA levels in humans. They call for more research to justify any restrictions on HAAs that would cause economic losses for the companies that make them.

Concerns about BPA, phthalates, and other HAAs highlight the difficulty of assessing the possible harmful health effects resulting from exposure to very low levels of various chemicals. Resolving these uncertainties will take decades of research. Some scientists argue that as a precaution during this period of research, we should sharply reduce our use of potentially harmful hormone disrupters, especially in products frequently used by pregnant women, infants, young children, and teenagers (Figure 14.10).

14.4 HOW CAN WE EVALUATE CHEMICAL HAZARDS?

CONCEPT 14.4A Scientists use live laboratory animals, case reports of poisonings, and epidemiological studies to estimate the toxicity of chemicals, but these methods have limitations.

CONCEPT 14.4B Many health scientists call for much greater emphasis on pollution prevention to reduce our exposure to potentially harmful chemicals.

Many Factors Determine the Harmful Health Effects of Chemicals

Toxicology is the study of the harmful effects of chemicals on humans and other organisms. **Toxicity** is a measure of the ability of a substance to cause injury, illness, or

death to a living organism. A basic principle of toxicology is that *any synthetic or natural chemical can be harmful if ingested or inhaled in a large enough quantity.* But the critical question is this: *At what level of exposure to a particular toxic chemical will the chemical cause harm?*

This is a difficult question to answer because of the many variables involved in estimating the effects of human exposure to chemicals. A key factor is the **dose**, the amount of a harmful chemical that a person has ingested, inhaled, or absorbed through the skin at any one time.

The effects of a particular chemical can also depend on the age of the person exposed to it. For example, toxic chemicals usually have a greater effect on fetuses, infants, and children than on adults. Current research suggests that exposure to chemical pollutants in the womb may be related to increasing rates of autism, childhood asthma, and learning disorders. Infants and young children are more susceptible to the effects of toxic substances than are adults, for three major reasons. *First*, they generally breathe more air, drink more water, and eat more food per unit of body weight than do adults. *Second*, they are exposed to toxins in dust and soil when they put their fingers, toys, and other objects in their mouths. *Third*, children usually have less well-developed immune systems and body detoxification processes than adults have.

The EPA has proposed that in determining any risk, regulators should assume that children have a 10-times higher risk factor than adults have. Some health scientists suggest that to be on the safe side, we should assume that this risk for children is 100 times the risk for adults.

Toxicity also depends on *genetic makeup*, which determines an individual's sensitivity to a particular toxin. Some people are sensitive to a number of toxins—a condition known as *multiple chemical sensitivity* (MCS). Another factor is how well the body's detoxification systems, including the liver, lungs, and kidneys, are working.

Several other variables can affect the level of harm caused by a chemical. One is its *solubility*. Water-soluble toxins (which are often inorganic compounds) can move throughout the environment and get into water supplies, as well as the aqueous solutions that surround the cells in our bodies. Oil- or fat-soluble toxins (which are usually organic compounds) can penetrate the membranes that surround our cells, because these membranes allow similar oil-soluble chemicals to pass through them. Thus, oil- or fat-soluble toxins can accumulate in body tissues and cells.

Another factor is a substance's *persistence*, or resistance to breaking down. Many chemicals, including DDT and PCBs, were used widely because they are not easily broken down in the environment. This means that they are more likely to remain in the body and have long-lasting harmful health effects.

Biological accumulation and magnification (see Figure 8.9, p. 171) can also play a role in toxicity. Animals that eat higher on the food chain are more susceptible to the effects of fat-soluble toxic chemicals because of the magnified concentrations of the toxins in their bodies. Examples of chemicals that can be biomagnified include DDT, PCBs, and methylmercury (**Core Case Study**).

The health damage resulting from exposure to a chemical is called the **response**. One type of response, an *acute effect,* is an immediate or rapid harmful reaction ranging from dizziness to death. By contrast, a *chronic effect* is a permanent or long-lasting consequence (kidney or liver damage, for example) of exposure to a single dose or to repeated lower doses of a harmful substance.

Some people have the mistaken idea that all natural chemicals are safe and all synthetic chemicals are harmful. In fact, many synthetic chemicals, including many of the medicines we take, are quite safe if used as intended, while many natural chemicals such as lead and mercury (**Core Case Study**) are deadly.

Scientists Use Live Laboratory Animals and Non-Animal Tests to Estimate Toxicity

The most widely used method for determining toxicity is to expose a population of live laboratory animals to measured doses of a specific substance under controlled conditions. Laboratory-bred mice and rats are widely used because, as mammals, their systems function, to some degree, similarly to human systems. Also, they are small and can reproduce rapidly under controlled laboratory conditions.

Animal tests typically take 2–5 years, involve hundreds to thousands of test animals, and can cost as much as $2 million per substance tested. Some of these tests can be painful to the test animals and can kill or harm them. Animal welfare groups want to limit or ban the use of test animals and, at the very least, want to ensure that they are treated in the most humane manner possible.

Scientists estimate the toxicity of a chemical by determining the effects of various doses of the chemical on test organisms and plotting the results in a **dose-response curve** (Figure 14.11). One approach is to determine the *lethal dose*—the dose that will kill an animal. A chemical's *median lethal dose (LD50)* is the dose that can kill 50% of the animals (usually rats and mice) in a test population within a given time period, usually expressed in milligrams of the chemical per kilogram of body weight (mg/kg). Then scientists use mathematical models to *extrapolate,* or estimate the effects of the chemical on humans, based on the results from testing the chemical on lab animals.

Chemicals vary widely in their toxicity (Table 14.1). Some poisons can cause serious harm or death after a single very low dose. Others cause such harm only at dos-

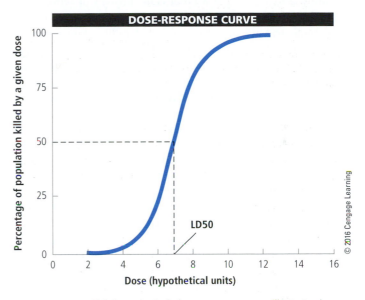

FIGURE 14.11 This hypothetical *dose-response curve* illustrates how scientists can estimate the LD50 for a chemical—the dosage that kills 50% of the animals in a test group. Toxicologists use this method to compare the toxicities of different chemicals.

TABLE 14.1 Toxicity Ratings and Average Lethal Doses for Humans

Toxicity Rating	LD50 (milligrams per kilogram of body weight)*	Average Lethal Dose†	Examples
Supertoxic	Less than 5	Less than 7 drops	Nerve gases, botulism toxin, mushroom toxin, dioxin (TCDD)
Extremely toxic	5–50	7 drops to 1 teaspoon	Potassium cyanide, heroin, atropine, parathion, nicotine
Very toxic	50–500	1 teaspoon to 1 ounce	Mercury salts, morphine, codeine
Moderately toxic	500–5,000	1 ounce to 1 pint	Lead salts, DDT, sodium hydroxide, sodium fluoride, sulfuric acid, caffeine, carbon tetrachloride
Slightly toxic	5,000–15,000	1 pint to 1 quart	Ethyl alcohol, household cleansers, soaps
Essentially nontoxic	15,000 or greater	More than 1 quart	Water, glycerin, table sugar

*Dosage that kills 50% of individuals exposed.

†Amounts of substances in liquid form at room temperature that are lethal when given to a 70-kilogram (150-pound) human.

ages so huge that it is nearly impossible to get enough into the body to cause injury or death. Most chemicals fall between these two extremes.

Some scientists challenge the validity of extrapolating data from test animals to humans, because human physiology and metabolism often differ from those of the test animals. Other scientists say that such tests and models work fairly well (especially for revealing cancer risks) when the correct experimental animal is chosen or when a chemical is toxic to several different test-animal species.

More humane methods for toxicity testing are available and are being used more often to replace testing on live animals. They include making computer simulations and using tissue cultures of cells and bacteria, chicken egg membranes, and individual animal cells, instead of whole, live animals. High-speed robot testing devices can now screen the biological activity of more than 1 million compounds a day to help determine their possible toxic effects.

The problems with estimating toxicities by using laboratory experiments get even more complicated (**Concept 14.4A**). In real life, each of us is exposed to a variety of chemicals, some of which can interact in ways that decrease or enhance their short- and long-term individual effects. Toxicologists already have great difficulty in estimating the toxicity of a single substance. Adding the problem of evaluating mixtures of potentially toxic substances, isolating the culprits, and determining how they can interact with one another is overwhelming from a scientific and economic standpoint. For example, just studying the interactions among 3 of the 500 most widely used industrial chemicals would take 20.7 million experiments—a physical and financial impossibility.

There Are Other Ways to Estimate the Harmful Effects of Chemicals

Scientists use several other methods to get information about the harmful effects of chemicals on human health. For example, *case reports*, usually made by physicians, provide information about people who have suffered from adverse health effects or died after exposure to a chemical. Such information often involves accidental or deliberate poisonings, drug overdoses, homicides, or suicide attempts.

Most case reports are not reliable sources for estimating toxicity because the actual dosage and the exposed person's health status are often unknown. But such reports can provide clues about environmental hazards and suggest the need for laboratory investigations.

Another source of information is *epidemiological studies*, which compare the health of people exposed to a particular chemical (the *experimental group*) with the health of a similar group of people not exposed to the agent (the *control group*). The goal is to determine whether the statistical association

between exposure to a toxic chemical and a health problem is strong, moderate, weak, or undetectable.

Four factors can limit the usefulness of epidemiological studies. *First,* in many cases, too few people have been exposed to high enough levels of a toxic agent to detect statistically significant differences. *Second,* the studies usually take a long time. *Third,* closely linking an observed effect with exposure to a particular chemical is difficult because people are exposed to many different toxic agents throughout their lives and can vary in their sensitivity to such chemicals. *Fourth,* we cannot use epidemiological studies to evaluate hazards from new technologies or chemicals to which people have not yet been exposed.

Are Trace Levels of Toxic Chemicals Harmful?

Almost everyone who lives in a more-developed country is now exposed to potentially harmful chemicals (Figure 14.12) that have built up to trace levels in their blood and in other parts of their bodies. A 2011 CDC study showed that the blood of an average American contains traces of 212 different chemicals, including potentially harmful chemicals such as arsenic and BPA.

Should we be concerned about trace amounts of various synthetic chemicals in our air, water, food, and bodies? In most cases, we simply do not know because there are too few data and because it is so difficult to determine the effects of exposures to low levels of these chemicals (**Concept 14.4A**).

Some scientists view exposures to trace amounts of synthetic chemicals with alarm, especially because of their potential long-term effects on the human immune, nervous, and endocrine systems. Others scientists view the threats from such exposures as minor. They point out that average life expectancy has been increasing in most countries, especially more-developed countries, for decades. Some scientists contend that the concentrations of such chemicals are so low that they are harmless. All agree that there is a need for much more research on the effects of trace levels of synthetic chemicals on human health.

Why Do We Know So Little about the Harmful Effects of Chemicals?

As we have seen, all methods for estimating toxicity levels and risks have serious limitations (**Concept 14.4A**), but they are all that we have. To take this uncertainty into account and to minimize harm, scientists and regulators typically set allowed levels of exposure to toxic substances at 1/100th or even 1/1,000th of the estimated harmful levels.

According to risk assessment expert Joseph V. Rodricks, "Toxicologists know a great deal about a few chemicals, a little about many, and next to nothing about most." The

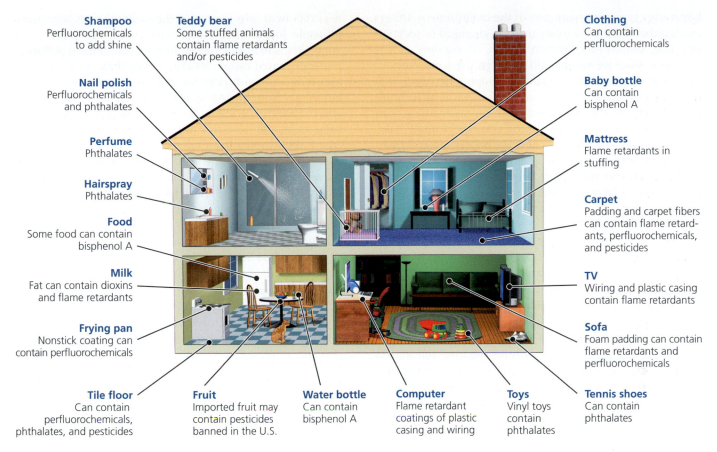

FIGURE 14.12 A number of potentially harmful chemicals are found in many homes. **Questions:** Does the fact that we do not know much about the long-term harmful effects of these chemicals make you more likely or less likely to minimize your exposure to them? Why?

Compiled by the authors using data from U.S. Environmental Protection Agency, Centers for Disease Control and Prevention, and New York State Department of Health.

U.S. National Academy of Sciences estimates that only 10% of the more than 85,000 registered synthetic chemicals in commercial use have been thoroughly screened for toxicity. Only 2% have been adequately tested to determine whether they are carcinogens, mutagens, or teratogens. Hardly any of the chemicals in commercial use have been screened for possible damage to the human nervous, endocrine, and immune systems. Because of insufficient data and the high costs of regulation, federal and state governments do not supervise the use of nearly 99.5% of the commercially available chemicals in the United States, and the problem is much worse in some less-developed countries.

How Far Should We Go in Using Pollution Prevention and the Precautionary Principle?

We know little about the potentially toxic chemicals around us and inside of us, and estimating their effects is very difficult, time-consuming, and expensive. So where does this leave us?

Some scientists and health officials, especially those in European Union countries, are pushing for much greater emphasis on *pollution prevention*. They say we should not release chemicals that we know or suspect can cause significant harm into the environment. This means looking for harmless or less harmful substitutes for toxic and hazardous chemicals. Another option is to recycle them within production processes to keep them from reaching the environment, as companies such as DuPont and 3M have been doing (see the Case Study that follows).

Pollution prevention is a strategy for implementing the **precautionary principle**. According to this principle, which can be applied to many other problems, when there is substantial preliminary evidence that an activity, technology, or chemical substance can harm humans, other organisms, or the environment, we should take precautionary measures to prevent or reduce such harm, rather than waiting for more conclusive scientific evidence (**Concept 14.4B**).

There is controversy over how far we should go in using the precautionary principle. Some argue that we should avoid using any chemical or technology that would harm any organism or any part of the environment. Others suggest that we should freely use any chemical or technology that can further human health or comfort or the economy. Most viewpoints fall between these extremes.

Those who favor a precautionary approach argue that those proposing to introduce a new chemical or technology should bear the burden of establishing its safety. This requires two major changes in the way we evaluate risks. *First,* we would assume that new chemicals and technologies could be harmful until scientific studies could show otherwise. *Second,* the existing chemicals and technologies that appear to have a strong chance of causing significant harm would be removed from the market until we could establish their safety.

Some movement is being made in the direction of applying the precautionary principle, especially in the European Union. In 2000, negotiators agreed to a global treaty that would ban or phase out the use of 12 of the most notorious *persistent organic pollutants (POPs),* also called the *dirty dozen.* These highly toxic chemicals have been shown to produce numerous harmful effects, including cancers, birth defects, compromised immune systems, and a 50% decline in sperm counts and sperm quality in men in a number of countries. The list includes DDT and eight other pesticides, PCBs, and dioxins. In 2009, nine more POPs were added, some of which are widely used in pesticides and in flame-retardants added to clothing, furniture, and other consumer goods. New chemicals will be added to the list when the harm they could potentially cause is seen as outweighing their usefulness. The POPs treaty went into effect in 2004 but has not been formally approved or implemented by the United States.

At U.S. Congressional hearings in 2009, experts testified that the current regulatory system in the United States makes it virtually impossible for the government to limit or ban the use of toxic chemicals. The hearings found that under this system, the EPA had required testing for only 200 of the more than 85,000 chemicals registered for use in the United States and had issued regulations to control fewer than 12 of those chemicals.

Manufacturers and businesses contend that widespread application of the much more precautionary approach used in the European Union would make it too expensive and almost impossible to introduce any new chemical or technology. It also does not take into account that there is some uncertainty in any scientific assessment of risk. This raises the question of what level of scientific uncertainty about a risk is needed to trigger the use of this principle.

Proponents of increased reliance on using pollution prevention agree that we can go too far. But they argue that we have an ethical responsibility to reduce known or potentially serious risks to human health, to the earth's life-support system, and to future generations, in keeping with the ethical **social science principle of sustainability**.

Proponents also point out that using the precautionary principle focuses the efforts and creativity of scientists, engineers, and businesses on finding solutions to pollution problems based on prevention rather than on cleanup. It also reduces health risks for employees and society, frees businesses from having to deal with pollution regulations, and reduces the threat of lawsuits from injured parties. In some cases, applying this principle helps companies to increase their profits from sales of safer products and innovative technologies, and it improves the public image of businesses operating in this manner.

In 2011, after a 35-year delay promoted by politically powerful coal companies and utilities that burn coal to produce electricity, the U.S. government took a step in this direction by issuing a rule to control emissions of mercury (**Core Case Study**) and harmful fine-particle pollution from older coal-burning plants in 28 states. Many eastern states have high depositions of mercury and harmful particles produced by coal-burning power and electric plants in the Midwest and blown eastward by prevailing winds (Figure 14.13). These new air pollution standards could prevent as many as 11,000 premature deaths, 200,000 non-fatal heart attacks, and 2.5 million asthma attacks. In 2014, the U.S. Supreme Court upheld these new EPA regulations.

In 2013, after 4 years of negotiations, representatives from many nations developed a U.N. treaty, known as the Minamata Convention, which seeks to curb or end most human-related inputs of mercury into the environment (**Core Case Study**). The overall goal is to reduce global mercury emissions by 15–35% in the next several decades. By January 2014, it had been signed by 140 countries and *ratified,* or formally approved, by some, including the United States. It will go into effect after 50 countries have ratified it. Once in effect, within 5 years, nations ratifying the treaty must require new coal-fired power plants, industrial plants, and smelters to use the best available mercury emission-control technologies. The treaty will also restrict the use of mercury in common household products such as cosmetics, light bulbs, switches, and batteries, and in measuring devices such as barometers, thermometers, and blood pressure monitors.

CASE STUDY

Pollution Prevention Pays

The 3M Company, based in the U.S. state of Minnesota, makes 60,000 different products in 100 manufacturing plants around the world. In 1975, 3M began a Pollution Prevention Pays (3P) program. Since then, it has reformulated some of its products, redesigned equipment and processes, reduced its use of hazardous raw materials, recycled and reused more waste materials, and sold some of its potentially hazardous but still useful wastes as raw materials to other companies. As of 2012, this program

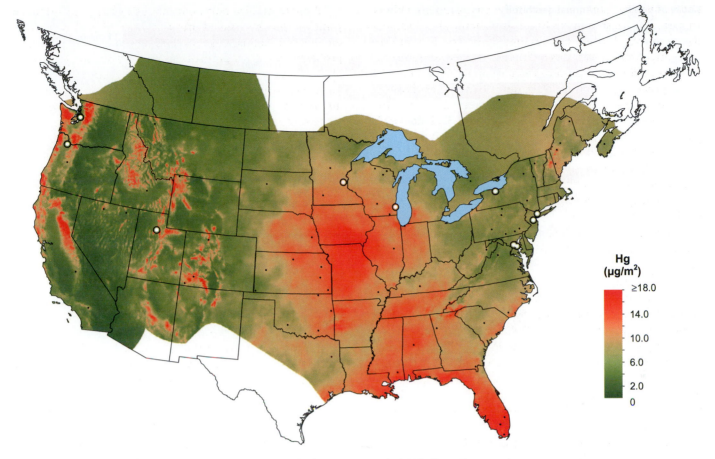

FIGURE 14.13 Atmospheric wet deposition of mercury in the lower 48 states in 2010. *Question:* Why do the highest levels occur mainly in the eastern half of the United States?

Source: Environmental Protection Agency and National Atmospheric Deposition Program.

had prevented more than 1.7 million metric tons (1.9 million tons) of pollutants from reaching the environment and saved the company $1.7 billion.

The 3M 3P program has been successful largely because employees are rewarded if projects they come up with eliminate or reduce a pollutant; reduce the amount of energy, materials, or other resources required in production; or save money through reduced pollution control costs, lower operating costs, or increased sales of new or existing products. Employees at 3M have now completed more than 10,000 3P projects.

Since 1990, a growing number of companies have adopted similar pollution and waste prevention programs that have led to cleaner production. (See the online Guest Essay by Peter Montague on cleaner production.) They are learning that, in addition to saving money by preventing pollution and reducing waste production, they have a much easier job of complying with pollution laws and regulations. Also, they find they can avoid lawsuits based on exposure to harmful chemicals, provide a safer environment for their workers (which can reduce their employee health insurance costs), and improve their public image.

14.5 HOW DO WE PERCEIVE RISKS AND HOW CAN WE AVOID THE WORST OF THEM?

CONCEPT 14.5 We can reduce the major risks we face by becoming informed, thinking critically about risks, and making careful choices.

The Greatest Health Risks Come from Poverty, Gender, and Lifestyle Choices

Risk analysis involves identifying hazards and evaluating their associated risks (*risk assessment;* Figure 14.2, left), ranking risks (*comparative risk analysis*), determining options and making decisions about reducing or eliminating risks (*risk management;* Figure 14.2, right), and informing decision makers and the public about risks (*risk communication*).

Statistical probabilities based on past experience, animal testing, and other assessments are used to estimate risks from older technologies and chemicals. To evaluate new technologies and products, risk evaluators use more

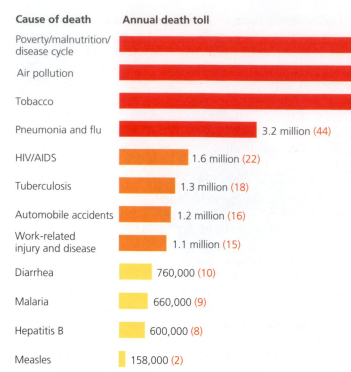

Cause of death **Annual death toll**

Cause of death	Annual death toll
Poverty/malnutrition/disease cycle	11 million (151)
Air pollution	7.0 million (96)
Tobacco	6.0 million (82)
Pneumonia and flu	3.2 million (44)
HIV/AIDS	1.6 million (22)
Tuberculosis	1.3 million (18)
Automobile accidents	1.2 million (16)
Work-related injury and disease	1.1 million (15)
Diarrhea	760,000 (10)
Malaria	660,000 (9)
Hepatitis B	600,000 (8)
Measles	158,000 (2)

FIGURE 14.14 Scientists have estimated the number of deaths per year in the world from various causes. Numbers in parentheses represent these death tolls in terms of the numbers of fully loaded 200-passenger jet airplanes crashing *every day of the year* with no survivors. **Question:** Which three of these causes are the most threatening to you?

Compiled by the authors using data from World Health Organization, Environmental Protection Agency, and U.S. Centers for Disease Control and Prevention.

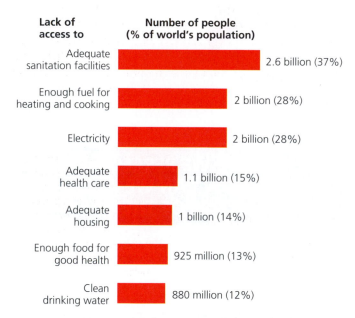

Lack of access to **Number of people (% of world's population)**

Lack of access to	Number of people (% of world's population)
Adequate sanitation facilities	2.6 billion (37%)
Enough fuel for heating and cooking	2 billion (28%)
Electricity	2 billion (28%)
Adequate health care	1.1 billion (15%)
Adequate housing	1 billion (14%)
Enough food for good health	925 million (13%)
Clean drinking water	880 million (12%)

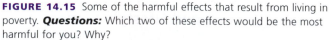

FIGURE 14.15 Some of the harmful effects that result from living in poverty. **Questions:** Which two of these effects would be the most harmful for you? Why?

Compiled by the authors using data from United Nations, World Bank, and World Health Organization.

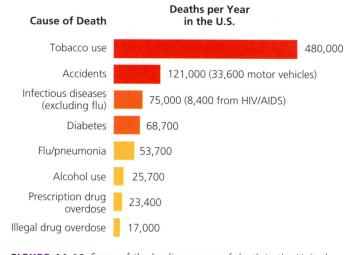

Cause of Death **Deaths per Year in the U.S.**

Cause of Death	Deaths per Year in the U.S.
Tobacco use	480,000
Accidents	121,000 (33,600 motor vehicles)
Infectious diseases (excluding flu)	75,000 (8,400 from HIV/AIDS)
Diabetes	68,700
Flu/pneumonia	53,700
Alcohol use	25,700
Prescription drug overdose	23,400
Illegal drug overdose	17,000

FIGURE 14.16 Some of the leading causes of death in the United States are preventable to the extent that they result from lifestyle choices. (Numbers are rounded off.)

Compiled by the authors using data from U.S. Centers for Disease Control and Prevention.

uncertain statistical probabilities, based on models rather than on actual experience and testing.

In terms of the number of deaths per year (Figure 14.14), *the greatest risk by far is poverty*. The high death toll ultimately resulting from poverty is caused by malnutrition, increased susceptibility to normally nonfatal infec-

tious diseases, and often-fatal infectious diseases transmitted by unsafe drinking water (Figure 14.15).

Studies indicate the four greatest risks in terms of shortened life spans are living in poverty, being born male, smoking (see the following Case Study), and being obese. Some of the greatest risks of premature death are illnesses that result primarily from lifestyle choices that people make (Figure 14.16) (**Concept 14.1**).

Cigarettes and E-Cigarettes

Cigarette smoking is the world's most preventable and largest cause of suffering and premature death among adults. The WHO estimates that tobacco use contributed to the deaths of 100 million people during the 20th century and could kill 1 billion people during this century unless individuals and governments act now to dramatically reduce smoking.

The WHO and a 2014 report by the U.S. Surgeon General estimated that each year, tobacco contributes to the premature deaths of about 6 million people resulting from 25 illnesses, including heart disease, stroke, type 2 diabetes, lung and other cancers, memory impairment, bronchitis, and emphysema (Figure 14.17). This amounts to an average of more than 16,400 deaths every day, or about one every 5 seconds.

According to the WHO, lifelong smokers reduce their life spans by an average of 10 years. By 2030, the annual death toll from smoking-related diseases is projected to reach more than 8 million—an average of 21,900 preventable deaths per day—according to the CDC and WHO. About 80% of these deaths are expected to occur in less-developed countries, especially China, where 30% of the world's smokers live. The annual death toll from smoking in China is about 1.2 million, an average of about 137 deaths every hour.

According to a 2014 CDC report, smoking kills more than 480,000 Americans per year—an average of 1,315 deaths per day, or nearly one every minute (Figure 14.17). This death toll is roughly equivalent to more than six fully loaded 200-passenger jet planes crashing *every day of the year* with no survivors. Smoking also causes about 8.6 million illnesses every year in the United States.

The overwhelming scientific consensus is that the nicotine inhaled in tobacco smoke is highly addictive. A British government study showed that adolescents who smoke more than one cigarette have an 85% chance of becoming long-term smokers.

Studies also show that breathing *secondhand smoke* poses health hazards. In 2010, the CDC and the California Environmental Protection Agency estimated that secondhand smoke annually contributes to 3,400 lung cancer deaths and 46,000 deaths from heart disease in the United States. A 2010 study by British researchers found that, globally, exposure to secondhand smoke contributes to about 600,000 deaths per year.

Studies indicate that cigarette smokers die, on average, 10 years earlier than nonsmokers, but that kicking the habit—even at 50 years of age—can cut such a risk in half. If people quit smoking by the age of 30, they can avoid nearly all the risk of dying prematurely.

In the United States, the percentage of adults who smoke dropped from 42% in 1964 to 18% in 2012, according to the CDC, and the goal is to reduce this to less than 10% by 2025. A 2012 study in the *Journal of the American Medical Association* estimated that this decline saved 8 million lives and added 2 years to the average U.S. citizen's life expectancy. This decline can be attributed to media coverage about the harmful health effects of smoking, sharp increases in cigarette taxes in many states, mandatory health warnings on cigarette packs, the ban on

GOOD NEWS

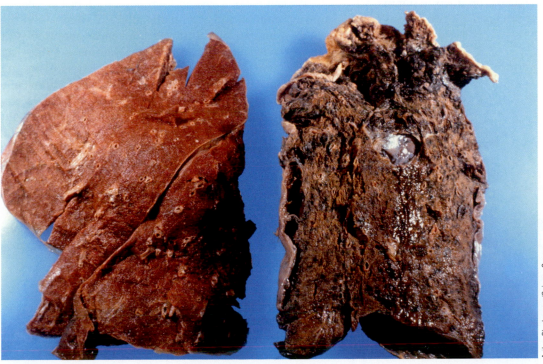

FIGURE 14.17 There is a startling difference between normal human lungs (left) and the lungs of a person who died of emphysema (right). The major causes of emphysema are prolonged smoking and exposure to air pollutants.

Arthur Glauberman/Science Source

sales to minors, and bans on smoking in workplaces, bars, restaurants, and public buildings.

Some people are using various forms of *electronic cigarettes* or *e-cigarettes* (Figure 14.18) as a substitute for tobacco cigarettes. These devices contain pure nicotine dissolved in a syrupy solvent with a flavoring to enhance taste and smell. A battery in the device heats the nicotine solution and converts it to a vapor as the user inhales. Smoking e-cigarettes is called *vaping*. E-cigarettes can be refilled with solutions that vary from 2% to 10% in their concentrations of nicotine.

Are e-cigarettes safe? No one knows, because they haven't been around long enough to be thoroughly evaluated. E-cigarettes reduce or eliminate the inhalation of tar and numerous other harmful chemicals found in regular cigarette smoke. But they still expose users to highly addictive nicotine, sometimes at levels of up to 5 times as high (10% nicotine) as that found in regular cigarettes (2% nicotine).

Preliminary research indicates that some e-cigarette vapors contain trace amounts of toxic cadmium, nickel, lead, and several substances that can cause cancer in test animals. Some of these toxins, not found in regular cigarette smoke, are nanoparticles that are small enough to get past the body's defense systems and travel deep into the lungs. A preliminary 2014 study by cancer researcher Maciej L. Goniewicz and his colleagues indicated that the high temperatures used to vaporize liquid nicotine could expose users to formaldehyde and a few other known or suspected carcinogens. However, it will take much more research for any direct link between e-cigarettes and cancer to be established.

The European Union (EU) has banned the advertising and sales of e-cigarettes and tobacco products to minors, as well as Internet sales of these products. EU regulations also limit the concentration of nicotine in e-cigarettes to 2% and require the disclosure of e-cigarette ingredients. And they require that these products have child-proof and tamper-proof packaging that carries graphic warnings on the harmful health effects of nicotine. In the United States, the FDA proposed similar guidelines in 2014, but they include little or no regulation of Internet sales or of nicotine content of e-cigarettes.

Estimating Risks from Technologies Is Not Easy

The more complex a technological system, and the more people needed to design and run it, the more difficult it is to estimate the risks of using the system. The overall *reliability* of such a system—the probability (expressed as a percentage) that the system will complete a task without failing—is the product of two factors:

System reliability (%) = Technology reliability (%) × Human reliability (%)

With careful design, quality control, maintenance, and monitoring, a highly complex system such as a nuclear power plant or a deep-sea oil-drilling rig can achieve a high degree of technological reliability. But human reliability usually is much lower than technological reliability and is almost impossible to predict.

Suppose the estimated technological reliability of a nuclear power plant is 95% (0.95) and human reliability is 75% (0.75). Then the overall system reliability is 71% ($0.95 \times 0.75 = 71\%$). Even if we could make the technology 100% reliable (1.0), the overall system reliability would still be only 75% ($1.0 \times 0.75 = 75\%$).

We can make a system more foolproof, or fail-safe, by moving more of the potentially fallible elements from the human side to the technological side. However, chance events such as a lightning strike can knock out an automatic control system, and no machine or computer program can completely replace human judgment. Also, the parts in any automated control system (such as the blowout protectors on the BP oil well that ruptured in the Gulf of Mexico in 2010) are manufactured, assembled, tested, certified, inspected, and maintained by fallible human beings. In addition, computer software programs used to monitor and control complex systems can be flawed because of human design error or can be deliberately sabotaged to cause their malfunction.

jps/Shutterstock.com

FIGURE 14.18 An e-cigarette that can be refilled with a solution of nicotine.

Most People Do a Poor Job of Evaluating Risks

Most of us are not good at assessing the relative risks from the hazards that we encounter. Many people deny or shrug off the high-risk chances of death (or injury) from the voluntary activities they enjoy such as *motorcycling* (with death by motorcycle crash being 30 times more likely than death by car crash), *smoking* (1 in 250 by age 70 for a pack-a-day smoker), *hang gliding* (1 in 1,250), and *driving* (1 in 3,300 without a seatbelt and 1 in 6,070 with a seatbelt). Indeed, the most dangerous thing that many people do each day is to drive or ride in a car.

Yet some of these same people may be terrified about their chances of being killed by *the flu* (a 1 in 130,000 chance), *a nuclear power plant accident* (1 in 200,000), *West Nile virus* (1 in 1 million), *lightning* (1 in 3 million), *a commercial airplane crash* (1 in 9 million), *snakebite* (1 in 36 million), or *shark attack* (1 in 281 million).

Five factors can cause people to see a technology or a product as being more or less risky than experts judge it to be. First is *fear*. Research shows that fear causes people to overestimate risks and to worry more about unusual risks than they do about common, everyday risks. Studies show that people tend to overestimate numbers of deaths caused by tornadoes, floods, fires, homicides, cancer, and terrorist attacks, and to underestimate death tolls from flu, diabetes, asthma, heart attack, stroke, and automobile accidents.

The second factor in our estimation of risk is the *degree of control* we have in a given situation. Most of us have a greater fear of things over which we do not have personal control. For example, some individuals feel safer driving their own car for long distances than traveling the same distance on a plane. But look at the numbers. The risk of dying in a car accident in the United States while using a seatbelt is 1 in 6,070, whereas the risk of dying in a U.S. commercial airliner crash is about 1 in 9 million.

The third factor is *whether a risk is catastrophic*, not chronic. We usually are more frightened by news of catastrophic accidents such as a plane crash than we are of a cause of death such as smoking, which has a much higher death toll spread out over time.

Fourth, some people suffer from *optimism bias*, the belief that risks that apply to other people do not apply to them. While people get upset when they see others driving erratically while talking on a cell phone or texting, they may believe that talking on the cell phone or texting does not impair their own driving ability.

A fifth factor is that many of the risky things we do are highly pleasurable and give *instant gratification*, while the potential harm from such activities comes later. Examples are smoking cigarettes, eating lots of ice cream, and getting a tan.

Certain Principles Can Help Us Evaluate and Reduce Risk

Here are four guidelines for evaluating and reducing risk (**Concept 14.5**):

- *Compare risks.* In evaluating a risk, the key question is not "Is it safe?" but rather "*How risky is it compared to other risks?*"

- *Determine how much risk you are willing to accept.* For most people, a 1 in 100,000 chance of dying or suffering serious harm from exposure to an environmental hazard is a threshold for changing their behavior. However, in establishing standards and reducing risk, the EPA generally assumes that a 1 in 1 million chance of dying from an environmental hazard is acceptable.

- *Evaluate the actual risk involved.* The news media usually exaggerate the daily risks we face in order to capture our interest and attract more readers, listeners, or television viewers. As a result, most people who are exposed to a daily diet of such exaggerated reports believe that the world is much more dangerous and risk-filled than it really is.

- *Concentrate on evaluating and carefully making important lifestyle choices.* When you worry about a risk, the most important question to ask is, "Do I have any control over this?" There is no point worrying about risks over which we have no control. But we do have control over major ways to reduce risks from heart attack, stroke, and certain forms of cancer, because we can decide whether to smoke, what to eat, and how much alcohol to drink. Other factors under our control are whether we practice safe sex, how much exercise we get, and how safely we drive.

BIG IDEAS

- We face significant hazards from infectious diseases such as flu, AIDS, tuberculosis, diarrheal diseases, and malaria, and from exposure to chemicals that can cause cancers and birth defects, as well as chemicals that can disrupt the human immune, nervous, and endocrine systems.

- Because of the difficulty of evaluating the harm caused by exposure to chemicals, many health scientists call for much greater emphasis on pollution prevention.

- By becoming informed, thinking critically about risks, and making careful choices, we can reduce the major risks we face.

Mercury's Toxic Effects and Sustainability

In the Core Case Study that opens this chapter, we saw that mercury (Hg) and its compounds that occur regularly in the environment can permanently damage the human nervous system, kidneys, and lungs and harm fetuses and cause birth defects. In this chapter, we also learned of many other chemical hazards, as well as biological, physical, cultural, and lifestyle hazards, in the environment. In addition, we saw how difficult it is to evaluate the nature and severity of threats from these various hazards.

One of the important facts discussed in this chapter is that on a global basis, the greatest threat to human health is poverty (often leading to malnutrition and disease), followed by the threats from smoking, air pollution, pneumonia and flu, and HIV/AIDS.

There are some threats that we can do little to avoid, but we can reduce other threats, partly by applying the three **scientific principles of sustainability** (see Supplement 7, p. S50). For example, we can greatly reduce our exposure to mercury and other pollutants by shifting from the use of nonrenewable fossil fuels (especially coal) to wider use of a diversity of renewable energy resources, including solar energy. We can reduce our exposure to harmful chemicals used in the manufacturing of various goods by cutting resource use and waste and by reusing and recycling material resources. We can also mimic biodiversity by using diverse strategies for solving environmental and health problems, and especially for reducing poverty and controlling population growth. In so doing, we also help to preserve the earth's biodiversity.

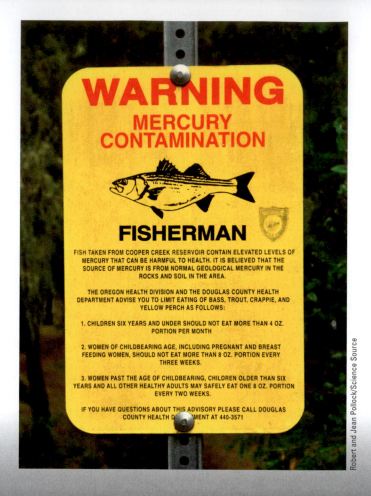

Robert and Jean Pollock/Science Source

Core Case Study

1. Describe the toxic effects of mercury and its compounds and explain how we are exposed to these toxins.

Section 14.1

2. What is the key concept for this section? Define and distinguish among **risk**, **risk assessment**, and **risk management**. Give an example of how scientists state probabilities. Give an example of a risk from each of the following: biological hazards, chemical hazards, natural hazards, cultural hazards, and lifestyle choices. What is a **pathogen**?

Section 14.2

3. What is the key concept for this section? Define **infectious disease**; define and distinguish among **bacteria**, **viruses**, and **parasites**; and give examples of diseases that each can cause. Define and distinguish between **transmissible disease** and **nontransmissible disease**, and give an example of each. List four ways in which infectious organisms can enter the

body. In terms of death rates, what are the world's four most serious infectious diseases? List five factors that have contributed to genetic resistance in bacteria to commonly used antibiotics. What is MRSA and why is it so threatening?

4. Describe the global threat from tuberculosis and list three factors that have helped it to spread. What is the biggest viral killer and how does it spread? Summarize the threat from the hepatitis B virus. What is the best way to reduce one's chances of getting an infectious disease? What is the focus of ecological medicine and what are some of its findings regarding the spread of diseases? Summarize the health threats from the global HIV/AIDS pandemic.

5. What is malaria and how does it spread? How much of the human population is subject to this threat? Explain how deforestation can promote the spread of malaria. List six major ways to reduce the global threat from infectious diseases.

Section 14.3

6. What is the key concept for this section? What is a **toxic chemical**? Define and distinguish among **carcinogens**, **mutagens**, and **teratogens**, and give an example of each. Describe the human immune, nervous, and endocrine systems, and for each of these systems, give an example of a chemical that can threaten it. What is a neurotoxin and why is methylmercury (**Core Case Study**) an especially dangerous one? What are six ways to prevent or control environmental inputs of mercury? What are hormonally active agents (HAAs), what risks do they pose, and how can we reduce those risks? Summarize health scientists' concerns about exposure to bisphenol A (BPA) and the controversy over what to do about exposure to this chemical. Summarize the concerns over exposure to phthalates. List six ways to reduce your exposure to HAAs.

Section 14.4

7. What are the two key concepts for this section? Define **toxicology**, **toxicity**, **dose**, and **response**.

What are three factors that affect the level of harm caused by a chemical? Give three reasons why children are especially vulnerable to harm from toxic chemicals. Describe how the toxicity of a substance can be estimated by testing laboratory animals, and explain the limitations of this approach. What is a **dose-response curve**? Explain how toxicities are estimated through use of case reports and epidemiological studies, and discuss the limitations of these approaches.

8. Summarize the controversy over the effects of trace levels of chemicals. Why do we know so little about the harmful effects of chemicals? What is the **precautionary principle**? Explain why the use of pollution prevention based on the precautionary principle to deal with health threats from chemicals is controversial. Describe some efforts to apply this principle on national and international levels. What is the U.N. Minamata Convention for reducing mercury pollution? How did pollution prevention pay off for the 3M Company?

Section 14.5

9. What is the key concept for this section? What is **risk analysis**? In terms of premature deaths, what are the three greatest threats that people face? What are six ways in which poverty can threaten one's health? Describe the health threats from smoking and how we can reduce these threats. Summarize our knowledge of the health effects of using e-cigarettes. How can we reduce the threats resulting from the use of various technologies? What are five factors that can cause people to misjudge risks? List four guidelines for evaluating and reducing risk.

10. What are this chapter's *three big ideas*? Explain how we can lessen the threats of harm from mercury in the environment by applying the three **scientific principles of sustainability**.

Note: Key terms are in bold type.

Critical Thinking

1. Assume that you are a national official with the power to set policy for controlling environmental mercury pollution from human sources (**Core Case Study**). List the goals of your policy and outline a plan for accomplishing those goals. List three or more possible problems that could result from implementing your policy.

2. What are three actions you would take to reduce the global threats to human health and life from each of the following: **(a)** tuberculosis, **(b)** HIV/AIDS, and **(c)** malaria?

3. Explain why you agree or disagree with each of the following statements:
   a. We should not worry much about exposure to toxic chemicals because almost any chemical, at a large enough dosage, can cause some harm.
   b. We should not worry much about exposure to toxic chemicals because, through genetic adaptation, we can develop immunities to such chemicals.
   c. We should not worry much about exposure to toxic chemicals because we can use genetic engineering to reduce our susceptibility to their effects.
   d. We should not worry about exposure to a chemical such as bisphenol A (BPA) because it has not been absolutely proven scientifically that BPA has killed anyone.

4. Should we ban the use of hormone mimics such as BPA in making products to be used by children younger than age 5? Should such a ban be extended to all products? Explain.

5. Workers in a number of industries are exposed to higher levels of various toxic substances than the general public is. Should we reduce the workplace levels allowed for such chemicals? What economic effects might this have?

6. Do you think there should be a ban on smoking indoors in all public places? Explain. Do you think that electronic cigarettes should be taxed and regulated like conventional cigarettes? Explain.

7. What are the three major risks you face from each of the following: (a) your lifestyle, (b) where you live, and (c) what you do for a living? Which of these risks are voluntary and which are involuntary? List three steps you could take to reduce each of these risks. Which of these steps do you already take or plan to take?

8. In deciding what to do about risks from chemicals in the area where you live, would you support legislation that requires the use of pollution prevention based on the precautionary principle and on the assumption that chemicals are potentially harmful until shown otherwise? Explain.

Doing Environmental Science

Pick a commonly used and potentially harmful chemical and use the library or Internet to learn about (a) what it is used for and how widely it is used, (b) its potential harm, (c) the scientific evidence for such claims, and (d) proposed solutions for dealing with this threat. Pick a study area, such as your dorm or apartment building, your block, or your city. In this area, try to determine the level of presence of the chemical you are studying. You could do this by finding four or five examples of items or locations containing the chemical and then estimating the total amount based on your sample. Write a report summarizing your findings.

Global Environment Watch Exercise

Search for *mercury pollution* and research the latest developments in studies of the harmful health effects of mercury (**Core Case Study**). Find an example of an effort to prevent or control mercury pollution and write a short report summarizing your findings. Try to find reports of two studies that reach different conclusions about how mercury should be regulated. Summarize the arguments for these conclusions on both sides. Based on what you have found, do you think that mercury pollution should be regulated more strictly in the state or country where you live? Explain your reasoning.

Data Analysis

The graph below shows the effects of AIDS on life expectancy at birth in Botswana, 1950–2000, and projects these effects to 2050. Study the graph and answer the questions that follow.

1. **(a)** By what percentage did life expectancy in Botswana increase between 1950 and 1995?
   (b) By what percentage was life expectancy in Botswana projected to decrease between 1995 and 2015?

2. **(a)** By what percentage is life expectancy in Botswana projected to increase between 2015 and 2050? **(b)** By what percentage was life expectancy in Botswana projected to decrease between 1995 and 2050?

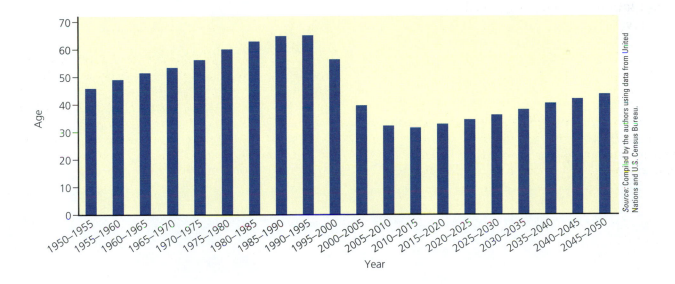

Source: Compiled by the authors using data from United Nations and U.S. Census Bureau.

CENGAGE**brain**.com To access course materials, including Aplia homework, please visit www.cengagebrain.com.

WWW.CENGAGEBRAIN.COM **385**

Nobody on this planet is
going to be untouched by
climate change.

RAJENDRA K. PACHAURI

15

AIR POLLUTION, CLIMATE
CHANGE, AND OZONE
DEPLETION

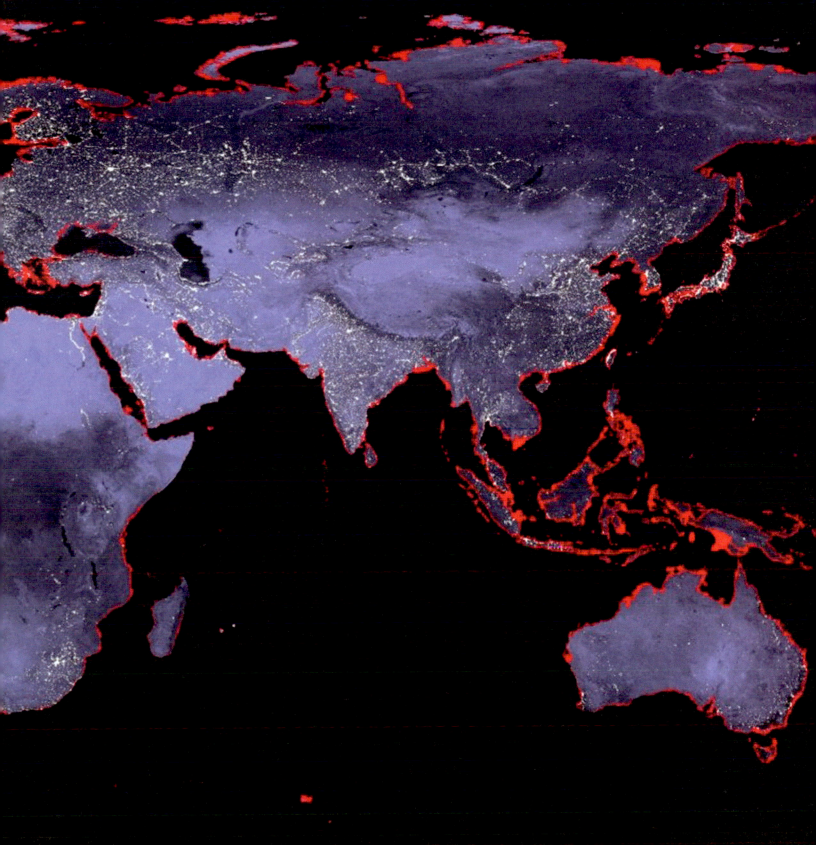

15.4 How might the earth's climate change in the future?

15.5 What are some possible effects of a warmer atmosphere?

15.6 What can we do to slow projected climate change?

15.7 How have we depleted ozone in the stratosphere and what can we do about it?

Areas that could be flooded by the end of this century (shown in red) by a 1-meter (3-foot) rise in sea level due to projected climate change.

NASA

Greenland is the world's largest island with a population of about 57,000 people. The ice that covers most of this mountainous island lies in glaciers that are as deep as 3.2 kilometers (2 miles).

Areas of the island's ice have been melting at an accelerating rate during Greenland's summers (Figure 15.1). Some of this ice is replaced by snow during winter months, but the annual net loss of Greenland's ice has increased during recent years.

So why should we care if ice in Greenland is melting? The answer is, considerable scientific evidence indicates that a key factor in this melting is *atmospheric warming*—the gradual rise of the average temperature of the atmosphere near the earth's land and water surfaces for 30 years or more. During this century, this warming is projected to increase significantly and to lead to *climate change*, which is very likely to have harmful effects on most of the earth's ecosystems, people, and economies.

Greenland's glaciers contain enough water to raise the global sea level by as much as 7 meters (23 feet) if all of it were to melt and drain into the sea. It is highly unlikely that this will happen, but even a moderate loss of this ice over one or several centuries would raise sea levels considerably (see chapter-opening photo). According to a 2014 study by earth scientist Michael Bevis, Greenland's ice loss has been responsible for nearly one-sixth of the global sea-level rise over the past 20 years. According to some scientists,

Greenland's melting ice is an early warning that human activities are very likely to disrupt the earth's climate in ways that could threaten life as we know it, especially during the latter half of this century.

In 1988, the World Meteorological Organization and the United Nations Environment Programme (UNEP) established the Intergovernmental Panel on Climate Change (IPCC) to document past climate changes and project future climate changes. The IPCC network includes more than 2,500 climate scientists and scientists in disciplines related to climate studies from more than 130 countries.

After reviewing tens of thousands of research studies for more than 25 years, the IPCC and most of the world's major scientific bodies,

including the U.S. National Academy of Sciences and the British Royal Society, have come to three major conclusions: **(1)** climate change is real and is happening now, **(2)** it is caused primarily by human activities such as the burning of fossil fuels and the clearing of forests, and **(3)** it is projected to accelerate, especially during the latter half of this century, unless we act now to slow it down and reduce its numerous harmful effects.

In this chapter, we examine the nature of the atmosphere, air pollution, the likely causes and effects of projected climate change, and depletion of ozone in the stratosphere. We also look at some possible ways to deal with these serious environmental, economic, and political challenges.

Compiled by the authors using data from Konrad Steffen, Russell Huff, and the National Aeronautics and Space Administration.

FIGURE 15.1 The total area of Greenland's glacial ice that melted by July 8, 2012 (red area in right image) was much greater than the amount that melted during the summer of 1982 (left).

15.1 WHAT IS THE NATURE OF THE ATMOSPHERE?

CONCEPT 15.1 The two innermost layers of the atmosphere are the *troposphere*, which supports life, and the *stratosphere*, which contains the protective ozone layer.

The Atmosphere Consists of Several Layers

We live under a thin blanket of gases surrounding the earth, called the **atmosphere**. It is divided into several spherical layers (Figure 15.2).

About 75–80% of the earth's air mass is found in the **troposphere**, the atmospheric layer closest to the earth's surface. This layer extends about 17 kilometers (11 miles) above sea level at the equator and 6 kilometers (4 miles) above sea level over the poles. If the earth were the size of an apple, this lower layer containing the air we breathe would be no thicker than the apple's skin.

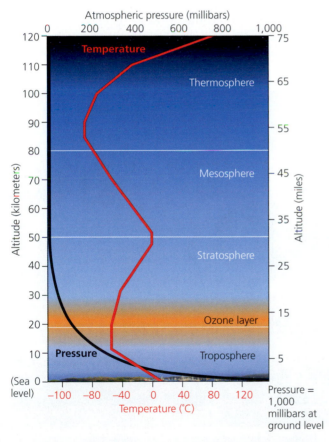

Atmospheric pressure (millibars)

FIGURE 15.2 **Natural capital:** The earth's atmosphere is a dynamic system that includes four layers. The average temperature of the atmosphere varies with altitude (red line) and with differences in the absorption of incoming solar energy. **Question:** Why do you think most of the planet's air is in the troposphere?

© 2016 Cengage Learning

Take a deep breath. About 99% of the volume of air you inhaled consists of two gases: nitrogen (78%) and oxygen (21%). The remainder is 0.93% argon (Ar), 0.040% carbon dioxide (CO_2), and small amounts of water vapor (H_2O, varying from the equator to the poles), dust and soot particles, and other gases, including methane (CH_4), ozone (O_3), and nitrous oxide (N_2O).

Several gases in the atmosphere, including H_2O, CO_2, CH_4, and N_2O, are called **greenhouse gases** because they absorb and release energy that warms the troposphere. Rising and falling air currents, winds, and concentrations of CO_2 and other greenhouse gases in the troposphere play major roles in the planet's short-term *weather* and long-term *climate*.

The atmosphere's second layer is the **stratosphere**, which extends from about 17 to about 48 kilometers (from 11 to 30 miles) above the earth's surface (Figure 15.2). Although the stratosphere contains less matter than the troposphere, its composition is similar, with two notable exceptions: its volume of water vapor is about 1/1,000th that of the troposphere and its concentration of ozone (O_3) is much higher.

Much of the atmosphere's small amount of ozone (O_3) is concentrated in a portion of the stratosphere called the **ozone layer**, found roughly 17–26 kilometers (11–16 miles) above sea level. Most stratospheric ozone is produced when some of the oxygen molecules in this layer interact with ultraviolet (UV) radiation emitted by the sun.

$$3 \ O_2 + UV \leftrightarrow 2 \ O_3$$

This UV filtering effect of ozone in the lower stratosphere acts as a "global sunscreen" that keeps about 95% of the sun's harmful UV radiation from reaching the earth's surface. It allows us and other life-forms to exist on land and helps to protect us from sunburn, skin and eye cancers, cataracts, and damage to our immune systems. It also prevents much of the oxygen in the troposphere from being converted to photochemical ozone, a harmful air pollutant when found near the ground. The stratospheric ozone layer is a vital part of the earth's natural capital that helps to keep us and other species alive, allows us to live on land, and supports our economies.

15.2 WHAT ARE THE MAJOR AIR POLLUTION PROBLEMS?

CONCEPT 15.2A Three major outdoor air pollution problems are *industrial smog* caused mostly by the burning of coal, *photochemical smog* caused by motor vehicle and industrial emissions, and *acid deposition* caused mainly by coal-burning power and industrial plants and motor vehicle emissions.

CONCEPT 15.2B The most threatening indoor air pollutants are smoke and soot from wood and coal fires (mostly in less-developed countries), cigarette smoke, and chemicals used in building materials and cleaning products.

Air Pollution Comes from Natural and Human Sources

Air pollution is the presence of chemicals in the atmosphere in concentrations high enough to harm organisms, ecosystems, or human-made materials, or to alter climate. Almost any chemical in the atmosphere can become a pollutant if it occurs in a high enough concentration. The effects of air pollution range from annoying to lethal.

Air pollutants come from natural and human sources. Natural sources include wind-blown dust, pollutants from wildfires and volcanic eruptions, and volatile organic chemicals released by some plants. Most natural air pollutants are spread out over the globe and diluted or are removed by chemical cycles, precipitation, and gravity. But in areas experiencing volcanic eruptions or forest fires, chemicals emitted by these sources can temporarily reach harmful levels.

Most human inputs of outdoor air pollutants occur in industrialized and urban areas where people, cars, and factories are concentrated. These pollutants are generated mostly by the burning of fossil fuels in power plants and industrial facilities *(stationary sources)* and in motor vehicles *(mobile sources)*.

Scientists classify outdoor air pollutants into two categories (Figure 15.3). **Primary pollutants** are chemicals or substances emitted directly into the air from natural processes and human activities at concentrations high enough to cause harm. While in the atmosphere, some primary pollutants react with one another and with other natural components of air to form new harmful chemicals, called **secondary pollutants**.

Outdoor air pollution was once a regional problem limited mostly to cities. Now it is a global problem, largely due to the sheer volume of pollutants produced by human activities. Prevailing winds can spread long-lived pollutants entering the atmosphere from India and China across the Pacific where they affect air pollution levels on the west coast of North America. There is no place on the planet's surface that has not been affected by air pollution from human activities.

Over the past 40 years, the quality of outdoor air in most of the more-developed countries has improved greatly. This has occurred primarily because grassroots pressure from citizens, especially during the 1970s, led governments in the United States and in most European countries to pass and enforce air-pollution-control laws.

On the other hand, according to the World Health Organization (WHO), more than 1.1 billion people (one of every six people on the earth) live in urban areas where outdoor air is unhealthy to breathe. Most of them live in densely populated cities in less-developed countries where air-pollution-control laws do not exist or are poorly enforced.

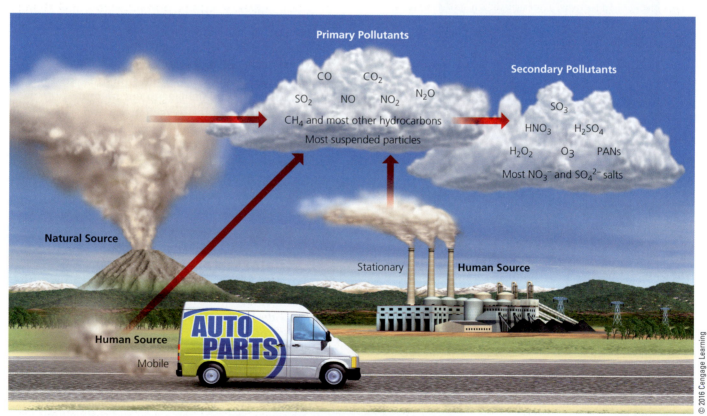

FIGURE 15.3 Human inputs of air pollutants come from *mobile sources* (such as cars) and *stationary sources* (such as industrial, power, and cement plants). Some *primary air pollutants* react with one another and with other chemicals in the air to form *secondary air pollutants*.

What Are the Major Outdoor Air Pollutants?

Carbon Oxides. *Carbon monoxide* (CO) is a colorless, odorless, and highly toxic gas that forms during the incomplete combustion of carbon-containing materials. Major sources are motor vehicle exhaust, the burning of forests and grasslands, the smokestacks of fossil fuel–burning power plants and industries, tobacco smoke, open fires, and inefficient stoves used for cooking or heating.

CO can combine with hemoglobin in red blood cells, which reduces the ability of blood to transport oxygen to body cells and tissues. Long-term exposure can trigger heart attacks and aggravate lung diseases such as asthma and emphysema. At high levels, CO can cause headache, nausea, drowsiness, confusion, collapse, coma, and death.

Carbon dioxide (CO_2) is a colorless, odorless gas. About 93% of the CO_2 in the atmosphere is the result of the natural carbon cycle (see Figure 3.14, p. 53), and the rest comes from human activities. There is considerable and growing scientific evidence that the rapid rise in atmospheric CO_2 levels since 1950 (see Figure 11, p. S43, in Supplement 5) is largely due to human activities, especially the burning of fossil fuels and the removal of CO_2-absorbing forests and grasslands. According to the IPCC, this increase in CO_2 levels is a major cause of atmospheric warming, which will very likely lead to increasingly harmful climate change during this century, unless we act now to sharply reduce our CO_2 emissions.

Nitrogen Oxides and Nitric Acid. *Nitric oxide* (NO) is a colorless gas that forms when nitrogen and oxygen gases react under high-combustion temperatures in automobile engines and coal-burning power and industrial plants. Lightning and certain bacteria in soil and water also produce NO as part of the nitrogen cycle (see Figure 3.15, p. 54).

In the air, NO reacts with oxygen to form *nitrogen dioxide* (NO_2), a reddish-brown gas. Collectively, NO and NO_2 are called *nitrogen oxides* (NO_x). Some of the NO_2 reacts with water vapor in the air to form *nitric acid* (HNO_3) and nitrate salts (NO_3^-), components of harmful *acid deposition,* which we discuss later in this chapter. Both NO and NO_2 play a role in the formation of *photochemical smog*—a mixture of chemicals formed under the influence of sunlight in cities with heavy traffic (discussed further below). *Nitrous oxide* (N_2O), a greenhouse gas, is emitted from fertilizers and animal wastes, and is produced by the burning of fossil fuels.

At high enough levels, nitrogen oxides can irritate the eyes, nose, and throat, aggravate lung ailments such as asthma and bronchitis, suppress plant growth, and reduce visibility when they are converted to nitric acid and nitrate salts.

Sulfur Dioxide and Sulfuric Acid. *Sulfur dioxide* (SO_2) is a colorless gas with an irritating odor. About one-third of the SO_2 in the atmosphere comes from natural sources such as volcanoes. The other two-thirds (and as much as 90% in some urban areas) come from human sources, mostly combustion of sulfur-containing coal in power and industrial plants, oil refining, and the smelting of sulfide ores.

In the atmosphere, SO_2 can be converted to *aerosols,* which consist of microscopic suspended droplets of *sulfuric acid* (H_2SO_4) and suspended particles of sulfate (SO_4^{2-}) salts that return to the earth as a component of acid deposition. Sulfur dioxide, sulfuric acid droplets, and sulfate particles reduce visibility and aggravate breathing problems. They can damage crops, trees, soils, and aquatic life in lakes. And they corrode metals and damage paint, paper, leather, and the stone used to build walls, statues, and monuments.

Particulates. *Suspended particulate matter* (SPM) consists of a variety of solid particles and liquid droplets that are small and light enough to remain suspended in the air for long periods. The U.S. Environmental Protection Agency (EPA) classifies particles as fine, or PM-10 (with diameters less than 10 micrometers), and ultrafine, or PM-2.5 (with diameters less than 2.5 micrometers). About 62% of the SPM in outdoor air comes from natural sources such as dust, wildfires, and sea salt. The remaining 38% comes from human sources such as coal-burning power and industrial plants, motor vehicles, wind erosion from exposed topsoil, and road construction.

These particles can irritate the nose and throat, damage the lungs, aggravate asthma and bronchitis, and shorten life spans. Toxic particulates such as lead and cadmium can cause genetic mutations, reproductive problems, and cancer. Particulates also reduce visibility, corrode metals, and discolor clothing and paints.

Ozone. One of the major ingredients of photochemical smog is *ozone* (O_3), a colorless and highly reactive gas. It can cause coughing and breathing problems, aggravate lung and heart diseases, reduce resistance to colds and pneumonia, and irritate the eyes, nose, and throat. It also damages plants, rubber in tires, fabrics, and paints. Scientific measurements indicate that some human activities have decreased the amount of beneficial ozone in the stratosphere and have increased the amount of harmful ozone in the troposphere near ground level—especially in some urban areas.

Volatile Organic Compounds (VOCs). Organic compounds that exist as gases in the atmosphere or that evaporate from sources on earth into the atmosphere are called *volatile organic compounds* (VOCs). Examples are hydrocarbons, emitted by the leaves of many plants, and *methane* (CH_4), a greenhouse gas that is 25 times more effective per molecule than CO_2 is at warming the atmosphere. About a third of global methane emissions come from natural sources, mostly plants, wetlands, and termites. The rest come from human sources such as rice paddies, landfills, natural gas wells and pipelines, and cows (mostly from their belching).

FIGURE 15.4 Industrial smog from an iron and steel factory in Czechoslovakia.

Other VOCs are liquids than can evaporate quickly into the atmosphere. Examples are benzene and other liquids used as industrial solvents, dry-cleaning fluids, and various components of gasoline, plastics, and other products.

Burning Coal Produces Industrial Smog

Seventy years ago, cities such as London, England, and the U.S. cities of Chicago, Illinois, and Pittsburgh, Pennsylvania, burned large amounts of coal in power plants and factories, and for heating homes and often for cooking food. People in such cities, especially during winter, were exposed to **industrial smog**, consisting mostly of an unhealthy mix of sulfur dioxide, suspended droplets of sulfuric acid, and a variety of suspended solid particles in outside air. Those who burned coal inside their homes were often exposed to dangerous levels of indoor air pollutants.

Today, urban industrial smog is rarely a problem in most of the more-developed countries where coal is burned only in large power and industrial plants with reasonably good pollution control. However, industrial smog remains a problem in industrialized urban areas of China, India, Ukraine, Czechoslovakia (Figure 15.4), Bulgaria, and Poland, where large quantities of coal are still burned in houses, power plants, and factories with inadequate pollution controls. Because of its heavy reliance on coal, China has some of the world's highest levels of industrial smog and 16 of the world's 20 most polluted cities. According to a 2014 Chinese government report, 92% of all Chinese cities did not meet the government's national outdoor air quality standards in 2013.

Sunlight Plus Cars Equals Photochemical Smog

A *photochemical reaction* is any chemical reaction activated by light. **Photochemical smog** is a mixture of primary and secondary pollutants formed under the influence of UV radiation from the sun. In greatly simplified terms,

$$\text{VOCs} + \text{NO}_x + \text{heat} + \text{sunlight} \rightarrow \begin{array}{l} \text{ground-level ozone (O}_3\text{)} \\ + \text{ other photochemical oxidants} \\ + \text{ aldehydes} \\ + \text{ other secondary pollutants} \end{array}$$

The formation of photochemical smog begins when exhaust from morning commuter traffic releases large

FIGURE 15.5 Photochemical smog is a serious problem in Los Angeles, California, although air pollution laws have helped to reduce the average number of severe smog days per year. **Question:** How serious is photochemical smog where you live?

amounts of NO and VOCs into the air over a city. The NO is converted to reddish-brown NO_2, which is why photochemical smog is sometimes called *brown-air smog*. When exposed to ultraviolet radiation from the sun, some of the NO_2 reacts in complex ways with VOCs released by certain trees (such as some oak species, sweet gums, and poplars), motor vehicles, and businesses (especially bakeries and dry cleaners). The resulting mixture of pollutants, dominated by ground-level ozone, usually builds up to peak levels by late morning, irritating people's eyes and respiratory tracts. Some of these pollutants, known as *photochemical oxidants,* can damage lung tissue.

All modern cities have some photochemical smog, but it is much more common in cities with sunny and warm climates, and a great number of motor vehicles. Examples are Los Angeles, California (Figure 15.5), and Salt Lake City, Utah, in the United States; Sydney, Australia; São Paulo, Brazil; Bangkok, Thailand; and Mexico City, Mexico.

Several Factors Can Decrease or Increase Outdoor Air Pollution

Five natural factors help *reduce* outdoor air pollution. First, *particles heavier than air* settle out as a result of gravitational attraction to the earth. Second, *rain and snow* partially cleanse the air of pollutants. Third, *salty sea spray from the oceans* washes out many pollutants from air that flows from land over the oceans. Fourth, *winds* sweep pollutants away and dilute them by mixing them with cleaner air. And fifth, some pollutants are removed by *chemical reactions*. For example, SO_2 can react with O_2 in the atmosphere to form SO_3, which reacts with water vapor to form droplets of H_2SO_4 that fall out of the atmosphere as acidic precipitation.

Six other factors can *increase* outdoor air pollution. First, *urban buildings* slow wind speed and reduce the dilution and removal of pollutants. Second, *hills and mountains* reduce the flow of air in valleys below them and allow pollutant levels to build up at ground level. Third, *high temperatures* promote the chemical reactions leading to the formation of photochemical smog. Fourth, *emissions of volatile organic compounds (VOCs)* from certain trees and plants in urban areas can promote the formation of photochemical smog.

The fifth factor—the so-called *grasshopper effect*—occurs when air pollutants are transported at high altitudes by evaporation and winds from tropical and temperate areas through the atmosphere to the earth's polar areas as part of the earth's global air circulation system (see Figure 7.3, p. 132). This happens mostly during winter. It explains why, for decades, pilots have reported seeing a reddish-brown haze over the Arctic. It also helps explain why polar bears, sharks, and native peoples in remote arctic areas have high levels of various harmful pollutants in their bodies.

The sixth factor has to do with the *vertical movement of air*. During daylight, the sun warms the air near the earth's surface. Normally, this warm air and most of the pollutants it contains rise to mix with the cooler air above and are dispersed. Under certain atmospheric conditions, however, a layer of warm air can temporarily lie atop a layer of cooler air nearer the ground, and this is called a **temperature inversion**. Because the cooler air near the surface is denser than the warmer air above, it does not rise and mix with the air above. If this condition persists, pollutants can build up to harmful and even lethal concentrations in the stagnant layer of cool air near the ground.

Acid Deposition Is a Serious Regional Air Pollution Problem

Most coal-burning power plants, metal ore smelters, oil refineries, and other industrial facilities emit sulfur dioxide (SO_2), suspended particles, and nitrogen oxides (NO_x). In more-developed countries, these facilities usually use tall smokestacks to vent their exhausts high into the atmo-

sphere where wind can dilute and disperse these pollutants. This reduces *local* air pollution, but it can increase *regional* air pollution, because prevailing winds can transport the SO_2 and NO_x pollutants as far as 1,000 kilometers (600 miles). During their trip, these compounds form secondary pollutants such as droplets of sulfuric acid (H_2SO_4), nitric acid vapor (HNO_3), and particles of acid-forming sulfate (SO_4^{2-}) and nitrate (NO_3^-) salts (Figure 15.3).

These acidic substances remain in the atmosphere for 2–14 days. During this period, they descend to the earth's surface in two forms: *wet deposition*, consisting of acidic rain, snow, fog, and cloud vapor, and *dry deposition*, consisting of acidic particles. The resulting mixture is called **acid deposition** (Figure 15.6)—often called *acid rain*. Most dry deposition occurs within 2–3 days of emission, relatively close to the industrial sources, whereas most wet deposition takes place within 4–14 days in more distant downwind areas.

Acid deposition is a problem in areas that lie downwind from coal-burning facilities and from urban areas with large numbers of cars (**Concept 15.2A**) (Figure 15.7). In some areas, soils contain compounds that can react with and help neutralize, or *buffer*, some inputs of acids. The

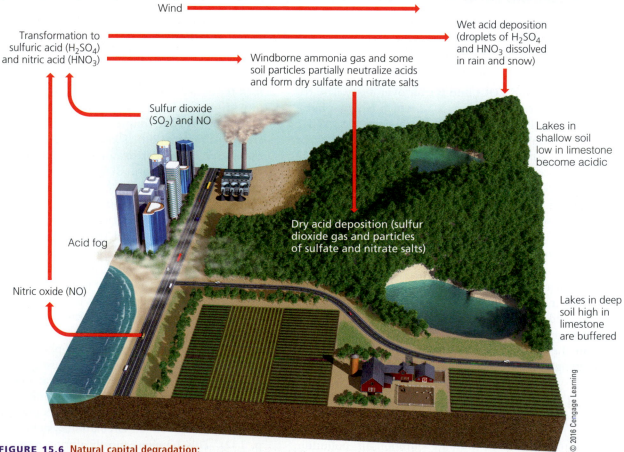

ANIMATED FIGURE 15.6 Natural capital degradation:
Acid deposition, which consists of rain, snow, dust, and other particles with a pH lower than 5.6, is commonly called acid rain (see Figure 5, p. S8, in Supplement 3).
Question: What are three ways in which your daily activities contribute to acid deposition?

areas most sensitive to acid deposition are those with thin, acidic soils that provide no such natural buffering (Figure 15.7, all green and most red areas) and those where the buffering capacity of soils has been depleted by decades of acid deposition.

Acid deposition (often along with other air pollutants such as ozone) can harm crops and reduce plant productivity. A combination of acid deposition and other air pollutants can affect forests by leaching essential plant nutrients such as calcium and magnesium from forest soils and by releasing from the soils ions of aluminum, lead, cadmium, and mercury, which are toxic to trees. These two effects rarely kill trees directly, but they can weaken them and leave them vulnerable to stresses such as severe cold, diseases, insect attacks, and drought.

Acid deposition damages statues and buildings, contributes to human respiratory diseases, and can leach toxic metals such as lead and mercury from soils and rocks into lakes used as sources of drinking water. These toxic metals can accumulate in the tissues of fish and of the animals that eat them, including people. This presents a serious health threat, especially for pregnant women (see Chapter 14, Core Case Study, p. 360). Because of excess acidity due to acid deposition, several thousand lakes in Norway and

Sweden, and 1,200 lakes in Ontario, Canada, contain few if any fish. In the United States, several hundred lakes (most of them in the Northeast) are similarly threatened.

In the United States, older coal-burning power and industrial plants without adequate pollution controls, especially in the Midwest, emit the largest quantities of SO_2, particulates, and other pollutants that cause acid deposition. Because of these emissions and those of other urban industries and motor vehicles, and because of the prevailing west-to-east winds, typical precipitation in parts of the eastern United States can be at least 10 times more acidic than natural precipitation is. Some mountaintop forests in the eastern United States and in areas to the east of large western U.S. cities are bathed in fog and dews that are about 1,000 times as acidic as normal precipitation.

This has also become an international problem wherever acid-producing emissions from one country are transported to other countries by prevailing winds. The worst acid deposition occurs in Asia, especially in China, which gets 68% of its total energy and 73% of its electricity from burning coal, according to the U.S. Energy Information Administration. Some of eastern Asia's emissions are traveling on strong winds all the way across the Pacific Ocean to the west coast of North America.

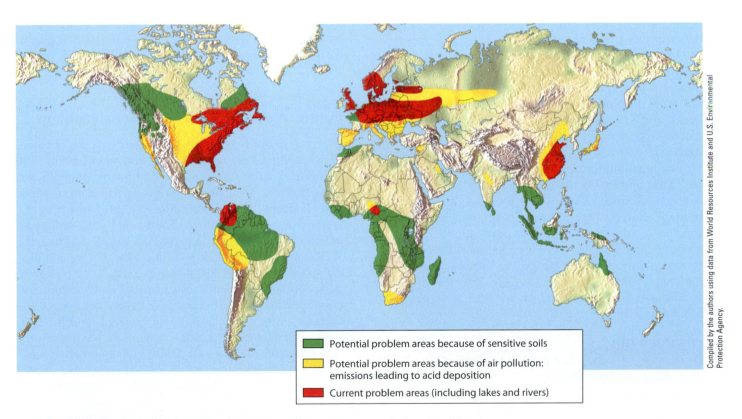

Compiled by the authors using data from World Resources Institute and U.S. Environmental Protection Agency.

Legend:
- Potential problem areas because of sensitive soils
- Potential problem areas because of air pollution: emissions leading to acid deposition
- Current problem areas (including lakes and rivers)

FIGURE 15.7 Regions where acid deposition is now a problem and regions with the potential to develop this problem. Such regions have large inputs of air pollution (mostly from power plants, industrial facilities, and ore smelters) or are sensitive areas with naturally acidic soils and bedrock that cannot neutralize (buffer) additional inputs of acidic compounds. **Question:** Do you live in or near an area that is affected by acid deposition or an area that is likely to be affected by acid deposition in the future?

Solutions

Acid Deposition

Prevention	Cleanup
Reduce coal use and burn only low-sulfur coal	Add lime to neutralize acidified lakes
Use natural gas and renewable energy resources in place of coal	Add phosphate fertilizer to neutralize acidified lakes
Remove SO_2 and NO_x from smokestack gases and remove NO_x from motor vehicular exhaust	Add lime to neutralize acidified soils
Tax SO_2 emissions	

FIGURE 15.8 Ways to reduce acid deposition and its damage. *Questions:* Which two of these solutions do you think are the best ones? Why?

Top: Brittany Courville/Shutterstock.com. Bottom: Yegor Korzh/Shutterstock.com.

We Know How to Reduce Acid Deposition

According to most scientists who study the acid rain problem, the best solutions are *preventive approaches* that reduce or eliminate emissions of sulfur dioxide (SO_2), nitrogen oxides (NO_x), and particulates. Since 1994, acid deposition has decreased sharply in the United States and especially in the eastern half of the country, partly because of significant reductions in SO_2 and NO_x emissions from coal-fired power and industrial plants under the 1990 amendments to the U.S. Clean Air Act. Figure 15.8 lists several ways to reduce acid deposition.

Implementing prevention solutions is politically difficult. One problem is that the people and ecosystems affected by acid deposition often are quite far downwind from the sources of the problem. Also, countries with large supplies of coal (such as China, India, Russia, Australia, and the United States) have a strong incentive to use it. However, in the United States, the increasing use of affordable wind (see Chapter 13, Core Case Study, p. 314) and cleaner-burning natural gas for generating electricity has reduced the use of coal to some extent.

CONSIDER THIS. . .

CONNECTIONS Low-Sulfur Coal, Atmospheric Warming, and Toxic Mercury

Some U.S. power plants have lowered SO_2 emissions by switching from high-sulfur to low-sulfur coals such as lignite (see Figure 13.11, p. 324). However, because low-sulfur coal has a lower heat value, more coal must be burned to generate a given amount of electricity, and this has led to increased CO_2 emissions, which contribute to atmospheric warming and climate change. Low-sulfur coal also has higher levels of toxic mercury and other trace metals, so by burning it, we emit more of these hazardous chemicals into the atmosphere.

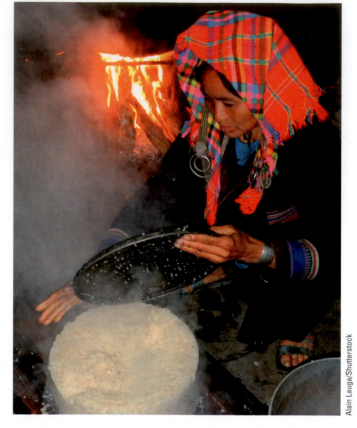

Alain Lauga/Shutterstock

FIGURE 15.9 By burning wood to cook food inside this dwelling in Laos, this woman is exposing herself and other occupants to dangerous levels of indoor air pollution.

Indoor Air Pollution Is a Serious Problem

In less-developed countries, the indoor burning of wood, charcoal, dung, crop residues, coal, and other fuels in open fires (Figure 15.9) or in unvented or poorly vented stoves exposes people to dangerous levels of particulate air pollution (**Concept 15.2B**). According to a 2014 report by the WHO, indoor air pollution is the world's most serious air pollution problem, especially for poor people. In 2012, the WHO estimated that such pollution killed 4.3 million people—an average of 11,780 deaths per day.

Indoor air pollution is also a serious problem in more-developed areas of all countries, mostly because of the chemicals used to make building materials and products such as furniture and paneling. Figure 15.10 shows some typical sources of indoor air pollution in a modern home.

EPA studies have revealed some alarming facts about indoor air pollution. *First,* levels of 11 common air pollutants generally are 2 to 5 times higher inside U.S. homes and commercial buildings than they are outdoors, and in some cases they are as much as 100 times higher. *Second,* pollution levels inside cars in traffic-clogged urban areas can be up to 18 times higher than outside levels. *Third,* the health risks from exposure to such chemicals are growing

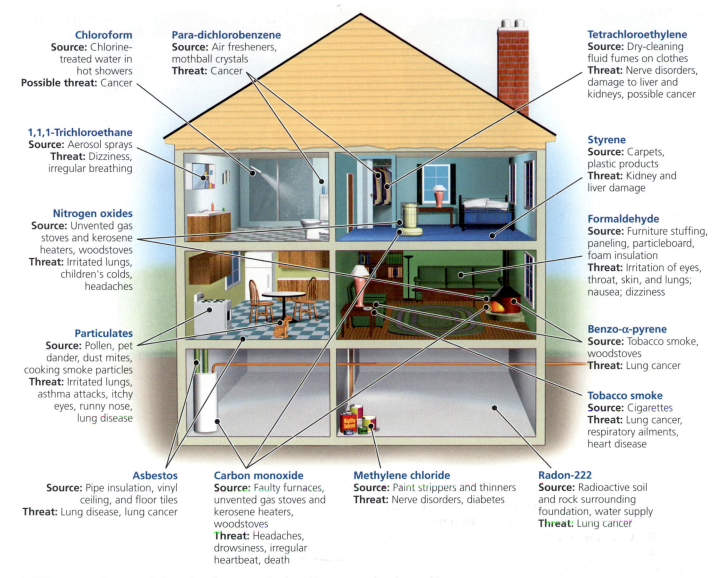

Chloroform
Source: Chlorine-treated water in hot showers
Possible threat: Cancer

Para-dichlorobenzene
Source: Air fresheners, mothball crystals
Threat: Cancer

Tetrachloroethylene
Source: Dry-cleaning fluid fumes on clothes
Threat: Nerve disorders, damage to liver and kidneys, possible cancer

1,1,1-Trichloroethane
Source: Aerosol sprays
Threat: Dizziness, irregular breathing

Styrene
Source: Carpets, plastic products
Threat: Kidney and liver damage

Nitrogen oxides
Source: Unvented gas stoves and kerosene heaters, woodstoves
Threat: Irritated lungs, children's colds, headaches

Formaldehyde
Source: Furniture stuffing, paneling, particleboard, foam insulation
Threat: Irritation of eyes, throat, skin, and lungs; nausea; dizziness

Particulates
Source: Pollen, pet dander, dust mites, cooking smoke particles
Threat: Irritated lungs, asthma attacks, itchy eyes, runny nose, lung disease

Benzo-α-pyrene
Source: Tobacco smoke, woodstoves
Threat: Lung cancer

Tobacco smoke
Source: Cigarettes
Threat: Lung cancer, respiratory ailments, heart disease

Asbestos
Source: Pipe insulation, vinyl ceiling, and floor tiles
Threat: Lung disease, lung cancer

Carbon monoxide
Source: Faulty furnaces, unvented gas stoves and kerosene heaters, woodstoves
Threat: Headaches, drowsiness, irregular heartbeat, death

Methylene chloride
Source: Paint strippers and thinners
Threat: Nerve disorders, diabetes

Radon-222
Source: Radioactive soil and rock surrounding foundation, water supply
Threat: Lung cancer

FIGURE 15.10 Numerous indoor air pollutants can be found in many modern homes (**Concept 15.2B**). *Question:* To which of these pollutants are you exposed?

Compiled by the authors using data from U.S. Environmental Protection Agency.

because most people in more-developed urban areas spend 70–98% of their time indoors or inside vehicles. Smokers, children younger than age 5, the elderly, the sick, pregnant women, people with respiratory or heart problems, and factory workers are especially at risk from indoor air pollution. **GREEN CAREER: Indoor air pollution specialist**

According to the EPA and public health officials, the four most dangerous indoor air pollutants in more-developed countries are *tobacco smoke* (see Chapter 14, Case Study, p. 379); *formaldehyde* emitted from many building materials and various household products (Figure 15.10); *radioactive radon-222 gas,* which can seep into houses from underground rock deposits; and *very small (ultrafine) particles* of various substances in emissions from motor vehicles, coal-burning facilities, wood fires, and forest and grass fires.

Air Pollution Is a Big Killer

Your respiratory system (Figure 15.11) helps to protect you from air pollution in various ways. Hairs in your nose filter out large particles. Sticky mucus in the lining of your upper respiratory tract captures smaller (but not the smallest) particles and dissolves some gaseous pollutants. Sneezing and coughing expel contaminated air and mucus when pollutants irritate your respiratory system.

In addition, hundreds of thousands of tiny, mucus-coated, hair-like structures, called *cilia*, line your upper respiratory tract. They continually move back and forth

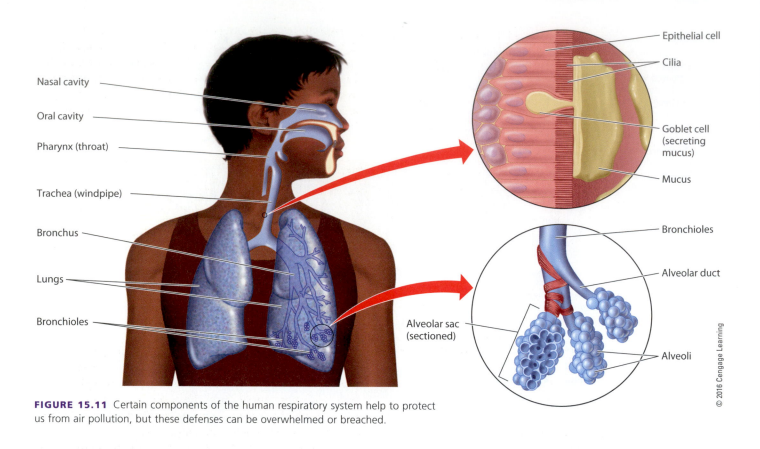

FIGURE 15.11 Certain components of the human respiratory system help to protect us from air pollution, but these defenses can be overwhelmed or breached.

FIGURE 15.12 Beijing, China, and a number of other major Chinese cities have very unhealthy levels of air pollution during parts of the year.

Air pollution has become a major health threat, especially for the residents of many growing cities.

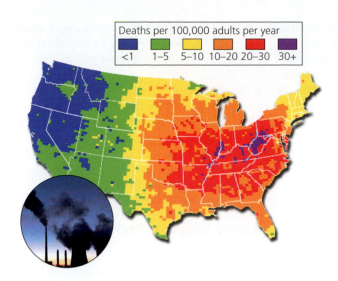

FIGURE 15.13 Distribution of premature deaths from air pollution in the United States, mostly from very small, fine, and ultrafine particles emitted into the atmosphere by coal-burning power plants. *Questions:* Why do the highest death rates occur in the eastern half of the United States? If you live in the United States, what is the risk at your home or where you go to school?

Compiled by the authors using data from U.S. Environmental Protection Agency.
Photo: Kodda/Shutterstock.com

and transport mucus and the pollutants it traps to your throat where they are swallowed or expelled.

Prolonged or acute exposure to air pollutants, including tobacco smoke, can overload or break down these natural defenses. Fine and ultrafine particulates get lodged deep in the lungs and contribute to lung cancer, asthma, heart attack, and stroke. Years of smoking or breathing polluted air can lead to other lung ailments such as chronic bronchitis and emphysema, which leads to acute shortness of breath.

In 2014, the WHO estimated that each year, indoor and outdoor air pollution kills about 7 million people. This averages out to about 800 deaths every hour. More than a third of these deaths occur in China (Figure 15.12) and India.

In 2013, Steven Barrett and other researchers at the Massachusetts Institute of Technology (MIT) estimated that outdoor air pollution, mostly fine-particle pollution, contributes to the deaths of roughly 200,000 Americans every year. About half of these deaths are blamed on car and truck exhaust and the other half on coal-burning power and industrial plants (Figure 15.13). This death toll is roughly equivalent to that of two fully loaded, 275-passenger airliners crashing every day of the year with no survivors. Millions more suffer from asthma attacks and other respiratory disorders brought on by indoor and outdoor air pollution.

According to EPA studies, each year, more than 125,000 Americans get cancer from breathing soot-laden diesel fumes emitted by buses, trucks, tractors, bulldozers and other construction equipment, trains, and ships.

A large diesel truck emits as much particulate matter as 150 cars, and according to a study led by Daniel Lack, the world's 100,000 or more diesel-powered oceangoing ships emit almost half as much particulate pollution as do the world's 1 billion cars. Thus, the largely unregulated shipping industry is one of the largest polluters of the atmosphere.

15.3 HOW SHOULD WE DEAL WITH AIR POLLUTION?

CONCEPT 15.3 Legal, economic, and technological tools can help us to clean up air pollution, but the best solution is to prevent it.

Laws and Regulations Can Reduce Outdoor Air Pollution

The United States provides an excellent example of how a regulatory approach can reduce air pollution (**Concept 15.3**). The U.S. Congress passed the Clean Air Acts in 1970, 1977, and 1990. With these laws, the federal government established air pollution regulations for key outdoor air pollutants that are enforced by states and major cities.

Congress directed the EPA to establish air quality standards for six major outdoor pollutants—carbon monoxide (CO), nitrogen dioxide (NO_2), sulfur dioxide (SO_2), suspended particulate matter (SPM, smaller than PM-10), ozone (O_3), and lead (Pb). Each standard specifies the maximum allowable level for a pollutant, averaged over a specific period. The EPA has also established national emission standards for more than 188 *hazardous air pollutants (HAPs)* that are thought to contribute to serious health and ecological problems.

According to a 2013 EPA report, the combined emissions of the six major outdoor air pollutants decreased by about 67% between 1980 and 2012, even with significant increases during the same period in gross domestic product (133%), vehicle miles traveled (97%), population (38%), and energy consumption (19%). Emissions during this period dropped by 99% for lead (Pb), 79% for SO_2, 72% for CO, 66% for PM-10, 59% for NO_x, 57% for volatile organic compounds, and 25% for ground-level ozone (O_3). *GOOD NEWS*

The reduction of outdoor air pollution in the United States since 1970 has been a remarkable success story, mostly because of two factors. *First,* during the 1970s, U.S. citizens insisted that laws be passed and enforced to improve air quality. Prior to 1970, when Congress passed the Clean Air Act, air-pollution-control equipment did not exist. *Second,* the country was affluent enough to afford such controls and improvements. For example, as a result of these factors, a new car today in the United States emits 75% less pollution than did a pre-1970 car.

Environmental scientists applaud this success, but they call for strengthening U.S. air pollution laws by:

- Putting much greater emphasis on air pollution prevention. With this approach, the question is not *What can we do about the air pollutants we produce?* but rather *How can we avoid producing these pollutants in the first place?* The power of prevention (**Concept 15.3**) was made clear by the 99% drop in atmospheric lead emissions after lead in gasoline was banned in 1976.

- Sharply reducing emissions from approximately 20,000 older coal-burning power and industrial plants, cement plants, oil refineries, and waste incinerators that have not been required to meet the air pollution standards for new facilities under the Clean Air Acts.

- Ramping up controls on atmospheric emissions of toxic pollutants such as mercury (see Figure 14.13, p. 377).

- Emphasizing reduction of emissions of air pollutants that blow across state boundaries (Figure 15.13).

- Continuing to improve fuel efficiency standards for motor vehicles, thereby also saving consumers money.

- Regulating more strictly the emissions from motorcycles and two-cycle gasoline engines used in devices such as chainsaws, lawnmowers, generators, scooters, and snowmobiles. The EPA estimates that running a typical gas-powered riding lawn mower for an hour creates as much air pollution as driving 34 cars for an hour.

- Setting much stricter air pollution regulations for airports and oceangoing ships.

- Sharply reducing indoor air pollution.

We Can Use the Marketplace to Reduce Outdoor Air Pollution

One approach to reducing pollutant emissions has been to allow producers of air pollutants to buy and sell government air pollution allotments in the marketplace. For example, with the goal of reducing SO_2 emissions, the Clean Air Act of 1990 authorized an *emissions trading,* or *cap-and-trade program,* which enables the 110 most polluting coal-fired power plants in 21 states to buy and sell SO_2 air pollution rights.

Under this system, each plant is annually given a number of pollution credits, which allow it to emit a certain amount of SO_2. A utility that emits less than its allotted amount has a surplus of pollution credits. That utility can use its credits to offset SO_2 emissions at its other plants, keep them for future plant expansions, or sell them to other utilities or private citizens or groups. Between 1990 and 2012, this emissions trading program helped to reduce SO_2 emissions from power plants in the United States by 76%, at a cost of less than one-tenth of the cost projected by the utility industry, according to the EPA.

Proponents of this market-based approach say it is cheaper and more efficient than government regulation of air pollution. Critics of this approach contend that it allows utilities with older, dirtier power plants to buy their way out of their environmental responsibilities and to continue to pollute.

The ultimate success of any emissions trading approach depends on two factors: how low the initial cap is set and how often it is lowered in order to promote continuing innovation in air pollution prevention and control. Without these two elements, emissions trading programs can shift pollution problems from one area to another without achieving an overall improvement in air quality.

There Are Many Ways to Reduce Outdoor Air Pollution

Figure 15.14 summarizes several ways to reduce emissions of sulfur oxides, nitrogen oxides, and particulate matter from stationary sources such as coal-burning power plants and industrial facilities—the primary contributors to industrial smog. Figure 15.15 lists several ways to prevent and reduce emissions from motor vehicles, the primary contributors to photochemical smog.

In more-developed countries, many of these solutions have been successful. However, the already poor air quality in urban areas of many less-developed countries is worsening as the numbers of motor vehicles in these nations rise. Over the next 10–20 years, technology could help all countries to clean up the air through improved

Solutions

Stationary Source Air Pollution

Prevention	Reduction or Dispersal
Burn low-sulfur coal or remove sulfur from coal	Disperse emissions using tall smokestacks (increases downwind pollution)
Convert coal to a liquid or gaseous fuel	Remove pollutants from smokestack gases
Switch from coal to natural gas and renewables	Tax each unit of pollution produced

FIGURE 15.14 Ways to prevent, reduce, or disperse emissions of sulfur oxides, nitrogen oxides, and particulate matter from stationary sources, especially coal-burning power plants and industrial facilities (**Concept 15.3**). *Questions:* Which two of these solutions do you think are the best ones? Why?

Top: Brittany Courville/Shutterstock.com. Bottom: Yegor Korzh/Shutterstock.com.

© 2016 Cengage Learning

engine and emission systems and hybrid-electric, plug-in hybrid, and all-electric vehicles (see Figure 13.23, p. 335).

GOOD NEWS

Reducing Indoor Air Pollution Should Be a Priority

Little effort has been devoted to reducing indoor air pollution even though it poses a much greater threat to human health than does outdoor air pollution (**Concept 15.2B**).

Air pollution experts suggest several ways to prevent or reduce indoor air pollution, as shown in Figure 15.16.

In less-developed countries, indoor air pollution from open fires (Figure 15.9) and inefficient stoves could be reduced if more people could use inexpensive clay or metal ovens that burn fuels more efficiently and vent their exhausts to the outside, or if they could use solar ovens and cookers (see Figure 13.32, p. 343).

Figure 15.17 lists some ways in which you can reduce your exposure to indoor air pollution.

Solutions

Motor Vehicle Air Pollution

Prevention

Walk or bike or use mass transit

Improve fuel efficiency

Get older, polluting cars off the road

Reduction

Require emission control devices

Inspect car exhaust systems twice a year

Set strict emission standards

© 2016 Cengage Learning

FIGURE 15.15 Ways to prevent or reduce emissions from motor vehicles (**Concept 15.3**). **Questions:** Which two of these solutions do you think are the best ones? Why?

Top: egd/Shutterstock.com. Bottom: Tyler Olson/Shutterstock.com.

Solutions

Indoor Air Pollution

Prevention

Ban indoor smoking

Set stricter formaldehyde emissions standards for carpet, furniture, and building materials

Prevent radon infiltration

Use naturally based cleaning agents, paints, and other products

Reduction and Dilution

Use adjustable fresh air vents for work spaces

Circulate air more frequently

Circulate a building's air through rooftop greenhouses

Use solar cookers and efficient, vented wood-burning stoves

© 2016 Cengage Learning

FIGURE 15.16 Ways to prevent or reduce indoor air pollution (**Concept 15.3**). **Questions:** Which two of these solutions do you think are the best ones? Why?

Top: Tribalium/Shutterstock.com. Bottom: Patstock/Age Fotostock.

What Can You Do?

Indoor Air Pollution

- Test for radon and formaldehyde inside your home and take corrective measures as needed

- Do not buy furniture and other products containing formaldehyde

- Test your home or workplace for asbestos fiber levels and check for any crumbling asbestos materials

- If you smoke, do it outside or in a closed room vented to the outside

- Make sure that wood-burning stoves, fireplaces, and kerosene- and gas-burning heaters are properly installed, vented, and maintained

- Install carbon monoxide detectors in all sleeping areas

- Use fans to circulate indoor air

- Grow house plants, the more, the better

- Do not store gasoline, solvents, or other volatile hazardous chemicals inside a home or attached garage

- Remove your shoes before entering your house to reduce inputs of dust, lead, and pesticides

© 2016 Cengage Learning

FIGURE 15.17 You can reduce your exposure to indoor air pollution. **Questions:** Which three of these actions do you think are the most important ones to take? Why?

15.4 HOW MIGHT THE EARTH'S CLIMATE CHANGE IN THE FUTURE?

CONCEPT 15.4 Considerable scientific evidence indicates that the earth's atmosphere is warming at a rate that is likely to lead to significant climate change.

Weather and Climate Are Not the Same

In thinking about climate change, it is very important to distinguish between weather and climate. **Weather** consists of short-term changes in atmospheric variables such as the temperature, precipitation, wind, and barometric pressure in a given area over a period of hours or days. By contrast, **climate** is determined by the *average* weather conditions of the earth or of a particular area, especially temperature and precipitation, over periods of at least three decades to thousands of years, according to the World Meteorological Society. Scientists have used long-term measurements of atmospheric temperature and precipitation to divide the earth into various climate zones (see Figure 7.2, p. 131).

During any period of 30 or more years, in a given area of the planet, there will often be hotter years and cooler years, and wetter years and drier years, as weather often fluctuates widely from day to day and from year to year. Climate scientists look at data on the normally fluctuating weather conditions for the earth as a whole and for particular areas of the earth to see if there has been a general rise or fall in measurements such as average temperature or precipitation over a period of at least 30 years. This is the only way they can determine how the climate of an area has changed, if at all. Thus, people who claim that an unusually cold winter, year, or even decade disproves atmospheric warming and climate change, and those who argue that an unusually warm summer, year, or decade demonstrates such warming and climate change, are making a common error by confusing weather with climate.

Atmospheric warming and climate change are often referred to as "global warming," which is a misleading term. It does not mean that all areas of the earth are getting warmer every year. Instead, as the earth's average atmospheric temperature rises, some areas get warmer at various times and some get cooler. However, when the *global average* atmospheric temperature rises or drops over a period of at least three decades, the earth's climate has changed.

Climate Change Is Not New but Recently Has Accelerated

Climate change is neither new nor unusual. Over the past 3.5 billion years, the planet's climate has been altered by factors such as volcanic eruptions, changes in solar input,

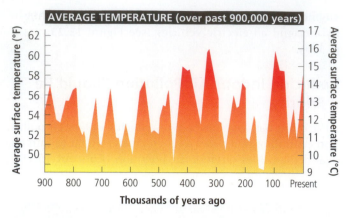

FIGURE 15.18 Over the past 900,000 years, the average global atmospheric temperature near the earth's surface has fluctuated widely. This graph is based on a body of scientific evidence that contains gaps, but the data clearly indicate general trends. **Question:** What are two conclusions you can draw from this graph?

Compiled by the authors using data from Goddard Institute for Space Studies, Intergovernmental Panel on Climate Change, National Academy of Sciences, National Aeronautics and Space Administration, National Center for Atmospheric Research, and National Oceanic and Atmospheric Administration.

continents moving slowly atop shifting tectonic plates (see Figure 4.D, p. 76), impacts by large meteors, the planet's wobbly orbit around the sun, and the fact that the earth is tilted in its orbital plane. The earth's climate is also affected by global air circulation patterns (see Figure 7.3, p. 132), large areas of ice (**Core Case Study**) that reflect incoming solar energy and help to cool the atmosphere, varying concentrations of the greenhouse gases found in the atmosphere, and occasional changes in ocean currents (see Figure 7.5, p. 133).

Over the past 900,000 years, the atmosphere has experienced climate change as a result of prolonged periods of considerable atmospheric warming and atmospheric cooling that led to ice ages (Figure 15.18). These alternating cycles of freezing and thawing are known as *glacial* and *interglacial* (between ice ages) *periods*.

For roughly 10,000 years, we have had the good fortune to live in an interglacial period characterized by a fairly stable climate based on a generally steady global average surface temperature. This important form of natural capital allowed the human population to grow as agriculture developed, and later as cities grew. For the past 1,000 years, the average temperature of the atmosphere near the earth's surface has remained fairly stable. But it began to rise during the 19th and 20th centuries when the expanding human population cleared large areas of forests and grasslands, which had been removing CO_2 from the atmosphere, and burned fossil fuels at steadily increasing rates, which added CO_2 to the atmosphere.

Most of the recent overall rapid rise in the global average atmospheric temperature on land has taken place since 1978 (Figure 15.19). According to the 2014 study on climate change by the American Association for the Advancement of Science (AAAS), evidence from numerous scientific

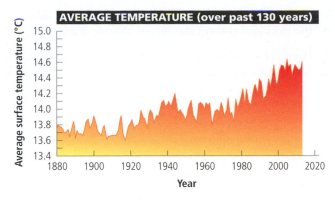

FIGURE 15.19 Average global atmospheric temperature near the earth's surface, 1880–2013. **Question:** What are two conclusions you can draw from this graph?

Compiled by the authors using data from Goddard Institute for Space Studies, Intergovernmental Panel on Climate Change, National Academy of Sciences, National Aeronautics and Space Administration, National Center for Atmospheric Research, and National Oceanic and Atmospheric Administration.

studies indicates that rising inputs of greenhouse gases from human activities are overwhelming the combined effects of natural factors that led to climate change in the past. Past temperature changes and the resulting changes in climate took place over periods of thousands to hundreds of thousands of years. Current climate change and that projected for the rest of this century are taking place within several

decades—many times faster than past climate changes caused by natural factors. Thus the problem we face is very rapid climate change caused mostly by human activities.

Scientists estimate past temperature changes such as those depicted in Figures 15.18 and 15.19 by analyzing many different types of evidence. These include radioisotopes in rocks and fossils; plankton and radioisotopes in ocean sediments; tiny bubbles, layers of soot, and other materials trapped in different layers of ancient air found in ice cores from glaciers (Figure 15.20); pollen from the bottoms of lakes and bogs; tree rings; and atmospheric temperature measurements taken regularly since 1861. These temperature measurements now include data from more than 40,000 measuring stations around the world, as well as from satellites.

Between 2007 and 2014, the world's leading scientific organizations—including the IPCC, U.S. National Academy of Sciences (NAS), British Royal Society, U.S. National Atmospheric and Oceanic Administration (NOAA), U.S. National Aeronautic and Space Administration (NASA), and AAAS—all reached the following three major conclusions, supported by the 2014 U.S. National Climate Assessment (NCA):

1. About 97% of the world's climate scientists agree that climate change is happening now, is caused mostly by human activities (especially deforestation and the burning of carbon-containing fossil fuels), and is a real

FIGURE 15.20 *Ice cores* are extracted from deep holes drilled into ancient glaciers at various sites such as this one near the South Pole in Antarctica. Analysis of ice cores yields information about the past composition of the lower atmosphere, temperature trends such as those shown in Figure 15.19, solar activity, snowfall, and forest fire frequency.

U.S. Geological Survey

threat that is very likely to get worse unless we act now to slow it. These conclusions are based on massive and overwhelming scientific evidence collected and evaluated over the past 25 years by thousands of the world's top scientists in this field. Thus, the popular view that there is significant scientific disagreement about atmospheric warming and climate change is a misleading myth that has delayed efforts to deal with this very serious threat to life as we know it and to human economies.

2. Immediate and sustained action to curb climate change is possible and affordable and would bring major benefits for human health and economies as well as for the environment.

3. The sooner we act to slow climate change, the lower the risks and costs of significant climate disruption.

Here are some of the thousands of pieces of scientific evidence, based on the analysis of tens of thousands of peer-reviewed scientific studies, that back up the conclusion that human-influenced climate change is happening now:

- Between 1906 and 2013, the earth's average global surface temperature rose by about 0.8C° (1.4F°), with much of this increase taking place since the mid-1970s (Figure 15.19).

- Nine of the ten warmest years on record since 1861 have taken place since 2000.

FIGURE 15.21 Between 1913 (top) and 2008 (bottom) much of the ice that covered Sperry Glacier in Montana's Glacier National Park melted.

Top: W. C. Alden/GNP Archives/US Geological Survey. Bottom: Lisa McKeon/US Geological Survey.

- In some parts of the world, glaciers that have existed for thousands of years are melting (Figure 15.21 and **Core Case Study**).

- In the Arctic, floating summer sea ice has been shrinking significantly.

- The melting of Greenland's ice sheets has accelerated (**Core Case Study**).

- In Alaska, glaciers and frozen ground (permafrost) are melting, loss of sea ice and rising sea levels are eating away at coastlines, and communities are being relocated inland.

- During the 20th century, the world's average sea level rose by 19 centimeters (7.6 inches)—mostly because of the expansion of ocean water as its temperature increased and because of increasing runoff from melting land-based ice, especially since 1975.

- Atmospheric levels of CO_2 and other greenhouse gases that warm the troposphere have been rising sharply (see Figure 11, p. S43, Supplement 5), mostly due to the burning of carbon-containing fossil fuels.

- As temperatures have risen, many terrestrial, marine, and freshwater species have migrated toward the poles and, on land, to cooler higher elevations (see Figure 7.9, p. 136). Species that cannot migrate face extinction.

The atmosphere continues to warm, although the rate of warming slowed down between 1998 and 2013. Scientific evidence indicates that this slowdown in the rate of atmospheric warming very likely is temporary and results from a combination of factors, including (1) short-term natural weather variations such as the El Niño–Southern Oscillation (see Figure 7.6, p. 134); (2) atmospheric ash from at least 17 volcanic eruptions since 2000 that have had a short-term cooling effect because sulfates in the volcanic emissions reflect sunlight; (3) increased air pollution from China and India that tend to cool the atmosphere similar to volcanic eruptions; (4) no net increase in energy from the sun since the 1970s; and (5) economic downturns such as the one between 2008 and 2012 that led to lower CO_2 emissions from the burning of fossil fuels.

In 2014, researchers Kevin Cowtan and Robert Way reviewed and found gaps in the worldwide atmospheric temperature data, as well as ways to fill some of those gaps by using satellite measurements. They concluded that atmospheric warming between 1997 and 2012 slowed by only about half as much as had been previously reported. In other words, warming during that period had been seriously underestimated. Other researchers are now examining their findings.

The Greenhouse Effect Plays a Key Role in Climate Change

The natural process called the **greenhouse effect** (see Figure 3.3, p. 44) plays a major role in determining the earth's climate. It occurs when some of the solar energy absorbed by the earth radiates into the atmosphere as infrared radiation (heat). As this radiation interacts with molecules in the air—especially the four major *greenhouse gases,* water vapor (H_2O), carbon dioxide (CO_2), methane (CH_4), and nitrous oxide (N_2O)—it increases the kinetic energy of air molecules and warms the lower atmosphere and the earth's surface.

Life on the earth and human economies are totally dependent on the greenhouse effect—one of the planet's most important forms of natural capital. It keeps the planet at a comfortable average temperature of around 15°C (58°F). Without it, the planet would be a frozen, uninhabitable place.

Since the beginning of the Industrial Revolution in the mid-1700s, human actions—mainly the burning of fossil fuels, deforestation, and agriculture—have led to significant increases in the concentrations of several greenhouse gases, especially CO_2, in the lower atmosphere (see Figure 11, p. S43, Supplement 5). The average atmospheric concentration of CO_2 rose by about 40% between 1880 and 2012 with more than half of the increase taking place since 1970. This is a long-lasting increase because CO_2

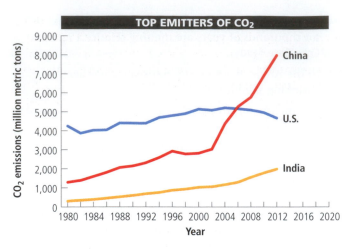

FIGURE 15.22 Carbon dioxide emissions from the burning of fossil fuels by the world's top three emitters, 1980 to 2012.

Compiled by the authors using data from Earth Policy Institute and T. A. Boden, G. Marland, and R. J. Andres, "Global, Regional, and National CO_2 Emissions," *Trends: A Compendium of Data on Global Change* (Oak Ridge, TN: Carbon Dioxide Information Analysis Center, 2012); BP, *Statistical Review of World Energy,* June 2013 (London: 2013).

typically remains in the atmosphere for 100 years or more. After oscillating between 180 and 280 parts per million (ppm) for 400,000 years, the concentration of CO_2 in the atmosphere averaged 396 ppm in 2013—higher than at any time in the last 800,000 years.

Also, since 1750, roughly the beginning of the Industrial Revolution, atmospheric methane levels have risen by 160%. According to a 2014 study by earth scientist Euan G. Nisbet and his colleagues, about two-thirds of the global emissions of methane are caused by human activities and the other third come from natural sources.

In 2012, the three largest emitters of energy-related CO_2 were China, the United States, and India, according to the Earth Policy Institute (Figure 15.22). However, in 2012, U.S. per capita emissions of CO_2 were several times higher than those of China and India. (See Figure 13 on p. S44 in Supplement 5 for data on per capita CO_2 emissions in high-, middle-, and low-income countries.) In comparing CO_2 emissions sources, scientists use the concept of a **carbon footprint**—the amount of CO_2 generated by an individual, an organization, a country, or any other entity over a given period of time. Thus, China has the largest national carbon footprint, and Americans have the largest per capita carbon footprints.

According to a 2014 study by Canadian researcher Damon Matthews and his colleagues, seven countries were responsible for 60% of the atmospheric warming taking place between 1906 and 2005. In order, they were the United States, China, Russia, Brazil, India, Germany, and the United Kingdom. The United States accounted for an estimated 22% of the temperature increase and China was responsible for 7%.

Climate models (Science Focus 15.1) project that rising levels of CO_2, water vapor, and atmospheric temperatures will likely bring about significant changes in the earth's climate that are likely to cause major ecological and economic disruption during this century. According to a 2014 NAS report, such changes are likely to last for at least a thousand years beyond the 21st century.

What Role Do the Oceans Play in Climate Change?

The world's oceans absorb CO_2 from the atmosphere as part of the carbon cycle and thus help to moderate the earth's average surface temperature and its climate. It is estimated that the oceans remove roughly 25% of the CO_2 pumped into the lower atmosphere by human activities. About 93% of it is then stored as carbon compounds in marine algae and vegetation and in coral reefs.

The oceans also absorb heat from the lower atmosphere. Indeed, an estimated 80–90% of the heat held in the lower atmosphere by greenhouse gases ends up in the ocean, according to NASA scientist Josh Willis. Then, partly driven by this heat, ocean currents slowly transfer some of the absorbed CO_2 to the deep ocean (see Figure 7.5, p. 133), where it is buried in carbon compounds in bottom sediments for several hundred million years. Thus, the average temperature of the oceans has also risen since 1970. However, because of their huge mass and volume, the oceans have warmed to a lesser degree and more slowly than the atmosphere has.

The uptake of CO_2 and heat by the world's oceans has helped to reduce the rate of climate change. However, this has resulted in the growing and serious problem of ocean acidification (see Science Focus 9.3, p. 206).

Cloud Cover Affects Atmospheric Warming

Warmer temperatures increase evaporation of surface water, which raises the relative humidity of the atmosphere in various parts of the world. This creates more clouds that can either cool or warm the atmosphere. An increase in thick and continuous *cumulus clouds* at low altitudes (Figure 15.23, left) could decrease surface warming by reflecting more sunlight back into space. But an increase in thin, wispy *cirrus clouds* at high altitudes (Figure 15.23, right) could cause more warming of the lower atmosphere by preventing some heat from escaping into space.

Climate modelers are working hard to understand more about the role of clouds in their climate models and the causes of cloud formation. According to the 2014 NAS report on climate, the latest scientific research indicates that the net global effect of cloud cover changes is likely to increase atmospheric warming. We need more research in order to evaluate this effect.

Tatiana Grozetskaya/Shutterstock.com

FIGURE 15.23 Cumulus clouds (left) are thick, relatively low-lying clouds that tend to decrease surface warming by reflecting some incoming solar radiation back into space. Cirrus clouds (right) are thin and float at high altitudes; they tend to warm the earth's surface by preventing some heat from flowing into space.

Cheryl Casey/Shutterstock.com

Outdoor Air Pollution Can Temporarily Slow Atmospheric Warming

According to the 2014 IPCC report, there is evidence that *aerosol* air pollution (suspended microscopic droplets and solid particles) from human activities has slowed the rate of atmospheric warming. These pollutants are released or formed in the troposphere by volcanic eruptions and by human activities (Figure 15.3). They can hinder or enhance both the greenhouse effect and cloud formation, depending on factors such as their size and reflectivity.

Most aerosols, such as light-colored sulfate particles produced by fossil fuel combustion, tend to reflect incoming sunlight and cool the lower atmosphere. However, black carbon particles, or *soot*, also released by such combustion, absorb solar energy and warm the lower atmosphere.

Climate scientists do not expect aerosols and soot particles to measurably counteract or enhance projected climate change resulting from greenhouse gas emissions in the next 50 years for two reasons. *First*, aerosols and soot fall back to the earth or are washed out of the lower atmosphere within weeks or months, whereas CO_2 typically remains in the lower atmosphere for 100 years or longer. *Second*, aerosol and soot inputs into the lower atmosphere are being reduced, especially in more-developed countries. The fact that they contribute to the deaths of large numbers of people has led to pollution prevention measures along with improvements in pollution-control technology.

USING MODELS TO PROJECT FUTURE CHANGES IN ATMOSPHERIC TEMPERATURES

To project the effects of increasing levels of greenhouse gases on future average global temperatures, scientists have developed complex *mathematical models* that simulate our current understanding of interactions among incoming sunlight, clouds, landmasses, oceans, ocean currents, concentrations of greenhouse gases and air pollutants, and other factors within the earth's complex climate system. They run these continually improving models on supercomputers and compare the results to known past climate changes, from which they project future changes in the earth's average atmospheric temperature. Figure 15.A gives a greatly simplified summary of some of the key interactions in the global climate system.

Such models provide *projections* of what is likely to happen to the average temperature of the lower atmosphere, based on available data and different assumptions about future changes such as CO_2 and CH_4 levels in the atmosphere. How well these projections match what actually happens in the real world depends on the validity of the assumptions, what variables are built into the models, and the accuracy of the data used.

Recall that while scientific research cannot give us absolute proof or certainty, it does provide us with varying levels of certainty.

When most experts in a particular scientific field generally agree on a level of 90% certainty about a set of measurements or model results, they say that their projections are *very likely* to be correct; when the level of certainty reaches 95% (a rarity in science), they contend that their projections are *extremely likely* to be correct.

In 1990, 1995, 2001, 2007, and 2014 the IPCC published reports on how global temperatures have changed in the past (Figures 15.18 and 15.19), how they are projected to change during this century, and how such changes are likely to affect the earth's climate. According to the 2014 IPCC report, based on analysis of past climate data and the use of more than two dozen climate models, it is extremely likely

FIGURE 15.A A simplified model of some major processes that interact to affect the earth's climate by determining the average temperature and greenhouse gas content of the lower atmosphere. Red arrows show processes that warm the atmosphere and blue arrows show those that cool it. **Question:** Why do you think a decrease in snow and ice cover is adding to the warming of the atmosphere?

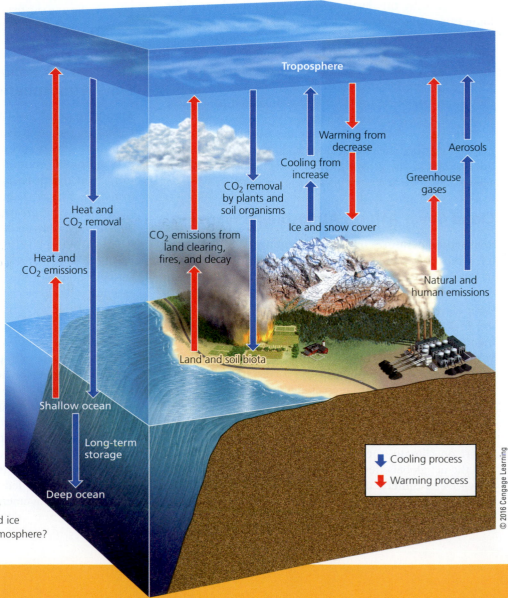

Sun
Troposphere
Heat and CO_2 removal
Heat and CO_2 emissions
CO_2 emissions from land clearing, fires, and decay
CO_2 removal by plants and soil organisms
Warming from decrease
Cooling from increase
Ice and snow cover
Aerosols
Greenhouse gases
Natural and human emissions
Land and soil biota
Shallow ocean
Long-term storage
Deep ocean

Cooling process
Warming process

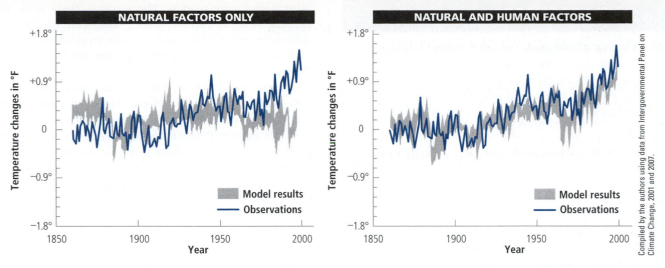

Compiled by the authors using data from Intergovernmental Panel on Climate Change, 2001 and 2007.

FIGURE 15.B Comparison of actual climate data with modeled projections for the period between 1860 and 2000 using natural factors only (left) and a combination of natural and human factors (right). Scientists have found that actual data match projections far more closely when human factors are included in the models.

(95% certainty) that human activities, especially the burning of fossil fuels, have played the dominant role in the observed atmospheric warming since 1975 (Figure 15.19). The researchers based this conclusion on the fact that, after thousands of times running the models, the only way they can get the model results to match actual measurements is by including the human activities factor (Figure 15.B).

The current results of more than two dozen climate models now in use suggest that it is *very likely* (with at least 90% certainty) that the earth's mean surface temperature will increase by 1.5–4.5C° (2.7–8.1F°) between 2013 and 2100 (Figure 15.C), unless we can sharply reduce deforestation along with our emissions of CO_2 and other greenhouse gases. In 2014, climate researcher Michael E.

Mann reported that the latest runs of key climate models indicated that the lower limit of the 2014 IPCC projected temperature increase should be raised from 1.5C° (2.7F°) to 2.5C° (4.5F°).

While there is an extremely high degree of certainty (95%) that the atmosphere has been warming primarily due to human activities, there is also a high degree of uncertainty in climate model results about projected future changes in the average atmospheric temperature. This is indicated by the wide range of projected temperature changes in Figure 15.C and the range of the projected rise in sea level (see Figure 16, p. S48, in Supplement 5). Climate experts are working to reduce such uncertainty by learning more about how the earth's climate system works and by improving climate data and models. Despite their limitations, these models are the best and only tools that we have for projecting likely average atmospheric temperatures in coming decades.

Critical Thinking

If the highest possible projected temperature increase shown in Figure 15.C takes place, what are three major ways in which this will likely affect your lifestyle and that of any children or grandchildren you eventually might have?

FIGURE 15.C Estimated changes (yellow area) and measured changes (black curve) in the average temperature of the atmosphere near the earth's surface between 1860 and 2013, and the projected range of temperature increase during the rest of this century.

Compiled by the authors using data from U.S. National Academy of Sciences, National Center for Atmospheric Research, Intergovernmental Panel on Climate Change, and Hadley Center for Climate Prediction and Research.

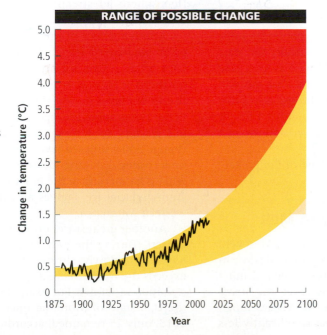

CONCEPT 15.5 The projected rapid change in the atmosphere's temperature could have severe and long-lasting consequences, including flooding, rising sea levels, shifts in the locations of croplands and wildlife habitats, and more extreme weather.

Rapid Atmospheric Warming Could Have Serious Consequences

Most historical changes in the temperature of the lower atmosphere took place over thousands of years (Figure 15.18). What makes the current problem urgent is that we face *a rapid projected increase in the average temperature of the lower atmosphere during your lifetime* (Figure 15.C). This, in turn, is very likely to change the fairly mild climate that we have had for the past 10,000 years. According to the 2014 AAAS report on climate change, "The rate of climate change now may be as fast as any extended warming period over the past 65 million years, and is projected to increase in coming years."

Climate models indicate that, in the *worst-case scenario*, rising atmospheric temperatures and concentrations of water vapor will likely lead to: rising sea levels that would flood many low-lying coastal cities (see chapter-opening photo); some forests being consumed in vast wildfires; some grasslands turning into dust bowls; some rivers drying up; certain ecosystems collapsing; the extinction of at least a fourth and perhaps half of the world's species; more intense and longer-lasting heat waves; and more destructive storms and flooding.

These effects will likely reduce food security and increase poverty and social conflict in many poorer nations, such as Bangladesh—countries that are typically the least responsible for atmospheric warming, as well as least able to protect themselves. The models indicate that we will have to deal simultaneously with many of these disruptive effects within this century—an incredibly short time to bring about a major shift in the way we live and interact with our life-support system.

Let's take a more in-depth look at some of these likely consequences.

More Ice and Snow Are Likely to Melt

Models project that climate change will be the most severe in the world's polar regions. Light-colored ice and snow in these regions help to cool the earth by reflecting incoming solar energy back into space—a process called the *albedo effect*. The melting of such ice and snow exposes much darker land and sea areas, which reflect significantly less sunlight and absorb more solar energy. This has warmed the atmosphere above the poles more and faster than the atmosphere is warming at lower latitudes, as projected by all major climate models. The result is likely to be more melting of snow and ice, which will cause further atmospheric warming above the poles in an escalating spiral of change as part of a positive, runaway feedback loop (see Figure 2.11, p. 36).

According to the 2014 IPCC report, arctic air temperatures have risen almost twice as fast as average temperatures in the rest of the world during the past 50 years, and they are now warmer than they have been in more than 44,000 years. Arctic ocean waters have also warmed. In addition, soot generated by North American, European, and Asian industries is darkening arctic ice and lessening its ability to reflect sunlight.

Mostly as a result of these factors, floating summer sea ice in the Arctic is disappearing faster than scientists thought it would only a few years ago (Figure 15.24). Measurements indicate that the melting of this ice is due to both warmer air above the ice and warmer water below. Because of changes in short-term weather conditions, summer arctic sea ice coverage is likely to fluctuate. But the overall projected long-term trends are for the Arctic to warm, for average summer sea ice coverage to decrease, and for the ice to become thinner.

If the current trend continues, summer arctic ice may be gone by 2050, according to the 2014 IPCC report. This would open these waters to shipping and would allow access to oil and mineral deposits in the arctic region. However, it could also lead to dramatic and long-lasting changes in weather and climate that could affect the whole planet. According to one hypothesis, this includes colder and snowier winters in Europe, eastern North America, and eastern Asia because of a slower Northern Hemisphere jet stream sinking southward and bringing cold air into those areas.

Another effect of arctic warming is faster melting of polar land-based ice, including that in Greenland (**Core Case Study**). This melting is adding freshwater to the northern seas, and is likely to contribute to a projected rise in sea level during this century. Glaciologist and National Geographic Emerging Explorer Erin Pettit is studying glaciers to better understand and project climate change and rising sea levels. She has pioneered the use of underwater listening devices (hydrophones) to record what is happening at the noisy boundary between disintegrating glacial ice shelves and the seas in Alaska and in western Antarctica.

Another great storehouse of ice is the earth's mountain glaciers. During the past 25 years, many of these glaciers have been slowly shrinking wherever summer melting exceeds the winter *snowpack*—the addition of ice from precipitation in winter. For example, Glacier National Park in the U.S. state of Montana once had 150 glaciers, but by 2013, only 25 remained, according to park officials.

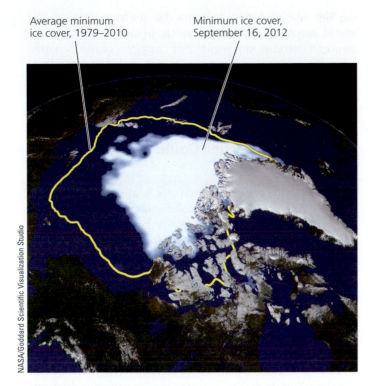

Average minimum ice cover, 1979–2010

Minimum ice cover, September 16, 2012

NASA/Goddard Scientific Visualization Studio

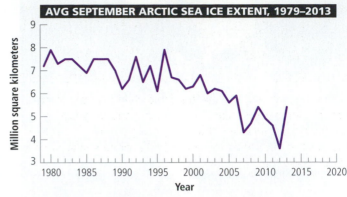

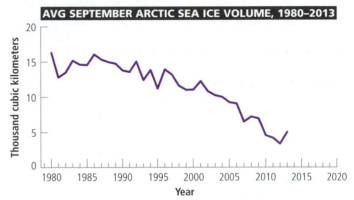

FIGURE 15.24 *The big melt:* Rising average atmospheric and ocean temperatures have caused more and more arctic sea ice to melt during the summer months. The yellow line added to this satellite image (left) shows the average summer minimum area of ice during the period 1979–2010, in contrast to the white, ice-covered summer minimum in 2012. The graphs (right) show that the sea ice melt generally has been increasing. *Question:* Which is declining faster, the summer sea ice area, or the summer sea ice volume?

Compiled by the authors using data from U.S. Goddard Space Flight Center, National Aeronautics and Space Administration, National Snow and Ice Data Center.

Mountain glaciers play a vital role in the water cycle (see Figure 3.13, p. 52) by storing water as ice during cold seasons and releasing it slowly to streams during warmer seasons. A prime example of high-elevation reservoirs is the glaciers of the Himalayan Mountains in Asia, which have been shrinking (Figure 15.21). They are a major source of water for large rivers such as the Ganges, which provides water for more than 400 million people in India and Bangladesh. They also feed China's Yangtze and Yellow Rivers, whose basins are home to more than 500 million people.

About 80% of the mountain glaciers in South America's Andes range are slowly shrinking. If this continues, 53 million people in Bolivia, Peru, and Ecuador who rely on meltwater from the glaciers for irrigation and hydropower could at some point face severe water, power, and food shortages. In the United States, according to climate models, people living in the Columbia, Sacramento, and Colorado River basins could face similar threats as the winter snowpack that feeds these rivers is projected to shrink by as much as 70% by 2050.

Permafrost Is Likely to Melt: Another Dangerous Scenario

Permafrost occurs in soils found beneath about 25% of the exposed land in Alaska, Canada, and Siberia in the northern hemisphere. Huge amounts of carbon are locked up in permafrost soils.

A 2012 report by the United Nations Environment Programme (UNEP) estimated that human-caused climate change is projected to thaw out significant amounts of permafrost, as is already happening in parts of Alaska and Siberia. If this trend continues, a great deal of organic material found below the permafrost will likely rot and release huge amounts of CH_4 and CO_2 into the atmosphere. This would accelerate projected atmospheric warming, which would in turn melt more permafrost and lead to more atmospheric warming in yet another worsening spiral of change as part of a positive feedback loop.

Some scientists are concerned about another methane source—a layer of permafrost on the Arctic Sea floor. Also, in 2012, aquatic ecologist and National Geographic Emerging Explorer Katey Walter Anthony found methane bubbling up from many arctic lake bottoms (Figure 15.25).

Sea Levels Are Rising

In 2014, the IPCC estimated that the average global sea level is likely to rise by 40–60 centimeters (1.3–2 feet) by the end of this century—about 10 times the rise that oc-

FIGURE 15.25 Scientists ignite a large bubble of methane gas released from an arctic lake in Alaska.

curred in the 20th century. Half to two-thirds of this rise will likely come from the melting of Greenland's ice (**Core Case Study**). However, accelerated melting could lead to seas rising by as much as 0.9–2 meters (3–7 feet), depending on how much of the land-based ice in Greenland and perhaps West Antarctica melt as the global temperature continues to rise. (See Figure 16 on p. S48 in Supplement 5 for a graph showing past and projected increases in the world's average sea level.)

According to the 2014 IPCC and NAS reports on climate change, a 1-meter (3-foot) rise in sea level during this century (excluding the additional effects of the resulting higher storm surges) could cause the following serious effects:

- Degradation or destruction of at least one-third of the world's coastal estuaries, wetlands, coral reefs, and deltas where much of the world's rice is grown.

- Disruption of many of the world's coastal fisheries.

- Flooding in large areas of low-lying countries such as Bangladesh, one of the world's poorest and most densely populated nations.

- Flooding and erosion of low-lying barrier islands and gently sloping coastlines, especially in U.S. coastal states such as Florida (Figure 15.26), Texas, Louisiana, New Jersey, South Carolina, and North Carolina.

- Flooding of some of the world's largest coastal cities such as Venice, London, and New Orleans (see red areas in chapter-opening photo and Figure 15.26), and displacement of at least 150 million people—an amount almost equal to half of the current U.S. population.

- Saltwater contamination of freshwater coastal aquifers resulting in degraded supplies of groundwater used as a source of water for drinking and irrigation.

- Submersion of low-lying island nations such as the Maldives and Fiji.

The sea-level rise will not be uniform around the world, according to climate models, because of factors such as ocean currents and winds. For example, coastal scientist John Pethick projects that by 2100, Bangladesh's sea level could rise by as much as 4 meters (13 feet), several times higher than the projected global average sea-level rise.

Severe Drought and Other Forms of Extreme Weather Could Become More Common

A 2005 study by National Center for Atmospheric Research scientist Aiguo Dai and his colleagues found that severe and prolonged drought was affecting at least 30% of the earth's land (excluding Antarctica)—an area the size of Asia. According to a 2007 study by climate researchers at NASA's Goddard Institute for Space Studies, by 2059, up to 45% of the world's land area could be experiencing extreme drought. Natural cyclical processes also cause extreme droughts so a specific drought cannot be tied to atmospheric warming. However, the extra heat energy in the atmosphere evaporates water from soils. According to a 2014 study by climate scientists Richard Seager and Martin Hoerling, this depletion of soil moisture prolongs droughts and makes them more severe, regardless of their causes.

Warming increases the kinetic energy in the atmosphere, and as a result, some areas will likely experience longer, more frequent, and more intense heat waves, which could raise the number of heat-related deaths, reduce crop production, and expand deserts. Since 1950, heat waves have become longer and more frequent. At the same time, because a warmer atmosphere can hold more moisture, other areas, such as the eastern half of the United States, will likely experience increased flooding, on average, from heavy and prolonged snow or rainfall.

FIGURE 15.26 If the average sea level rises by 1 meter (3 feet), the areas shown here in red in the U.S. state of Florida will be flooded.

FIGURE 15.27 With warmer winters, populations of mountain pine beetles have exploded and killed large numbers of trees (orange areas in photo) in this lodgepole pine forest in the Canadian province of British Columbia.

Reproduced with permission from Natural Resources Canada, Canadian Forest Service, 2014

Large areas of forest are now seeing the effects of climate change.

In some areas, global atmospheric warming will likely lead to colder winter weather, according to climate models, largely because of changes in global air circulation patterns due to the warming. (See Figure 15 on p. S47 in Supplement 5 for graphs showing global changes in atmospheric temperature, atmospheric moisture, heat waves, and extreme rainfalls for the past several decades.)

In 2010, a World Meteorological Organization panel of experts concluded that projected atmospheric warming is likely to lead to fewer but stronger hurricanes and typhoons that could cause more damage in coastal areas where urban populations have grown rapidly (see chapter-opening photo). However, climate scientists point out that while climate change is likely to increase the overall chances of extreme weather events over periods of 30 years or longer, there is not sufficient evidence to link any specific extreme weather event such as a heat wave, storm, hurricane, or typhoon to climate change.

Climate Change Threatens Biodiversity

According to the 2007 and 2014 IPCC reports, projected climate change is likely to alter ecosystems and take a toll on biodiversity in areas of every continent. For example, up to 85% of the Amazon rain forest—one of the world's major centers of biodiversity—could be lost and converted to tropical savanna if the global atmospheric temperature rises by the highest projected amount (Figure 15.C), according to a 2009 study led by Chris Jones.

As the atmosphere warms, 25–50% of the world's species could face extinction by 2100. The hardest hit will be cold-climate plant and animal species, including the polar bear in the Arctic and penguins in Antarctica; species that live at higher elevations; species with limited ranges such as some amphibians (see Chapter 4 Core Case Study, p. 64); and those with limited tolerance for temperature change, such as corals. The primary cause of such extinctions would be loss of habitat. On the other hand, the populations of plant and animal species that thrive in warmer climates could grow.

Research indicates that the most vulnerable ecosystems are coral reefs, polar seas, coastal wetlands, high-elevation forests, and alpine and arctic tundra. Primarily because of drier conditions, forest fires will likely become more frequent and intense in some areas such as the southeastern and western United States. A warmer climate could also greatly increase populations of insects and fungi that damage trees, especially in the absence of winter weather cold enough to control their populations. According to scientists, this helps to explain the recent severe damage to pine forests in Canada (Figure 15.27) and in the American West. Climate change is also threatening many existing state and national parks, wildlife reserves, wilderness areas, and wetlands, along with much of the biodiversity they contain.

Food Production Could Decline

Farmers will face dramatic changes due to shifting climates and an intensified hydrologic cycle, if the atmosphere keeps warming as projected (Figure 15.C). According to the IPCC, crop productivity is projected to increase slightly at middle to high latitudes in areas such as midwestern Canada, Russia, and Ukraine, with moderate warming, but it will decrease if warming goes too far. However, the projected rise in crop productivity might be limited because the soils in these northern regions generally lack sufficient plant nutrients.

Climate change models project a decline in agricultural productivity and food security in tropical and subtropical regions, especially in Southeast Asia and Central America.

Also, the flooding of river deltas due to rising sea levels could reduce crop production, partly because some aquifers that supply irrigation water will be infiltrated by salt water. This flooding could also affect fish production in coastal aquaculture ponds. Food production could also drop in farm regions that are dependent on rivers fed by melting glaciers, and in arid and semiarid areas where droughts become more prolonged.

According to the IPCC, food is likely to be plentiful for a while in a warmer world, because of the longer growing season in northern regions. But IPCC scientists warn that during the latter half of this century, several hundred million of the world's poorest and most vulnerable people could face starvation and malnutrition due to a drop in food production caused by projected climate change.

Climate Change Will Likely Threaten Human Health, National Security, and Economies

According to IPCC and other reports, more frequent and prolonged heat waves in some areas will raise the numbers of deaths and illnesses, especially among older people, people in poor health, and the urban poor who cannot afford air conditioning. On the other hand, fewer people will die from cold weather. However, a study led by Mercedes Medina-Ramon of the Harvard University School of Public Health suggests that during the latter half of this century, the projected rise in the number of heat-related deaths will likely exceed the projected drop in the number of cold-related deaths.

A warmer and more CO_2-rich atmosphere will likely favor rapidly multiplying insects, including mosquitoes and ticks that transmit diseases such as West Nile virus and Lyme disease. Warming will also favor microbes, toxic molds, and fungi, as well as some plants that produce pollens that cause allergies and asthma attacks. Also, insect pests and weeds are likely to multiply, spread, and reduce crop yields. Microbes that cause infectious tropical diseases such as dengue fever and yellow fever could expand or at least shift their ranges and numbers if the mosquitoes that carry them spread to warmer temperate and higher elevation areas. But we need more research in order to establish this connection.

It is also likely that higher atmospheric temperatures and higher levels of water vapor in urban areas will contribute to heavier photochemical smog in such areas. This in turn will likely cause more pollution-related deaths and illnesses due to heart ailments and respiratory problems.

Recent studies by the U.S. Department of Defense and the National Academy of Sciences warn that climate change could affect U.S. national security. Likely effects include global geopolitical impacts that could contribute to increased food and water scarcity, poverty, environmental degradation, social unrest, mass migration of environmental refugees, political instability, and the weakening of fragile governments.

Climate change is also likely to take a toll on human economies, although there is disagreement over the likely impact. Various economists estimate that the cost of projected climate change, if current trends continue, would be 5% to 20% per year of the world's economic activity as measured by gross domestic output. Other researchers say the cost could be higher, because most models used to calculate such costs have not adequately accounted for economic losses from projected ocean acidification (see Science Focus 9.3, p. 206), rising sea levels, and biodiversity losses. Also, a growing number of economists and major multinational companies are recognizing that more drought, flooding, and extreme weather events are contributing to lower economic productivity, disruption of supply chains, higher costs for food and other commodities, and higher levels of financial risk for companies and investors.

15.6 WHAT CAN WE DO TO SLOW PROJECTED CLIMATE CHANGE?

CONCEPT 15.6 We can reduce greenhouse gas emissions and the threat of climate change while saving money and improving human health if we cut energy waste and rely more on cleaner renewable energy resources.

Dealing with Projected Climate Change Is Difficult

Many climate scientists and other analysts believe that reducing the threat of projected climate change is one of the most urgent scientific, political, economic, and ethical issues that humanity faces. But the following characteristics of this complex problem make it difficult to tackle:

- *The problem is global.* Dealing with this threat will require unprecedented and prolonged international cooperation.

- *The problem is a long-term political issue.* Climate change is now happening and is already having harmful impacts, but it is not viewed as an urgent problem by most voters and elected officials. And most of the people who will likely suffer the most serious harm from projected climate change during the latter half of this century have not yet been born.

- *The projected harmful and beneficial impacts of climate change are not spread evenly.* For example, higher-latitude nations such as Canada, Russia, and New Zealand may temporarily have higher crop yields, fewer deaths in winter, and lower heating bills. But other, mostly poor nations such as Bangladesh could see more flooding and higher death tolls.

- *Proposed solutions, such as sharply reducing or phasing out the use of fossil fuels, are controversial* because they could disrupt economies and lifestyles and threaten the profits of the economically and politically powerful fossil-fuel companies.

- *The projected effects are uncertain.* Current climate models lead to a wide range in the projected temperature increase (Figure 15.C) and sea-level rise (Figure 16, p. S48, in Supplement 5). Thus, there is considerable uncertainty over whether the harmful changes will be moderate or catastrophic. This makes it difficult to plan for avoiding or managing risk, and it highlights the urgent need for more scientific research to reduce the uncertainty in climate models.

There are two basic approaches to dealing with the projected harmful effects of global climate change. One, called *mitigation,* is to try to slow it down in order to avoid its most harmful effects. The other approach, called *adaptation,* is to recognize that our current failure to sharply reduce our contribution to atmospheric warming means that some climate change is unavoidable and that we will now have to try to adapt to some of its harmful effects. Most analysts call for a combination of both approaches.

Regardless of how we approach climate change, most climate scientists argue that our most urgent priority is to avoid any and all **climate change tipping points**—those estimated thresholds beyond which natural systems could change irreversibly. Figure 15.28 lists some of the possible tipping points that climate scientists are most concerned about. For example, if we continue to add CO_2 to the atmosphere at the current rate, we will likely exceed the estimated tipping point marked by 450 ppm of atmospheric CO_2 within a couple of decades. This would likely lock in severe climate change for hundreds or perhaps thousands of years and increase the likelihood of exceeding many of the other tipping points shown in Figure 15.28.

We Can Prevent and Control Greenhouse Gas Emissions

The world's nations and energy companies combined hold reserves of fossil fuels that, if burned, would emit nearly five times the amount of CO_2 that climate scientists say we can safely emit. Currently, the economic well-being of the politically powerful fossil-fuel companies and most of the world's national economies depend on using all or most of these reserves of fossil fuels.

Even if we could stop burning all fossil fuels tomorrow, the amount of CO_2 already in the atmosphere will very likely continue to warm the earth for many decades because it will remain in the atmosphere for at least 100 years. In other words, our fossil fuel–driven civilization has very likely (with at least 90% certainty) irreversibly committed future generations to a warmer world with rising seas and other harmful consequences—a serious viola-

- Atmospheric carbon level of 450 ppm
- Melting of all arctic summer sea ice
- Collapse and melting of the Greenland ice sheet
- Collapse and melting of most of the western Antarctic ice sheet
- Massive release of methane from thawing arctic permafrost and from the arctic seafloor
- Collapse of part of the Gulf Stream
- Severe ocean acidification, collapse of phytoplankton populations, and a sharp drop in the ability of the oceans to absorb CO_2
- Massive loss of coral reefs
- Severe shrinkage or collapse of Amazon rain forest

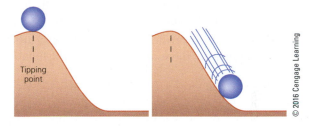

© 2016 Cengage Learning

FIGURE 15.28 Climate scientists have come up with this list of possible climate change tipping points.

tion of the ethical **principle of sustainability** (see Supplement 7, p. S51), which calls for us to leave the world in a condition as good as or better than what we inherited.

A growing number of scientists and other analysts recognize that shifting away from our dependence on fossil fuels, especially abundant coal and oil, over the next 4 to 5 decades will be difficult but can be done. During the history of human energy use, we shifted our dependence first from wood to coal, then to oil, and now to natural gas with each shift taking about 50 years. We now have the knowledge and ability to shift to a reliance on energy efficiency and renewable energy over the next 50 years. All we need is the political and ethical will to commit to and complete such a shift.

Some scientists contend that we could get a jump on slowing atmospheric warming by focusing on black carbon (a component of soot) and methane emissions, because these chemicals are short-lived in the atmosphere, compared to CO_2 emissions, and we have the technologies to accomplish such cuts fairly quickly. Although methane emissions account for only 9% of U.S. greenhouse gas emissions, methane molecules warm the air 25 times more effectively than CO_2 molecules do. According to a 2011 UNEP study, major cuts in these greenhouse gas emissions could potentially cut the rate of atmospheric warming in half by 2050.

Figure 15.29 lists a number of ways to slow the rate and degree of atmospheric warming and some of the projected resulting climate change caused by human activities (**Concept 15.6**). Among these solutions are four major prevention strategies that, according to a 2010 NAS study and the 2014 IPCC report, could reduce human greenhouse gas emissions by 57–83% by 2050:

1. Improve energy efficiency to reduce fossil fuel use, especially the use of coal. This would also save consumers money.

2. Shift from carbon-based fossil fuels to a mix of low-carbon renewable energy resources based on local and regional availability.

3. Stop cutting down tropical forests and plant trees to help remove more CO_2 from the atmosphere.

4. Shift to more sustainable and climate-friendly food production (see Figure 10.27, p. 242).

5. Work on cutting greenhouse gas emissions in urban areas and making them more adaptable to climate change.

According to the 2014 IPCC report, there is much good news related to dealing with the threat of climate change:

- We know how to engineer zero-carbon buildings and how to reduce the carbon footprints of existing buildings.

- If we increase the use of hybrid, plug-in hybrid, and electric cars, while charging their batteries only with electricity produced from renewable energy sources, we will greatly reduce the carbon footprints of motor vehicles.

- The shift to renewable energy is accelerating as the prices for electricity from low-carbon wind turbines and solar cells are falling very rapidly, and investments in these technologies are growing.

- If we act now, we have inexpensive ways to keep atmospheric concentrations of CO_2 below 450 ppm and the atmospheric temperature increase to no more than 2°C (3.6°F) above the preindustrial level. This would require reducing global consumption of goods and services during this century, but only by 0.06% per year, according to IPCC estimates. And it could result in major and widespread economic benefits in the form of new technologies, markets, and jobs.

Solutions

Slowing Climate Change

Prevention	Cleanup
Cut fossil fuel use (especially coal)	Sequester CO_2 by planting trees and preserving forests and wetlands
Shift from coal to natural gas	
Repair leaky natural gas pipelines and facilities	Sequester carbon in soil
Improve energy efficiency	Sequester CO_2 deep underground (with no leaks allowed)
Shift to renewable energy resources	
Reduce deforestation	Sequester CO_2 in the deep ocean (with no leaks allowed)
Use more sustainable agriculture and forestry	
Put a price on greenhouse gas emissions	Remove CO_2 from smokestack and vehicle emissions

© 2016 Cengage Learning

FIGURE 15.29 Some ways to slow atmospheric warming and the resulting climate change projected for this century (**Concept 15.6**). *Questions:* Which five of these solutions do you think are the best ones? Why?

Top: Mark Smith/Shutterstock.com. Center: Yegor Korzh/Shutterstock.com. Bottom: pedrosala/Shutterstock.com.

Some Promote Cleanup and Geoengineering Schemes to Ward Off Climate Change

Some scientists and engineers want to use cleanup strategies for removing some of the CO_2 from the atmosphere or from smokestacks, and storing (sequestering) it in other parts of the environment. One such strategy, known as **carbon capture and storage (CCS)**, would remove some of the CO_2 gas from smokestack emissions of coal-burning power and industrial plants or from the atmosphere, convert it to a liquid, and force it under pressure into underground storage sites (Figure 15.30).

One problem with CCS schemes is that, with current technology, they can remove only part of the CO_2 from smokestack emissions and the atmosphere, at great cost, and they would not address the massive amounts of CO_2 from motor vehicle exhausts, food production, and the burning of forests. CCS schemes also require a lot of energy, which could lead to greater use of fossil fuels and higher emissions of CO_2 and other air pollutants. In addition, the CO_2 that is removed would have to remain sealed from the atmosphere forever. Any large-scale leaks or a number of smaller continuous leaks could dramatically increase atmospheric warming in a very short time.

So far, the experimental projects for capturing CO_2 have not been very effective and have been quite costly. For example, the first large-scale plant designed to convert coal to gas and to capture 65% of the CO_2 produced has

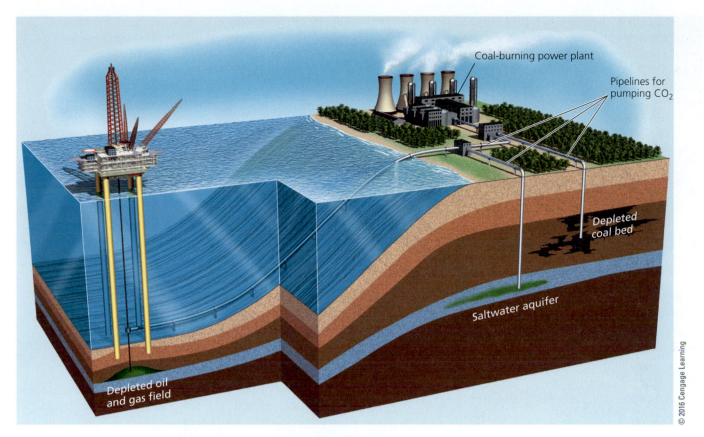

FIGURE 15.30 Some proposed carbon capture and storage (CCS) schemes for removing some of the carbon dioxide from smokestack emissions and from the atmosphere and storing (sequestering) it in soil, plants, deep underground reservoirs, and sediments beneath the ocean floor. **Questions:** Which of these proposed strategies do you think would work best and which would be the least effective? Why?

experienced large cost overruns and is projected to cost $5.2 billion. It will be one of the most expensive fossil-fuel plants ever built in the United States. Such high costs are making U.S. utilities reluctant to build CCS plants.

Other approaches to cleaning up carbon, or sequestering it, include planting large areas of trees and fertilizing the ocean with iron pellets to boost populations of phytoplankton (a strategy that, preliminary experiments indicate, may not work very well and could disrupt marine ecosystems). These and other such strategies would have the aim of removing more CO_2 from the atmosphere.

Other strategies fall under the umbrella of **geoengineering**, or trying to manipulate certain natural conditions to help counter our enhancement of the greenhouse effect. Some of these proposals are shown in Figure 15.31. One such proposal calls for injecting sulfate particles into the stratosphere to reflect some of the incoming sunlight into space in order to slow the warming of the troposphere. Other scientists have called for placing a series of giant mirrors in orbit above the earth to reflect incoming sunlight for the same purpose. Another scheme is to deploy a large fleet of computer-controlled ocean-going ships to inject saltwater high into the sky in order to make clouds whiter and more reflective.

Some scientists reject the idea of launching sulfates into the stratosphere as being too risky because of our limited knowledge about possible unknown effects. For example, if the sulfates reflected too much sunlight, they could reduce evaporation enough to alter global rainfall patterns and worsen the already dangerous droughts in Asia and Africa. Also, a 2008 study by atmospheric scientist Simone Tilmes indicated that chlorine released by reactions involved in this scheme could speed up the thinning of the earth's vital ozone layer (a problem that we discuss in Section 15.7).

According to some scientists, a major problem with most of these technological fixes is that, if they succeed, they could be used to justify the continued rampant use of fossil fuels, and this would allow CO_2 levels in the lower atmosphere to continue building and adding to the serious problem of ocean acidification. In the long run, they might not succeed in slowing atmospheric warming, in which case, atmospheric temperatures will likely soar at a rapid rate and essentially ensure severe and irreversible climate change, as well as other possible unexpected harmful side effects.

In addition, thinking that we can use geoengineering schemes to slow or prevent climate change could seriously delay a shift from relying on nonrenewable fossil fuels to using improved energy efficiency and a mix of renewable

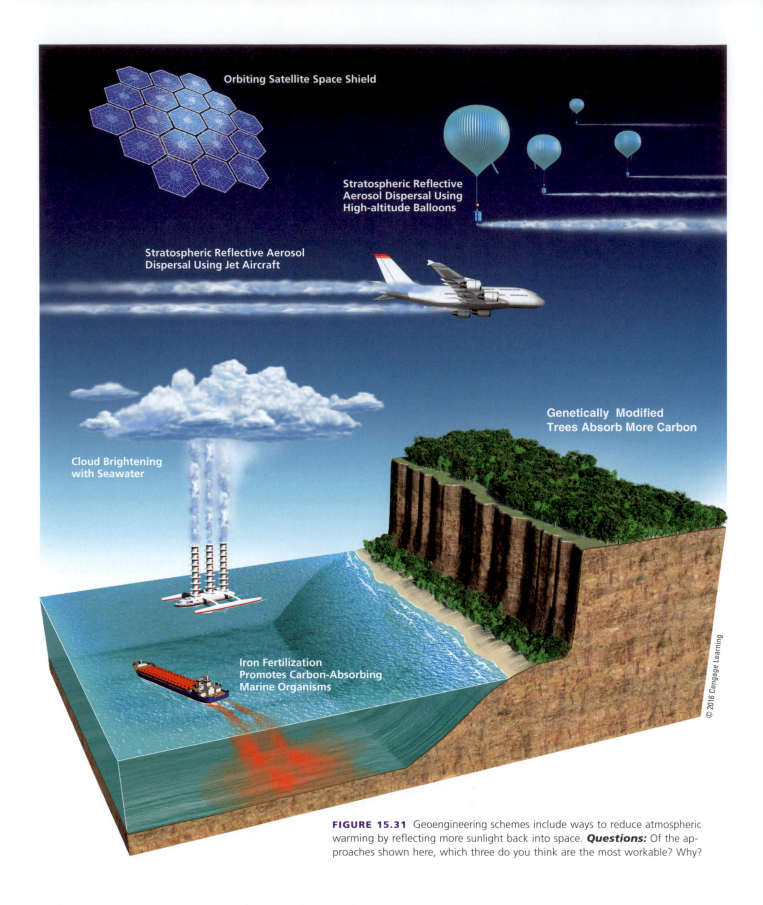

FIGURE 15.31 Geoengineering schemes include ways to reduce atmospheric warming by reflecting more sunlight back into space. **Questions:** Of the approaches shown here, which three do you think are the most workable? Why?

Orbiting Satellite Space Shield

Stratospheric Reflective Aerosol Dispersal Using High-altitude Balloons

Stratospheric Reflective Aerosol Dispersal Using Jet Aircraft

Genetically Modified Trees Absorb More Carbon

Cloud Brightening with Seawater

Iron Fertilization Promotes Carbon-Absorbing Marine Organisms

© 2016 Cengage Learning

energy resources. Many scientists say we cannot afford such a delay.

Governments Can Help to Reduce the Threat of Climate Change

Governments can use seven major strategies to promote the solutions listed in Figure 15.29 (**Concept 15.6**). They can:

- Strictly regulate carbon dioxide (CO_2) and methane (CH_4) as climate-changing pollutants that can harm public health.

- Phase out the most polluting coal-burning power plants over the next 50 years and replace them with cleaner natural gas and renewable energy alternatives such as wind power and solar power. Use of next-generation nuclear power plants, but only those that meet the criteria listed in Figure 13.20 (p. 331), could help to further reduce power plant CO_2 emissions.

- Put a price on carbon emissions by phasing in taxes on each unit of CO_2 or CH_4 emitted, or phasing in energy taxes on each unit of any fossil fuel burned (Figure 15.32). These tax increases could be offset by reductions in taxes on income, wages, and profits. In other words, *tax pollution, not payrolls and profits*.

- Use a cap-and-trade system (see p. 400 and Figure 15.33), which would use the marketplace to help reduce emissions of CO_2 and CH_4.

- Phase out government subsidies and tax breaks for the fossil-fuel industry and unsustainable industrialized food production, and phase in such subsidies and tax breaks for energy-efficient technologies, low-carbon renewable energy development, and more sustainable food production.

- Focus research and development efforts on innovations that lower the cost of clean energy alternatives, so that they can compete more favorably with fossil fuels that have benefited from government (taxpayer) subsidized research and development for over 50 years.

- Work out agreements to finance and monitor efforts to reduce deforestation—which accounts for 12% to 17% of global greenhouse gas emissions—and to promote global tree-planting efforts (see Chapter 9, Core Case Study, p. 186).

Environmental economists and a growing number of business leaders, along with the President of the World Bank, call for putting a price on carbon emissions as the best way to curb them before it is too late to avoid the projected catastrophic effects of climate change. This would help us to include the estimated harmful environmental and health costs of using fossil fuels in fuel prices, in keeping with the full-cost pricing **principle of sustainability**. However, establishing laws and regulations that raise the price of fossil fuels is politically difficult because of the immense political and economic power of the fossil fuel lobby and because reliance on fossil fuels is widespread and well-entrenched among consumers and businesses.

Governments have also entered into international climate negotiations. In December 1997, delegates from 161 nations met in Kyoto, Japan, to negotiate a treaty to slow atmospheric warming and projected climate change. The first phase of the resulting *Kyoto Protocol* went into effect in February 2005 with 187 of the world's 194 countries (not including the United States) ratifying the agreement by late 2009.

The 36 participating more-developed countries agreed to cut their emissions of CO_2, CH_4, and N_2O to certain levels by 2012, when the treaty was to expire, but they failed to do so. Less-developed countries, including China, were excused from this requirement, because such reductions would curb their economic growth. In 2005, participating countries began negotiating a second phase of the treaty,

Trade-Offs

Carbon and Energy Taxes

Advantages	Disadvantages
Simple to administer	Tax laws can get complex
Clear price on carbon	Vulnerable to loopholes
Covers all emitters	Doesn't guarantee lower emissions
Predictable revenues	Politically unpopular

© 2016 Cengage Learning

FIGURE 15.32 Using carbon and energy taxes or fees to help reduce greenhouse gas emissions has advantages and disadvantages. *Question:* Which two advantages and which two disadvantages do you think are the most important and why?

Trade-Offs

Cap-and-Trade Policies

Advantages	Disadvantages
Clear legal limit on emissions	Revenues not predictable
Rewards cuts in emissions	Vulnerable to cheating
Record of success	Rich polluters can keep polluting
Low expense for consumers	Puts variable price on carbon

© 2016 Cengage Learning

FIGURE 15.33 Using a cap-and-trade policy to help reduce greenhouse gas emissions has advantages and disadvantages. *Question:* Which two advantages and which two disadvantages do you think are the most important and why?

which was supposed to go into effect in 2012, but these negotiations have failed to extend the original agreement. Negotiators hope to reach an agreement in 2015 that would go into effect in 2020. Some analysts see top-down international climate agreements as weak, slow, and ineffective responses to an urgent global problem.

Some Countries, States, Cities, and Private Companies Are Leading the Way in Reducing Carbon Footprints

Some nations are leading others in facing the challenges of projected climate change. Costa Rica aims to be the first country to become *carbon neutral* by cutting its net carbon emissions to zero by 2030. The country generates 78% of its electricity with renewable hydroelectric power and another 18% from renewable wind and geothermal energy.

Some U.S. state and local governments are moving ahead in dealing with climate change, not waiting for the U.S. federal government, which has acted slowly in this arena. By 2014, at least 30 U.S. states had set goals for reducing greenhouse gas emissions. California plans to get 33% of its electricity from low-carbon renewable energy sources by 2030. That state is showing that it is possible to implement policies that cut carbon emissions and create jobs. Since 1990, local governments in more than 650 cities around the world (including more than 450 U.S. cities) have established programs to reduce their greenhouse gas emissions.

Leaders of some big U.S. companies, including Alcoa, DuPont, Ford Motor Company, General Electric, Johnson & Johnson, PepsiCo, and Shell Oil, have joined with leading environmental organizations to form the U.S. Climate Action Partnership. Each is working on reducing its carbon footprint, and such actions will reduce their contributions to climate change while saving them money. For example, between 1990 and 2006, DuPont slashed its energy usage and cut its greenhouse gas emissions by 72% and saved $3 billion, while the company grew its business by almost a third. These companies have called on the government to enact strong national climate change legislation, saying, "In our view, the climate change challenge will create more economic opportunities than risks for the U.S. economy."

Colleges and Universities Are Reducing Their Carbon Footprints

Many colleges and universities are also taking action. Arizona State University (ASU) boasts the largest collection of solar panels of any U.S. university. ASU was also the first U.S. university to establish a School of Sustainability. The College of the Atlantic in Bar Harbor, Maine, has been carbon neutral since 2007. It gets all of its electricity from renewable hydropower, and many of its buildings are heated with the use of renewable wood pellets. Students there built a wind turbine that powers a nearby organic farm, which offers organic produce to the campus, to local schools, and to food banks.

Students at the University of Washington in Seattle agreed to an increase in their fees to help the school buy electricity from renewable energy sources. At Florida's University of Miami, drivers of hybrid cars get a 50% parking discount. At Texas A & M University, used cooking oil from the cafeteria is converted to biofuel for the school's delivery trucks.

EARTH University in Costa Rica has a mission to promote sustainable development in tropical countries. Its sustainable agriculture degree program has attracted students from more than 20 different countries. And a growing number of campus groups are urging the administrators at their schools to help slow climate change by ending their endowment fund investments in fossil fuel companies.

CONSIDER THIS. . .

THINKING ABOUT What Your School Can Do

What are three steps that you think your school should take to help reduce its carbon footprint?

Every Individual Choice Makes a Difference

Each of us will play a part in the atmospheric warming and climate change projected to occur during this century. Whenever we use energy generated by fossil fuels, for example, we add a certain amount of CO_2 to the atmosphere. Each use of energy adds to an individual's carbon foot-

What Can You Do?

Reducing CO_2 Emissions

- Calculate your carbon footprint (there are several helpful websites)
- Drive a fuel-efficient car, walk, bike, carpool, and use mass transit
- Reduce garbage by reducing consumption, recycling, and reusing more items
- Use energy-efficient appliances and compact fluorescent or LED lightbulbs
- Wash clothes in warm or cold water and hang them up to dry
- Close window curtains to keep heat in or out
- Use a low-flow showerhead
- Eat less meat or no meat
- Heavily insulate your house and seal all air leaks
- Use energy-efficient windows
- Set your hot-water heater to 49°C (120°F)
- Plant trees
- Buy from businesses working to reduce their emissions

© 2016 Cengage Learning

FIGURE 15.34 You can reduce your annual emissions of CO_2. **Question:** Which of these steps, if any, do you take now or plan to take in the future?

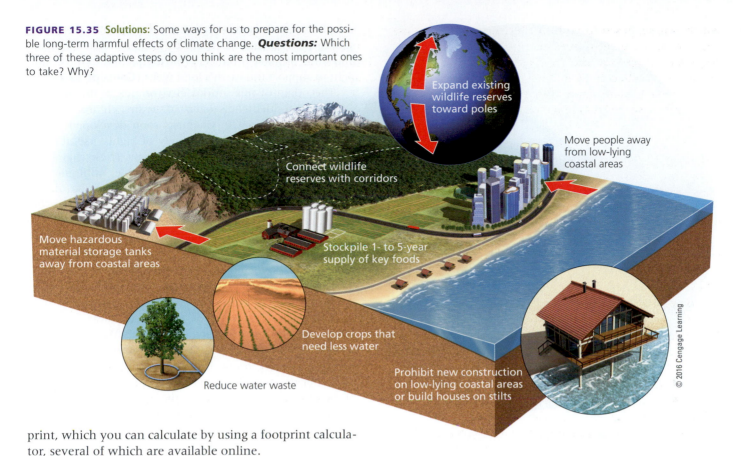

FIGURE 15.35 Solutions: Some ways for us to prepare for the possible long-term harmful effects of climate change. **Questions:** Which three of these adaptive steps do you think are the most important ones to take? Why?

Labels within figure:
- Expand existing wildlife reserves toward poles
- Connect wildlife reserves with corridors
- Move people away from low-lying coastal areas
- Move hazardous material storage tanks away from coastal areas
- Stockpile 1- to 5-year supply of key foods
- Develop crops that need less water
- Reduce water waste
- Prohibit new construction on low-lying coastal areas or build houses on stilts
- © 2016 Cengage Learning

print, which you can calculate by using a footprint calculator, several of which are available online.

With this sort of information, you can make adjustments in your daily living to reduce your carbon footprint or to *offset* part of it by finding ways to reduce CO_2 in the atmosphere. For example, the band Coldplay offset the amount of CO_2 emitted for the production of one of their CDs by providing 10,000 CO_2-absorbing mango trees to some villages in India. Figure 15.34 lists some ways in which you can cut your CO_2 emissions while increasing your beneficial environmental impact.

We Can Prepare for Climate Change

According to global climate models, the world needs to make a 50–85% cut in emissions of greenhouse gases by 2050 to stabilize concentrations of these gases in the atmosphere. This would help to prevent the atmosphere from warming by more than 2C° (3.6F°) and it would head off rapid changes in the world's climate along with the harmful environmental, economic, and health effects that are projected to result.

However, because of the political difficulty of making such large reductions, many analysts believe that while we work to slash greenhouse gas emissions, we should also begin to prepare for the likely harmful effects of projected climate change. Figure 15.35 shows some ways to do so.

For example, relief organizations such as the International Red Cross are turning their attention to projects such as expanding mangrove forests as buffers against storm surges, building shelters on high ground, and planting trees on slopes to help prevent landslides in the face of

projected higher levels of precipitation and rising levels. The U.S. state of Alaska, which is warming quickly, has plans to relocate coastal villages at risk from rising sea levels, coastal erosion, and higher storm surges.

Some coastal communities in the United States now require that new houses and other new construction be built high enough off of the ground or further back from the current shoreline to survive such hazards. In anticipation of rising sea levels, Boston has elevated one of its sewage treatment plants. And some cities plan to establish cooling centers to shelter residents during increasingly intense heat waves. In the low-lying Netherlands, people are dealing with the threat of a rising sea level by building houses that float.

15.7 HOW HAVE WE DEPLETED OZONE IN THE STRATOSPHERE AND WHAT CAN WE DO ABOUT IT?

CONCEPT 15.7A Widespread use of certain chemicals has reduced ozone levels in the stratosphere and allowed more harmful ultraviolet (UV) radiation to reach the earth's surface.

CONCEPT 15.7B To reverse ozone depletion, we must stop producing ozone-depleting chemicals and adhere to the international treaties that ban such chemicals.

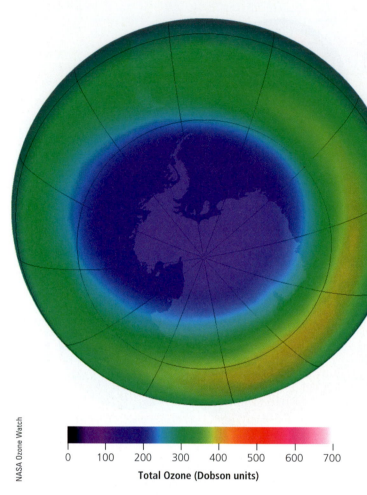

NASA Ozone Watch

0   100   200   300   400   500   600   700
Total Ozone (Dobson units)

FIGURE 15.36 **Natural capital degradation:** This colorized satellite image shows ozone thinning over Antarctica during September of 2013. Ozone depletion of 50% or more occurred in the dark blue and purple areas.

Our Use of Certain Chemicals Threatens the Ozone Layer

A layer of ozone in the lower stratosphere (Figure 15.2) keeps about 95% of the sun's harmful ultraviolet (UV-A and UV-B) radiation from reaching the earth's surface and harming us and a number of other species. In other words, the ozone layer in the stratosphere is a vital form of natural capital that supports all life on land and in shallow aquatic environments.

However, measurements taken by researchers show a considerable seasonal depletion, or thinning, of ozone concentrations in the stratosphere above Antarctica (Figure 15.36) and above the Arctic since the 1970s. Similar measurements reveal more slight overall ozone thinning everywhere except over the tropics. The loss of ozone over Antarctica has been called an *ozone hole*. A more accurate term is *ozone thinning* because the ozone depletion varies with altitude and location.

Based on these measurements and on mathematical and chemical models, the overwhelming consensus of re-

searchers in this field is that ozone depletion in the stratosphere poses a serious threat to humans, other animals, and some primary producers (mostly plants) that use sunlight to support the earth's food webs (**Concept 15.7A**).

The origin of this serious environmental threat began with the accidental discovery of the first chlorofluorocarbon (CFC) in 1930. Chemists soon developed similar compounds to create a family of highly useful CFCs, known by their trade name Freons. These chemically unreactive, odorless, nonflammable, nontoxic, and noncorrosive compounds seemed to be dream chemicals. Inexpensive to manufacture, they became popular as coolants in air conditioners and refrigerators, propellants in aerosol spray cans, cleansers for electronic parts such as computer chips, fumigants for granaries and ships' cargo holds, and gases used to make insulation and packaging.

It turned out that CFCs were too good to be true. Starting in 1974 with the work of chemists Sherwood Rowland and Mario Molina (Individuals Matter 15.1), scientists demonstrated that CFCs are persistent chemicals that destroy protective ozone in the stratosphere. Satellite data and other measurements and models indicate that 75–85% of the observed ozone losses in the stratosphere since 1976 resulted from people releasing CFCs and other ozone-depleting chemicals into the troposphere beginning in the 1950s. After entering the troposphere these long-lived chemicals eventually reached the stratosphere, where they began destroying ozone faster than it was being formed by the interaction of stratospheric oxygen and incoming solar radiation—a disruption of one of the earth's most important forms of natural capital. While in the troposphere, CFCs also act as greenhouse gases that help to warm the lower atmosphere.

Why Should We Worry about Ozone Depletion?

Why is ozone depletion something that should concern us? Figure 15.37 lists some of the demonstrated effects of stratospheric ozone thinning. One effect is that more biologically damaging UV-A and UV-B radiation will reach the earth's surface (**Concept 15.7A**) and is a likely contributor to rising numbers of eye cataracts, damaging sunburns, and skin cancers. Figure 15.38 lists ways in which you can protect yourself from harmful UV radiation.

Another serious threat from ozone depletion and the resulting increase in UV radiation reaching the planet's surface is the possible impairment or destruction of phytoplankton, especially in Antarctic waters (see Figure 3.10, p. 49). These tiny marine plants play a key role in removing CO_2 from the atmosphere and they form the base of many ocean food webs. By destroying them, we would be eliminating the vital ecological services they provide. This could also accelerate projected atmospheric warming by reducing the capacity of the oceans for removal of CO_2 from the atmosphere.

HAL GARB/AFP/Getty Images

Donna Coveney/MIT

Sherwood Rowland and Mario Molina—A Scientific Story of Expertise, Courage, and Persistence

In 1974, calculations by the late Sherwood Rowland (top photo) and Mario Molina (bottom photo), chemists at the University of California–Irvine, indicated that chlorofluorocarbons (CFCs) were lowering the average concentration of ozone in the stratosphere. These scientists decided they had an ethical obligation to go public with the results of their research. They shocked both the scientific community and the $28-billion-per-year CFC industry by calling for an immediate ban of CFCs in spray cans, for which substitutes were available.

The research of these two scientists led them to four major conclusions. *First*, once CFCs are put into the atmosphere, these persistent chemicals remain there for a long time.

Second, over 11–20 years, these compounds rise into the stratosphere through convection, random drift, and the turbulent mixing of air in the lower atmosphere.

Third, once they reach the stratosphere, the CFC molecules break down under the influence of high-energy UV radiation. This releases highly reactive chlorine atoms (Cl), as well as atoms of fluorine (F) and bromine (Br), all of which accelerate the breakdown of ozone (O_3) into O_2 and O in a cyclic chain of chemical reactions. As a consequence, ozone is destroyed faster than it forms in some parts of the stratosphere.

Fourth, each CFC molecule can last in the stratosphere for 65–385 years, depending on its type. During that time, each chlorine atom released during the breakdown of CFCs can break down hundreds of O_3 molecules.

The CFC industry (led by DuPont) was a powerful, well-funded adversary with a lot of profits and jobs at stake. It attacked Rowland's and Molina's calculations and conclusions, but the two researchers held their ground, expanded their research, and explained their results to other scientists, elected officials, and the media. After 14 years of delaying tactics, DuPont officials acknowledged in 1988 that CFCs were depleting the ozone layer, and they agreed to stop producing them and to sell higher-priced alternatives that their chemists had developed.

In 1995, Rowland and Molina received the Nobel Prize in chemistry for their work on CFCs.

Background photo: NASA

We Can Reverse Stratospheric Ozone Depletion

According to researchers in this field, we should immediately stop producing all ozone-depleting chemicals (**Concept 15.7B**). However, models and measurements indicate that even with immediate and sustained action, it will take at least 60 years for the earth's ozone layer to recover the levels of ozone it had in 1980, and it could take about 100 years for it to recover to pre-1950 levels.

In 1987, representatives of 36 nations met in Montreal, Canada, and developed the *Montreal Protocol*. This treaty's goal was to cut emissions of CFCs (but no other ozone-depleting chemicals) by about 35% between 1989 and 2000. After hearing more bad news about seasonal ozone thinning above Antarctica in 1989, representatives of 93 countries had more meetings and in 1992 adopted the *Copenhagen Amendment*, which accelerated the phase-out of CFCs and added some other key ozone-depleting chemicals to the agreement.

These international agreements set an important precedent because nations and companies worked together, using a *prevention approach* to try to solve a serious environmental problem. This approach worked for three reasons. *First*, there was convincing and dramatic scientific evidence of a serious problem. *Second*, CFCs were produced by a small number of international companies and this meant there was less corporate resistance to finding a solution. *Third*, the certainty that CFC sales would decline over a period of years because of government bans un-

Natural Capital Degradation

Effects of Ozone Depletion

Human Health and Structures

- Worse sunburns
- More eye cataracts and skin cancers
- Immune system suppression

Food and Forests

- Reduced yields for some crops
- Reduced seafood supplies due to smaller phytoplankton populations
- Decreased forest productivity for UV-sensitive tree species

Wildlife

- More eye cataracts in some species
- Shrinking populations of aquatic species sensitive to UV radiation
- Disruption of aquatic food webs due to shrinking phytoplankton populations

Air Pollution and Climate Change

- Increased acid deposition
- Increased photochemical smog
- Degradation of outdoor painted surfaces, plastics, and building materials
- While in troposphere, CFCs act as greenhouse gases

© 2016 Cengage Learning

FIGURE 15.37 Decreased levels of ozone in the stratosphere can have a number of harmful effects (**Concept 15.7A**). *Questions:* Which three of these effects do you think are the most threatening? Why?

What Can You Do?

Reducing Exposure to UV Radiation

- Stay out of the sun, especially between 10 A.M. and 3 P.M.
- Do not use tanning parlors or sunlamps
- When in the sun, wear clothing and sunglasses that protect against UV-A and UV-B radiation
- Be aware that overcast skies do not protect you
- Do not expose yourself to the sun if you are taking antibiotics or birth control pills
- When in the sun, use a sunscreen with a protection factor of at least 15

© 2016 Cengage Learning

FIGURE 15.38 You can reduce your exposure to harmful UV radiation. *Question:* Which of these precautions do you already take?

leashed the economic and creative resources of the private sector to find even more profitable substitute chemicals.

Progress in reducing ozone depletion could be set back by projected climate change. Scientists from Johns Hopkins University reported in 2009 that tropospheric warming makes the stratosphere cooler, which slows down its rate of ozone repair. Thus, as the troposphere warms and the global climate changes, ozone levels may take much longer to return to where they were before ozone thinning began, or they may never recover completely. In 2012, Harvard University chemist James G. Anderson and his colleagues suggested that updrafts from large and more intense thunderstorms resulting from projected climate change could inject water vapor into the stratosphere (which is normally drier than a desert). This could set off chemical reactions that could accelerate ozone depletion.

On the other hand, according to a 2013 study by a team of scientists led by Francisco Estrada, measures taken to comply with the ozone reduction agreements have lowered the global average temperature by about 0.2°C (0.4°F) by reducing emissions of CFC greenhouse gases.

GOOD NEWS

At any rate, the landmark international agreements on stratospheric ozone, now signed by all 196 of the world's countries, are important examples of successful global cooperation in response to a serious global environmental problem (**Concept 15.7B**).

BIG IDEAS

- We need to give top priority status to the prevention of outdoor and indoor air pollution and the reduction of stratospheric ozone depletion.

- Reducing the projected harmful effects of rapid climate change during this century requires emergency action to increase energy efficiency, sharply reduce greenhouse gas emissions, and rely more on renewable energy resources.

- While we can prepare for some climate change that is now happening, we could realize important economic, ecological, and health benefits by drastically reducing greenhouse gas emissions with the goal of slowing climate change.

Melting Ice in Greenland and Sustainability

In this chapter, we have seen that human activities such as the burning of fossil fuels and the widespread clearing and burning of forests for planting crops have contributed to higher levels of greenhouse gases in the atmosphere. This has contributed to increased atmospheric warming, which in turn is projected to result in possibly disruptive climate change during this century.

The effects of climate change could include rapid melting of land-based ice (**Core Case Study**) and arctic sea ice, worsening drought, rising sea levels, declining biodiversity, and severe threats to human health and economies. Many of these effects could further accelerate climate change in worsening spirals of change. We have also seen how the widespread use of certain chemicals has thinned the stratospheric ozone layer, which has led to plant and animal health problems.

We can apply the six **principles of sustainability** (see Supplement 7, pp. S50–S51) to help reduce the harmful effects of air pollution, projected climate change, and stratospheric ozone depletion. We can reduce emissions of pollutants, greenhouse gases, and ozone-depleting chemicals by relying more on direct and indirect forms of solar energy than on fossil fuels; recycling and reusing matter

resources much more widely than we do now; and mimicking biodiversity by using a variety of often locally available low-carbon renewable energy resources in place of fossil fuels and nuclear power. We can advance toward these goals by including the harmful environmental and health costs of fossil fuel use in market prices; seeking win-win solutions that will benefit both the economy and the environment; and giving high priority to passing along to future generations an environment in which they too can thrive.

Jan Martin Will/Shutterstock.com

Chapter Review

Core Case Study

1. Summarize the story of Greenland's melting glaciers and the possible effects of this process. Explain how it fits into projections about climate change during this century.

Section 15.1

2. What is the key concept for this section? Define and distinguish among **atmosphere**, **troposphere**, **stratosphere**, and **ozone layer**. Define and give two examples of **greenhouse gases**. Why is the ozone layer important?

Section 15.2

3. What are the two key concepts for this section? What is **air pollution**? Distinguish between **primary pollutants** and **secondary pollutants** and give an example of each. List the major outdoor air pollutants and their harmful effects. Distinguish

between **industrial smog** and **photochemical smog**. List and briefly explain five natural factors that help to reduce outdoor air pollution and six natural factors that help to worsen it. What is a **temperature inversion** and how can it affect outdoor air pollution levels? What is **acid deposition**, how does it form, and what are its major environmental impacts on vegetation, lakes, human-built structures, and human health? List three major ways to reduce acid deposition.

4. What is the most threatening indoor air pollutant in many less-developed countries? What are the four most dangerous indoor air pollutants in more-developed countries? Briefly describe the human body's defenses against air pollution, how they can be overwhelmed, and illnesses that can result. In the world and in the United States, about how many people die prematurely from air pollution each year?

Section 15.3

5. What is the key concept for this section? Summarize the use of air-pollution-control laws in the United States and how they could be improved. List the

advantages and disadvantages of using an emissions trading program to reduce outdoor air pollution. List the major ways to reduce emissions from power plants and motor vehicles. What are four ways to reduce indoor air pollution?

Section 15.4

6. What is the key concept for this section? Define and distinguish between **weather** and **climate**. Why is "global warming" a misleading term? Summarize the trends in atmospheric warming and cooling during the past 900,000 years. What has happened to the atmosphere's temperature near the earth's surface since 1975? How do scientists get information about past temperatures and climates? List three major conclusions of the IPCC and other scientific bodies regarding changes in the temperature of the earth's atmosphere. List nine pieces of scientific evidence that support the conclusion that human-influenced climate change is happening now.

7. What is the **greenhouse effect** and why is it so important to life on the earth? What role do CO_2 emissions play in atmospheric warming and what are two major sources of these emissions? Define **carbon footprint**. Explain how scientists use models to project future changes in atmospheric temperatures. Describe how each of the following might affect average atmospheric temperatures and projected climate change during this century: **(a)** the oceans, **(b)** cloud cover, and **(c)** outdoor air pollution.

Section 15.5

8. What is the key concept for this section? Summarize the projections of scientists on how climate change is likely to affect each of the following: ice and snow cover, permafrost, sea levels, severe drought, extreme weather events, biodiversity, food production, human health and economies, and national security.

Section 15.6

9. What is the key concept for this section? What are five reasons why dealing with projected climate change is a difficult problem? Define **climate change tipping point**, and list five possible examples. What are two basic approaches to dealing with climate change? List five major prevention strategies and four cleanup approaches for slowing projected climate change. List seven pieces of good news related to dealing with climate change. What is **carbon capture and storage (CCS)**? List three problems associated with capturing and storing carbon dioxide emissions. Define **geoengineering** and describe two proposed geoengineering strategies for dealing with the threat of climate change. What are the main potential problems with relying on such strategies? List seven things that governments can do to help slow projected climate change. What are the major advantages and disadvantages of using **(a)** carbon or energy taxes and **(b)** a cap-and-trade system for reducing the threat of climate change? What is the *Kyoto Protocol* and how effective has it been in dealing with the threat of climate change? List five ways in which you can reduce your carbon footprint. List five ways in which we can prepare for the possible long-term harmful effects of projected climate change.

Section 15.7

10. What are the two key concepts for this section? How did stratospheric ozone depletion occur, and what are its effects? Explain how scientists Sherwood Roland and Mario Molina helped to awaken the world to this threat. What has the world done to reduce the threat of ozone depletion in the stratosphere? What are the *three big ideas* for this chapter? Explain how we can apply the six **principles of sustainability** to the problems of air pollution, climate change, and ozone depletion.

Note: Key terms are in bold type.

Critical Thinking

1. If you had convincing evidence that at least half of Greenland's glaciers (**Core Case Study**) were sure to melt during this century, would you argue for taking serious actions now to slow projected climate change? Summarize your arguments for or against such actions.

2. Suppose someone tells you that carbon dioxide (CO_2) should not be classified as an air pollutant because it is a natural chemical that we add to the atmosphere every time we exhale. Would you consider this to be faulty reasoning? Explain.

3. China's burning of coal has caused major and growing air pollution problems for the country and for its neighboring nations, and it has contributed to projected climate change. In addition, air pollution generated in China now sometimes spreads across the Pacific Ocean to the west coast of North America. Do you think China is justified in developing its coal resource aggressively as other countries, including the United States, have done? Explain. If you think that

China should sharply reduce its dependence on coal, would you also call for the United States to sharply reduce its use of coal? Explain.

4. If you live in the United States, list three important ways in which your life would be different if citizen-led actions during the 1970s and 1980s had not led to the Clean Air Acts of 1970, 1977, and 1990. Which one or more of the six **principles of sustainability** (see Supplement 7, pp. S50–S51) were applied by the passage of these acts? Explain.

5. Explain why you agree or disagree with IPCC scientists and most of the world's climate scientists that **(a)** climate change is happening now, **(b)** human activities are the dominant cause of this climate change, and **(c)** only action by humans can slow down the rate of climate change and avert or delay its projected harmful environmental, health, and economic effects.

6. Explain why you would support or oppose each of the strategies listed in Figure 15.29 for slowing projected climate change caused by atmospheric warming.

7. Some scientists have suggested that, in order to help cool the warming atmosphere, we could annually inject huge quantities of sulfate particles into the stratosphere. This might have the effect of reflecting some incoming sunlight back into space. Explain why you would support or oppose this geoengineering scheme.

8. What are three consumption patterns or other aspects of your lifestyle that directly add greenhouse gases to the atmosphere? Which, if any, of these habits would you be willing to give up in order to help slow projected climate change?

Doing Environmental Science

Gather data on the trends in average annual temperatures and average annual precipitation over the past 30 years in the area where you live. (Possible sources include weather sites on the Internet, your school library, local TV and radio meteorologists, and local or regional weather bureaus.) Try to find data for as many of these years as possible, and plot these data to determine whether the average temperature and precipitation during this period has increased, decreased, or stayed about the same. Write a report summarizing your search for data, your results, and your conclusions.

Global Environment Watch Exercise

Within the GREENR database, use the World Map feature, and under "Browse," select *Climate Change* and click on the pin for Greenland. Find the latest information on the melting of glaciers in Greenland (**Core Case Study**). What is the rate of melting of the glaciers and how might this affect sea levels during this century? Briefly summarize the evidence used by scientists to support their statements about melting ice in Greenland. Summarize any information you find about ongoing studies of Greenland's ice.

Data Analysis

Coal often contains sulfur (S) as an impurity that is released as gaseous SO_2 during combustion, and SO_2 is one of six primary air pollutants monitored by the EPA. The U.S. Clean Air Act limits sulfur emissions from large coal-fired boilers to 0.54 kilograms (1.2 pounds) of sulfur per million Btu (British thermal units) of heat generated. (1 metric ton = 1,000 kilograms = 2,200 pounds = 1.1 ton; 1 kilogram = 2.2 pounds.)

1. Given that coal burned in power plants has a heating value of 27.5 million Btus per metric ton (25 million Btus per ton), determine the number of kilograms (and pounds) of coal needed to produce 1 million Btus of heat.

2. Assuming that all of the sulfur in the coal is released to the atmosphere during combustion, what is the maximum percentage of sulfur that the coal can contain and still allow the utility to meet the standards of the Clean Air Act?

CENGAGE**brain**.com To access course materials, including Aplia homework, please visit www.cengagebrain.com.

16

SOLID AND HAZARDOUS WASTE

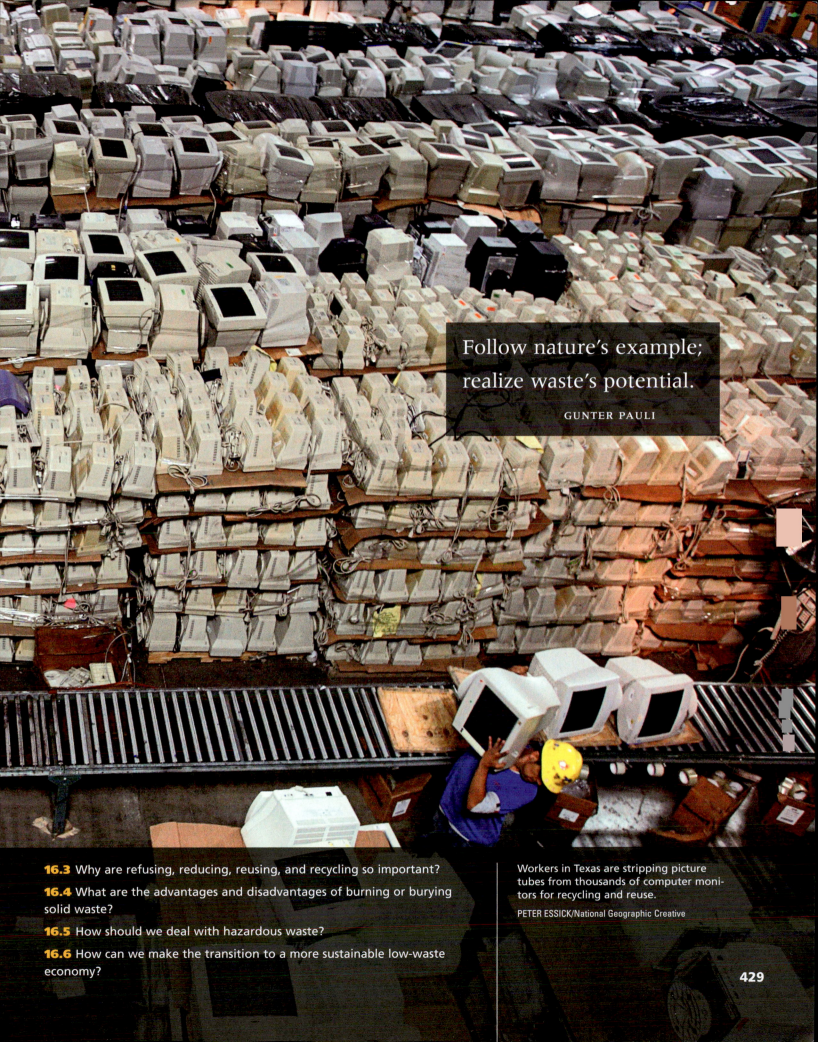

> Follow nature's example;
> realize waste's potential.
>
> GUNTER PAULI

16.3 Why are refusing, reducing, reusing, and recycling so important?

16.4 What are the advantages and disadvantages of burning or burying solid waste?

16.5 How should we deal with hazardous waste?

16.6 How can we make the transition to a more sustainable low-waste economy?

Workers in Texas are stripping picture tubes from thousands of computer monitors for recycling and reuse.

PETER ESSICK/National Geographic Creative

E-Waste—An Exploding Problem

Electronic waste, or *e-waste*, consists of discarded television sets, cell phones, computers, and other electronic devices (see chapter-opening photo). It is the fastest-growing solid waste problem in the United States and in the world.

Only about 14% of all U.S. e-waste is recycled. Most of the rest of it ends up in landfills and incinerators, even though many of the components in electronic devices contain gold, rare earths (see Chapter 12, Core Case Study, p. 290), and other valuable materials that could be recycled or reused. E-waste is also a source of toxic and hazardous chemicals that can contaminate air, surface water, groundwater, and soil and cause human health problems.

Much of the e-waste in the United States that is not buried or incinerated is shipped to China (Figure 16.1), India, and other Asian and African countries where labor is cheap and environmental regulations are weak. Workers there—many of them children—dismantle, burn, and treat e-waste products with acids to recover valuable metals and reusable parts. As they do this, they are exposed to toxic metals such as lead and mercury and other harmful chemicals. The remaining scrap is dumped into waterways and fields or burned in open fires, exposing many people to highly toxic chemicals called dioxins. So by discarding a cell phone or other electronic device in the United States, one can contribute to health problems for people in China, India, or Africa.

Transfer of such hazardous waste from more-developed to less-developed countries was banned by the International Basel Convention, a treaty designed to reduce and control the movement of hazardous waste across international borders. Despite this ban, much of the world's e-waste is not officially classified as hazardous waste, or it is illegally smuggled out of some countries. The United States can export its e-waste legally because it is the only industrialized nation that has not ratified the Basel Convention.

The European Union (EU) has led the way in dealing with e-waste by requiring manufacturers to take back electronic products at the end of their useful lives and repair, remanufacture, or recycle them. In the EU, e-waste is banned from landfills and incinerators. Japan is also taking this approach. To cover the costs of these programs, consumers pay a recycling tax on electronic products, an example of implementing the full-cost pricing **principle of sustainability**. However, because of the high costs involved in complying with these laws, much of this waste that is supposed to be recycled domestically is shipped illegally to other countries.

Electronic waste is just one of many types of solid and hazardous waste that we discuss in this chapter.

Peter Essick/National Geographic Creative

FIGURE 16.1 Workers in Taizhou City in China's Zhejiang Province are recovering valuable materials from scrapped computers shipped from the United States. **Questions:** Have you disposed of an electronic device lately? If so, how did you dispose of it?

WHAT ARE THE PROBLEMS RELATED TO SOLID AND HAZARDOUS WASTES?

CONCEPT 16.1A Solid waste contributes to pollution and includes valuable resources that could be reused or recycled.

CONCEPT 16.1B Hazardous waste contributes to pollution, as well as to natural capital degradation, health problems, and premature deaths.

We Throw Away Huge Amounts of Useful Things

In the natural world, wherever humans are not dominant, there is essentially no waste, because the wastes of one organism become nutrients or raw materials for others (see Figure 3.10, p. 49). This natural cycling of nutrients is the basis for one of the three **scientific principles of sustainability**. We violate this principle by producing huge amounts of waste materials that go into landfills or litter and pollute the environment. Studies and experience indicate that by mimicking nature, we could reduce this waste of potential resources and the resulting environmental harm by up to 80%.

In their 2013 book, *The Upcycle*, William McDonough and Michael Braungart call for us to think of solid wastes and pollution as potentially useful and economically valuable materials and chemicals. In other words, instead of asking "How do I get rid of these wastes?" they say we need to ask: "How much money can I get for these resources?" and "How can I design products that don't end up as wastes or pollutants?" This means designing products so that they can be recycled or reused, much like nutrients in the biosphere, in keeping with the chemical cycling **principle of sustainability**. With this approach we would think of trash cans and garbage trucks as resource containers. This is an important way to increase our beneficial environmental impact.

One major category of waste is **solid waste**—any unwanted or discarded material we produce that is not a liquid or a gas. Solid waste can be divided into two types. One type is **industrial solid waste** produced by mines (see Figure 12.8, p. 297), farms, and industries that supply people with goods and services. The other is **municipal solid waste (MSW)**, often called *garbage* or *trash*, which consists of the combined solid wastes produced by homes and workplaces other than factories. Examples include paper and cardboard, food wastes, cans, bottles, yard wastes, furniture, plastics, metals, glass, wood, and e-waste (**Core Case Study**). Some resource experts suggest that we change the name of the trash we produce from MSW to MWR—mostly wasted resources.

In more-developed countries, most MSW is buried in landfills or burned in incinerators. In many less-developed countries, much of it ends up in open dumps, where poor people eke out a living finding items they can use or sell. The United States is the world's largest producer of MSW (see the following Case Study).

CASE STUDY

Solid Waste in the United States

The United States leads the world in total solid waste production and in solid waste per person. With only 4.5% of the world's people, the United States produces about 25% of the world's solid waste. According to the U.S. Environmental Protection Agency (EPA), about 98.5% of all solid waste produced in the United States is industrial solid waste from mining (76%), agriculture (13%), and industry (9.5%). The remaining 1.5% of U.S. solid waste is MSW.

Every year, the United States generates enough MSW to fill a bumper-to-bumper convoy of garbage trucks long enough to encircle the earth's equator almost six times. About 67% of it is dumped in landfills and about 9% is incinerated, but much of it ends up as litter (Figure 16.2).

Consider some of the solid wastes that consumers throw away each year, on average, in the high-waste economy of the United States:

- Enough tires to encircle the earth's equator almost three times.

- An amount of disposable diapers that, if linked end to end, would reach to the moon and back seven times.

- Enough carpet to cover the U.S. state of Delaware.

- Enough nonreturnable plastic bottles to form a stack that would reach from the earth to the moon and back about six times.

- About 100 billion plastic shopping bags, or 274 million per day, an average of nearly 3,200 every second.

- Enough office paper to build a wall 3.5 meters (11 feet) high across the country from New York City to San Francisco, California.

- 25 billion plastic foam cups—enough, if lined up end to end, to circle the earth's equator 95 times.

Most of these wastes break down very slowly, if at all. Lead, mercury, glass, plastic foam, and most plastic bottles do not break down; an aluminum can takes 500 years to disintegrate; plastic bags, 400 to 1,000 years; and a plastic six-pack holder, 100 years.

Hazardous Waste Is a Serious and Growing Problem

Another major category of waste is **hazardous**, or **toxic, waste**—any discarded material or substance that threatens human health or the environment because it is poisonous, dangerously chemically reactive, corrosive, or flammable. Examples include industrial solvents, hospital

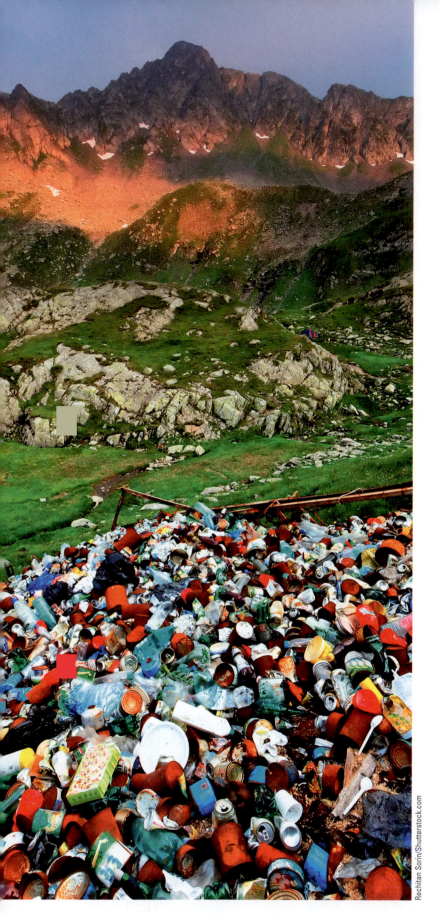

FIGURE 16.2 Various types of solid waste have been dumped in this isolated mountain area.

What Harmful Chemical Are in Your Home?

Cleaning
Disinfectants
Drain, toilet, and window cleaners
Spot removers
Septic tank cleaners

Paint Products
Paints, stains, varnishes, and lacquers
Paint thinners, solvents, and strippers
Wood preservatives
Artist paints and inks

General
Dry-cell batteries (mercury and cadmium)
Glues and cements

Gardening
Pesticides
Weed killers
Ant and rodent killers
Flea powders

Automotive
Gasoline
Used motor oil
Antifreeze
Battery acid
Brake and transmission fluid

FIGURE 16.3 Harmful chemicals are found in many homes. The U.S. Congress has exempted the disposal of many of these household chemicals and other items from government regulation. **Question:** Which of these chemicals could you find in the place where you live?

Top: tuulijumala/Shutterstock.com. Center: Katrina Outland/Shutterstock.com. Bottom: Karramba Production/Shutterstock.com.

medical waste, car batteries (containing lead and acids), household pesticide products, dry-cell batteries (containing mercury and cadmium), and ash and sludge from incinerators and coal-burning power and industrial plants. Figure 16.3 lists some of the harmful chemicals found in many homes.

Another form of extremely hazardous waste is the highly radioactive waste produced by nuclear power plants and nuclear weapons facilities (see Chapter 13, pp. 329–330). Such wastes must be stored safely for 10,000 to 240,000 years, depending on what radioactive isotopes they contain. After 60 years of research, scientists and governments have not found a scientifically and politically acceptable way to safely isolate these dangerous wastes for such long periods of time.

According to the U.N. Environment Programme (UNEP), more-developed countries produce 80–90% of the world's hazardous wastes, and the United States is the largest producer. However, as China continues to industrialize rapidly, largely without adequate pollution controls, it may soon take over the number-one spot.

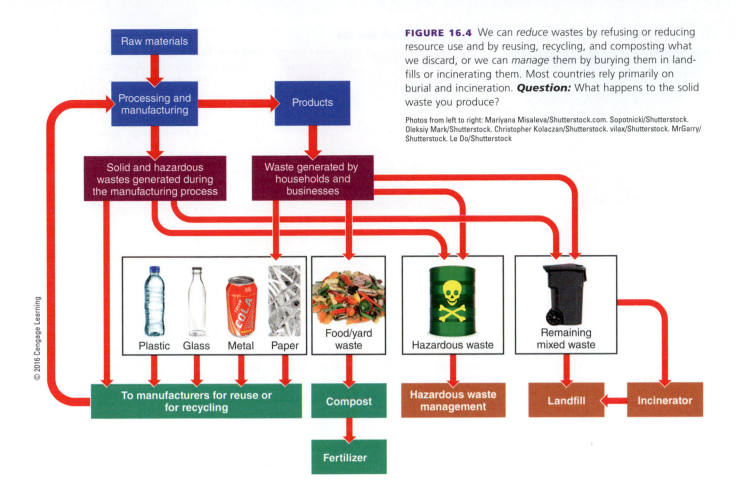

© 2016 Cengage Learning

FIGURE 16.4 We can *reduce* wastes by refusing or reducing resource use and by reusing, recycling, and composting what we discard, or we can *manage* them by burying them in landfills or incinerating them. Most countries rely primarily on burial and incineration. **Question:** What happens to the solid waste you produce?

Photos from left to right: Mariyana Misaleva/Shutterstock.com. Sopotnicki/Shutterstock. Oleksiy Mark/Shutterstock. Christopher Kolaczan/Shutterstock. vilax/Shutterstock. MrGarry/Shutterstock. Le Do/Shutterstock

16.2 HOW SHOULD WE DEAL WITH SOLID WASTE?

CONCEPT 16.2 A sustainable approach to solid waste is first to produce less of it, then to reuse or recycle it, and finally to safely dispose of what is left.

We Can Burn, Bury, or Recycle Solid Waste or Produce Less of It

We can deal with the solid wastes we create in two ways. One method is **waste management**, in which we attempt to control wastes in ways that reduce their environmental harm without seriously trying to reduce the amount of waste produced. This approach begins with the question, "What do we do with solid waste?" It typically involves mixing wastes together and then usually burying them, burning them, or shipping them to another location. The second approach is **waste reduction**, in which we produce much less solid waste, and the wastes we do produce are considered to be potential resources that we can reuse, recycle, or compost (**Concept 16.2**). This waste prevention approach begins with the question, "How can we avoid producing so much solid waste?"

Most analysts call for using **integrated waste management**—a variety of coordinated strategies for both waste disposal and waste reduction (Figure 16.4). Figure 16.5 compares the science-based waste management goals of the EPA and National Academy of Sciences with waste management trends based on an analysis of actual waste data by Columbia University and *BioCycle* researchers.

Some scientists and economists estimate that we could eliminate up to 80% of the solid waste we produce by rigorously applying the strategies shown in Figure 16.5 (left), thereby mimicking the earth's chemical cycling **principle of sustainability**. Let us look more closely at these options in the order of priorities suggested by scientists.

We Can Cut Solid Wastes by Refusing, Reducing, Reusing, and Recycling

Waste reduction (**Concept 16.2**) is based on the four Rs of resource use:

- **Refuse:** Don't use it.
- **Reduce:** Use less.
- **Reuse:** Use it over and over.
- **Recycle:** Convert used resources to useful items and buy products made from recycled materials.

What We Should Do

Reduce
Reuse
Recycle/Compost
Incinerate
Bury

What We Do

Bury (67%)
Recycle/Compost (23.7%)
Incinerate (9%)
Reuse (0.2%)
Reduce (<0.1%)

FIGURE 16.5 Priorities recommended by the U.S. National Academy of Sciences for dealing with municipal solid waste (left) compared with actual waste-handling practices in the United States (right). **Question:** Why do you think most countries do not follow most of the scientific-based priorities listed on the left?

Compiled by the authors using data from U.S. Environmental Protection Agency, U.S. National Academy of Sciences, Columbia University, and *BioCycle*.

An important form of recycling is **composting**, which mimics nature by using bacteria to decompose yard trimmings, vegetable food scraps, and other biodegradable organic wastes into materials than can be used to improve soil fertility.

From an environmental standpoint, the first three Rs are preferred because they are *input*, or *waste prevention*, approaches that tackle the problem of waste production at the front end—before it occurs. Recycling is important but it deals with wastes after they have been produced. By refusing, reducing, reusing, recycling, and composting, we save matter and energy resources, reduce pollution (including greenhouse gas emissions), help protect biodiversity, and save money. Figure 16.6 lists some ways in which you can use the four Rs of waste reduction to reduce your output of solid waste.

Here are six strategies that industries and communities have used to reduce resource use, waste, and pollution.

First, *change industrial processes to eliminate or reduce the use of harmful chemicals*. Since 1975, the 3M Company has taken this approach and, in the process, saved $1.2 billion (see Chapter 14, Case Study, p. 376).

Second, *redesign manufacturing processes and products to use less material and energy*. For example, the weight of a typical car has been reduced by about one-fourth since the 1960s through the use of lighter steel, aluminum, and lightweight plastics and composite materials.

Third, *develop products that are easy to repair, reuse, remanufacture, compost, or recycle*. For example, Xerox photocopiers that are leased by businesses are made of reusable or recyclable parts that allow for easy remanufacturing and are projected to save the company $1 billion in manufacturing costs.

Fourth, *eliminate or reduce unnecessary packaging*. Use the following hierarchy for product packaging: no packaging, reusable packaging, and recyclable packaging.

Fifth, *use fee-per-bag waste collection systems* that charge consumers for the amount of waste they throw away but provide free pickup of recyclable and reusable items.

Sixth, *pass laws* that require companies to take back various consumer products such as electronic equipment (**Core Case Study**), appliances, and motor vehicles, as Japan and many European countries do, for recycling or remanufacturing.

What Can You Do?

Solid Waste

- Follow the four Rs of resource use: Refuse, Reduce, Reuse, Recycle
- Ask yourself whether you really need what you're buying and refuse packaging wherever possible
- Rent, borrow, or barter goods and services when you can, buy secondhand, and donate or sell unused items
- Buy things that are reusable, recyclable, or compostable, and be sure to reuse, recycle, and compost them
- Buy products with little or no packaging and recycle any packaging as much as possible
- Avoid disposables such as paper and plastic bags, plates, cups, and utensils, disposable diapers, and disposable razors whenever reusable versions are available
- Cook with whole, fresh foods, avoid heavily packaged processed foods, and buy products in bulk whenever possible
- Discontinue junk mail as much as possible and read online newspapers and magazines and e-books

FIGURE 16.6 Individuals matter: You can save resources and money by reducing your output of solid waste and pollution. **Questions:** Which three of these steps do you think are the most important ones to take? Why? Which of these things do you already do?

© 2016 Cengage Learning

16.3 WHY ARE REFUSING, REDUCING, REUSING, AND RECYCLING SO IMPORTANT?

CONCEPT 16.3 By refusing and reducing resource use and by reusing and recycling what we use, we decrease our consumption of matter and energy resources, reduce pollution and natural capital degradation, and save money.

There Are Alternatives to the Throwaway Economy

In today's industrialized societies, we have increasingly substituted throwaway items for reusable ones, which has resulted in growing masses of solid waste. However, by applying the four Rs, we can slow or stop this trend.

Refusing to buy some things, especially those with a significant environmental impact, is one way to use resources more sustainably. It begins by asking important questions such as: "Do I really need this piece of clothing or this electronic gadget? Do I really need to replace my mobile phone, computer, or other electronic device just because a newer model has come out?"

Reducing is the next level of using resources more sustainably. It means buying less, and involves asking questions such as: "How many pairs of shoes or pants or how many blouses or skirts do I need? How many electronic devices do I need?"

The third level of more sustainable resource use is *reuse*. It involves cleaning and using items such as containers (Figure 16.7), dishes, utensils, and razor blades over and over, and thus increasing the typical life span of such products. This form of waste reduction decreases the use of matter and energy resources, cuts waste and pollution (including greenhouse gas emissions), creates local jobs, and helps us to save money and expand our beneficial environmental impacts (**Concept 16.3**).

Finally, the fourth level of more sustainable resource use, *recycling and composting*, involves converting used items and materials to other useful forms. It requires more energy and cuts waste and pollution to a lesser extent than does reuse of used items.

Reuse is on the rise. For example, many coffee shops are reducing waste and cutting their costs by discounting their hot drink prices for customers who bring their own refillable containers (Figure 16.7, right). Denmark, Finland, and the Canadian province of Prince Edward Island have banned all beverage containers that cannot be reused. In Finland, 95% of all soft drink, beer, wine, and spirits containers are refillable. The use of rechargeable batteries is cutting toxic waste by reducing the amount of conventional batteries that are thrown away. The newest rechargeable batteries come fully charged, can

Levent Konuk/Shutterstock

ffolas/Shutterstock.com

FIGURE 16.7 *Reuse:* Instead of using throwaway paper (left) or plastic cups, many people take their own refillable cups (right) to workplaces, coffee shops, restaurants, and school cafeterias. **Question:** What are some other reusable items that you could use?

hold a charge for up to 2 years when not in use, and can be recharged in about 15 minutes.

The single-use plastic bag is a highly visible symbol of the throwaway mentality. Globally, about 2 million of such bags are used every minute, on average. Most of them end up littering the land or surface waters and endangering wildlife. Laid end-to-end, the 100 billion plastic bags used in the United States each year would reach to the moon and back about 60 times.

Instead of using throwaway paper or plastic bags to carry home groceries and other purchased items, many people have switched to reusable cloth bags. Both paper and plastic bags are environmentally harmful, and the question of which is more damaging has no clear-cut answer. In many countries, the landscape is littered with plastic bags. They can take 400 to 1,000 years to break down and can kill animals that try to eat them or become ensnared in them. Huge quantities of plastic bags and other plastic products and solid wastes end up in the ocean (see Case Study that follows).

To encourage people to carry reusable bags, the governments of Denmark, Ireland, Taiwan, and the Netherlands tax plastic shopping bags. In Ireland, a tax of about 25¢ per bag helped to cut plastic bag litter by 90% as people switched to reusable bags. Kenya, Uganda, Rwanda, South Africa, Bangladesh, Bhutan, Taiwan, China, Australia, France, Italy, and parts of India also have banned the use of all or most types of plastic shopping bags. In 2014, the European Union passed a directive aimed at cutting the use of single-use plastic bags by 80%.

In 2014, Hawaii became the first U.S. state to ban such bags. They have also been banned in 133 U.S. cities or counties, including San Francisco and Los Angeles, California; Dallas, Texas; Portland, Oregon; Seattle, Washington; Chicago, Illinois; and Washington, D.C.

What Can You Do?

Reuse

- Buy beverages in refillable glass containers

- Use reusable lunch containers

- Store refrigerated food in reusable containers

- Use rechargeable batteries and recycle them when their useful life is over

- When eating out, bring your own reusable container for leftovers

- Carry groceries and other items in a reusable basket or cloth bag

- Buy used furniture, cars, and other items, whenever possible

© 2016 Cengage Learning

FIGURE 16.8 **Individuals matter:** Ways to reuse the items we purchase. **Question:** Which of these suggestions have you tried and how did they work for you?

Brenda Carson/Shutterstock.com

There are many other ways to reuse various items, some of which are listed in Figure 16.8.

CASE STUDY

Ocean Garbage Patches: There Is No Away

In 1997, ocean researcher Charles Moore discovered two gigantic, slowly rotating masses of plastic and other solid wastes in the middle of the North Pacific Ocean near the Hawaiian Islands. These wastes, mostly small particles floating on or just beneath the ocean's surface and known as the Great Pacific Garbage Patch, are trapped there by a vortex where four rotating ocean currents called *gyres* meet (Figure 16.9).

Roughly 80% of this trash (much of it plastic bags and bottles) comes from the land—washed or blown off beaches, pouring out of storm drains, and floating down streams and rivers that empty into the sea from the west coast of North America, the east coast of Asia, and hundreds of Pacific islands. Most of the rest is dumped into the ocean from cargo and cruise ships.

According to some estimates, the North Pacific Garbage Patch occupies an area at least the size of Texas. However, such estimates are difficult to verify because this plastic-laden soup consists mostly of small particles of plastic sus-

pended slightly underwater, and thus it is difficult to see and measure.

Research shows that the tiny plastic particles can be harmful to marine mammals, to some seabirds such as albatrosses, and to some fishes and other aquatic species that mistake them for food and swallow them. Because these animals cannot digest the plastic, it can cause them to die from starvation, poisoning, or choking.

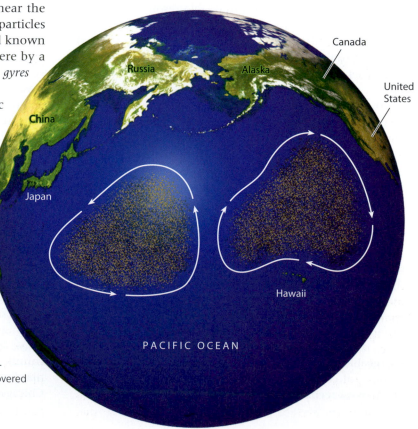

FIGURE 16.9 The Great Pacific Garbage Patch consists of two vast slowly swirling subsurface pools of small plastic particles. Four additional huge garbage patches have been discovered in the world's other major oceans.

© 2016 Cengage Learning

Mike Biddle's Contribution to Plastics Recycling

In 1994, Mike Biddle, a former engineer with Dow Chemical, and his business partner Trip Allen founded MBA Polymers, Inc. Their goal was to develop a commercial process for recycling high-value plastics from complex streams of manufactured goods such as computers, electronics, appliances, and automobiles. They succeeded by designing a 16-step automated process that separates plastics from nonplastic items in mixed waste streams, and then separates plastics from each other by type and grade. The process then converts them to pellets that can be sold and used to make new products.

The pellets are cheaper than virgin plastics because the company's process uses 90% less energy than that needed to make a new plastic and because the raw material is discarded junk that is cheap or free. In addition, greenhouse gas emissions from this recycling process are much lower than those from the process of making virgin plastics. And recycling plastic wastes reduces the need to incinerate them or bury them in landfills.

MBA Polymers has been selected by *Inc.* magazine as one of "America's Most Innovative Companies." Biddle has been named a Technology Pioneer by the World Economic Forum and has received some of the world's most important environmental awards.

Mike Biddle

Background photo: justasc/Shutterstock.com

Who is to blame for the garbage patches? Each of us risks making a contribution every time we use and discard a plastic item. We may think we have simply thrown it away. But there is no away.

Unfortunately, there is no practical or affordable way to clean up this mess that we have created through throwaway habits. All we can do is to prevent the garbage patches from growing larger, and the only way to do this is to reduce our output of wastes by practicing the four Rs of resource use.

There Is Great Potential for Recycling

Households and workplaces produce five major types of materials that can be recycled: paper products, glass, aluminum, steel, and some plastics. We can reprocess such materials into new, useful products in two ways. In **primary**, or **closed-loop**, **recycling**, materials such as aluminum cans are recycled into new products of the same type. In **secondary recycling**, waste materials are converted to different products.

Recycling involves three steps: the collection of materials for recycling, the conversion of recycled materials to new products, and the selling and buying of products that contain recycled material. Recycling does not work unless all three of these steps are taken consistently.

According to Columbia University researchers and *Biocycle* magazine, the United States recycles or composts about 24% of its MSW (lower than the EPA estimate of about 34%). This includes the recycling of about 96% of all lead-acid batteries, 72% of all newspapers, directories, and newspaper inserts, 67% of all steel cans, and 50% of all discarded aluminum cans. Other categories such as food waste and plastics have much lower recycling rates. Experts say that with education and proper incentives, Americans could recycle and compost at least 80% of their MSW, in keeping with the chemical cycling **principle of sustainability**.

Currently, only about 7% by weight of all plastic wastes in the United States (and 13% of plastic containers and packaging) is recycled. These percentages are low because there are many different types of plastic resins, which are difficult to separate from products that contain them. However, progress is being made in the recycling of plastics (Individuals Matter 16.1) and in the development of more degradable bioplastics (Science Focus 16.1).

Composting is another form of recycling that mimics nature's recycling of nutrients. It involves using bacteria to

BIOPLASTICS

Most of today's plastics are made from organic polymers produced from petroleum-based chemicals. This may change as scientists shift to developing plastics made from biologically based chemicals.

Henry Ford, who developed the first Ford car and founded Ford Motor Company, supported research on the development of a bioplastic made from soybeans and another made from hemp. A 1914 photograph shows him using an ax to strike the body of a Ford car made from soy bioplastic to demonstrate its strength and resistance to denting.

However, as oil became widely available, petrochemical plastics took over the market. Now, with projected climate change and other environmental problems associated with the use of oil, chemists are stepping up efforts to make biodegradable and more environmentally sustainable plastics. These *bioplastics* can be made from corn, soy, sugarcane, switchgrass, chicken feathers, and some components of garbage.

Compared to conventional oil-based plastics, properly designed bioplastics are lighter, stronger, and cheaper, and the process of making them requires less energy and produces less pollution per unit of weight. Instead of being sent to landfills, some packaging made from bioplastics can be composted to produce a soil conditioner, in keeping with the chemical cycling **principle of sustainability**.

Critical Thinking

What might be some disadvantages of bioplastics? Do you think they outweigh the advantages?

---

decompose yard trimmings, vegetable food scraps, and other biodegradable organic wastes. The resulting organic material can be added to soil to supply plant nutrients, slow soil erosion, retain water, and improve crop yields. Homeowners can compost such wastes in simple backyard containers, in composting piles that must be turned over occasionally, or in small composting drums that can be rotated to mix the wastes and speed up the decomposition process.

Some cities in Canada and in many European Union countries collect and compost 85% or more of their biodegradable wastes in centralized community facilities. In the United States in 2013, at least 25 states had some version of a ban on yard wastes in sanitary landfills, and about 3,000 municipal composting programs were recycling about 60% of the yard wastes in the country's MSW (Figure 16.10). However, such programs must exclude toxic materials that can contaminate the compost and make it unsafe for fertilizing crops and lawns. While roughly a third of all food produced in the United States is thrown away uneaten, only about 2.5% of this wasted food is composted, according to the EPA.

Recycling Has Advantages and Disadvantages

Figure 16.11 lists the advantages and disadvantages of recycling (**Concept 16.3**). Whether recycling makes economic sense depends on how we look at its economic and environmental benefits and costs.

Critics of recycling programs argue that recycling is costly and adds to the taxpayer burden in communities where recycling is funded through taxation. They say that recycling may make economic sense for valuable and easy-to-recycle materials in MSW such as paper and paperboard, steel, and aluminum, but probably not for cheap or plentiful resources such as glass.

Proponents of recycling point to studies showing that the net economic, health, and environmental benefits of recycling (Figure 16.11, left) far outweigh the costs. For example, the EPA estimates that recycling and composting in the United States in 2010 reduced emissions of climate-changing carbon dioxide by an amount roughly equal to that emitted by 36 million passenger vehicles. Recycling, reuse, and composting create 6–10 times as many jobs as landfills and waste incineration. The U.S. recycling industry employs about 1.1 million people, and doubling the U.S. recycling rate would create about 1 million new jobs.

Cities that make money by recycling and that have higher recycling rates tend to use a *single-pickup system* for both recyclable and nonrecyclable materials, instead of a more expensive dual-pickup system. Successful systems also tend to use a pay-as-you-throw approach that charges for picking up trash but not for picking up recyclable or reusable materials, and they require citizens and businesses to sort their trash and recyclables by type. San Francisco, California, uses such a system, and in 2012, the city recycled, composted, or reused 78% of its MSW.

Composting turns wastes into a useful resource.

imging/Shutterstock.com

FIGURE 16.10 Large-scale municipal composting site.

Trade-Offs

Recycling

Advantages

Reduces energy and mineral use and air and water pollution

Reduces greenhouse gas emissions

Reduces solid waste

Disadvantages

Can cost more than burying in areas with ample landfill space

Reduces profits for landfill and incinerator owners

Inconvenient for some

© 2016 Cengage Learning

FIGURE 16.11 Recycling solid waste has advantages and disadvantages (**Concept 16.3**). *Questions:* Which single advantage and which single disadvantage do you think are the most important? Why?

Photo: Jacqui Martin/Shutterstock.com

16.4 WHAT ARE THE ADVANTAGES AND DISADVANTAGES OF BURNING OR BURYING SOLID WASTE?

CONCEPT 16.4 Technologies for burning and burying solid wastes are well developed, but burning can contribute to air and water pollution and greenhouse gas emissions, and buried wastes can contribute to water pollution.

Burning Solid Waste Has Advantages and Disadvantages

Globally, MSW is burned in more than 800 large *waste-to-energy incinerators,* which use the heat they generate to boil water and make steam for heating water or interior spaces,

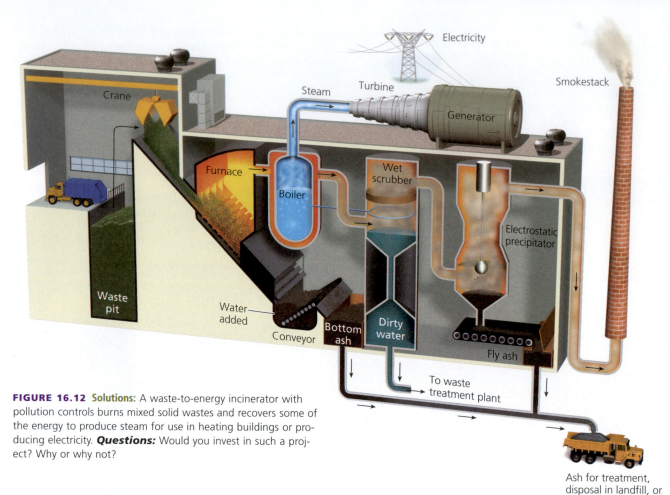

Electricity

Turbine

Smokestack

Steam

Generator

Crane

Furnace

Boiler

Wet
scrubber

Electrostatic
precipitator

Waste
pit

Water
added

Conveyor

Bottom
ash

Dirty
water

Fly ash

To waste
treatment plant

Ash for treatment,
disposal in landfill, or
use as landfill cover

© 2016 Cengage Learning

FIGURE 16.12 Solutions: A waste-to-energy incinerator with pollution controls burns mixed solid wastes and recovers some of the energy to produce steam for use in heating buildings or producing electricity. **Questions:** Would you invest in such a project? Why or why not?

Trade-Offs

Waste-to-Energy Incineration

Advantages	Disadvantages
Reduces trash volume	Expensive to build
Produces energy	Produces a hazardous waste
Concentrates hazardous substances into ash for burial	Emits some CO_2 and other air pollutants
Sale of energy reduces cost	Encourages waste production

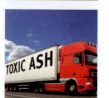

© 2016 Cengage Learning

FIGURE 16.13 Incinerating solid waste has advantages and disadvantages (**Concept 16.4**). These trade-offs also apply to the incineration of hazardous waste. **Questions:** Which single advantage and which single disadvantage do you think are the most important? Why?

Top: Ulrich Mueller/Shutterstock.com. Bottom: Dmitry Kalinovsky/Shutterstock.com.

or for producing electricity (Figure 16.12). There are 115 of these incinerators in the United States, according to Columbia University and *BioCycle* researchers.

The United States incinerates about 9% of its MSW. One reason for this fairly low percentage is that incineration has a bad reputation stemming from past use of highly polluting and poorly regulated incinerators. Also, incineration competes with an abundance of low-cost landfills in many parts of the country. By contrast, Denmark incinerates 54% of its MSW in state-of-the-art waste-to-energy incinerators that exceed European air pollution standards by a factor of 10. A 2009 EPA study concluded that landfills emit more air pollutants than modern waste-to-energy incinerators. On the other hand, the resulting incinerator ash contains toxic chemicals that still have to be stored somewhere.

Many U.S. citizens, local governments, and environmental scientists oppose waste incineration because it undermines efforts to increase reuse and recycling by creating a demand for burnable wastes and making it easier for consumers to discard reusable and recyclable items. Figure 16.13 lists the advantages and disadvantages of using incinerators to burn solid waste.

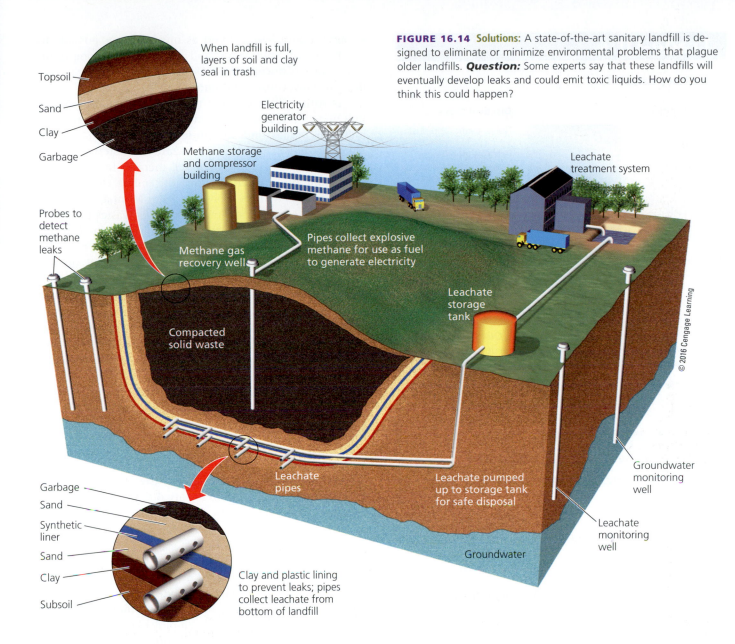

Topsoil
Sand
Clay
Garbage

When landfill is full, layers of soil and clay seal in trash

FIGURE 16.14 Solutions: A state-of-the-art sanitary landfill is designed to eliminate or minimize environmental problems that plague older landfills. **Question:** Some experts say that these landfills will eventually develop leaks and could emit toxic liquids. How do you think this could happen?

Electricity generator building

Methane storage and compressor building

Probes to detect methane leaks

Methane gas recovery well

Pipes collect explosive methane for use as fuel to generate electricity

Leachate treatment system

Leachate storage tank

Compacted solid waste

© 2016 Cengage Learning

Garbage
Sand
Synthetic liner
Sand
Clay
Subsoil

Leachate pipes

Clay and plastic lining to prevent leaks; pipes collect leachate from bottom of landfill

Leachate pumped up to storage tank for safe disposal

Groundwater

Groundwater monitoring well

Leachate monitoring well

Burying Solid Waste Has Advantages and Disadvantages

In the United States, about 67% of all MSW, by weight, is buried in sanitary landfills, compared to 80% in Canada, 15% in Japan, and 4% in Denmark. There are two types of landfills. In newer landfills, called **sanitary landfills** (Figure 16.14), solid wastes are spread out in thin layers, compacted, and covered daily with a fresh layer of clay or plastic foam. This process helps to keep the material dry, reduces leakage of contaminated water (leachate) from the landfill, lessens the risk of fire, decreases odors, and reduces accessibility to the landfill for rats and other vermin.

The bottoms and sides of well-designed sanitary landfills have strong double liners and containment systems that collect the liquids leaching from them. Some landfills also have systems for collecting methane, the potent greenhouse gas that is produced when the wastes decompose in the absence of oxygen, and burning it as a fuel. Figure 16.15 lists the advantages and disadvantages of using sanitary landfills to dispose of solid waste.

The second type of landfill is an **open dump**, essentially a field or large pit where garbage is deposited and sometimes burned. Open dumps are rare in more-developed countries, but are widely used near major cities in many less-developed countries. China disposes of much of its rapidly growing mountains of solid waste mostly in rural open dumps or in poorly designed and poorly regulated landfills that do not have most of the features of sanitary landfills.

Trade-Offs

Sanitary Landfills

Advantages	Disadvantages
Low operating costs	Noise, traffic, and dust
Can handle large amounts of waste	Releases greenhouse gases (methane and CO_2) unless they are collected
Filled land can be used for other purposes	Output approach that encourages waste production
No shortage of landfill space in many areas	Eventually leaks and can contaminate groundwater

© 2016 Cengage Learning

FIGURE 16.15 Using sanitary landfills to dispose of solid waste has advantages and disadvantages (**Concept 16.4**). *Questions:* Which single advantage and which single disadvantage do you think are the most important? Why?

Photo: Pedro Miguel Sousa/Shutterstock.com

16.5 HOW SHOULD WE DEAL WITH HAZARDOUS WASTE?

CONCEPT 16.5 A more sustainable approach to hazardous waste is first to produce less of it, then to reuse or recycle it, then to convert it to less-hazardous materials, and finally to safely store what is left.

We Can Use Integrated Management of Hazardous Waste

Figure 16.16 shows an integrated management approach suggested by the U.S. National Academy of Sciences that establishes three priority levels for dealing with hazardous waste: produce less; convert as much of it as possible to less-hazardous substances; and put the rest in long-term, safe storage (**Concept 16.5**). Denmark follows these priorities, but most countries do not.

As with solid waste, the top priority should be pollution prevention and waste reduction. With this approach, industries try to find substitutes for toxic or hazardous materials, reuse or recycle the hazardous materials within industrial processes, or use or sell them as raw materials for making other products. (See the online Guest Essays on this subject by Lois Gibbs and Peter Montague.)

At least 33% of industrial hazardous wastes produced in the European Union are exchanged through clearinghouses where they are sold as raw materials for use by other industries. The producers of these wastes do not have to pay for their disposal and recipients get low-cost raw materials. About 10% of the hazardous waste in the United States is exchanged through such clearinghouses, a figure that could be raised significantly.

We can also recycle e-waste (**Core Case Study**). However, most e-waste recycling can create health hazards, especially for workers in some less-developed countries (see the following Case Study).

CASE STUDY

Recycling E-Waste

In some countries, workers in e-waste recycling operations—many of them children—are often exposed to toxic chemicals as they dismantle the electronic trash to extract its valuable metals or other parts that can be sold for reuse or recycling (**Core Case Study**).

According to the United Nations, more than 70% of the world's e-waste is shipped to China. A center for such waste is the small port city of Guiyu, where the air reeks of burning plastic and acid fumes. There, more than 5,500 small-scale e-waste businesses employ over 30,000 people (including some children). They work for very low wages in dangerous conditions to extract valuable metals like gold, silver, copper, and various rare earth metals (see

Produce Less Hazardous Waste	Convert to Less Hazardous or Nonhazardous Substances	Put in Perpetual Storage
■ Change industrial processes to reduce or eliminate hazardous waste production ■ Recycle and reuse hazardous waste	■ Natural decomposition ■ Incineration ■ Thermal treatment ■ Chemical, physical, and biological treatment ■ Dilution in air or water	■ Landfill ■ Underground injection wells ■ Surface impoundments ■ Underground salt formations

© 2016 Cengage Learning

FIGURE 16.16 *Integrated hazardous waste management:* The U.S. National Academy of Sciences has suggested these priorities for dealing with hazardous waste (**Concept 16.5**). *Question:* Why do you think most countries do not follow these priorities?

Chapter 12, Core Case Study, p. 290) from millions of discarded computers, television sets, and cell phones.

These workers usually wear no masks or gloves, often work in rooms with no ventilation, and are usually exposed to a cocktail of toxic chemicals. They carry out dangerous activities such as smashing TV picture tubes with large hammers to recover certain components—a method that releases large amounts of toxic lead dust into the air. They also burn computer wires to expose copper, melt circuit boards in metal pots over coal fires to extract lead and other metals, and douse the boards with strong acid to extract gold. After the valuable metals are removed, leftover parts are burned or dumped into rivers or onto the land. An estimated 82% of the Guiyu area's children younger than age 6 suffer from lead poisoning.

The United States produces roughly 50% of the world's e-waste and recycles only about 14% of it, according to the Consumer Electronics Association. However, by 2013, at least 20 states had banned the disposal of computers and TV sets in landfills and incinerators, and these measures set the stage for an emerging, highly profitable *e-cycling* industry. In 2013, 13 states along with New York City made manufacturers responsible for recycling most electronic devices.

Some have called for a U.S. federal law that would make manufacturers responsible for taking back all electronic devices they produce and recycling them domestically. It could be similar to laws in the European Union, where a recycling fee typically covers the costs of such programs. Without such a law there is little incentive for recycling e-waste and plastics. Mike Biddle (Individuals Matter 16.1, p. 437) chose to open his plastics recycling plants in China, England, and Austria, where markets for his product are stronger, instead of in the United States.

Recycling probably will not keep up with the explosive growth of e-waste, and there is money to be made from illegally sending such materials to other countries. The only real long-term solution is a *prevention* approach through which electrical and electronic products would be designed to be produced and easily repaired, remanufactured, or recycled, without the use of toxic materials.

We Can Detoxify Hazardous Wastes

The first step in dealing with hazardous wastes is to collect them. In Denmark, all hazardous and toxic wastes from industries and households are delivered to any of 21 transfer stations throughout the country. From there they are taken to a large processing facility, where three-fourths of the waste is detoxified by physical, chemical, and biological methods. The rest is buried in a carefully designed and monitored landfill.

Physical methods for detoxifying hazardous wastes include using charcoal or resins to filter out harmful solids, distilling liquid wastes to separate out harmful chemicals, and precipitating such chemicals from solution. Especially deadly wastes can be encapsulated in glass, cement, or ceramics and then put in secure storage sites. *Chemical methods* are used to convert hazardous chemicals to harmless or less harmful chemicals through chemical reactions.

Some scientists and engineers consider *biological methods* for treatment of hazardous waste to be the wave of the future. One such approach is *bioremediation*, in which bacteria and enzymes help to destroy toxic or hazardous substances or convert them to harmless compounds. Bioremediation takes a little longer to work than most physical and chemical methods, but it costs much less. (See the online Guest Essay by John Pichtel on this topic.)

Another approach is *phytoremediation*, which involves using natural or genetically engineered plants to absorb, filter, and remove contaminants from polluted soil and water. Various plants have been identified as "pollution sponges" that can be used to help clean up soil and water contaminated with chemicals such as pesticides, organic solvents, and radioactive or toxic metals. Phytoremediation is still being evaluated and is slow, compared to other alternatives.

Incineration of hazardous wastes has the same combination of advantages and disadvantages as does the burning of solid wastes (Figure 16.13). However, incinerating hazardous waste without effective and expensive air pollution controls can release air pollutants such as highly toxic dioxins, and it produces an extremely toxic ash that must be safely and permanently stored in a landfill or vault especially designed for this hazardous end product.

We can also detoxify hazardous wastes by using *plasma gasification*, a technology that uses arcs of electrical energy to produce very high temperatures in order to vaporize trash in the absence of oxygen. This process reduces the volume of a given amount of waste by 99%, produces a synthetic gaseous fuel, and encapsulates toxic metals and other materials in glassy lumps of rock. So why are we not making widespread use of the plasma arc torch to detoxify hazardous wastes? The main reason is its high cost. Plasma arc companies are working to bring the cost down.

We Can Store Some Forms of Hazardous Waste

Ideally, we should use burial on land or long-term storage of hazardous and toxic wastes in secure vaults only as the third and last resort after the first two priorities have been exhausted (Figure 16.16 and **Concept 16.5**). Currently, however, burial on land is the most widely used method in the United States and in most countries, largely because of its lower cost.

The most common form of burial is *deep-well disposal*, in which liquid hazardous wastes are pumped under high pressure through a pipe into dry, porous rock formations far beneath aquifers that are tapped for drinking and irrigation water (see Figure 11.26, p. 272). Theoretically, these liquids soak into the porous rock material and are isolated from overlying groundwater by essentially imper-

Trade-Offs

Deep-Well Disposal

Advantages	Disadvantages
Safe if sites are chosen carefully	Leaks can occur from corrosion of well casing
Wastes can often be retrieved	Emits CO_2 and other air pollutants
Low cost	Output approach that encourages waste production

© 2016 Cengage Learning

FIGURE 16.17 Injecting liquid hazardous wastes into deep underground wells has advantages and disadvantages. ***Questions:*** Which single advantage and which single disadvantage do you think are the most important? Why?

Trade-Offs

Surface Impoundments

Advantages	Disadvantages
Low cost	Water pollution from leaking liners and overflows
Wastes can often be retrieved	Air pollution from volatile organic compounds
Can store wastes indefinitely with secure double liners	Output approach that encourages waste production

© 2016 Cengage Learning

FIGURE 16.18 Storing liquid hazardous wastes in surface impoundments has advantages and disadvantages. ***Questions:*** Which single advantage and which single disadvantage do you think are the most important? Why?

Photo: Jim West/Alamy

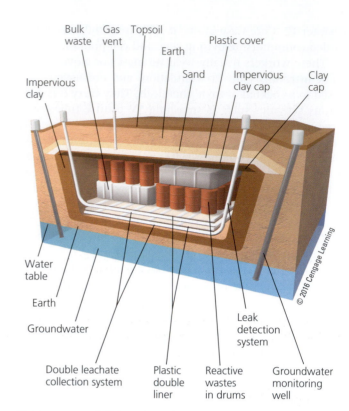

© 2016 Cengage Learning

FIGURE 16.19 **Solutions:** Hazardous wastes can be isolated and stored in a secure hazardous waste landfill.

meable layers of clay and rock. The cost is low and the wastes can often be retrieved if problems develop. However, there are a limited number of such sites and limited space within them. Sometimes the wastes can leak into groundwater from the well shaft or migrate into groundwater in unexpected ways. Also, this is an output approach that encourages the production of hazardous wastes.

In the United States, almost two-thirds of all liquid hazardous wastes are injected into deep disposal wells. And this amount will increase sharply as the country relies more on fracking to produce natural gas and oil trapped in shale rock (see Science Focus 13.1, p. 318). Many scientists argue that current regulations for deep-well disposal in the United States are inadequate and should be improved (see the Case Study that follows). Figure 16.17 lists the advantages and disadvantages of using deep-well disposal of liquid hazardous wastes.

Some liquid hazardous wastes are stored in lined ponds, pits, or lagoons, called *surface impoundments*. Sometimes they include liners that help to contain the waste. As the water evaporates, the waste settles and becomes more concentrated. However, where liners are not used and wherever the liners leak, such wastes can percolate into groundwater. Also, because these impoundments are not covered, volatile harmful chemicals can evaporate, polluting the air. In addition, powerful storms can cause such ponds to overflow. Figure 16.18 lists the advantages and disadvantages of using this method.

EPA studies have found that 70% of all U.S. hazardous waste storage ponds have no liners and could threaten groundwater supplies. According to the EPA, eventually, all impoundment liners are likely to leak and could contaminate groundwater.

Sometimes both liquid and solid hazardous wastes are put into drums or other containers and buried in carefully designed and monitored *secure hazardous waste landfills* (Figure 16.19). This is the least-used method because of the expense involved.

Figure 16.20 lists some ways in which you can reduce your output of hazardous waste—the first step in dealing with it.

What Can You Do?

Hazardous Waste

- Avoid using pesticides and other hazardous chemicals, or use them in the smallest amounts possible

- Use less harmful substances instead of commercial household cleaners. For example, use vinegar to polish metals, clean surfaces, and remove stains and mildew, and baking soda to clean utensils and to deodorize and remove stains.

- Do not dump pesticides, paints, solvents, oil, antifreeze, or other hazardous chemicals down the toilet, down the drain, into the ground, into the garbage, or down storm drains. Free hazardous waste disposal services are available in many cities.

- Do not throw old fluorescent lightbulbs (which contain mercury) into regular trash. Many communities and home product retailers offer free recycling of these bulbs.

FIGURE 16.20 **Individuals matter:** You can reduce your output of hazardous wastes (**Concept 16.5**). **Questions:** Which two of these measures do you think are the most important ones to take? Why?

CASE STUDY

Hazardous Waste Regulation in the United States

About 5% of all hazardous waste produced in the United States is regulated under the Resource Conservation and Recovery Act (RCRA, pronounced "RICK-ra"), passed by the U.S. Congress in 1976 and amended in 1984. Under this act, the EPA sets standards for the management of several types of hazardous waste and issues permits to companies that allow them to produce and dispose of a certain amount of those wastes by approved methods. Permit holders must use a *cradle-to-grave* system to keep track of waste they transfer from a point of generation (cradle) to an approved off-site disposal facility (grave), and they must submit proof of this disposal to the EPA.

RCRA is a good start, but about 95% of the hazardous and toxic wastes, including e-waste, produced in the United States is not regulated. In most other countries, especially less-developed countries, the amount of regulated waste is even smaller.

CONSIDER THIS. . .

THINKING ABOUT Hazardous Waste

Why is it that 95% of the hazardous waste, including the growing amounts of e-waste (**Core Case Study**) produced in the United States, is not regulated? Do you favor regulating such wastes? What do you think would be the economic consequences of doing so?

In 1976, the U.S. Congress passed the Toxic Substances Control Act, which was intended to regulate and ensure the safety of the thousands of chemicals used in the manufacture of many products and contained in many products. Under this law, companies must notify the EPA before introducing a new chemical into the marketplace, but they are not required to provide any data about its safety. In other words, any new chemical is viewed as safe unless the EPA can show that it is harmful.

Since 1976, the EPA, with a very limited budget, has used this act to ban only 5 of the roughly 80,000 chemicals in use. Environmental and health scientists call for Congress to reform this almost 40-year-old law by requiring manufacturers to provide data showing that a chemical or product containing a certain chemical is safe before it can be sold in the marketplace.

In 1980, the U.S. Congress passed the *Comprehensive Environmental Response, Compensation, and Liability Act,* commonly known as the *CERCLA* or *Superfund* program, supervised by the EPA. Its goals are to identify sites, commonly called Superfund sites, where hazardous wastes have contaminated the environment and to clean them up, using EPA-approved methods, on a priority basis. The worst sites—those that represent an immediate and severe threat to human health—are put on a *National Priorities List* and scheduled for cleanup.

By May 2014, there were 1,326 sites on the Superfund list, along with 51 proposed new sites, and 375 sites had been cleaned up and removed from the list. The Waste Management Research Institute estimates that at least 10,000 sites should be on the priority list and that cleanup of these sites would cost about $1.7 trillion, not including legal fees. This is a glaring example of the economic and environmental value of emphasizing waste reduction and pollution prevention over the *end-of-pipe* cleanup approach that the United States and most countries rely on.

In 1984, Congress amended the Superfund Act to give citizens the right to know what toxic chemicals are being stored or released in their communities. This required 23,800 large manufacturing facilities to report their annual releases of any of nearly 650 toxic chemicals. If you live in the United States, you can find out what toxic chemicals are being stored and released in your neighborhood by going to the EPA's *Toxic Release Inventory* website.

The Superfund Act, designed to make polluters pay for cleaning up abandoned hazardous waste sites, greatly reduced the number of illegal dumpsites around the country. It also forced waste producers who were fearful of liability claims to reduce their production of such waste and to recycle or reuse much more of it. However, under pressure from polluters, the U.S. Congress refused to renew the tax on oil and chemical companies that had financed the Superfund legislation after it expired in 1995. The Superfund is now broke, and taxpayers, not polluters, are footing the bill for cleanups when the responsible parties cannot be found. As a result, the pace of cleanup has slowed.

16.6 HOW CAN WE MAKE THE TRANSITION TO A MORE SUSTAINABLE LOW-WASTE ECONOMY?

CONCEPT 16.6 Shifting to a low-waste economy will require individuals and businesses to reduce resource use and to reuse and recycle most solid and hazardous wastes at local, national, and global levels.

Grassroots Action Has Led to Better Solid and Hazardous Waste Management

In the United States, individuals have organized grassroots (bottom-up) citizen movements to prevent the construction of hundreds of incinerators, landfills, treatment plants for hazardous and radioactive wastes, and polluting chemical plants in or near their communities. Health risks from incinerators and landfills, when averaged over the entire country, are quite low, but the risks for people living near such facilities are much higher.

Manufacturers and waste industry officials point out that something must be done with the toxic and hazardous wastes created in the production of certain goods and services. They contend that even if local citizens adopt a "not in my back yard" (NIMBY) approach, the waste will always end up in someone's back yard.

Many citizens do not accept this argument. To them, the best way to deal with most toxic and hazardous waste is to produce much less of it, as suggested by the U.S. National Academy of Sciences (Figure 16.16). For such materials, they argue that the goal should be "not in anyone's back yard" (NIABY) or "not on planet Earth" (NOPE), which calls for drastically reducing production of such wastes by emphasizing pollution and waste prevention.

We Can Encourage Reuse and Recycling

Three factors hinder reuse and recycling. *First,* the market prices of almost all products do not include the harmful environmental and health costs associated with producing, using, and discarding them—a violation of the full-cost pricing **principle of sustainability**.

Second, the economic playing field is uneven, because in most countries, resource extraction industries receive more government tax breaks and subsidies than reuse and recycling industries get.

Third, the demand and thus the price paid for recycled materials fluctuates, mostly because it is not a high priority for most governments, businesses, and individuals to buy goods made of recycled materials.

How can we encourage reuse and recycling? Proponents say that leveling the economic playing field is the best way to start. Governments can *increase* subsidies and tax breaks for reusing and recycling materials, and *decrease* subsidies and tax breaks for making items from virgin resources.

Another strategy is to greatly increase use of the fee-per-bag waste collection system that charges households for the trash they throw away but not for their recyclable and reusable wastes. When the U.S. city of Fort Worth, Texas, instituted such a program, the proportion of households recycling their trash went from 21% to 85%. The city went from losing $600,000 in its recycling program to making $1 million a year because of increased sales of recycled materials to industries.

Governments can also pass laws requiring companies to take back and recycle or reuse packaging and electronic waste discarded by consumers (**Core Case Study**), as is done in Japan and some European Union countries. Another important strategy is to encourage or require government purchases of recycled products to help increase demand for and lower prices of these products. Also, citizens can pressure governments to require product labeling that lists the recycled content of products, as well as the types and amounts of any hazardous materials they contain.

Reuse, Recycling, and Composting Present Economic Opportunities

A growing number of people are saving money through reuse, regularly going to yard sales, flea markets, second-hand stores, and online sites such as eBay and craigslist. Another such site, the Freecycle Network, links people who want to give away their unused household belongings to people who want or need them.

For many, recycling has become a business opportunity. In particular, *upcycling,* or recycling materials into products of a higher value, is a growing field. For example, a British company called Worn Again is converting discarded textiles, such as old hot-air balloons and worn-out seat covers, into windbreaker jackets and other products. And researcher Na Lu at the University of North Carolina, Charlotte, has found a way to upcycle plastic bottles to make a building material that could outperform composite lumber and wood lumber. Entrepreneurs see upcycling as an area of great economic opportunity, and it represents a golden opportunity for those interested in creating or expanding their positive environmental impact.

International Treaties Have Reduced Hazardous Waste

For decades, some countries regularly shipped hazardous wastes to other countries for disposal or processing. However, since 1992, an international treaty known as the Basel Convention has banned participating countries from shipping hazardous waste (including e-waste; see **Core Case Study**) to or through other countries without their permission. By 2013, this agreement had been ratified (formally approved and implemented) by 179 countries. The United States has signed but has not ratified the convention. In 1995, the treaty was amended to outlaw all transfers of hazardous wastes from industrial countries to less-developed countries. This ban is likely to be ratified by enough countries to go into effect in the next few years.

This ban will help, but it will not wipe out the very profitable illegal shipping of hazardous wastes. Hazardous waste smugglers evade the laws by using an array of tactics, including bribes, false permits, and mislabeling of hazardous wastes as recyclable materials.

In 2000, delegates from 122 countries completed a global treaty known as the Stockholm Convention on Persistent Organic Pollutants (POPs). It regulates the use of 12 widely used persistent organic pollutants that can accumulate in the fatty tissues of humans and other animals that occupy high trophic levels in food webs. At such levels, these hazardous chemicals can reach levels hundreds of thousands of times higher than their levels in the general environment (see Figure 8.9, p. 171). Because they persist in the environment, POPs can also be transported long distances by wind and water.

The original list of 12 hazardous chemicals, called the *dirty dozen*, includes DDT and eight other chlorine-containing persistent pesticides, PCBs, dioxins, and furans. Using blood tests and statistical sampling, medical researchers at New York City's Mount Sinai School of Medicine found that it is likely that nearly every person on earth has detectable levels of POPs in their bodies. The long-term health effects of this involuntary global chemical experiment are largely unknown.

By 2014, 179 countries had ratified a strengthened version of the POPs treaty that seeks to ban or phase out the use of these hazardous chemicals and to detoxify or isolate existing stockpiles. It allows 25 countries to continue using DDT to combat malaria until safer alternatives are found. As of 2014, the United States had not yet ratified this treaty. The list of regulated POPs is expected to grow.

In 2000, the Swedish Parliament enacted a law that, by 2020, will ban all potentially hazardous chemicals that are persistent in the environment and can accumulate in living tissue. This law also requires industries to perform risk assessments on the chemicals they use and to show that these chemicals are safe to use, as opposed to requiring the government to show that they are dangerous. In other words, chemicals are assumed to be guilty until proven innocent— the reverse of the current policy in the United States and most other countries. There is strong opposition to this approach in the United States, especially from most of the industries that produce and use potentially hazardous chemicals.

We Can Make the Transition to Low-Waste Economies

According to physicist Albert Einstein, "A clever person solves a problem; a wise person avoids it." Many people are taking these words seriously. The governments of Norway, Austria, and the Netherlands have committed to re- ducing their resource waste by 75%. Many school cafeterias, restaurants, national parks, and corporations are participating in a rapidly growing "zero waste" movement to reduce, reuse, and recycle, and some have lowered their waste outputs by up to 80%, with the ultimate goal of eliminating their waste outputs. They are applying nature's chemical cycling **principle of sustainability** (see the following Case Study}.

CASE STUDY

Industrial Ecosystems: Copying Nature

An important goal for a more sustainable society is to make its industrial manufacturing processes cleaner and more sustainable by redesigning them to mimic the way nature deals with wastes—an approach called *biomimicry*. In nature, according to the chemical cycling **principle of sustainability**, the waste outputs of one organism become the nutrient inputs of another organism, so that all of the earth's nutrients are endlessly recycled. This explains why there is essentially no waste in undisturbed ecosystems.

One way for industries to mimic nature is to reuse or recycle most of the minerals and chemicals they use, instead of burying or burning them or shipping them somewhere. Another method that industries can use to mimic nature is to interact with each other through *resource exchange webs* in which the wastes of one manufacturer become the raw materials for another—similar to food webs in natural ecosystems (see Figure 3.10, p. 49).

This is happening in Kalundborg, Denmark, where an electric power plant and nearby industries, farms, and homes are collaborating to save money and to reduce their outputs of waste and pollution within what is called an *ecoindustrial park*, or *industrial ecosystem*. They exchange waste outputs and convert them into resources, as shown in Figure 16.21. This cuts pollution and waste and reduces the flow of nonrenewable mineral and energy resources through the local economy.

Today, more than 40 ecoindustrial parks (18 of them in the United States) operate in various places around the world, and more are being built or planned. A number of people who work in the rapidly growing field of industrial ecology are focusing on developing a global network of industrial ecosystems over the next few decades, and this could lead to an important *ecoindustrial revolution*. **GREEN CAREER: Industrial ecology**

Ecoindustrial parks provide many economic benefits for businesses. By encouraging recycling and waste reduction prevention, they reduce the costs of managing solid wastes, controlling pollution, and complying with pollution regulations. They also reduce a company's chances of being sued because of damages, to people or to the environment, caused by their actions. In addition, companies improve the health and safety of workers by reducing their exposure to toxic and hazardous materials, thereby reducing

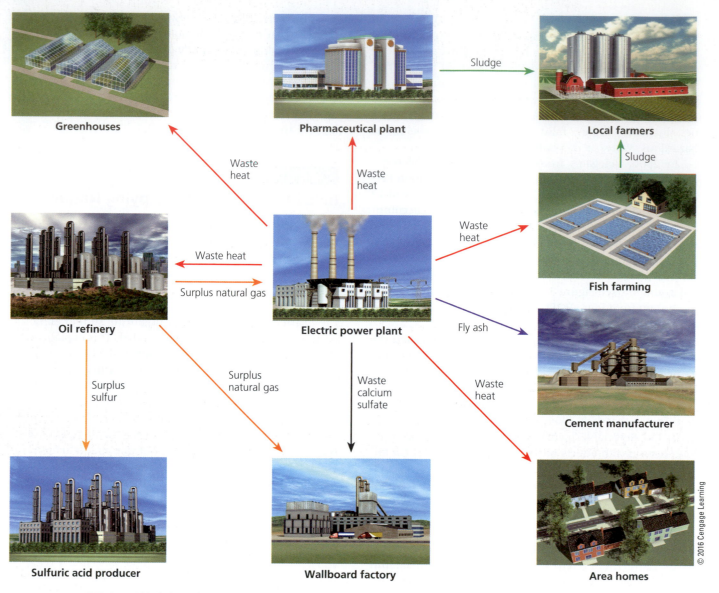

FIGURE 16.21 Solutions: This *industrial ecosystem* in Kalundborg, Denmark, reduces waste production by mimicking a natural ecosystem's food web. The wastes of one business become the raw materials for another, thus mimicking the way nature recycles chemicals. **Questions:** Is there an industrial ecosystem near where you live or go to school? If not, think about where and how such a system could be set up.

Labels in figure:
- Greenhouses
- Pharmaceutical plant
- Local farmers
- Sludge
- Sludge
- Waste heat
- Waste heat
- Waste heat
- Waste heat
- Oil refinery
- Surplus natural gas
- Electric power plant
- Fly ash
- Fish farming
- Surplus sulfur
- Surplus natural gas
- Waste calcium sulfate
- Waste heat
- Cement manufacturer
- Sulfuric acid producer
- Wallboard factory
- Area homes

© 2016 Cengage Learning

company health insurance costs. Biomimicry also encourages companies to come up with new, environmentally beneficial, and less resource-intensive chemicals, processes, and products that they can sell worldwide.

Biomimicry involves two major steps. The first is to study how natural systems have responded to changes in environmental conditions over many millions of years. The second step is to try to copy or adapt these responses within human systems in order to help us deal with various environmental challenges. In the case of solid and hazardous wastes, the food web serves as a natural model for responding to the growing problem of these wastes. This is in keeping with the three **scientific principles of sustainability** that nature has used for billions of years.

BIG IDEAS

- The order of priorities for dealing with solid waste should be first to minimize production of it, then to reuse and recycle as much of it as possible, and finally to safely burn or bury what is left.

- The order of priorities for dealing with hazardous waste should be first to minimize production of it, to reuse or recycle it, to convert it to less-hazardous material, and to safely store what is left.

- We can view solid wastes as wasted resources, and hazardous wastes as materials that we want to avoid producing in the first place.

E-Waste and Sustainability

One of the problems of maintaining a high-waste society is the growing mass of e-waste (**Core Case Study**) and other types of solid and hazardous waste discussed in this chapter. The challenge is to make the transition from an unsustainable high-waste, throwaway economy to a more sustainable low-waste, reducing–reusing–recycling economy as soon as possible.

Such a transition will require applying the six **principles of sustainability**. We can reduce our outputs of solid and hazardous waste by relying much less on fossil fuels and nuclear power (which produces long-lived, hazardous radioactive wastes) while relying much more on renewable energy from the sun, wind, and flowing water. We can mimic nature's chemical cycling processes by reusing and recycling materials as much as possible. Integrated waste management, which uses a diversity of approaches and emphasizes waste reduction and pollution prevention, is a way to mimic nature's use of biodiversity.

By including more of the harmful environmental and health costs of the consumer economy in market prices,

we would be applying the full-cost pricing **principle of sustainability** while encouraging people to refuse, reduce, reuse, and recycle. In doing so, we would benefit the environment, create new jobs and businesses capitalizing on the four Rs, and gain health and environmental benefits for us, thus finding win-win solutions. This could also lead to lower levels of resource use per person, and thus a lower demand for materials that eventually become solid and hazardous wastes. All these measures together would help us to pass along to future generations a world that is at least as livable or more so than the one we have enjoyed.

Bakalusha/Shutterstock.com

Chapter Review

Core Case Study

1. Explain how and why electronic waste (e-waste) has become a growing solid waste problem.

Section 16.1

2. What are the two key concepts for this section? Distinguish among **solid waste**, **industrial solid waste**, **municipal solid waste (MSW)**, and **hazardous (toxic) waste**, and give an example of each. Summarize the types and sources of solid waste generated in the United States and explain what happens to it.

Section 16.2

3. What is the key concept for this section? Distinguish among **waste management**, **waste reduction**, and **integrated waste management**. Summarize the priorities that prominent scientists suggest we should use for dealing with solid waste and compare them to actual practices in the United States.

Distinguish among **refusing**, **reducing**, **reusing**, and **recycling** in dealing with the solid wastes we produce. Why are the first three Rs preferred from an environmental standpoint? What is **composting**? List six ways in which industries and communities can reduce resource use, waste, and pollution.

Section 16.3

4. What is the key concept for this section? Explain why refusing, reducing, reusing, and recycling are so important and give examples of each. List five ways to reuse various items. What is the Great Pacific Garbage Patch and how did it come to be? How does it harm marine life and how can the growth of such patches be prevented?

5. Distinguish between **primary (closed-loop) recycling** and **secondary recycling**. What are three important steps that must occur for any recycling program to work? What are some benefits of composting? Explain how some plastics are being recycled and describe Mike Biddle's contributions to doing this. What are bioplastics? What are the major advantages and disadvantages of recycling?

Section 16.4

6. What is the key concept for this section? What are the major advantages and disadvantages of using incinerators to burn solid and hazardous waste? Distinguish between **sanitary landfills** and **open dumps**. What are the major advantages and disadvantages of burying solid waste in sanitary landfills?

Section 16.5

7. What is the key concept for this section? What are the priorities that scientists suggest we should use in dealing with hazardous waste? Summarize the problems involved in sending e-wastes to less-developed countries for recycling. Describe three ways to detoxify hazardous wastes. What is bioremediation? What is phytoremediation? What are the major advantages and disadvantages of incinerating hazardous wastes? How can we use plasma gasification to detoxify hazardous wastes?

8. What are the major advantages and disadvantages of storing liquid hazardous wastes in deep underground wells and in surface impoundments? What is a secure hazardous waste landfill? List four ways to reduce your output of hazardous waste. Summarize the story of regulation of hazardous wastes in the United States.

Section 16.6

9. What is the key concept for this section? How has grassroots action led to improved solid and hazardous waste management in the United States? What are three factors that discourage recycling? What are three ways to encourage recycling and reuse? Give three examples of how people are saving or making money through reuse, recycling, and composting. Describe regulation of hazardous wastes at the global level through the Basel Convention and the treaty to control persistent organic pollutants (POPs). What is biomimicry? What is an industrial ecosystem?

10. What are this chapter's three big ideas? Explain how we could deal with the growing problems of e-waste and other wastes (**Core Case Study**) by applying the six **principles of sustainability**.

Note: Key terms are in bold type.

Critical Thinking

1. Do you think that manufacturers of computers, television sets, cell phones, and other electronic products (**Core Case Study**) should be required to take their products back at the end of their useful lives for repair, remanufacture, or recycling in a manner that is environmentally responsible and that does not threaten the health of recycling workers? Explain. Would you be willing to pay more for these products to cover the costs of such a take-back program? If so, what percentage more per purchase would you be willing to pay for these products?

2. Think of three items that you regularly use once and then throw away. Are there reusable items that you could use in place of these disposable items?

3. Do you think that you could consume less by refusing to buy some of the things you regularly buy? If so, what are three of those things? Do you think that this is something you ought to do? Explain.

4. A company called Changing World Technologies has built a pilot plant to test a process it has developed for converting a mixture of discarded computers, old tires, turkey bones and feathers, and other wastes into oil by mimicking and speeding up natural processes for converting biomass into oil. Explain how this recycling process, if it turns out to be technologically and economically feasible, could lead to increased waste production.

5. Would you oppose having **(a)** a sanitary landfill, **(b)** a hazardous waste surface impoundment, **(c)** a hazardous waste deep-injection well, or **(d)** a solid waste incinerator in your community? For each of these facilities, explain your answer. If you oppose having such facilities in your community, how do you think the solid and hazardous wastes generated in your community should be managed?

6. How does your school dispose of its solid and hazardous wastes? Does it have a recycling program? How well does it work? Does your school encourage reuse? If so, how? Does it have a hazardous waste collection system? If so, describe it. List three ways in which you would improve your school's waste reduction and management systems.

7. List three ways in which you could apply **Concept 16.6** to making your lifestyle more environmentally sustainable.

8. Congratulations! You are in charge of the world. List the three most important components of your strategy for dealing with **(a)** solid waste and **(b)** hazardous waste.

Doing Environmental Science

Collect the trash (excluding food waste) that you generate in a typical week. Measure its total weight and volume. Sort it into major categories such as paper, plastic, metal, and glass. Then weigh each category and calculate its percentage by weight of the total amount of trash that you have measured. What percentage by weight of this waste consists of materials that could be recycled? What percentage consists of materials for which you could have used a reusable substitute, such as a coffee mug instead of a disposable cup? What percentage by weight of the items could you have done without? Compare your answers to these questions with those of your classmates. Together with your classmates, combine all the results and do the same analysis for the entire class. Use these results to estimate the same values for the entire student population at your school.

Global Environment Watch Exercise

Within the GREENR database, go to the *E-Waste* topic portal. Research and find statistics on how rapidly the world's production of e-waste (**Core Case Study**) is growing and how rapidly e-waste production is growing in the United States. Write a brief report on what the United States and one other country of your choice are doing to deal with this growing waste problem. Include statistics on how much e-waste is generated in each country, on how much of it is recycled, and on how much of it goes to landfills. Compare the two approaches in terms of how successful they are.

Ecological Footprint Analysis

Researchers estimate that the average daily municipal solid waste production per person in the United States is 3.2 kilograms (7 pounds). Use the data in the pie chart below to get an idea of a typical annual MSW ecological footprint for each American by calculating the total weight in kilograms (and pounds) for each category generated during 1 year (1 kilogram = 2.20 pounds). Use the table (below, right) to enter your answers.

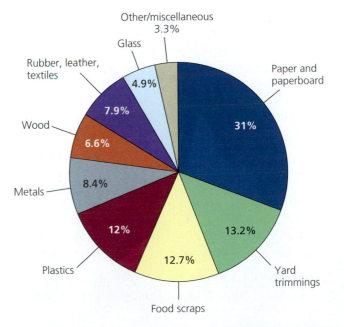

Composition of a typical sample of U.S. municipal solid waste, 2010.
Compiled by the authors using data from the U.S. Environmental Protection Agency.

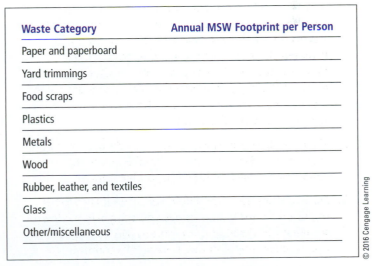

Waste Category	Annual MSW Footprint per Person
Paper and paperboard	
Yard trimmings	
Food scraps	
Plastics	
Metals	
Wood	
Rubber, leather, and textiles	
Glass	
Other/miscellaneous	

© 2016 Cengage Learning

When it is asked how much it will cost to protect the environment, one more question should be asked: How much will it cost our civilization if we do not?

GAYLORD NELSON

17

ENVIRONMENTAL ECONOMICS, POLITICS, AND WORLDVIEWS

17.3 How can we implement more sustainable and just environmental policies?

17.4 What are some major environmental worldviews?

17.5 How can we live more sustainably?

This mountain environment hosts many different kinds of natural capital, which help to sustain all life and economies.

Pichugin Dmitry/Shutterstock.com

The United States, China, and Sustainability

The greatest challenge we face is to learn how to live more sustainability during the next few decades. Meeting this challenge depends largely on the decisions and actions of the United States and China—the two countries that lead the world in resource consumption and production of wastes and pollutants.

From 1940 to 1970, the United States experienced rapid economic growth along with severe pollution and degradation of air, water, and land. By 1970, public awareness of these problems had grown and spurred an environmental movement made up of millions of citizens who demanded an end to this environmental degradation. This prompted the U.S. Congress to pass a number of environmental laws that have led to improvements in the nation's environmental quality.

Despite this important progress, the United States, with the third largest population, has the world's largest ecological footprint (Figure 17.1, left), mostly because it uses far more resources per person than any other country. That means it also has the world's largest per capita

ecological footprint (Figure 17.1, left). If everyone in the world used resources equal to what the average American uses, we would need about five planet Earths to support them (Figure 17.1, right), according to the World Wildlife Fund (WWF) and the Global Footprint Network. And the U.S. population growth rate is the highest of any industrialized country.

China has the world's largest population and the second largest economy. China's economy grew rapidly for three decades, beginning in the late 1970s, after the country shifted partially from a centrally planned economy to more of a market-based economy. As in the U.S. case, China's economic growth has resulted in severe environmental problems, which have contributed to its total ecological footprint—the second largest in the world (Figure 17.1, left). Also, China's consumer middle class, living

mostly in its large cities, has grown to roughly 300 million—a number almost equal to the U.S. population. Still, the large majority of Chinese citizens are poor and as a result, the per capita ecological footprint in China is about one-sixth that of the United States (Figure 17.1, left).

Since the 1960s, China has cut its birth rate in half and its population is now growing at a rate slower than that of the United States. However, if its middle class continues to grow and consume more resources as projected, China could have the world's largest per capita and total ecological footprints within a decade or two.

This chapter is about the economic, political, and ethical aspects of environmental problems and solutions. As you will see, all three aspects are illustrated in a comparison of economic growth in the United States and China, along with consideration of the future of such growth and what it will mean to global sustainability.

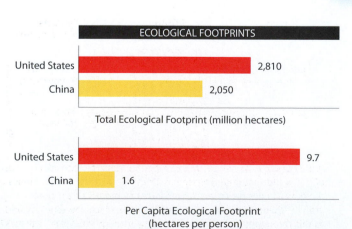

ECOLOGICAL FOOTPRINTS

United States	2,810
China	2,050

Total Ecological Footprint (million hectares)

United States	9.7
China	1.6

Per Capita Ecological Footprint (hectares per person)

FIGURE 17.1 Comparison of total and per capita ecological footprints of the United States and China. The image to the right shows the number of planet Earths that would be needed to indefinitely support the global population today (top), in 2030 (center), and today if everyone were to have the same ecological footprint as the average American now has (bottom).

© 2016 Cengage Learning

Natural Capital + Manufactured Capital + Human Capital = Goods and Services

© 2016 Cengage Learning

FIGURE 17.2 We use three types of resources to produce goods and services.

Center: Elena Elisseeva/Shutterstock.com. Right center: Michael Shake/Shutterstock.com. Right: Yuri Arcurs/Shutterstock.com.

17.1 HOW ARE ECONOMIC SYSTEMS RELATED TO THE BIOSPHERE?

CONCEPT 17.1 Ecological economists and most sustainability experts regard human economic systems as subsystems of the biosphere.

Economic Systems Depend on Natural Capital

Economics is the social science that deals with the production, distribution, and consumption of goods and services to satisfy people's needs and wants. In a market-based economic system, buyers and sellers interact competitively to make economic decisions about how goods and services are produced, distributed, and consumed. In a truly *free-market* economic system, all economic decisions are governed solely by the competitive interactions of *supply* (the amount of a good or service that is available), *demand* (the amount of a good or service that people want), and *price* (the market value of a good or service). If the demand for a good or service is greater than the supply, its price rises, and when supply exceeds demand, the price falls.

Also, in a truly *free-market* economic system, **(1)** no company or small group of companies could control the prices of any goods or services; **(2)** the market prices of goods and services would include all of their direct and indirect costs *(full-cost pricing)*; and **(3)** consumers would have full information about the beneficial and harmful environmental and health effects of the goods and services available to them. The economies of the world's countries are not truly free-market economies, because the combined actions of their governments and many of their corporations result in major violations of these three principles.

Most economic systems use three types of *capital*, or resources, to produce goods and services (Figure 17.2). **Natural capital** (see Figure 1.3, p. 7) includes resources and ecosystem services produced by the earth's natural processes, which support all life and all economies.

Human capital includes the physical and mental talents of the people who provide labor, organizational and management skills, and innovation. **Manufactured capital** includes the machinery, materials, and factories that people create using natural resources.

There Is Controversy over the Sustainability of Economic Growth

Economic growth is an increase in the capacity of a nation, state, city, or company to provide goods and services to people. Today, a typical industrialized country depends on a **high-throughput economy**, which attempts to boost economic growth by increasing the flow of natural matter and energy resources through the economic system to produce more goods and services (Figure 17.3). Such an economy produces valuable goods and services, but it also converts large quantities of high-quality matter and energy resources into waste, pollution, and low-quality heat, which tend to flow into planetary *sinks* (air, water, soil, and organisms).

Economic development is any set of efforts focused on creating economies that can serve to improve human well-being by meeting basic human needs for items such as food, shelter, physical and economic security, and good health. The world's countries vary greatly in their levels of

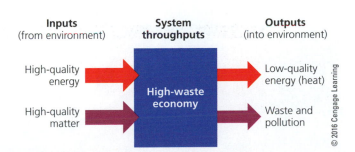

Inputs (from environment) — System throughputs — Outputs (into environment)

High-quality energy → **High-waste economy** → Low-quality energy (heat)

High-quality matter → Waste and pollution

© 2016 Cengage Learning

ANIMATED FIGURE 17.3 The *high-throughput economies* of most of the world's more-developed countries rely on continually increasing the flow of energy and matter resources to promote economic growth. *Question:* What are three ways in which you regularly add to this throughput of matter and energy through your daily activities?

economic growth and development. (See Figure 17.1 and Figure 3, p. S18, in Supplement 4 to compare the world's countries in terms of per capita income.)

For more than 200 years, there has been a debate over whether there are limits to economic growth. *Neoclassical economists,* such as the late Alfred Marshall (1842–1924) and the late Milton Friedman (1912–2006), assume that the potential for economic growth is essentially unlimited and is necessary for providing profits for businesses and jobs for workers. They also consider natural capital to be important but not indispensable because they believe we can find substitutes for essentially any resource that we might deplete or degrade.

Ecological economists such as Herman Daly (see his on-line Guest Essay on this topic) and Robert Costanza disagree with the neoclassical model. They point out that there are no substitutes for many vital natural resources, such as clean air, clean water, fertile soil, and biodiversity, or for crucial ecosystem services such as climate control, air and water purification, pest control, pollination, topsoil renewal, and nutrient cycling. In contrast to neoclassical economists, they view human economic systems as subsystems of the biosphere that depend heavily on the earth's irreplaceable natural resources and ecosystem services (Figure 17.4).

Closely related to the school of ecological economics is that of *environmental economics.* Economists in this school favor adjusting existing economic policies and tools to be more environmentally beneficial over inventing all-new policies and tools.

Now the debate among these economists is shifting to questions about what kinds of economic growth and development we should encourage. Economic growth does not necessarily mean more pollution and environmental degradation. It depends on the kinds of economic growth, the technologies involved, and whether nations pass and enforce laws that regulate inputs of harmful chemicals into the environment (see Table 1.1, p. 13).

Ecological and environmental economists call for recognizing the relationship between our economic systems and the earth's natural capital shown in Figure 17.4, and they contend that, without regulation of harmful inputs into the environment, conventional economic growth eventually will become unsustainable. Therefore, they argue, we should promote **environmentally sustainable economic development**, an approach that uses political

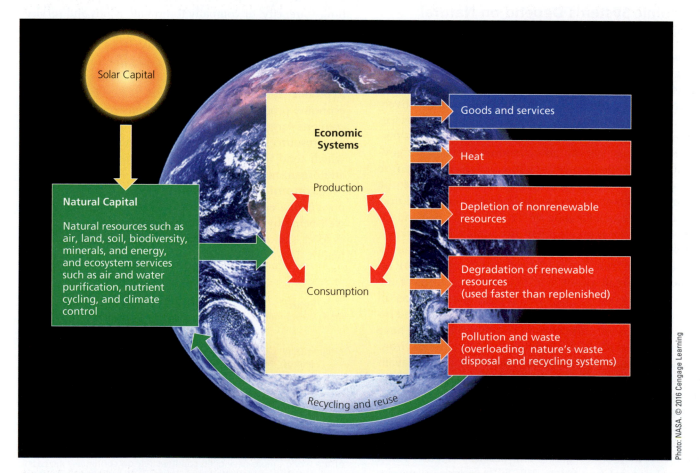

Photo: NASA. © 2016 Cengage Learning

ANIMATED FIGURE 17.4 *Ecological economists* see all human economies as subsystems of the biosphere that depend on natural resources and services provided by the sun and earth. ***Question:*** Do you agree or disagree with this model? Explain.

and economic systems to encourage environmentally beneficial and more sustainable forms of economic growth and to discourage environmentally harmful forms of economic growth that degrade natural capital.

CONCEPT 17.2 We can use resources more sustainably by including the harmful environmental and health costs of producing goods and services in their market prices *(full-cost pricing)*, by subsidizing environmentally beneficial goods and services, and by taxing pollution and waste instead of wages and profits.

We Can Apply the Principle of Full-Cost Pricing

The *market price,* or *direct price,* that we pay for a product or service usually does not include all of the *indirect,* or *external, costs* of harm to the environment and human health associated with its production and use. Such costs are often called *hidden costs.*

For example, if we buy a car, the price we pay includes the *direct,* or *internal, costs* of raw materials, labor, shipping, and a markup for dealer profit. In using the car, we pay additional direct costs for gasoline, maintenance, repairs, and insurance. However, in order to extract and process raw materials to make a car, manufacturers use energy and mineral resources, produce solid and hazardous wastes, disturb land, pollute the air and water, and release greenhouse gases into the atmosphere. These are the hidden external costs that can have short- or long-term harmful effects on us, on future generations, on our economies, and on the earth's life-support systems.

Because these harmful external costs are not included in the market price of a car, most people do not connect them with car ownership. Still, the car buyer and other people in a society pay these hidden costs sooner or later, in the forms of poorer health, higher expenses for health care and insurance, higher taxes for pollution control, and degradation of natural capital.

Many economists and environmental experts call for including these harmful external costs in the market prices of goods—a practice called **full-cost pricing**, and the basis for one of the **social science principles of sustainability**. They cite this failure to include the harmful environmental costs in the market prices of goods and services as one of the major causes of the environmental problems we face (see Figure 1.11, p. 14).

According to its proponents, full-cost pricing would reduce resource waste, pollution, and environmental degradation and improve human health by encouraging producers to invent more resource-efficient and less-polluting methods of production. It would also enable consumers to make more informed decisions about the goods and services they buy. For example, if the harmful environmental and health costs of mining and burning coal to produce electricity (Figure 17.5) were included in the market prices of coal and coal-fired electricity, coal would be much more expensive and likely would be replaced by less environmentally harmful resources such as natural gas and solar and wind power.

Implementation of full-cost pricing would result in some industries and businesses disappearing or remaking themselves, while new ones would appear—a normal and revitalizing process in a dynamic and creative capitalist economy. If we were to phase in a shift to full-cost pricing over a decade or two, some environmentally harmful businesses would have time to transform themselves into profitable, environmentally beneficial businesses.

Full-cost pricing seems to make a lot of sense. Why, then, is it not used more widely? *First,* most producers of harmful products and services would have to charge more for them, and some would go out of business. Naturally, these producers oppose such pricing. *Second,* it is difficult to estimate many environmental and health costs, and to know how they might change in the future. *Third,* many environmentally harmful businesses have used their political and economic power to obtain government **subsidies**, or payments and protections of various forms that help them to stay in business, and this helps them to avoid true free-market competition and to retain their economic advantage.

We Can Shift from Environmentally Harmful to Environmentally Beneficial Subsidies

Governments can use subsidies to encourage producers to work toward full-cost pricing. However, some subsidies enable businesses to operate in such a way that they do damage to the environment or to human health. According to environmental scientist Norman Myers, such *perverse subsidies* cost the world's governments (taxpayers) at least $2 trillion a year—an average $3.8 million a minute! Myers also estimates that perverse subsidies cost the average American taxpayer $2,000 per year. (See Myers' online Guest Essay on this topic.)

Perverse subsidies can distort the economic playing field and create a huge economic incentive for resource depletion and environmental degradation. Examples include depletion subsidies and tax breaks for extracting minerals and fossil fuels, cutting timber on public lands, irrigating with low-cost water, and overfishing commercially valuable aquatic species.

A number of environmental scientists and economists call for phasing out environmentally perverse subsidies and tax breaks. They also call for phasing in subsidies and

The price of coal does not include the harmful environmental costs of producing and burning it.

Andreas Reinhold/Shutterstock

FIGURE 17.5 Most of the harmful environmental effects of strip-mining coal and burning it to produce electricity are not included in the cost of electricity.

tax breaks for environmentally beneficial businesses such as those involved in pollution prevention, waste prevention, sustainable forestry and agriculture, conservation of water supplies, energy efficiency improvements, renewable energy use, and measures to slow projected climate change. In other words, over the next few decades we could make a *subsidy shift* from environmentally harmful to environmentally beneficial subsidies, thereby increasing our beneficial environmental impact.

Making such a shift would not be easy because the powerful interests that receive perverse subsidies spend a lot of time and money *lobbying,* or trying to influence governments to continue their subsidies. They also work to deny subsidies to their more environmentally beneficial competitors. But countries including Japan, France, and Belgium have made such shifts.

CONSIDER THIS. . .

THINKING ABOUT Subsidies

Do you favor phasing out environmentally harmful government subsidies and tax breaks, and phasing in environmentally beneficial ones? Explain. What are three things you could you do to help bring this about? How might such subsidy shifting affect your lifestyle?

Environmental Indicators Can Help Us Live More Sustainably

Economic growth is usually measured by the percentage of change per year in a country's **gross domestic product (GDP)**: the annual market value of all goods and services produced by all firms and organizations, foreign and domestic, operating within a country. A country's economic growth per person is measured by changes in the **per capita GDP**: the GDP divided by the country's total population at midyear.

GDP and per capita GDP indicators provide a standardized, useful method for measuring and comparing the economic outputs of nations. However, the GDP was deliberately designed to measure such outputs without taking into account their beneficial or harmful environmental impacts. Environmental and ecological economists and environmental scientists call for the development and widespread use of new indicators—called *environmental indicators*—to help monitor environmental quality and human well-being.

One such indicator is the **genuine progress indicator (GPI)**—the GDP plus the estimated value of beneficial transactions that meet basic needs, minus the estimated harmful environmental, health, and social costs of all transactions. Examples of beneficial transactions included in the GPI are unpaid volunteer work, health care provided by family members, child care, and housework. Harmful costs that are subtracted to arrive at the GPI include the costs of pollution, resource depletion and degradation, and crime. The GPI was developed by environmental economists with the hope that governments would adopt it. However, it has not yet been implemented by any of the world's economies.

Another set of environmental indicators, developed by the United Nations, compares the world's countries in terms of environmental quality. It uses various measures, including CO$_2$ emissions, marine and terrestrial protected areas, forest cover, water supplies, and municipal waste collection. Figure 17.6 shows one example of the sort of data used for these indicators.

Researchers led by Kelly Cain at the St. Croix Institute for Sustainable Community Development have developed a computer model that makes use of sustainable community development indicators. For example, the model estimates the amount of money and other resources that leave any community that imports most of its food, usually through large grocery stores. Cain argues that such a community can save large amounts of money and shrink its ecological footprint by learning how to produce much more of its own food and energy.

These and other environmental indicators now being developed are far from perfect. However, without such indicators, it will be impossible to monitor the overall effects of human activities on human health, on the environment, and on the planet's natural capital. Such indicators are also helpful for finding the best ways to improve environmental quality and life satisfaction.

We Can Tax Pollution and Wastes Instead of Wages and Profits

Another way to discourage pollution and resource waste is to tax it (**Concept 17.2**). *Green taxes* could be levied on a per-unit basis on the amount of pollution and hazardous waste produced by a farm, business, or industry, and on the use of fossil fuels, nitrogen fertilizer, timber, minerals, water, and other resources. This approach would help us to implement the full-cost pricing **principle of sustainability** and increase our beneficial environmental impact.

To many analysts, the tax systems in most countries are backward. They *discourage* what we want more of— jobs, income, and profit-driven innovation—and *encourage* what we want less of—pollution, resource waste, and environmental degradation. A more environmentally sustainable economic and political system would *lower* taxes on labor, income, and wealth, and *raise* taxes on environmental activities that produce pollution, wastes, and environmental degradation. Some 2,500 economists, including eight Nobel Prize winners in economics, have endorsed this *tax-shifting* concept.

Proponents point out three requirements for the successful implementation of green taxes. *First,* they would have to be phased in over 10–20 years to allow businesses to plan for the future. *Second,* income, payroll, or other taxes would have to be reduced by an amount equal to that of the green tax so that there would be no net increase in taxes. *Third,* the poor and lower-middle class would need a safety net to reduce the regressive nature of any new taxes on essentials such as fuel, water, electricity, and food. Figure 17.7 lists some of the advantages and disadvantages of using green taxes.

In Europe and the United States, polls indicate that once such tax shifting is explained to voters, 70% of them support the idea. Germany's green tax on fossil fuels, introduced in 1999, has reduced pollution and greenhouse gas emissions, helped to create up to 250,000 new jobs, lowered taxes on wages, and greatly increased the use of renewable energy resources. Costa Rica, Sweden, Denmark, Spain, and the Netherlands have raised taxes on several environmentally harmful activities while cutting taxes on income, wages, or both.

The U.S. Congress has not enacted green taxes, mostly because economically and politically powerful industries, including the automobile, fossil fuel, mining, and chemical industries, claim that such taxes will reduce their competitiveness and harm the economy and consumers by forcing producers to raise the prices of their goods and services. In addition, most voters have been conditioned to oppose any new taxes and have not been educated about the economic and environmental benefits of a tax-shifting

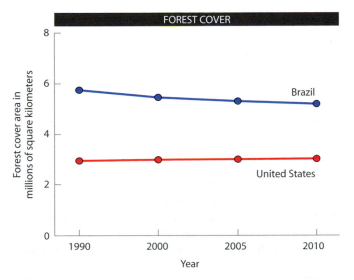

FIGURE 17.6 *Monitoring environmental progress:* Comparison of the United States and Brazil in terms of changes in forest cover. The UN's environmental indicators are based on several measures of environmental quality, including growth or shrinkage of forest cover.

Compiled by the authors using data from United Nations Statistics Division.

Trade-Offs

Environmental Taxes and Fees

Advantages	Disadvantages
Help bring about full-cost pricing	Low-income groups are penalized unless safety nets are provided
Encourage businesses to develop environmentally beneficial technologies and goods	Hard to determine optimal level for taxes and fees
Easily administered by existing tax agencies	If set too low, wealthy polluters can absorb taxes as costs

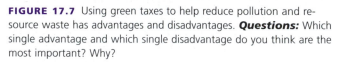

© 2016 Cengage Learning

FIGURE 17.7 Using green taxes to help reduce pollution and resource waste has advantages and disadvantages. *Questions:* Which single advantage and which single disadvantage do you think are the most important? Why?

Top: chuong/Shutterstock.com. Bottom: EduardSV/Shutterstock.com.

approach that would improve environmental quality with no net increase in taxes.

In 2014, the Sustainable Public Procurement Programme was launched. The goal of this global program, sponsored by a number of organizations including the U.N. Environment Programme (UNEP), is to encourage and assist governments in devoting more of their spending to purchases of goods and services that are sustainably produced and provided.

Environmental Laws and Regulations Can Discourage or Encourage Innovation

Environmental regulation is a form of government intervention in the marketplace that is widely used to help control or prevent pollution and environmental degradation and to encourage more efficient use of resources. It involves enacting and enforcing laws that set pollution standards, regulate the release of toxic chemicals into the environment, and protect certain slowly replenished resources such as public forests, parks, and wilderness areas from unsustainable use.

Such regulation is another way to help implement the full-cost pricing **principle of sustainability**, because it forces companies to include more of the costs of pollution control and other regulated aspects in the prices of their products. For this reason, opponents of regulation claim that it hurts business, especially in markets where

they are competing with China and other countries whose regulations are not as strong as those of the United States (**Core Case Study**).

However, proponents of regulation point to the results of China's lax environmental regulations. While that country's economy has been growing rapidly since 1980, its environmental problems have also multiplied dramatically. Now, according to the Chinese Academy of Sciences, its major cities suffer from serious air pollution; about 57% of its urban groundwater, used for drinking water for hundreds of millions of people, and 43% of its surface water is too polluted to use; its topsoil is severely polluted; and some of its food is tainted with harmful chemicals. These problems are leading to civil unrest in China as well as to a less favorable standing in the global marketplace.

So far, most environmental regulation in the United States and in many other countries has involved passing laws that are typically enforced through a *command-and-control* approach. Critics say that this strategy can unnecessarily increase costs and discourage innovation, because many of these government regulations concentrate on cleanup instead of prevention. Some regulations also set compliance deadlines that are often too short to allow companies to find innovative solutions to pollution and waste.

A different approach favored by many economists, as well as environmental and business leaders, is to use *incentive-based environmental regulations*. Rather than to require all companies in a particular market to follow the same fixed procedures or use the same technologies, this approach uses the economic forces of the marketplace to encourage businesses to be innovative in reducing pollution and resource waste.

Several European nations use such *innovation-friendly environmental regulation,* which involves setting goals, establishing heavy penalties for not meeting the goals, freeing industries to meet the goals in any way that works, and allowing enough time for innovation. This has motivated several companies to develop green products and industrial processes that have created jobs. It has also helped some companies to boost their profits while becoming more competitive in national and international markets.

GOOD NEWS

We Can Use the Marketplace to Reduce Pollution and Resource Waste

In one incentive-based regulation system, the government decides on acceptable levels of total pollution or resource use, sets limits, or *caps,* to maintain these levels, and gives or sells companies a certain number of *tradable pollution* or *resource-use permits* governed by the caps.

With this *cap-and-trade* approach, a permit holder that does not use its entire allocation can save credits for future expansion, use them in other parts of its operation, or sell them to other companies. The United States has used this approach to reduce the emissions of sulfur dioxide (see

Chapter 15, p. 400) and several other air pollutants. Tradable rights could also be established among countries to help preserve biodiversity and reduce emissions of greenhouse gases and other regional and global pollutants.

Figure 17.8 lists the advantages and disadvantages of using tradable pollution and resource-use permits. The effectiveness of such programs depends on how high or low the initial cap is set and on the rate at which the cap is regularly reduced to encourage further innovation.

We Can Reduce Pollution and Resource Waste by Selling Services Instead of Things

One approach to working toward more environmentally beneficial economies is to sell certain services in place of the products that provide those services. With this approach, a manufacturer or service provider makes more money if the production of its product involves minimal material use and pollution, and if the product lasts, is energy efficient, produces as little pollution as possible while in use, and is easy to maintain, repair, reuse, or recycle.

Such an economic shift is under way in some businesses. Since 1992, Xerox has been leasing most of its copy machines as part of its mission to provide *document services* instead of selling photocopiers. When a customer's service contract expires, Xerox takes the machine back for reuse or remanufacture. It has a goal of sending no material to landfills or incinerators. To save money, Xerox designs machines to have the fewest possible parts, be energy ef-

ficient, and emit as little noise, heat, ozone, and chemical waste as possible.

In Europe, Carrier has begun shifting from selling heating and air conditioning equipment to providing indoor heating and cooling services. The company makes higher profits by leasing and installing energy-efficient equipment that is durable and easy to rebuild or recycle. Carrier also makes money through helping clients to save energy by adding insulation, eliminating heat losses, and boosting energy efficiency in their offices and homes.

Reducing Poverty Would Help Us to Deal with Environmental Problems

Poverty is defined as the condition under which people cannot meet their basic economic needs. According to the World Bank, poverty is the way of life for nearly half of the world's people who have to live on incomes equivalent to less than $2.25 per day. One fifth of the world's people live in extreme poverty (Figure 17.9), struggling to survive on incomes of less than $1.25 a day or on no income at all.

Some analysts are alarmed at the widening gap between rich and poor countries and between super-rich individuals and the rest of the world. In 2014, Oxfam reported that the richest 85 people in the world had as much wealth as the poorest 3.5 billion people—half the world's

James P. Blair/National Geographic Creative

FIGURE 17.9 This 3-year-old girl was sleeping in her family's shack in a slum in Port-Au-Prince, Haiti.

population. Some economists say that part of this wealth will trickle down to the poor and middle class. Others point out that for almost three decades, instead of trickling down, most wealth has been flowing up to rich individuals, corporations, and countries.

Poverty can have severely harmful health effects (see Figures 14.14 and 14.15, p. 378) and has been identified as one of the five major causes of the environmental problems we face. To reduce poverty and its harmful effects, governments, businesses, international lending agencies, and wealthy individuals could undertake the following:

- Mount a massive global effort to combat malnutrition and the infectious diseases that kill millions of people prematurely (Figure 14.15, p. 378).

- Provide universal primary school education for all children and for the world's nearly 800 million illiterate adults.

- Provide assistance to help less-developed countries reduce their population growth, mostly by investing in family planning, reducing poverty, and elevating the social and economic status of women.

- Focus on sharply reducing the total and per capita ecological footprints of more-developed countries such as the United States and rapidly growing less-developed countries such as China (**Core Case Study**) and India.

- Make large investments in small-scale infrastructure such as solar-cell power facilities for rural villages and sustainable agriculture projects to help less-developed nations work toward more energy-efficient and environmentally beneficial economies.

- Encourage lending agencies to make small loans to poor people who want to increase their income (see Case Study that follows).

Ecologist and National Geographic Emerging Explorer Sasha Kramer has been working in the impoverished and ecologically degraded nation of Haiti to attack the problems of hunger, soil depletion, and water pollution all at once. Her nonprofit organization has distributed waterless composting toilets throughout the country to collect human wastes and transform them into compost, which Haitian farmers can use to rebuild depleted soil and boost food production. This process also keeps human wastes out of Haiti's water supply and reduces the dangerous threat of waterborne infectious diseases.

CASE STUDY

Microlending

Most of the world's poor people want to work and earn enough to climb out of poverty and make a better life for themselves and their families. But few of them have credit records or assets that they could use as collateral to secure the loans that they could use to buy what they would need to start farming or to start small businesses.

For over three decades, an innovation called *microlending*, or *microfinance*, has helped a number of people living in poverty to deal with this problem. In 1983, economist Muhammad Yunus started the Grameen (Village) Bank in Bangladesh, a country with a high poverty rate and a rapidly growing population. Unlike commercial banks, the Grameen Bank is essentially owned and run by borrowers and by the Bangladeshi government. Since it was founded, the bank has provided more than $8 billion in microloans of $50 to $500 at low interest rates to 7.6 million impoverished people in Bangladesh who do not qualify for loans at traditional banks.

About 97% of these loans have been used by women, mostly to start small businesses, to plant crops, to buy small irrigation pumps, to buy cows and chickens for producing and selling milk and eggs, or to buy bicycles for transportation. Grameen Bank microloans are also being used to develop day-care centers, health-care clinics, reforestation projects, drinking water supply projects, literacy programs, and small-scale solar- and wind-power systems in rural villages.

To promote loan repayment, the bank puts borrowers into groups of five. If a group member fails to make a weekly payment, other members must pay it. The average repayment rate on its microloans has been 95% or higher—nearly twice the average repayment rate for loans by conventional commercial banks—and the bank consistently has made a profit. Typically, about half of Grameen's borrowers move above the poverty line within 5 years of receiving their loans. In addition, birth rates are lower among most of the borrowers, a majority of whom are women.

In 2006, Yunus and his colleagues at the bank jointly won the Nobel Peace Prize for their pioneering use of microcredit loans that change people's lives. He has said that "Unleashing the energy and creativity in each human being is the answer to poverty." Banks based on the Grameen microcredit model have spread to 58 countries (including the United States) with an estimated 500 million participants.

We Can Work Toward Achieving the Millennium Development Goals

In 2000, the world's nations set goals—called *Millennium Development Goals*—for sharply reducing hunger and poverty, improving health care, achieving universal primary education, empowering women, and moving toward environmental sustainability by 2015. More-developed countries pledged to donate 0.7%—or $7 of every $1,000—of their annual national income to less-developed countries to help them in achieving these goals. But so far, only five countries—Denmark, Luxembourg, Sweden, Norway, and the Netherlands—have donated what they had promised.

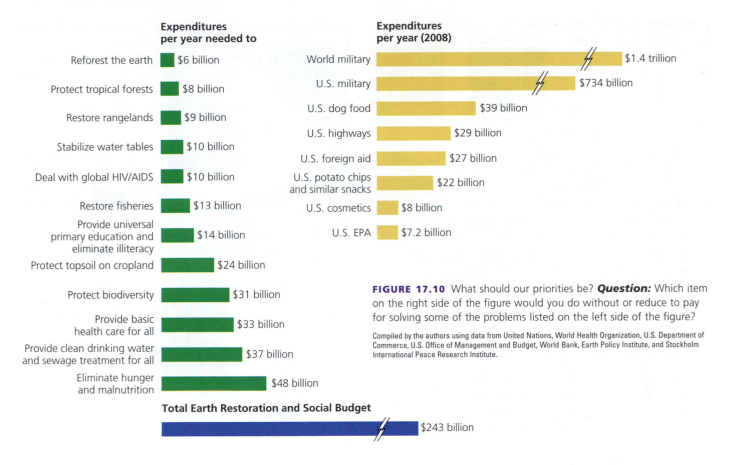

Expenditures per year needed to		Expenditures per year (2008)	
Reforest the earth	$6 billion	World military	$1.4 trillion
Protect tropical forests	$8 billion	U.S. military	$734 billion
Restore rangelands	$9 billion	U.S. dog food	$39 billion
Stabilize water tables	$10 billion	U.S. highways	$29 billion
Deal with global HIV/AIDS	$10 billion	U.S. foreign aid	$27 billion
Restore fisheries	$13 billion	U.S. potato chips and similar snacks	$22 billion
Provide universal primary education and eliminate illiteracy	$14 billion	U.S. cosmetics	$8 billion
Protect topsoil on cropland	$24 billion	U.S. EPA	$7.2 billion
Protect biodiversity	$31 billion		
Provide basic health care for all	$33 billion		
Provide clean drinking water and sewage treatment for all	$37 billion		
Eliminate hunger and malnutrition	$48 billion		

Total Earth Restoration and Social Budget $243 billion

FIGURE 17.10 What should our priorities be? *Question:* Which item on the right side of the figure would you do without or reduce to pay for solving some of the problems listed on the left side of the figure?

Compiled by the authors using data from United Nations, World Health Organization, U.S. Department of Commerce, U.S. Office of Management and Budget, World Bank, Earth Policy Institute, and Stockholm International Peace Research Institute.

In fact, the average amount donated in most years has been 0.25% of national income. The United States—the world's richest country—gives only 0.16% of its national income to help poor countries; Japan, another wealthy country, gives only 0.18% compared with the 0.9% given by Sweden. For any country, deciding whether or not to commit 0.7% of annual national income toward the Millennium Development Goals is an ethical issue that requires individuals and nations to evaluate their priorities (Figure 17.10). The United States and China could set an example by increasing their nonmilitary aid to poor countries (**Core Case Study**).

CONSIDER THIS. . .

THINKING ABOUT The Millennium Development Goals

Do you think the country where you live should devote at least 0.7% of its annual national income toward achieving the Millennium Development Goals? Explain.

We Can Use Lessons from Nature to Shift to More Environmentally Sustainable Economies

The three scientific laws governing matter and energy changes (see Chapter 2, pp. 33 and 35–37) and the six **principles of sustainability** (see Supplement 7, pp. S50–S51) suggest that the best long-term solution to our environmental and resource problems is to shift away from a high-throughput (high-waste) economy based on ever-increasing matter and energy flow. Our goal would be to develop a **low-throughput (low-waste) economy** that would work with nature to reduce excessive throughput, inefficient use of matter and energy resources, and the resulting pollution and wastes. Another goal could be to reuse, recycle, and compost most of our solid wastes (Figure 17.11).

The drive to improve environmental quality and to work toward environmental sustainability has created new major growth industries along with profits and large numbers of new *green jobs* (Figure 17.12). While the United States was once a leader in the green economy, other countries including Germany and China (**Core Case Study**) have taken the lead in some areas, such as wind and solar energy development. However, according to Ethan Pollack of the Economic Policy Institute, by 2010, there were more than 3.1 million green jobs in the United States and that number has been rising.

Making the shift to more sustainable economies will require governments and industries to greatly increase their spending on research and development—especially in the areas of energy efficiency and renewable energy—as Germany has done in recent years. The shift toward sustainability will also require business leaders to understand why such a shift is important ecologically and economically (Individuals Matter 17.1).

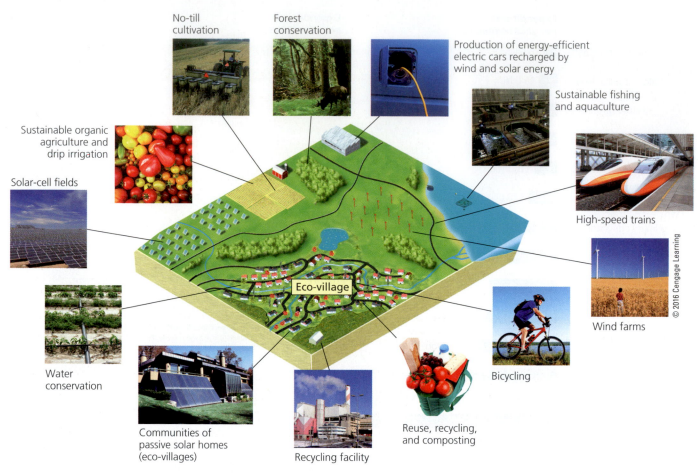

No-till cultivation

Forest conservation

Production of energy-efficient electric cars recharged by wind and solar energy

Sustainable fishing and aquaculture

Sustainable organic agriculture and drip irrigation

Solar-cell fields

High-speed trains

Eco-village

Wind farms

Water conservation

Bicycling

Communities of passive solar homes (eco-villages)

Recycling facility

Reuse, recycling, and composting

© 2016 Cengage Learning

FIGURE 17.11 Solutions: Some of the components of more environmentally sustainable economic development. **Question:** What are three new types of jobs that could be generated by such an economy?

Photos going clockwise starting at "No-till cultivation": Jeff Vanuga/National Resource Conservation Service. Natalia Bratslavsky/Shutterstock. Pi-Lens/Shutterstock.com. DAVID DOUBILET/National Geographic Creative. hxdbzxy/Shutterstock.com. Varina and Jay Patel/Shutterstock.com. Kalmatsuy Tatyana/Shutterstock.com. Brenda Carson/Shutterstock.com. Alexander Chaikin/Shutterstock.com. National Renewable Energy Laboratory. Anhong/Dreamstime. pedrosala/Shutterstock.com. Robert Kneschke/Shutterstock.com.

Environmentally Sustainable Businesses and Careers

Aquaculture	Environmental design and architecture	 	Environmental law
Biodiversity protection			Environmental nanotechnology
Biofuels	Environmental economics		Fuel-cell technology
Climate change research	Environmental education		Geographic information systems (GIS)
Conservation biology	Environmental engineering		Geothermal geologist
Ecotourism management	Environmental entrepreneur		Hydrogen energy
Energy-efficient product design	Environmental health		Hydrologist
Environmental chemistry	Environmental writer		Marine science
			Pollution prevention
			Reuse and recycling

Selling services in place of products

Solar-cell technology

Sustainable agriculture

Sustainable forestry

Urban gardening

Urban planning

Waste reduction

Watershed hydrologist

Water conservation

Wind energy

© 2016 Cengage Learning

FIGURE 17.12 *Green careers:* Some key environmental businesses and careers are expected to flourish during this century, while environmentally harmful, or sunset, businesses are expected to decline. See the website for this book for more information on various environmental careers. **Question:** How could some of these careers help you to apply any of the **principles of sustainability**?

Top: Goodluz/Shutterstock.com. Second from top: Goodluz/Shutterstock.com. Second from bottom: Dusit/Shutterstock.com. Bottom: Corepics VOF/Shutterstock.com.

Ray Anderson

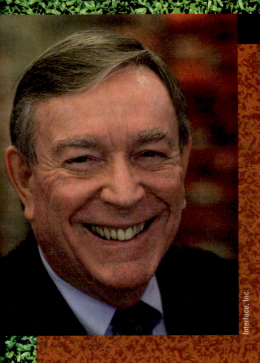

Ray Anderson (1934–2011), founder of the American company Interface, was one of the world's most respected and effective leaders in the movement to make businesses more sustainable. The company is the world's largest commercial manufacturer of carpet tiles, with 25 factories in 7 countries, and customers in 110 countries.

Anderson changed the way he viewed the world—and his business—after reading Paul Hawken's book *The Ecology of Commerce*. In 1994, he announced plans to develop the nation's first totally sustainable corporation. Within 16 years, Interface had cut its water usage by 74%, its net greenhouse gas emissions by 32%, its solid waste by 63%, its fossil fuel use by 60%, and its energy use by 44%. The company now gets 31% of its total energy from renewable resources. These efforts have saved Interface more than $433 million.

Anderson also sent his carpet design team into the forest and told them to learn how nature would design floor covering. The team observed that there were no regularly repeating patterns on the forest floor and that instead, there was disorder and diversity. They had discovered nature's biodiversity principle of sustainability. With this in mind, the design team created a line of carpet tiles, no two of which had the same design. Within 18 months after it was introduced, this new product line was the company's top-selling design.

Interface also applies nature's chemical cycling principle of sustainability by making its carpet tiles from recycled fibers. And the carpet tile factory runs partly on solar energy, thus applying another of nature's sustainability principles. In addition, the company invented a new carpet recycling process that does not emit carbon dioxide to the atmosphere.

Under Anderson's leadership, Interface became the world's largest seller of carpet tiles and profits tripled. Anderson also created a new consulting group as part of Interface to help other businesses start on the path toward becoming more sustainable. He was an exemplary leader in the practice of upcycling—working not only to reduce his company's ecological footprint but to create a large and expanding positive environmental impact.

17.3 HOW CAN WE IMPLEMENT MORE SUSTAINABLE AND JUST ENVIRONMENTAL POLICIES?

CONCEPT 17.3 Individuals can work together to take part in political processes that influence how environmental policies are made and implemented.

Developing Environmental Policy Is Not Easy

The roles played by a government are determined largely by its **policies**—the set of laws and regulations it enforces and the programs it funds. **Politics** is the process by which individuals and groups try to influence or control the policies and actions of governments at local, state, national, and international levels. One important application of this process is the development of **environmental policy**—environmental laws, regulations, and programs that are designed, implemented, and enforced by one or more government agencies.

Here, we focus primarily on how policies are developed within a **representative democracy**—a government run by the people through elected officials and representatives. This form of government has often been compromised by groups or organizations that gain enough wealth and power to have more influence over government policies than the average citizen can have. However, the ideals for representative democracy are usually embodied in a document called a *constitution,* which provides the basis of government authority and, in most cases, limits government power by mandating free elections and guaranteeing the right of free speech. Another name for such a government is *constitutional democracy.*

Political institutions in most constitutional democracies are designed to allow gradual change that ensures economic and political stability. In the United States, for example, rapid and destabilizing change is curbed by a system of checks and balances that distributes power among three branches of government—*legislative, executive,* and *judicial*—and among federal, state, and local governments.

The major function of government in democratic countries is to develop and implement policies for dealing with various issues. A typical *policy life cycle* consists of four stages:

- *Problem recognition.* A problem is identified by members of the public or by a lawmaker or executive.

- *Policy formulation.* A cause or causes of the problem are identified and a solution such as a law or program to help deal with the problem is proposed and developed.

- *Policy implementation.* A law is passed to put the solution into effect. A legislative body then provides funds for implementing and enforcing the new law, and the appropriate government department or agency must draw up regulations or rules for implementing it.

- *Policy adjustment.* The new program is monitored, evaluated, and adjusted as necessary.

In passing laws, developing budgets, and formulating regulations, elected and appointed government officials must deal with pressure from many competing *special-interest groups.* Each of these groups advocates for the passage of certain laws, for receipt of subsidies or tax breaks, and for regulations favorable to its cause. Such groups also seek to weaken or repeal laws, subsidies, tax breaks, and regulations that are unfavorable to their positions. Important examples of special-interest groups include *profit-making organizations,* such as corporations; *nongovernmental organizations (NGOs),* most of which are nonprofit organizations such as environmental groups; *labor unions,* representing the interests of workers; and *trade associations,* representing various industries.

As they have evolved, most democratic governments have taken on features that hinder their ability to deal with environmental problems (see the Case Study that follows). Such problems are usually invisible or difficult to understand, as well as intertwined in complex ways, and they can have long-lasting effects and require integrated, long-term solutions. However, most politicians spend much of their time seeking reelection and tend to focus on short-term, isolated issues rather than on long-term, complex problems. Also, political leaders, with hundreds of issues to deal with, usually have little or no understanding of how the earth's natural systems work and how those systems support all life, economies, and societies.

By comparison, even though it has shifted partially to a market-based economy, China has a far more centralized, less democratic government, without the checks and balances of Western democracies. Change there can occur more quickly, as policies are developed by the central government with little public input. This has helped the Chinese economy to grow rapidly, but it has also led to major environmental problems (**Core Case Study**).

Because of its strong central government, public protest in China has been limited. However, as China's rapidly expanding middle class gains more economic power, it is putting pressure on the government to do something about increasingly intolerable environmental conditions. To reduce the growing threat of civil unrest in this rapidly urbanizing country, the government will have to put much greater emphasis on dealing with its serious environmental problems, as the United States did in the 1970s.

CASE STUDY

Managing Public Lands in the United States—Politics in Action

No nation has set aside as much of its land for public use, resource extraction, enjoyment, or wildlife habitat as has the United States. The federal government manages roughly 35% of the country's land, which is jointly owned by all U.S. citizens. About three-fourths of this federal public land is in Alaska and another fifth is in the western states (Figure 17.13).

Some federal public lands are used for many purposes. For example, the *National Forest System* consists of 155 national forests and 22 national grasslands. These lands, managed by the U.S. Forest Service (USFS), are used for logging, mining, livestock grazing, farming, oil and gas extraction, recreation, and conservation of watershed, soil, and wildlife resources.

The Bureau of Land Management (BLM) manages large areas of land—40% of all land managed by the federal government and 13% of the total U.S. land surface—mostly in the western states and Alaska. These lands are used primarily for mining, oil and gas extraction, and livestock grazing.

The U.S. Fish and Wildlife Service (USFWS) manages 560 *National Wildlife Refuges.* Most refuges protect habitats and breeding areas for waterfowl and big game to provide a harvestable supply for hunters. Permitted activities in most refuges include hunting, trapping, fishing, oil and gas development, mining, logging, grazing, farming, and some military activities.

The uses of some other public lands are more restricted. The *National Park System,* managed by the National Park Service (NPS), includes 59 major parks (Figure 17.14) and 342 national recreation areas, monuments, memorials, battlefields, historic sites, parkways, trails, rivers, seashores, and lakeshores. Only camping, hiking, sport fishing, and boating can take place in the national parks, whereas sport hunting, mining, and oil and gas drilling are allowed in national recreation areas.

The most restricted public lands are 756 roadless areas that make up the *National Wilderness Preservation System.* These areas lie within the other public lands and are

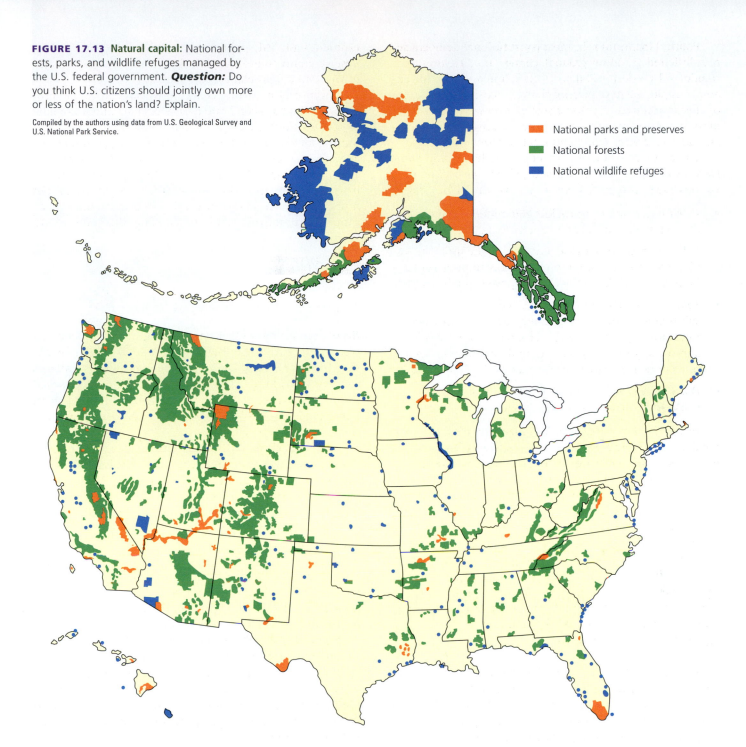

FIGURE 17.13 Natural capital: National forests, parks, and wildlife refuges managed by the U.S. federal government. **Question:** Do you think U.S. citizens should jointly own more or less of the nation's land? Explain.

Compiled by the authors using data from U.S. Geological Survey and U.S. National Park Service.

- ■ National parks and preserves
- ■ National forests
- ■ National wildlife refuges

managed by the agencies in charge of those surrounding lands. Most of these areas are open only for recreational activities such as hiking, sport fishing, camping, and non-motorized boating.

Many federal public lands contain valuable oil, natural gas, coal, geothermal, timber, and mineral resources. Since the 1800s, there has been intense controversy over how to use and manage the resources on these public lands.

Most conservation biologists, environmental economists, and many free-market economists believe that four principles should govern use of public lands:

1. They should be used primarily for protecting biodiversity, wildlife habitats, and ecosystems as something like a national eco-insurance policy.

2. No one should receive government subsidies or tax breaks for using or extracting resources on public lands.

3. The American people deserve fair compensation for the use of their property.

4. All users or extractors of resources on public lands should be fully responsible for any environmental damage they cause.

FIGURE 17.14 Gibbon River in Yellowstone National Park, Wyoming.

There is strong and effective opposition to these ideas. Developers, resource extractors, many economists, and many citizens tend to view public lands in terms of their usefulness in providing mineral, timber, and other resources and increasing short-term economic growth. They have succeeded in blocking implementation of the four principles listed above. For example, in recent years, analyses of budgets and appropriations reveal that the government has spent an average of $1 billion a year—an average of $2.7 million a day—on subsidies and tax breaks for privately owned interests that use U.S. public lands for activities such as mining, fossil fuel extraction, logging, and livestock grazing.

Some developers and resource extractors have sought to go further in opening up more federal lands for economic development. Here are five of the ideas that such interests have proposed to Congress:

1. Sell public lands or their resources to corporations or individuals, usually at prices that are less than market value, or turn over their management to state and local governments.

2. Slash federal funding for the administration of regulations related to public lands.

3. Cut diverse old-growth forests in the national forestlands for timber and for making biofuels, and replace them with tree plantations to be harvested for the same purposes.

4. Open national parks, national wildlife refuges, and wilderness areas to oil drilling, mining, off-road vehicles, and commercial development.

5. Eliminate or take regulatory control away from the National Park Service and launch a 20-year construction program in the parks to build new concessions and theme parks that would be run by private firms.

CONSIDER THIS. . .

THINKING ABOUT U.S. Public Lands

Explain why you would support or oppose each of the five proposals for changing the use and management of U.S. public lands, listed above.

Environmental Justice Is an Important Priority for Policy Makers

Environmental justice is an ideal whereby every person is entitled to protection from environmental hazards regardless of race, gender, age, national origin, income, social class, or any political factor. (See the online Guest Essay on this subject by Robert Bullard, one of the pioneers of the environmental justice movement in the United States.)

Studies have shown that a lopsided share of polluting factories, hazardous waste dumps, incinerators, and landfills in the United States are located in communities populated mostly by African Americans, Asian Americans, Latinos, and Native Americans. Other research has shown that, in general, toxic waste sites in white communities have been cleaned up faster and more completely than similar sites in African American and Latino communities have. In China, this problem is worse (Core Case Study). Because of intolerable pollution in urban areas, many factories in China are being moved to the countryside, and rates of cancer and other serious human illnesses are rising sharply in these rural areas.

Such environmental discrimination in many parts of the world has led to a growing grassroots effort known as the *environmental justice movement*. Supporters of this movement have pressured governments, businesses, and environmental organizations to become aware of environmental injustice and to act to prevent it. They have made some progress toward their goals, but have a long way to go.

Some politicians and business representatives suggest that economics should be the main factor in decisions about where to locate power plants, freeways, landfills, incinerators, and other such potentially disruptive facilities, arguing that they can cut costs by locating them in less desirable areas. Often, however, such areas are home to low-income residents who have much less political power than do developers and corporations. Many analysts argue that an ethical principle of environmental justice should carry as much weight as economic factors do in such decisions. This is a political struggle that is still unresolved in many areas of the world.

CONSIDER THIS. . .

THINKING ABOUT Environmental Justice

Do you think that the principles of environmental justice should get equal weight, more weight, or less weight in political decisions about where to locate potentially environmentally harmful facilities such as incinerators? Explain.

Certain Principles Can Guide Us in Making Environmental Policy

Analysts suggest that when evaluating existing or proposed environmental policies, legislators and individuals should be guided by seven principles designed to reduce environmental harm:

- *The reversibility principle:* Avoid making decisions that cannot be reversed later if they turn out to be harmful. For example, two essentially irreversible actions are the production of indestructible, toxic coal ash in coal-burning power plants, and the production of deadly radioactive wastes throughout the nuclear power fuel cycle. In both cases, the hazardous wastes must be stored safely for thousands of years.

- *The net energy principle:* Avoid the widespread use of energy resources and technologies with low or

negative net energy yields (see Figure 13.2, p. 315), which cannot compete in the open marketplace without government subsidies and tax breaks. Examples of such energy alternatives include nuclear power (considering the whole fuel cycle), tar sands, shale oil, ethanol made from corn, and hydrogen fuel, as discussed in Chapter 13.

- *The precautionary principle:* When substantial evidence indicates that an activity threatens human health or the environment, take precautionary measures to prevent or reduce such harm, even if some of the cause-and-effect relationships are not well established, scientifically.

- *The prevention principle:* Make decisions that help to prevent a problem from occurring or becoming worse.

- *The polluter-pays principle:* Develop regulations and use economic tools such as green taxes to ensure that polluters bear the costs of dealing with the pollutants and wastes they produce in accordance with the full-cost pricing **principle of sustainability**. This also stimulates the development of innovative ways to reduce and prevent pollution and wastes.

- *The environmental justice principle:* In implementing environmental policy, no group of people should bear an unfair share of the burden created by pollution, environmental degradation, or the execution of environmental laws. This ethical principle also addresses environmental injustices committed by one generation and affecting future generations, and thus it echoes the future generations **principle of sustainability**.

- *The holistic principle:* Recognize that the environmental and other problems we face are connected. Thus, we need to focus on long-term solutions that address root causes of such interconnected problems instead of focusing on short-term and often ineffective fixes that treat each problem separately.

Implementing such principles is not easy and requires policy makers throughout the world to become more environmentally literate. It also requires robust debate among politicians and citizens, mutual respect for diverse beliefs, and a dedication to implementing the win-win **principle of sustainability**. This would replace the more polarized I-win-you-lose approach that fails to recognize that the solutions to urgent environmental problems will require openness, inclusiveness, innovation, and compromise among political players and other people with divergent views.

CONSIDER THIS. . .

THINKING ABOUT Environmental Political Principles

Which three of the seven principles listed here do you think are the most important? Why? Which ones do you think could influence legislators in your city, state, or country? Why?

Grassroots political action has led to improvements in environmental quality in many countries.

David Grossman/Alamy

FIGURE 17.15 Marchers in New York City in 2011 protesting against natural gas fracking.

Individuals and Environmental Leaders Working Together Can Make a Difference

A major theme of this book is that *individuals matter*. History shows that significant change usually comes from the *bottom up* when individuals join together to bring about change. Without previous bottom-up, or grassroots, political action by millions of individual citizens and organized citizen groups (Figure 17.15), the air that millions of people breathe today and the water they drink would be much more polluted, and much more of the earth's biodiversity would have disappeared (**Concept 17.3**). (See the Case Study that follows.)

With the growth of the Internet, digital technology, and social media, individuals have become more empowered. Partly because of this social networking, the number of citizens' groups, national and global action networks, and NGOs focused on environmental and other problems has grown rapidly. In some instances, what people have done to improve environmental quality in their own neighborhoods, schools, or workplaces has inspired actions on regional, national, and global stages. Figure 17.16 lists ways in which individuals living in constitutional democracies have influenced government policies.

Each of us can provide environmental leadership in several different ways. First, we can *lead by example*, using our own lifestyles and values to show others that change is possible and can be beneficial. For example, we can use fewer disposable products, eat food that has been sustainably produced (see Figure 10.27, p. 242), and walk, bike, or take mass transit to work or school. We can reuse and recycle many items, and we can reduce our matter and energy consumption by thinking more about whether we should limit more of our purchases to things we really need. In addition to setting a good example, we can also save money and expand our beneficial environmental impacts by taking such steps.

Second, we can *work within existing economic and political systems to bring about environmental improvement* by campaigning and voting for informed and ecologically literate candidates, and by communicating with elected officials. As environmental writer and activist Bill McKibben says, "First change your politicians, then worry about your lightbulbs." We can also send a message to companies that we feel are harming the environment through their production processes or products by *voting with our wallets*—not buying their products or services—and letting them know why. Another way to work within the system is to choose one of the many rapidly growing green careers highlighted throughout this book and described in Figure 17.12 and on this book's companion website.

Third, we can *run for some sort of local office*. Look in the mirror. Maybe you are one of those who can make a difference as an officeholder.

Fourth, we can *propose and work for better solutions to environmental problems*. Leadership is much more than just

What Can You Do?

Influencing Environmental Policy

- Become informed on issues

- Make your views known at public hearings

- Make your views known to elected representatives and understand their positions on environmental issues

- Contribute money and time to candidates who support your views

- Vote

- Run for office

- Form or join nongovernment organizations (NGOs) seeking change

- Support reform of election campaign financing that reduces undue influence by corporations and wealthy individuals

© 2016 Cengage Learning

FIGURE 17.16 **Individuals matter:** Some ways in which you can influence environmental policy. **Questions:** Which three of these actions do you think are the most important? Which ones, if any, have you taken?

taking a stand for or against something. It also involves coming up with solutions to problems and persuading people to work together to achieve them.

Some environmentally active citizens and leaders are motivated by two important findings. *First,* research by social scientists indicates that social change requires active support by only 5–10% of the population, which is often enough to lead to a political tipping point. *Second,* experience has shown that reaching such a critical mass can bring about social change much faster than most people think. **GOOD NEWS**

CASE STUDY

U.S. Environmental Laws and Regulations

Concerned citizens have persuaded the U.S. Congress to enact a number of important federal environmental and resource protection laws. Most of them were enacted in the 1970s (Figure 17.17).

U.S. environmental laws generally fit into five categories. The first type *requires evaluation of the environmental*

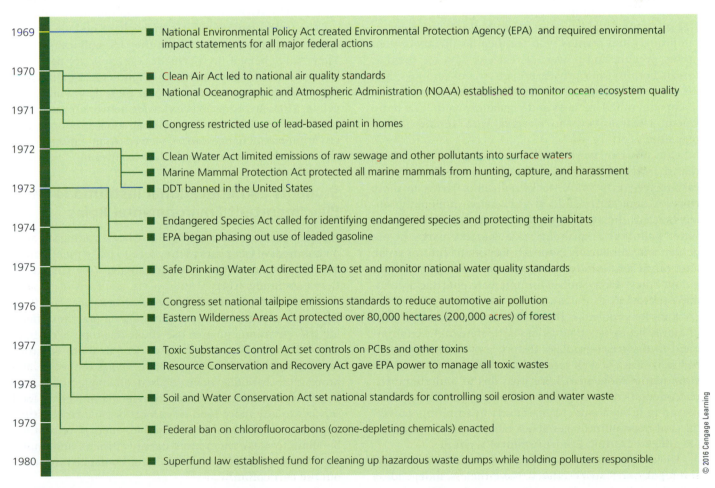

1969	■ National Environmental Policy Act created Environmental Protection Agency (EPA) and required environmental impact statements for all major federal actions
1970	■ Clean Air Act led to national air quality standards
	■ National Oceanographic and Atmospheric Administration (NOAA) established to monitor ocean ecosystem quality
1971	■ Congress restricted use of lead-based paint in homes
1972	■ Clean Water Act limited emissions of raw sewage and other pollutants into surface waters
	■ Marine Mammal Protection Act protected all marine mammals from hunting, capture, and harassment
1973	■ DDT banned in the United States
1974	■ Endangered Species Act called for identifying endangered species and protecting their habitats
	■ EPA began phasing out use of leaded gasoline
1975	■ Safe Drinking Water Act directed EPA to set and monitor national water quality standards
1976	■ Congress set national tailpipe emissions standards to reduce automotive air pollution
	■ Eastern Wilderness Areas Act protected over 80,000 hectares (200,000 acres) of forest
1977	■ Toxic Substances Control Act set controls on PCBs and other toxins
	■ Resource Conservation and Recovery Act gave EPA power to manage all toxic wastes
1978	■ Soil and Water Conservation Act set national standards for controlling soil erosion and water waste
1979	■ Federal ban on chlorofluorocarbons (ozone-depleting chemicals) enacted
1980	■ Superfund law established fund for cleaning up hazardous waste dumps while holding polluters responsible

© 2016 Cengage Learning

FIGURE 17.17 Some of the major environmental laws and their amended versions enacted in the United States from 1969 to 1980.

impacts of certain human activities. It is represented by one of the first and most far-reaching federal environmental laws, the National Environmental Policy Act, or NEPA, passed in 1970. Under NEPA, an *environmental impact statement (EIS)* must be developed for every major federal project likely to have an effect on environmental quality. The EIS must identify, and suggest ways to lessen, any harmful impacts, and it must present an evaluation of alternatives to the project.

The second major type of environmental legislation *sets standards for pollution levels* (as in the Clean Air Acts, see Chapter 15, p. 399). A third type *sets aside or protects certain species, resources, and ecosystems* (the Endangered Species Act, see Chapter 8, p. 177, and the Wilderness Act, see Chapter 9, p. 200). A fourth type *screens new substances for safety and sets standards* (as in the Safe Drinking Water Act). And a fifth type *encourages resource conservation* (the Resource Conservation and Recovery Act, see Chapter 16, p. 445).

U.S. environmental laws have been highly effective, especially in controlling some forms of pollution. However, since 1980, a well-organized and well-funded movement has mounted a strong campaign to weaken or repeal existing U.S. environmental laws and regulations and to change the ways in which public lands (Figure 17.14) are used.

Three major groups are strongly opposed to various U.S. environmental laws and regulations: **(1)** some corporate leaders and owners who see them as threats to their profits, wealth, and power; **(2)** citizens who see them as threats both to their private property rights and jobs; and **(3)** state and local government officials who resent having to implement federal laws and regulations with little or no federal funding, or who disagree with certain federal regulations.

One problem working against effective regulations is that the focus of environmental issues has shifted from easy-to-see, dirty smokestacks and filthy rivers to more complex, long-term, and less visible environmental problems such as climate change, biodiversity loss, and groundwater pollution. Explaining such complex issues to the public and mobilizing support for often controversial, long-range solutions to such problems is difficult. (See the online Guest Essay on environmental reporting by Andrew C. Revkin.)

Another problem is that some environmentalists have primarily brought bad news about the state of the environment to the general public. History shows that bearers of bad news are not well received, and opponents of the environmental movement have used this to undermine environmental concerns. History also shows that people are moved to bring about change mostly by inspiring, powerful, and hopeful messages about what the world could be like. **GREEN CAREER: Environmental writer**

Since 2000, efforts to weaken U.S. environmental laws and regulations have escalated. Nevertheless, independent polls show that more than 80% of the U.S. public strongly support environmental laws and regulations and do not want them weakened. However, polls also show that less than 10% of the U.S. public (and in economic downturns only about 2–3%) considers the environment to be one of the nation's most pressing problems. As a result, environmental concerns often do not get transferred to the ballot box or to personal spending decisions.

Citizen Environmental Groups Play Important Roles

The spearheads of the global conservation, environmental, and environmental justice movements are the tens of thousands of nonprofit NGOs working at the international, national, state, and local levels. The growing influence of these organizations is one of the most important factors in the forging of environmental decisions and policies.

NGOs range in size from grassroots groups with just a few members to *mainline* organizations such as the World Wildlife Fund (WWF), a 5-million-member global conservation organization that operates in 100 countries, with 1.2 million members in the United States. Other international groups with large memberships include Greenpeace, the Nature Conservancy, and Conservation International. In the United States, more than 8 million citizens belong to more than 30,000 NGOs that deal with environmental issues.

The largest environmental groups have become powerful and important forces within the U.S. political system. They have helped to persuade Congress to pass and strengthen environmental laws (Figure 17.17), and they fight attempts to weaken or repeal these laws. Taken together, a loosely connected worldwide network of grassroots NGOs working for bottom-up political, social, economic, and environmental change can be viewed as an emerging citizen-based *global sustainability movement*.

GOOD NEWS

Students and Educational Institutions Can Play Important Environmental Roles

Campus environmental groups have been leading the way on hundreds of campuses as they seek to make their schools and local communities more sustainable (see Chapter 1 Core Case Study, p. 4).

Many student groups make *environmental audits* of their campuses or schools. They gather data on practices affecting the environment and use them to propose changes that will make their campuses or schools more environmentally sustainable while usually saving money in the process. Such audits have focused on implementing or improving recycling programs, convincing university food services to buy more food from local organic farms, shifting from fossil fuels to renewable energy, retrofitting buildings to make them more energy efficient, and implementing concepts of environmental sustainability throughout the curriculum.

Other students have focused on institutional investments. In 2012, more than 300 student-led campaigns were pressuring colleges and universities to stop investing their endowment funds in environmentally harmful industries, such as coal-fired electricity production, and to increase their investments in renewable energy and other environmentally beneficial businesses.

At Northland College in Ashland, Wisconsin, students helped to design a green living and learning center (Figure 17.18) that houses 150 students and features a wind turbine, solar panels, furniture made of recycled materials, and waterless (composting) toilets. Northland students voted to impose a *green fee* of $40 per semester on themselves to help finance the college's sustainability programs.

Dickinson College in Carlisle, Pennsylvania, integrates sustainability throughout its curriculum and uses wind power to offset all of its electricity use. Since 1990, De Anza Community College in Cupertino, California, has been integrating sustainability concepts into its curriculum. In addition, a team of students, faculty, administrators, and members of the local community worked together to build the LEED-platinum-certified Kirsch Center for Environmental Studies. The University of Connecticut—cited in 2013 by the Sierra Club as the greenest U.S. university—offers almost 600 sustainability classes and 40% of its faculty members carry out research related to environmental sustainability.

CONSIDER THIS. . .

THINKING ABOUT The Greening of Your Campus

What major steps is your school taking to increase its own environmental sustainability and to educate its students about environmental sustainability?

Environmental Security Will Be Increasingly Important

Countries are legitimately concerned with *military security* and *economic security*. However, ecologists and many economists point out that all economies are supported by the earth's natural capital (Figure 17.4). Thus, environmental security, economic security, and national security are interrelated.

According to environmental scientist Norman Myers:

If a nation's environmental foundations are degraded or depleted, its economy may well decline, its social fabric deteriorate, and its political structure become destabilized as growing numbers of people seek to sustain themselves from declining resource stocks. Thus, national security is no longer about fighting forces and weaponry alone. It relates increasingly to watersheds, croplands, forests, genetic resources, climate, and other factors that, taken together, are as crucial to a nation's security as are military factors.

Myers and other analysts call for all countries to make environmental security a major focus of diplomacy and government policy at all levels. (See Myers' online Guest Essay on this subject.)

A 2014 article by *Foreign Policy in Focus* and TheNation.com warned that the United States does not have ade-

Courtesy of Northland College

FIGURE 17.18 The Environmental Living and Learning Center is a residence hall and meeting space at Northland College in Ashland, Wisconsin. Northland students had a major role in designing the building to be more sustainable than conventional buildings.

quate safety rules and inspection and maintenance programs for protecting many of the dangerous and environmentally harmful materials it has produced and facilities it has built since 1950. This includes toxic chemical dumps, nuclear weapons, nuclear power plants, radioactive waste storage facilities (see Figure 13.19, p. 329), oil and gas pipelines, oil refineries, and offshore oil rigs. This is partly because not enough citizens or lobbyists have pressured Congress to pass regulations or to designate funds for improving security around these facilities.

A number of international environmental organizations help to shape and set global environmental policy and improve environmental security and sustainability. Perhaps the most influential is the United Nations, which houses a large family of organizations including the U.N. Environment Programme (UNEP), the World Health Organization (WHO), the U.N. Development Programme (UNDP), and the Food and Agriculture Organization (FAO).

Other organizations that make or influence environmental decisions are the World Bank, the Global Environment Facility (GEF), and the International Union for the Conservation of Nature (IUCN). Despite their limited funding, these and other international organizations have played important roles in:

- expanding the understanding of environmental issues,

- gathering and evaluating environmental data,

- developing and monitoring international environmental treaties,

- providing grants and loans for sustainable economic development and reduction of poverty, and

- helping more than 100 nations to develop environmental laws and institutions.

17.4 WHAT ARE SOME MAJOR ENVIRONMENTAL WORLDVIEWS?

CONCEPT 17.4 Major environmental worldviews differ on which is more important—human needs and wants, or the overall health of ecosystems and the biosphere.

Environmental Worldviews Differ in Important Ways

People disagree on how serious our environmental problems are, as well as on what we should do about them. These conflicts arise mostly out of differing **environmental worldviews**—ways of thinking about how the world works and beliefs that people hold about their roles in the natural world. People with differing environmental worldviews can take the same data, be logically consistent in their analysis of those data, and arrive at quite different conclusions because they start with different assumptions and values.

An environmental worldview is determined partly by a person's **environmental ethics**—what one believes about what is right and what is wrong in our behavior toward the environment. According to environmental ethicist Robert Cahn:

> The main ingredients of an environmental ethic are caring about the planet and all of its inhabitants, allowing unselfishness to control the immediate self-interest that harms others, and living each day so as to leave the lightest possible footprints on the planet.

Environmental ethics are playing an increasingly important role in many political and economic decisions that affect the environment. For example, China and the United States both have abundant reserves of coal (**Core Case Study**). People with one ethical view argue that China has the right to exploit its coal resource to help drive further economic growth, much in the way that the United States has done. However, those with a different ethical view contend that China should not burn all of its coal, because the resulting CO_2 emissions could push the world past a dangerous climate change tipping point (see Figure 15.28, p. 415). Others also argue that the United States and other countries with large coal reserves should not burn all of their coal. Ethical issues such as this one present difficult challenges to the world's decision makers.
GREEN CAREER: Environmental ethicist

Some Environmental Worldviews Are Human-Centered

A **human-centered (anthropocentric) worldview** is one that is focused primarily on the needs and wants of people. One such worldview, called the *planetary manage-* *ment worldview*, sees humans as the planet's most important species—the one that can and should dominate and manage the earth, mostly for its own benefit. This view holds that other species and parts of nature are to be evaluated according to how useful they are to humans.

Another human-centered environmental worldview is the *stewardship worldview*. It assumes that we have an ethical responsibility to be caring and responsible managers, or *stewards*, of the earth. According to this view, when we use the earth's natural capital, we are borrowing from the earth and from future generations. Thus, we have an ethical responsibility to pay this debt by leaving the earth's life-support system in as good a condition as what we now enjoy or better.

Some people contend that any human-centered worldview will eventually fail because it wrongly assumes we now have or can gain enough knowledge to become effective managers or stewards of the earth. Critics of human-centered worldviews point out that we do not even know how many plant and animal species live on the earth, much less what their roles are, how they interact with one another and their nonliving environment, and how they support our lives and economies. As biologist David Ehrenfeld puts it, "In no important instance have we been able to demonstrate comprehensive successful management of the world, nor do we understand it well enough to manage it even in theory." This position is illustrated by the failure of the Biosphere 2 project (Science Focus 17.1).

Some Environmental Worldviews Are Life-Centered and Others Are Earth-Centered

Life-centered (biocentric) worldviews hold that all forms of life have value as participating members of the biosphere, regardless of their potential or actual use to humans. **Earth-centered worldviews** also take this view and expand it to include the entire biosphere, especially ecosystems.

Eventually, all species become extinct. However, most people who hold a life-centered worldview believe we have an ethical responsibility to avoid hastening the extinction of any species through our activities, for two reasons. *First*, each species is a unique package of genetic information that should be respected and protected because it is part of a genetic storehouse that will help the earth's life to continue by changing in response to changes in environmental conditions. *Second*, every species has the potential for providing economic benefits directly or indirectly through its participation in providing ecosystem services.

People with earth-centered worldviews believe we have an ethical responsibility to take a wider view and to try to preserve the earth's biodiversity and the functioning of its life-support systems for the benefit of all forms of life, now and in the future. They argue that humans are not in charge of the world and that human economies and other systems are subsystems of the biosphere (Figure 17.4) and

BIOSPHERE 2—A LESSON IN HUMILITY

FIGURE 17.A Biosphere 2, constructed near Tucson, Arizona, was designed to be a self-sustaining life-support system.

PRNewsFoto/Huron Valley Travel/AP Images

In 1991, eight scientists (four men and four women) were sealed inside Biosphere 2, a $200 million glass and steel enclosure designed to be a self-sustaining life-support system (Figure 17.A) that would increase our understanding of Biosphere 1—the *earth's* life-support system.

This sealed system of interconnected domes was built in the desert near Tucson, Arizona. It contained artificial ecosystems including a tropical rain forest, a savanna, a desert, a lake, streams, freshwater and saltwater wetlands, and a mini-ocean with a coral reef.

Biosphere 2 was designed to mimic the earth's natural chemical cycling systems. Water evaporated from its ocean and other aquatic systems and then condensed to provide rainfall over the tropical rain forest. The precipitation trickled through soil filters into the marshes and back into the mini-ocean before beginning the cycle again.

The facility was stocked with more than 4,000 species of plants and animals, including small primates, chickens, cats, and insects, selected to help maintain life-support functions. Human and animal excrement and other wastes were treated and recycled to help support plant growth. Sunlight and external natural gas–powered generators provided energy. The Biospherians were to be isolated for 2 years and to raise their own food using intensive organic agriculture. They were to breathe air recirculated by plants and to drink water cleansed by natural chemical recycling processes.

From the beginning, many unexpected problems cropped up and the life-support system began to unravel. The level of oxygen in the air declined with soil organisms converting it to carbon dioxide. Additional oxygen had to be pumped in from the outside to keep the Biospherians from suffocating.

Tropical birds died after the first freeze. An ant species got into the enclosure, proliferated, and killed off most of the system's original insect species. In total, 19 of the Biosphere's 25 small animal species became extinct. Before the 2-year period was over, all plant-pollinating insects became extinct, thereby dooming to extinction most of the plant species.

Despite many problems, the facility's waste and wastewater were recycled. With much hard work, the Biospherians were also able to produce 80% of their food supply. However, they suffered from persistent hunger and weight loss.

In the end, an expenditure of $200 million failed to maintain a life-support system for eight people for 2 years. Ecologists Joel Cohen and David Tilman, who evaluated the project, concluded, "No one yet knows how to engineer systems that provide humans with life-supporting services that natural ecosystems provide for free."

Critical Thinking

Some analysts argue that the problems with Biosphere 2 resulted mostly from inadequate design and that a better team of scientists and engineers could make it work. Explain why you agree or disagree with this view.

---

dependent on natural capital. They also contend that preventing the depletion and degradation of this natural capital and expanding our beneficial environmental impacts are vital priorities. This worldview holds that the sustainability of our species, civilizations, and economies depends on the sustainability of the biosphere, of which we are just one part.

The earth-centered worldview holds that, while some talk about saving the earth, it has existed for billions of years and does not need saving. Rather, they argue, what we need to save is our own species and cultures—which have been around for less than an eye-blink in the 3.8-billion-year history of life on the earth—along with the numerous other species that may become extinct because of our activities. (See the online Guest Essay on this topic by Lester W. Milbrath.) This begins by understanding that we are part of, not apart from, the earth, and that we are utterly dependent on the earth's natural capital.

17.5 HOW CAN WE LIVE MORE SUSTAINABLY?

CONCEPT 17.5 We can live more sustainably by becoming environmentally literate, learning from nature, living more simply and lightly on the earth, and becoming active environmental citizens.

We Can Become More Environmentally Literate

There is widespread scientific evidence and agreement that we are a species in the process of degrading our own life-support system and that, during this century, this behavior will very likely threaten human civilization and the existence of up to half of the world's other species. Part of the problem stems from our incomplete understanding of how the earth's life-support system works, how our actions affect its life-sustaining systems, and how we can change our behavior toward the earth and thus toward ourselves. Improving this understanding begins by grasping three important ideas that form the foundation of environmental literacy:

1. Natural capital matters because it supports the earth's life and our economies.

2. Our ecological footprints are immense and are expanding rapidly.

3. We should not exceed any of the estimated planetary boundaries or ecological tipping points (see Figure 3.A, p. 58, and Figure 15.28, p. 415). Once we cross such points, neither wealth nor technology can save us from the resulting consequences that could last for hundreds to thousands of years.

Learning how to live more sustainably requires a foundation of environmental education aimed at producing environmentally literate citizens. We hope that, with this course and textbook, you have begun to build your foundation and we encourage you to keep learning about the environment throughout your life.

We Can Learn from the Earth

Formal environmental education is important, but is it enough? Many analysts say *no*. They call for us to appreciate not only the economic value of nature, but also its ecological, aesthetic, and spiritual values. To these analysts, the problem is not just a lack of environmental literacy but also, for many people, a lack of intimate contact with nature and an incomplete understanding of how nature works and sustains us. This can reduce our ability to act more responsibly toward the earth and thus toward ourselves.

A growing chorus of analysts suggest that we have much to learn from nature. They call for us to acquire a sense of awe, wonder, mystery, excitement, and humility by standing under the stars, exploring a forest, enjoying a beautiful scene in nature (Figure 17.19), or taking in the majesty and power of the sea. We might pick up a handful of topsoil and try to sense the teeming microscopic life within it that helps to keep us alive by supporting food production. We might look at a tree, a mountain (see chapter-opening photo), a rock, or a bee, or listen to the sound of a bird and try to sense how each of them is connected to us and we to them, through the earth's life-sustaining processes.

Such direct experiences with nature can reveal parts of the complex web of life that cannot be bought, re-created with technology, or reproduced with genetic engineering. Understanding and directly experiencing the precious and free gifts we receive from nature can help us to make an ethical commitment to live more sustainably on this earth.

Some psychologists and other thinkers argue that such direct experience and understanding of nature is necessary for healthy living. Journalist Richard Louv, who has specialized in studying relationships among family, community, and nature, coined the term **nature-deficit disorder** to describe a wide range of problems, including anxiety, depression, and attention-deficit disorders that might be resulting from a lack of contact with nature. Louv argues that the problem is especially apparent among children who play mostly indoors and at best view the natural world digitally—something new in the history of humankind (see About the Cover, p. ii).

As more than half of the world's people now live in rapidly growing urban areas and seldom or never see natural settings, this disorder is thought to be on the rise. Louv and others urge us to reconnect with nature, as many people do through birdwatching, taking photographs of what they enjoy in nature, hiking or biking through natural areas, camping, canoeing, snorkeling (see this book's cover photo), and gardening.

We Can Live More Simply and Lightly on the Earth

On a time scale of hundreds of thousands to millions of years, the earth is very resilient and can survive the environmental wounds we are inflicting. However, scientists warn that if we continue on our current path, within your lifetime, it is very likely that we will be living on a changed planet with a warmer and sometimes harsher climate, less dependable supplies of water, more acidic oceans, extensive soil degradation, a sixth mass extinction underway, degradation of key ecosystem services, and widespread ecological and economic disruption.

Some analysts urge people who have a habit of consuming excessively to learn how to live more simply and lightly. Seeking happiness through the pursuit of material things is considered folly by almost every major religion and philosophy. Yet, today's avalanche of advertising messages encourage people to buy more and more things to fill a growing list of wants as a way to achieve happiness. As American humorist and writer Mark Twain (1835–1910)

FIGURE 17.19 Experiencing nature can help us to understand the need to protect the earth's natural capital and to live more sustainably.

observed: "Civilization is the limitless multiplication of unnecessary necessities."

Some people are now adopting a lifestyle of *voluntary simplicity*. This can involve learning to live with fewer possessions, using products and services that have smaller harmful environmental impacts, and even creating beneficial environmental impacts (**Concept 17.5**). Many of these people view voluntary simplicity not as a sacrifice but as a way to have a more fulfilling and satisfying life. Instead of working longer to pay for bigger vehicles and houses, they are spending more time with their loved ones, friends, and neighbors.

Practicing voluntary simplicity is a way to apply the Indian philosopher and leader Mahatma Gandhi's *principle of enoughness:* "The earth provides enough to satisfy every person's need but not every person's greed. . . . When we take more than we need, we are simply taking from each other, borrowing from the future, or destroying the environment and other species." Most of the world's major religions have similar teachings.

Living more lightly starts with asking the question: How much is enough? Similarly, one can ask: What do I really

■ Avoid buying something just because a friend has bought it

■ Go on an ad diet by not watching or reading advertisements

■ Avoid shopping for recreation and buying on impulse

■ Stop using credit and buy only with cash to avoid overspending

■ Borrow and share things like books, tools, and other consumer goods

FIGURE 17.20 Five ways to withdraw from an addiction to buying more and more stuff.

© 2016 Cengage Learning

need? These are not easy questions to answer, because people in affluent societies are conditioned to want more and more material possessions and to view them as needs instead of wants. As a result, many people are addicted to buying more and more stuff as a way to find meaning in their lives. Figure 17.20 lists five steps that some psychologists have advised people to take to help them withdraw from this addiction.

Concept 17.5   **477**

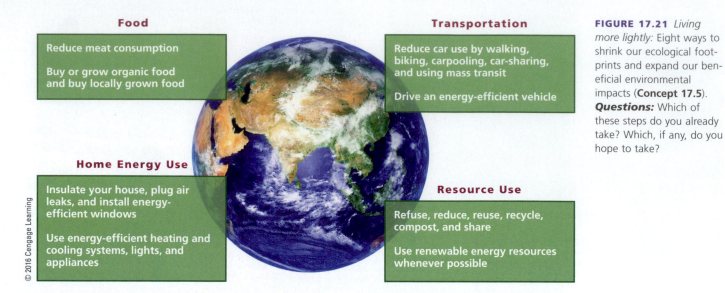

Food

Reduce meat consumption

Buy or grow organic food and buy locally grown food

Transportation

Reduce car use by walking, biking, carpooling, car-sharing, and using mass transit

Drive an energy-efficient vehicle

Home Energy Use

Insulate your house, plug air leaks, and install energy-efficient windows

Use energy-efficient heating and cooling systems, lights, and appliances

Resource Use

Refuse, reduce, reuse, recycle, compost, and share

Use renewable energy resources whenever possible

© 2016 Cengage Learning

FIGURE 17.21 *Living more lightly:* Eight ways to shrink our ecological footprints and expand our beneficial environmental impacts (**Concept 17.5**). *Questions:* Which of these steps do you already take? Which, if any, do you hope to take?

Environmental ethicists have developed a number of ethical guidelines for living more sustainably by converting environmental concerns, literacy, and wisdom into environmentally responsible actions. Here are five of those guidelines—all of them in keeping with all six **principles of sustainability** (see Figure 1.2, p. 6, and Figure 1.5, p. 9, or Supplement 7, pp. S50–S51):

1. Mimic the ways in which nature sustains itself and consider the effects that certain activities will have on other people and other life forms, present and future.

2. Protect the earth's natural capital and repair ecological damage caused by human activities.

3. Use matter and energy resources as efficiently as possible.

4. Celebrate and protect biodiversity.

5. Leave the earth in a condition that is as good as what we have enjoyed, or better.

Throughout this text, you have encountered lists of steps we can take to live more lightly by reducing the size and impact of our ecological footprints on the earth. It would be difficult for most of us to do all or even most of these things. So which ones are the most important? The human activities that have the greatest harmful impacts on the environment are food production, transportation, home energy use, and our overall resource use. Based on this analysis, Figure 17.21 lists eight key ways in which some people are choosing to live more simply to try to make beneficial environmental impacts.

It is important to recognize that there is no single correct or best solution to each of the environmental problems we face. Indeed, one of nature's three **scientific principles of sustainability** holds that preserving diversity—in this case, being flexible and adaptable in trying a variety of cultural, economic, political, and technological solutions to our problems—is the best way to adapt to the earth's largely unpredictable, ever-changing environmental conditions. Finally, we should laugh every day and enjoy and celebrate nature, connectedness, friendship, and love.

We Can Bring about a Sustainability Revolution in Your Lifetime

The Industrial Revolution, which began around the mid-18th century, has been a remarkable global transformation. Now, in this century, environmental leaders say it is time for another global transformation—a *sustainability revolution*. Here is what business entrepreneur and environmental writer Paul Hawken told the 2009 graduating class at the University of Portland:

When asked if I am pessimistic or optimistic about the future, my answer is always the same: If you look at the science about what is happening on the earth and aren't pessimistic, you don't understand the data. But if you meet the people who are working to restore this earth and the lives of the poor, and you aren't optimistic, you haven't got a pulse. . . . You join a multitude of caring people.

Figure 17.22 lists some of the major cultural shifts in emphasis that could help to bring about a sustainability revolution. One of the leaders in the movement to develop and promote detailed plans for making the shift to more sustainable ways of living is Lester R. Brown (Individuals Matter 17.2).

We know what needs to be done and we can change the way we treat the earth and thus our life-support system. History also shows that we can bring about change faster than we might think, once we have the courage to leave behind ideas and practices that no longer work and to nurture new positive trends such as the rapidly growing seedlings of sustainability listed in Figure 17.23.

Lester R. Brown: Champion of Sustainability

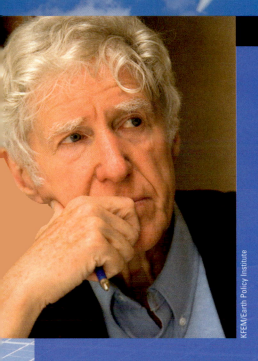

KFEM/Earth Policy Institute

Lester R. Brown is president of the Earth Policy Institute, which he founded in 2001. The purpose of this nonprofit, interdisciplinary research organization is to provide a plan for a more sustainable future and a roadmap showing how we could get there. In 1974, he founded the Worldwatch Institute, which publishes annual reports on environmental issues and has helped to educate citizens and national leaders throughout the world about the serious environmental problems we face.

Brown is an interdisciplinary thinker and one of the pioneers of the global sustainability movement. For decades, he has been researching and describing the complex and interconnected environmental issues we face and proposing concrete strategies for dealing with them. The *Washington Post* has called him "one of the world's most influential thinkers," and in 2011, *Foreign Policy* named him one of the Top Global Thinkers.

Brown's Plan B for shifting to a more environmentally and economically sustainable future has four main goals: (1) stabilize population growth, (2) stabilize climate change, (3) eradicate poverty, and (4) restore the earth's natural support systems.

Brown has written or coauthored more than 50 books, which have been translated into more than 40 languages. He has received numerous prizes and awards, including 25 honorary degrees, the United Nations Environment Prize, and Japan's Blue Planet Prize. In 2012, he was inducted into the Earth Hall of Fame in Kyoto, Japan. He also holds three honorary professorships in China.

Despite the serious environmental challenges we face, Brown sees reasons for hope. They include his understanding that social change can sometimes occur very quickly. He is also encouraged by improvements in fuel efficiency, the emerging shift from using coal to using solar and wind energy to produce electricity, and a growing public understanding of our need to live more sustainably.

Background photo: Petr Kopka/Shutterstock.com

Current Emphasis

Sustainability Emphasis

Current Emphasis	Sustainability Emphasis
Energy and Climate	
Fossil fuels	→ Direct and indirect solar energy
Energy waste	→ Energy efficiency
Climate disruption	→ Climate stabilization
Matter	
High resource use and waste	→ Less resource use and waste
Consume and throw away	→ Reduce, reuse, and recycle
Waste disposal and pollution control	→ Waste prevention and pollution prevention
Life	
Deplete and degrade natural capital	→ Protect natural capital
Reduce biodiversity	→ Protect biodiversity
Population growth	→ Population stabilization

© 2016 Cengage Learning

FIGURE 17.22 Solutions: Some of the cultural shifts in emphasis that scientists say will be necessary to bring about a sustainability revolution. *Questions:* Which of these shifts do you think are most important? Why?

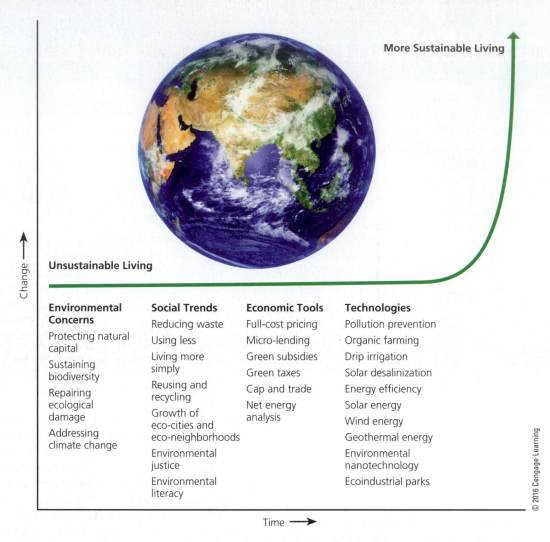

More Sustainable Living

Change →

Unsustainable Living

Environmental Concerns	Social Trends	Economic Tools	Technologies
Protecting natural capital	Reducing waste	Full-cost pricing	Pollution prevention
Sustaining biodiversity	Using less	Micro-lending	Organic farming
Repairing ecological damage	Living more simply	Green subsidies	Drip irrigation
Addressing climate change	Reusing and recycling	Green taxes	Solar desalinization
	Growth of eco-cities and eco-neighborhoods	Cap and trade	Energy efficiency
	Environmental justice	Net energy analysis	Solar energy
	Environmental literacy		Wind energy
			Geothermal energy
			Environmental nanotechnology
			Ecoindustrial parks

Time →

© 2016 Cengage Learning

FIGURE 17.23 The concerns, trends, tools, and technologies listed under this curve of exponential growth could all be parts of a major shift toward sustainability. These agents of change are growing slowly, but at some point, some or all of them could take off, round the J-curve of exponential growth, and help to bring about a sustainability revolution within your lifetime. *Questions:* Which two items in each of these four categories do you think are the most important to promote? What other items would you add to this list?

While some skeptics say the idea of a sustainability revolution is idealistic and unrealistic, entrepreneur Paul Hawken has argued that "the most unrealistic person in the world is the cynic, not the dreamer." And according to the late Steve Jobs, cofounder of Apple Inc., "The people who are crazy enough to think they can change the world are the ones who do." If these and other individuals had not had the courage to forge ahead with ideas that others called idealistic and unrealistic, very few of the achievements that we now celebrate would have come to pass.

The key to a sustainability revolution is our understanding that individuals matter—that each and every one of our choices and actions makes a difference. Virtually all of the environmental progress we have made during the last few decades occurred because individuals banded together to insist that we can do better. This is an exciting time to be alive as we learn how to live more sustainably on our planetary home.

BIG IDEAS

• A more sustainable economic system would include in market prices the harmful environmental and health costs of producing and using goods and services, subsidize environmentally beneficial goods and services, tax pollution and waste instead of wages and profits, and reduce poverty.

• Individuals can work together to become part of the political processes that influence how environmental policies are made and implemented.

• Living more sustainably means becoming environmentally literate, learning from nature, living more simply, and becoming active environmental citizens.

The United States, China, and Sustainability

As the world's two largest economies, the United States and China will play key roles in determining whether or not the world can make a transition to a more sustainable future during this century (**Core Case Study**). The governments and citizens of these countries can help to make such a transition by applying the **scientific principles of sustainability** (see Figure 1.2, p. 6)—relying much more on solar energy and other renewable energy sources, reusing and recycling much more of what is produced, and celebrating, restoring, and protecting as much as possible of the biodiversity that supports our lives and economies.

U.S. and Chinese policy makers and business leaders could also be guided by the three **social science principles of sustainability** (see Figure 1.5, p. 9). In the political arena, they

Ragnarrock | Dreamstime.com

will have to try harder to find win-win solutions that benefit the largest numbers of people while also benefiting the environment. Such solutions will likely include internalizing the harmful environmental and health costs of producing and using goods and services (full-cost pricing). And these decision makers, as well as all the rest of us, will need to make each decision with future generations in mind—seeking to leave the world in as good a condition as that which we now enjoy, if not better.

Chapter Review

Core Case Study

1. Explain why the decisions and actions of the United States and China will play a major role in determining whether the world can make a shift to a more sustainable future, and why this chapter's Core Case Study has economic, political, and ethical aspects.

Section 17.1

2. What is the key concept for this section? What is **economics**? Distinguish among **natural capital**, **human capital**, and **manufactured capital**. Define **economic growth**. What is a **high-throughput economy**? Define **economic development** and **environmentally sustainable economic development**. Explain how the views of neoclassical economists differ from those of ecological and environmental economists.

Section 17.2

3. What is the key concept for this section? Why do products and services actually cost more than most people think? What is **full-cost pricing** and what are some benefits of using it to determine the market values of goods and services? Give three reasons why it is not widely used. Define **subsidies**. What are perverse subsidies and how do they contribute to environmental problems? Define and distinguish between

gross domestic product (GDP) and **per capita GDP**. What is the **genuine progress indicator (GPI)** and how does it differ from the GDP? Why are environmental indicators important? What are the major advantages and disadvantages of green taxes? Why do some analysts find most tax systems to be backward? What are three suggested requirements for the successful implementation of green taxes?

4. Distinguish between command-and-control and incentive-based, or innovation-friendly, government regulations and list the advantages of the second approach. What are the major advantages and disadvantages of using the cap-and-trade approach to implementing environmental regulations for controlling pollution and resource use? What are some environmental benefits of selling services instead of goods? Give two examples of this approach.

5. What is **poverty** and how is it related to population growth and environmental degradation? List six ways in which governments, businesses, lenders, and individuals can help to reduce poverty. What is *microlending* and how can it benefit the poor and the environment? What are the Millennium Development Goals? What is a **low-throughput (low-waste) economy**? List six components of more environmentally sustainable economic development. Name five green businesses or careers that would be important in more environmentally sustainable economies. Summarize the story of how the late Ray Anderson worked to develop a more environmentally sustainable business.

Section 17.3

6. What is the key concept for this section? Define **policies**, **politics**, **environmental policy**, and **representative democracy**. List the four components of the policy life cycle. What are special interest groups? Give four examples. What are four major types of public lands in the United States? Summarize the political controversy over managing these lands. What is **environmental justice** and what kinds of environmental injustice occur in the United States and in China? List seven principles that decision makers can use in making environmental policy. What are four ways in which individuals in democracies can help to influence environmental policy? What are three ways to provide environmental leadership?

7. What are five major types of U.S. environmental laws? What are two problems that hinder effective environmental regulation? Describe the roles of grassroots and mainstream environmental organizations. Give two examples of how students and educational institutions have led the way in shifting to more sustainable ways of living and working. Explain the importance of environmental security, relative to economic and military security.

Section 17.4

8. What is the key concept for this section? What is an **environmental worldview**? What are **environmental ethics**? What is a **human-centered**

(**anthropocentric) worldview**? Summarize the debate over whether we can effectively manage the earth. Summarize the ecological lessons learned from the failure of the Biosphere 2 project. Define **life-centered (biocentric) worldview** and **earth-centered worldview**.

Section 17.5

9. What is the key concept for this section? What are three basic principles of environmental literacy? Give two examples of how we can learn from the earth. What is **nature-deficit disorder** and why is it important? What does it mean to adopt a lifestyle of voluntary simplicity? What is the first step to living more lightly? List five guidelines for living more sustainably. List eight important steps that individuals can take to help make the transition to living more sustainably. List six important cultural shifts that could be part of a sustainability revolution. Describe Lester R. Brown's role in helping the world to move toward a more environmentally sustainable future.

10. What are this chapter's *three big ideas*? Explain how the United States and China could apply the **principles of sustainability** in making the shift to a more sustainable future.

Note: Key terms are in bold type.

Critical Thinking

1. In making a shift to a more sustainable future, what do you think are the three most important things that **(a)** the United States needs to do, **(b)** China needs to do (**Core Case Study**), and **(c)** you need to do?

2. Suppose that over the next 20 years, the environmental and health costs of goods and services are gradually added to their market prices until those prices more closely reflect their total costs. What harmful effects and what beneficial effects might such a *full-cost pricing* process have on your lifestyle and on the lives of any children, grandchildren, and great grandchildren you eventually might have?

3. Which three of the components of more sustainable economic development shown in Figure 17.11 do you think are the most important ones to promote? Which three do you think are the least important? Assume you are the advisor to the leaders of your country and write an explanation of your reasoning.

4. Explain why you agree or disagree with each of the seven principles listed on p. 469, which are recommended by some analysts for use in making

environmental policy decisions. Which three of these principles do you think are the most important? Why?

5. Explain why you agree or disagree with **(a)** each of the four principles that biologists and some economists have suggested for using public lands in the United States (p. 467), and **(b)** each of the five suggestions made by developers and resource extractors for managing and using U.S. public lands (p. 468).

6. This chapter summarized several different environmental worldviews. Go through these worldviews and find the beliefs you agree with, and then describe your own environmental worldview. Which of your beliefs, if any, were added or modified as a result of taking this course? Compare your answer with those of your classmates.

7. Explain why you agree or disagree with the following statements: **(a)** everyone has the right to have as many children as they want; **(b)** all people have a right to use as many resources as they want; **(c)** individuals should have the right to do whatever they want with land they own, regardless of whether such actions harm the environment, their neighbors, or the local community; **(d)** other species exist to be used by humans; **(e)** all forms of life have an intrinsic value and therefore have a right to exist. Are your

answers consistent with the beliefs making up your environmental worldview, which you described in question 6? If not, explain.

8. Do you think we have a reasonable chance of bringing about a sustainability revolution within your lifetime? Explain. If you are nearing the end of this course, is your view of the future more hopeful or less hopeful than it was when you began this course? Compare your answers with those of your classmates.

Doing Environmental Science

Choose an environmental issue, such as water pollution, climate change, population growth, biodiversity loss, or any other issue that you have studied in this course. Conduct a poll of students, faculty, staff, and local residents by asking them the questions that follow relating to your particular environmental issue. Poll as many people as you can in order to get a large sample. Create categories. For example, note whether each respondent is male or female. By creating such categories, you are placing each person into a *respondent pool*. You can add other questions about age, political leaning, and other factors to refine your pools as you see fit.

Poll Questions

Question 1   On a scale of 1 to 10, how knowledgeable are you about environmental issue X?

Question 2   On a scale of 1 to 10, how aware are you of ways in which you, as an individual, impact environmental issue X?

Question 3   On a scale of 1 to 10, how important is it for you to learn more about environmental issue X?

Question 4   On a scale of 1 to 10, how sure are you that an individual can have a positive influence on environmental issue X?

Question 5   On a scale of 1 to 10, how sure are you that the government is providing the appropriate level of leadership with regard to environmental issue X?

1. Collect your data and analyze your findings to measure any differences among the respondent pools.

2. List any major conclusions you would draw from the data.

3. Publicize your findings on your school's website or in the local newspaper.

Global Environment Watch Exercise

Within the GREENR database, use the World Map feature, and under "Browse," select *Sustainability*. Click on the pins for the United States, Japan, China, India, and one other country of your choice and research what each of them is doing to try to become more sustainable. Which of these sustainability programs are working well? Which ones are not working well? Write a report comparing these programs and include details on what is working and what is not working in each of the programs. Also, include your thoughts about why some are not working well and what could be done to make them more successful.

Ecological Footprint Analysis

Working with classmates, conduct an ecological footprint analysis of your campus. Work with a partner, or in small groups, to research and investigate one or more aspects of your school, such as recycling or composting; water use; food service practices; energy use; building management and energy conservation; transportation for both on- and off-campus trips; or grounds maintenance. Depending on your school and its location, you may want to add more areas to the investigation. You can also decide to study the campus as a whole, or to break it down into smaller research areas, such as dorms, administrative buildings, classroom buildings, grounds, and other areas.

1. After deciding on your group's research area, conduct your analysis. As part of your analysis, develop a list of questions that will help to determine the ecological impact related to your chosen topic. For example, with regard to water use, you might ask how much water is used, what is the estimated amount that is lost through leaking pipes and faucets, and what is the average monthly water bill for the school, among other questions. Use such questions as a basis for your research.

2. Analyze your results and share them with the class to determine what can be done to shrink the ecological footprint of your school within the area you have chosen.

3. Arrange a meeting with school officials to share your action plan with them.

CENGAGE**brain**.com To access course materials, including Aplia homework, please visit www.cengagebrain.com.

Length

Metric

1 kilometer (km) = 1,000 meters (m)
1 meter (m) = 100 centimeters (cm)
1 meter (m) = 1,000 millimeters (mm)
1 centimeter (cm) = 0.01 meter (m)
1 millimeter (mm) = 0.001 meter (m)

English

1 foot (ft) = 12 inches (in)
1 yard (yd) = 3 feet (ft)
1 mile (mi) = 5,280 feet (ft)
1 nautical mile = 1.15 miles

Metric–English

1 kilometer (km) = 0.621 mile (mi)
1 meter (m) = 39.4 inches (in)
1 inch (in) = 2.54 centimeters (cm)
1 foot (ft) = 0.305 meter (m)
1 yard (yd) = 0.914 meter (m)
1 nautical mile = 1.85 kilometers (km)

Area

Metric

1 square kilometer (km^2) = 1,000,000 square meters (m^2)
1 square meter (m^2) = 1,000,000 square millimeters (mm^2)
1 square meter (m^2) = 10,000 square centimeters (cm^2)
1 hectare (ha) = 10,000 square meters (m^2)
1 hectare (ha) = 0.01 square kilometer (km^2)

English

1 square foot (ft^2) = 144 square inches (in^2)
1 square yard (yd^2) = 9 square feet (ft^2)
1 square mile (mi^2) = 27,880,000 square feet (ft^2)
1 acre (ac) = 43,560 square feet (ft^2)

Metric–English

1 hectare (ha) = 2.471 acres (ac)
1 square kilometer (km^2) = 0.386 square mile (mi^2)
1 square meter (m^2) = 1.196 square yards (yd^2)
1 square meter (m^2) = 10.76 square feet (ft^2)
1 square centimeter (cm^2) = 0.155 square inch (in^2)

Volume

Metric

1 cubic kilometer (km^3) = 1,000,000,000 cubic meters (m^3)
1 cubic meter (m^3) = 1,000,000 cubic centimeters (cm^3)
1 liter (L) = 1,000 milliliters (mL) = 1,000 cubic centimeters (cm^3)
1 cubic meter (m^3) = 1,000 liters (L)
1 milliliter (mL) = 0.001 liter (L)
1 milliliter (mL) = 1 cubic centimeter (cm^3)

English

1 gallon (gal) = 4 quarts (qt)
1 quart (qt) = 2 pints (pt)

Metric–English

1 liter (L) = 0.265 gallon (gal)
1 liter (L) = 1.06 quarts (qt)
1 liter (L) = 0.0353 cubic foot (ft^3)
1 cubic meter (m^3) = 35.3 cubic feet (ft^3)
1 cubic meter (m^3) = 1.30 cubic yards (yd^3)
1 cubic kilometer (km^3) = 0.24 cubic mile (mi^3)
1 barrel (bbl) = 159 liters (L)
1 barrel (bbl) = 42 U.S. gallons (gal)

Mass

Metric

1 kilogram (kg) = 1,000 grams (g)
1 gram (g) = 1,000 milligrams (mg)
1 gram (g) = 1,000,000 micrograms (μg)
1 milligram (mg) = 0.001 gram (g)
1 microgram (μg) = 0.000001 gram (g)
1 metric ton (mt) = 1,000 kilograms (kg)

English

1 ton (t) = 2,000 pounds (lb)
1 pound (lb) = 16 ounces (oz)

Metric–English

1 metric ton (mt) = 2,200 pounds (lb) = 1.1 tons (t)
1 kilogram (kg) = 2.20 pounds (lb)
1 pound (lb) = 454 grams (g)
1 gram (g) = 0.035 ounce (oz)

Energy and Power

Metric

1 kilojoule (kJ) = 1,000 joules (J)
1 kilocalorie (kcal) = 1,000 calories (cal)
1 calorie (cal) = 4.184 joules (J)

Metric–English

1 kilojoule (kJ) = 0.949 British thermal unit (Btu)
1 kilojoule (kJ) = 0.000278 kilowatt-hour (kW-h)
1 kilocalorie (kcal) = 3.97 British thermal units (Btu)
1 kilocalorie (kcal) = 0.00116 kilowatt-hour (kW-h)
1 kilowatt-hour (kW-h) = 860 kilocalories (kcal)
1 kilowatt-hour (kW-h) = 3,400 British thermal units (Btu)
1 quad (Q) = 1,050,000,000,000,000 kilojoules (kJ)
1 quad (Q) = 293,000,000,000 kilowatt-hours (kW-h)

Temperature Conversions

Fahrenheit (°F) to Celsius (°C): °C = (°F − 32.0) ÷ 1.80
Celsius (°C) to Fahrenheit (°F): °F = (°C × 1.80) + 32.0

Graphs and Maps Are Important Visual Tools

A **graph** is a tool for conveying information that we can summarize numerically by illustrating that information in a visual format. This information, called *data*, is collected in experiments, surveys, and other information-gathering activities. Graphing can be a powerful tool for summarizing and conveying complex information.

In this textbook and the accompanying web-based Active Graphing exercises, we use three major types of graphs: *line graphs, bar graphs,* and *pie graphs.* Here, you will explore each of these types of graphs and learn how to read them. In the web-based Active Graphing exercises, you can try your hand at creating some graphs.

An important visual tool used to summarize data that vary over small or large areas is a **map**. We discuss some aspects of reading maps relating to environmental science at the end of this supplement.

Line Graphs

Line graphs usually represent data that fall in some sort of sequence such as a series of measurements over time or distance. In most such cases, units of time or distance lie on the horizontal *x-axis*. The possible measurements of some quantity or variable such as temperature or oil use that changes over time or distance usually lie on the vertical *y-axis*.

In Figure 1, the x-axis shows the years between 1965 and 2020, and the y-axis displays a range of possible values for the annual amounts of oil consumed worldwide. Usually, the y-axis appears on the left end of the x-axis, although y-axes can appear on the right end, in the middle, or on both ends of the x-axis.

The curving line on a line graph represents the measurements taken at certain time or distance intervals. In Figure 1, the curve represents changes in annual global oil consumption between 1965 and 2012. To find the average annual global oil consumption for any year, find that year on the x-axis (a point called the *abscissa*) and run a vertical line from the axis to the curve. At the point where your line intersects the curve, run a horizontal line to the y-axis. The value at that point on the y-axis, called the *ordinate*, is the amount you are seeking. You can go through the same process in reverse to find a year in which global oil consumption was at a certain point.

Questions

1. About how many metric tons of oil were consumed in the world in 2012?
2. Roughly how many times more oil was consumed in 2012 than in 1985? About how many times more oil was consumed in 2012 than in 1965?

Line graphs have several important uses. One of the most common applications is to compare two or more variables. Figure 2 compares two variables: monthly temperature and precipitation (rain and snowfall) during a typical year in a temperate deciduous forest. However, in this case the variables are measured on two different scales, so there are two y-axes. The y-axis on the left end of the graph shows a Centigrade temperature scale, while the y-axis on the right shows the range of precipitation measurements in millimeters. The x-axis displays the first letters of each of the 12 months' names.

Questions

1. In which month does most precipitation fall? Which is the driest month of the year? Which is the hottest month?
2. If the temperature curve were almost flat, running throughout the year at roughly its highest point of about 30°C, how do you think this forest would change

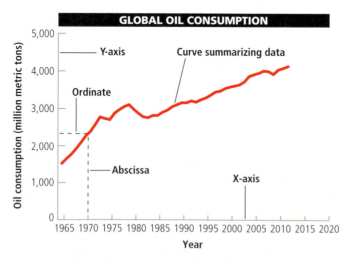

FIGURE 1 Global oil consumption, 1965–2012.

Compiled by the authors using data from U.S. Energy Information Administration, International Energy Agency, British Petroleum, and United Nations.

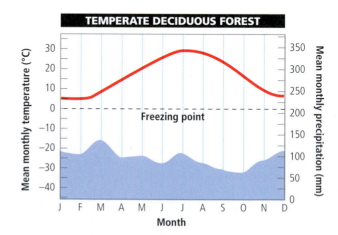

FIGURE 2 Typical variations in annual temperature (red) and precipitation (blue) in a temperate deciduous forest.

© 2016 Cengage Learning

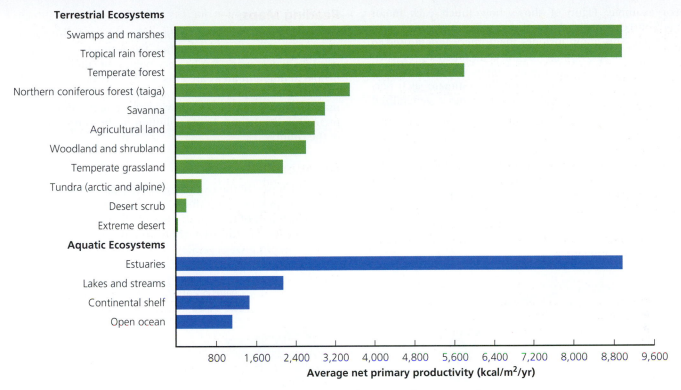

FIGURE 3 Estimated average annual *net primary productivity* in major life zones and ecosystems, in kilocalories of energy produced per square meter per year (kcal/m²/yr).

Compiled by the authors using data from R. H. Whittaker, *Communities and Ecosystems*, 2nd ed., New York: Macmillan, 1975.

from what it is now (see Figure 7.14, center, p. 142)? If the annual precipitation suddenly dropped and remained under 25 centimeters all year, what do you think would eventually happen to this forest?

Bar Graphs

The *bar graph* is used to compare measurements for one or more variables across categories. Unlike the line graph, a bar graph typically does not involve a sequence of measurements over time or distance. The measurements compared on a bar graph usually represent data collected at some point in time or during a well-defined period. For instance, we can compare the *net primary productivity (NPP),* a measure of chemical energy produced by plants in an ecosystem, for different ecosystems, as represented in Figure 3.

In most bar graphs, the categories to be compared are laid out on the x-axis, while the range of measurements for the variable under consideration lies along the y-axis. In Figure 3, the categories (ecosystems) are on the y-axis, and the variable range (NPP) lies on the x-axis. In either case, reading the graph is straightforward. Simply run a line perpendicular to the bar you are reading from the top of that bar (or the right or left end, if it lies horizontally) to the variable value axis. In Figure 3, you can see that the NPP for the continental shelf, for example, is close to 1,600 kcal/m²/yr.

Questions

1. What are the two terrestrial ecosystems that are closest in NPP value of all pairs of such ecosystems? About how many times greater is the NPP in a tropical rain forest than the NPP in a savanna?
2. Which is the most productive of the aquatic ecosystems shown here? Which is the least productive?

An important application of the bar graph used in this book is the *age-structure diagram* (see Figure 6.7, p. 108), which describes a population by showing the numbers of males and females in certain age groups (see Chapter 6, pp. 108–110).

Pie Graphs

Like bar graphs, *pie graphs,* or *pie charts,* illustrate numerical values for two or more categories. But in addition to that, they can also show each category's proportion of the total of all measurements. The categories are usually ordered on the graph from largest to smallest, for ease of comparison, although this is not always the case. Also, as with bar graphs, pie graphs are generally snapshots of a data set at a point in time or during a defined time period. Unlike line graphs, one pie graph cannot show changes over time.

For example, Figure 4 shows how much each major energy source contributes to the world's total amount of energy used. This graph includes the numerical data used to construct it—the percentages of the total taken up by each part of the pie. But we can use pie graphs without including the numerical data and we can roughly estimate such percentages. The pie graph in that case provides a generalized picture of the composition of a data set.

Questions

1. How many times bigger was coal use than nuclear energy use in 2012?
2. How many times bigger was oil use than hydropower use in 2012?

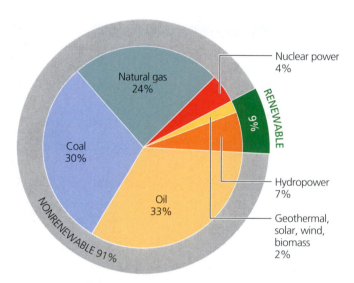

FIGURE 4 Global energy use by source in 2012.

Compiled by the authors using data from British Petroleum, U.S. Energy Information Administration, and International Energy Agency.

Reading Maps

We can use maps for considerably more than showing where places are relative to one another. For example, in environmental science, maps can be very helpful in comparing how people in different areas are affected by environmental problems such as air pollution and acid deposition. Figure 5 is a map of the United States showing the relative numbers of premature deaths due to air pollution in the various regions of the country.

Questions

1. Which part of the country generally has the lowest level of premature deaths due to air pollution?
2. Which part of the country has the highest level? What is the level in the area where you live or go to school?

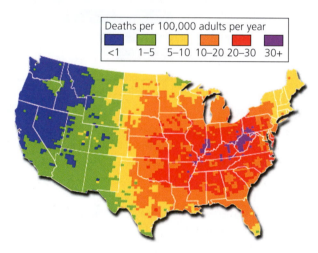

FIGURE 5 Distribution of premature deaths from air pollution in the United States, mostly from very small, fine, and ultrafine particles added to the atmosphere by coal-burning power plants.

Compiled by the authors using data from U.S. Environmental Protection Agency.

Chemists Use the Periodic Table to Classify Elements on the Basis of Their Chemical Properties

The basic unit of each element is a unique *atom* that is different from the atoms of all other elements. Each atom consists of an extremely small and dense center called its *nucleus,* which contains one or more protons and, in most cases, one or more neutrons, as well as one or more electrons moving rapidly somewhere around the nucleus (see Figure 2.4, p. 30). Each atom has equal numbers of positively charged protons and negatively charged electrons. Because these electrical charges cancel one another, *atoms as a whole, in their simplest form, have no net electrical charge.*

We cannot determine the exact location of the electrons around any nucleus. Instead, we can estimate the *probability* that they will be found at various locations outside the nucleus—sometimes called an *electron probability cloud.* This is somewhat like saying that there are six birds flying around inside a cloud. We do not know their exact location, but the cloud represents an area in which we can probably find them.

Matter consists of elements and compounds (see Chapter 2, pp. 29–30). Chemists have developed a way to classify the elements according to their chemical behavior in what is called the *Periodic Table of Elements* (Figure 1). Each horizontal row in the table is called a *period.* Each vertical column lists elements with similar chemical properties and is called a *group.*

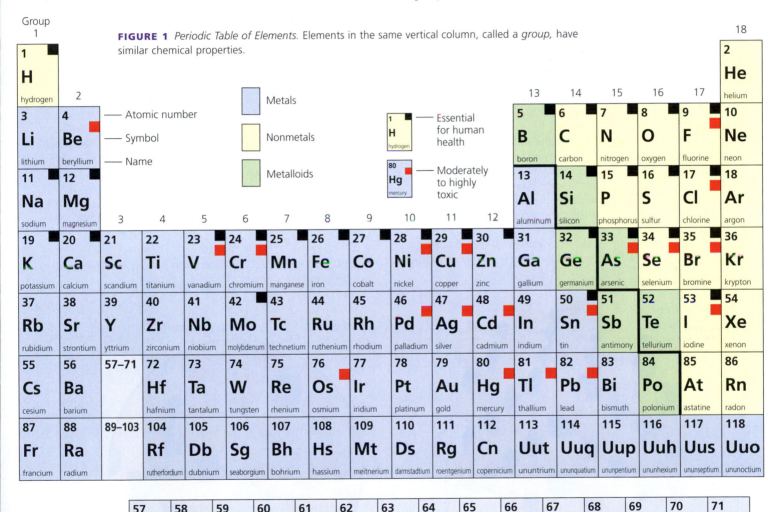

FIGURE 1 *Periodic Table of Elements.* Elements in the same vertical column, called a *group,* have similar chemical properties.

© 2016 Cengage Learning

The Periodic Table in Figure 1 shows how the elements can be classified as *metals, nonmetals,* and *metalloids.* Examples of metals are sodium (Na), calcium (Ca), aluminum (Al), iron (Fe), lead (Pb), silver (Ag), and mercury (Hg).

Atoms of *metals* tend to lose one or more of their electrons to form positively charged ions such as Na^+, Ca^{2+}, and Al^{3+}. For example, an atom of the metallic element sodium (Na, atomic number 11) with 11 positively charged protons and 11 negatively charged electrons can lose one of its electrons. It then becomes a sodium ion with a positive charge of 1 (Na^+) because it now has 11 positive charges (protons) but only 10 negative charges (electrons).

Examples of *nonmetals* are hydrogen (H), carbon (C), nitrogen (N), oxygen (O), phosphorus (P), sulfur (S), chlorine (Cl), and fluorine (F). Atoms of some nonmetals such as chlorine, oxygen, and sulfur tend to gain one or more electrons lost by metallic atoms to form negatively charged ions such as O^{2-}, S^{2-}, and Cl^-. For example, an atom of the nonmetallic element chlorine (Cl, atomic number 17) can gain an electron and become a chlorine ion. The ion has a negative charge of 1 (Cl^-) because it has 17 positively charged protons and 18 negatively charged electrons. Atoms of nonmetals can also combine with one another to form molecules in which they share one or more pairs of their electrons. Hydrogen, a nonmetal, is placed by itself above the center of the table because it does not fit very well into any of the groups.

The elements arranged in a diagonal staircase pattern between the metals and nonmetals have a mixture of metallic and nonmetallic properties and are called *metalloids.* Examples are germanium (Ge) and arsenic (As).

Figure 1 also identifies the elements required as *nutrients* (marked by small black squares) for all or some forms of life, and elements that are moderately or highly toxic (marked by small red squares) to all or most forms of life. Note that some elements such as copper (Cu) serve as nutrients, but can also be toxic at high enough doses. Six nonmetallic elements—carbon (C), oxygen (O), hydrogen (H), nitrogen (N), sulfur (S), and phosphorus (P)—make up about 99% of the atoms of all living things.

CONSIDER THIS

THINKING ABOUT The Periodic Table

Use the Periodic Table to identify, by name and symbol, two elements that should have chemical properties similar to those of **(a)** Ca, **(b)** potassium, **(c)** S, **(d)** lead.

Ionic and Covalent Bonds Hold Compounds Together

Sodium chloride (NaCl) consists of a three-dimensional network of oppositely charged *ions* (Na^+ and Cl^-) held together by the forces of attraction between opposite charges (Figure 2). The strong forces of attraction between such oppositely charged ions are called *ionic bonds.* They are formed when an electron is transferred from a metallic atom such as sodium (Na) to a nonmetallic element such as chlorine (Cl). Because ionic compounds consist of ions formed from atoms of metallic elements (positive ions) and nonmetallic elements (negative ions), they can be described as *metal–nonmetal compounds.*

Sodium chloride and many other ionic compounds tend to dissolve in water and break apart into their individual ions (Figure 3).

NaCl	$\rightarrow$	Na^+	+	Cl^-
sodium chloride (in water)		sodium ion		chloride ion

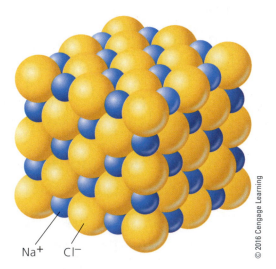

Na⁺   Cl⁻

FIGURE 2 A solid crystal of an ionic compound such as sodium chloride consists of a three-dimensional array of oppositely charged ions held together by *ionic bonds* that result from the strong forces of attraction between opposite electrical charges. Such a bond is formed when an electron is transferred from a metallic atom such as sodium (Na) to a nonmetallic element such as chlorine (Cl).

© 2016 Cengage Learning

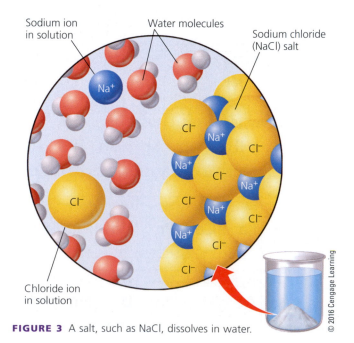

Sodium ion in solution

Water molecules

Sodium chloride (NaCl) salt

Chloride ion in solution

FIGURE 3 A salt, such as NaCl, dissolves in water.

© 2016 Cengage Learning

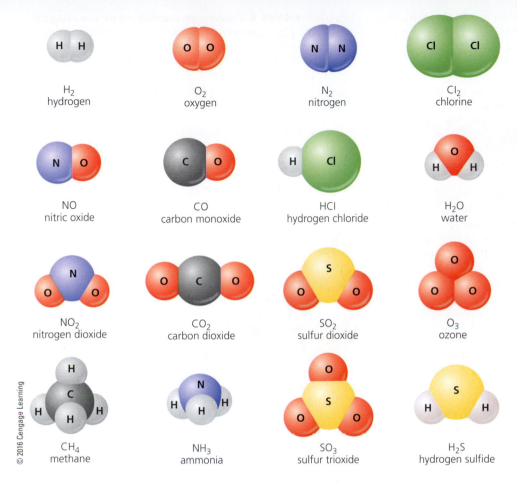

© 2016 Cengage Learning

FIGURE 4 Chemical formulas and shapes for some *covalent compounds* formed when atoms of one or more nonmetallic elements combine with one another. The bonds between the atoms in such molecules are called *covalent bonds*.

H_2
hydrogen

O_2
oxygen

N_2
nitrogen

Cl_2
chlorine

NO
nitric oxide

CO
carbon monoxide

HCl
hydrogen chloride

H_2O
water

NO_2
nitrogen dioxide

CO_2
carbon dioxide

SO_2
sulfur dioxide

O_3
ozone

CH_4
methane

NH_3
ammonia

SO_3
sulfur trioxide

H_2S
hydrogen sulfide

Scientists use **pH** as a measure of the acidity of a solution based on its concentration of hydrogen ions (H^+). By definition, a neutral solution has a pH of 7, an acidic solution has a pH less than 7, and a basic solution has a pH greater than 7.

Each single unit change in pH represents a tenfold increase or decrease in the concentration of hydrogen ions per liter. For example, an acidic solution with a pH of 3 is 10 times more acidic than a solution with a pH of 4. Figure 5 shows the approximate pH and hydrogen ion concentration per liter of solutions for various common substances.

Water, a *covalent compound,* consists of molecules made up of uncharged atoms of hydrogen (H) and oxygen (O). Each water molecule consists of two hydrogen atoms chemically bonded to an oxygen atom, yielding H_2O molecules. The bonds between the atoms in such molecules are called *covalent bonds* and form when the atoms in the molecule share one or more pairs of their electrons. Because they are formed from atoms of nonmetallic elements (Figure 1), covalent compounds can be described as *nonmetal–nonmetal compounds.* Figure 4 shows the chemical formulas and shapes of the molecules that are the building blocks for several common *covalent compounds.*

What Makes Solutions Acidic? Hydrogen Ions and pH

The *concentration,* or number of hydrogen ions (H^+) in a specified volume of a solution (typically a liter), is a measure of its acidity. Pure water (not tap water or rainwater) has an equal number of hydrogen (H^+) and hydroxide (OH^-) ions. It is called a **neutral solution**. An **acidic solution** has more hydrogen ions than hydroxide ions per liter. A **basic solution** has more hydroxide ions than hydrogen ions per liter.

CONSIDER THIS

THINKING ABOUT pH

A solution has a pH of 2. How many times more acidic is this solution than one with a pH of 6?

The measurement of acidity is important in the study of environmental science, as environmental changes involving acidity can have serious environmental impacts. For example, when coal and oil are burned, they give off acidic compounds that can return to the earth as *acid deposition* (see Figure 15.6, p. 394, and Figure 15.7, p. 395), which has become a major environmental problem.

There Are Weak Forces of Attraction between Some Molecules

Ionic and covalent bonds form between the ions or atoms *within* a compound. There are also weaker forces of attraction *between* the molecules of covalent compounds (such as water) resulting from an unequal sharing of electrons by two atoms.

For example, an oxygen atom has a much greater attraction for electrons than does a hydrogen atom. Thus, the electrons shared between the oxygen atom and its two hydrogen

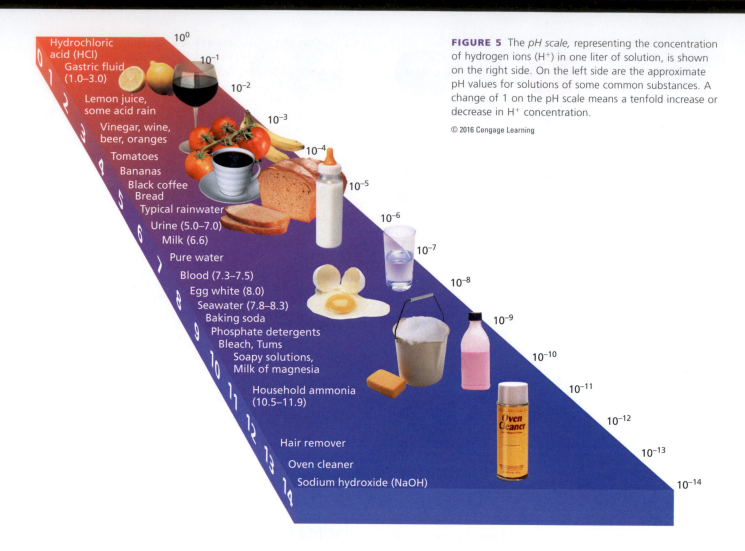

FIGURE 5 The *pH scale,* representing the concentration of hydrogen ions (H^+) in one liter of solution, is shown on the right side. On the left side are the approximate pH values for solutions of some common substances. A change of 1 on the pH scale means a tenfold increase or decrease in H^+ concentration.

© 2016 Cengage Learning

Hydrochloric acid (HCl) — 10^0

Gastric fluid (1.0–3.0) — 10^{-1}

Lemon juice, some acid rain — 10^{-2}

Vinegar, wine, beer, oranges — 10^{-3}

Tomatoes

Bananas — 10^{-4}

Black coffee
Bread — 10^{-5}

Typical rainwater

Urine (5.0–7.0) — 10^{-6}

Milk (6.6)

Pure water — 10^{-7}

Blood (7.3–7.5)

Egg white (8.0) — 10^{-8}

Seawater (7.8–8.3)

Baking soda

Phosphate detergents — 10^{-9}
Bleach, Tums

Soapy solutions,
Milk of magnesia — 10^{-10}

Household ammonia (10.5–11.9) — 10^{-11}

— 10^{-12}

Hair remover

Oven cleaner — 10^{-13}

Sodium hydroxide (NaOH) — 10^{-14}

atoms in a water molecule are pulled closer to the oxygen atom, but not actually transferred to the oxygen atom. As a result, the oxygen atom in a water molecule has a slightly negative partial charge and its two hydrogen atoms have a slightly positive partial charge.

The slightly positive hydrogen atoms in one water molecule are then attracted to the slightly negative oxygen atoms in another water molecule. These forces of attraction *between* water molecules are called *hydrogen bonds* (Figure 6). They account for many of water's unique properties (see Science

FIGURE 6 *Hydrogen bond:* The slightly unequal sharing of electrons in the water molecule creates a molecule with a slightly negatively charged end and a slightly positively charged end. Because of this electrical polarity, the hydrogen atoms of one water molecule are attracted to the oxygen atoms in other water molecules. These fairly weak forces of attraction *between* molecules (represented by the dashed lines) are called *hydrogen bonds.*

© 2016 Cengage Learning

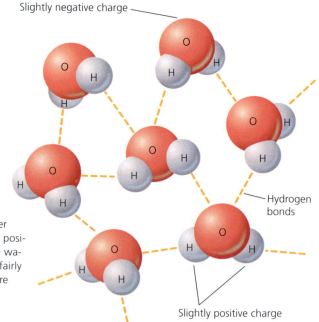

Slightly negative charge

Hydrogen bonds

Slightly positive charge

Focus 3.2, p. 51). Hydrogen bonds also form between other covalent molecules or between portions of such molecules containing hydrogen and nonmetallic atoms with a strong ability to attract electrons.

Four Types of Large Organic Compounds Are the Molecular Building Blocks of Life

Larger and more complex organic compounds, called *polymers,* consist of a number of basic structural or molecular units *(monomers)* linked by chemical bonds, somewhat like rail cars linked in a freight train. Four types of macromol-ecules—complex carbohydrates, proteins, nucleic acids, and lipids—are the molecular building blocks of life.

Complex carbohydrates consist of two or more monomers of *simple sugars* (such as glucose, Figure 7) linked together. One example is the starches that plants use to store energy and also to provide energy for animals that feed on plants. Another is cellulose, the earth's most abundant organic compound, which is found in the cell walls of bark, leaves, stems, and roots.

Proteins are large polymer molecules formed by linking together long chains of monomers called *amino acids* (Figure 8). Living organisms use about 20 different amino acid

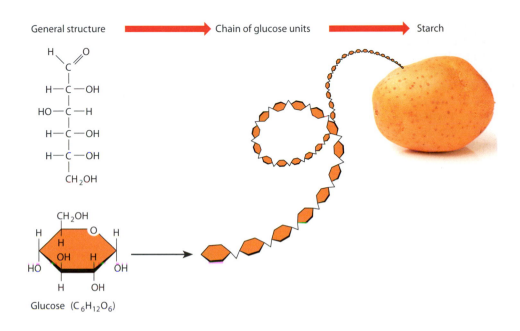

FIGURE 7 Straight-chain and ring structural formulas of glucose, a simple sugar that can be used to build long chains of complex carbohydrates such as starch and cellulose.

© 2016 Cengage Learning
Photo: JIANG HONGYAN/Shutterstock

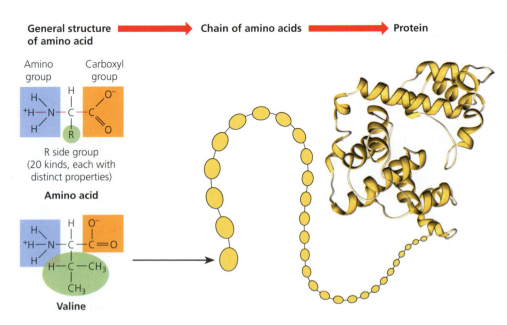

FIGURE 8 General structural formula of amino acids (upper left) and a specific structural formula of one of the 20 different amino acid molecules (lower left) that can be linked together in chains to form proteins that fold up into more complex shapes.

© 2016 Cengage Learning

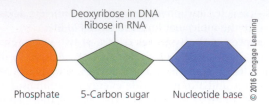

Deoxyribose in DNA
Ribose in RNA

Phosphate   5-Carbon sugar   Nucleotide base

© 2016 Cengage Learning

FIGURE 9 Generalized structures of the nucleotide molecules linked in various numbers and sequences to form large nucleic acid molecules such as various types of DNA (deoxyribonucleic acid) and RNA (ribonucleic acid). In DNA, the five-carbon sugar in each nucleotide is deoxyribose; in RNA it is ribose.

molecules to build a variety of proteins, which play different roles. Some help to store energy. Some are components of the *immune system* that protects the body against diseases and harmful substances by forming antibodies that make invading agents harmless. Others are *hormones* that are used as chemical messengers in the bloodstreams of animals to turn various bodily functions on or off. In animals, proteins are also components of hair, skin, muscle, and tendons. In addition, some proteins act as *enzymes* that catalyze or speed up certain chemical reactions.

Nucleic acids are large polymer molecules made by linking hundreds to thousands of four types of monomers called *nucleotides*. Two nucleic acids—DNA (**d**eoxyribo**n**ucleic **a**cid) and RNA (**r**ibo**n**ucleic **a**cid)—participate in the building of proteins and carry hereditary information used to pass traits from parent to offspring. Each nucleotide consists of a *phosphate group*, a *sugar molecule* containing five carbon atoms (deoxyribose in DNA molecules and ribose in RNA molecules), and one of four different *nucleotide bases* (represented by A, G, C, and T, the first letter in each of their names, or A, G, C, and U in RNA) (Figure 9). The four basic nucleotides used to make various forms of DNA molecules differ in the types of nucleotide bases they contain—adenine (A), guanine (G), cytosine (C), and thymine (T). (Uracil, labeled U, occurs instead of thymine in RNA.) In the cells of living organisms, these nucleotide units combine in different numbers and sequences to form *nucleic acids* such as various types of DNA and RNA (Figure 10).

Hydrogen bonds formed between parts of the four nucleotides in DNA hold two DNA strands together like a spiral staircase, forming a double helix (Figure 10). DNA molecules can unwind and replicate themselves.

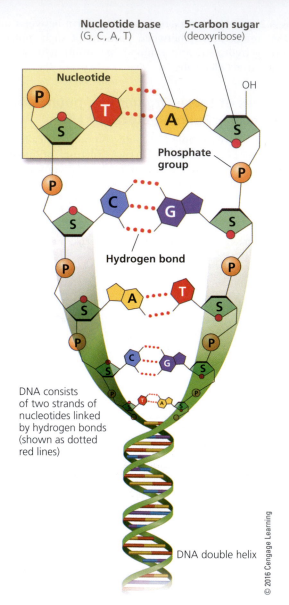

DNA consists of two strands of nucleotides linked by hydrogen bonds (shown as dotted red lines)

DNA double helix

© 2016 Cengage Learning

FIGURE 10 A portion of the double helix of a DNA molecule. The double helix is composed of two spiral (helical) strands of nucleotides. Each nucleotide contains a unit of phosphate (P), deoxyribose (S), and one of four nucleotide bases: adenine (A), guanine (G), cytosine (C), and thymine (T). The two strands are held together by hydrogen bonds formed between various pairs of the nucleotide bases. Guanine (G) bonds with cytosine (C), and adenine (A) with thymine (T).

Fatty acid
(lipid)

Fat molecule
(triglyceride)

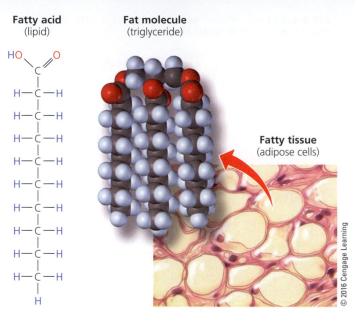

HO — C = O

H — C — H

H — C — H

H — C — H

H — C — H

H — C — H

H — C — H

H — C — H

H — C — H

H — C — H

H

Fatty tissue
(adipose cells)

© 2016 Cengage Learning

FIGURE 11 Structural formula of fatty acid (one form of lipid, left). Fatty acids are converted into more complex fat molecules (center) that are stored in adipose cells (right).

The total weight of the DNA needed to reproduce all of the world's people is only about 50 milligrams—the weight of a small match. If the DNA coiled in your body were unwound, it would stretch about 960 million kilometers (600 million miles)—more than six times the distance between the sun and the earth.

The different molecules of DNA that make up the millions of species found on the earth are like a vast and diverse genetic library. Each species is a unique book in that library. The *genome* of a species is made up of the entire sequence of DNA "letters" or base pairs that combine to "spell out" the chromosomes in typical members of each species. Scientists have been able to map out the genome for the human species by using powerful computers to help them analyze the 3.1 billion base sequences in human DNA.

Lipids, a fourth building block of life, are a chemically diverse group of large organic compounds that do not dissolve in water. Examples are *fats and oils* for storing energy (Figure 11), *waxes* for structure, and *steroids* for producing hormones.

Figure 12 shows the relative sizes of simple and complex molecules, cells, and multicelled organisms.

1 centimeter	(cm)	= 1/100 meter, or 0.4 inch
1 millimeter	(mm)	= 1/1,000 meter
1 micrometer	(μm)	= 1/1,000,000 meter
1 nanometer	(nm)	= 1/1,000,000,000 meter

$1 \text{ meter} = 10^{2} \text{ cm} = 10^{3} \text{ mm} = 10^{6} \text{ μm} = 10^{9} \text{ nm}$

FIGURE 12 Comparison of the relative sizes of simple molecules, complex molecules, cells, and multicellular organisms. Scale is exponential, not linear. Each unit of measure is 10 times larger than the unit preceding it.
Photo: STILLFX/Shutterstock

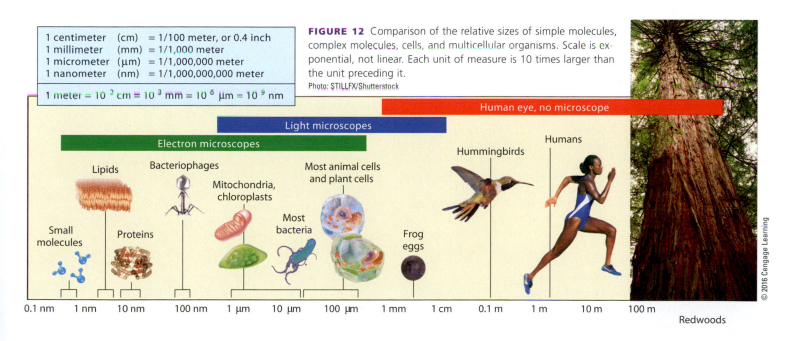

© 2016 Cengage Learning

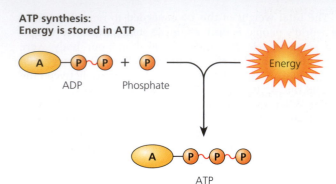

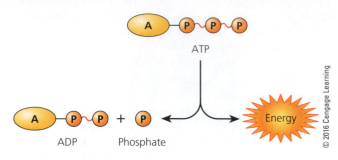

© 2016 Cengage Learning

FIGURE 13 Models showing energy storage (left) and release (right) in cells.

Certain Molecules Store and Release Energy in Cells

Chemical reactions occurring in plant cells during photosynthesis (see Chapter 3, p. 44) release energy that is absorbed by adenosine diphosphate (ADP) molecules and stored as chemical energy in adenosine triphosphate (ATP) molecules (Figure 13, left). When cellular processes require energy, ATP molecules release it to form ADP molecules (Figure 13, right).

Chemists Balance Chemical Equations to Keep Track of Atoms

Chemists use a shorthand system, or equation, to represent any chemical reaction. Chemical equations are also used as an accounting system to verify that no atoms are created or destroyed in a chemical reaction as required by the law of

conservation of matter (see Chapter 2, p. 33). As a consequence, each side of a chemical equation must have the same number of atoms or ions of each element involved. Ensuring that this condition is met leads to what chemists call a *balanced chemical equation*. The equation for the burning of carbon ($C + O_2 \rightarrow CO_2$) is balanced because one atom of carbon and two atoms of oxygen are on both sides of the equation.

Consider the following chemical reaction: When electricity passes through water (H_2O), the latter can be broken down into hydrogen (H_2) and oxygen (O_2), as represented by the following equation:

$$H_2O \quad \rightarrow \quad H_2 \quad + \quad O_2$$

H_2O	H_2	O_2
2 H atoms	2 H atoms	2 O atoms
1 O atom		

This equation is unbalanced because one atom of oxygen is on the left side of the equation but two oxygen atoms are on the right side. We cannot change the subscripts of any of the formulas to balance this equation because that would change the arrangements of the atoms, leading to different substances. Instead, we must use different numbers of the molecules involved to balance the equation. For example, we could use two water molecules:

$$2\ H_2O \quad \rightarrow \quad H_2 \quad + \quad O_2$$
4 H atoms   2 H atoms   2 O atoms
2 O atoms

This equation is still unbalanced. Although the numbers of oxygen atoms on both sides of the equation are now equal, the numbers of hydrogen atoms are not. We can correct this problem by having the reaction produce two hydrogen molecules:

$$2\ H_2O \quad \rightarrow \quad 2\ H_2 \quad + \quad O_2$$
4 H atoms   4 H atoms   2 O atoms
2 O atoms

Now the equation is balanced, and the law of conservation of matter has been observed. For every two molecules of water through which we pass electricity, two hydrogen molecules and one oxygen molecule are produced.

CONSIDER THIS

THINKING ABOUT Chemical Equations

Try to balance the chemical equation for the reaction that combines nitrogen gas (N_2) and hydrogen gas (H_2) to form ammonia gas (NH_3).

Matter Can Undergo Nuclear Changes

In addition to physical and chemical changes, matter can undergo three types of nuclear change (see Chapter 2, pp. 32–33), as summarized in Figure 14.

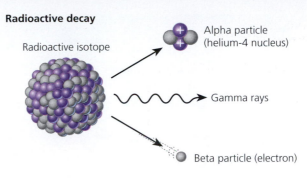

Radioactive decay

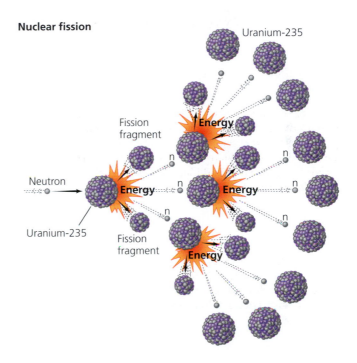

Nuclear fission

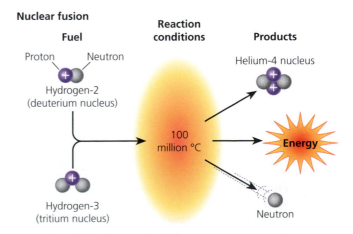

Nuclear fusion

FIGURE 14 There are three types of nuclear changes: natural radioactive decay (top), nuclear fission (center), and nuclear fusion (bottom).

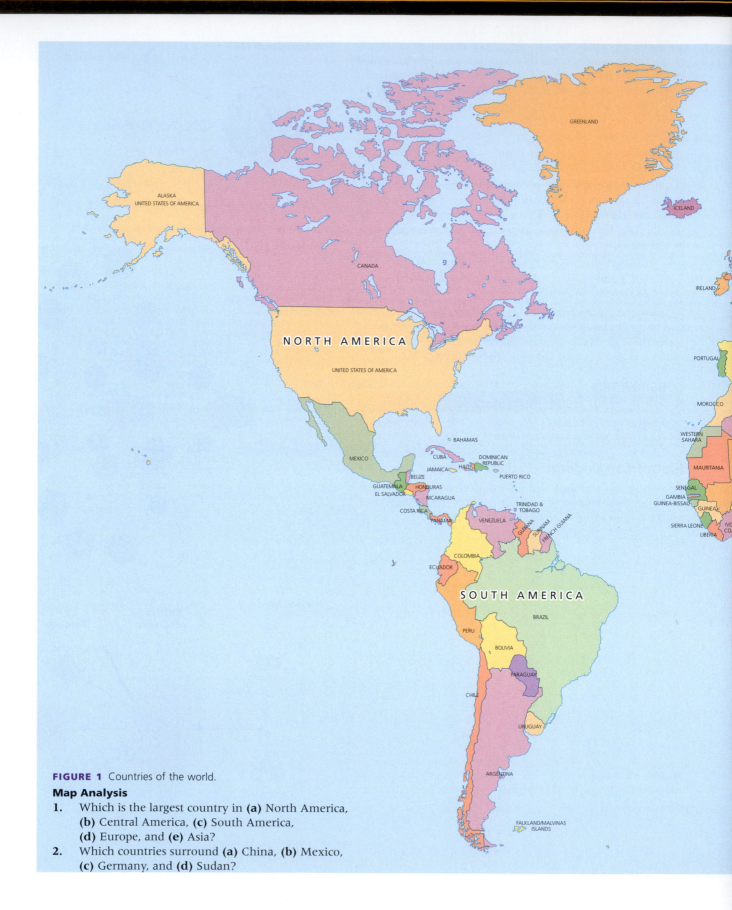

FIGURE 1 Countries of the world.

Map Analysis

1. Which is the largest country in **(a)** North America,
    (b) Central America, **(c)** South America,
    (d) Europe, and **(e)** Asia?
2. Which countries surround **(a)** China, **(b)** Mexico,
    (c) Germany, and **(d)** Sudan?

FIGURE 2 Composite satellite view of the earth showing its major terrestrial and aquatic features.

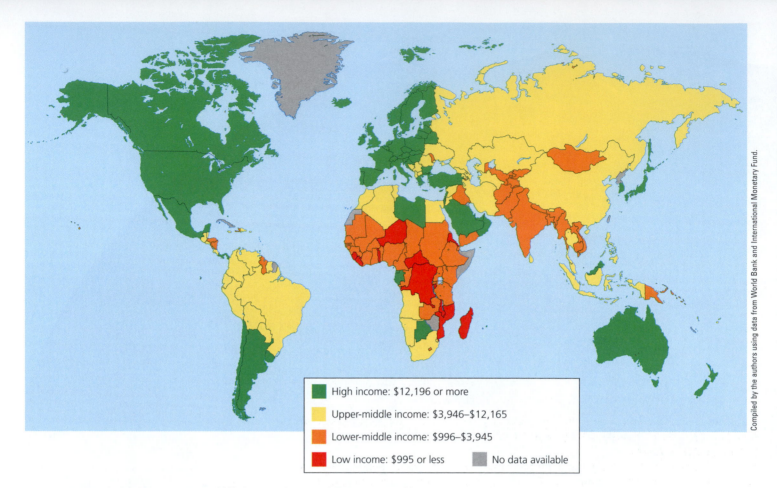

Compiled by the authors using data from World Bank and International Monetary Fund.

High income: $12,196 or more

Upper-middle income: $3,946–$12,165

Lower-middle income: $996–$3,945

Low income: $995 or less   No data available

FIGURE 3 High-income, upper-middle-income, lower-middle-income, and low-income countries in terms of gross national income (GNI) purchasing power parity (PPP) per capita (U.S. dollars) in 2010.

Data and Map Analysis

1. In how many countries is the per capita average income $995 or less? Look at Figure 1 and find the names of three of these countries.

2. In how many instances does a lower-middle- or low-income country share a border with a high-income country? Look at Figure 1 and find the names of the countries that reflect three examples of this situation.

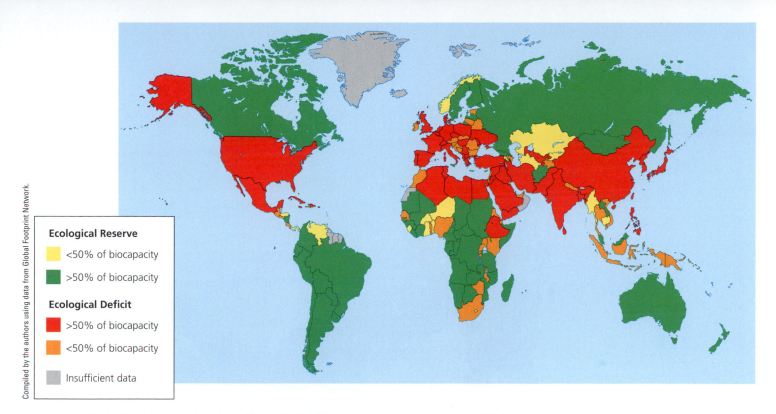

Ecological Reserve

<50% of biocapacity

>50% of biocapacity

Ecological Deficit

>50% of biocapacity

<50% of biocapacity

Insufficient data

FIGURE 4 *Ecological debtors and creditors:* The ecological footprints of some countries exceed their biocapacity, while other countries still have ecological reserves.

Data and Map Analysis

1. List five countries, including the three largest, in which the ecological deficit is greater than 50% of biocapacity. (See Figure 1 of this supplement for country names.)

2. On which two continents does land with ecological reserves of more than 50% of biocapacity occupy the largest percentage of total land area? (See Figure 1 of this supplement for continent names.)

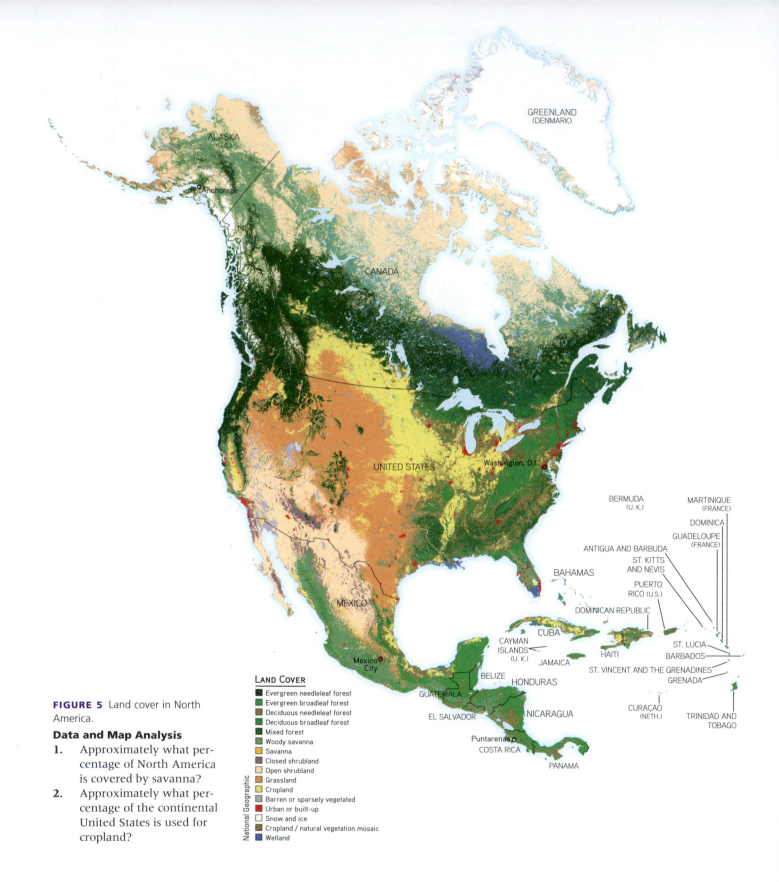

FIGURE 5 Land cover in North America.

Data and Map Analysis

1. Approximately what percentage of North America is covered by savanna?
2. Approximately what percentage of the continental United States is used for cropland?

LAND COVER

- Evergreen needleleaf forest
- Evergreen broadleaf forest
- Deciduous needleleaf forest
- Deciduous broadleaf forest
- Mixed forest
- Woody savanna
- Savanna
- Closed shrubland
- Open shrubland
- Grassland
- Cropland
- Barren or sparsely vegetated
- Urban or built-up
- Snow and ice
- Cropland / natural vegetation mosaic
- Wetland

National Geographic

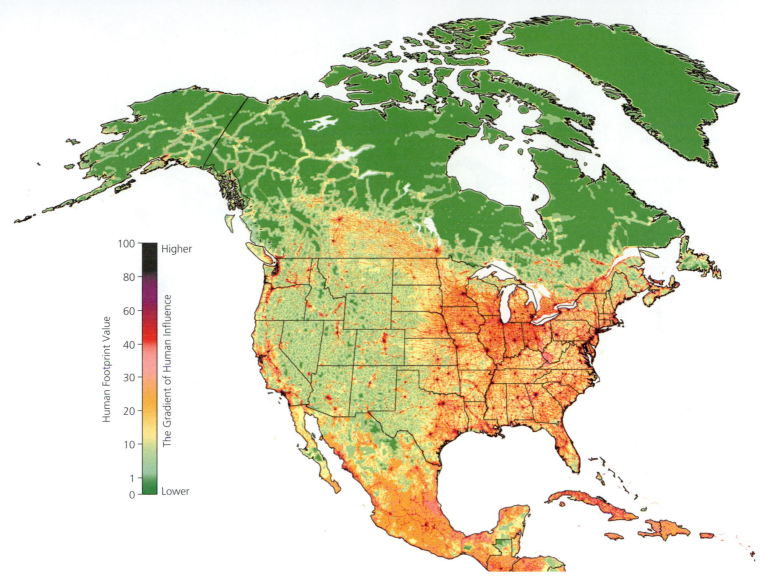

FIGURE 6 **Natural capital degradation:** The human ecological footprint in North America. Colors represent the percentage of each area influenced by human activities.

Compiled by the authors using data from Wildlife Conservation Society and Center for International Earth Science Information Network at Columbia University.

Data and Map Analysis
1. Which general area of the United States has the highest human footprint values?
2. What is the relative value of the human ecological footprint in the area where you live or go to school?

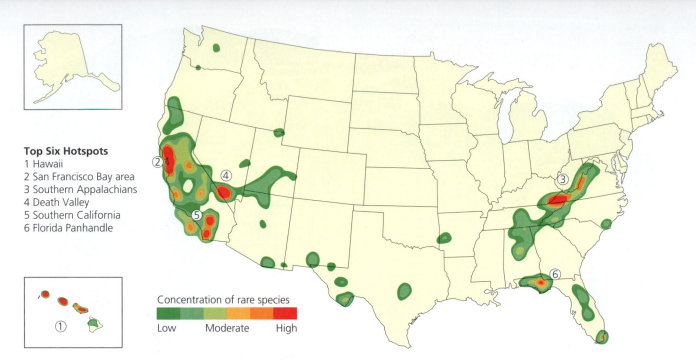

Top Six Hotspots
1 Hawaii
2 San Francisco Bay area
3 Southern Appalachians
4 Death Valley
5 Southern California
6 Florida Panhandle

Concentration of rare species

Low   Moderate   High

FIGURE 7 *Endangered natural capital:* Major biodiversity hotspots in the United States that need emergency protection. The shaded areas contain the largest concentrations of rare and potentially endangered species.

Compiled by the authors using data from State Natural Heritage Programs, Nature Conservancy, and Association for Biodiversity Information.

Data and Map Analysis

1. If you live in the United States, which of the top six hotspots is closest to where you live or go to school?

2. Which general part of the country has the highest overall concentration of rare species? Which part has the second-highest concentration?

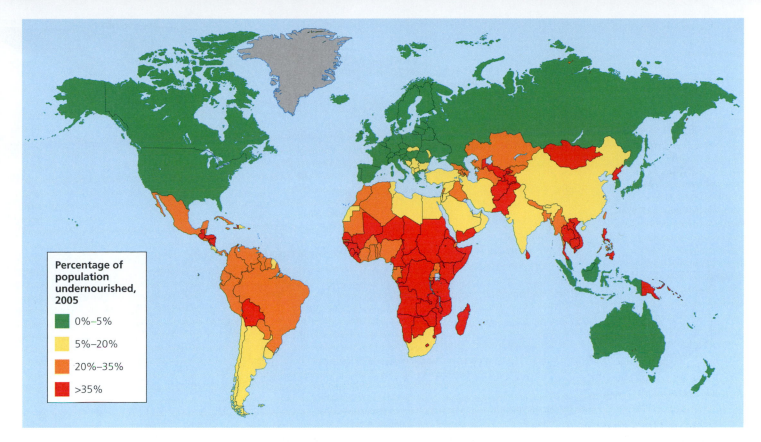

FIGURE 8 World hunger expressed as percentages of populations suffering from chronic hunger and malnutrition.

Compiled by the authors using data from UN Food and Agriculture Organization, World Health Organization, and U.S. Department of Agriculture.

Data and Map Analysis

1. List the continents in order, starting with the one that has the highest percentage of undernourished people and ending with the one that has the lowest such percentage. (See Figure 1 of this supplement for continent names.)

2. On which continent is the largest block of countries that suffer the highest levels of undernourishment? List five of these countries. (See Figure 1 of this supplement for country names.)

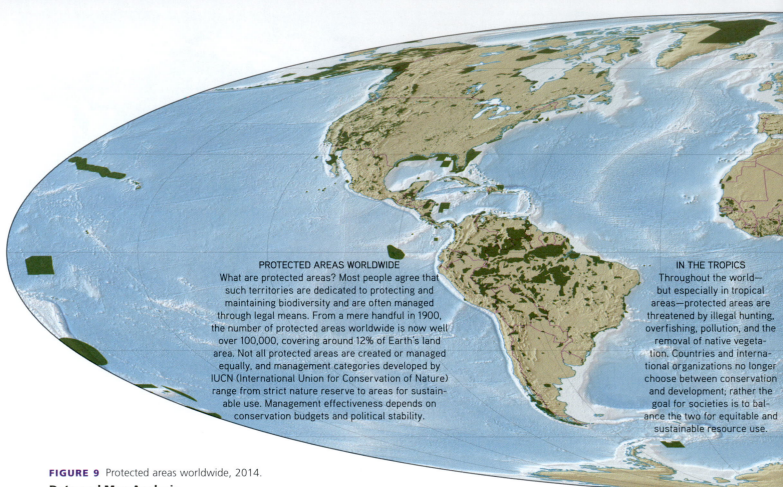

PROTECTED AREAS WORLDWIDE
What are protected areas? Most people agree that such territories are dedicated to protecting and maintaining biodiversity and are often managed through legal means. From a mere handful in 1900, the number of protected areas worldwide is now well over 100,000, covering around 12% of Earth's land area. Not all protected areas are created or managed equally, and management categories developed by IUCN (International Union for Conservation of Nature) range from strict nature reserve to areas for sustainable use. Management effectiveness depends on conservation budgets and political stability.

IN THE TROPICS
Throughout the world—but especially in tropical areas—protected areas are threatened by illegal hunting, overfishing, pollution, and the removal of native vegetation. Countries and international organizations no longer choose between conservation and development; rather the goal for societies is to balance the two for equitable and sustainable resource use.

FIGURE 9 Protected areas worldwide, 2014.

Data and Map Analysis

1. Describe the locations of three relatively large protected areas. (See Figure 1 of this supplement for country and continent names.)
2. What continent appears to have the largest percentage of protected land? What continent appears to have the smallest percentage of protected land?

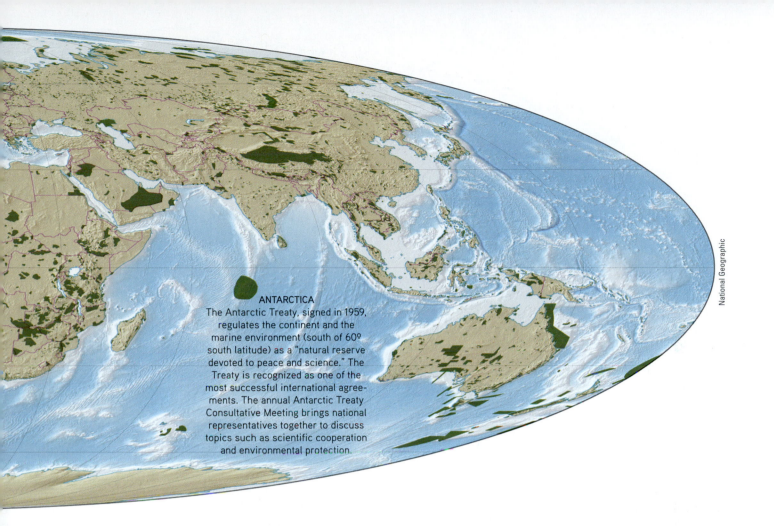

ANTARCTICA
The Antarctic Treaty, signed in 1959, regulates the continent and the marine environment (south of 60º south latitude) as a "natural reserve devoted to peace and science." The Treaty is recognized as one of the most successful international agreements. The annual Antarctic Treaty Consultative Meeting brings national representatives together to discuss topics such as scientific cooperation and environmental protection.

National Geographic

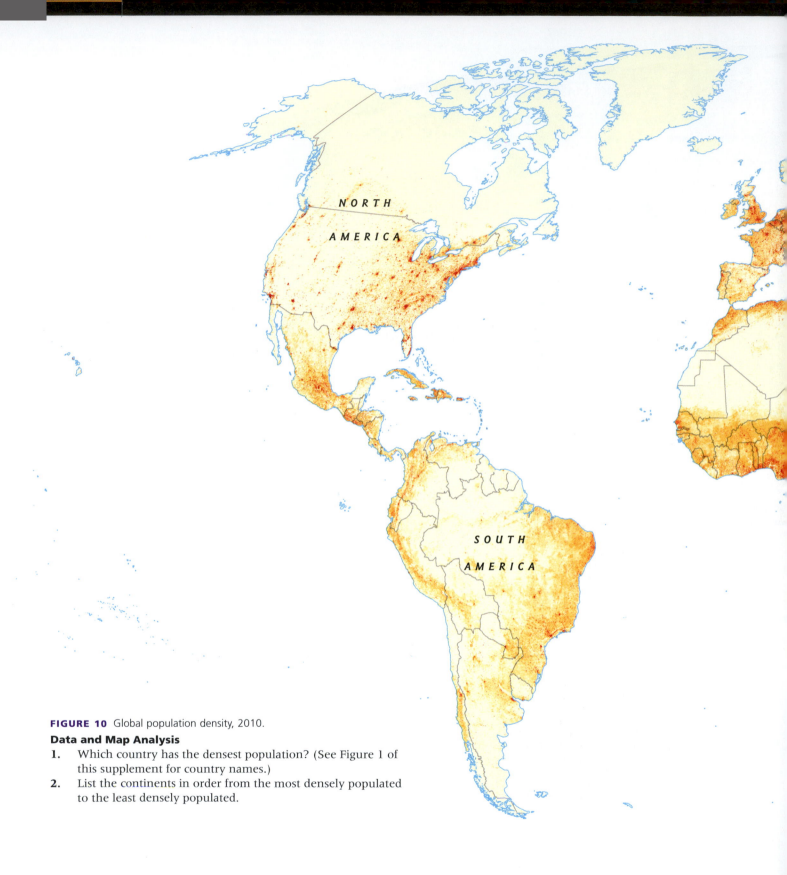

FIGURE 10 Global population density, 2010.

Data and Map Analysis

1. Which country has the densest population? (See Figure 1 of this supplement for country names.)
2. List the continents in order from the most densely populated to the least densely populated.

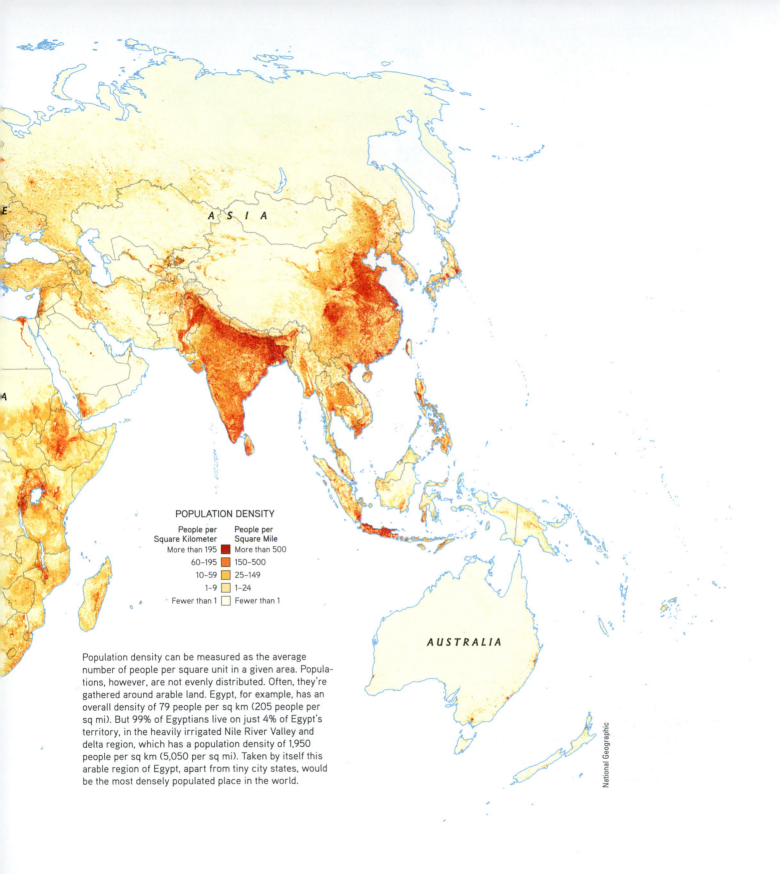

POPULATION DENSITY

People per Square Kilometer		People per Square Mile
More than 195		More than 500
60–195		150–500
10–59		25–149
1–9		1–24
Fewer than 1		Fewer than 1

A S I A

AUSTRALIA

Population density can be measured as the average number of people per square unit in a given area. Populations, however, are not evenly distributed. Often, they're gathered around arable land. Egypt, for example, has an overall density of 79 people per sq km (205 people per sq mi). But 99% of Egyptians live on just 4% of Egypt's territory, in the heavily irrigated Nile River Valley and delta region, which has a population density of 1,950 people per sq km (5,050 per sq mi). Taken by itself this arable region of Egypt, apart from tiny city states, would be the most densely populated place in the world.

National Geographic

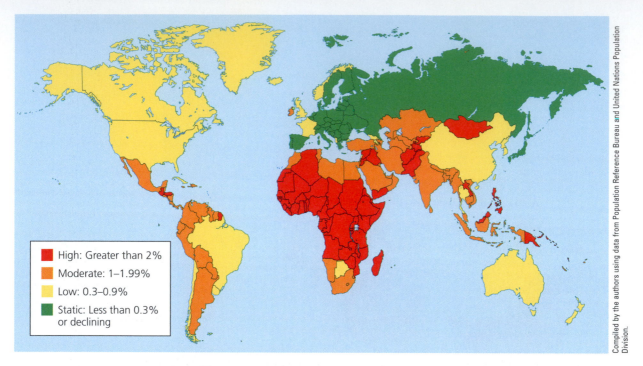

Compiled by the authors using data from Population Reference Bureau and United Nations Population Division.

FIGURE 11 Comparative rates of population growth (%) throughout the world in 2013.

Data and Map Analysis

1. Which continent has the greatest number of countries with high rates of population growth? Which continent has the greatest number of countries with static rates? (See Figure 1 of this supplement for continent names.)

2. Name three countries from each population growth rate category on this map (see Figure 1 for country names).

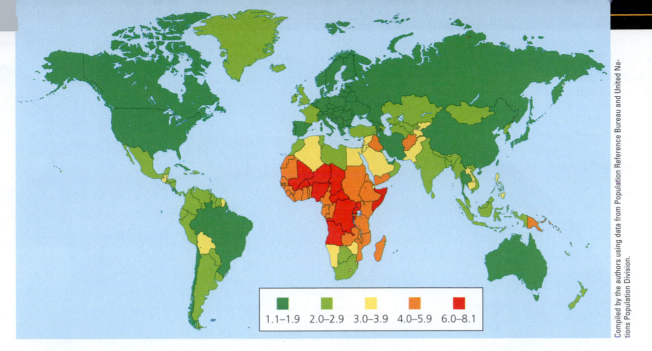

Compiled by the authors using data from Population Reference Bureau and United Nations Population Division.

FIGURE 12 Comparison of total fertility rates (TFRs)—the average number of children born to the women in any population throughout their lifetimes—as measured in 2013.

Data and Map Analysis

1. Which country in the middle TFR category borders two countries in the lowest TFR category? What are those two countries? (See Figure 1 of this supplement for country names.)
2. Describe two geographic patterns that you see on this map.

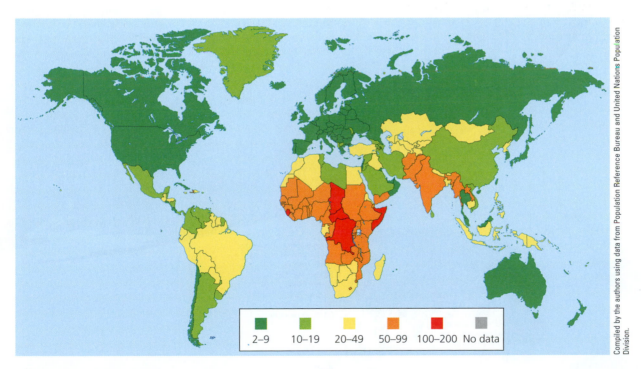

Compiled by the authors using data from Population Reference Bureau and United Nations Population Division.

FIGURE 13 Comparison of the infant mortality rates of countries around the world in 2013.

Data and Map Analysis

1. Describe a geographic pattern that you can see on this map.
2. Describe any similarities that you see in geographic patterns between this map and the one in Figure 12.

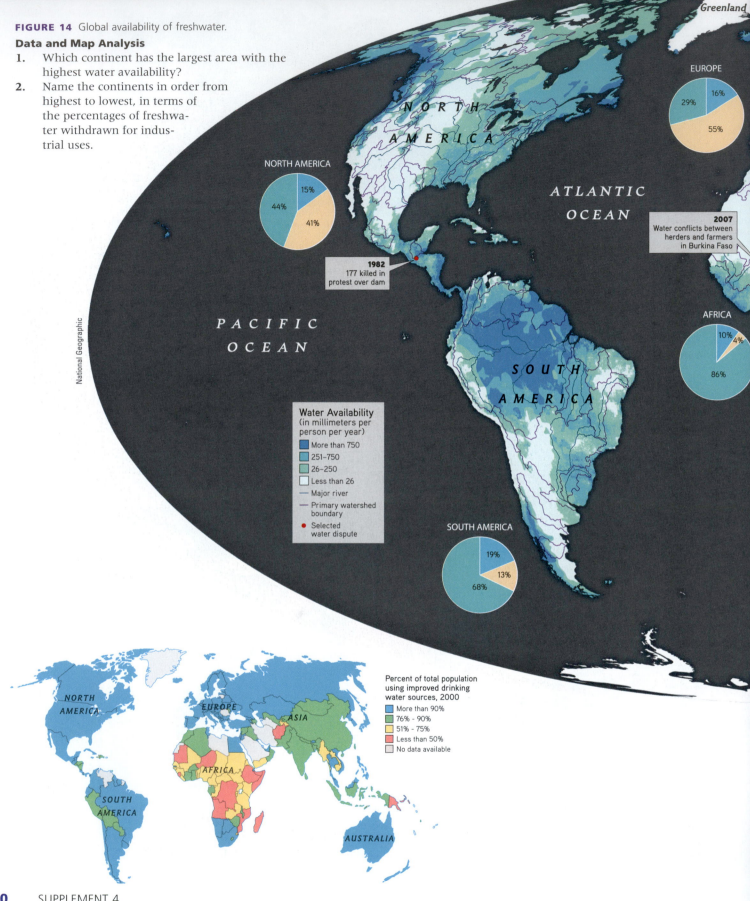

FIGURE 14 Global availability of freshwater.

Data and Map Analysis

1. Which continent has the largest area with the highest water availability?
2. Name the continents in order from highest to lowest, in terms of the percentages of freshwater withdrawn for industrial uses.

National Geographic

NORTH AMERICA

EUROPE
- 16%
- 29%
- 55%

1982
177 killed in protest over dam

2007
Water conflicts between herders and farmers in Burkina Faso

NORTH AMERICA
- 15%
- 44%
- 41%

AFRICA
- 10%
- 4%
- 86%

ATLANTIC OCEAN

PACIFIC OCEAN

Water Availability
(in millimeters per person per year)
- More than 750
- 251–750
- 26–250
- Less than 26
- Major river
- Primary watershed boundary
- Selected water dispute

SOUTH AMERICA
- 19%
- 13%
- 68%

Percent of total population using improved drinking water sources, 2000
- More than 90%
- 76% - 90%
- 51% - 75%
- Less than 50%
- No data available

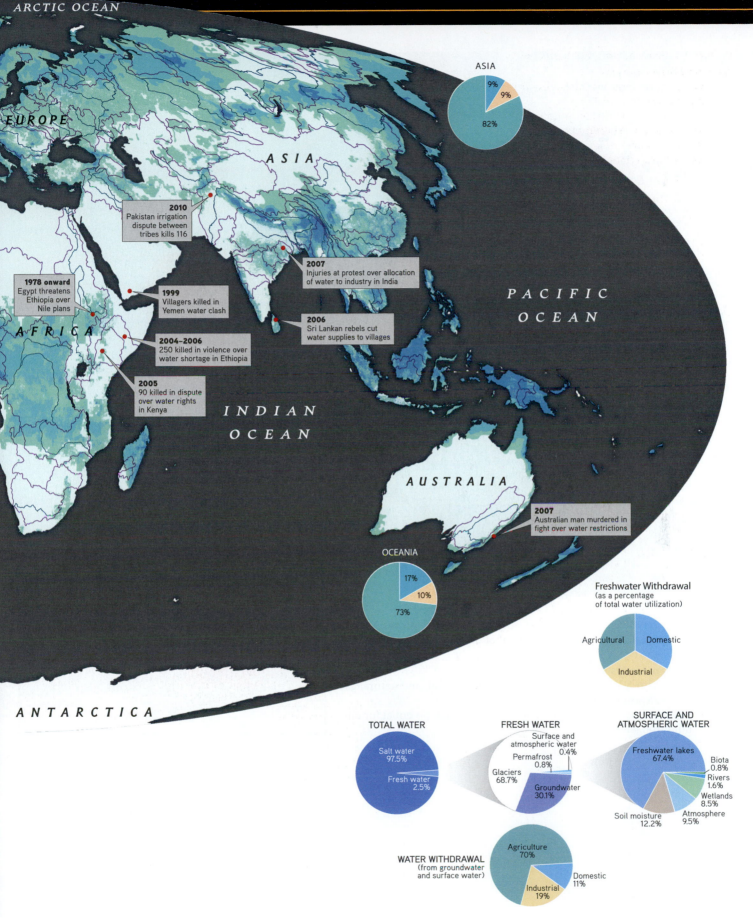

ARCTIC OCEAN

EUROPE

ASIA

AFRICA

INDIAN OCEAN

PACIFIC OCEAN

AUSTRALIA

ANTARCTICA

2010
Pakistan irrigation dispute between tribes kills 116

1978 onward
Egypt threatens Ethiopia over Nile plans

1999
Villagers killed in Yemen water clash

2007
Injuries at protest over allocation of water to industry in India

2006
Sri Lankan rebels cut water supplies to villages

2004–2006
250 killed in violence over water shortage in Ethiopia

2005
90 killed in dispute over water rights in Kenya

2007
Australian man murdered in fight over water restrictions

ASIA
9%
9%
82%

OCEANIA
17%
10%
73%

Freshwater Withdrawal
(as a percentage of total water utilization)
Agricultural
Domestic
Industrial

TOTAL WATER
Salt water 97.5%
Fresh water 2.5%

FRESH WATER
Surface and atmospheric water 0.4%
Permafrost 0.8%
Glaciers 68.7%
Groundwater 30.1%

SURFACE AND ATMOSPHERIC WATER
Freshwater lakes 67.4%
Biota 0.8%
Rivers 1.6%
Wetlands 8.5%
Atmosphere 9.5%
Soil moisture 12.2%

WATER WITHDRAWAL
(from groundwater and surface water)
Agriculture 70%
Domestic 11%
Industrial 19%

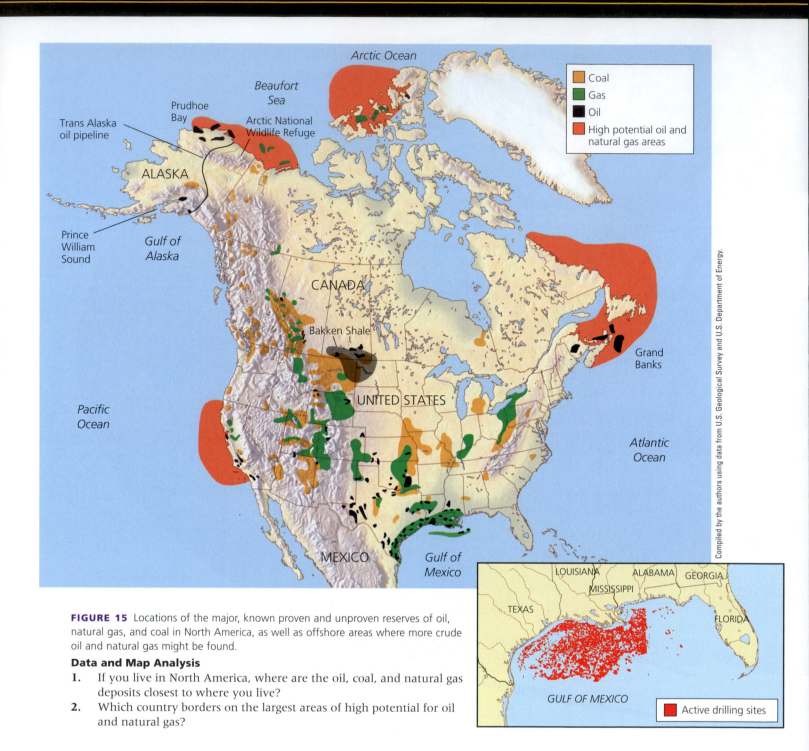

FIGURE 15 Locations of the major, known proven and unproven reserves of oil, natural gas, and coal in North America, as well as offshore areas where more crude oil and natural gas might be found.

Data and Map Analysis

1. If you live in North America, where are the oil, coal, and natural gas deposits closest to where you live?
2. Which country borders on the largest areas of high potential for oil and natural gas?

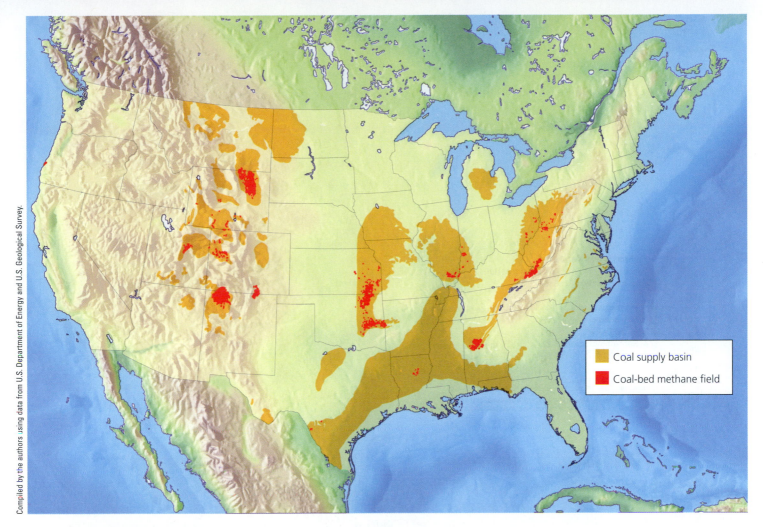

Compiled by the authors using data from U.S. Department of Energy and U.S. Geological Survey.

Coal supply basin

Coal-bed methane field

FIGURE 16 Major coal supply basins and coal-bed methane fields in the lower 48 states of the United States.

Data and Map Analysis

1. If you live in the United States, where are the coal-bed methane deposits closest to where you live?
2. Removing these deposits requires lots of water. Compare the locations of the major deposits of coal-bed methane with water-deficit areas shown in Figure 11.5, p. 253, and Figure 11.6, p. 254.

FIGURE 17 Major natural gas shale deposits in North America.

Data and Map Analysis

1. What state has the largest area of natural gas shale deposits?
2. Name three areas where two states or two countries share a border over a natural gas shale deposit.

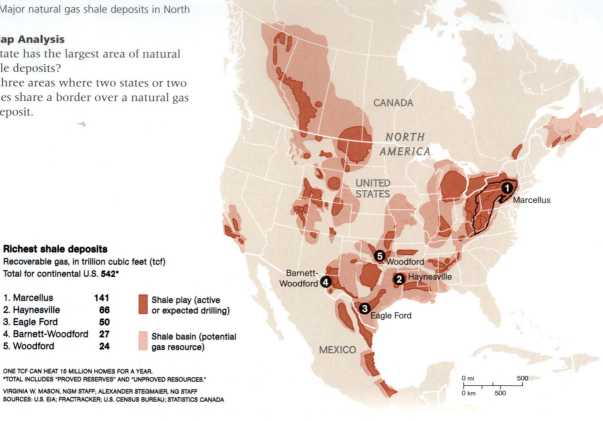

Richest shale deposits

Recoverable gas, in trillion cubic feet (tcf)
Total for continental U.S. **542***

1. Marcellus	**141**	
2. Haynesville	**66**	▮ Shale play (active or expected drilling)
3. Eagle Ford	**50**	
4. Barnett-Woodford	**27**	▮ Shale basin (potential gas resource)
5. Woodford	**24**	

ONE TCF CAN HEAT 15 MILLION HOMES FOR A YEAR.
*TOTAL INCLUDES "PROVED RESERVES" AND "UNPROVED RESOURCES."

VIRGINIA W. MASON, NGM STAFF; ALEXANDER STEGMAIER, NG STAFF
SOURCES: U.S. EIA; FRACTRACKER; U.S. CENSUS BUREAU; STATISTICS CANADA

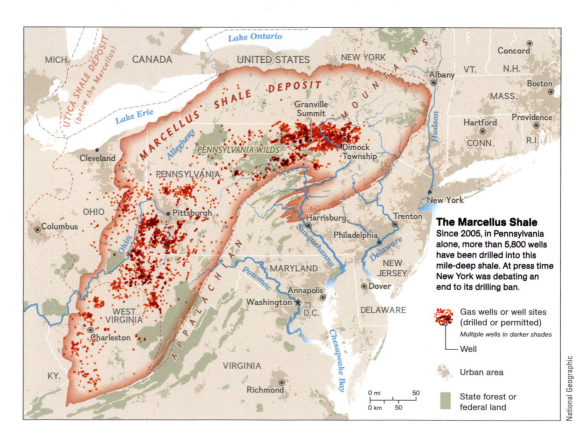

The Marcellus Shale

Since 2005, in Pennsylvania alone, more than 5,800 wells have been drilled into this mile-deep shale. At press time New York was debating an end to its drilling ban.

Gas wells or well sites (drilled or permitted)
Multiple wells in darker shades
Well

Urban area

State forest or federal land

National Geographic

FIGURE 18 Locations of the 100 commercial nuclear power reactors in the United States.

Compiled by the authors using data from U.S. Nuclear Regulatory Commission and U.S. Department of Energy.

Data and Map Analysis

1. Name five states that have more than three commercial nuclear power reactors.
2. Which state has the largest number of commercial nuclear power reactors?

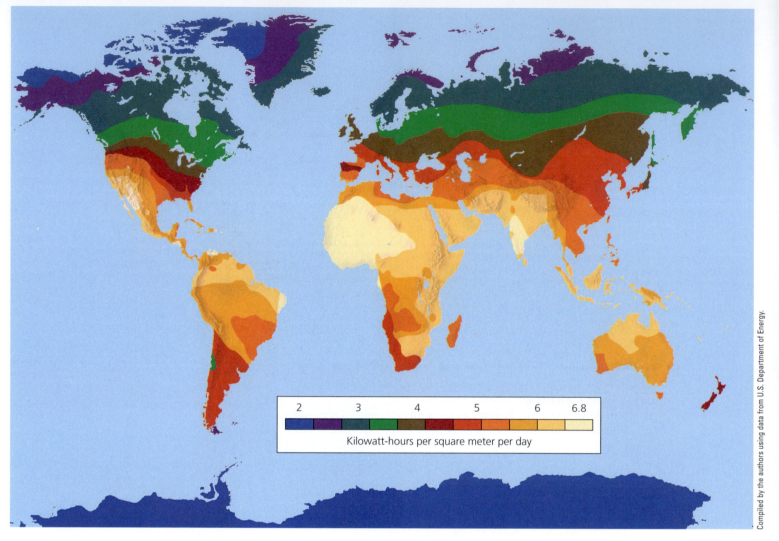

Compiled by the authors using data from U.S. Department of Energy.

FIGURE 19 Global availability of direct solar energy. Areas with more than 3.5 kilowatt-hours per square meter per day (see scale) are good candidates for passive and active solar heating systems and use of solar cells to produce electricity.

Data and Map Analysis

1. What is the potential for making greater use of solar energy to provide heat and produce electricity (with solar cells) where you live or go to school?

2. List the continents in order of overall availability of direct solar energy, from those with the highest to those with the lowest. (See Figure 1 of this supplement for continent names.)

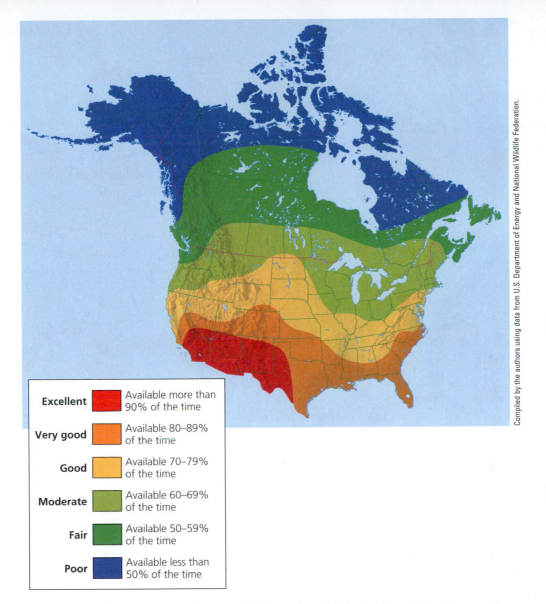

Compiled by the authors using data from U.S. Department of Energy and National Wildlife Federation.

Excellent	🟥	Available more than 90% of the time
Very good	🟧	Available 80–89% of the time
Good	🟨	Available 70–79% of the time
Moderate	🟩	Available 60–69% of the time
Fair	🟩	Available 50–59% of the time
Poor	🟦	Available less than 50% of the time

FIGURE 20 Availability of direct solar energy in the continental United States and Canada.

Data and Map Analysis

1. If you live in the United States or Canada, what is the potential for making increased use of solar energy to provide heat and electricity (with solar cells) where you live or go to school?

2. How many states and Canadian provinces have areas with excellent, very good, or good availability of direct solar energy?

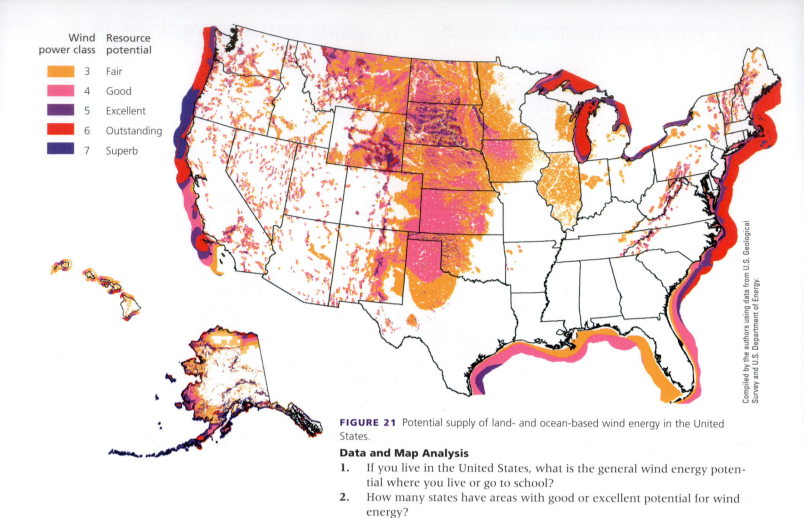

Wind
power class | Resource
potential
3 | Fair
4 | Good
5 | Excellent
6 | Outstanding
7 | Superb

Compiled by the authors using data from U.S. Geological Survey and U.S. Department of Energy.

FIGURE 21 Potential supply of land- and ocean-based wind energy in the United States.

Data and Map Analysis

1. If you live in the United States, what is the general wind energy potential where you live or go to school?

2. How many states have areas with good or excellent potential for wind energy?

FIGURE 22 Potential geothermal energy resources in the continental United States.

Data and Map Analysis

1. If you live in the United States, what is the potential for using geothermal energy to provide heat or to produce electricity where you live or go to school?

2. How many states have areas with very good or excellent potential for using geothermal energy?

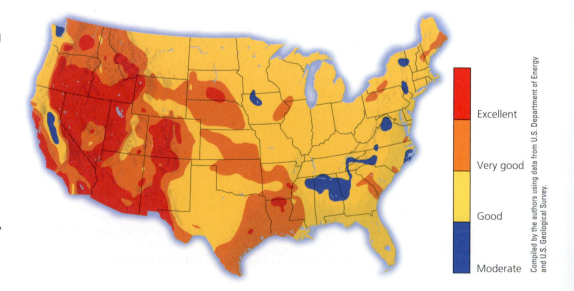

Excellent

Very good

Good

Moderate

Compiled by the authors using data from U.S. Department of Energy and U.S. Geological Survey.

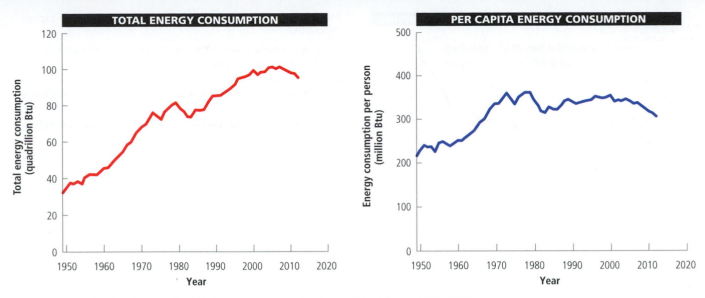

FIGURE 1 Total (left) and per capita (right) energy consumption in the United States, 1950–2012.

Compiled by the authors using data from U.S. Energy Information Administration and U.S. Census Bureau.

Data and Graph Analysis

1. In what year or years did total U.S. energy consumption reach 80 quadrillion Btus?
2. In what year did energy consumption per person reach its highest level shown on this graph, and about what was that level of consumption?

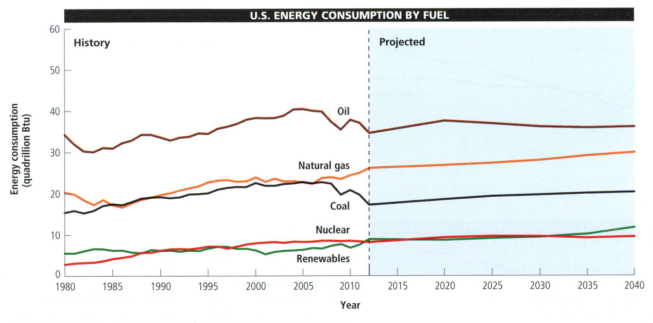

FIGURE 2 Energy consumption by fuel in the United States, 1980–2012, with projections to 2040.

Compiled by the authors using data from U.S. Energy Information Administration Primary Energy Consumption Estimates by Source, 1949–2012, and Annual Energy Outlook 2013.

Data and Graph Analysis

1. Usage of which energy source grew the most between 1990 and 2012? How much did it grow (in quadrillion Btus)?
2. For which two energy sources is usage expected to grow the fastest between 2012 and 2040?

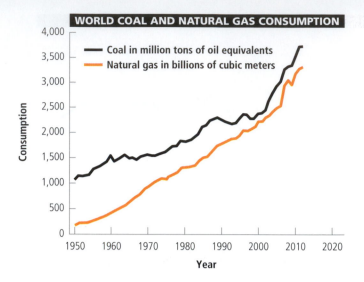

WORLD COAL AND NATURAL GAS CONSUMPTION

Coal in million tons of oil equivalents
Natural gas in billions of cubic meters

Consumption

Year

FIGURE 3 World coal and natural gas consumption for the period 1950–2012.

Compiled by the authors using data from British Petroleum and International Energy Agency.

Data and Graph Analysis

1. Which energy source has grown more steadily—coal or natural gas? In what years has coal use grown most sharply?
2. In what year did coal use reach a level twice as high as it was in 1960?

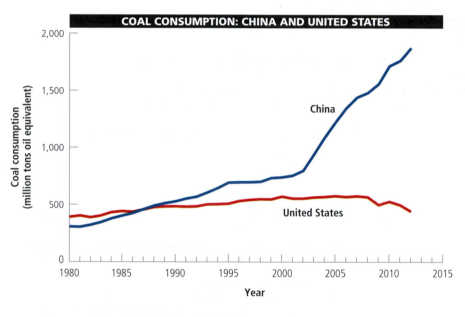

COAL CONSUMPTION: CHINA AND UNITED STATES

Coal consumption
(million tons oil equivalent)

China

United States

Year

FIGURE 4 Coal consumption in China and the United States, 1980–2012.

Compiled by the authors using data from Earth Policy Institute and British Petroleum.

Data and Graph Analysis

1. By what percentage did coal consumption increase in China between 1980 and 2012?
2. By what percentage did coal consumption decrease in the United States between 2005 and 2012?

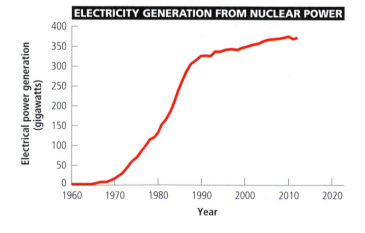

ELECTRICITY GENERATION FROM NUCLEAR POWER

Electrical power generation
(gigawatts)

Year

FIGURE 5 Global capacity for generation of electricity from nuclear power, 1960–2012.

Compiled by the authors using data from International Energy Agency, Worldwatch Institute, and Earth Policy Institute.

Data and Graph Analysis

1. After 1980, how long did it take to double that year's generating capacity?
2. Considering the decades of the 1970s, 1980s, 1990s, and 2000s, which decade saw the sharpest growth in generating capacity? During which decade did this growth level off?

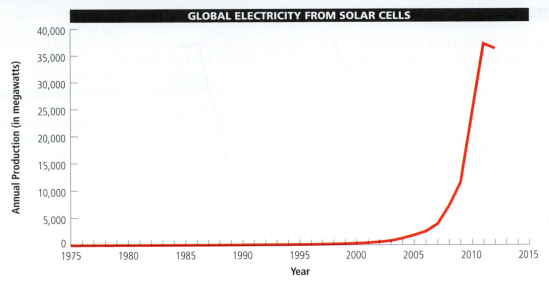

FIGURE 6 Global electricity production from solar cells, 1975–2012.

Compiled by the authors using data from U.S. Energy Information Administration, International Energy Agency, Worldwatch Institute, and Earth Policy Institute.

Data and Graph Analysis

1. About how many times more electricity was produced by solar cells in 2012 than in 2000?
2. How long did it take the world to go **(a)** from 2 to 5,000 megawatts of electricity produced in a year by solar cells and **(b)** from 5,000 to 30,000 megawatts of same?

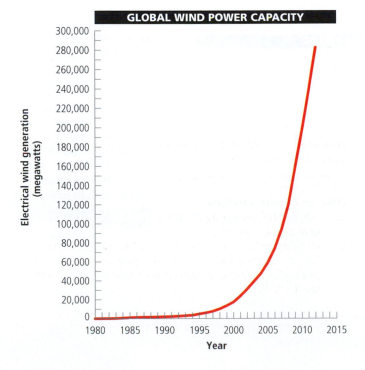

FIGURE 7 Global installed capacity for generation of electricity by wind energy, 1980–2012.

Compiled by the authors using data from Global Wind Energy Council, European Wind Energy Association, American Wind Energy Association, Worldwatch Institute, World Wind Energy Association, and Earth Policy Institute.

Data and Graph Analysis

1. How long did it take for the world to go from zero to 100,000 megawatts of installed capacity for wind-generated electricity?
2. In 2012, the world's installed capacity for generating electricity by wind power was about how many times more than it was in 1995?

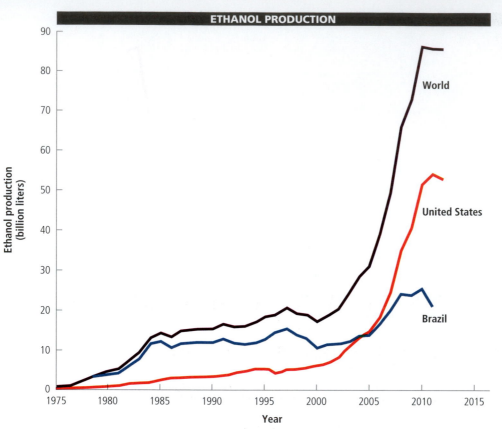

FIGURE 8 Production of ethanol motor fuel in the world, in Brazil, and in the United States, 1975–2012.

Compiled by the authors using data from USDA, U.S. Energy Information Administration, Worldwatch Institute, and Earth Policy Institute.

Data and Graph Analysis

1. By roughly what percentage did global ethanol production increase between 2000 and 2012?

2. In what years did ethanol production peak in **(a)** the United States, **(b)** Brazil, and **(c)** the world?

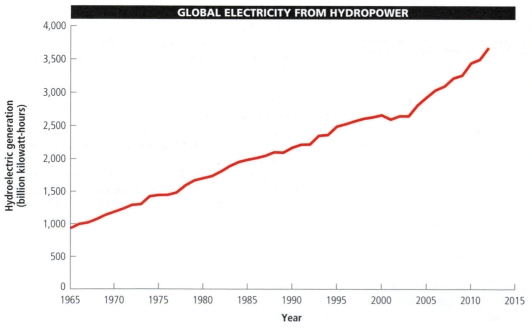

FIGURE 9 World hydroelectric generation, 1965–2012.

Compiled by the authors using data from International Energy Agency, British Petroleum, Worldwatch Institute, and Earth Policy Institute.

Data and Graph Analysis

1. After 1965, how many years did it take for the world to double its hydroelectric capacity?

2. Between what years was growth in hydroelectric capacity the sharpest?

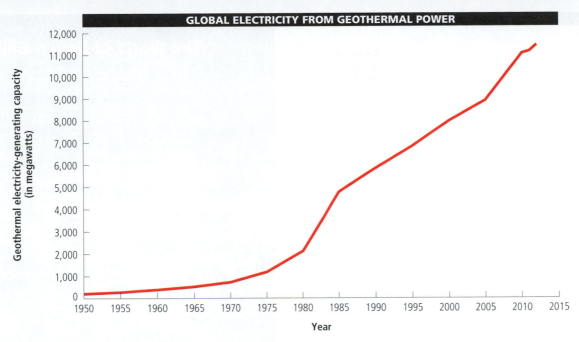

GLOBAL ELECTRICITY FROM GEOTHERMAL POWER

FIGURE 10 Global cumulative installed geothermal electricity-generating capacity, 1950–2012.

Compiled by the authors using data from International Energy Agency, Worldwatch Institute, Earth Policy Institute, and BP.

Data and Graph Analysis

1. About how many times more geothermal electricity-generating capacity was available in 2012 than in 1965?
2. Between what years was growth in this power source the sharpest?

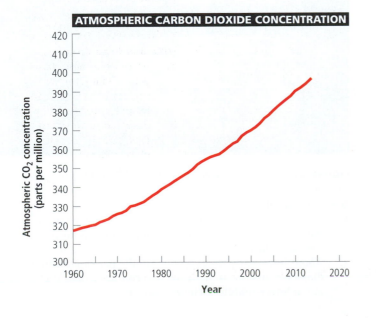

ATMOSPHERIC CARBON DIOXIDE CONCENTRATION

FIGURE 11 Atmospheric concentration of carbon dioxide (CO_2) measured at a major atmospheric research center in Mauna Loa, Hawaii, 1960–2013.

Compiled by the authors using data from Scripps Institute of Oceanography, U.S. Energy Information Agency, and Earth Policy Institute.

Data and Graph Analysis

1. By how much did atmospheric CO_2 concentrations grow between 1960 and 2013 **(a)** in parts per million and **(b)** in percentage?
2. Assuming that atmospheric CO_2 concentrations continue growing as is reflected on this graph, estimate the year in which such concentrations will reach 450 parts per million.

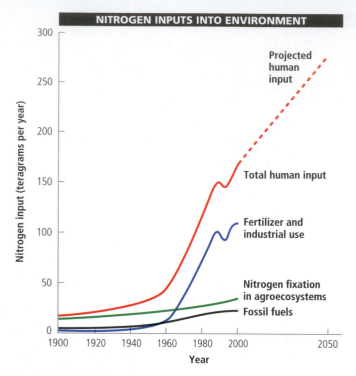

NITROGEN INPUTS INTO ENVIRONMENT

Nitrogen input (teragrams per year)

Projected human input

Total human input

Fertilizer and industrial use

Nitrogen fixation in agroecosystems

Fossil fuels

Year

FIGURE 12 Global trends in the annual inputs of nitrogen into the environment from human activities, with projections to 2050.

Compiled by the authors using data from Millennium Ecosystem Assessment and International Fertilizer Industry Association.

Data and Graph Analysis
1. About how many times higher than the total human input of nitrogen to the environment in 1900 was the human input in 2000?
2. If nitrogen inputs continue as projected, about how many times higher will the total inputs be in 2050 than they were in 1980?

The World of Seven Billion

Population
Most future population growth will happen in the less developed countries, where birthrates remain highest.

Life expectancy at birth
Improved health care and nutrition have raised life expectancy from a global average of 52 years in 1960 to 69 years today.

Deaths under age five (per 1,000 live births)
Worldwide there has been remarkable improvement. Since 1960, the number of children who die before age five has fallen by more than half.

Access to improved sanitation (percent)
The UN defines this as access to toilets–even simple pit toilets– that keep excrement away from humans, animals, and insects.

Deaths caused by infectious disease (percent)
The top five causes of death by infectious disease are acute respiratory infections (such as pneumonia), HIV/AIDS, diarrhea, TB, and malaria.

Years of education
Increases in education affect not only economic development but population: The more education a woman receives, the fewer children she is likely to bear.

Literacy rate (percent)
Global literacy is 82 percent. But for those who live where printed materials, even signs or product boxes, are rare, reading is a "use it or lose it" skill.

Fertility rate (children per woman)
In most of the world, the fertility rate has fallen. Among the reasons: decline in infant mortality, economic improvements, and education of women.

Rate of natural population increase (percent)
A country's annual natural growth rate is measured by subtracting the number of deaths from the number of births. It does not include migration, in or out.

Net migration rate (per 1,000 people)
More than 200 million people–over 3 percent of the world's population–live outside the country in which they were born.

Urban population (percent)
As of 2008, the world's population has shifted from mainly rural to more than 50 percent urban. Most urbanites live in cities of fewer than 500,000 people.

Carbon dioxide emissions (per capita, in metric tons)
Energy demand, largely for fossil fuels, continues to rise. China has surpassed the U.S. in total CO_2 emissions, but, per capita, U.S. emissions are four times higher.

Data for each category above were first compiled for all countries in each income level. The data were then averaged, accounting for differences in population.

National Geographic

FIGURE 13 Some characteristics of low-, middle-, and high-income countries.

Data and Graph Analysis
1. For every child under age 5 in a high-income country who dies, how many children under 5 in low-income countries die?
2. About how many times higher are the per capita carbon dioxide emissions in high-income countries than they are in lower-middle-income countries?

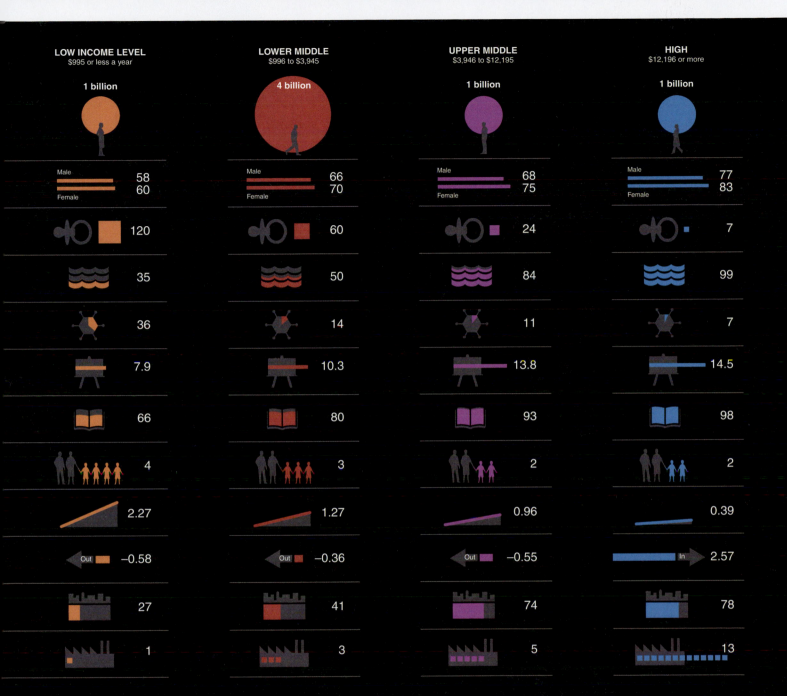

LOW INCOME LEVEL $995 or less a year	LOWER MIDDLE $996 to $3,945	UPPER MIDDLE $3,946 to $12,195	HIGH $12,196 or more
1 billion	4 billion	1 billion	1 billion
Male 58 / Female 60	Male 66 / Female 70	Male 68 / Female 75	Male 77 / Female 83
120	60	24	7
35	50	84	99
36	14	11	7
7.9	10.3	13.8	14.5
66	80	93	98
4	3	2	2
2.27	1.27	0.96	0.39
Out −0.58	Out −0.36	Out −0.55	In 2.57
27	41	74	78
1	3	5	13

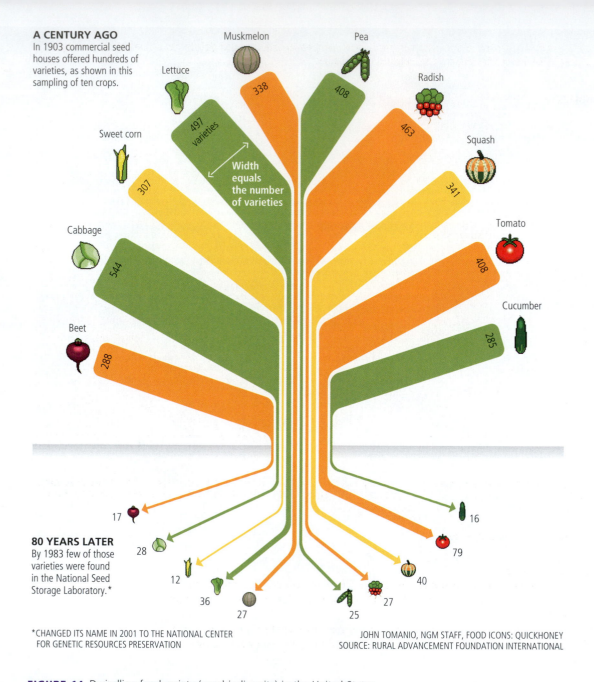

*CHANGED ITS NAME IN 2001 TO THE NATIONAL CENTER
 FOR GENETIC RESOURCES PRESERVATION

JOHN TOMANIO, NGM STAFF, FOOD ICONS: QUICKHONEY
SOURCE: RURAL ADVANCEMENT FOUNDATION INTERNATIONAL

FIGURE 14 Dwindling food variety (agrobiodiversity) in the United States.

National Geographic

Data and Graph Analysis

1. What percentage of the lettuce varieties available in 1903 were available in 1983? Answer the same question for tomato varieties.
2. Which three of the varieties shown here had the greatest declines in numbers of varieties available?

WHY SO WILD?

The atmosphere is getting warmer and wetter. Those two trends, which are clear in data averaged globally and annually, are increasing the chances of heat waves, heavy rains, and perhaps other extreme weather.

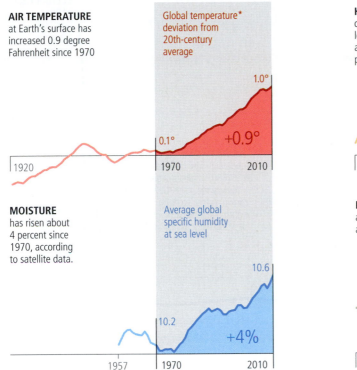

AIR TEMPERATURE at Earth's surface has increased 0.9 degree Fahrenheit since 1970

Global temperature* deviation from 20th-century average

1.0°

0.1° +0.9°

1920   1970   2010

MOISTURE has risen about 4 percent since 1970, according to satellite data.

Average global specific humidity at sea level

10.6

10.2 +4%

1957   1970   2010

HEAT WAVES— of which nighttime lows are one indicator— are striking a growing portion of the U.S.

Percentage of U.S. experiencing summer minimum temperatures much above normal

35%

4% +31%

1920   1970   2010

EXTREME RAINFALLS are now affecting larger areas of the U.S. as well.

Percentage of U.S. getting an elevated portion of precipitation from extreme events

16%

9% +7%

1920   1970   2010

GRAPHS ARE SMOOTHED USING A TEN-YEAR MOVING AVERAGE.
*AVERAGE TEMPERATURE OVER LAND AND OCEAN

JOHN TOMANIO, NGM STAFF, ROBERT THOMASON. SOURCES: JEFF MASTERS, WEATHER UNDERGROUND; NATIONAL CLIMATIC DATA CENTER (TEMPERATURE, HEAT WAVES, AND RAINFALL); NOAA (HUMIDITY)

FIGURE 15 Indicators of climate change.

National Geographic

Data and Graph Analysis

1. Compared to the percentage of the United States that experienced summer minimum temperatures much above normal in 1970, how much larger was that percentage in 2010?

2. For every 0.1 degree (F) increase in average air temperature that occurred between 1970 and 2010, how much did moisture in the atmosphere rise (on average and by percent)?

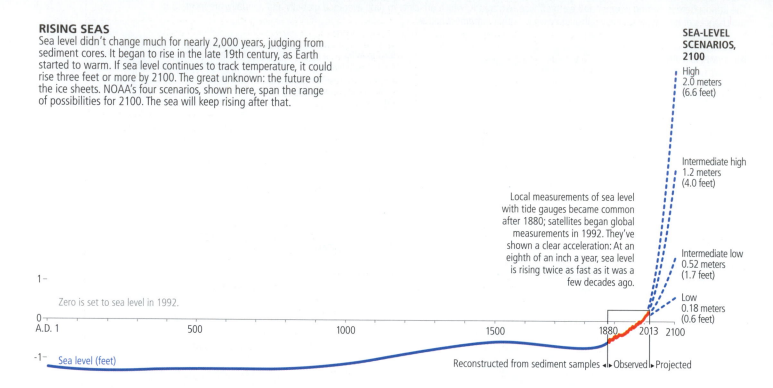

RISING SEAS

Sea level didn't change much for nearly 2,000 years, judging from sediment cores. It began to rise in the late 19th century, as Earth started to warm. If sea level continues to track temperature, it could rise three feet or more by 2100. The great unknown: the future of the ice sheets. NOAA's four scenarios, shown here, span the range of possibilities for 2100. The sea will keep rising after that.

SEA-LEVEL SCENARIOS, 2100

High
2.0 meters
(6.6 feet)

Intermediate high
1.2 meters
(4.0 feet)

Intermediate low
0.52 meters
(1.7 feet)

Low
0.18 meters
(0.6 feet)

Local measurements of sea level with tide gauges became common after 1880; satellites began global measurements in 1992. They've shown a clear acceleration: At an eighth of an inch a year, sea level is rising twice as fast as it was a few decades ago.

Zero is set to sea level in 1992.

Sea level (feet)

Reconstructed from sediment samples ◀▶ Observed ▶ Projected

FIGURE 16 *Rising seas.* Past and projected changes in sea level.

National Geographic

Data and Graph Analysis

1. Estimate the amount by which the sea level rose (in meters and feet) during the observed period between 1880 and 2013. (Hint: the red part of the curve above the x-axis represents a rise of 0.18 meters, or 0.6 feet.)

2. How many times higher than the low estimate for the rise in sea level by 2100 is the high estimate?

Era	Period	Time (millions of years ago)	Major Events (approximate time in millions of years ago, in parentheses)
Cenozoic (Age of Mammals)	Quaternary	1.6 – present	Likely beginning of new mass extinction (now) Human civilization develops (0.01 to now) Modern humans (*Homo sapiens sapiens*) (0.2) First humans (1.2)
	Tertiary	6.5 – 1.6	Oldest human ancestors (4.4) Grasses diversify and spread Mammals diversify and spread
Mesozoic (Age of Reptiles)	Cretaceous	146 – 6.5	Mass extinction (75% of species, including dinosaurs) (66) First primates First flowering plants
	Jurassic	208 – 146	Mass extinction (75% of species) (200) First birds Dinosaurs diversify and spread
	Triassic	245 – 208	First dinosaurs First mammals
Paleozoic (Age of Fishes)	Permian	290 – 245	Mass extinction (90–96% of species) (251) Reptiles diversify and spread
	Pennsylvanian	322 – 290	First reptiles
	Mississippian	362 – 322	Coal deposits form
	Devonian	408 – 362	Mass extinction (70% of species) (375) First land animals (amphibians) Fish diversify and spread First forests
	Silurian	439 – 408	First land plants and corals
	Ordovician	510 – 439	Mass extinction (60–70% of species) (450) First fish
	Cambrian	545 – 510	First shellfish Ozone layer forms Oxygen increases in atmosphere Photosynthetic organisms proliferate
Precambrian	Proterozoic	2,500 – 545	First animals in sea (jellyfish) First multicellular organisms
	Archean	4,600 – 2,500	First photosynthesis and oxygen in atmosphere (2,800) First plants in sea (algae) (3,200) Atmospheric water vapor condenses to oceans (3,700) First rocks (3,800) Likely origin of life (first one-celled organisms) (3,800) Earth forms (4,600)

Dates are approximate.

Three scientific principles of sustainability can guide us in making a shift to a more sustainable future.

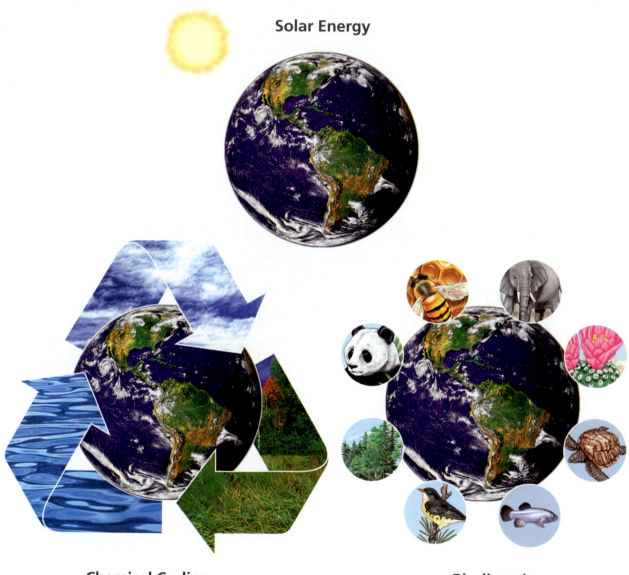

Solar Energy

Chemical Cycling

Biodiversity

Three social science principles of sustainability can guide us in making a shift to a more sustainable future.

ECONOMICS
Full-cost pricing

POLITICS
Win-win results

ETHICS
Responsibility to
future generations

acid deposition The falling of acids and acid-forming compounds from the atmosphere to the earth's surface. Acid deposition is commonly known as *acid rain*, a term that refers to the wet deposition of droplets of acids and acid-forming compounds.

acidic solution Any water solution that has more hydrogen ions (H^+) than hydroxide ions (OH^-); any water solution with a pH less than 7. Compare *basic solution, neutral solution*.

acidity Chemical characteristic that helps determine how a substance dissolved in water (a solution) will interact with and affect its environment; based on the comparative amounts of hydrogen ions (H^+) and hydroxide ions (OH^-) contained in a particular volume of the solution. See *pH*.

acid rain See *acid deposition*.

active solar heating system System that uses solar collectors to capture energy from the sun and store it as heat for space heating and water heating. Liquid or air pumped through the collectors transfers the captured heat to a storage system such as an insulated water tank or rock bed. Pumps or fans then distribute the stored heat or hot water throughout a dwelling as needed. Compare *passive solar heating system*.

adaptation Any genetically controlled structural, physiological, or behavioral characteristic that helps an organism survive and reproduce under a given set of environmental conditions. It usually results from a beneficial mutation. See *biological evolution, mutation, natural selection*.

adaptive trait See *adaptation*.

aerobic respiration Complex process that occurs in the cells of most living organisms, in which nutrient organic molecules such as glucose ($C_6H_{12}O_6$) combine with oxygen (O_2) to produce carbon dioxide (CO_2), water (H_2O), and energy. Compare *photosynthesis*.

affluence Wealth that results in high levels of consumption and unnecessary waste of resources, driven by the assumption that buying more and more material goods will bring fulfillment and happiness.

age structure Percentage of the population (or number of people of each gender) at each age level in a population.

agrobiodiversity The genetic variety of animal and plant species used on farms to produce food.

air pollution One or more chemicals in high enough concentrations in the air to harm humans, other animals, vegetation, or materials. Such chemicals or physical conditions are called air pollutants. See *primary pollutant, secondary pollutant*.

alley cropping Planting of crops in strips with rows of trees or shrubs on each side.

anaerobic respiration Form of cellular respiration in which some decomposers get the energy they need through the breakdown of glucose (or other nutrients) in the absence of oxygen. Compare *aerobic respiration*.

animal manure Dung and urine of animals used as a form of organic fertilizer. Compare *green manure*.

Anthropocene New era in which humans have become major agents of change in the functioning of the earth's life-support system as their ecological footprints have spread over the earth. Compare *Holocene*.

anthropocentric Human-centered. Compare *biocentric*.

anthropocentric environmental worldview See *human-centered environmental worldview*.

aquaculture Growing and harvesting of fish and shellfish for human use in freshwater ponds, irrigation ditches, and lakes, or in cages or fenced-in areas of coastal lagoons and estuaries, or in the open ocean.

aquatic life zone Marine and freshwater portions of the biosphere. Examples include freshwater life zones (such as lakes and streams) and ocean or marine life zones (such as estuaries, coastlines, coral reefs, and the open ocean).

aquifer Porous, water-saturated layers of sand, gravel, or bedrock that can yield an economically significant amount of water.

area strip mining Type of surface mining used where the terrain is flat. An earthmover strips away the overburden and a power shovel digs a trench to remove the mineral deposit. The trench is then filled with overburden, and a new trench is made parallel to the previous one. The process is repeated over the entire site. Compare *mountaintop removal mining, open-pit mining, subsurface mining*.

artificial selection Process by which humans select one or more desirable genetic traits in the population of a plant or animal species and then use selective breeding to produce populations containing many individuals with the desired traits. Compare *genetic engineering, natural selection*.

asthenosphere Zone within the earth's mantle made up of hot, partly melted rock that flows and can be deformed like soft plastic.

atmosphere The whole mass of air surrounding the earth. See *stratosphere, troposphere*. Compare *biosphere, geosphere, hydrosphere*.

atom The smallest unit of an element that can exist and still have the unique characteristics of that element; made of subatomic particles, it is the basic building block of all chemical elements and thus all matter. Compare *ion, molecule*.

atomic number Number of protons in the nucleus of an atom. Compare *mass number*.

atomic theory Idea that all elements are made up of atoms; the most widely accepted scientific theory in chemistry.

autotroph See *producer*.

background extinction rate Normal extinction rate of various species as a result of changes in local environmental conditions; usually refers to the extinction rate that existed before the human population grew exponentially. Compare *mass extinction*.

bacteria Prokaryotic, one-celled organisms, some of which transmit diseases. Most act as decomposers and get the nutrients they need by breaking down complex organic compounds in the tissues of living or dead organisms into simpler inorganic nutrient compounds.

basic solution Water solution with more hydroxide ions (OH^-) than hydrogen ions (H^+); water solution with a pH greater than 7. Compare *acidic solution, neutral solution*.

bioaccumulation An increase in the concentration of a chemical in specific organs or tissues at a level higher than would normally be expected. Compare *biomagnification*.

biocentric Life-centered. Compare *anthropocentric*.

biodegradable pollutant Material that can be broken down into simpler substances (elements and compounds) by bacteria or other decomposers. Paper and most organic wastes such as animal manure are biodegradable but can take decades to biodegrade in modern landfills. Compare *nondegradable pollutant*.

biodiversity Variety of different species (*species diversity*), genetic variability among individuals within each species (*genetic diversity*), variety of ecosystems (*ecological diversity*), and variety of functions such as energy flow and matter cycling that are part of the life-support system of the biosphere (*functional diversity*).

biodiversity hotspot An area especially rich in plant species that are found nowhere else and are in great danger of extinction. Such areas suffer serious ecological disruption, mostly because of rapid human population growth and the resulting pressure on natural resources.

biofuel Gas (such as methane) or liquid fuel (such as ethyl alcohol or biodiesel) made from plant material (biomass).

biogeochemical cycle Natural processes that recycle nutrients in various chemical forms from the nonliving environment to living organisms and then back to the nonliving environment. Examples include the carbon, oxygen, nitrogen, phosphorus, sulfur, and hydrologic cycles. See *nutrient cycling*.

biological community See *community*.

biological diversity See *biodiversity*.

biological evolution through natural selection Change in the genetic makeup of a population of a species in successive generations. If continued long enough, it can lead to the formation of a new species. Note that populations, not individuals, evolve. See also *adaptation, natural selection, theory of evolution*.

biological extinction Complete disappearance of a species from the earth. It happens when a species cannot adapt and successfully reproduce under new environmental conditions, when a species evolves into one or more new species, or when all the species' individuals are killed off by forces in the environment. Compare *speciation*. See also *endangered species, mass extinction, threatened species*.

biomagnification Increase in concentration of slowly degradable, fat-soluble chemicals such as DDT and PCBs in organisms at successively higher trophic levels of a food chain or web. Compare *bioaccumulation*.

biomass Organic matter produced by plants and other photosynthetic producers; total dry weight of all living organisms that can be supported at each trophic level in a food chain or web; dry weight of all organic matter in plants and animals in an ecosystem; plant materials and animal wastes used as fuel.

biome A terrestrial region distinguished by the predominance of certain types of vegetation and other forms of life. Examples include various types of deserts, grasslands, and forests.

biomimicry Process of observing certain changes in nature, studying how natural systems have responded to such changing conditions over many millions of years, and applying what is learned to dealing with some environmental challenge.

biosphere Zone of the earth where life is found. It consists of parts of the atmosphere (the troposphere), hydrosphere (mostly surface water and groundwater), and lithosphere (mostly soil and surface rocks and sediments on the bottoms of oceans and other bodies of water). Compare *atmosphere, geosphere, hydrosphere*.

birth rate See *crude birth rate*.

broadleaf deciduous plants Plants such as oak and maple trees that survive drought and cold by shedding their leaves and becoming dormant. Compare *broadleaf evergreen plants, coniferous evergreen plants*.

broadleaf evergreen plants Plants that keep most of their broad leaves year-round. An example is the trees found in the canopies of tropical rain forests. Compare *broadleaf deciduous plants, coniferous evergreen plants*.

buffer Substance that can react with hydrogen ions in a solution and thus hold the acidity or pH of a solution fairly constant. See *pH*.

carbon capture and storage (CCS) Process of removing carbon dioxide gas from coal-burning power and industrial plants and storing it somewhere (usually underground or under the seabed). To be effective, it must be stored so that it cannot be released into the atmosphere, essentially forever.

carbon cycle Cyclic movement of carbon in different chemical forms from the environment to organisms and then back to the environment.

carbon footprint The amount of carbon dioxide generated by an individual, an organization, or a geographically or politically defined area (such as a city, state, or country) in a given period of time.

carbon oxides Collective term for the compounds carbon monoxide and carbon dioxide, which can act as pollutants.

carcinogen Chemicals, ionizing radiation, and viruses that cause or promote the development of cancer. Compare *mutagen, teratogen*.

carnivore Animal that feeds on other animals. Compare *herbivore, omnivore*.

carrying capacity (K) Maximum population of a particular species that a given habitat can support over a given period.

CCD See *colony collapse disorder*.

CCS See *carbon capture and storage*.

cell Smallest living unit of an organism. Each cell is encased in an outer membrane or wall and contains genetic material (DNA) and other substances that enable it to perform its life function. Organisms such as bacteria consist of only one cell, but most organisms contain many cells.

cell theory The idea that all living things are composed of cells; the most widely accepted scientific theory in biology.

CFCs See *chlorofluorocarbons*.

chemical change Interaction between chemicals in which the chemical composition of the elements or compounds involved changes. Compare *nuclear change, physical change*.

chemical cycling The circulation of chemicals from the environment (mostly from soil and water) through organisms and back to the environment. See *nutrient cycling*.

chemical formula Shorthand way to show the number of atoms (or ions) in the basic structural unit of a compound. Examples include H_2O, $NaCl$, and $C_6H_{12}O_6$.

chemical reaction See *chemical change*.

chemosynthesis Process in which certain organisms (mostly specialized bacteria) extract inorganic compounds from their environment and convert them into organic nutrient compounds without the presence of sunlight. Compare *photosynthesis*.

chlorofluorocarbons (CFCs) Organic compounds, made up of atoms of carbon, chlorine, and fluorine, that can deplete the ozone layer when they slowly rise into the stratosphere and react with ozone molecules.

CHP (combined heat and power) See *cogeneration*.

chromosome A grouping of genes and associated proteins in plant and animal cells that carry certain types of genetic information. See *genes*.

chronic malnutrition Faulty nutrition, caused by a diet that does not supply an individual with enough protein, essential fats, vitamins, minerals, and other nutrients needed for good health. Compare *overnutrition, chronic undernutrition*.

chronic undernutrition Condition suffered by people who cannot grow or buy enough food to meet their basic energy needs. Most chronically undernourished children live in developing countries and are likely to suffer from mental retardation and stunted growth, and to die from infectious diseases. Compare *chronic malnutrition, overnutrition*.

clear-cutting Method of timber harvesting in which all trees in a forested area are removed in a single cutting. Compare *selective cutting, strip cutting*.

climate General pattern of atmospheric conditions in a given area over periods ranging from at least 30 years to thousands of years. The two main factors determining an area's climate are its average temperature, with its seasonal variations, and the average amount and distribution of precipitation. Compare *weather*.

climate change tipping point Point at which an environmental problem reaches a threshold level where scientists fear it could cause irreversible climate disruption.

closed-loop recycling See *primary recycling*.

coal Solid, combustible mixture of organic compounds with 30–98% carbon by weight, mixed with various amounts of water and small amounts of sulfur and nitrogen compounds. It forms in several stages as the remains of plants are subjected to heat and pressure over millions of years.

coal gasification Conversion of solid coal to synthetic natural gas (SNG).

coal liquefaction Conversion of solid coal to a liquid hydrocarbon fuel such as synthetic gasoline or methanol.

coastal wetland Land along a coastline, extending inland from an estuary that is covered with salt water all or part of the year. Examples include marshes, tidal flats, and mangrove swamps. Compare *inland wetland*.

coastal zone Warm, nutrient-rich, shallow part of the ocean that extends from the high-tide mark on land to the edge of a shelf-like extension of continental land masses known as the continental shelf. Compare *open sea*.

coevolution Evolution in which two or more species interact and exert selective pressures on each other that can lead each species to undergo adaptations. See *evolution, natural selection*.

cogeneration Production of two useful forms of energy, such as high-temperature heat or steam and electricity, from the same fuel source.

cold front Leading edge of an advancing mass of cold air. Compare *warm front*.

colony collapse disorder (CCD) The occurrence of very large losses of European honeybee colonies in the United States and in parts of Europe; the bees have been disappearing from their colonies during the winter and not returning as expected in the spring.

combined heat and power (CHP) production See *cogeneration*.

commensalism An interaction between organisms of different species in which one type of organism benefits and the other type is neither helped nor harmed to any great degree. Compare *mutualism*.

commercial forest See *tree plantation*.

commercial inorganic fertilizer Commercially prepared mixture of inorganic plant nutrients such as nitrates, phosphates, and potassium applied to the soil to restore fertility and increase crop yields. Compare *organic fertilizer*.

common-property resource Resource that is owned jointly by a large group of individuals, such as land that belongs to a whole village and that can be used by anyone for grazing cows or sheep. Compare *open access renewable resource*. See *tragedy of the commons*.

community Populations of all species living and interacting in an area at a particular time.

complex carbohydrate Two or more monomers of simple sugars (such as glucose) linked together. One example is the starches that plants use to store energy and also to provide energy for animals that feed on plants. Another is cellulose, the earth's most abundant organic compound, which is found in the cell walls of bark, leaves, stems, and roots.

compost Partially decomposed organic plant and animal matter used as a soil conditioner or fertilizer.

composting A form of recycling that mimics nature by using bacteria to decompose yard trimmings, vegetable food scraps, and other biodegradable organic wastes into materials that can be used to increase soil fertility.

compound Combination of atoms, or oppositely charged ions, of two or more elements held together by attractive forces called chemical bonds. Examples are NaCl, CO_2, and $C_6H_{12}O_6$. Compare *element*.

coniferous evergreen plants Cone-bearing plants (such as cedars, spruces, pines, and firs) that keep most of their leaves all year. Most of these tree species have narrow, pointed leaves (needles). Compare *broadleaf deciduous plants*, *broadleaf evergreen plants*.

coniferous trees Cone-bearing trees, mostly evergreens, that have needle-shaped or scale-like leaves. Examples are pines, firs, cedars, and spruce trees. They produce wood known commercially as softwood. Compare *deciduous plants*.

conservation biology Multidisciplinary science created to deal with the crisis of accelerating losses of the genes, species, communities, and ecosystems that make up earth's biological diversity. Its goals are to investigate human impacts on biodiversity and to develop practical approaches to preserving it.

conservation-tillage farming Crop cultivation in which the soil is disturbed little (minimum-tillage farming) or not at all (no-till farming) in an effort to reduce soil erosion, lower labor costs, and save energy. Compare *conventional-tillage farming*.

consumer Organism that cannot synthesize the organic nutrients it needs and gets its organic nutrients by feeding on the tissues of producers or of other consumers; generally divided into *primary consumers* (herbivores), *secondary consumers* (carnivores), *tertiary (higher-level) consumers*, *omnivores*, and *detritivores* (decomposers and detritus feeders). In economics, one who uses economic goods. Compare *producer*.

contour farming Plowing and planting across the changing slope of land, rather than in straight lines, to help retain water and reduce soil erosion.

contour strip mining Form of surface mining used on hilly or mountainous terrain. A power shovel cuts a series of terraces into the side of a hill. An earthmover removes the overburden, and a power shovel extracts the coal. The overburden from each new terrace is dumped onto the one below. Compare *area strip mining, mountaintop removal mining, open-pit mining, subsurface mining*.

conventional-tillage farming Crop cultivation method in which a planting surface is made by plowing land, breaking up the exposed soil, and then smoothing the surface. Compare *conservation-tillage farming*.

coral reef Formation produced by massive colonies containing billions of tiny coral animals, called polyps, that secrete a stony substance (calcium carbonate) around themselves for protection. When the corals die, their empty outer skeletons form layers; reefs grow with the accumulation of such layers. They are found in the coastal zones of warm tropical and subtropical oceans.

core Inner zone of the earth that consists of a solid inner core and a fluid outer core. Compare *crust, mantle*.

corrective feedback loop See *negative feedback loop*.

crop rotation Planting a field, or an area of a field, with different crops from year to year to reduce soil nutrient depletion. A plant such as corn, tobacco, or cotton, which removes large amounts of nitrogen from the soil, is planted one year. The next year a legume such as soybeans, which adds nitrogen to the soil, is planted.

crown fire Extremely hot forest fire that burns ground vegetation and treetops. Compare *ground fire, surface fire*.

crude birth rate Annual number of live births per 1,000 people in the population of a geographic area at the midpoint of a given year. Compare *crude death rate*.

crude death rate Annual number of deaths per 1,000 people in the population of a geographic area at the midpoint of a given year. Compare *crude birth rate*.

crude oil Gooey liquid consisting mostly of hydrocarbon compounds and small amounts of compounds containing oxygen, sulfur, and nitrogen. Extracted from underground accumulations, it is sent to oil refineries, where it is converted to heating oil, diesel fuel, gasoline, tar, and other materials.

crust Solid outer zone of the earth. It consists of oceanic crust and continental crust. Compare *core, mantle*.

cultural eutrophication Overnourishment of aquatic ecosystems with plant nutrients (mostly nitrates and phosphates) resulting from human activities such as agriculture, urbanization, and discharges from sewage treatment plants. See *eutrophication*.

dam A structure built across a river to control the river's flow or to create a reservoir. See *reservoir*.

data Factual information collected by scientists.

death rate See *crude death rate*.

deciduous plants Trees, such as oaks and maples, and other plants that survive during dry or cold seasons by shedding their leaves. Compare *coniferous trees*.

decomposer Organism that digests parts of dead organisms, as well as cast-off fragments and wastes of living organisms, by breaking down the complex organic molecules in those materials into simpler inorganic compounds and then absorbing the soluble nutrients. The balance of these broken down materials return to the soil and water for reuse. Decomposers primarily consist of various bacteria and fungi. Compare *consumer, detritivore, producer*.

deforestation Removal of trees from a forested area.

delta An area at the mouth of a river built up by deposits of river sediments, often containing estuaries and coastal wetlands.

demographic transition Hypothesis that countries, as they become industrialized, have declines in death rates followed by declines in birth rates.

depletion time The time it takes to use a certain fraction (usually 80%) of the known or estimated supply of a nonrenewable resource at an assumed rate of use. Finding and extracting the remaining 20% usually costs more than it is worth. See *economic depletion*.

desalination Purification of salt water or brackish (slightly salty) water by removal of dissolved salts.

desert Biome in which evaporation exceeds precipitation and the average amount of precipitation is less than 25 centimeters (10 inches) per year. Such areas have little vegetation or have widely spaced, mostly low vegetation. Compare *forest, grassland*.

desertification Conversion of rangeland, rain-fed cropland, or irrigated cropland to desert-like land, with a drop in agricultural productivity of 10% or more. It usually is caused by a combination of overgrazing, soil erosion, prolonged drought, and climate change.

detritivore Consumer organism that feeds on detritus—parts of dead organisms and cast-off fragments and wastes of living organisms. Examples include earthworms, termites, and crabs. Compare *decomposer*.

detritus Parts of dead organisms and cast-off fragments and wastes of living organisms.

detritus feeder See *detritivore*.

dissolved oxygen (DO) content Amount of oxygen gas (O_2) dissolved in a given volume of water at a particular temperature and pressure, often expressed as a concentration in parts of oxygen per million parts of water.

dose Amount of a potentially harmful substance an individual ingests, inhales, or absorbs through the skin. Compare *response*. See *dose-response curve*.

dose-response curve Plot of data showing the effects of various doses of a toxic agent on a group of test organisms. See *dose, response*.

drainage basin See *watershed*.

drought Condition in which an area does not get enough water because of lower-than-normal precipitation or higher-than-normal temperatures that increase evaporation.

earth-centered environmental worldview Worldview holding that we are part of, and dependent on, nature; that the earth's life-support system exists for all species, not just for us; that our economic success and the long-term survival of our cultures and our species depend on learning how the earth has sustained itself for billions of years and integrating such lessons from nature into the ways we think and act. Compare *human-centered environmental worldview* and *life-centered environmental worldview*. See *environmental wisdom worldview*.

earthquake Shaking of the ground resulting from the fracturing and displacement of subsurface rock, which produces a fault, or from subsequent movement along the fault.

ecological diversity The variety of forests, deserts, grasslands, oceans, streams, lakes, and other biological communities. See *biodiversity*. Compare *functional diversity, genetic diversity, species diversity*.

ecological footprint Amount of biologically productive land and water needed to supply a population with the renewable resources it uses and to absorb or dispose of the wastes from such resource use. It is a measure of the average environmental impact of populations in different countries and areas. See *per capita ecological footprint*.

ecological niche Total way of life, or role of a species in an ecosystem. It includes all physical, chemical, and biological conditions that a species needs in order to live and reproduce in an ecosystem.

ecological restoration Deliberate alteration of a degraded habitat or ecosystem to restore as much of its ecological structure and function as possible.

ecological succession Process in which communities of plant and animal species in a particular area are replaced over time by a series of different and often more complex communities. See *primary ecological succession, secondary ecological succession*.

ecological tipping point Point at which an environmental problem reaches a threshold level, which causes an often irreversible shift in the behavior of a natural system.

ecologist Biological scientist who studies relationships between living organisms and their environment.

ecology Biological science that studies the relationships between living organisms and their environment; study of the structure and functions of nature.

economic depletion Exhaustion of 80% of the estimated supply of a nonrenewable resource. Finding, extracting, and processing the remaining 20% usually costs more than it is worth. May also apply to the depletion of a renewable resource, such as a fish or tree species. See *depletion time*.

economic development Improvement of human living standards by economic growth. Compare *economic growth*.

economic growth Increase in the capacity to provide people with goods and services; an increase in gross domestic product (GDP). Compare *economic development*. See *gross domestic product*.

economic resources Natural resources, capital goods, and labor used in an economy to produce material goods and services. See *natural resources*.

economics Social science that deals with the production, distribution, and consumption of goods and services to satisfy people's needs and wants.

ecosystem One or more communities of different species interacting with one another and with the chemical and physical factors making up their nonliving environment.

ecosystem services Natural services that support life on the earth and are essential to the quality of human life and the functioning of the

world's economies; an important form of natural capital. Examples are the chemical cycles, natural pest control, and natural purification of air and water. See *natural capital, natural resources*.

edge effect Tendency for a transition zone between two different ecosystems to have greater species diversity and a higher density of organisms than are found in either of the individual ecosystems.

electromagnetic radiation Forms of kinetic energy traveling as electromagnetic waves. Examples include radio waves, TV waves, microwaves, infrared radiation, visible light, ultraviolet radiation, X-rays, and gamma rays.

electron (e) Tiny particle moving around outside the nucleus of an atom. Each electron has one unit of negative charge and almost no mass. Compare *neutron, proton*.

element Chemical, such as hydrogen (H), iron (Fe), sodium (Na), carbon (C), nitrogen (N), or oxygen (O), whose distinctly different atoms serve as the basic building blocks of all matter. Two or more elements combine to form the compounds that make up most of the world's matter. Compare *compound*.

endangered species Wild species with so few individual survivors that the species could soon become extinct in all or most of its natural range. Compare *threatened species*.

endemic species Species that is found in only one area. Such species are especially vulnerable to extinction.

energy Capacity to do work or to transfer heat. Can involve mechanical, physical, chemical, or electrical tasks or heat transfers between objects at different temperatures.

energy efficiency Percentage of the total energy input that does useful work and is not converted into low-quality, generally useless heat in an energy conversion system or process. See *energy quality, net energy yield*.

energy quality Ability of a form of energy to do useful work. High-temperature heat and the chemical energy in fossil fuels are examples of concentrated high-quality energy. Low-quality energy such as low-temperature heat is dispersed or diluted and cannot do much useful work. See *high-quality energy, low-quality energy*.

environment All external conditions, factors, matter, and energy, living and nonliving, that affect any living organism or other specified system.

environmental degradation Depletion or destruction of a potentially renewable resource such as soil, grassland, forest, or wildlife that is used faster than it is naturally replenished. If such use continues, the resource becomes nonrenewable (on a human time scale) or nonexistent (extinct). See also *natural capital degradation, sustainable yield*.

environmental ethics Human beliefs about what is right or wrong with how we treat the environment.

environmentalism Social movement dedicated to protecting the earth's life-support systems for us and other species.

environmental justice Fair treatment and meaningful involvement of all people, regard-

less of race, color, sex, national origin, or income, with respect to the development, implementation, and enforcement of environmental laws, regulations, and policies.

environmentally sustainable economic development An approach that uses political and economic systems to encourage environmentally beneficial and more sustainable forms of economic growth and to discourage environmentally harmful forms of economic growth that degrade natural capital.

environmental movement Citizens organized to demand that political leaders enact laws and develop policies to curtail pollution, clean up polluted environments, and protect unspoiled areas from environmental degradation.

environmental policy Laws, rules, and regulations related to environmental issues that are developed, implemented, and enforced by a particular government body or agency.

environmental resistance All of the limiting factors that act together to limit the growth of a population. See *limiting factor*.

environmental science Interdisciplinary study that uses information and ideas from the physical sciences (such as biology, chemistry, and geology) as well as those from the social sciences and humanities (such as economics, politics, and ethics) to learn how nature works, how we interact with the environment, and how we can deal with environmental problems.

environmental scientist Scientist who uses information from the physical sciences and social sciences to learn how nature works and how humans interact with the environment and to develop solutions to environmental problems. See *environmental science*.

environmental wisdom worldview Earth-centered worldview holding that humans are part of and totally dependent on nature and that nature exists for all species, not just for us. Our success depends on learning how the earth sustains itself and integrating such environmental wisdom into the ways we think and act. Compare *frontier environmental worldview, planetary management worldview, stewardship worldview*. See *earth-centered environmental worldview, human-centered environmental worldview, life-centered environmental worldview*.

environmental worldview Set of assumptions and beliefs about how nature works and how humans relate to the environment; includes an environmental ethic, or a belief about what is right or wrong with how we treat the environment. See *earth-centered environmental worldview, human-centered environmental worldview, life-centered environmental worldview, environmental wisdom worldview, frontier environmental worldview, planetary management worldview, stewardship worldview*.

EPA U.S. Environmental Protection Agency; responsible for managing U.S. government efforts to control air and water pollution, radiation and pesticide hazards, environmental research, hazardous waste, and solid waste disposal.

epidemiology Study of the patterns of disease or other harmful effects from exposure to toxins and diseases caused by pathogens within defined groups of people to find out why some people get sick and some do not.

erosion Process or group of processes by which loose or consolidated earth materials, especially topsoil, are dissolved, loosened, or worn away and removed from one place and deposited in another.

estuary Partially enclosed coastal area at the mouth of a river from which freshwater, carrying fertile silt and runoff from the land, mixes with salty seawater.

eutrophication Physical, chemical, and biological changes that take place after a lake, estuary, or slow-flowing stream receives inputs of plant nutrients—mostly nitrates and phosphates—from natural erosion and runoff from the surrounding land basin. See *cultural eutrophication*.

eutrophic lake Lake with a large or excessive supply of plant nutrients, mostly nitrates and phosphates. Compare *oligotrophic lake*.

evaporation Conversion of a liquid into a gas.

evergreen plants Plants that keep some of their leaves or needles throughout the year. Examples include cone-bearing trees (conifers) such as firs, spruces, pines, redwoods, and sequoias. Compare *deciduous plants*.

evolution See *biological evolution*.

exhaustible resource See *nonrenewable resource*.

experiment Procedure a scientist uses to study some phenomenon under known conditions. Scientists conduct some experiments in the laboratory and others in nature. The resulting scientific data or facts must be verified or confirmed by repeated observations and measurements, ideally by several different investigators.

exponential growth Growth in which some quantity, such as population size or economic output, increases at a constant rate per unit of time. An example is the growth sequence 2, 4, 8, 16, 32, 64, and so on, which increases by 100% at each interval. When the increase in quantity over time is plotted, this type of growth yields a curve shaped like the letter J.

external cost Harmful environmental, economic, or social effect of producing and using an economic good that is not included in the market price of the good. Compare *internal cost*. See *full-cost pricing*.

extinction See *biological extinction*.

extinction rate Percentage or number of species that go extinct within a certain period of time such as a year.

family planning Providing information, clinical services, and contraceptives to help people choose the number and spacing of children they want to have.

feedback Any system output of matter, energy, or information that, when fed back into the system, increases or decreases a change to the system. See *feedback loop, negative feedback loop*, and *positive feedback loop*.

feedback loop The process that occurs when an output of matter, energy, or information is fed back into the system as an input and leads to changes in that system. See *feedback*. Compare *negative feedback loop* and *positive feedback loop*.

feedlot Confined outdoor or indoor space used to raise hundreds to thousands of domesticated livestock.

first law of thermodynamics Whenever energy is converted from one form to another in a physical or chemical change, no energy is created or destroyed, but energy can be changed from one form to another; you cannot get more energy out of something than you put in; in terms of energy quantity, you cannot get something for nothing. This law does not apply to nuclear changes, in which large amounts of energy can be produced from small amounts of matter. See *second law of thermodynamics*.

fishery Concentration of particular aquatic species suitable for commercial harvesting in a given ocean area or inland body of water.

fish farming See *aquaculture*.

fishprint Area of ocean needed to sustain the consumption of an average person, a nation, or the world. Compare *ecological footprint*.

floodplain Flat valley floor next to a stream channel. For legal purposes, the term often applies to any low area that has the potential for flooding, including certain coastal areas.

flows See *throughputs*.

food chain Series of organisms in which each organism is eaten or decomposed by the next one in the series. Compare *food web*.

food insecurity Condition under which people live with chronic hunger and malnutrition that threatens their ability to lead healthy and productive lives. Compare *food security*.

food security Condition under which every person in a given area has daily access to enough nutritious food to have an active and healthy life. Compare *food insecurity*.

food web Complex network of many interconnected food chains and feeding relationships. Compare *food chain*.

forest Biome with enough average annual precipitation to support the growth of tree species and smaller forms of vegetation. Compare *desert, grassland*.

fossil fuel Product of partial or complete decomposition of plants and animals; occurs as crude oil, coal, natural gas, or heavy oil as a result of exposure to heat and pressure in the earth's crust over millions of years. See *coal, crude oil, natural gas*.

fossils Skeletons, bones, shells, body parts, leaves, seeds, or impressions of such items that provide recognizable evidence of organisms that lived long ago.

fracking See *hydraulic fracturing*.

freshwater Relatively pure water containing few dissolved salts.

freshwater life zones Aquatic systems where water with a dissolved salt concentration of less than 1% by volume accumulates on or flows through the surfaces of terrestrial biomes. Examples include standing (lentic) bodies of freshwater such as lakes, ponds, and inland wetlands and flowing (lotic) systems such as streams and rivers. Compare *biome*.

front The boundary between two air masses with different temperatures and densities. See *cold front, warm front*.

frontier environmental worldview Human-centered worldview held by European colonists settling North America in the 1600s, holding that the continent had vast resources and was a wilderness to be conquered, cleared, and cultivated by settlers. See *earth-centered environmental worldview, human-centered environmental worldview, life-centered environmental worldview*.

frontier science See *tentative science*.

full-cost pricing Setting market prices of goods and services to include all the harmful environmental and health costs of producing and using them. See *external cost, internal cost*.

functional diversity Biological and chemical processes or functions such as energy flow and matter cycling that are parts of the life-support system of the biosphere. See *biodiversity, ecological diversity, genetic diversity, species diversity*.

GDP See *gross domestic product*.

generalist species Species with a broad ecological niche. They can live in many different places, eat a variety of foods, and tolerate a wide range of environmental conditions. Examples include flies, cockroaches, mice, rats, and humans. Compare *specialist species*.

genes Coded units of information about specific traits that are passed from parents to offspring during reproduction. They consist of segments of DNA molecules found in chromosomes.

genetically modified organism (GMO) Organism whose genetic makeup has been altered by genetic engineering.

genetic diversity Variability in the genetic makeup among individuals within a single species. See *biodiversity*. Compare *ecological diversity, functional diversity, species diversity*.

genetic engineering Insertion of an alien gene into an organism to give it a beneficial genetic trait. Compare *artificial selection, natural selection*.

genuine progress indicator (GPI) GDP plus the estimated value of beneficial transactions that meet basic needs, but in which no money changes hands, minus the estimated harmful environmental, health, and social costs of all transactions. Compare *gross domestic product*.

geoengineering Any use of technology to try to manipulate certain natural conditions in order to counter our enhancement of the greenhouse effect.

geographic isolation Separation of populations of a species into different areas for long periods of time.

geology Study of the earth's dynamic history, as recorded in its rocks, and of the features and processes of the earth's interior and surface.

geosphere Earth's intensely hot core, thick mantle composed mostly of rock, and thin outer crust that contains most of the earth's rock, soil, and sediment. Compare *atmosphere, biosphere, hydrosphere*.

geothermal energy Heat transferred from the earth's underground concentrations of dry steam (steam with no water droplets), wet steam (a mixture of steam and water droplets), or hot water trapped in fractured or porous rock.

global warming Warming of the earth's lower atmosphere (troposphere) because of increases in the concentrations of one or more greenhouse gases. It can result in climate change that can last for decades to thousands of years. See *greenhouse effect, greenhouse gases*.

GMO See *genetically modified organism*.

GPI See *genuine progress indicator*.

GPP See *gross primary productivity*.

graph A tool for conveying information that we can summarize numerically by illustrating that information in a visual format.

grassland Biome found in regions with enough annual average precipitation to support the growth of grass and small plants but not enough to support large stands of trees. Compare *desert, forest*.

greenhouse effect Natural effect that releases heat in the atmosphere near the earth's surface. Water vapor, carbon dioxide, ozone, and other gases in the lower atmosphere (troposphere) absorb some of the infrared radiation (heat) radiated by the earth's surface. Their molecules vibrate and transform the absorbed energy into longer-wavelength infrared radiation in the troposphere. If the atmospheric concentrations of these greenhouse gases increase and other natural processes do not remove them, the average temperature of the lower atmosphere will increase.

greenhouse gases Gases in the earth's lower atmosphere (troposphere) that cause the greenhouse effect. Examples include carbon dioxide, chlorofluorocarbons, ozone, methane, water vapor, and nitrous oxide.

green manure Freshly cut or still-growing green vegetation that is plowed into the soil to increase the organic matter and humus available to support crop growth. Compare *animal manure*.

green revolution Popular term for the introduction of scientifically bred or selected varieties of grain (rice, wheat, corn) that, with adequate inputs of fertilizer and water, can greatly increase crop yields.

gross domestic product (GDP) Annual market value of all goods and services produced by all firms and organizations, foreign and domestic, operating within a country. See *per capita GDP*. Compare *genuine progress indicator (GPI)*.

gross primary productivity (GPP) Rate at which an ecosystem's producers capture and store a given amount of chemical energy as biomass in a given length of time. Compare *net primary productivity*.

ground fire Fire that burns decayed leaves or peat deep below the ground's surface. Compare *crown fire, surface fire*.

groundwater Water that sinks into the soil and is stored in slowly flowing and slowly renewed underground reservoirs called *aquifers*; underground water in the zone of saturation, below the water table. Compare *runoff, surface water*.

habitat Place or type of place where an organism or population of organisms lives. Compare *ecological niche*.

habitat fragmentation Breakup of a larger habitat area into smaller areas, usually as a result of human activities.

hazardous chemical Chemical that can cause harm because it is flammable or explosive, can irritate or damage the skin or lungs (such as strong acidic or alkaline substances), or can cause allergic reactions of the immune system (allergens). Also referred to as a *toxic chemical*.

hazardous waste Any solid, liquid, or containerized gas that can catch fire easily, is corrosive to skin tissue or metals, is unstable and can explode or release toxic fumes, or has harmful concentrations of one or more toxic materials that can leach out. These substances are usually by-products of manufacturing processes. Sometimes called *toxic waste*.

heat Total kinetic energy of all randomly moving atoms, ions, or molecules within a given substance, excluding the overall motion of the whole object. Heat always flows spontaneously from a warmer sample of matter to a colder sample of matter. This is one way to state the *second law of thermodynamics*.

herbivore Plant-eating organism. Examples include deer, sheep, grasshoppers, and zooplankton. Compare *carnivore, omnivore*.

heterotroph See *consumer*.

high Air mass with a high pressure. Compare *low*.

high-grade ore Ore containing a large amount of a desired mineral. Compare *low-grade ore*.

high-input agriculture See *industrialized agriculture*.

high-quality energy Energy that is concentrated and has great ability to perform useful work. Examples include high-temperature heat and the energy in electricity, coal, oil, gasoline, sunlight, and nuclei of uranium-235. Compare *low-quality energy*.

high-quality matter Matter that is concentrated and contains a high concentration of a useful resource. Compare *low-quality matter*.

high-throughput economy Economic system in most advanced industrialized countries, in which ever-increasing economic growth is sustained by maximizing the rate at which matter and energy resources are used, with little emphasis on pollution prevention, recycling, reuse, reduction of unnecessary waste, and other forms of resource conservation. Compare *low-throughput economy*.

high-waste economy See *high-throughput economy*.

HIPPCO Acronym used by conservation biologists for the six most important secondary causes of premature extinction: **H**abitat destruction, degradation, and fragmentation; **I**nvasive (nonnative) species; **P**opulation growth (too many people consuming too many resources); **P**ollution; **C**limate change; and **O**verexploitation.

Holocene Period of relatively stable climate and other environmental conditions that followed the long period of glaciation; it has allowed the human population to grow, develop agriculture, and take over a large and growing share of the earth's land and other resources. Compare *Anthropocene*.

horizontal drilling A method of drilling first vertically to a certain point, then bending the flexible well bore and drilling horizontally to gain access to oil and natural gas deposits trapped within layers of shale or other rock deposits.

human capital People's physical and mental talents that provide labor, innovation, culture, and organization. Compare *manufactured capital, natural capital*.

human-centered (anthropocentric) environmental worldview Worldview that sees the natural world primarily as a support system for human life. Examples are the *planetary management worldview*, which holds that humans are separate from and in charge of nature and that we can manage the earth mostly for our benefit, into the distant future; and the *stewardship worldview*, which holds that we can and should manage the earth for our benefit, but we have an ethical responsibility to be caring and responsible managers, or *stewards*, of the earth. Compare *earth-centered environmental worldview* and *life-centered environmental worldview*.

human resources See *human capital*.

hunger See *chronic undernutrition*.

hydraulic fracturing Process of injecting water mixed with sand and some toxic chemicals underground through horizontal natural gas wells and then using explosives and high pressure to fracture the deep rock and free up the natural gas stored there. The gas flows out of the well along with much of the water and a mix of compounds pulled from the rocks that can include salts, toxic heavy metals, and naturally occurring radioactive materials. Commonly referred to as *fracking*.

hydroelectric power plant Structure in which the energy of falling or flowing water spins a turbine generator to produce electricity.

hydrologic cycle Biogeochemical cycle that collects, purifies, and distributes the earth's fixed supply of water from the environment to living organisms and then back to the environment.

hydropower Electrical energy produced by falling or flowing water. See *hydroelectric power plant*.

hydrosphere The sum total of the earth's liquid water (oceans, lakes, other bodies of surface water, and underground water), frozen water (polar ice caps, floating ice caps, and ice in soil, known as *permafrost*), and water vapor in the atmosphere. See also *hydrologic cycle*. Compare *atmosphere, biosphere, geosphere*.

igneous rock Rock formed when molten rock material (magma) wells up from the earth's interior, cools, and solidifies into rock masses. Compare *metamorphic rock, sedimentary rock*. See *rock cycle*.

immigration Migration of people into a country or area to take up permanent residence.

indicator species Species whose decline serves as early warnings that a community or ecosystem is being degraded. Compare *keystone species, native species, nonnative species*.

industrialized agriculture (high-input agriculture) Production of large quantities of crops and livestock for sale to major markets; involves use of large inputs of energy from fossil fuels (especially oil and natural gas), water, fertilizer, and pesticides.

industrial smog Type of air pollution consisting mostly of a mixture of sulfur dioxide, suspended droplets of sulfuric acid formed from some of the sulfur dioxide, and suspended solid particles. Compare *photochemical smog*.

industrial solid waste Solid waste produced by mines, factories, refineries, food growers, and businesses that supply people with goods and services. Compare *municipal solid waste*.

inertia (persistence) The ability of a living system such as a grassland or a forest to survive moderate disturbances.

inexhaustible resource Resource such as sunlight that is expected to last essentially forever, in human terms. See *nonrenewable resource, renewable resource*.

infant mortality rate Number of babies out of every 1,000 born each year who die before their first birthday.

infectious disease Disease caused when a pathogen such as a bacterium, virus, or parasite invades the body and multiplies in its cells and tissues. Examples are flu, HIV, malaria, tuberculosis, and measles. See *transmissible disease*. Compare *nontransmissible disease*.

inland wetland Land away from the coast, such as a swamp, marsh, or bog, that is covered all or part of the time with freshwater. Compare *coastal wetland*.

inorganic compounds All compounds not classified as organic compounds. See *organic compounds*.

inputs Matter and energy from the environment that is put into a living system. See *system*. Compare *outputs* and *throughputs*.

instrumental value Value of an organism, species, ecosystem, or the earth's biodiversity based on its usefulness to humans. Compare *intrinsic value*.

integrated pest management (IPM) Combined use of biological, chemical, and cultivation methods in proper sequence and timing to keep the size of a pest population below the level that causes economically unacceptable loss of a crop or livestock animal.

integrated waste management Variety of strategies for both waste reduction and waste management designed to deal with the solid wastes we produce.

internal cost Direct cost paid by the producer and the buyer of an economic good. Compare *external cost*. See *full-cost pricing*.

interspecific competition Attempts by members of two or more species to use the same limited resources in an ecosystem.

intrinsic value Value of an organism, species, ecosystem, or the earth's biodiversity based on its existence, regardless of whether it has any usefulness to humans. Compare *instrumental value*.

invasive species See *nonnative species*.

inversion See *temperature inversion*.

ion Atom or group of atoms with one or more positive (+) or negative (−) electrical charges. Examples are Na^+ and Cl^-. Compare *atom, molecule*.

IPM See *integrated pest management*.

irrigation Supplying water to crops by artificial means rather than by relying on natural rainfall.

isotopes Two or more forms of a chemical element that have the same number of protons but different mass numbers because they have different numbers of neutrons in their nuclei.

keystone species Species that play roles affecting many other organisms in an ecosystem; these roles often have the effect of supporting or maintaining an ecosystem. Compare *indicator species, native species, nonnative species*.

kinetic energy Energy that matter has because of its mass and speed, or velocity. Compare *potential energy*.

K-selected species Species that tend to do well in competitive conditions when their population size is near the carrying capacity (K) of their environment. They tend to reproduce later in life and have a small number of offspring with fairly long life spans. Examples include elephants, whales, humans, birds of prey, saguaro cactus, and most tropical rain forest trees. Compare *r-selected species*.

lake Large natural body of standing freshwater formed when water from precipitation, land runoff, or groundwater flow fills a depression in the earth created by glaciation, earth movement, volcanic activity, or a giant meteorite. See *eutrophic lake, oligotrophic lake*.

land degradation Decrease in the ability of land to support crops, livestock, or wild species in the future as a result of natural or human-induced processes.

landfill See *sanitary landfill*.

land-use planning Planning to determine the best present and future uses of each parcel of land.

law of conservation of energy See *first law of thermodynamics*.

law of conservation of matter In any physical or chemical change, matter is neither created nor destroyed but merely changed from one form to another; in physical and chemical changes, existing atoms are rearranged into different spatial patterns (physical changes) or different combinations (chemical changes).

law of nature See *scientific law*.

LD50 See *median lethal dose*.

less-developed country Country that has low to moderate industrialization and low to moderate per capita GDP. Most are located in Africa, Asia, and Latin America. Compare *more-developed country*.

life-centered environmental worldview Worldview holding that all species have value as participating members of the biosphere, regardless of their potential or actual use to humans; includes the belief that we have an ethical responsibility to avoid hastening the extinction of species through our activities. Compare *earth-centered environmental worldview* and *human-centered environmental worldview*.

life expectancy Average number of years a newborn infant can be expected to live.

limiting factor Single factor that limits the growth, abundance, or distribution of the population of a species in an ecosystem. See *limiting factor principle*.

limiting factor principle Too much or too little of any abiotic factor can limit or prevent growth of a population of a species in an ecosystem, even if all other factors are at or near the optimal range of tolerance for the species.

lipids A chemically diverse group of large organic compounds that do not dissolve in water. Examples are fats and oils for storing energy, waxes for structure, and steroids for producing hormones.

liquefied natural gas (LNG) Natural gas converted to liquid form by cooling it to a very low temperature.

liquefied petroleum gas (LPG) Mixture of liquefied propane (C_3H_8) and butane (C_4H_{10}) gas removed from natural gas and used as a fuel.

lithosphere Outer shell of the earth, composed of the crust and the rigid, outermost part of the mantle outside the asthenosphere. Compare *crust, geosphere, mantle*.

LNG See *liquefied natural gas*.

logistic growth Pattern in which exponential population growth occurs when the population is small, and population growth decreases steadily with time as the population approaches the carrying capacity.

low Air mass with a low pressure. Compare *high*.

low-grade ore Ore containing a small amount of a desired mineral. Compare *high-grade ore*.

low-quality energy Energy that is dispersed and has little ability to do useful work. An example is low-temperature heat. Compare *high-quality energy*.

low-quality matter Matter that is dilute or dispersed or contains a low concentration of a useful resource. Compare *high-quality matter*.

low-throughput (low-waste) economy Economy based on working with nature by recycling and reusing discarded matter; preventing pollution; conserving matter and energy resources by reducing unnecessary waste and use; and building things that are easy to recycle, reuse, and repair. Compare *high-throughput economy*.

low-waste economy See *low-throughput economy*.

LPG See *liquefied petroleum gas*.

malnutrition See *chronic malnutrition*.

mangrove forest Ecosystem, found on some coastlines in warm tropical climates, that may contain any of 69 species of trees and shrubs that can live partly submerged in the salty waters of coastal swamps.

mantle Zone of the earth's interior between its core and its crust. Compare *core, crust*. See *geosphere, lithosphere*.

manufactured capital See *manufactured resources*.

manufactured inorganic fertilizer Commercially prepared mixture of inorganic plant nutrients such as nitrates, phosphates, and potassium applied to the soil to restore fertility and increase crop yields. Compare *organic fertilizer*.

manufactured resources Manufactured items made from natural resources and used to produce and distribute economic goods and services bought by consumers. They include tools, machinery, equipment, factory buildings, and transportation and distribution facilities. Compare *human capital, natural resources*.

map An important visual tool used to summarize data that vary over small or large areas in 2- or 3-dimensional representations of those areas.

marine life zones See *saltwater life zones*.

mass extinction Catastrophic, widespread, often global event in which large numbers of species become extinct over a short time period, compared with normal (background) extinctions. Compare *background extinction rate*.

mass number Sum of the number of neutrons and the number of protons in the nucleus of an atom. It gives the approximate mass of that atom. Compare *atomic number*.

matter Anything that has mass (the amount of material in an object) and takes up space. On the earth, where gravity is present, we weigh an object to determine its mass.

matter quality Measure of how useful a matter resource is, based on its availability and concentration. See *high-quality matter, low-quality matter*.

median lethal dose (LD50) Amount of a toxic substance per unit of test animal body weight that kills half the test animal population within a certain time period.

megacity City with 10 million or more people.

metamorphic rock Rock produced when a preexisting rock is subjected to high temperatures (which may cause it to melt partially), high pressures, chemically active fluids, or a combination of these agents. Compare *igneous rock, sedimentary rock*. See *rock cycle*.

microorganisms Organisms such as bacteria that are so small that it takes a microscope to see them.

migration Movement of any species into and out of a specific geographic area.

mineral Any naturally occurring inorganic substance found in the earth's crust as a crystalline solid. See *mineral resource*.

mineral resource Concentration of mineral material in or on the earth's crust in a form and amount such that extracting and converting it into useful materials or items is currently or potentially profitable. Mineral resources are classified as metallic (such as iron and tin ores) or nonmetallic (such as sand and salt).

minimum-tillage farming See *conservation-tillage farming*.

model Approximate representation or simulation of a system being studied.

molecule Combination of two or more atoms of the same chemical element (such as O_2) or different chemical elements (such as H_2O) held together by chemical bonds. Compare *atom, ion*.

monoculture Cultivation of a single crop, usually on a large area of land. Compare *polyculture*.

more-developed country Country that is highly industrialized and has a high per capita GDP. Compare *less-developed country*.

mountaintop removal mining Type of surface mining that uses explosives, massive power shovels, and large machines called draglines to remove the top of a mountain and expose seams of coal underneath the removed soil and rocks, which are pushed into valleys below, causing multiple harmful environmental effects. Compare *area strip mining, contour strip mining*.

MSW See *municipal solid waste*.

municipal solid waste (MSW) Solid materials discarded by homes and businesses in or near urban areas. See *solid waste*. Compare *industrial solid waste*.

mutagen Agent such as a chemical or form of radiation that increases the frequency of mutations in the DNA molecules found in cells. See *carcinogen, mutation, teratogen*.

mutation Random change in the DNA molecules that make up genes; such changes can alter anatomy, physiology, or behavior in the affected organism's offspring. See *mutagen*.

mutualism Type of species interaction in which both participating species generally benefit. Compare *commensalism*.

nanotechnology Uses science and engineering to manipulate and create materials out of atoms and molecules at the ultrasmall scale of less than 100 nanometers.

native species Species that normally live and thrive in a particular ecosystem. Compare *indicator species, keystone species, nonnative species*.

natural capital Natural resources and ecosystem services that keep us and other species alive and support our economies. See *ecosystem services, natural resources*.

natural capital degradation The waste, depletion, or destruction of any of the earth's natural capital. See *environmental degradation*.

natural gas Underground deposits of gases consisting of 50–90% by weight methane gas (CH_4) and small amounts of heavier gaseous hydrocarbon compounds such as propane (C_3H_8) and butane (C_4H_{10}).

natural income Renewable resources such as plants, animals, and soil provided by natural capital.

natural resources Forms of matter in nature, such as air, water, and soil, and forms of energy, such as solar and wind energy, that are essential or useful to humans. See *natural capital*.

natural selection Process by which a particular beneficial gene (or set of genes) is reproduced in succeeding generations more than other genes. The result of natural selection tends to be a population that contains a greater proportion of organisms better adapted to certain environmental conditions. See *adaptation, biological evolution, mutation*.

natural services See *ecosystem services*.

nature-deficit disorder A wide range of problems, including anxiety, depression, and attention-deficit and behavior disorders, that might be resulting at least partially from a lack of contact with nature.

negative feedback loop The process that occurs when an output of matter, energy, or information is fed back into the system as an input and slows or stops a change occurring to the system or causes the system to change in the opposite direction. See *system*. Compare *positive feedback loop*.

net energy yield Total amount of useful energy available from an energy resource or energy system over its lifetime, minus the amount of energy *used* (the first energy law), *automatically wasted* (the second energy law), and *unnecessarily wasted* in finding, processing, concentrating, and transporting it to users.

net primary productivity (NPP) Rate at which all the plants in an ecosystem produce net useful chemical energy; it is equal to the difference between the rate at which the plants in an ecosystem produce useful chemical energy (gross primary productivity) and the rate at which they use some of that energy through cellular respiration. Compare *gross primary productivity*.

neutral solution Water solution containing an equal number of hydrogen ions (H^+) and hydroxide ions (OH^-); water solution with a pH of 7. Compare *acidic solution, basic solution*.

neutron (n) Elementary particle in the nuclei of all atoms (except hydrogen-1). It has a relative mass of 1 and no electric charge. Compare *electron, proton*.

niche See *ecological niche*.

nitric oxide (NO) Colorless gas that forms when nitrogen and oxygen gas in air react at the high-combustion temperatures in automobile engines and coal-burning plants. Lightning and certain bacteria in soil and water also produce NO as part of the *nitrogen cycle*.

nitrogen cycle Cyclic movement of nitrogen in different chemical forms from the environment to organisms and then back to the environment.

nitrogen dioxide (NO_2) Reddish-brown gas formed when nitrogen oxide reacts with oxygen in the air.

nitrogen oxides (NO_x) The collective term for nitric oxide and nitrogen dioxide. See *nitric oxide* and *nitrogen dioxide*.

noise pollution Any unwanted, disturbing, or harmful sound that impairs or interferes with hearing.

nondegradable pollutant Material that is not broken down by natural processes. Examples include the toxic elements lead and mercury. Compare *biodegradable pollutant*.

nonnative species Species that migrate into an ecosystem or are deliberately or accidentally introduced into an ecosystem by humans. Compare *native species*.

nonpoint source Broad and diffuse area, rather than a specific point, from which pollutants enter bodies of surface water or air. Examples include runoff of chemicals and sediments from cropland, livestock feedlots, logged forests, urban streets, parking lots, lawns, and golf courses. Compare *point source*.

nonrenewable resource Resource that exists in a fixed amount (stock) in the earth's crust and has the potential for renewal by geological, physical, and chemical processes taking place over hundreds of millions to billions of years. Examples include copper, aluminum, coal, and oil. We classify these resources as exhaustible because we are extracting and using them at a much faster rate than they are formed. Compare *inexhaustible resource, renewable resource*.

nontransmissible disease Disease that is not caused by living organisms and does not spread from one person to another. Examples include most cancers, diabetes, cardiovascular disease, and malnutrition. Compare *transmissible disease*.

no-till farming See *conservation-tillage farming*.

NPP See *net primary productivity*.

nuclear change Process in which nuclei of certain isotopes spontaneously change, or are forced to change, into one or more different isotopes. The three principal types of nuclear change are radioactive decay, nuclear fission, and nuclear fusion. Compare *chemical change, physical change*.

nuclear fission Nuclear change in which the nuclei of certain isotopes with large mass numbers (such as uranium-235 and plutonium-239) are split apart into lighter nuclei when struck by a neutron. This process releases more neutrons and a large amount of energy. Compare *nuclear fusion*.

nuclear fuel cycle Includes the mining of uranium, processing and enriching the uranium to make fuel, using it in the reactor, safely storing the resulting highly radioactive wastes for thousands of years until their radioactivity falls to safe levels, and retiring the highly radioactive plant by taking it apart and storing its high- and moderate-level radioactive material safely for thousands of years.

nuclear fusion Nuclear change in which two nuclei of isotopes of elements with a low mass number (such as hydrogen-2 and hydrogen-3) are forced together at extremely high temperatures until they fuse to form a heavier nucleus (such as helium-4). This process releases a large amount of energy. Compare *nuclear fission*.

nucleic acid Large polymer molecule made by linking hundreds to thousands of four types of monomers called nucleotides. Two nucleic acids—DNA (deoxyribonucleic acid) and RNA (ribonucleic acid)—participate in the building of proteins and carry hereditary information used to pass traits from parent to offspring.

nucleus Extremely tiny center of an atom, making up most of the atom's mass. It contains one or more positively charged protons and one or more neutrons with no electrical charge (except for a hydrogen-1 atom, which has one proton and no neutrons in its nucleus).

nutrient Any chemical that an organism must ingest in order to live, grow, or reproduce.

nutrient cycle See *biogeochemical cycle*.

nutrient cycling The circulation of chemicals necessary for life, from the environment (mostly from soil and water) through organisms and back to the environment.

ocean acidification The rising overall levels of acidity in ocean waters, occurring because the oceans absorb about 25% of the CO_2 emitted into the atmosphere by human activities, especially the burning of carbon-containing fossil fuels. The CO_2 reacts with ocean water to form a weak acid, which is having harmful effects on various aquatic species.

ocean currents Mass movements of surface water produced by prevailing winds blowing over the oceans and shaped largely by continental land forms.

oil sand See *tar sand*.

old-growth (primary) forest Virgin and old, second-growth forests containing trees that are often hundreds—sometimes thousands—of years old. Examples include forests of Douglas fir, western hemlock, giant sequoia, and coastal redwoods in the western United States.

oligotrophic lake Lake with a low supply of plant nutrients. Compare *eutrophic lake*.

omnivore Animal that can use both plants and other animals as food sources. Examples include pigs, rats, cockroaches, and humans. Compare *carnivore, herbivore*.

open access renewable resource Renewable resource owned by no one and available for use by anyone at little or no charge. Examples include clean air, groundwater, the open ocean and its fish, and the ozone layer. Compare *common-property resource*.

open dump Fields or holes in the ground where garbage is deposited and sometimes covered with soil. They are rare in more-developed countries, but are widely used in many less-developed countries, especially to handle wastes from megacities. Compare *sanitary landfill*.

open-pit mining Removing minerals such as gravel, sand, and metal ores by digging them out of the earth's surface and leaving an open pit behind. Compare *area strip mining, contour strip mining, mountaintop removal mining, subsurface mining*.

open sea Part of any ocean that lies beyond the continental shelf. Compare *coastal zone*.

ore Part of a metal-yielding material that can be economically extracted from a mineral; typically contains two parts: the ore mineral, which contains the desired metal, and waste mineral material (gangue). See *high-grade ore, low-grade ore*.

organic agriculture Growing crops with limited or no use of synthetic pesticides and synthetic fertilizers and no use of genetically modified crops; raising livestock without use of synthetic growth regulators and feed additives; and using organic fertilizer (manure, legumes, compost) and natural pest controls (bugs that eat harmful bugs, plants that repel bugs, and environmental controls such as crop rotation).

organic compounds Compounds containing carbon atoms combined with each other and with atoms of one or more other elements such as hydrogen, oxygen, nitrogen, sulfur, phosphorus, chlorine, and fluorine. All other compounds are called *inorganic compounds*.

organic farming See *organic agriculture* and *sustainable agriculture*.

organic fertilizer Organic material such as animal manure, green manure, and compost applied to cropland as a source of plant nutrients. Compare *manufactured inorganic fertilizer*.

organism Any form of life.

outputs Matter and energy that leaves a living system and enters the environment. See *system*. Compare *inputs* and *throughputs*.

overburden Layer of soil and rock overlying a mineral deposit. Surface mining removes this layer.

overgrazing Destruction of vegetation when too many grazing animals feed too long on a specific area of pasture or rangeland and exceed the carrying capacity of a rangeland or pasture area.

overnutrition Diet so high in calories, saturated (animal) fats, salt, sugar, and processed foods and so low in vegetables and fruits that the consumer runs a high risk of developing diabetes, hypertension, heart disease, and other health hazards. Compare *chronic malnutrition* and *chronic undernutrition*.

ozone (O_3) Colorless and highly reactive gas and a major component of photochemical smog. Also found in the ozone layer in the stratosphere. See *ozone layer, photochemical smog*.

ozone depletion Decrease in concentration of ozone (O_3) in the stratosphere. See *ozone layer*.

ozone layer Layer of gaseous ozone (O_3) in the stratosphere that protects life on earth by filtering out most harmful ultraviolet radiation from the sun.

parasite Consumer organism that lives on or in, and feeds on, a living plant or animal, known as the host, over an extended period. The parasite draws nourishment from and gradually weakens its host; it may or may not kill the host. See *parasitism*.

parasitism Interaction between species in which one organism, called the parasite, preys on another organism, called the host, by living on or in the host.

particulates Also known as suspended particulate matter (SPM); variety of solid particles and liquid droplets small and light enough to remain suspended in the air for long periods. About 62% of the SPM in outdoor air comes from natural sources such as dust, wild fires, and sea salt. The remaining 38% comes from human sources such as coal-burning electric power and industrial plants, motor vehicles, plowed fields, road construction, unpaved roads, and tobacco smoke.

passive solar heating system System that, without the use of mechanical devices, captures sunlight directly within a structure and converts it to low-temperature heat for space heating or for heating water for domestic use. Compare *active solar heating system*.

pasture Managed grassland or enclosed meadow that usually is planted with domesticated grasses or other forage to be grazed by livestock.

pathogen Living organism that can cause disease in another organism. Examples include bacteria, viruses, and parasites.

PCBs See *polychlorinated biphenyls*.

peak production Point in time when the pressure in an oil well drops and its rate of conventional crude oil production starts declining, usually a decade or so; for a group of wells or for a nation, the point at which all wells on average have passed peak production.

peer review Process of scientists reporting details of the methods and models they used, the results of their experiments, and the reasoning behind their hypotheses for other scientists working in the same field (their peers) to examine and criticize.

per capita ecological footprint Amount of biologically productive land and water needed to supply each person in a population with the renewable resources he or she uses and to absorb or dispose of the wastes from such resource use. It measures the average environmental impact of individuals in populations in different countries and areas. Compare *ecological footprint*.

per capita GDP Annual gross domestic product (GDP) of a country divided by its total population at midyear. See *gross domestic product*. Compare *genuine progress indicator (GPI)*.

perennial Plant that can live for more than 2 years.

permafrost Perennially frozen layer of the soil that forms when the water in the soil freezes. It is found in arctic tundra.

perpetual resource See *inexhaustible resource*.

persistence See *inertia*.

pest Unwanted organism that directly or indirectly interferes with human activities.

pesticide Any chemical designed to kill or inhibit the growth of a pest.

petrochemicals Chemicals obtained by refining (distilling) crude oil. They are used as raw materials in manufacturing most industrial chemicals, fertilizers, pesticides, plastics, synthetic fibers, paints, medicines, and many other products.

petroleum See *crude oil*.

pH Numeric value that indicates the relative acidity or alkalinity of a substance on a scale of 0 to 14, with the neutral point at 7. Acid solutions have pH values lower than 7; basic or alkaline solutions have pH values greater than 7.

phosphorus cycle Cyclic movement of phosphorus in different chemical forms from the environment to organisms and then back to the environment.

photochemical smog Complex mixture of air pollutants produced in the lower atmosphere by the reaction of hydrocarbons and nitrogen oxides under the influence of sunlight. Especially harmful components include ozone, peroxyacyl nitrates (PANs), and various aldehydes. Compare *industrial smog*.

photosynthesis Complex process that takes place in cells of green plants. Radiant energy from the sun is used to combine carbon dioxide (CO_2) and water (H_2O) to produce oxygen (O_2), carbohydrates (such as glucose, $C_6H_{12}O_6$), and other nutrient molecules. Compare *aerobic respiration*.

photovoltaic (PV) cell Device that converts solar energy directly into electrical energy. Also called a solar cell.

physical change Process that alters one or more physical properties of an element or a compound without changing its chemical composition. Examples include changing the size and shape of a sample of matter (crushing ice and cutting aluminum foil) and changing a sample of matter from one physical state to another (boiling and freezing water). Compare *chemical change, nuclear change*.

phytoplankton Small, drifting plants, mostly algae and bacteria, found in aquatic ecosystems. Compare *plankton, zooplankton*.

pioneer species First hardy species—often microbes, mosses, and lichens—that begin colonizing a site as the first stage of ecological succession. See *ecological succession*.

planetary management worldview A human-centered worldview holding that humans are separate from nature, that nature exists mainly to meet our needs and increasing wants, and that we can use our ingenuity and technology to manage the earth's life-support systems, mostly for our benefit. It assumes that economic growth is unlimited. Compare *environmental wisdom worldview, stewardship worldview*. See *earth-centered environmental worldview, human-centered environmental worldview, life-centered environmental worldview*.

plankton Small plant organisms (phytoplankton) and animal organisms (zooplankton) that float in aquatic ecosystems. See *phytoplankton, zooplankton*.

plantation agriculture A form of industrialized agriculture used primarily in tropical less-developed countries to grow cash crops such as bananas, coffee, vegetables, soybeans, sugarcane, and palm oil. These crops are grown on large monoculture plantations, mostly for export to more-developed countries.

plate tectonics Theory of geophysical processes that explains the movements of lithospheric plates and the processes that occur at their boundaries. See *lithosphere, tectonic plates*.

point source Single identifiable source that discharges pollutants into the environment. Examples include the smokestack of a power plant or an industrial plant, drainpipe of a meatpacking plant, chimney of a house, or exhaust pipe of an automobile. Compare *nonpoint source*.

policies Programs, and the laws and regulations through which they are enacted, that a government enforces and funds.

politics Process through which individuals and groups try to influence or control government policies and actions that affect local, state, national, and international communities.

pollutant Particular chemical or form of energy that can adversely affect the health, survival, or activities of humans or other living organisms. See *pollution*.

pollution Undesirable change in the physical, chemical, or biological characteristics of air, water, soil, or food that can adversely affect the health, survival, or activities of humans or other living organisms.

pollution cleanup Device, process, or strategy used to remove or reduce the level of a pollutant after it has been produced. Examples include automobile emission control devices and sewage treatment plants. Compare *pollution prevention*.

pollution prevention Device, process, or strategy used to prevent a potential pollutant from forming or entering the environment or to sharply reduce the amount entering the environment. Compare *pollution cleanup*.

polychlorinated biphenyls (PCBs) Group of 209 toxic, oily, synthetic chlorinated hydrocarbon compounds that can be biologically amplified in food chains and webs.

polyculture Complex form of intercropping in which a large number of different plants maturing at different times are planted together. Compare *monoculture*.

population Group of individual organisms of the same species living in a particular area.

population change Increase or decrease in the size of a population. It is equal to (Births + Immigration) − (Deaths + Emigration).

population crash Dieback of a population that has depleted its supply of resources, exceeding the carrying capacity of its environment. See *carrying capacity*.

population density Number of organisms in a particular population found in a specified area of land or volume of water or air.

positive feedback loop The process that occurs when an output of matter, energy, or information is fed back into the system as an input and causes the system to change further in the same direction. See *system*. Compare *negative feedback loop*.

potential energy Energy stored and potentially available for use. Examples include a rock held in your hand, the water in a reservoir behind a dam, and the chemical energy stored in the carbon atoms of coal or in molecules of food. Compare *kinetic energy*.

poverty Inability of people to meet their basic needs for food, clothing, and shelter.

prairie See *grassland*.

precautionary principle When there is significant scientific uncertainty about potentially serious harm from chemicals or technologies, decision makers should act to prevent harm to humans and to the environment. See *pollution prevention*.

precipitation Water in the form of rain, sleet, hail, and snow that falls from the atmosphere onto land and bodies of water.

predation Interaction in which an organism of one species (the predator) captures and feeds on some or all parts of an organism of another species (the prey).

predator Organism that captures and feeds on some or all parts of an organism of another species (the prey).

predator–prey relationship Relationship that has evolved between two organisms, in which one organism has become the prey for the other, the latter called the predator. See *predator, prey*.

prey Organism that is killed by an organism of another species (the predator) and serves as its source of food.

primary (closed-loop) recycling Process in which materials are recycled into new products of the same type—turning used aluminum cans into new aluminum cans, for example.

primary consumer Organism that feeds on some or all parts of plants (herbivore) or on other producers. Compare *detritivore, omnivore, secondary consumer*.

primary ecological succession Ecological succession in an area without soil or bottom sediments. See *ecological succession*. Compare *secondary ecological succession*.

primary pollutant Chemical that has been added directly to the air by natural events or human activities and occurs in a harmful concentration. Compare *secondary pollutant*.

primary productivity See *gross primary productivity, net primary productivity*.

primary sewage treatment Mechanical sewage treatment in which large solids are filtered out by screens and suspended solids settle out as sludge in a sedimentation tank. Compare *secondary sewage treatment*.

principles of sustainability See *scientific principles of sustainability, social science principles of sustainability*.

probability Mathematical statement about how likely it is that something will happen.

producer Organism that uses solar energy (green plants) or chemical energy (some bacteria) to manufacture the organic compounds it needs as nutrients from simple inorganic compounds obtained from its environment. Compare *consumer, decomposer*.

protein Large polymer molecules formed by linking together long chains of monomers called amino acids.

proton (p) Positively charged particle in the nuclei of all atoms. Each proton has a relative mass of 1 and a single positive charge. Compare *electron, neutron*.

proven oil reserves Identified deposits from which conventional crude oil can be extracted profitably at current prices with current technology.

PV cell See *photovoltaic cell*.

pyramid of energy flow Diagram representing the flow of energy through each trophic level in a food chain or food web. With each energy transfer, only a small part (typically 10%) of the usable energy entering one trophic level is transferred to the organisms at the next trophic level.

radioactive decay Change of a radioisotope to a different isotope by the emission of radioactivity.

radioactive waste Waste products of nuclear power plants, research, medicine, weapons production, and other processes involving nuclear reactions. See *radioactivity*.

radioactivity Nuclear change in which unstable nuclei of atoms spontaneously shoot out "chunks" of mass, energy, or both at a fixed rate. The three principal types of radioactive products are gamma rays and fast-moving alpha particles and beta particles.

rain shadow effect Effects of prolonged low precipitation on the leeward side of a high mountain range when prevailing winds flow up and over these mountains, dropping their moisture on the windward side and creating semiarid and arid conditions on their leeward side.

rangeland Land that supplies forage or vegetation (grasses, grass-like plants, and shrubs) for grazing and browsing animals. Compare *pasture*.

range of tolerance Range of variations in any physical or chemical environmental factor under which a population can survive. See *limiting factor*.

reconciliation ecology Science of inventing, establishing, and maintaining habitats to conserve species diversity in places where people live, work, or play.

recycle To collect and reprocess discarded materials so that they can be made into new products; one of the four Rs of resource use. An example is collecting aluminum cans, melting them down, and using the aluminum to make new cans or other aluminum products. See *primary recycling, secondary recycling*. Compare *reduce, refuse,* and *reuse*.

reduce To consume less and live a simpler lifestyle; one of the four Rs of resource use. Compare *recycle, refuse,* and *reuse*.

refining A complex process of heating crude oil to separate it into various fuels and other components with different boiling points.

refuse To choose not to consume a particular product or service in order to shrink one's impact; one of the four Rs of resource use. Compare *recycle, reduce,* and *reuse*.

reliable science Concepts and ideas that are widely accepted by experts in a particular field of the natural or social sciences. Compare *tentative science, unreliable science*.

reliable surface runoff Surface runoff of water that generally can be counted on as a stable source of water from year to year. See *runoff*.

renewable resource Resource that can be replenished rapidly (in hours to several decades) through natural processes as long as it is not used up faster than it is replaced. Examples include trees in forests, grasses in grasslands, wild animals, fresh surface water in lakes and streams, most groundwater, fresh air, and fertile soil. If such a resource is used faster than it is replenished, it can be depleted. Compare *nonrenewable resource* and *inexhaustible resource*. See also *environmental degradation*.

replacement-level fertility Average number of children a couple must bear to replace themselves. The average for a country or the world usually is slightly higher than two children per couple (2.1 in the United States and 2.5 in some less-developed countries) mostly because some children die before reaching their reproductive years. See also *total fertility rate*.

representative democracy A government run by the people through elected officials and representatives.

reproductive isolation Long-term geographic separation of members of a particular sexually reproducing species.

reserves Resources that have been identified and from which a usable mineral can be extracted profitably at present prices with current mining or extraction technology.

reservoir Artificial lake created when a stream is dammed. See *dam*.

resilience The ability of a living system to be restored through secondary succession after a more severe disturbance.

resource Anything obtained from the environment to meet human needs and wants. It can also apply to other species.

resource partitioning Process of dividing up resources in an ecosystem so that species with similar needs (overlapping ecological niches) use the same scarce resources at different times, in different ways, or in different places. See *ecological niche*.

response Amount of health damage caused by exposure to a certain dose of a harmful substance or form of radiation. See *dose, dose-response curve*.

restoration ecology Research and scientific study devoted to restoring, repairing, and reconstructing damaged ecosystems.

reuse To use a product over and over again in the same form. An example is collecting, washing, and refilling glass beverage bottles. One of the four Rs. Compare *reduce, refuse,* and *recycle*.

riparian zone A thin strip or patch of vegetation that borders a stream. These zones provide habitats and resources for wildlife.

risk Probability that something undesirable will result from deliberate or accidental exposure to a hazard. See *risk analysis, risk assessment, risk management*.

risk analysis Identifying hazards, evaluating the nature and severity of risks associated with the hazards *(risk assessment)*, ranking risks, and using this and other information to determine options and make decisions about reducing or eliminating risks *(risk management)*.

risk assessment Process of gathering data and making assumptions to estimate short- and long-term harmful effects on human health or the environment from exposure to hazards associated with the use of a particular product or technology. See *risk, risk analysis, risk management*.

risk management Use of risk assessment and other information to determine options and make decisions about reducing or eliminating risks. See *risk, risk analysis, risk assessment*.

rock Any solid material that makes up a large, natural, continuous part of the earth's crust. See *mineral*.

rock cycle Largest and slowest of the earth's cycles, consisting of geologic, physical, and chemical processes that form and modify rocks and soil in the earth's crust over millions of years.

r-selected species Species that have a capacity for a high rate of population growth (r). They tend to have many, usually small, offspring and to give them little or no parental care or protection. Examples include algae, bacteria, and most insects with short life spans. Compare *K-selected species*.

runoff Freshwater from precipitation and melting ice that flows on the earth's surface into streams, lakes, wetlands, and reservoirs. See *reliable surface runoff, surface runoff, surface water*. Compare *groundwater*.

salinity Concentration of various salts dissolved in a given volume of water.

salinization Accumulation of salts in soil that can eventually make the soil unable to support plant growth.

saltwater life zones Aquatic life zones that depend on saltwater, including oceans and their accompanying bays, estuaries, coastal wetlands, shorelines, coral reefs, and mangrove forests; also called *marine life zones*.

sanitary landfill Waste disposal site on land in which waste is spread in thin layers, compacted, and covered with a fresh layer of clay or plastic foam each day. Compare *open dump*.

science Broad field of study involving efforts to discover order in nature and to use that knowledge to make projections about what is likely to happen in nature. See *data, reliable science, scientific hypothesis, scientific law, scientific model, scientific theory, tentative science, unreliable science*.

scientific hypothesis An educated guess aimed at explaining a scientific law or certain scientific observations. Compare *scientific law, scientific model, scientific theory*.

scientific law Description of what scientists find happening in nature repeatedly in the same way, without known exception. See *first law of thermodynamics, law of conservation of matter, second law of thermodynamics*. Compare *scientific hypothesis, scientific model, scientific theory*.

scientific model A simulation of a complex process or system. Many are mathematical models that are run and tested using computers. Compare *scientific hypothesis, scientific law, scientific theory*.

scientific principles of sustainability Principles by which nature has sustained itself for billions of years by relying on solar energy, biodiversity, and nutrient recycling.

scientific theory A well-tested and widely accepted scientific hypothesis. Compare *scientific hypothesis, scientific law, scientific model*.

secondary consumer Organism that feeds only on primary consumers. Compare *detritivore, omnivore, primary consumer*.

secondary ecological succession Ecological succession in an area in which natural vegetation has been removed or destroyed but the soil or bottom sediment has not been destroyed. See *ecological succession*. Compare *primary ecological succession*.

secondary pollutant Harmful chemical formed in the atmosphere when a primary air pollutant reacts with normal air components or other air pollutants. Compare *primary pollutant*.

secondary recycling A process in which waste materials are converted into different products; for example, used tires can be shredded and turned into rubberized road surfacing. Compare *primary (closed-loop) recycling*.

secondary sewage treatment Second step in most waste treatment systems in which aerobic bacteria decompose as much as 90% of degradable, oxygen-demanding organic wastes in wastewater. Compare *primary sewage treatment*.

second-growth forest A stand of trees resulting from secondary ecological succession; these forests develop after the trees in an area have been removed by human activities, such as clear-cutting for timber or cropland, or by natural forces such as fire, hurricanes, or volcanic eruptions.

second law of thermodynamics Whenever energy is converted from one form to another in a physical or chemical change, we end up with lower-quality or less usable energy than we started with. In any conversion of heat energy to useful work, some of the initial energy input is always degraded to lower-quality, more dispersed, less useful energy—usually low-temperature heat that flows into the environment. See *first law of thermodynamics*.

sedimentary rock Rock that forms from the accumulated products of erosion and in some cases from the compacted shells, skeletons, and other remains of dead organisms. Compare *igneous rock, metamorphic rock*. See *rock cycle*.

selective cutting Cutting of intermediate-aged, mature, or diseased trees in an uneven-aged forest stand, either singly or in small groups. This encourages the growth of younger trees and maintains an uneven-aged stand. Compare *clear-cutting, strip cutting*.

septic tank Underground tank for treating wastewater from a home, used in rural and suburban areas. Bacteria in the tank decompose organic wastes, and the sludge settles to the bottom of the tank. The effluent flows out of the tank into the ground through a field of drainpipes.

shale oil Slow-flowing, dark brown, heavy oil obtained when kerogen in oil shale is vaporized at high temperatures and then condensed. Shale oil can be refined to yield gasoline, heating oil, and other petroleum products.

smart growth Form of urban planning that recognizes that urban growth will occur but uses zoning laws and other tools to prevent sprawl, direct growth to certain areas, protect ecologically sensitive and important lands and waterways, and develop urban areas that are more environmentally sustainable and more enjoyable places to live.

smelting Process in which a mineral ore is heated in order to separate a desired metal from the other elements in the ore.

SNG See *synthetic natural gas*.

social science principles of sustainability Derived from studies of economics, political science, and ethics, these principles could help guide us toward a more sustainable future. Full-cost pricing: the harmful environmental and health costs of producing and using goods and services should be included in their market prices; win-win politics: government policies should be beneficial to humans and to the environment; responsibility to future generations: we should leave the planet's life-support systems in at least as good a condition as that which we now enjoy, if not better, for future generations.

soil Complex mixture of inorganic minerals (clay, silt, pebbles, and sand), decaying organic matter, water, air, and living organisms underlying most terrestrial ecosystems.

soil conservation Methods used to reduce soil erosion, prevent depletion of soil nutrients, and restore nutrients previously lost by erosion, leaching, and excessive crop harvesting.

soil erosion Movement of soil components, especially topsoil, from one place to another, usually by wind, flowing water, or both. This natural process can be greatly accelerated by human activities that remove vegetation from soil. Compare *soil conservation*.

soil horizons Horizontal zones, or layers, that make up a particular mature soil. Each horizon has a distinct texture and composition that vary with different types of soils. See *soil profile*.

soil profile Cross-sectional view of the horizons in a soil. See *soil horizons*.

solar cell See *photovoltaic cell*.

solar energy Direct radiant energy from the sun and a number of indirect forms of energy produced by the direct input of such radiant energy. Principal indirect forms of solar energy include wind, falling and flowing water (hydropower), and biomass (solar energy converted into chemical energy stored in the chemical bonds of organic compounds in trees and other plants)—none of which would exist without direct solar energy.

solar thermal system System also known as *concentrated solar power (CSP)* that uses one of several different methods to collect and concentrate solar energy in order to boil water and produce steam for generating electricity.

solid waste Any unwanted or discarded material that is not a liquid or a gas. See *industrial solid waste, municipal solid waste*.

specialist species Species with a narrow ecological niche. They may be able to live in only one type of habitat, tolerate only a narrow range of climatic and other environmental conditions, or use only one type or a few types of food. Compare *generalist species*.

speciation Formation of two species from one species because of divergent natural selection in response to changes in environmental conditions; usually takes thousands of years. Compare *extinction*.

species Group of organisms with certain defining characteristics in common, and for sexually reproducing organisms, a set of individuals that can mate and produce fertile offspring. Every organism is a member of a certain species.

species diversity The degree of variety of species in a given area; often measured by the number of different species (species richness) combined with the relative abundance of individuals within each of those species (species evenness) in a given area. See *biodiversity, species evenness, species richness*. Compare *ecological diversity, genetic diversity*.

species evenness Degree to which comparative numbers of individuals of each of the species present in a community are similar. See *species diversity*. Compare *species richness*.

species richness The number of different species contained in a community. See *species diversity*. Compare *species evenness*.

spoils Unwanted rock and other waste materials produced when a material is removed from the earth's surface or subsurface by mining, dredging, quarrying, or excavation.

stewardship worldview Human-centered worldview holding that we can manage the earth for our benefit but that we have an ethical responsibility to be caring and responsible managers, or stewards, of the earth. It calls for encouraging environmentally beneficial forms of economic growth and discouraging environmentally harmful forms. Compare *environmental wisdom worldview, frontier environmental worldview, planetary management worldview.* See *earth-centered environmental worldview, human-centered environmental worldview, life-centered environmental worldview.*

stratosphere Second layer of the atmosphere, extending about 17–48 kilometers (11–30 miles) above the earth's surface. It contains a layer of gaseous ozone (O_3), which filters out about 95% of the incoming harmful ultraviolet radiation emitted by the sun. Compare *troposphere.* See *ozone layer.*

strip-cropping Planting regular crops and close-growing plants, such as hay or nitrogen-fixing legumes, in alternating rows or bands to help reduce depletion of soil nutrients.

strip cutting Variation of clear-cutting in which a strip of trees is clear-cut along the contour of the land, with the corridor being narrow enough to allow natural regeneration within a few years. After regeneration, another strip is cut above the first, and so on. Compare *clear-cutting, selective cutting.*

strip mining Form of surface mining in which bulldozers, power shovels, or stripping wheels remove large areas of soil and vegetation to access a mineral deposit beneath a land surface. See *area strip mining, contour strip mining, surface mining.* Compare *subsurface mining.*

subsidence Slow or rapid sinking of part of the earth's crust often caused by removal of structural material under the surface by activities such as subsurface mining.

subsidies Payments and protections of various forms that help businesses and industries to survive and thrive.

subsurface mining Extraction of a metal ore or fuel resource such as coal from a deep underground deposit. Compare *surface mining.*

sulfur cycle Cyclic movement of sulfur in various chemical forms from the environment to organisms and then back to the environment.

sulfur dioxide (SO_2) Colorless gas with an irritating odor. About one-third of the SO_2 in the atmosphere comes from natural sources as part of the sulfur cycle. The other two-thirds come from human sources, mostly combustion of sulfur-containing coal in electric power and industrial plants and from oil refining and smelting of sulfide ores.

sulfuric acid Compound containing hydrogen, sulfur, and oxygen; a hazardous chemical that is often a component of acid precipitation. See *acid deposition.*

surface fire Forest fire that burns only undergrowth and leaf litter on the forest floor. Compare *crown fire, ground fire.*

surface mining Removing soil, subsoil, and other strata and then extracting a mineral deposit found fairly close to the earth's surface. See *area strip mining, contour strip mining,* *mountaintop removal mining, open-pit mining, strip mining.* Compare *subsurface mining.*

surface runoff Water flowing off the land into bodies of surface water. See *reliable surface runoff.*

surface water Precipitation that does not infiltrate the ground or return to the atmosphere by evaporation or transpiration. Found in streams, rivers, lakes, and wetlands. See *surface runoff.* Compare *groundwater.*

survivorship curve Line graph that shows the percentages of the members of a population surviving at different ages. There are three generalized types of survivorship curves: *late loss, early loss,* and *constant loss.*

suspended particulate matter See *particulates.*

sustainability Ability of earth's various systems, including human cultural systems and economies, to survive and adapt to changing environmental conditions indefinitely.

sustainable agriculture Method of growing crops and raising livestock by relying on organic fertilizers, soil conservation, water conservation, biological pest control, and minimal use of nonrenewable fossil-fuel energy.

sustainable yield Highest rate at which a potentially renewable resource can be used indefinitely without reducing its available supply. See also *environmental degradation.*

synfuels Synthetic gaseous and liquid fuels produced from solid coal or sources other than natural gas or crude oil.

synthetic biology A technology that enables scientists to make new sequences of DNA and to use such genetic information to design and create new cells, tissues, organisms, and devices, and to redesign existing natural biological systems.

synthetic natural gas (SNG) Gaseous fuel containing mostly methane produced from solid coal.

system A set of components that function and interact in some regular way. Most living systems have inputs, throughputs, and outputs of matter and energy from the environment. See *inputs, outputs, and throughputs.*

tailings Rock and other waste materials removed as impurities when waste mineral material is separated from the metal in an ore.

tar sand Deposit of a mixture of clay, sand, water, and varying amounts of a tar-like heavy oil known as bitumen. Bitumen can be extracted from tar sand by heating. It is then purified and upgraded to synthetic crude oil.

tectonic plates Various-sized pieces of the earth's lithosphere that move slowly around atop the mantle's flowing asthenosphere. Most earthquakes and volcanoes occur near the boundaries of these plates. See *lithosphere.*

temperature inversion Layer of dense, cool air trapped under a layer of less dense, warm air. It prevents upward-flowing air currents from developing. In a prolonged inversion, air pollution in the trapped layer may build up to harmful levels.

tentative science Preliminary scientific data, hypotheses, and models that have not been widely tested and accepted. Compare *reliable science, unreliable science.*

teratogen Chemical, ionizing agent, or virus that causes birth defects. Compare *carcinogen, mutagen.*

terracing Planting crops across a long, steep slope that has been converted into a series of broad, nearly level terraces with short vertical drops from one to another that run along the contour of the land to retain water and reduce soil erosion.

tertiary (higher-level) consumers Animals that feed on carnivores at high trophic levels in food chains and webs. Examples include hawks, lions, bass, and sharks. See *carnivore.* Compare *detritivore, primary consumer, secondary consumer.*

theory of evolution Widely accepted scientific theory that all life forms developed from earlier life forms. It is the way most biologists explain how life has changed over the past 3.8 billion years and why it is so diverse today.

theory of island biogeography Widely accepted scientific theory holding that the number of different species (species richness) found on an island is determined by the interactions of two factors: the rate at which new species immigrate to the island and the rate at which species become extinct, or cease to exist, on the island. See *species richness.*

thermal inversion See *temperature inversion.*

threatened species Wild species that is still abundant in its natural range but is likely to become endangered because of a decline in numbers. Compare *endangered species.*

throughputs Matter and energy flowing through a living system. See *system.* Compare *inputs* and *outputs.*

tipping point Threshold level at which an environmental problem causes a fundamental and irreversible shift in the behavior of a system. See *climate change tipping point, ecological tipping point.*

topsoil The fertile top layer of many soils; it is a vital component of natural capital, because it stores the water and nutrients needed by plants.

total fertility rate (TFR) Estimate of the average number of children that the women in a given population will have during their childbearing years.

toxic chemical Chemical that can cause harm to an organism. See *carcinogen, mutagen, teratogen.*

toxicity Measure of the harmfulness of a substance.

toxicology Study of the adverse health effects of chemicals.

toxic waste See *hazardous waste.*

traditional intensive agriculture Mode of traditional (non-industrialized) farming aimed at producing enough food for a farm family to survive and to have a surplus that can be sold. This type of agriculture uses higher inputs of labor, fertilizer, and water than traditional subsistence agriculture. Compare *industrialized agriculture, traditional subsistence agriculture.*

traditional subsistence agriculture Mode of traditional (non-industrialized) farming aimed at producing enough food for a farm family to survive and to put aside some food for harder times. Compare *industrialized agriculture, traditional intensive agriculture*.

tragedy of the commons Depletion or degradation of a potentially renewable resource to which a large number of people have free and unmanaged access. An example is the depletion of commercially desirable fish species in the open ocean beyond areas controlled by coastal countries. See *common-property resource, open access renewable resource*.

trait Characteristic passed on from parents to offspring during reproduction in an animal or plant.

transform fault Area where the earth's lithospheric plates move in opposite but parallel directions along a fracture (fault) in the lithosphere.

transmissible disease Disease that is caused by living organisms (such as bacteria, viruses, and parasitic worms) and can spread from one person to another through air, water, food, or body fluids, or in some cases through insects or other organisms. Compare *nontransmissible disease*.

transpiration Process in which water is absorbed by the root systems of plants, moves up through the plants, passes through pores (stomata) in their leaves or other parts, and evaporates into the atmosphere as water vapor.

tree farm See *tree plantation*.

tree plantation Site planted with one or only a few tree species that are all the same age. When the stand matures it is usually harvested by clear-cutting and then replanted. These farms normally raise rapidly growing tree species for fuelwood, timber, or pulpwood. Compare *old-growth forest, second-growth forest*.

trophic level The set of all organisms that are the same number of energy transfers away from the original source of energy (usually sunlight) that enters an ecosystem. For example, all producers belong to the first trophic level and all herbivores belong to the second trophic level in a food chain or a food web.

troposphere Innermost layer of the atmosphere. It contains about 75% of the mass of earth's air and extends about 17 kilometers (11 miles) above sea level. Compare *stratosphere*.

tsunami Series of large waves generated when part of the ocean floor suddenly rises or drops, typically due to an earthquake.

undernutrition See *chronic undernutrition*.

unreliable science Scientific results or hypotheses presented as reliable science without having undergone the rigors of the peer review process. Compare *reliable science, tentative science*.

upwelling Movement of nutrient-rich bottom water to the ocean's surface. It can occur far from shore but usually takes place along certain steep coastal areas where the warm surface layer of ocean water flows away from shore and is replaced by cold, nutrient-rich bottom water.

urban area Geographic area containing a community with a population of 2,500 or more. The minimum number of people used in this definition varies among countries, from 2,500 to 50,000.

urban growth Rate of growth of an urban population.

urbanization Creation or growth of urban areas, or cities, and their surrounding developed land. See *urban area*.

urban sprawl Growth of low-density development on the edges of cities and towns. See *smart growth*.

virtual water Water that is not directly consumed but is used to produce food and other products.

virus Infectious agent that is smaller than a bacterium; it works by invading a cell and taking over its genetic machinery to copy itself. It then multiplies and spreads throughout one's body, causing a viral disease such as flu or AIDS.

volatile organic compounds (VOCs) Organic compounds that exist as gases in the atmosphere and act as pollutants, some of which are hazardous.

volcano Vent or fissure in the earth's surface through which magma, liquid lava, and gases are released into the environment.

warm front Boundary between an advancing warm air mass and the cooler one it encounters. Because warm air is less dense than cool air, an advancing warm front rises over a mass of cool air. Compare *cold front*.

waste management Managing wastes to reduce their environmental harm without trying to reduce the amount of waste produced. See *integrated waste management*. Compare *waste reduction*.

waste reduction Reducing the amount of waste produced; wastes that are produced are viewed as potential resources that can be reused, recycled, or composted. See *integrated waste management*. Compare *waste management*.

wastewater Water that contains sewage and other wastes from homes, farms, and industries.

water cycle See *hydrologic cycle*.

water footprint A rough measure of the volume of water used directly and indirectly to keep a person or group alive and to support their lifestyles.

waterlogging Saturation of soil with irrigation water or excessive precipitation so that the water table rises close to the surface.

water pollution Any physical or chemical change in surface water or groundwater that can harm living organisms or make water unfit for certain uses.

watershed (drainage basin) Land area that delivers water, sediment, and dissolved substances via small streams to a larger stream, lake, or wetland.

water table Upper surface of the zone of saturation, in which all available pores in the soil and rock in the earth's crust are filled with water. See *zone of aeration, zone of saturation*.

weather Short-term changes in the temperature, barometric pressure, humidity, precipitation, sunshine, cloud cover, wind direction and speed, and other conditions in the troposphere at a given place and time. Compare *climate*.

wetland Land that is covered all or part of the time with saltwater or freshwater, excluding streams, lakes, and the open ocean. See *coastal wetland, inland wetland*.

wilderness Area where the earth and its ecosystems have not been seriously disturbed by humans and where humans are only temporary visitors; in law, an area that is protected from all harmful human activities.

windbreak Row of trees or hedges planted to partially block wind flow and reduce soil erosion on cultivated land.

wind farm Cluster of wind turbines in a windy area on land or at sea, built to capture wind energy and convert it to electrical energy.

worldview A set of beliefs about how the world works and what one's role in the world should be. See *earth-centered environmental worldview, environmental wisdom worldview, frontier environmental worldview, human-centered environmental worldview, life-centered environmental worldview, planetary management worldview, stewardship worldview*.

zone of aeration Zone in soil that is not saturated with water and that lies above the water table. See *water table, zone of saturation*.

zone of saturation Zone below the water table where all available pores in soil and rock in the earth's crust are filled by water. See *water table, zone of aeration*.

zoning Designating parcels of land for particular types of use.

zooplankton Animal plankton; small floating herbivores that feed on plant plankton (phytoplankton). Compare *phytoplankton*.